A First Course in Magnetohydrodynamics

This text introduces readers to magnetohydrodynamics (MHD), the physics of ionised fluids. Traditionally, MHD is taught as part of a graduate curriculum in plasma physics. By contrast, this text – one of a very few – teaches MHD exclusively from a fluid dynamics perspective, making it uniquely accessible to senior undergraduate students. Part I of the text uses the MHD Riemann problem as a focus to introduce the fundamentals of MHD: Alfvén's theorem, waves, shocks, rarefaction fans, and so on. Part II builds upon this with presentations of broader areas of MHD: fluid instabilities, viscid hydrodynamics, steady-state MHD, and non-ideal MHD. Throughout the text, more than 125 problems and several projects (with solutions available to instructors) reinforce the main ideas. In addition, large-font lesson plans for a "flipped-style" class are available free of charge to instructors who use this text as required reading for their course. This book is suitable for advanced undergraduate and beginning graduate students of physics, requiring no previous knowledge of fluid dynamics or plasma physics.

David Clarke is a retired professor of astronomy and physics from Saint Mary's University in Halifax, Nova Scotia. Over his thirty-year career, he has taught numerous courses in physics and astronomy at the undergraduate and graduate levels, including courses in fluid dynamics and MHD that inspired this text. He is co-developer of the original *ZEUS* MHD code and currently the primary developer of *ZEUS-3D* that he uses for his research in astrophysical jets and has made available open source to hundreds of investigators worldwide.

Professor Hannes Olof Gösta Alfvén (1908–1995), father of magnetohydrodynamics, 1970 Nobel Prize laureate for physics. Portrait created in 1972 by Benno Movin-Hermes (1902–1977) using a *three-colour foil method* developed by the artist. Reproduced with permission from I. Movin and the Moderna Museet, Stockholm.

A First Course in
Magnetohydrodynamics

David Alan Clarke
Saint Mary's University, Halifax NS Canada

Shaftesbury Road, Cambridge CB2 8EA, United Kingdom

One Liberty Plaza, 20th Floor, New York, NY 10006, USA

477 Williamstown Road, Port Melbourne, VIC 3207, Australia

314–321, 3rd Floor, Plot 3, Splendor Forum, Jasola District Centre, New Delhi – 110025, India

103 Penang Road, #05–06/07, Visioncrest Commercial, Singapore 238467

Cambridge University Press is part of Cambridge University Press & Assessment, a department of the University of Cambridge.

We share the University's mission to contribute to society through the pursuit of education, learning and research at the highest international levels of excellence.

www.cambridge.org
Information on this title: www.cambridge.org/9781009381475

DOI: 10.1017/9781009381468

© David Alan Clarke 2025

This publication is in copyright. Subject to statutory exception and to the provisions of relevant collective licensing agreements, no reproduction of any part may take place without the written permission of Cambridge University Press & Assessment.

When citing this work, please include a reference to the DOI 10.1017/9781009381468

First published 2025

A catalogue record for this publication is available from the British Library

Library of Congress Cataloging-in-Publication Data
Names: Clarke, David Alan, 1958– author.
Title: A first course in magnetohydrodynamics / David Alan Clarke, Saint Mary's University, Nova Scotia.
Description: Cambridge, United Kingdom ; New York, NY : Cambridge University Press, 2025. | Includes bibliographical references and index.
Identifiers: LCCN 2024014429 | ISBN 9781009381475 (hardback) | ISBN 9781009381468 (ebook)
Subjects: LCSH: Magnetohydrodynamics – Textbooks.
Classification: LCC QC718.5.M36 C53 2025 | DDC 538/.6–dc23/eng/20240629
LC record available at https://lccn.loc.gov/2024014429

ISBN 978-1-009-38147-5 Hardback

Cambridge University Press & Assessment has no responsibility for the persistence or accuracy of URLs for external or third-party internet websites referred to in this publication and does not guarantee that any content on such websites is, or will remain, accurate or appropriate.

For Jodi, my life-long love.

Contents

Preface   xiii

Introduction   1

Part I   1-D MHD in Ten Weeks   5

1   The Fundamentals of Hydrodynamics   7
1.1   Definition of a fluid   7
1.2   A quick review of kinetic theory   9
1.3   The equations of ideal hydrodynamics   12
1.4   The internal energy density   18
1.5   Primitive, integral, and conservative form   20
Problem Set 1   22

2   Selected Applications of Hydrodynamics   26
2.1   Sound waves   26
     2.1.1   Wave equation approach   26
     2.1.2   Eigenvalue approach   29
2.2   Rankine–Hugoniot jump conditions   33
     2.2.1   Case 1: Trivial solution   35
     2.2.2   Case 2: Tangential discontinuity   35
     2.2.3   Case 3: Shock   36
2.3   Bores and hydraulic jumps (optional)   42
     2.3.1   Bores in the "lab frame"   46
2.4   Bernoulli's theorem   47
Problem Set 2   56

3   The Hydrodynamical Riemann Problem   69
3.1   Eulerian and Lagrangian frames of reference   70
3.2   The three characteristics of hydrodynamics   72
3.3   Characteristic paths and space-time diagrams   74
3.4   The MoC and the Riemann problem   78
3.5   Non-linear hydrodynamical waves   81
     3.5.1   Hyperbolic system of equations   81

		3.5.2	Left and right eigenvectors (optional)	82
	3.6	3.5.3	Hydrodynamical rarefaction fans	85
	3.6	Solution to the HD Riemann problem		90
	Problem Set 3			94

4 The Fundamentals of Magnetohydrodynamics — 99
- 4.1 A brief introduction to MHD 99
- 4.2 The ideal induction equation 101
- 4.3 Alfvén's theorem 103
- 4.4 Modifications to the momentum equation 106
- 4.5 The MHD Poynting power density 107
- 4.6 Modifications to the total energy equation 107
- 4.7 The equations of ideal MHD 109
- 4.8 Vector potential and magnetic helicity (optional) 112
  - 4.8.1 Magnetic topology 114
- Problem Set 4 117

5 MHD Waves and Discontinuities — 123
- 5.1 Primitive and conservative equations of MHD 124
- 5.2 MHD wave families 127
  - 5.2.1 (Shear) Alfvén waves 130
  - 5.2.2 Fast and slow magnetosonic waves 139
  - 5.2.3 Summary of MHD waves 146
- 5.3 The MHD Rankine–Hugoniot jump conditions 149
  - 5.3.1 Case 1: Trivial solution 154
  - 5.3.2 Case 2: Contact discontinuity 154
  - 5.3.3 Case 3: Tangential discontinuity 154
  - 5.3.4 Case 4: Rotational discontinuity 155
  - 5.3.5 Case 5: Fast, slow, and intermediate shocks 155
- Problem Set 5 172

6 The MHD Riemann Problem — 183
- 6.1 Overview 183
- 6.2 Non-linear MHD waves 185
  - 6.2.1 Fast and slow eigenkets 186
  - 6.2.2 Fast and slow rarefaction fans 192
- 6.3 Space-time diagrams 202
- 6.4 An MHD Riemann solver 204
  - 6.4.1 Problem parameters 205
  - 6.4.2 Strategy for the Riemann solver 208
  - 6.4.3 Algorithm for an exact MHD Riemann solver 210
  - 6.4.4 Sample problems 221
- Problem Set 6 235

Part II Additional Topics in (M)HD — 241

7 Fluid Instabilities — 243
- 7.1 Kelvin–Helmholtz instability — 245
  - 7.1.1 Normal mode analysis of the KHI — 246
  - 7.1.2 The development of the KHI — 251
  - 7.1.3 The KHI in nature — 252
  - 7.1.4 Numerical analysis of the KHI (optional) — 254
- 7.2 Rayleigh–Taylor instability — 256
  - 7.2.1 Normal mode analysis of the RTI — 257
  - 7.2.2 Numerical analysis of the RTI (optional) — 262
  - 7.2.3 Kruskal–Schwarzschild instability — 267
- 7.3 Magneto-rotational instability — 268
  - 7.3.1 Mathematical model of the MRI — 269
  - 7.3.2 Physical model of the MRI — 275
  - 7.3.3 Angular momentum transport — 278
  - 7.3.4 Numerical analysis of the MRI (optional) — 283
- 7.4 Parker instability — 286
  - 7.4.1 A qualitative description — 286
  - 7.4.2 A quantitative description (optional) — 291
- Problem Set 7 — 305

8 Viscid Hydrodynamics — 309
- 8.1 Introduction — 309
- 8.2 The stress tensor — 310
  - 8.2.1 The trace of the stress tensor — 312
  - 8.2.2 Viscosity and Newtonian fluids — 314
  - 8.2.3 Non-isotropic "pressure" — 315
- 8.3 The Navier–Stokes equation — 317
- 8.4 The viscid energy equation — 320
  - 8.4.1 Viscous dissipation — 321
- 8.5 The Reynolds number — 322
- 8.6 Applications — 326
  - 8.6.1 Plane laminar viscous flow — 327
  - 8.6.2 Forced flow between parallel plates — 328
  - 8.6.3 Open channel flow — 329
  - 8.6.4 Hagen–Poiseuille flow — 333
  - 8.6.5 Couette flow — 334
- Problem Set 8 — 337

9 Steady-State MHD — 341
- 9.1 The Weber–Davis constants — 342
- 9.2 The MHD Bernoulli function — 349
  - 9.2.1 Critical launching angle — 354

9.3	Stellar winds	357
	9.3.1 Critical points	359
	9.3.2 The Weber–Davis Model	361
9.4	Astrophysical jets (optional)	366
	Problem Set 9	371

10 Non-ideal MHD — 378

10.1	Introducing non-ideal MHD	378
10.2	The three players	380
	10.2.1 A weakly ionised, isothermal, one-fluid model	380
	10.2.2 Relative importance of the non-ideal terms	387
10.3	Resistive dissipation	391
	10.3.1 The resistive induction equation	391
	10.3.2 Dissipation of magnetic energy	392
	10.3.3 Magnetic diffusion and reconnection	393
	10.3.4 Dynamo theory (optional)	399
10.4	The Hall effect	406
	10.4.1 The case of a completely ionised fluid	408
	10.4.2 Magnetic reconnection, revisited	411
10.5	Ambipolar diffusion	414
	10.5.1 Overview and motivation	414
	10.5.2 A two-fluid, non-isothermal model for AD	416
	Problem Set 10	428

Appendices — 441

A Essentials of Vector Calculus — 443

A.1	Vector identities	443
	A.1.1 Identities involving dyadics	444
	A.1.2 Vector derivatives of $\vec{r}$	445
A.2	Theorems of vector calculus	445
A.3	Orthogonal coordinate systems	446
A.4	Euler's and the momentum equations	448
A.5	The Lorentz force	451

B Essentials of Electrodynamics — 453

B.1	Maxwell's equations	453
B.2	Electric energy density	454
B.3	Magnetic energy density	455
B.4	Resistive energy density	457
B.5	The Poynting vector	458

C The "Conics" of PDEs — 459

Contents

D The Secant Method — **462**
    D.1 Univariate root finder 462
    D.2 Multivariate root finder 465

E Roots of a Cubic — **467**

F Sixth-Order Runge–Kutta — **468**

G Coriolis' Theorem — **476**

H The Diffusion Equation — **484**

Variable Glossary — **486**

References — **493**

Index — **500**

Preface

WHENEVER a university professor stares down the barrel of a new course preparation, the first question invariably asked is *Is there a text?* For the most part in undergraduate physics, the answer is usually *Yes, and plenty to choose from*. But for a senior undergraduate or beginning graduate course in *magnetohydrodynamics* (MHD), the selection is much narrower.

MHD is a relatively new branch of physics. Developed by Hannes Alfvén during the 1940s, it didn't gain wide acceptance among physicists writ large until the late 1950s culminating in Alfvén's Nobel Prize in 1970.[1] As such, MHD is often touted as a "classical afterthought", the only branch of classical physics introduced *after* quantum mechanics with many of its fundamentals – the Riemann problem, magneto-rotational instability, and non-ideal effects – still being worked out in the 1990s and early aughts.

Thus, MHD has not had as long a history as other branches of physics in which textbooks could accumulate, particularly at the undergraduate level. Indeed, MHD has largely been considered a graduate-level subject and, because of this, the vast majority of existing texts specialise in areas such as fusion physics, solar physics, and planetary discs, many written for students already with some familiarity of plasma physics or fluid dynamics.

Another extenuating circumstance is MHD is a divided field whose practitioners – largely plasma physicists and fluid dynamicists – approach the subject in two very different ways. From a plasma physicist's point of view (PoV), an MHD system is the isotropic limit of an ensemble of charged particles – a plasma – governed by velocity moments of the *Vlasov–Boltzmann equation*, a 6-D inhomogeneous partial differential equation (PDE) at the heart of plasma physics. From a fluid dynamicist's PoV, an MHD system is never considered as an ensemble of particles, but rather as a continuous medium governed by simple conservation rules that *any* undergraduate physics student can understand. This leads to a hyperbolic set of equations that can be analysed entirely in terms of *waves*. The two approaches couldn't be more different.

After thirty years of teaching graduate and undergraduate (astro)physics, it is my considered opinion that for MHD to be approachable by undergraduates, it needs to be taught from the fluids PoV. Wave mechanics – so fundamental in classical mechanics, electrodynamics, and quantum mechanics – is already ingrained in the mind of a fourth-year student. On the other hand, velocity moments of a six-

---

[1] On pages 127–128, there's an amusing anecdote on what – or who – changed the physics community's collective mind on MHD, and a link to Anthony Peratt's short biography on Alfvén.

dimensional PDE are not. Further, most texts taking the plasma PoV are focused on laboratory plasmas and fusion physics, and the wave nature of MHD is often overlooked. To my taste, MHD is *the* prototype for teaching and reinforcing *wave mechanics*, and this is precisely the approach I take in this text.

While there are plenty of textbooks on MHD from the plasma PoV, precious few exist from the fluids PoV. In the survey of texts I did as part of my proposal for this text, I found more than 100 books written over the past six decades from the plasma PoV focused on plasma physics with a substantial portion devoted to the "MHD limit". Indeed, a dozen or so of these textbooks include MHD in their titles.[2] By contrast, I found just *two* texts on MHD written entirely from the fluids PoV and directed to senior undergraduates: Kendall & Plumpton's *Magnetohydrodynamics with Hydrodynamics* (1964); and Galtier's *Introduction to Modern Magnetohydrodynamics* (2016).

This text offers a third. My approach focuses on the *fundamentals* of the subject and teaches MHD for its own sake rather than dwelling on directed applications and current areas of research; these, I argue, are better suited for graduate texts of which there are plenty. I do provide numerous examples from the literature, but these are selected to emphasise certain ideas (*e.g.*, planetary discs with non-ideal MHD, stellar winds with steady-state MHD, astrophysical jets with Bernoulli's principle, *etc.*) and none should distract the reader from the current discussion. Once endowed with the fundamentals, I contend, students can carry these forward to further their study at the graduate level, should they choose.

In keeping with the undergraduate theme, the first part – *1-D MHD in Ten Weeks* – is designed around a single goal: solving the 1-D MHD Riemann problem. I also assert that to understand MHD, one first has to understand ordinary hydrodynamics (HD) which is, after all, just the zero-field limit of MHD. To these ends, Chap. 1 introduces the student to the fundamentals of HD that includes a novel and simple derivation of the three ideal HD equations. Chapter 2 focuses on 1-D applications of HD including sound waves, shocks, bores, and Bernoulli's principle while Chap. 3 develops a semi-analytic solution to the hydrodynamical Riemann problem. In so doing, students learn how the equations of HD lead to a wave equation, and are shown three ways to extract information about hydrodynamical waves: direct solution of the wave equation; normal mode analysis using linear algebra; and via Riemann invariants and their characteristic paths. In my experience, introducing students to these methods – particularly the latter two – for the relatively simple case of HD is critical for them to understand how they apply to the much more complicated MHD case.

The magnetic induction, $\vec{B}$, doesn't appear until Chap. 4 where the ideal induction equation and the Lorentz force are introduced, along with Alfvén's theorem, magnetic helicity, and flux linking. Chapter 5 examines the MHD equations in 1-D to uncover all three types of waves (slow, Alfvén, fast) and all discontinuities (tan-

---

[2]The most ambitious and a very recent example of this is Goedbloed, Keppens, & Poedts' *Magnetohydrodynamics of Laboratory and Astrophysical Plasmas* (2019). This is a comprehensive tour de force which could support at least three graduate-level courses.

gential, rotational, shocks) including all three shock subtypes (slow, intermediate, fast). Chapter 6 introduces slow and fast rarefaction fans, and then brings it all together to show students how an exact MHD Riemann solver can be assembled. As this is a semi-analytical solution, students learn or have reinforced semi-analytical techniques including Runge–Kutta methods, multivariate secant root finders, methods for maintaining machine accuracy, and the list goes on.

Part I is designed to be completed in twenty-five hours of instruction (ten weeks at most Canadian universities). The four chapters in Part II, *Additional Topics in (M)HD*, are independent from each other, depend only on material from Part I, and give the instructor options to complete the semester. These include (M)HD instabilities in Chap. 7, viscid HD (Navier–Stokes equation) in Chap. 8, steady-state MHD in Chap. 9, and non-ideal MHD in Chap. 10. In the interest of expediency, sections designated as "optional" can be omitted without loss of continuity.

Parts of the text may come across as "mathematically dense"; this is deliberate. As an undergraduate, I always found it frustrating and distracting when I was unable to fill in the large gaps left between lines of logic in the texts my professors chose, and I was not going to produce a text that did the same. That said, the densest parts of the mathematics can largely be skimmed on first read and certainly don't need to be covered in detail in class, as the main results from which the physics is extracted are always boxed within each development. For the student like I was who needs to know how the derivations are done, the gaps between the mathematical steps are small enough that a careful second read should suffice.

More than 125 problems – many exploiting "teaching moments" – and several computer projects are distributed amongst the ten chapters' problem sections, each generally digestible by a senior undergraduate or first-year graduate student. Problems without an asterisk can and should be done in a page or less (and often a few lines), one asterisk indicates a two-page solution, two asterisks indicate a three- to four-page solution, while three asterisks indicate a more involved problem, generally requiring more than five pages and/or a substantive computer program to solve. A complete solution set including the computer projects is available to the instructor upon request.

The eight appendices are designed to remind students of particularly critical material prerequisite to this text. Students who do not recognise, recall, or know how to use any of this material are encouraged to review the relevant material from previous courses. Following the appendices is a glossary of symbols used throughout the text, a list of references, and finally an extensive index.

While this text assumes no previous knowledge of (M)HD, students should have had second- and third-year courses in mechanics, electrodynamics, and thermodynamics. On the math side, students should be fluent in vector calculus (at the level of App. A), adept at solving differential equations including PDEs such as the wave equation, and thoroughly familiar with linear algebra and, in particular, *eigen*algebra. In addition, some experience in scientific computing (algorithm and code development) would be beneficial.

Finally, an acknowledgement of the biases of the author is in order. While this text includes numerous astrophysical applications, astronomy is by no means a prerequisite, nor is this text designed just for budding astrophysicists. I would like to think *any* physicist interested in learning about fundamental MHD will find this book useful. As for units, I follow the bulk of the physics community (but not astronomy!) and use mks exclusively. Lastly, all program listings in this text are in *FORTRAN 77*, the *only* computer language – this old programmer would assert – a computational scientist really needs to know!

Like many of my contemporaries, I learned MHD "at the knee of my advisors", by reading select chapters in certain texts, by going through journal articles, and talking to experts. As a student and post-doctoral fellow, it always struck me as a bit unfair that all other branches of physics seemed to be taught in more systematic and accessible ways – dedicated courses, self-contained textbooks, problem sets at appropriate levels – but somehow not MHD. Granted, MHD doesn't enjoy the same "critical mass" of students as other areas of physics, and perhaps this inaccessibility is part of the reason why.

This text – almost two decades in the making – is the textbook I wish I had access to forty years ago when I started out in this game. And now as I enter my retirement, it is my profound hope that within these pages, new students of MHD will find a self-contained introduction to the subject that will help launch them into a fascinating, life-long adventure as I have enjoyed!

Each chapter in Part II benefitted from written projects or theses submitted by Saint Mary's graduate and undergraduate students taking my (M)HD course and/or working with me as a research student in years past. For these efforts, I thank Joel Tanner, Patrick Rogers, Jonathan Ramsey, Nicholas MacDonald, Michael Power, and Christopher MacMackin.

I thank my editors Nicholas Gibbons, Sarah Armstrong, Stephanie Windows, and Jane Chan at Cambridge University Press for their capable and patient guidance of this first-time author. It definitely made my job a lot less daunting! A big thank-you goes to Patricia Langille at Saint Mary's Patrick Power Library for doing *all* the heavy lifting in getting permissions for the copyrighted material used in this text; she gets the first signed copy! I would also like to acknowledge the academic freedom afforded to me over the past three decades by Saint Mary's University that made long-term projects like this possible. Thank you all.

This text was typeset using Donald Knuth's TeX and Leslie Lamport's LaTeX. Many of the figures were created using Xfig developed by Supoj Sutanthavibul, Ken Yap, Brian V. Smith, and others. Figures from *ZEUS-3D* simulations were created using PSPlot developed by Kevin E. Kohler. Countless members of the scientific community are indebted to these people for placing their software into the public domain.

And then there are the magnificent villages of Ménerbes and Saint-Pierre-

Toirac, France. The most difficult sections of this text for me to write were inspired and completed during my long *séjours* there and, other than my home province of Nova Scotia, I can't think of anywhere else I'd rather spend months at a time than in a small French village. So much about France charms me including, of all things, their speed-limit signs! It's not enough to tell you what the speed limit is, the French also feel the need to tell you you're being *reminded* of what the speed limit is! So in homage to my second country, all footnotes throughout the text serving as a "reminder to the reader" are heralded by the French translation *Rappel*.

And finally to my wife Jodi (MEL) to whom this text is dedicated. Words can't express my love and gratitude for sticking by me these nearly forty years and for helping create such a wonderful and supportive home for our family here in Halifax. You're the best!

Any constructive feedback on what works and doesn't work in this text, as well as any error reports, omissions, redundancies, *etc.* are welcome and can be sent to me directly at `AfciMHD@gmail.com`. Instructors of courses based on this text may download a solution set to the problems and a fully-developed set of course lecture notes from CUP's website, `www.cambridge.org/9781009381475`.

David Clarke
Halifax, Nova Scotia
`www.ap.smu.ca/~dclarke`

Introduction

At last, some remarks are made about the transfer of momentum from the sun to the planets, which is fundamental to the theory. The importance of magnetohydrodynamic waves in this respect are [sic] *pointed out.*

> First published mention of the term *magnetohydrodynamic*, from "On the cosmogony of the solar system III" by Hannes Alfvén, 1942, *Stockholm's Observatoriums Annaler*, v. 14, 9.1–9.29.

THE ANCIENT GREEKS knew the universe to be made up of the four elements: earth; water; wind; and fire. Today, we know these as the four states of matter: solid; liquid; gas; and plasma, three of which fall into the realm of *fluid dynamics*. Indeed, more than 99.9% of "ordinary matter" in the universe is in the fluid state and, in particular, the plasma (magnetohydrodynamical) state.[1]

Yet, as a pure science, fluid dynamics has often been omitted from many university undergraduate physics curricula. In fact, if you want to find regularly offered courses in fluid dynamics in a university calendar, you're more likely to find them among the engineering or applied mathematics offerings than physics.

One could come up with a number of reasons for this:

- areas of physics such as classical mechanics, electrodynamics, and quantum mechanics are deemed more "fundamental" and courses such as fluid dynamics get relegated as "optional", if offered at all;

- analytical progress generally requires mathematics not typically understood by most undergraduate students of physics until their fourth year; and

- historically, the really interesting problems required the use of major laboratory facilities (such as those available in a large engineering department) or theorems of advanced applied mathematics.

An alternative to expensive laboratories or a degree in Applied Mathematics is computing. While supercomputers capable of solving interesting problems in fluid dynamics have been available since the mid 1980s, it is only since the turn of the 21st century that *cheap* supercomputing has become widely available so that "ordinary" physicists and astrophysicists can once again do interesting problems in the subject.

Indeed, many of the more "interesting" problems in astrophysics such as those

---

[1] www.plasma-universe.com/99-999-plasma/.

in star formation, planetary discs, stellar evolution, the interstellar medium, formation of galaxies, galactic and extragalactic outflows and accretion, the early universe, cosmology, even the Big Bang itself have awaited this "promised land" of cheap supercomputing. Now that it has "arrived", more and more of the literature in astrophysics is being devoted to applications of fluid dynamics and, in particular, magnetohydrodynamics. More than for any other practitioner of physics, astrophysicists are finding the role of fluid dynamics is becoming *increasingly* important with time, not less. For this reason alone, I would argue, university physics curricula should be offering more courses in fluid dynamics, lest the discipline be taken over completely by the engineers and applied mathematicians!

Before we start, let us agree on some basic terms and their uses.

1. A *fluid* is a state of matter that can flow. A liquid is an *incompressible* fluid, while gas and plasma (ionised gas) are *compressible* fluids. A more technical definition of a fluid involves the notion of *granularity*, where the *mean free path* (or *collision length* defined as the distance a particle in the fluid can travel, on average, before colliding with another particle), δl, is much less than any measurable scale length of interest ($\mathcal{L}$). When $\delta l \ll \mathcal{L}$, a fluid can be treated as a *continuum* rather than as an *ensemble of particles* which simplifies the governing equations enormously.

2. *Fluid Dynamics*, a term which is interchangeable with *hydrodynamics* (HD), is the physics of fluid flow (compressible or incompressible), and involves the concepts of mass and energy conservation, Newton's second law, and an *equation of state*.

3. *Fluid Mechanics* has come to refer to fluid dynamics from an engineering vantage point, with more emphasis on experimentation than on theory. Typically (but not always), a text entitled *Fluid Mechanics* will be an engineering text, while a text entitled *Fluid Dynamics* will be a physics text. A notable exception is Landau and Lifshitz' classic text *Fluid Mechanics*, which, in many ways, is the definitive treatment of the subject from a theoretical physicist's perspective.

4. *Gas Dynamics* is compressible fluid dynamics in which all the fluid particles are neutral.

5. *Magnetohydrodynamics (MHD)* is compressible or incompressible fluid in which an appreciable fraction of the particles are charged (ionised) and where charge neutrality is observed at all length scales of interest. Thus, within any volume element however small, there must be as many negative charges as positive. In an MHD fluid, circulation of charged particles at the sub-fluid length scale implies a current and thus a magnetic field which, in turn, interacts with ionised particles on the post-fluid length scale. Note that an MHD fluid need not be 100% ionised for the equations of MHD to apply (*e.g.*, Chap. 10 on

non-ideal MHD). Neutrals in a partially (even a few percent) ionised fluid can couple to the magnetic field via collisions with charged particles. By contrast, a completely neutral gas can neither generate nor interact with a magnetic field.

An MHD fluid can be created from an HD fluid by increasing the ionisation fraction. For a gas, this can be done by increasing its temperature and thus compressible MHD fluids are plasmas. For a liquid such as water, the ionisation fraction can be increased by dissolving salts. While the earth's oceans permeated by the earth's magnetic field technically constitutes an MHD fluid, the weakness of the earth's magnetic field ($\sim 4 \times 10^{-5}$ T, $\beta \sim 10^9$ defined in §5.2)[2] and the extremely low fraction of particles that are ionised renders the MHD effects just about immeasurable.

6. *Plasma Physics* is the study of the collective behaviour of an ensemble of charged particles at length scales smaller than the fluid length scale thereby rendering the MHD equations inapplicable. Plasma physics is generally described by the *Vlasov–Boltzmann equation* which can account for non-fluid-like behaviour such as charge separation and plasma oscillations. An MHD fluid can be described as a plasma in which charge neutrality is observed at all length scales of interest, and thus MHD is an important special case of plasma physics. An excellent first text on plasma physics, which is beyond the scope of this text, is Volume 1 of Francis Chen's now-classic text *Plasma Physics and Controlled Fusion* (1984).

The equations of MHD reduce to the equations of HD when the magnetic induction ($\vec{B}$) is set to zero. As we shall see, HD becomes MHD by adding the Lorentz force to the hydrodynamic version of Newton's second law, and by introducing Faraday's law of induction that governs how the magnetic induction evolves. These modifications, which will seem rather elementary when first introduced, belie the incredible complexity magnetism provides an ionised fluid. For example, while a hydrodynamical fluid can support compressive waves only (and thus, much of HD can be understood in one dimension), the tension along lines of magnetic induction allow a magnetohydrodynamical fluid to support transverse waves as well, thus requiring all three dimensions to describe.

To understand MHD is to understand wave mechanics, and much of this text is devoted to building the students' mathematical skills and physical intuition in this area. By the end of Part I, the student will be able to solve the most complex MHD problem one can do exactly (albeit, semi-analytically), namely the MHD Riemann problem. And while the development of a general, multidimensional computer code

---

[2]Strictly speaking, it is the magnetic *induction*, $\vec{B}$, that has units tesla while the magnetic *field*, $\vec{H} = \vec{B}/\mu$ (App. B), has units ampere/metre. Thus, the earth's magnetic *field* is about 30 A/m. In this book, I attempt to be consistent with this distinction by using the term *magnetic field* when referring to magnetism generically, and *magnetic induction* when reference is to $\vec{B}$ specifically although, for the most part and especially in astrophysics, this difference is largely academic since all that separates them is the constant μ_0.

to solve more complex problems in MHD is beyond the scope of this book, the 1-D Riemann problem and the ideas upon which it is based are at the core of virtually every general computer program written and with which a whole host of interesting (astro)physical problems become accessible.

PART I

1-D MHD IN TEN WEEKS

1 The Fundamentals of Hydrodynamics

Everything flows and nothing abides; everything gives way and nothing stays fixed.

Heraclitus (c. 535–c. 475 BCE)

1.1 Definition of a fluid

THE PHYSICS of hydrodynamics (HD), namely conservation of mass, conservation of energy, and Newton's second law, are all concepts familiar to first-year undergraduate students, though the mathematics to solve the relevant equations is not. Consider an *ensemble of particles* within some volume V, and let these particles interact with each other via elastic collisions. We can let V remain fixed (in which case we allow the particles to collide elastically with the walls of the container too), or we can let V increase or decrease as the particles move apart or come together; it does not matter. If the mass, total energy, and momentum of the ensemble of particles are M, E_T, and $\vec{S}$ respectively, then we have:

$$\frac{dM}{dt} = 0, \qquad \text{conservation of mass;} \qquad (1.1)$$

$$\frac{dE_T}{dt} = \sum \mathcal{P}_{\text{app}}, \qquad \text{conservation of total energy;} \qquad (1.2)$$

$$\frac{d\vec{S}}{dt} = \sum \vec{F}_{\text{ext}}, \qquad \text{Newton's second law.} \qquad (1.3)$$

Here, $\sum \mathcal{P}_{\text{app}}$ is the rate at which work is done (power) by all forces *applied* to the ensemble of particles, and $\sum \vec{F}_{\text{ext}}$ are all forces *external* to and acting on the ensemble of particles. Note that the applied forces – normally just collisions from neighbouring ensembles of particles – are typically a subset of the external forces, which include collisions from neighbouring particles *plus* forces arising from gravity, magnetism, radiation, *etc*. This is because in addition to the thermal and kinetic energies, the *total energy*, E_T, includes gravitational, magnetic, radiative, and possibly other energies as well.

It is how we model the collisional forces from neighbouring ensembles of particles that defines both what constitutes a fluid and how Eq. (1.1)–(1.3) are further developed. Consider a small cube with volume $\Delta V = (\Delta l)^3$ as shown in Fig. 1.1a.

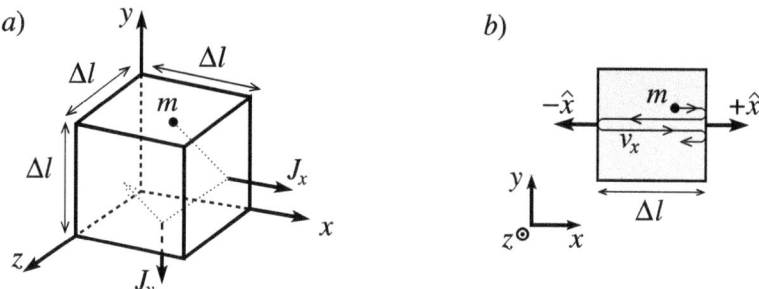

Figure 1.1. *a*) A single particle bounces elastically from the walls of a cube of edge length, Δl, imparting impulses J_x, J_y, etc. *b*) An x–y cut through the cube in panel *a* showing one particle whose motion is entirely in the x-direction.

Let the walls of the cube be perfectly reflecting and let there be just one particle inside the cube moving at some speed v in an arbitrary direction.

When the particle collides with the wall, both the particle and cube suffer a change in momentum in a direction normal to the surface of the cube. Moments later, the particle collides with a different wall, and the particle and cube suffer changes in momentum in a direction normal to that wall. A change in momentum is an impulse, J, which when multiplied by the time over which the collision occurs, Δt, constitutes the average force. Thus formally, the "pressure", p, the collision exerts on the wall of the box is this average force divided by the area of the wall:

$$p \sim \frac{J \Delta t}{(\Delta l)^2}.$$

In this scenario, the "pressure" is highly variable in time, and by no means could the "pressure" be construed as isotropic. At a given time, the "pressure" one wall feels will have nothing to do with the "pressures" felt by the other walls.

However, by arbitrarily increasing the number of particles, $\mathcal{N}$, inside our small volume, ΔV, the number of collisions with a given wall, n, occurring in a time Δt will be the same at each wall to within some arbitrarily small variance, Δn. Put another way, averaged over Δt, particle collisions exert the same "pressure" on each wall to within a variance made as small as we please by making $\mathcal{N}$ as large as we please. Thus, we have rendered the particle "pressure" inside the cube *isotropic* because each wall now feels the same force.

There is a contrived exception to this picture. If all the particles were to be placed initially on the mid-plane of the cube and all were launched with the same speed towards one wall of the cube, then it is only with this and the opposite wall that particles would ever collide, and they would do so in a highly ordered, periodic fashion. The remaining four walls would, in principle, never feel any collisions, and thus the "pressure" in the cube would not be isotropic even with $\mathcal{N}$ chosen arbitrarily large. Such a well-ordered and well-directed ensemble of particles is said to be *streaming* and, as $\mathcal{N}$ is made larger, it becomes increasingly difficult in practice to maintain streaming motion. Small perturbations will eventually cause one particle to collide with another which in turn collide with others, and the ensuing chain reaction quickly reduces the streaming motion to chaos. Isotropic "pressure" (the same "pressure" measured on each of the six walls) is once again the result.

We can now state the key criterion for an ensemble of particles to be treated as a fluid. If there is a sufficient number of particles inside our box (volume element) of dimension Δl so that the motion of particles within the volume element can always be considered isotropic, then the effect of the collisions of particles against the walls of the volume element (which may be rigid walls, or "soft" walls of neighbouring ensembles of particles) is to exert an isotropic "pressure" against all walls. Since isotropy is maintained by particle–particle collisions within the volume element, we may "mathematise" this criterion as,

$$\delta l \ll \Delta l < \mathcal{L}, \tag{1.4}$$

where δl is the mean free path (collision length) of the particles, Δl is the length of one side of our cubic volume element containing an arbitrarily large number of particles, and $\mathcal{L}$ is the smallest length scale of interest in our physical problem. If Ineq. (1.4) holds, we say the ensemble of particles behaves as a *fluid* or a *continuum*. This assumption is an important one; it allows us to treat the applied forces resulting from collisions – which otherwise could be *extremely* difficult to deal with – in a very simple way, namely as an isotropic "pressure".

1.2 A quick review of kinetic theory

To now, I have been enclosing the word *pressure* in quotation marks. This is because I haven't yet made the logical connection between particle collisions (and more specifically, the momentum transferred by particle collisions) and what we commonly think of as *pressure*, such as the *barometric* pressure of the air. So, before we examine how Eq. (1.1)–(1.3) become the equations of hydrodynamics (HD) under the assumption that the ensemble of particles behaves as a fluid (when Ineq. 1.4 is valid), let us review how the "pressure" and the "temperature" of a fluid relate to properties of the ensemble of particles. These ideas form the basis of *kinetic theory*, often exposed to students for the first time in a first-year physics course.[1]

Consider a cube whose edges of length Δl are aligned with the x-, y-, and z-axes of a Cartesian coordinate system, as depicted in Fig. 1.1. Returning to our example in the previous section, suppose a single point particle of mass m moves inside the cube with velocity $v_x \hat{x}$ and collides with the wall whose normal is $+\hat{x}$. If collisions are all elastic, then the particle reflects from the wall with a velocity $-v_x \hat{x}$ and thus suffers a change in momentum of $\Delta S_x = -2mv_x$. Conservation of momentum then demands that an impulse of $+2mv_x$ be imparted against the wall. At a time $\Delta t = 2\Delta l / v_x$ later, the same particle again collides with the wall, imparting another impulse of $+2mv_x$ against it. Thus, the rate at which momentum is delivered to the

---

[1] For example, Halliday, Resnick, & Walker (2003).

wall by a single particle is given by,

$$\frac{\Delta S_x}{\Delta t} = \frac{2mv_x}{2\Delta l/v_x} = \frac{mv_x^2}{\Delta l} = \langle F \rangle,$$

where $\langle F \rangle$ is the average force felt by the wall. Thus, the average pressure exerted by this one particle, defined as force per unit area, is given by,

$$\langle p \rangle = \frac{\langle F \rangle}{(\Delta l)^2} = \frac{mv_x^2}{V},$$

where $V = (\Delta l)^3$ is the volume of the cube. For $\mathcal{N}$ particles, we simply add over all particles:

$$p \equiv \sum_{i=1}^{\mathcal{N}} \langle p_i \rangle = \sum_{i=1}^{\mathcal{N}} \frac{mv_{x,i}^2}{V} = \frac{m}{V} \sum_{i=1}^{\mathcal{N}} v_{x,i}^2 = \frac{m\mathcal{N}}{V} \langle v_x^2 \rangle, \tag{1.5}$$

where each point particle is assumed to have the same mass, m, and where $\langle v_x^2 \rangle = \sum v_{x,i}^2/\mathcal{N}$ is the arithmetic mean of the squares of the particle velocities.

For any given particle, $v^2 = v_x^2 + v_y^2 + v_z^2$ and, for large $\mathcal{N}$, one would expect $\langle v_x^2 \rangle = \langle v_y^2 \rangle = \langle v_z^2 \rangle$ since one Cartesian direction shouldn't be favoured over another. Thus,

$$\langle v^2 \rangle = \langle v_x^2 \rangle + \langle v_y^2 \rangle + \langle v_z^2 \rangle = 3\langle v_x^2 \rangle, \tag{1.6}$$

and Eq. (1.5) becomes,

$$p = \frac{\mathcal{N} m v_{\text{rms}}^2}{3V}, \tag{1.7}$$

where,

$$v_{\text{rms}} \equiv \sqrt{\langle v^2 \rangle},$$

is the *root-mean-square* (rms) speed of the particles in the volume V. Comparing Eq. (1.7) with the *ideal gas law*:

$$p = \frac{\mathcal{N} k_{\text{B}} T}{V}, \tag{1.8}$$

(where $k_{\text{B}} = 1.3807 \times 10^{-23}$ J K^{-1} is the Boltzmann constant) yields:

$$T = \frac{mv_{\text{rms}}^2}{3k_{\text{B}}} \quad \Rightarrow \quad \frac{3}{2} k_{\text{B}} T = \frac{1}{2} m v_{\text{rms}}^2 = \langle K \rangle, \tag{1.9}$$

where $\langle K \rangle$ is the average kinetic energy per point particle. Thus, while the *pressure*, p, is a measure of the rate at which momentum is transferred from the particles of the fluid (gas) to, for example, the diaphragm of the measuring device (barometer), the *temperature* (or more precisely $3k_{\text{B}}T/2$) is a measure of the average kinetic energy of the particles.

The randomly directed kinetic energy of a system of $\mathcal{N}$ particles is called its *internal energy*, E, and, for the point particles under discussion, is given by,

$$E = \mathcal{N} \langle K \rangle = \frac{3}{2} \mathcal{N} k_{\text{B}} T.$$

The factor 3/2 is significant and warrants comment. A point particle, as may be found exclusively in a monatomic gas, has three *degrees of freedom* of motion, namely *translation* in each of the three Cartesian directions (Fig. 1.2, left).

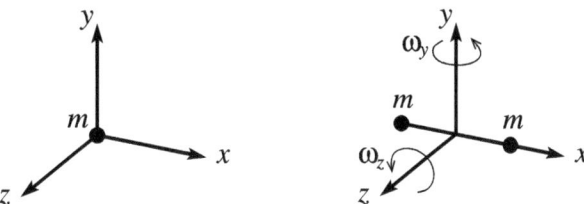

Figure 1.2. A point particle (left) has three degrees of freedom for movement, while a "dumb-bell" (right) has five.

From Eq. (1.6), we have $\langle v_i^2 \rangle = \langle v^2 \rangle / 3$ for $i = x, y, z$, and thus to each (translational) degree of freedom we can associate an internal energy $E_i = \mathcal{N} k_B T / 2$, where $E = E_x + E_y + E_z = 3 E_i$.

Now, a diatomic molecule (essentially two point masses connected by a massless rod) has the same three translational degrees of freedom as a monatomic particle *plus* two *rotational* degrees of freedom, namely rotation about each of the two principle axes orthogonal to its own axis (the x-axis in Fig. 1.2, right), for a total of five degrees of freedom.[2] Note that spinning about the x-axis itself does not constitute a degree of freedom as the moment of inertia about this axis is essentially zero. Because of the *principle of equipartition*,[3] each degree of freedom stores the same amount of kinetic energy, and the internal energy of a diatomic gas must be,

$$E = \frac{5}{2} \mathcal{N} k_B T.$$

Thus, in general, we write,

$$E = \frac{1}{\gamma - 1} \mathcal{N} k_B T, \qquad (1.10)$$

where $\gamma = 5/3$ for a monatomic gas, $\gamma = 7/5$ for a diatomic gas, and $4/3 \leq \gamma < 7/5$ for molecules more complex than diatomic.[4] One can show that $\gamma = C_P / C_V$, the ratio of specific heats of the gas, and that for an adiabatic gas (where heat is neither lost nor gained from the system), $p \propto \rho^\gamma$, where ρ is the *mass density* of the gas.

Dividing Eq. (1.10) by the volume of the sample and using Eq. (1.8) gives an expression for the *internal energy density*, e:

$$e = \frac{E}{V} = \frac{1}{\gamma - 1} \frac{\mathcal{N} k_B T}{V} = \frac{p}{\gamma - 1}.$$

Thus, an alternate form of the ideal gas law, and the form most frequently used in

---

[2] In principle, there are also two vibrational degrees of freedom which, at "ordinary temperatures", statistical mechanics tells us are insignificant.

[3] Left to their own devices, systems will distribute the available energy equally among all possible ways energy can be stored. Thus, for a large number of diatomic molecules randomly colliding with each other and the walls of their container, one would not expect $m \langle v_x^2 \rangle$ to differ significantly from $m \langle v_y^2 \rangle$ or $m \langle v_z^2 \rangle$ any more than it should differ from $I_y \langle \omega_y^2 \rangle$ or $I_z \langle \omega_z^2 \rangle$, where I_y and I_z are the moments of inertia about the y- and z-axes respectively.

[4] Polyatomic molecules are significantly more complex than diatomic molecules, and the full power of statistical mechanics along with a tensor treatment of its moment of inertia are required to explain the value of γ for any individual molecule.

hydrodynamics, is,

$$p = (\gamma - 1)e, \quad (1.11)$$

which states that the rate at which momentum is transferred via collisions is proportional to the average kinetic energy density (*i.e.*, per unit volume) of the random particle motion.

Possibly the second most frequently used form of the ideal gas law in hydrodynamics is,

$$p = \frac{\rho k_B T}{m}, \quad (1.12)$$

which follows directly from Eq. (1.8) noting that $\rho = \mathcal{N}m/V$. Finally, from Eq. (1.9) [and replacing the '3' with $2/(\gamma-1)$], we find:

$$v_{\text{rms}} = \sqrt{\frac{2k_B T}{(\gamma-1)m}} = \sqrt{\frac{2p}{(\gamma-1)\rho}}. \quad (1.13)$$

Thus, the rms speed goes as the square root of the temperature. We shall encounter another characteristic speed of the gas proportional to the square root of the temperature in §2.1.1, namely the sound speed, c_s. Indeed, c_s and v_{rms} arise from essentially the same physics, as will be explained when the sound speed is properly introduced.

1.3 The equations of ideal hydrodynamics

In hydrodynamics, the adjective *ideal* means that internal dissipative forces such as viscosity are ignored. A fluid without (with) viscosity is said to be *inviscid* (*viscid*). In this chapter, our discussion is exclusively restricted to inviscid flow. Viscid flow is more the realm of terrestrial HD (though there are important applications for astrophysical fluids as well), and is covered in some depth in Chap. 8.

We begin our discussion by defining the adjectives *extensive* and *intensive*. Variables such as mass, volume, and energy which are *proportional* to the amount of substance being measured are *extensive* quantities, while mass density (often just referred to as density), energy density, and temperature are *independent* of the amount of substance being studied and are examples of *intensive* quantities.

To give a precise relationship between extensive and intensive quantities, consider a small sample of substance with volume ΔV. For every extensive quantity, $Q(V,t)$, of that sample, we can define a corresponding intensive quantity, $q(\vec{r},t)$, such that,

$$q(\vec{r},t) = \lim_{\Delta V \to 0} \frac{\Delta Q(V,t)}{\Delta V} = \frac{\partial Q(V,t)}{\partial V}. \quad (1.14)$$

This is a *microscopic* description of the system; q may well change from point to point. A *macroscopic* description of the system can be obtained by integrating Eq. (1.14) over a finite volume, V, to recover Q:

$$Q(V,t) = \int_V q(\vec{r},t)\, dV. \quad (1.15)$$

Note that Eq. (1.15) requires that q be an *integrable* function of the coordinates over the volume V, and thus q can be discontinuous and have poles of order less than unity. On the other hand, Eq. (1.14) requires that Q be a *differentiable* function of V, and thus it must be both continuous and free from any poles of any order. Evidently, differentiability is a more restrictive requirement than integrability, and this observation will have important consequences as we develop the theory further.

We're now ready to introduce and prove a theorem that provides a particularly simple way to derive the equations of hydrodynamics from the conservation laws of Eq. (1.1)–(1.3).

Theorem 1.1. Theorem of hydrodynamics.[5] *If the time dependence of an extensive quantity, Q, is given by:*

$$\frac{dQ}{dt} = \Sigma, \tag{1.16}$$

where Σ represents the possibly time-dependent "source terms" (reasons for Q not being "conserved"), then the evolution equation for the corresponding intensive quantity, $q(\vec{r}, t)$, is given by,

$$\frac{\partial q}{\partial t} + \nabla \cdot (q\vec{v}) = \sigma, \tag{1.17}$$

where $\vec{v} = d\vec{r}/dt$, $Q = \int_V q \, dV$, $\Sigma = \int_V \sigma \, dV$, and where the product $q\vec{v}$ must be a differentiable function of the coordinates.

Proof:

$$\frac{dQ}{dt} = \Sigma \quad \Rightarrow \quad \frac{d}{dt} \int_V q \, dV = \int_V \sigma \, dV,$$

where, in general, the volume element $V = V(t)$ also varies in time. Thus, using the standard definition of the derivative,

$$\frac{d}{dt} \int_{V(t)} q \, dV = \lim_{\Delta t \to 0} \frac{1}{\Delta t} \left[\int_{V(t+\Delta t)} q(\vec{r}, t+\Delta t) dV - \int_{V(t)} q(\vec{r}, t) dV \right]$$

$$= \lim_{\Delta t \to 0} \frac{1}{\Delta t} \left[\int_{V(t+\Delta t) - V(t)} q(\vec{r}, t+\Delta t) dV \right.$$

$$\left. + \int_{V(t)} q(\vec{r}, t+\Delta t) dV - \int_{V(t)} q(\vec{r}, t) dV \right]$$

$$= \lim_{\Delta t \to 0} \frac{1}{\Delta t} \int_{\Delta V} q(\vec{r}, t+\Delta t) dV$$

$$+ \lim_{\Delta t \to 0} \frac{1}{\Delta t} \int_{V(t)} [q(\vec{r}, t+\Delta t) - q(\vec{r}, t)] dV,$$

---

[5] This theorem is a variant of *Reynolds' transport theorem*, a volume-integral application of the Leibniz formula for the derivative of an integral.

where, as shown in the inset, performing the volume integral over the difference in volumes, $\Delta V = V(t+\Delta t) - V(t)$, is the same as integrating over the closed surface, ∂V, using a volume differential given by $dV = (\vec{v}\Delta t) \cdot (\hat{n} dA)$. Thus,

$$\frac{d}{dt}\int_{V(t)} q\, dV = \lim_{\Delta t \to 0} \frac{1}{\Delta t} \oint_{\partial V} q(\vec{r}, t+\Delta t)(\vec{v}\Delta t) \cdot (\hat{n} dA)$$

$$+ \int_{V(t)} \lim_{\Delta t \to 0} \frac{q(\vec{r}, t+\Delta t) - q(\vec{r}, t)}{\Delta t} dV$$

$$= \oint_{\partial V} q(\vec{r}, t)\vec{v} \cdot \hat{n}\, dA + \int_{V(t)} \frac{\partial q(\vec{r}, t)}{\partial t} dV$$

$$= \int_{V(t)} \nabla \cdot (q(\vec{r}, t)\vec{v}) dV + \int_{V(t)} \frac{\partial q(\vec{r}, t)}{\partial t} dV \quad \text{(Gauss; Eq. A.30)}$$

$$= \int_{V(t)} \left(\frac{\partial q(\vec{r}, t)}{\partial t} + \nabla \cdot (q(\vec{r}, t)\vec{v}) \right) dV = \int_{V(t)} \sigma(\vec{r}, t) dV$$

$$\Rightarrow \int_V \left(\frac{\partial q}{\partial t} + \nabla \cdot (q\vec{v}) - \sigma \right) dV = 0.$$

As this is true for any V, the integrand must be zero, proving the theorem. □

Note that q is not the conserved quantity, Q is (at least to within a known source term, Σ). However, since Q is the volume-integral of q, we'll refer to q as a *volume-conserved quantity*.

The quantity $q\vec{v} \equiv \vec{f}_Q$ is the *advective flux density* of Q whose units are those of Q times m^{-2} s^{-1}; this will require a little unpacking. The flux,[6] $\mathcal{F}_Q$, of a vector field, $\vec{f}_Q$, is a measure of how much $\vec{f}_Q$ "passes through" a given surface area with arbitrary normal, $\hat{n}$. Mathematically,

$$\mathcal{F}_Q = \oint_S \vec{f}_Q \cdot \hat{n}\, dA \quad \text{or} \quad \mathcal{F}_Q = \int_\Sigma \vec{f}_Q \cdot \hat{n}\, dA, \quad (1.18)$$

depending on whether the surface is closed (S) or open (Σ) respectively. Thus, the units of $\vec{f}_Q$ are those of $\mathcal{F}_Q$ per unit area, and $\vec{f}_Q$ can also be interpreted as a *flux density* of $\mathcal{F}_Q$. And so, $\mathcal{F}_Q$ is the flux of $\vec{f}_Q$ while $\vec{f}_Q$ is the flux *density* of $\mathcal{F}_Q$.

An *advective* flux density is more specific to fluid dynamics and refers to some quantity, Q, being *advected* (*i.e.*, transported) by the flow across a surface at a certain rate. Thus, while $\vec{f}_Q$ is the *flux density* of $\mathcal{F}_Q$ with units of $\mathcal{F}_Q$ per unit area, $\vec{f}_Q = q\vec{v}$ is also the *advective* flux density of Q – the volume integral of q – with units of Q per unit area per unit time. It is the "per unit time" part that triggers the adjective *advective*.

Evidently, we have four different types of "fluxes" to keep straight (flux, flux

---

[6] From the Latin *fluxus* or "flow", this term was introduced to physics by Sir Isaac Newton.

density, advective flux, advective flux density) and the literature seems to blur all four; often you'll find any or all of these terms used interchangeably. In this text, while I maintain the distinction between *flux* and *flux density*, I've chosen to drop the adjective *advective* to simplify the language a bit, relying instead on context. If a particular flux/flux density has a "per unit time" aspect to it, it is an *advective* flux/flux density; otherwise just flux/flux density.

Last point before getting to the equations of HD: Eq. (1.16) is an *integral* equation (Q and Σ both being volume integrals of intensive quantities, q and σ), and thus represents a *global* statement (valid over a finite sample of the fluid) on the conservation of the extensive quantity, Q. On the other hand, Eq. (1.17) is a *differential* equation (often referred to as the *differential form* of Eq. 1.16) and thus represents a *local* statement (valid at a point) on the conservation of Q, involving the corresponding intensive quantity, q. *Global and local forms of an equation are not identical.* Because differential equations require the functions to be differentiable, solutions of the differential form of the equations can be more restrictive than those of the integral form where functions need only be integrable. More on this in §1.5.

Example 1.1. Let $Q = M$, the mass of the sample of fluid. Find the evolution equation for the corresponding intensive quantity, $q = \rho$ (mass density).

Solution: From Eq. (1.1), $\Sigma = 0 \Rightarrow \sigma = 0$, and Theorem 1.1 requires that:

$$\boxed{\frac{\partial \rho}{\partial t} + \nabla \cdot (\rho \vec{v}) = 0.} \tag{1.19}$$

This is the *continuity equation*; the first equation of HD. □

Example 1.2. Let $Q = E_T$, the total energy of the fluid sample[7] of mass M:

$$E_T = E + \frac{1}{2}Mv^2 + M\phi,$$

where E is the internal (thermal) energy and ϕ is the gravitational potential. Find the evolution equation for the corresponding intensive quantity, the *total energy density*, namely,

$$e_T = e + \frac{1}{2}\rho v^2 + \rho\phi, \tag{1.20}$$

where once again, e is the internal energy density, whose units $\mathrm{J\,m^{-3}} = \mathrm{N\,m^{-2}}$ are the same as those for pressure, as expected from Eq. (1.11).

Solution: From Eq. (1.2), $\Sigma = \mathcal{P}_{\mathrm{app}} \Rightarrow \sigma = p_{\mathrm{app}}$, the *applied power density* interpreted as the rate at which work is done on a unit volume of the fluid sample by all *applied* forces. Thus, Theorem 1.1 implies:

$$\frac{\partial e_T}{\partial t} + \nabla \cdot (e_T \vec{v}) = p_{\mathrm{app}}. \tag{1.21}$$

---
[7]When we introduce magnetism in Chap. 4, we'll add a magnetic term to E_T.

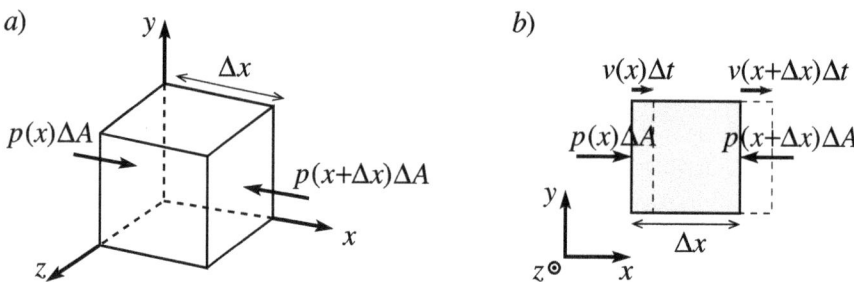

Figure 1.3. *a)* A cube of edge length Δx with external pressure forces acting on the x-faces indicated. *b)* An x–y cut through the cube in panel (*a*) showing both the pressure forces on and motion of the x-faces.

As discussed in §1.1, applied forces are collisions of external particles with the fluid sample. Thus, the applied power is the rate at which work is done on the fluid sample by the external fluid as the former expands or contracts within the latter.

To find an expression for the applied power, $\mathcal{P}_{\text{app}}$, consider a small cube of fluid with dimension Δx in the x-direction and cross-sectional area $\Delta A = \Delta V/\Delta x$ (Fig. 1.3). The pressure force exerted on the left face of the cube is $F(x) = +p(x)\Delta A$ and, in time Δt, the left face is displaced by $v_x(x)\Delta t$. Thus, the work done on the left face by the external fluid is $\Delta W_{\text{L}} = +p(x)\,v_x(x)\Delta t\,\Delta A$. Similarly, the work done on the right face is $\Delta W_{\text{R}} = -p(x+\Delta x)\,v_x(x+\Delta x)\Delta t\,\Delta A$ [since $p(x+\Delta x)$ and $v(x+\Delta x)$ are oppositely directed; Fig. 1.3*b*], and the net work done on the fluid cube is:

$$\Delta W = \Delta W_{\text{L}} + \Delta W_{\text{R}} = p(x)\,v_x(x)\Delta t\,\Delta A - p(x+\Delta x)\,v_x(x+\Delta x)\Delta t\,\Delta A$$
$$= p(x)\,\delta V(x) - p(x+\Delta x)\,\delta V(x+\Delta x),$$

where $\delta V(x)$ [$\delta V(x+\Delta x)$] is the small volume change on the left [right] face of the cubic sample of volume ΔV by virtue of the motion of the left [right] face. Because of its form, this work is frequently referred to as the "pdV term".

Dividing ΔW by Δt gives us the applied power,

$$\mathcal{P}_{\text{app}} = \frac{\Delta W}{\Delta t} = -\Delta A\,\Delta x\,\frac{p(x+\Delta x)\,v_x(x+\Delta x) - p(x)\,v_x(x)}{\Delta x} = -\Delta V\,\frac{\Delta(pv_x)}{\Delta x},$$

and thus the applied power density is given by:

$$p_{\text{app}} = \frac{\mathcal{P}_{\text{app}}}{\Delta V} = -\frac{\Delta(pv_x)}{\Delta x}.$$

Taking into account similar terms in the y- and z-directions, and letting $\Delta \to \partial$, we have:

$$p_{\text{app}} = -\nabla \cdot (p\vec{v}). \tag{1.22}$$

Substituting Eq. (1.22) into Eq. (1.21) yields:

$$\frac{\partial e_{\text{T}}}{\partial t} + \nabla \cdot (e_{\text{T}}\vec{v}) = -\nabla \cdot (p\vec{v}),$$

$$\Rightarrow \boxed{\frac{\partial e_T}{\partial t} + \nabla \cdot \left((e_T + p)\vec{v}\right) = 0,} \tag{1.23}$$

the *total energy equation* and the second equation of ideal HD. □

Example 1.3. Let $Q = \vec{S}$, the *total momentum* of the fluid sample. Find the evolution equation for the corresponding intensive quantity, $q = \vec{s} = \rho\vec{v}$ (the *momentum density*).

Solution: From Eq. (1.3), $\Sigma = \sum \vec{F}_{\text{ext}} \Rightarrow \sigma = \sum \vec{f}_{\text{ext}}$, the *external force densities*. Thus, Theorem 1.1 requires that:

$$\frac{\partial \vec{s}}{\partial t} + \nabla \cdot (\vec{s}\vec{v}) = \sum \vec{f}_{\text{ext}}, \tag{1.24}$$

where the Cartesian representation of the divergence term is:

$$\nabla \cdot (\vec{s}\vec{v}) = \left(\nabla \cdot (s_x \vec{v}), \nabla \cdot (s_y \vec{v}), \nabla \cdot (s_z \vec{v})\right).$$

(See §A.4 for other orthogonal coordinate systems.)

For now, we will limit the external force densities to terms arising from pressure gradients and gravity. In Chap. 4, we'll add the Lorentz force, in Chap. 8 viscous stress, and in Chap. 10, drag forces exerted between ions and neutral particles. Starting with the pressure gradient, consider once again the small cube of fluid with edge length Δx and face area ΔA in Fig. 1.3a. If the pressure at the left and right sides of the cube are respectively $p(x)$ and $p(x + \Delta x)$, then the net pressure force acting on the cube in the x-direction is given by:

$$F(x + \Delta x) + F(x) = -p(x + \Delta x)\Delta A + p(x)\Delta A = -\frac{\Delta p}{\Delta x}\Delta A \Delta x = -\frac{\Delta p}{\Delta x}\Delta V.$$

Thus, the pressure force density in the x-direction is:

$$f_x = \frac{\Delta F_x}{\Delta V} = -\frac{\Delta p}{\Delta x} \to -\frac{\partial p}{\partial x} \quad \text{as } \Delta x \to 0.$$

Accounting for all three components,

$$\vec{f}_p = -\nabla p. \tag{1.25}$$

The gravitational force density, $\vec{f}_\phi$, is even simpler to derive. If the fluid sample has mass ΔM, then the gravitational force on ΔM is $-\Delta M \nabla \phi$, where ϕ is the local gravitational potential arising from all external masses, including other regions of fluid and distant or embedded point masses (*e.g.*, stars). Thus, $\vec{f}_\phi$ is given by:

$$\vec{f}_\phi = -\frac{\Delta M \nabla \phi}{\Delta V} \to -\rho \nabla \phi \quad \text{as } \Delta V \to 0. \tag{1.26}$$

Substituting both Eq. (1.25) and (1.26) into Eq. (1.24) yields the *momentum equation*, the third and final equation of ideal HD:

$$\boxed{\frac{\partial \vec{s}}{\partial t} + \nabla \cdot (\vec{s}\vec{v}) = -\nabla p - \rho \nabla \phi.} \quad \square \tag{1.27}$$

Summary of §1.3: Equations (1.19), (1.23), and (1.27) constitute two scalar equations and one vector equation which, when combined with Eq. (1.11), (1.20), and $\vec{s} = \rho\vec{v}$ (the *constitutive equations*), provide a closed system of equations for the *fluid flow variables*, namely the volume-conserved quantities ρ, $\vec{s}$, and e_T. This suite of equations comprises our first set of equations of ideal hydrodynamics:

Equation Set 1:

$$\frac{\partial \rho}{\partial t} + \nabla \cdot (\rho \vec{v}) = 0; \quad \text{continuity}$$

$$\frac{\partial e_T}{\partial t} + \nabla \cdot \big((e_T + p)\vec{v}\big) = 0; \quad \text{total energy equation}$$

$$\frac{\partial \vec{s}}{\partial t} + \nabla \cdot (\vec{s}\vec{v}) = -\nabla p - \rho \nabla \phi; \quad \text{momentum equation}$$

$$e_T = e + \frac{1}{2}\rho v^2 + \rho\phi; \quad \text{constitutive equation 1}$$

$$p = (\gamma - 1)e; \quad \text{constitutive equation 2}$$

$$\vec{s} = \rho\vec{v}. \quad \text{constitutive equation 3}$$

The gravitational potential, ϕ, is computed by adding up all the potentials of the contributing point masses, and/or by computing the self-gravitational potential of the gas from the density distribution from Poisson's equation:

$$\nabla^2 \phi = 4\pi G \rho. \tag{1.28}$$

As a PDE, Poisson's equation is qualitatively different from the equations of hydrodynamics. It has no time derivative, spatial derivatives are second order, and Poisson's equation is an example of an *elliptical* PDE rather than the *hyperbolic* PDEs of HD (App. C). Analytical methods for solving Poisson's equation can be found in any intermediate or advanced text on electrodynamics (*e.g.*, Paris & Hurd, 1969; Lorrain & Corson, 1970; Jackson, 1975 to suggest a few), while numerical treatments can be found in widely available resources such as *Numerical Recipes* (Press *et al.*, 1992). We shall not address such methods in this text.

1.4 The internal energy density

Equation (1.23) governs the evolution of the total energy density, e_T. We can eliminate the need for the first constitutive equation by finding an evolution equation for the internal energy density, e, alone, and our approach shall be via thermodynamics.

The combined first and second law of thermodynamics is:

$$TdS = dE + p\,dV, \tag{1.29}$$

where the only new variable being introduced is S, the total *entropy* of the fluid

sample.[8] If the mass of the sample, $M = \rho V$, is fixed, then:
$$dV = -\frac{M}{\rho^2}\,d\rho,$$
and Eq. (1.29) becomes:
$$T\,dS = dE - \frac{Mp}{\rho^2}\,d\rho. \tag{1.30}$$

Define $s \equiv S/M$ and $\varepsilon \equiv E/M$ to be the *specific* entropy and *specific* internal energy respectively. As the term *density* is used to connote *per unit volume*, so the term *specific* is used to connote *per unit mass*. Note, for example, that the specific internal energy and internal energy density are related by $e = \rho\varepsilon$, and while $e \propto p$, the pressure (Eq. 1.11), $\varepsilon \propto T$, the temperature.

With these definitions, Eq. (1.30) becomes,
$$\frac{d\varepsilon}{dt} - \frac{p}{\rho^2}\frac{d\rho}{dt} = T\frac{ds}{dt}, \tag{1.31}$$
where I've divided through by the differential dt to obtain an expression relating time derivatives.

Now, because $\varepsilon = e/\rho$, we have:
$$\frac{d\varepsilon}{dt} = \frac{1}{\rho}\frac{de}{dt} - \frac{e}{\rho^2}\frac{d\rho}{dt}.$$

Further, from continuity (Eq. 1.19) and use of the chain rule for partial derivatives, we have:
$$\frac{\partial \rho}{\partial t} + \vec{v}\cdot\nabla\rho + \rho\nabla\cdot\vec{v} = \frac{d\rho}{dt} + \rho\nabla\cdot\vec{v} = 0 \quad\Rightarrow\quad \frac{d\rho}{dt} = -\rho\nabla\cdot\vec{v}.$$

Substituting these into Eq. (1.31) yields:
$$\frac{1}{\rho}\frac{de}{dt} + \frac{p+e}{\rho}\nabla\cdot\vec{v} = T\frac{ds}{dt}. \tag{1.32}$$

Another invocation of the chain rule gives us:
$$\frac{de}{dt} = \frac{\partial e}{\partial t} + \vec{v}\cdot\nabla e,$$
and thus, upon multiplying through by ρ, Eq. (1.32) becomes:
$$\frac{\partial e}{\partial t} + \vec{v}\cdot\nabla e + e\nabla\cdot\vec{v} = -p\nabla\cdot\vec{v} + \rho T\frac{ds}{dt}$$
$$\Rightarrow\quad \frac{\partial e}{\partial t} + \nabla\cdot(e\vec{v}) = -p\nabla\cdot\vec{v} + p\frac{m}{k_{\rm B}}\frac{ds}{dt},$$
where the ideal gas law (1.12) has been used to replace ρT with $pm/k_{\rm B}$. Finally, by defining the *unitless entropy per particle*, $\mathcal{S} \equiv ms/k_{\rm B}$, we obtain:
$$\frac{\partial e}{\partial t} + \nabla\cdot(e\vec{v}) = -p\left(\nabla\cdot\vec{v} - \frac{d\mathcal{S}}{dt}\right). \tag{1.33}$$

---
[8]Unavoidably, S is an over-used symbol. It has already been defined and indeed is used throughout the text as the closed surface of integration. Here, it is being used to represent the total entropy of the fluid (an extensive quantity), while the vector $\vec{S}$ represents the total momentum.

For an adiabatic process, the entropy per particle remains constant, and we arrive at our final form for the evolution equation for internal energy density:

$$\boxed{\frac{\partial e}{\partial t} + \nabla \cdot (e \vec{v}) = -p \nabla \cdot \vec{v}.} \tag{1.34}$$

Note that Eq. (1.34) can replace Eq. (1.20) and (1.23), thus giving rise to a somewhat simpler set of hydrodynamical equations:

Equation Set 2:

$$\frac{\partial \rho}{\partial t} + \nabla \cdot (\rho \vec{v}) = 0; \qquad \text{continuity}$$

$$\frac{\partial e}{\partial t} + \nabla \cdot (e \vec{v}) = -p \nabla \cdot \vec{v}; \qquad \text{internal energy equation}$$

$$\frac{\partial \vec{s}}{\partial t} + \nabla \cdot (\vec{s}\vec{v}) = -\nabla p - \rho \nabla \phi; \qquad \text{momentum equation}$$

$$p = (\gamma - 1)e; \qquad \text{constitutive equation 2}$$

$$\vec{s} = \rho \vec{v}. \qquad \text{constitutive equation 3}$$

1.5 Primitive, integral, and conservative form

For a so-called *barotropic* gas (where p is a function of ρ only; both adiabatic and isothermal gases are examples of *barotropes*), it is left to Problem 1.5 to derive the so-called *pressure equation*:

$$\boxed{\frac{\partial p}{\partial t} + \vec{v} \cdot \nabla p = -\rho \frac{dp}{d\rho} \nabla \cdot \vec{v}.} \tag{1.35}$$

It is further left to Problem 1.2 to show that the continuity equation, (Eq. 1.19), and the momentum equation, (Eq. 1.27), combine to yield an evolution equation for the velocity:

$$\boxed{\frac{\partial \vec{v}}{\partial t} + (\vec{v} \cdot \nabla) \vec{v} = -\frac{1}{\rho} \nabla p - \nabla \phi,} \tag{1.36}$$

where, in Cartesian coordinates, we have:

$$(\vec{v} \cdot \nabla) \vec{v} = (\vec{v} \cdot \nabla v_x, \vec{v} \cdot \nabla v_y, \vec{v} \cdot \nabla v_z).$$

(See §A.4 for other orthogonal coordinate systems.) Equation (1.36) is known as *Euler's equation* named for Leonhard Euler (1707–1783), the Swiss mathematician and physicist often described as *the* most prolific mathematician of all time.[9]

---

[9] www.wikipedia.org/wiki/Leonhard_Euler

Collecting Eq. (1.35) and (1.36) with the continuity equation, (Eq. 1.19), gives us our third set of HD equations:

Equation Set 3:

$$\frac{\partial \rho}{\partial t} + \nabla \cdot (\rho \vec{v}) = 0; \qquad \text{continuity}$$

$$\frac{\partial p}{\partial t} + \vec{v} \cdot \nabla p = -\rho \frac{dp}{d\rho} \nabla \cdot \vec{v}; \qquad \text{pressure equation}$$

$$\frac{\partial \vec{v}}{\partial t} + (\vec{v} \cdot \nabla) \vec{v} = -\frac{1}{\rho} \nabla p - \nabla \phi. \qquad \text{Euler's equation}$$

These three equations form a closed set; no constitutive equations are necessary. Equation Set 3 is said to be in *primitive form* because it governs the time evolution of the three so-called *primitive variables* ρ, p, and $\vec{v}$.

Finally, one can write down the equations of ideal HD in *integral form* by performing volume integrals on each term of Eq. Set 1, this time setting $\phi = 0$. This yields a set of *integro-differential* equations highly reminiscent of the fundamental conservation laws (Eq. 1.1–1.3) upon which our current discussion is based. Accordingly, I have designated these as *Equation Set 0*:

Equation Set 0: Integral Equations of Ideal HD

$$\frac{\partial M}{\partial t} + \oint_S \rho \vec{v} \cdot \hat{n} \, d\sigma = 0; \qquad (1.37)$$

$$\frac{\partial E_\mathrm{T}}{\partial t} + \oint_S (e_\mathrm{T} + p) \vec{v} \cdot \hat{n} \, d\sigma = 0; \qquad (1.38)$$

$$\frac{\partial \vec{s}}{\partial t} + \oint_S \rho(\vec{v}\vec{v}) \cdot \hat{n} \, d\sigma = -\oint_S p\hat{n} \, d\sigma, \qquad (1.39)$$

---

with constitutive equations:

$$M = \int_V \rho \, dV; \quad E_\mathrm{T} = \int_V e_\mathrm{T} \, dV; \quad e_\mathrm{T} = \frac{p}{\gamma - 1} + \frac{\rho v^2}{2}; \quad \text{and} \quad \vec{s} = \int_V \rho \vec{v} \, dV,$$

and where S is the surface (not entropy!) enclosing the volume element V. Because all spatial derivatives in Eq. Set 1 are either perfect divergences or perfect gradients, their volume integrals can be replaced with surface integrals by the use of Gauss' theorem (Eq. A.30 and A.31). *Note that the same cannot be done with the internal energy equation in Eq. Set 2 and the pressure and Euler's equations in Eq. Set 3 because of the imperfect divergences and gradients in these equations.*

The integral form in Eq. Set 0 completely exposes the three conservation laws upon which fluid dynamics is based. For each equation, the time rate of change of the extensive quantity within a given volume element, V, however large or small is determined completely by the (advective) *flux density* of that quantity (integrand of the surface integral) passing through the closed surface, S.[10] Since Eq. Set 0 follows so directly from Eq. Set 1, Eq. Set 1 is said to be in *conservative form*.

The distinction between conservative and primitive forms is more than semantic. To be valid, the primitive equations require that the primitive variables p, $\vec{v}$, and thus ρ be individually differentiable – and therefore continuous – everywhere. This necessarily precludes discontinuities in ρ, p, and $\vec{v}$ and thus the primitive equations are valid only for *smooth flow*. On the other hand, the conservative equations only require that the functions (flux densities) $\rho\vec{v}$, $\vec{v}(e_T + p)$, and $\rho\vec{v}\vec{v} + p\mathsf{I}$[11] be continuous,[12] and not necessarily the primitive variables individually. Thus, the conservative equations in terms of the conservative variables $(\rho, e_T, \vec{s})$ can, in principle, admit solutions with discontinuities in ρ, p, and $\vec{v}$ (*i.e., discontinuous flow*) so long as these discontinuities combine to yield continuous flux densities. We shall exploit this observation when we write down the *Rankine–Hugoniot jump conditions* in §2.2.3, and then again for MHD in §5.3.

Of course, in addition to discontinuous solutions, the conservative equations also admit all smooth solutions admitted by the primitive equations, and thus the conservative set of equations is the more general of the two. Still, there are times when use of the primitive equations is far more convenient, as we shall see when we discuss the all-important Riemann problem in Chap. 3 and 6.

Problem Set 1

1.1 On a cold winter afternoon, you enter your winter cabin (which has not been heated for weeks) freezing cold. You light a roaring fire in the hearth and after an hour, the cabin is warm enough to take off your winter clothing.

- a) Does the air in your cabin contain more, less, or the same total internal energy, E, now that it is warm than when it was cold? Explain.

- b) If you conclude that the air contains less or the same internal energy after being heated as before, where does all the energy from the fire go?

1.2 Derive Euler's equation (Eq. 1.36 in the text) from the continuity and momentum equations (Eq. 1.19 and 1.27). Your proof should be valid for all coordinate systems, not just Cartesian.

Hint: Vector identity (A.21) from App. A should be particularly helpful.

---

[10] Note that in this picture, the pressure p contributes to the flux densities of both $\vec{S}$ and E_T.

[11] $\vec{v}\vec{v}$ is the *dyadic product* of $\vec{v}$ with itself creating a rank 2 tensor (matrix; see Eq. A.16), while I is the "identity tensor", which you can think of as the identity matrix.

[12] To see how one arrives at the conclusion that the momentum equation, (1.27), only requires that $\rho\vec{v}\vec{v} + p\mathsf{I}$ be continuous, it is instructive to note that formally, $\nabla p = \nabla \cdot (p\mathsf{I})$. Try it!

Problem Set 1

1.3 Derive the internal energy equation for an adiabatic gas (Eq. 1.34 in the text) from the hydrodynamical equations alone by substituting the definition for e_T (Eq. 1.20) into the total energy equation (Eq. 1.23) and then by using the continuity equation (Eq. 1.19) and Euler's equation (Eq. 1.36) to simplify.

Hint: The gravitational potential, ϕ, solves Poisson's equation (Eq. 1.28) and, as such, has no explicit time dependence. Thus, you can set $\partial \phi / \partial t = 0$.

1.4*

a) Equation (1.34) in the text is the evolution equation for the internal energy of an adiabatic gas. Show that the analogous equation for an *isothermal* gas is:
$$\frac{\partial e}{\partial t} + \nabla \cdot (e\,\vec{v}) = 0. \qquad (1.40)$$
Physically, what do you suppose is happening in an isothermal gas to maintain its isothermality?

b) For an adiabatic gas, we argued that the unitless entropy per particle, $\mathcal{S}$, remains constant in time. Find $d\mathcal{S}/dt$ for an isothermal gas.

c) We can model a real gas by an equation of state of the form $p = \kappa \rho^n$ where, in principle, both κ and the power-law index n could vary from point to point. For an adiabatic gas, $n = \gamma$, while for an isothermal gas, $n = 1$ (why?). Argue that for a real gas, $1 < n < \gamma$ and thus the isothermal and adiabatic conditions represent limits in between which a given real gas should be found.

1.5* A *barotropic* equation of state is one where the pressure depends only on the density, that is $p = p(\rho)$.

a) Starting with the internal energy density equation for an adiabatic gas, Eq. (1.34) in the text, show that:
$$\frac{\partial p}{\partial t} + \vec{v} \cdot \nabla p = -\gamma p \nabla \cdot \vec{v}. \qquad (1.41)$$

b) Starting with the continuity equation and assuming a barotropic equation of state, derive the "pressure equation", Eq. (1.35).

c) Show that for an adiabatic gas where $p \propto \rho^\gamma$, Eq. (1.35) reduces to Eq. (1.41).

1.6 The *vorticity* is defined as $\vec{\omega} = \nabla \times \vec{v}$, and is a measure of fluid *circulation*.

a) Starting from either Eq. (1.27) or (1.36) in the text and assuming the fluid to be barotropic (as defined in Problem 1.5), show that the evolution equation for the vorticity is given by:

$$\frac{\partial \vec{\omega}}{\partial t} = \nabla \times (\vec{v} \times \vec{\omega}). \tag{1.42}$$

Hint: Vector identity (A.15) in App. A *might* be of help.

b) If the fluid is not barotropic [*e.g.*, $p = p(\rho, e)$], show that Eq. (1.42) is still valid if the fluid is *incompressible*, that is where the density may be taken as constant in both space and time and thus the continuity equation (Eq. 1.19) reduces to $\nabla \cdot \vec{v} = 0$.

1.7 Define the *circulation*, Γ, of a fluid about a closed loop, C, to be:

$$\Gamma = \oint_C \vec{v} \cdot d\vec{l}.$$

By inspection, Γ is non-zero only if there is net circulation about the loop, whence its name.

a) Show that:

$$\Gamma = \int_\Sigma \vec{\omega} \cdot d\vec{\sigma}, \tag{1.43}$$

where Σ is the open surface enclosed by the closed loop, C, and $\vec{\omega} = \nabla \times \vec{v}$ is the vorticity defined in Problem 1.6. This should be a one-liner. Thus, the circulation, Γ, can also be interpreted as the "vorticity flux" passing through a closed loop.

b) Prove that for a barotropic (Problem 1.5) or incompressible (ρ = constant) fluid,

$$\frac{d\Gamma}{dt} = 0.$$

This is *Kelvin's circulation theorem*, and asserts that vorticity flux is a conserved quantity for inviscid barotropic flow.

Hint: Start with Eq. (1.43) and examine $d\Gamma/dt$, noting that the surface over which the integral is performed, Σ, is also time-dependent; this must somehow be taken into account in taking the time derivative. If this doesn't seem like a familiar problem, review the proof of the theorem of hydrodynamics (Theorem 1.1 in the text). Finally, you should come to a point where Eq. (1.42) from Problem 1.6 would be useful; feel free to use it!

Discussion: As we shall see in Chap. 4, lines of magnetic induction and vortex lines share many properties since both $\vec{\omega}$ and $\vec{B}$ are solenoidal ($\nabla \cdot \vec{\omega} = \nabla \cdot \vec{B} = 0$), and both are governed by an "induction equation" (*cf.* Eq. 1.42 and 4.4). Given that magnetic flux is a conserved quantity, it should then come as no surprise that vorticity flux is also conserved.

An immediate consequence of Kelvin's circulation theorem is that if a barotropic or incompressible fluid starts off with zero vorticity (and thus zero circulation everywhere), it must develop in a such way to maintain zero vorticity. If it didn't, then one could find a patch of area through which $\Gamma \neq 0$, violating Kelvin's theorem. Note that dissipative encounters with walls or introduction of viscosity (numerical or physical) into the fluid, which are not present in Euler's equation used to prove Kelvin's theorem, could cause an initially irrotational fluid to develop vorticity. Otherwise, Kelvin's theorem essentially states that for an inviscid fluid, "once irrotational, always irrotational".

If one can establish that $\vec{\omega} = \nabla \times \vec{v} = 0$ for all time, then the velocity field can be expressed as the gradient of a scalar; $\vec{v} = \nabla \psi$. Such a *velocity potential* can be useful, particularly for incompressible fluids where $\nabla \cdot \vec{v} = 0$ since this means the velocity potential will satisfy Laplace's equation, $\nabla^2 \psi = 0$. In this case, all the mathematics used in problems in electrostatics and, in particular, in *potential theory* can be brought to bear on solving Laplace's equation instead of the much more difficult Euler's equation.

2 Selected Applications of Hydrodynamics

I chatter over stony ways, in little sharps and trebles;
I bubble into eddying bays, I babble on the pebbles...
And out again I curve and flow to join the brimming river;
For men may come and men may go, but I go on forever.

Lord Alfred Tennyson (1809–1892)
from *The Song of the Brook*, 1842

2.1 Sound waves

SO MANY PHENOMENA in physics can be described in terms of waves, and Tennyson's babbling brook is a prime example. Gravity waves[1] ripple the interface between the water and air – two fluids of very different density – and within each fluid, sound waves propagate. For most of us, sound waves are an important method of communication and gaining information about the environment. If one does not see the on-coming truck, for example, its sound will be a clue to get out of the way! Likewise for a fluid, the propagation of sound waves is how one region of the fluid "knows" what the next is doing, and to react accordingly.

In this section, we identify sound waves by finding the wave equations hidden within the fluid equations derived in the previous chapter. We then look at these waves in two very useful ways: first, as solutions to the wave equation itself; then second as an algebraic "eigen-problem", where the eigenvalues and eigenvectors of the governing equations provide all the information we need to understand how waves propagate through the fluid.

2.1.1 Wave equation approach

Consider the barotropic pressure equation (Eq. 1.35) and Euler's equation (Eq. 1.36) in the absence of gravity ($\phi = 0$):

$$\frac{\partial p}{\partial t} + \vec{v} \cdot \nabla p = -\rho p'(\rho) \nabla \cdot \vec{v}; \tag{2.1}$$

$$\frac{\partial \vec{v}}{\partial t} + (\vec{v} \cdot \nabla)\vec{v} = -\frac{1}{\rho}\nabla p, \tag{2.2}$$

---
[1] Not *gravitational waves* which are rather different!

where $p' = dp/d\rho$. Now consider *small perturbations* in the co-moving frame of the fluid to what is otherwise a static state. Thus, let:

$$\rho = \rho_0 + \epsilon\rho_p; \qquad \vec{v} = \epsilon\vec{v}_p; \qquad p = p_0 + \epsilon p_p; \qquad p' = p'_0 + \epsilon p'_p, \qquad (2.3)$$

where ρ_0 and p_0 are constants,[2] where quantities with the subscript 'p' are the perturbations, and where p'_0 is p' evaluated at ρ_0. Thus, $\rho_p \ll \rho_0$, $p_p \ll p_0$, $p'_p \ll p'_0$, and $v_p = |\vec{v}_p| \ll c_s$, where c_s is the sound speed to be introduced later in this subsection. Following common practice in perturbation theory, ϵ is just a label (whose formal value is 1) used to identify second and higher order terms which are systematically ignored in a first-order perturbation analysis.

Substituting Eq. (2.3) into Eq. (2.1) and (2.2) gives:

$$\epsilon\frac{\partial p_p}{\partial t} + \epsilon^2 \vec{v}_p \cdot \nabla p_p = -\epsilon(\rho_0 + \epsilon\rho_p)(p'_0 + \epsilon p'_p)\nabla \cdot \vec{v}_p; \qquad (2.4)$$

$$\epsilon\frac{\partial \vec{v}_p}{\partial t} + \epsilon^2 (\vec{v}_p \cdot \nabla)\vec{v}_p = -\frac{\epsilon}{\rho_0 + \epsilon\rho_p}\nabla p_p, \qquad (2.5)$$

where the fact that ρ_0 and p_0 are constant in both time and space has been used to simplify the derivatives. Further, a binomial expansion yields:

$$\frac{\epsilon}{\rho_0 + \epsilon\rho_p} = \frac{\epsilon}{\rho_0}\left(1 + \epsilon\frac{\rho_p}{\rho_0}\right)^{-1} = \frac{\epsilon}{\rho_0}\left(1 - \epsilon\frac{\rho_p}{\rho_0} + \ldots\right). \qquad (2.6)$$

Substituting Eq. (2.6) into Eq. (2.5), and then dropping all terms of order ϵ^2 in Eq. (2.4) and (2.5), we get:

$$\frac{\partial p}{\partial t} = -\rho_0 p'_0 \nabla \cdot \vec{v}; \qquad (2.7)$$

$$\frac{\partial \vec{v}}{\partial t} = -\frac{1}{\rho_0}\nabla p, \qquad (2.8)$$

setting all remaining factors of ϵ to 1 and dropping all subscripts 'p' (since $\nabla p = \nabla p_p$, *etc.*). Equations (2.7) and (2.8) are, respectively, the *linearised* pressure and Euler equations in the co-moving frame of the fluid. The adjective "linearised" refers to the fact that by restricting the solutions to small perturbations, the one and only non-linear term, $(\vec{v} \cdot \nabla)\vec{v}$, is eliminated.[3]

Taking the time-derivative of Eq. (2.7) and then using Eq. (2.8), we get:

$$\frac{\partial^2 p}{\partial t^2} = -\rho_0 p'_0 \frac{\partial}{\partial t}(\nabla \cdot \vec{v}) = -\rho_0 p'_0 \nabla \cdot \left(\frac{\partial \vec{v}}{\partial t}\right) = -\rho_0 p'_0 \nabla \cdot \left(-\frac{1}{\rho_0}\nabla p\right) = p'_0 \nabla^2 p,$$

resulting in a *wave equation*:

$$\frac{\partial^2 p}{\partial t^2} = c_s^2 \nabla^2 p, \qquad (2.9)$$

in which the speed of wave propagation in the co-moving frame of the fluid is,

$$c_s \equiv \sqrt{p'_0} = \sqrt{\left.\frac{dp}{d\rho}\right|_{\rho_0}}. \qquad (2.10)$$

---

[2] There is no $\vec{v}_0$ since we are in the *co-moving* frame of the fluid.

[3] In the "lab frame", this term becomes $(\vec{v}_0 \cdot \nabla)\vec{v}$ and, while not eliminated, is rendered linear because of the leading $\vec{v}_0$ instead of $\vec{v}$.

These small perturbations are *sound waves* and, accordingly, c_s is the *sound speed*. For the special cases of an adiabatic ($p \propto \rho^\gamma$) and isothermal ($p \propto \rho$) equations of state for a gas, we have:

$$c_s = \begin{cases} \sqrt{\dfrac{\gamma p}{\rho}} = \sqrt{\dfrac{\gamma k_B T}{m}} & \text{adiabatic sound speed;} \\ \sqrt{\dfrac{p}{\rho}} = \sqrt{\dfrac{k_B T}{m}} & \text{isothermal sound speed,} \end{cases} \quad (2.11)$$

using the ideal gas law, Eq. (1.12), and where m is the *average* mass per particle.[4]

Now, the solution to the 1-D wave equation ($\nabla^2 \to \partial^2/\partial x^2$) is:

$$p(x,t) = f(kx - \omega t) + g(kx + \omega t), \quad (2.12)$$

where f (g) is an arbitrary function describing a wave or a pulse moving in the $+\hat{x}$ ($-\hat{x}$) direction, and where $p(x,0) = f(kx) + g(kx)$ is the pressure distribution at $t = 0$. Here, $k = 2\pi/\lambda$ is the *wave number*, λ is the *wavelength*, $\omega = 2\pi\nu$ is the *angular frequency* (rad s^{-1}), and ν is the *frequency* (Hz). Evidently, the wave speed is $c_s = \lambda\nu = \omega/k$, and thus,

$$\omega = k c_s, \quad (2.13)$$

is the so-called *dispersion relation*, giving ω in terms of k. The reader may recall from classical mechanics that when the *group velocity*, $v_g = d\omega/dk$, is equal to the *phase velocity*, $v_p = \omega/k$, the medium is *non-dispersive*; that is, all wavelengths travel at the same propagation speed. Since from Eq. (2.13), $v_g = v_p = c_s$, an ordinary gas is an example of a non-dispersive medium.

One can easily verify that Eq. (2.12) solves the 1-D wave equation as follows. Let $\xi_\pm \equiv x \pm c_s t$. Then $p(x,t) = f(k\xi_-) + g(k\xi_+)$ and we have by the chain rule,

$$\frac{\partial p}{\partial t} = \frac{df}{d(k\xi_-)} \frac{\partial (k\xi_-)}{\partial t} + \frac{dg}{d(k\xi_+)} \frac{\partial (k\xi_+)}{\partial t} = -f' k c_s + g' k c_s$$

$$\Rightarrow \quad \frac{\partial^2 p}{\partial t^2} = c_s^2 k^2 (f'' + g''),$$

where $'$ indicates differentiation with respect to the argument. Similarly,

$$\frac{\partial^2 p}{\partial x^2} = k^2 (f'' + g'') \quad \Rightarrow \quad \frac{\partial^2 p}{\partial t^2} = c_s^2 \frac{\partial^2 p}{\partial x^2},$$

and $p(x,t)$ in Eq. (2.12) solves the 1-D wave equation.

To summarise what we've just shown, *any function whose dependence upon the independent variables, x and t, has the form $\xi_i = x - u_i t$ where u_i is the wave speed, is a solution to the wave equation*, a fact we'll use frequently throughout this text. In the same vein, Problem 2.1 shows that,

$$p(\vec{r}, t) = f(\vec{k} \cdot \vec{r} - \omega t) + g(\vec{k} \cdot \vec{r} + \omega t), \quad (2.14)$$

solves the 3-D wave equation, Eq. (2.9), where $\vec{k}$ is the *wave vector* whose magnitude is the wave number and whose direction is the direction of wave propagation.

---

[4] *Rappel*: k_B is the Boltzmann constant.

Problem 2.2 shows that the velocity function, $v(x,t)$, also obeys a wave equation with wave speed $c_s = \sqrt{p_0'}$ and thus has a solution similar in form to Eq. (2.12). Therefore, the two quantities set into oscillation by the passage of a sound wave are pressure (transporting potential energy density since $p \propto e$) and v_x (transporting kinetic energy density, $\frac{1}{2}\rho v_x^2$). Since, in 1-D, v_x is in the direction of wave propagation, sound waves are said to be *longitudinal* as opposed to *transverse*, such as those propagating along a taut wire.

We conclude this subsection with a comment on the nature of the sound speed itself. Regardless of whether the gas is isothermal or adiabatic, Eq. (2.11) tells us that $c_s \propto \sqrt{T}$. Now, whatever the thermodynamics of the gas may be, sound waves generally propagate adiabatically simply because heat flow is not normally responsive enough to maintain strict isothermality within the small packets of fluid undergoing the rapid oscillations caused by the passage of a sound wave.

Thus, let's set some "yardsticks" by considering the adiabatic sound speed (first of Eq. 2.11), namely,

$$c_s = \sqrt{\frac{\gamma k_B T}{m}}. \tag{2.15}$$

For dry air at STP, $m = 4.81 \times 10^{-26}$ kg (average mass of the atmospheric constituents) and $\gamma = 7/5$ (99% of the atmosphere is diatomic). Thus the speed of sound at the surface of the earth is $c_s \sim 20.0\sqrt{T} \sim 331$ m s^{-1} (for $T = 273$ K), or about 1,200 km/hr. On the other hand, an astrophysical fluid is typically monatomic and consists mostly of hydrogen (75%) and helium (25%). Therefore, the average mass per particle is $7m_p/4 \sim 2.9 \times 10^{-27}$ kg in a neutral astrophysical gas, and $7m_p/9 \sim 1.3 \times 10^{-27}$ kg in a fully ionised gas. Thus, for a neutral gas, $c_s \sim 90\sqrt{T}$ (e.g., ~ 3 km/s for $T \sim 1{,}000$ K), and for a completely ionised gas, $c_s \sim 130\sqrt{T}$ (e.g., ~ 40 km/s for $T \sim 10^5$ K).

Finally, on comparing Eq. (2.15) and (1.13), we find:

$$\frac{c_s}{v_{\text{rms}}} = \sqrt{\frac{2}{\gamma(\gamma-1)}},$$

which is about 1.34 (1.89) for a monatomic (diatomic) gas. Thus, the sound speed is less than, but on the order of, the rms speed of the gas particles. That the two speeds should be so closely related makes sense since, at the particle level, it is only through particle–particle collisions that information of the passage of a wave may be propagated.

2.1.2 Eigenvalue approach

Most authors introduce sound waves by developing the wave equation as done in the previous subsection but, to my taste, this is not the most physically transparent approach. Consider again the linearised pressure and Euler equations, this time in 1-D and in the *lab frame* from which the fluid is in motion (*i.e.* $v_0 \neq 0$):

$$\frac{\partial p}{\partial t} + v_0 \frac{\partial p}{\partial x} = -\rho_0 c_s^2 \frac{\partial v}{\partial x} \quad \Rightarrow \quad \frac{\partial p}{\partial t} + \frac{\partial}{\partial x}(v_0 p + \rho_0 c_s^2 v) = 0;$$

$$\frac{\partial v}{\partial t} + v_0 \frac{\partial v}{\partial x} = -\frac{1}{\rho_0}\frac{\partial p}{\partial x} \quad \Rightarrow \quad \frac{\partial v}{\partial t} + \frac{\partial}{\partial x}\left(\frac{p}{\rho_0} + v_0 v\right) = 0,$$

where $c_s^2 = p_0'$ is taken to be constant. In "matrix" form, these equations can be written as,

$$\frac{\partial}{\partial t}\begin{bmatrix} p \\ v \end{bmatrix} + \frac{\partial}{\partial x}\begin{bmatrix} v_0 p + \rho_0 c_s^2 v \\ p/\rho_0 + v_0 v \end{bmatrix} = 0 \qquad (2.16)$$

$$\Rightarrow \quad \frac{\partial}{\partial t}\begin{bmatrix} p \\ v \end{bmatrix} + \begin{bmatrix} v_0 & \rho_0 c_s^2 \\ 1/\rho_0 & v_0 \end{bmatrix}\frac{\partial}{\partial x}\begin{bmatrix} p \\ v \end{bmatrix} = 0. \qquad (2.17)$$

If we define:

$$|q\rangle \equiv \begin{bmatrix} p \\ v \end{bmatrix} \quad \text{and} \quad |f\rangle \equiv \begin{bmatrix} v_0 p + \rho_0 c_s^2 v \\ p/\rho_0 + v_0 v \end{bmatrix},$$

where $|q\rangle$ is the "column vector", or *ket*,[5] of primitive variables p and v[6] and $|f\rangle$ is the ket of (advective) flux densities, then Eq. (2.16) and (2.17) may be written more compactly as:

$$\frac{\partial |q\rangle}{\partial t} + \frac{\partial |f\rangle}{\partial x} = 0 \quad \Rightarrow \quad \frac{\partial |q\rangle}{\partial t} + \mathsf{J}\frac{\partial |q\rangle}{\partial x} = 0, \qquad (2.18)$$

where:

$$\mathsf{J} = \begin{bmatrix} v_0 & \rho_0 c_s^2 \\ 1/\rho_0 & v_0 \end{bmatrix}, \qquad (2.19)$$

is the so-called *Jacobian matrix*, whose $(i,j)^{\text{th}}$ matrix element is given by:

$$J_{ij} = \frac{\partial f_i}{\partial q_j}. \qquad (2.20)$$

As we saw in the previous subsection, both p and v satisfy the wave equation and therefore so must $|q\rangle$. From our conclusion at the end of §2.1.1, $|q\rangle$ can then be written in the form,[7]

$$|q(x,t)\rangle = |\tilde{q}(\xi_i)\rangle, \qquad (2.21)$$

where $\xi_i = x - u_i t$, $i = 1, 2$, is the coordinate co-moving with the wave whose wave speed in the rest frame of the fluid is u_i. Eq. (2.21) is known as a *normal mode solution*, where "normal mode" – a term from classical mechanics – refers to the fact that all points of interest in a system oscillate at the same frequency. The *general solution* is then a linear combination of all normal mode solutions with coefficients determined by applying boundary conditions.

---

[5] The designation of a column vector as $|a\rangle$ ("ket") and a row vector as $\langle b|$ ("bra" as in "hat", not "bra" as in "hot") is due to Paul Dirac. This notation is most commonly used in quantum mechanics, but equally applicable here. In this convention, the inner ("dot") product of two vectors is always the matrix product of a "bra" (a row vector $\equiv$ a $1 \times n$ matrix) and a "ket" (a column vector $\equiv$ an $n \times 1$ matrix). This product, known as a "bra-ket", or *bracket*, generates a scalar and is written $\langle a|b\rangle$ (cf., $\vec{a}\cdot\vec{b}$ in "normal" vector notation). In this book, where it is useful to indicate whether a vector is a row or a column, we shall use Dirac's bra-ket notation liberally.

[6] *Rappel*: In the last chapter, the *primitive* variables were defined as the set $(\rho, p, \vec{v})$, whereas the *conservative* variables were defined as the set $(\rho, e_{\text{T}}, \vec{s})$.

[7] The "tilde" (˜) over the q recognises that formally, while the functions $q(x,t)$ and $\tilde{q}(\xi_i)$ may be equal numerically, they are not the same functions of their argument(s) and thus should be represented by different names.

To this end, start by differentiating $|q(x,t)\rangle$ in Eq. (2.21) with respect to t and x:

$$\frac{\partial |q\rangle}{\partial t} = \frac{\partial \xi_i}{\partial t}\frac{d|\tilde{q}\rangle}{d\xi_i} = -u_i|\tilde{q}'\rangle \quad \text{and} \quad \frac{\partial |q\rangle}{\partial x} = \frac{\partial \xi_i}{\partial x}\frac{d|\tilde{q}\rangle}{d\xi_i} = |\tilde{q}'\rangle, \quad (2.22)$$

where $'$ indicates differentiation with respect to ξ_i. Substituting Eq. (2.22) directly into the second of Eq. (2.18) yields:

$$\mathsf{J}|\tilde{q}'\rangle = u_i|\tilde{q}'\rangle, \quad (2.23)$$

which must be true for the normal mode solution to be valid. Since J is a matrix, $|\tilde{q}'\rangle$ a column vector, and u_i are scalars, Eq. (2.23) is an "eigen-equation" where the *eigenvalues* of J, namely u_i, are the allowed wave speeds, and where $|\tilde{q}'\rangle$ are proportional to the associated *eigenvectors* or *eigenkets*, $|r_i\rangle$.

To find the eigenvalues of J, we rewrite Eq. (2.23) as,

$$(\mathsf{J} - u_i\mathsf{I})|\tilde{q}'\rangle \equiv \mathsf{A}|\tilde{q}'\rangle = 0, \quad (2.24)$$

where I is the identity matrix. Now, if the matrix A had an inverse, we could immediately write,

$$\mathsf{A}^{-1}\mathsf{A}|\tilde{q}'\rangle = \mathsf{A}^{-1} \times 0 \quad \Rightarrow \quad |\tilde{q}'\rangle = 0,$$

and the *trivial* solution would be the *only* solution to Eq. (2.24). Therefore, for there to be *non-trivial* solutions, A must be *singular* (possess no inverse) and thus its determinant must be zero. This condition leads to the so-called *secular* or *characteristic* equation; that is,

$$\det \mathsf{A} = \det(\mathsf{J} - u_i\mathsf{I}) = \begin{vmatrix} v_0 - u_i & \rho_0 c_s^2 \\ 1/\rho_0 & v_0 - u_i \end{vmatrix}$$

$$= (v_0 - u_i)^2 - c_s^2 = (v_0 - u_i + c_s)(v_0 - u_i - c_s) = 0$$

$$\Rightarrow \quad u_1 = v_0 - c_s \quad \text{and} \quad u_2 = v_0 + c_s, \quad (2.25)$$

are the two eigenvalues and thus wave speeds. These correspond to sound waves moving in the $-\hat{x}$ (left, u_1), and $+\hat{x}$ (right, u_2) directions within a fluid moving at speed v_0 relative to the lab frame.

Next, let the eigenket associated with $u_1 = v_0 - c_s$ be $|r_1\rangle \propto |\tilde{q}'\rangle$. To find it, we write the matrix equation (2.23) as,

$$(\mathsf{J} - u_1\mathsf{I})|r_1\rangle = 0 \quad \Rightarrow \quad \begin{bmatrix} c_s & \rho_0 c_s^2 \\ 1/\rho_0 & c_s \end{bmatrix}\begin{bmatrix} r_{11} \\ r_{12} \end{bmatrix} = 0.$$

Breaking this up into ordinary algebraic equations, we get,

$$c_s r_{11} + \rho_0 c_s^2 r_{12} = 0 \quad \text{and} \quad \frac{1}{\rho_0}r_{11} + c_s r_{12} = 0,$$

where, as usual, one of the equations is redundant for correctly determined eigenvalues. That is, we can evaluate only one of the components of the eigenket $|r_1\rangle$ in terms of the other, in which case we find,

$$r_{11} = -c_s \rho_0 r_{12} \equiv -Z_0 r_{12},$$

where Z_0 is the *impedance* of the fluid. Thus, the eigenket associated with u_1 is:

$$|r_1\rangle = \begin{bmatrix} r_{11} \\ r_{12} \end{bmatrix} = \begin{bmatrix} -Z_0 \\ 1 \end{bmatrix} r_{12} = \begin{bmatrix} -Z_0 \\ 1 \end{bmatrix}, \quad (2.26)$$

choosing $r_{12} = 1$ as our "normalisation".[8]

From Eq. (2.23), we have $|\tilde{q}_1'\rangle \propto |r_1\rangle$, and we write:

$$|\tilde{q}_1'(\xi_1)\rangle = w_1'(\xi_1)|r_1\rangle, \quad (2.27)$$

where, for convenience, the proportionality factor, w_1', is expressed as the first derivative of some function, w_1, of the co-moving coordinate $\xi_1 = x - (v_0 - c_s)t$. Integrating with respect to ξ_1 (and with respect to which the eigenkets are constant), we get:

$$|q_1(x,t)\rangle = |\tilde{q}_1(\xi_1)\rangle = w_1(\xi_1)|r_1\rangle,$$

since $|q_1(x,t)\rangle = |\tilde{q}(\xi_1)\rangle$. In a similar manner, associated with the second eigenvalue, $u_1 = v_0 + c_s$, is the eigenket:

$$|r_2\rangle = \begin{bmatrix} Z_0 \\ 1 \end{bmatrix}, \quad (2.28)$$

and thus the solution for the *right*-moving wave is:

$$|q_2(x,t)\rangle = |\tilde{q}_2(\xi_2)\rangle = w_2(\xi_2)|r_2\rangle,$$

where w_2 is an arbitrary function of $\xi_2 = x - (v_0 + c_s)t$. Therefore, the general solution is the superposition of the left- and right-moving waves, and we have:

$$|q(x,t)\rangle = |q_1(x,t)\rangle + |q_2(x,t)\rangle = w_1(\xi_1)|r_1\rangle + w_2(\xi_2)|r_2\rangle. \quad (2.29)$$

To find the functions (coefficients) w_1 and w_2, we apply initial conditions. At $t = 0$, $|q(x,0)\rangle = |\tilde{q}(x)\rangle$ – presumably a known function – and we've got,

$$|\tilde{q}(x)\rangle = \begin{bmatrix} \tilde{p}(x) \\ \tilde{v}(x) \end{bmatrix} = w_1(x)|r_1\rangle + w_2(x)|r_2\rangle = w_1(x)\begin{bmatrix} -Z_0 \\ 1 \end{bmatrix} + w_2(x)\begin{bmatrix} Z_0 \\ 1 \end{bmatrix}$$

$$\Rightarrow \quad \tilde{p}(x) = -w_1(x)Z_0 + w_2(x)Z_0 \quad \text{and} \quad \tilde{v}(x) = w_1(x) + w_2(x),$$

which, when solved for $w_1(x)$ and $w_2(x)$, yield,

$$w_1(x) = \frac{1}{2}\left(\tilde{v}(x) - \frac{1}{Z_0}\tilde{p}(x)\right) \quad \text{and} \quad w_2(x) = \frac{1}{2}\left(\tilde{v}(x) + \frac{1}{Z_0}\tilde{p}(x)\right).$$

Substituting these expressions into Eq. (2.29) yields,

$$|q(x,t)\rangle = \begin{bmatrix} p(x,t) \\ v(x,t) \end{bmatrix} = w_1(\xi_1)\begin{bmatrix} -Z_0 \\ 1 \end{bmatrix} + w_2(\xi_2)\begin{bmatrix} Z_0 \\ 1 \end{bmatrix}$$

$$= \frac{1}{2}\begin{bmatrix} \tilde{p}(\xi_2) + \tilde{p}(\xi_1) + Z_0(\tilde{v}(\xi_2) - \tilde{v}(\xi_1)) \\ \tilde{v}(\xi_2) + \tilde{v}(\xi_1) + \frac{1}{Z_0}(\tilde{p}(\xi_2) - \tilde{p}(\xi_1)) \end{bmatrix}. \quad (2.30)$$

---

[8]Since the eigenket components have different units, the term "normalisation" seems out of place given what is "normalised" won't have magnitude unity. For fluid eigenkets, a better term might be "scaled for convenience", although "normalised" is so well established in the vernacular, that we seem to be stuck with it.

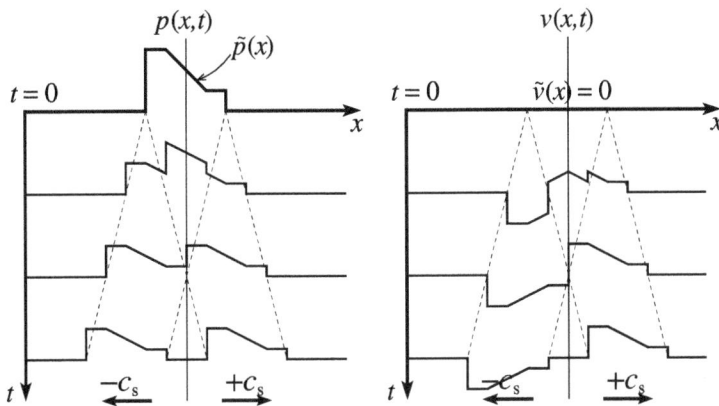

Figure 2.1. A perturbation is initially applied to the pressure profile [$\tilde{p}(x)$], but not to the velocity profile [$\tilde{v}(x) = 0$]. As time progresses, half of the pressure pulse moves off in each direction, while waves closely related to the pressure waves develop in the velocity profile. These profiles were determined by direct application of Eq. (2.30) which the reader is encouraged to verify.

Note that Eq. (2.30) is specific enough to construct a plot of the pressure and velocity profiles for all time given the initial profiles, $\tilde{p}(x)$ and $\tilde{v}(x)$. An example of this is shown in Fig. 2.1. Problems 2.3–2.5 give the reader the opportunity to study Fig. 2.1 further, and to use Eq. (2.30) to determine the waveforms launched by a given initial perturbation.

In this subsection, we've introduced a normal mode, linear algebraic method to analyse the wave nature of the fluid equations in the particularly simple setting of 1-D linearised hydrodynamics. In so doing, we found a systematic way to identify the various waves the system can support, to calculate their wave speeds from the reference frame of the fluid or the lab, and to evaluate the eigenkets which, as Eq. (2.30) and Fig. 2.1 exemplify, determine the profiles of the flow variables as a function of position and time. In due course, we shall apply these methods to the more complicated cases of non-linear hydrodynamics (§3.5) where we'll discover another type of wave known as a *rarefaction fan*, and to 1-D MHD (§5.2) where we'll find that the 7 × 7 Jacobian matrix with variable matrix elements supports *seven* waves rather than the two found here.

2.2 Rankine–Hugoniot jump conditions

Figure 2.2 depicts what is sometimes called a *shock tube*: a *steady-state*[9] flow of gas along the x-axis with all transverse derivatives ($\partial/\partial y$, $\partial/\partial z$) set to zero. Without

---

[9] "Steady state" means all *explicit* time derivatives, *e.g.* $\partial\rho/\partial t$, are zero. Conversely, *implicit* time dependence, *e.g.* $d\rho/dt = \partial\rho/\partial t + \vec{v}\cdot\nabla\rho = \vec{v}\cdot\nabla\rho$, need not be zero.

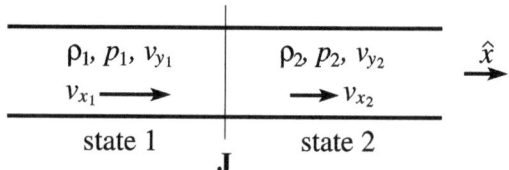

Figure 2.2. $1\frac{1}{2}$-D fluid flow, as viewed from the reference frame **J**.

loss of generality, we shall assume flow is from left to right and $v_z = 0$ (e.g., align $\hat{y}$ with the component of $\vec{v}$ perpendicular to $\hat{x}$). Thus, even though we assume no transverse derivatives, we allow for a transverse velocity;[10] a situation often referred to as "$1\frac{1}{2}$-D flow".

Let **J** be a fixed reference line dividing the fluid into two "states": an *upwind* state 1 (from **J** to state 1 in Fig. 2.2, one must go *against* the flow, *i.e.* "upwind") and a *downwind* state 2 (from **J** to state 2, one must go *with* the flow, *i.e.* "downwind"). Note that **J** is not a physical barrier but a fixed reference line with respect to which flow velocities are measured.

Let the primitive variables be $(\rho_1, p_1, v_{x_1}, v_{y_1})$ and $(\rho_2, p_2, v_{x_2}, v_{y_2})$ in states 1 and 2 respectively. As depicted in Fig. 2.2, the two states are in contact with and possibly in motion relative to each other (*i.e.*, $\vec{v}_1 \neq \vec{v}_2$) at **J**. Now, it is quite conceivable **J** corresponds to nothing significant at all, in which case the upwind and downwind states are identical. On the other hand, **J** could be parked on top of a stationary feature in the flow – a "jump", if you will, in some of the variables – that renders the two states different. Thus, what we really want to know is: How do the downwind fluid variables depend upon the upwind values?

To answer this, we begin with the differential equations in conservative form (Eq. Set 1) so that discontinuities in the primitive variables at **J** may be admitted, and impose upon them the assumptions of steady state ($\partial/\partial t = 0$) and zero transverse derivatives. Thus, $\partial/\partial x \to d/dx$ and we have:

$$\frac{d}{dx}(\rho v_x) = 0;$$

$$\frac{d}{dx}\left(\frac{1}{2}\rho v^2 v_x + \frac{\gamma}{\gamma - 1}p v_x\right) = 0;$$

$$\frac{d}{dx}(\rho v_x^2 + p) = 0;$$

$$\frac{d}{dx}(\rho v_y v_x) = 0,$$

after substituting the constitutive equations into the total energy and momentum equations, and where $v^2 = v_x^2 + v_y^2$. These ODEs are trivial to integrate and, when applied to the conditions of Fig. 2.2, lead to the so-called *Rankine–Hugoniot jump*

---

[10]Strictly speaking, in an actual laboratory shock tube, $v_y = 0$ too.

conditions:[11]

$$\rho_2 v_{x_2} = \rho_1 v_{x_1}; \tag{2.31}$$

$$\tfrac{1}{2}\rho_2 v_2^2 v_{x_2} + \frac{\gamma}{\gamma-1} p_2 v_{x_2} = \tfrac{1}{2}\rho_1 v_1^2 v_{x_1} + \frac{\gamma}{\gamma-1} p_1 v_{x_1}; \tag{2.32}$$

$$\rho_2 v_{x_2}^2 + p_2 = \rho_1 v_{x_1}^2 + p_1; \tag{2.33}$$

$$\rho_2 v_{y_2} v_{x_2} = \rho_1 v_{y_1} v_{x_1}. \tag{2.34}$$

2.2.1 Case 1: Trivial solution

Obviously, $\rho_2 = \rho_1$, $p_2 = p_1$, $v_{x_2} = v_{x_1}$, and $v_{y_2} = v_{y_1}$ solve Eq. (2.31)–(2.34), and such a solution corresponds to the possibility we already guessed, namely **J** corresponds to nothing significant in the flow. This solution is referred to as the *trivial solution*, and is consistent with a uniform gas flowing at a constant velocity, $\vec{v}_1 = \vec{v}_2$.

2.2.2 Case 2: Tangential discontinuity

Suppose $v_{x_1} = v_{x_2} = 0$. Then Eq. (2.31), (2.32), and (2.34) are satisfied trivially, while Eq. (2.33) requires $p_2 = p_1$. No constraints are placed on ρ nor v_y, and thus these variables may jump by an arbitrary amount across **J**. This phenomenon is known as a *tangential discontinuity*. Because the pressure is constant across a tangential discontinuity, the ideal gas law, Eq. (1.12), requires that a density jump be accompanied by a temperature jump equal to its inverse. Thus:

$$\frac{T_2}{T_1} = \frac{\rho_1}{\rho_2}.$$

Note that for a given pressure, the gas with the higher temperature will also have a higher specific entropy, and thus there is an entropy jump as well.

A meteorological example of a tangential discontinuity that may or may not include a density jump is *wind shear*, which can play havoc with air travellers particularly near take-off or landing. Another example which may or may not include a jump in v_y is a *weather front*. A *cold front* is where cool dry air displaces warm moist air while a *warm front* is the reverse. Typically, both fronts can result in significant precipitation as the hot, moist, and less-dense air[12] rises over the cool, dry, denser air, cools, and releases its moisture as rain or snow.

Finally, a tangential discontinuity across which v_y is constant (and thus can

---

[11] Named for Scottish physicist William Rankine (1820–1872) and French engineer Pierre Hugoniot (1851–1887) who pioneered the theoretical explanation of shock waves.

[12] Notwithstanding our physiological impression that hot, muggy air is "heavier" (denser) than cool, dry air, the reverse is actually true. At STP, every 22.4 litres of air, dry or humid, contains one mole of particles. For humid air, a good number of molecules such as N_2, O_2, CO_2, and Ar have been displaced by significantly less massive H_2O molecules, and thus humid air is less dense than dry air.

be set to zero with an appropriate Galilean transformation) is called a *contact discontinuity* (CD), or simply a *contact*. Note that in a strictly 1-D hydrodynamical flow, there can be no transverse accelerations (*e.g.*, in the y-direction) and thus a shear cannot develop spontaneously. Therefore, a contact is the only type of tangential discontinuity to arise in 1-D HD. By contrast, in an MHD fluid gradients in the x-direction can give rise to a Lorentz force in the y-direction and tangential discontinuities with a velocity shear can arise spontaneously in 1-D MHD, a subject we shall revisit in §5.3.

2.2.3 Case 3: Shock

Suppose now that $v_{x_1}, v_{x_2} \neq 0$. Right away we see that Eq. (2.31) and (2.34) require $v_{y_2} = v_{y_1}$ which can be set to zero by applying an appropriate Galilean transformation. Therefore, let us simplify the notation slightly by setting $v_1 = v_{x_1}$ and $v_2 = v_{x_2}$. Then, substituting Eq. (2.31) into Eq. (2.33) yields:

$$p_2 = \rho_1 v_1^2 + p_1 - \rho_1 v_1 v_2, \qquad (2.35)$$

while dividing Eq. (2.31) into Eq. (2.32) yields:

$$\frac{v_2^2}{2} + \frac{\gamma}{\gamma-1}\frac{p_2 v_2}{\rho_2 v_2} = \frac{v_1^2}{2} + \frac{\gamma}{\gamma-1}\frac{p_1}{\rho_1}. \qquad (2.36)$$

Next, substitute Eq. (2.31) and (2.35) into Eq. (2.36) to get:

$$\frac{v_2^2}{2} + \frac{\gamma}{\gamma-1}\frac{v_2}{\rho_1 v_1}(\rho_1 v_1^2 + p_1 - \rho_1 v_1 v_2) = \frac{v_1^2}{2} + \frac{\gamma}{\gamma-1}\frac{p_1}{\rho_1},$$

and, after some straight-forward algebra, we obtain a quadratic in v_2/v_1:

$$\frac{\gamma+1}{2}\left(\frac{v_2}{v_1}\right)^2 - \left(\gamma + \frac{c_1^2}{v_1^2}\right)\frac{v_2}{v_1} + \frac{\gamma-1}{2} + \frac{c_1^2}{v_1^2} = 0, \qquad (2.37)$$

where $c_1^2 = \gamma p_1/\rho_1$ is the square of the sound speed (introduced in §2.1) in state 1. Using the quadratic formula on Eq. (2.37), one finds:

$$\frac{v_2}{v_1} = \frac{\gamma + M_1^{-2} \pm (1 - M_1^{-2})}{\gamma+1}, \qquad (2.38)$$

where $M_1 \equiv v_1/c_1$ is the *Mach number*, named for Ernst Mach (1838–1916), the Austrian (Czech) mathematician, physicist, and philosopher best known for the *Mach principle* (relating all motion with the inertia of all else) who also happened to be the first to describe supersonic motion in a gas.[13]

The $+$root of Eq. (2.38) yields $v_2/v_1 = 1$ (thus recovering case 1), while the $-$root yields:

$$\boxed{\frac{v_2}{v_1} = \frac{M_1^2(\gamma-1) + 2}{M_1^2(\gamma+1)} = 1 - 2\frac{M_1^2-1}{M_1^2(\gamma+1)}.} \qquad (2.39)$$

---

[13] www.wikipedia.org/wiki/Ernst_Mach

Thus, from Eq. (2.31) we have:

$$\boxed{\frac{\rho_2}{\rho_1} = \frac{v_1}{v_2} = \frac{M_1^2(\gamma+1)}{M_1^2(\gamma-1)+2} = 1 + 2\frac{M_1^2-1}{M_1^2(\gamma-1)+2}.} \quad (2.40)$$

Next, rearranging Eq. (2.35) gives us:

$$p_2 = p_1 + \rho_1 v_1^2 \left(1 - \frac{v_2}{v_1}\right),$$

which, after substituting $\rho_1 = \gamma p_1/c_1^2$ and Eq. (2.39), yields:

$$\boxed{\frac{p_2}{p_1} = \frac{2\gamma M_1^2 - \gamma + 1}{\gamma + 1} = 1 + 2\gamma \frac{M_1^2 - 1}{\gamma + 1}.} \quad (2.41)$$

Equations (2.39)–(2.41) are solutions to the Rankine–Hugoniot jump conditions (Eq. 2.31–2.34) describing simultaneous *jumps* in the primitive variables ρ, p, and v across the reference line **J** in Fig. 2.2. When such conditions exist in nature, it is called a *shock wave*, or simply a *shock*.

Other aspects of a shock are self-evident from the present discussion. For example, the temperature jump is given by the ideal gas law, Eq. (1.12), with Eq. (2.39) and (2.41):

$$\frac{T_2}{T_1} = \frac{p_2}{p_1}\frac{\rho_1}{\rho_2} = \frac{p_2}{p_1}\frac{v_2}{v_1} = \left(1 + 2\gamma\frac{M_1^2-1}{\gamma+1}\right)\left(1 - 2\frac{M_1^2-1}{M_1^2(\gamma+1)}\right),$$

which, after a little manipulation, can be expressed as:

$$\frac{T_2}{T_1} = 1 + \frac{2(\gamma-1)}{(\gamma+1)^2}\left(\gamma + \frac{1}{M_1^2}\right)(M_1^2 - 1). \quad (2.42)$$

Finally, the downwind Mach number, M_2, is given by:

$$M_2^2 = \left(\frac{v_2}{c_2}\right)^2 = \frac{v_2^2 \rho_2}{\gamma p_2} = \frac{v_2 v_1 \rho_1}{\gamma p_2} = \frac{v_1^2 \rho_1}{\gamma p_1}\frac{p_1}{p_2}\frac{v_2}{v_1}$$

$$= M_1^2 \frac{\gamma+1}{2\gamma M_1^2 - \gamma + 1}\frac{M_1^2(\gamma-1)+2}{M_1^2(\gamma+1)}$$

$$\Rightarrow M_2^2 = \frac{M_1^2(\gamma-1)+2}{2\gamma M_1^2 - \gamma + 1} = 1 - \frac{(\gamma+1)(M_1^2-1)}{2\gamma M_1^2 - \gamma + 1}. \quad (2.43)$$

Note that all jumps and the downwind Mach number have been expressed in terms of γ (embodying the thermodynamics of the gas) and the upwind Mach number M_1^2 (embodying the kinematics of the gas). These dependencies are depicted in Fig. 2.3.

Evidently, M_1 is *the* critical parameter to determine the nature of a shock. For *transonic* flow ($M_1 = 1$), the right-hand sides of Eq. (2.39)–(2.43) reduce to unity. Thus, $M_1 = 1$ corresponds to the trivial case of §2.2.1.

For *hypersonic* flow ($M_1 \gg 1$), Eq. (2.39)–(2.43) have the following limits:

$$\lim_{M_1^2 \to \infty} \frac{v_2}{v_1} \to \frac{\gamma-1}{\gamma+1} < 1; \quad (2.44)$$

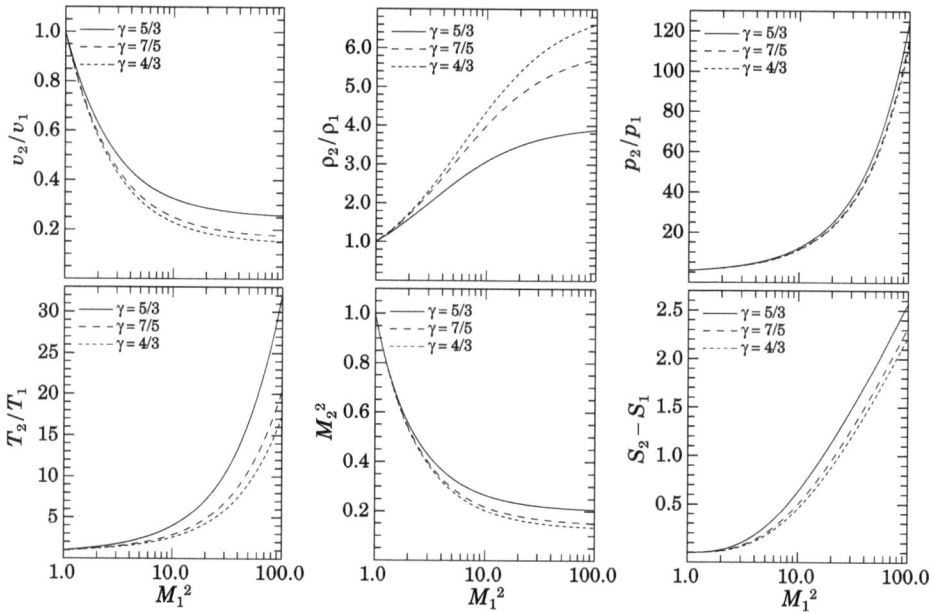

Figure 2.3. Graphical representations of Eq. (2.39)–(2.43) and (2.53) for various values of γ, with all variables defined in the text. All velocities and Mach numbers are relative to the reference frame of the shock. As shown in the text, a necessary condition for a shock to exist is $M_1 > 1$, and thus all plots start at $M_1 = 1$.

$$\lim_{M_1^2 \to \infty} \frac{\rho_2}{\rho_1} \to \frac{\gamma+1}{\gamma-1} > 1; \qquad (2.45)$$

$$\lim_{M_1^2 \to \infty} \frac{p_2}{p_1} \to \frac{2\gamma}{\gamma+1} M_1^2 \to \infty; \qquad (2.46)$$

$$\lim_{M_1^2 \to \infty} \frac{T_2}{T_1} \to \frac{2\gamma(\gamma-1)}{(\gamma+1)^2} M_1^2 \to \infty; \qquad (2.47)$$

$$\lim_{M_1^2 \to \infty} M_2^2 \to \frac{\gamma-1}{2\gamma} < 1. \qquad (2.48)$$

Table 2.1 summarises these limits for various values of γ. In particular, across a hypersonic shock, the speed drops while the density, pressure, and temperature rise. Meanwhile, $M_2 < 1$ indicates that flow downwind of the shock is subsonic (relative to the shock). As we'll see, these results are general and hold for all values of $M_1^2 > 1$, not just hypersonic flow.

It turns out that it is impossible to attain the shock jump conditions for *subsonic* flow ($M_1 < 1$). To see this, we must evaluate the entropy jump across a shock, and then invoke the combined first and second law of thermodynamics (Eq. 1.29):

$$S_2 - S_1 = \Delta S = \int_1^2 dS = \int_1^2 \frac{dE}{T} + \int_1^2 \frac{p\,dV}{T}, \qquad (2.49)$$

where ΔS is the change in entropy across the shock. To this end, from Eq. (1.10)

γ	v_2/v_1	ρ_2/ρ_1	p_2/p_1	T_2/T_1	M_2^2
5/3	0.250	4	$1.250\,M_1^2$	$0.313\,M_1^2$	0.200
7/5	0.167	6	$1.167\,M_1^2$	$0.194\,M_1^2$	0.143
4/3	0.143	7	$1.143\,M_1^2$	$0.163\,M_1^2$	0.125

Table 2.1. Entries show the limits of Eq. (2.39)–(2.43) for the given value of γ in the limit as $M_1^2 \to \infty$.

we get:
$$dE = \frac{1}{\gamma - 1} \mathcal{N} k_B dT, \tag{2.50}$$

and from the ideal gas law, $pV = \mathcal{N} k_B T$, we have,
$$p\,dV + V\,dp = \mathcal{N} k_B dT$$
$$\Rightarrow\quad p\,dV = \mathcal{N} k_B dT - V\,dp = \mathcal{N} k_B dT - \mathcal{N} k_B T \frac{dp}{p}. \tag{2.51}$$

Substituting Eq. (2.50) and (2.51) into Eq. (2.49) yields:
$$\Delta S = \frac{\mathcal{N} k_B}{\gamma - 1} \int_1^2 \frac{dT}{T} + \mathcal{N} k_B \int_1^2 \frac{dT}{T} - \mathcal{N} k_B \int_1^2 \frac{dp}{p}. \tag{2.52}$$

As before (page 19), define the unitless entropy per particle,[14] $\mathcal{S} = (\gamma - 1)S/\mathcal{N} k_B$. Then, on performing the integrations in Eq. (2.52), we get,

$$\Delta \mathcal{S} = \gamma \ln \frac{T_2}{T_1} - (\gamma - 1) \ln \frac{p_2}{p_1} = \ln \left[\left(\frac{p_2 \rho_1}{p_1 \rho_2} \right)^\gamma \left(\frac{p_2}{p_1} \right)^{1-\gamma} \right]$$
$$= \ln \left[\frac{p_2}{p_1} \left(\frac{\rho_1}{\rho_2} \right)^\gamma \right] = \ln \left(\frac{p_2/\rho_2^\gamma}{p_1/\rho_1^\gamma} \right) = \ln \frac{\kappa_2}{\kappa_1}, \tag{2.53}$$

where κ is the "constant" in the adiabatic equation of state, namely $p = \kappa \rho^\gamma$. Equation (2.53) gives us a very simple and practical result: The entropy jump across a shock is proportional to the logarithm of the ratio of the kappas.

We can now express $\Delta \mathcal{S}$ in terms of M_1^2 and γ. Substituting Eq. (2.40) and (2.41) into Eq. (2.53) yields:

$$\Delta \mathcal{S} = \ln \left(\frac{2\gamma M_1^2 - \gamma + 1}{\gamma + 1} \right) + \gamma \ln \left(\frac{M_1^2(\gamma - 1) + 2}{M_1^2(\gamma + 1)} \right). \tag{2.54}$$

To find the critical points, if any, of $\Delta \mathcal{S}(M_1^2)$ ($\Delta \mathcal{S}$ expressed as a function of M_1^2), we set its first derivative to zero. After a little algebra, we find:

$$\frac{d(\Delta \mathcal{S})}{d(M_1^2)} = \frac{(M_1^2 - 1)^2 (\gamma - 1)}{M_1^2 [2\gamma M_1^2 - (\gamma - 1)][M_1^2(\gamma - 1) + 2]} = 0,$$

---

[14] Here, and out of convenience, $\mathcal{S}$ is defined with an extra factor of $(\gamma - 1)$.

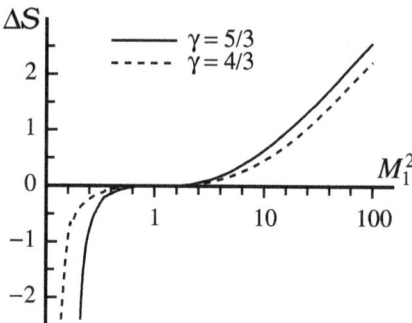

Figure 2.4. Equation (2.54) showing graphically that $\Delta S > 0 \Rightarrow M_1^2 > 1$.

which, by inspection, yields just one critical point, namely $M_1^2 = 1$. A little additional algebra shows that the second derivative is zero at $M_1^2 = 1$, and thus $M_1^2 = 1$ is an inflection point of $\Delta S(M_1^2)$. Thus, there are no extrema and $\Delta S(M_1^2)$ is a monotonic function of $M_1^2 \in (0, \infty)$. Evaluating Eq. (2.54) at $M_1^2 = 1$, we find $\Delta S = 0$ meaning there is no entropy change for the transonic, trivial case. Further, as $M_1^2 \to \infty$, we find,

$$\lim_{M_1^2 \to \infty} \Delta S \to \ln M_1^2 > 0.$$

Thus, $\Delta S(M_1^2)$ is a monotonically *increasing* function with a root and inflection point at $M_1^2 = 1$, and therefore positive definite for $M_1^2 > 1$ and negative definite for $M_1^2 < 1$.[15] Figure 2.4 is a graphical representation of Eq. (2.54) for $\gamma = 5/3$ and $4/3$ showing clearly that $\Delta S > 0 \Rightarrow M_1^2 > 1$.

As the second law of thermodynamics states that all spontaneous processes in an isolated system will result in $\Delta S \geq 0$, we may finally conclude that the shock solution to the Rankine–Hugoniot jump conditions is possible only for $M_1^2 \geq 1$. Since $M_1^2 = 1$ corresponds to the trivial case, we may state that $M_1^2 > 1$ for shocks, which are therefore inherently supersonic phenomena.

Having established that $M_1^2 > 1$ for shocks, we can determine whether the flow variables increase or decrease across any shock – not just hypersonic ones – simply by inspecting Eq. (2.39)–(2.43). This leads us to the following general result:

$$v_2 < v_1; \quad \rho_2 > \rho_1; \quad p_2 > p_1; \quad T_2 > T_1; \quad M_2^2 < 1, \qquad (2.55)$$

known as Zemplén's theorem (or the *entropy condition*).

So far, all discussion has considered the shock from its own frame of reference. However, setting up a "standing shock" in the laboratory may not be as practical as a moving shock, such as that illustrated in Fig. 2.5. In this case, supersonic fluid moves along a 1-D tube and collides with the closed end, exciting a shock wave that moves in the upwind direction *with an unknown speed*. It would, therefore, be useful to express the jump conditions in terms of the upwind Mach number *relative*

---

[15]In fact, upon inspection of the first term in Eq. (2.54), we see that $\Delta S \notin \mathbb{R}$ for $0 \leq M_1^2 < \frac{\gamma-1}{2\gamma}$.

Rankine–Hugoniot jump conditions

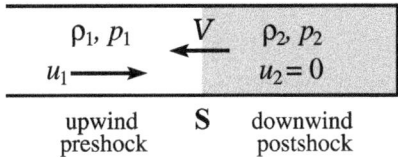

Figure 2.5. A 1-D shock tube, as viewed from the reference frame of the post-shock gas (the lab frame), shown here in grey. **S** indicates the shock which is moving to the left at a speed $\mathcal{V}$ relative to the lab frame.

to the closed end of the tube ($\mathcal{M}_1$) instead of the upwind Mach number relative to the shock itself (M_1).

To this end, let $u_1 > 0$ (left to right) be the velocity of the preshock gas in the "lab frame", and let $\mathcal{V} < 0$ (right to left) be the propagation velocity of the shock (Fig. 2.5). The shocked gas is at rest relative to the end of the tube, and is therefore in the lab frame. Thus the speed of the shocked gas relative to the lab is $u_2 = 0$, and the Galilean transformation between the shock frame **J** in Fig. 2.2 and the lab frame must be:

$$v_1 = u_1 - \mathcal{V}; \qquad v_2 = -\mathcal{V}. \tag{2.56}$$

From Eq. (2.39), we have:

$$\frac{v_2}{v_1} = \frac{v_1^2(\gamma - 1) + 2c_1^2}{v_1^2(\gamma + 1)}$$

$$\Rightarrow \quad v_2 v_1 (\gamma + 1) = v_1^2(\gamma - 1) + 2c_1^2. \tag{2.57}$$

Substituting the transformation Eq. (2.56) into Eq. (2.57) yields:

$$(u_1 - \mathcal{V})(-\mathcal{V})(\gamma + 1) = (u_1 - \mathcal{V})^2 (\gamma - 1) + 2c_1^2$$

$$\Rightarrow \quad \frac{\mathcal{V}^2}{u_1^2} - \frac{\mathcal{V}}{u_1} \frac{3 - \gamma}{2} - \frac{\gamma - 1}{2} - \frac{c_1^2}{u_1^2} = 0.$$

The roots of the quadratic are:

$$\frac{\mathcal{V}}{u_1} = \frac{3 - \gamma}{4} \pm \sqrt{\left(\frac{\gamma + 1}{4}\right)^2 + \frac{1}{\mathcal{M}_1^2}},$$

where $\mathcal{M}_1 = u_1/c_1$ is the Mach number of the preshock gas relative to the lab frame.

Only one of these roots is physical. The +root requires $\mathcal{V} > 0$ which is contrary to our assumption and thus discarded. For $\gamma > 1$, $\gamma + 1 > 3 - \gamma$, and the −root assures $\mathcal{V} < 0$, as desired. Thus, we find:

$$\frac{\mathcal{V}}{u_1} = 1 - \mathcal{Q} \quad \Rightarrow \quad \frac{\mathcal{M}_S}{\mathcal{M}_1} = -\frac{\mathcal{V}/c_1}{u_1/c_1} = \mathcal{Q} - 1,$$

where,

$$\mathcal{Q} \equiv \sqrt{\left(\frac{\gamma + 1}{4}\right)^2 + \frac{1}{\mathcal{M}_1^2}} + \frac{\gamma + 1}{4} > \frac{\gamma + 1}{2} > 1,$$

Figure 2.6. Naturally occurring tidal bores propagating along the Peticodiac River in New Brunswick. The left image (photo credit David Milligan) shows an undulating bore, while the right image (photo credit Charles LeGresley) shows a foaming bore, both propagating along the river from right to left. Note that the ends of the undulating bore are actually foaming; why? (Answer in the text.)

and where $\mathcal{M}_S$ is the Mach number of the shock (relative to the sound speed in the upwind gas) as measured in the lab frame. Note that $\mathcal{M}_S$ is taken to be positive, even though $\mathcal{V}$ is negative.

It is left to Problem 2.8 to show that the jump conditions in terms of speeds (Mach numbers) measured in the lab frame (where the shocked gas is stagnant) are given by:

$$\frac{\rho_2}{\rho_1} = \frac{\mathcal{Q}}{\mathcal{Q}-1}; \qquad \frac{p_2}{p_1} = 1 + \gamma \mathcal{M}_1^2 \, \mathcal{Q}; \qquad \frac{T_2}{T_1} = (\mathcal{Q}-1)\left(\frac{1}{\mathcal{Q}} + \gamma \mathcal{M}_1^2\right). \quad (2.58)$$

2.3 Bores and hydraulic jumps (optional)

The provinces of New Brunswick and Nova Scotia are separated by the Bay of Fundy, in which the greatest difference in sea level between low and high tides can be found on the planet.[16] Whilst the average tidal range around the world is between two and three metres, the tidal range in the Minas Basin, one of the inlets off the Bay of Fundy, can be *seventeen* metres during the new and full moon.

The physical reasons for these extraordinarily high tides makes for a fun physics problem in its own right (and so, see Problem 2.12). However, it's not the tides themselves that interest us here, but one of their many consequences. As the tides rush in and push into the many rivers that empty into the bay, the oppositely flowing waters trigger some of the most impressive *tidal bores* one can find anywhere.

As discussed below, bores come in various flavours including *undulating bores*,

---

[16]Incidentally, another Canadian body of water, Ungava Bay in Québec's far north, exhibits a tidal range very similar to the Fundy tides. Indeed, local proponents of each site engage in a colourful argument over whose site actually has the highest tides, with many experts (but not all!) suggesting the true answer lies somewhere within the statistical uncertainties.

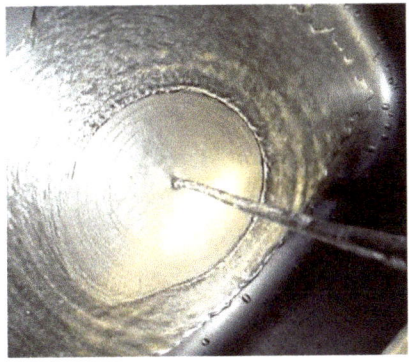 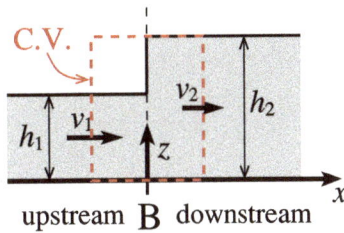

Figure 2.7. (left) A laminar stream of water from a kitchen faucet strikes the bottom of a sink, setting up a common example of a standing bore or hydraulic jump. (Photo credit: Petr Vita and his kitchen sink!) (right) A schematic of a standing bore, showing the two depths of water and two flow speeds on either side of the discontinuity, **B**, that is established as a consequence of the applicable conservation laws, and the control volume (C.V.) used to analyse it.

foaming bores, and *standing bores*[17] (a.k.a., *hydraulic jumps*). Figure 2.6 shows examples of the former two moving up the Peticodiac River from the Bay of Fundy towards Moncton, New Brunswick. The left image shows an undulating bore, while the right image shows a more energetic foaming bore used to advantage in this photograph by several surfers being pushed up-river! Indeed, a thriving "white water rafting" industry is sustained by the foaming tidal bores each province experiences twice daily.

While shocks do propagate within "incompressible" fluids, they do so by exploiting the very little bit of compressibility incompressible fluids actually have. In a strictly incompressible fluid, bores are mathematically the most similar phenomenon to shocks that propagate, though they are much more properly described as *gravity waves*, relying as they do on the restoring force the earth's background gravitational field provides.

An example of a standing bore can easily be set up in your kitchen sink (left panel of Fig. 2.7), something most people have probably seen so many times that they've long since ceased to notice. In the schematic of a standing bore shown in the right panel of Fig. 2.7, the water depth jumps at **B** from h_1 upstream of **B** to $h_2 > h_1$ downstream. *Question*: What upstream and downstream water speeds relative to **B**, v_1 and v_2, are necessary to sustain the observed depths, h_1 and h_2?

A useful construct for such problems is a *control volume* (C.V.), depicted in red in the right panel of Fig. 2.7. With the control volume fixed to the reference frame of **B** (the bore), the flow into and out of the C.V. is in *steady state* (no dependence on time). Then, for incompressible flow ($\rho =$ constant), the continuity equation (Eq. 1.19) becomes:

$$\cancelto{0}{\partial_t \rho} + \nabla \cdot (\rho \vec{v}) = \rho \nabla \cdot \vec{v} = 0 \quad \Rightarrow \quad \nabla \cdot \vec{v} = 0, \tag{2.59}$$

---

[17] Not to be confused with your professor!

which, integrated over the C.V. gives:

$$\int_V \nabla \cdot \vec{v}\, dV = \oint_S \vec{v} \cdot \hat{n}\, d\sigma = 0, \tag{2.60}$$

using Gauss' theorem.

Note that $\vec{v} = v_x \hat{x}$ is perpendicular to the surface normal, $\hat{n}$, at both the top and bottom of the C.V. Further, $\vec{v} = 0$ along the portions of the C.V. not in contact with water. Thus, the only parts of the C.V. that contribute to the surface integral in Eq. (2.60) are the portions of the vertical sides immersed in water, and Eq. (2.60) becomes:

$$-v_1 h_1 w + v_2 h_2 w = 0 \quad \Rightarrow \quad v_1 h_1 = v_2 h_2 \quad \text{(continuity)}, \tag{2.61}$$

where w is some arbitrary width into the page.

Next, in steady state, the x-component of the momentum equation (Eq. 1.27) is:

$$\cancelto{0}{\partial_t s_x} + \nabla \cdot (s_x \vec{v} + p\hat{x}) = 0,$$

which, when integrated over the C.V., gives:

$$\int_V \nabla \cdot (s_x \vec{v} + p\hat{x})\, dV = \oint_S (\rho v_x^2 + p)\hat{x} \cdot \hat{n}\, d\sigma = 0, \tag{2.62}$$

since $s_x = \rho v_x$ and $\vec{v} = v_x \hat{x}$. Once again, only the ends of the C.V. (where $\hat{n} = \pm\hat{x}$) contribute to the surface integral, and Eq. (2.62) becomes:

$$-\rho v_1^2 \cancel{w} h_1 - \cancel{w} \int_0^{h_2} p_{\text{L}}(z)dz + \rho v_2^2 \cancel{w} h_2 + \cancel{w} \int_0^{h_2} p_{\text{R}}(z)dz = 0, \tag{2.63}$$

where $p_{\text{L/R}}(z)$ is the pressure profile at the left/right side of the C.V. which, as the following argument shows, can be determined using standard freshman physics.

Consider the column of water delineated by the dashed line in the inset whose cross-sectional area is A and depth is d. Evidently, the atmospheric pressure pushes down on the column with a force $p_{\text{atm}} A$, and the weight of the column measured at P is $\rho A g d$, where ρ is the density of water and $g = 9.81\,\text{m}\,\text{s}^{-2}$ is the usual acceleration of gravity. Thus, the water pressure at P must be:

$$p(d) = \frac{1}{A}(p_{\text{atm}} A + \rho A g d) = p_{\text{atm}} + \rho g d,$$

where the depth, d, is known as the *pressure head*. Thus, for depths $d_{1,2} = h_{1,2} - z$, where z is the vertical coordinate depicted in the right panel of Fig. 2.7, we have:

$$\left.\begin{aligned} p_{\text{L}}(z) &= \begin{cases} p_{\text{atm}} + \rho g(h_1 - z), & 0 < z < h_1; \\ p_{\text{atm}}, & h_1 < z < h_2; \end{cases} \\ p_{\text{R}}(z) &= p_{\text{atm}} + \rho g(h_2 - z), \quad 0 < z < h_2. \end{aligned}\right\} \tag{2.64}$$

Bores and hydraulic jumps (optional)

This allows us to evaluate the integrals in Eq. (2.63):

$$\int_0^{h_2} p_L(z)\,dz = p_{\text{atm}} z\Big|_0^{h_2} + \rho g\left(h_1 z - \tfrac{1}{2}z^2\right)_0^{h_1} = p_{\text{atm}} h_2 + \rho g \frac{h_1^2}{2};$$

$$\int_0^{h_2} p_R(z)\,dz = p_{\text{atm}} z\Big|_0^{h_2} + \rho g\left(h_2 z - \tfrac{1}{2}z^2\right)_0^{h_2} = p_{\text{atm}} h_2 + \rho g \frac{h_2^2}{2},$$

which, when substituted into Eq. (2.63) gives:

$$\rho v_2^2 h_2 - \rho v_1^2 h_1 + \cancel{p_{\text{atm}} h_2} + \rho g \frac{h_2^2}{2} - \cancel{p_{\text{atm}} h_2} - \rho g \frac{h_1^2}{2} = 0$$

$$\Rightarrow \quad g\frac{h_2^2 - h_1^2}{2} = v_1^2 h_1 \frac{h_2}{h_2} - v_2^2 h_2 \frac{h_2}{h_2} = v_1^2 \frac{h_1}{h_2}(h_2 - h_1),$$

using Eq. (2.61). Note how all terms involving p_{atm} cancel out.

Solving for v_1, then using Eq. (2.61) to find v_2, we get our final result:

$$\boxed{v_1 = \sqrt{\frac{gh_2(h_1+h_2)}{2h_1}};} \quad \text{and} \quad \boxed{v_2 = v_1 \frac{h_1}{h_2} = \sqrt{\frac{gh_1(h_1+h_2)}{2h_2}},} \qquad (2.65)$$

the unique speeds that support the given upstream and downstream depths, h_1 and h_2 that conserve both momentum and mass conservation.

Having considered the continuity and momentum equations, it is reasonable to ask why the energy equation wasn't needed. It is left to Problem 2.13 to show that for an incompressible fluid, one cannot conserve mass, momentum, and energy simultaneously at **B**. Thus energy is not conserved across a bore, but rather is dissipated in one of two ways:

1. For $h_2/h_1 \lesssim 1.3$, waves are excited downstream of the jump (deeper side), transporting energy away from **B**. This is an *undulating bore* shown in the left panel of Fig. 2.6.

2. For $h_2/h_1 \gtrsim 1.3$, the bore becomes turbulent downstream, and energy is dissipated as heat, noise, and waves. This is a *foaming bore* shown in the right panel of Fig. 2.6.

Tidal bores, such as those shown in Fig. 2.6, are triggered when the incoming tide meets outgoing flow from a river. At the river mouth where the velocity differential is the greatest, the bore has the greatest height and energy and, if sufficiently high, begins its journey up the river as a foaming bore. As it advances, its energy is dissipated, height differential diminished, and the bore evolves into an undulating bore, starting in the centre of the river where h_1 is the greatest, and eventually including the edges where h_1 is the least. The wavelength of undulations is evidently determined by factors such as g, $h_2 - h_1$, and properties of the water such as its density and viscosity. For water, the critical point where a bore switches from foaming to undulating is at $h_2/h_1 \sim 1.3$, a semi-empirical result that also depends upon the intrinsic properties of the fluid.

2.3.1 Bores in the "lab frame"

An advancing tidal bore is typically observed from the riverbank which, given the relative speeds of the river flow and the advancing bore, is approximately the frame of reference of the shallower water, *i.e.*, the river as it flows into the ocean. Let us refer to this reference frame as **L** – the "lab frame" – in analogy with the situation depicted in Fig. 2.5 where we similarly considered a shock wave. Thus, and as illustrated in the inset, taking the upstream flow-speed, u_1 (relative to **L**), to be zero, we have:

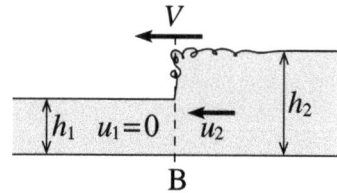

$$u_1 = 0; \quad u_2 = v_2 - v_1; \quad \mathcal{V} = -v_1,$$

where, as before, v_1 and v_2 are respectively the upstream and downstream flow speeds relative to the bore, **B**, and $\mathcal{V}$ is the speed of the left-moving bore relative to **L**. Thus, from Eq. (2.65),

$$\mathcal{V} = -\sqrt{\frac{gh_2(h_1 + h_2)}{2h_1}}; \tag{2.66}$$

and $\quad u_2 = \left(1 - \frac{h_1}{h_2}\right)v_1 = -(h_2 - h_1)\sqrt{\frac{g(h_1 + h_2)}{2h_1 h_2}}, \tag{2.67}$

where for both $\mathcal{V}$ and u_2, the leading negative sign means motion relative to **L** is leftward.

Example 2.1. In five hours, a tidal bore travels at a constant speed along a 36-km inlet of average depth $h_1 = 0.2$ m.

a) Find the average height (above the undisturbed water) of the bore.

b) Is the bore undulating or foaming?

Solution: a) Assuming the undisturbed (shallow) end of the bore is at rest relative to the observer, the bore speed is given by Eq. (2.66):

$$\mathcal{V} = -\sqrt{\frac{gh_2(h_1 + h_2)}{2h_1}} \Rightarrow gh_2^2 + gh_1 h_2 - 2h_1 \mathcal{V}^2 = 0$$

$$\Rightarrow h_2 = \frac{-h_1 + \sqrt{h_1^2 + 8h_1 \mathcal{V}^2/g}}{2}, \tag{2.68}$$

discarding the $-$ root to keep $h_2 > 0$.

From the problem description, $\mathcal{V} = (36\,\text{km})/(5\,\text{hr}) = 2\,\text{m s}^{-1}$. Thus, with $g = 9.81\,\text{m s}^{-2}$ and $\langle h_1 \rangle = 0.2\,\text{m}$, Eq. (2.68) $\Rightarrow \langle h_2 \rangle = 0.316\,\text{m}$, and the average height of the bore above the undisturbed water is:

$$\langle h \rangle = \langle h_2 \rangle - \langle h_1 \rangle = \underline{\underline{0.116\,\text{m}}}.$$

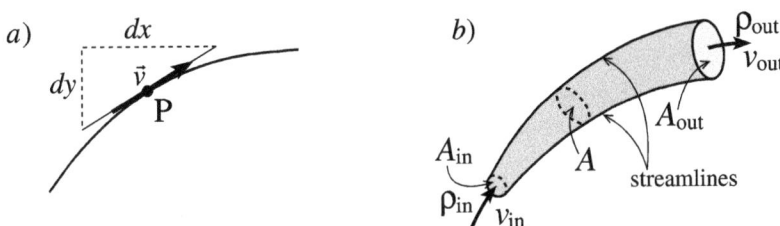

Figure 2.8. *a*) A streamline is always tangential to the local velocity vector. *b*) A streamtube is a "pencil of fluid" bounded on all sides by streamlines.

b) To determine the nature of the bore,

$$\frac{\langle h_2 \rangle}{\langle h_1 \rangle} = \frac{0.316}{0.2} = 1.58 > 1.3 \quad \Rightarrow \quad \underline{\underline{\text{foaming bore.}}}$$

2.4 Bernoulli's theorem

We begin our discussion of Bernoulli's theorem with some definitions:

Streamline: A streamline is a path in a fluid whose tangents are everywhere parallel to the local velocity vector. This definition can easily be mathematised. If, as in Fig. 2.8a, the tangent to the streamline at point P is characterised by a "rise" of dy over a "run" of dx, then for the tangent to be parallel to the local velocity vector, (v_x, v_y), we have in 2-D:

$$\frac{dy}{dx} = \frac{v_y}{v_x},$$

which, in principle, can be solved to yield $y(x)$, the path describing the streamline. An entirely analogous but slightly more complicated procedure can be followed to find a streamline in 3-D.

Streakline: A streakline is a line in the fluid created by the convection of a tracer (*e.g.*, smoke, dye) that is continually released into the fluid at a fixed location. When people think of a streamline, they are often thinking, in fact, of a streakline (*e.g.*, the smoke streaks created by an engineer in a wind tunnel that pass over the surface of, say, a vehicle being examined for its aerodynamic properties). In general, streaklines and streamlines are different.

Particle path: A particle path is the trajectory of an individual point being *convected* with the fluid. Thus, a particle path is the locus of points at different times representing the position of the same particle, whereas a streakline is the locus of points representing the positions of different particles at the same time.

Flowline: In a *steady-state* fluid (one characterised by setting all the $\partial/\partial t$ terms to zero), streamlines, streaklines, and particle paths are all the same, and are referred to collectively as flowlines. If the flow is *not* in steady state, then all three paths are, in general, different.

Streamtube: A streamtube (or in steady state, a *flowtube*) is a "pencil of fluid" bounded by streamlines (flowlines), as depicted in Fig. 2.8b.

Since, by definition, flow does not cross streamlines, the "walls" of a streamtube are as impervious to the fluid as though they were physical barriers, and the only matter entering or leaving a streamtube will be through its "ends" (Fig. 2.8b). We may formalise this idea by considering the integral form of the continuity equation (Eq. 1.37) in the steady state:

$$\oint_S \rho \vec{v} \cdot \hat{n} \, d\sigma = 0.$$

If S is the closed surface of a flowtube, then only the "ends" of the flowtube (A_{in} and A_{out} in Fig. 2.8b) contribute to the integral, since $\vec{v} \perp \hat{n}$ everywhere else. Therefore we have,

$$\oint_S \rho \vec{v} \cdot \hat{n} \, d\sigma = \rho_{\text{out}} v_{\text{out}} A_{\text{out}} - \rho_{\text{in}} v_{\text{in}} A_{\text{in}} = 0,$$

where $\rho_{\text{in/out}}$ and $v_{\text{in/out}}$ are mean values across $A_{\text{in/out}}$, and where the velocity is the component parallel to the area normal, $\hat{n}$. The negative sign arises because $\vec{v}_{\text{in}}$ and $\vec{A}_{\text{in}}$ are antiparallel. Thus we can write:

$$\rho_{\text{out}} v_{\text{out}} A_{\text{out}} = \rho_{\text{in}} v_{\text{in}} A_{\text{in}}, \tag{2.69}$$

which, loosely translated, means "what goes in must come out". Note that $\rho v A$, with units kg s^{-1}, is a *mass flux* and Eq. (2.69) is a statement of mass flux conservation. Thus, for an arbitrary flowtube cross-section, A, whose normal is parallel to the flow, we have:

$$\boxed{\rho v A = \text{constant.}} \tag{2.70}$$

Note that if mass flux weren't conserved, the mass in any given volume would either accumulate or disappear over time, contradicting the steady-state assumption.

Bernoulli's theorem identifies another useful constant along streamlines for fluid flow in the steady state, and can be derived in a number of ways. Here we begin with the total energy equation (Eq. 1.23) setting $\partial e_T / \partial t = 0$:

$$\nabla \cdot (e_T + p) \vec{v} = \nabla \cdot \left(\frac{e_T + p}{\rho} \rho \vec{v} \right) = \rho \vec{v} \cdot \nabla \left(\frac{e_T + p}{\rho} \right) + \frac{e_T + p}{\rho} \underbrace{\nabla \cdot (\rho \vec{v})}_{= 0} = 0,$$

where the continuity equation (Eq. 1.19) in steady state allows us to set the last term to zero. Thus, on dividing through by $\rho \neq 0$, we have,

$$\vec{v} \cdot \nabla \left(\frac{e_T + p}{\rho} \right) = 0,$$

and $\vec{v}$ is everywhere perpendicular to the gradient of the function $\mathcal{B} \equiv (e_T + p)/\rho$. This is equivalent to saying $\vec{v}$ – which maps out streamlines – is everywhere parallel to lines of constant $\mathcal{B}$ (or tangential to surfaces of constant $\mathcal{B}$ in 3-D), and we have,

$$\boxed{\mathcal{B} = \frac{v^2}{2} + \frac{e+p}{\rho} + \phi = \text{constant along streamlines,}} \qquad (2.71)$$

where Eq. (1.20) has been used to replace the total energy density, e_T. This is Bernoulli's theorem, named for Daniel Bernoulli (1700–1782), the Dutch/Swiss physicist and a contemporary of Johan Euler. The Bernoulli family was a bit of a mathematical dynasty in the 18th century, including his father, Johann (a professor of mathematics at Basel whose most famous pupil was Euler himself), his brother, Nicolas Bernoulli, and his uncle, Jacob Bernoulli. Daniel's interests took him mostly into applied mathematics and physics.[18]

Bernoulli's theorem takes on a variety of forms depending on the equation of state one uses for e. The two studied in this text are for ideal gases (compressible or incompressible), and ideal liquids (strictly incompressible). The former is obtained by using the ideal gas law, Eq. (1.11), to substitute $e = p/(\gamma-1)$ into Eq. (2.71):

$$\mathcal{B}_{\text{gas}} = \frac{v^2}{2} + \frac{\gamma}{\gamma-1}\frac{p}{\rho} + \phi = \frac{v^2}{2} + \hbar + \phi = \text{constant along streamlines,} \qquad (2.72)$$

where,

$$\hbar = \frac{\gamma p}{(\gamma-1)\rho} = \frac{c_s^2}{\gamma-1}, \qquad (2.73)$$

is the *enthalpy* of the gas. As for liquids (which do not obey the ideal gas law), the internal energy density is essentially independent of the pressure, and depends almost exclusively on the temperature. Indeed, the pressure in a liquid can vary dramatically even if e and ρ remain constant to within one part in $10^{>6}$ and we shall therefore take ρ as constant. Further, if the liquid is isothermal, e will be constant in which case, Bernoulli's theorem takes on the form:

$$\mathcal{B}_{\text{liq}} \equiv \mathcal{B} - \frac{e}{\rho} = \frac{v^2}{2} + \frac{p}{\rho} + \phi = \text{constant along streamlines.} \qquad (2.74)$$

Note that for terrestrial examples (gases and liquids), one takes $\phi = gz$, where g is the acceleration of gravity, and z is the vertical distance from a reference height.

Bernoulli's theorem is a powerful tool for analysing some fluid problems, but it is worth remembering the restrictions under which Bernoulli's theorem applies. *For Eq. (2.71) to be valid, the flow must be in a steady state, and one must be considering events along the same streamline.* While these restrictions may seem severe, it turns out they're not as restrictive as one may at first think as the examples below and several of the problems in the problem set illustrate.

Broad-crested weir[19]

In certain parts of the world where sudden rushes of water, particularly near mountain bases, are a risk to lives and property, *arroyos* (as they are known in south-

---

[18] www.wikipedia.org/wiki/Daniel_Bernoulli
[19] Adapted from an example in Tom Faber's text, *Fluid Dynamics for Physicists* (1995).

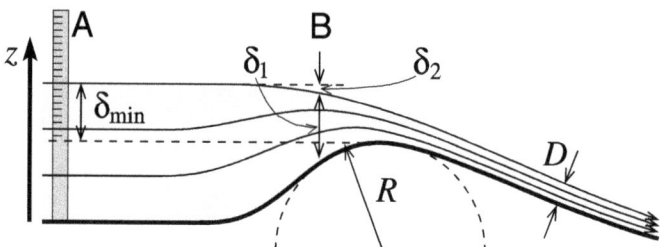

Figure 2.9. Cross section of a "broad-crested weir" used to measure flow rate in an "arroyo". The adjective "broad-crested" refers to the fact that $R \gg D$, where R is the radius of curvature of the weir and D is the depth of water passing over the weir.

western U.S.) are constructed to redirect the flow away from homes, roads, *etc.* These arroyos are often built with a *broad-crested weir* (Fig. 2.9) which is not meant to restrict the flow, but to provide an astonishingly simple way of measuring the flow rate (*e.g.*, m^3 s^{-1}) through the arroyo.

As shown in Fig. 2.9, an asymmetric mound, or *weir*, is built across an arroyo so that its "blunt-end" faces upstream, while the downstream side tapers gradually back to the level of the arroyo floor. Somewhat upstream of the weir is a vertical pole with markings indicating the height of water *above* the crest of the weir.

Question: How does the flow rate through the arroyo depend upon water depth?

One can clearly imagine such a flow to be (very nearly) in steady state, isothermal, and exhibit easily identified streamlines. For example, shown in Fig. 2.9 are three streamlines, the top corresponding to the surface of the water. Thus, this problem is a natural application of Bernoulli's theorem for an isothermal liquid.

Consider two points along the surface, A and B. Point A is chosen to be far enough back from the crest of the weir so that the water surface at A is approximately level, and point B is some arbitrary point between A and the crest of the weir. As shown in Fig. 2.9, δ_1 is the vertical distance between the arroyo floor and the water level at point B, while δ_2 is the vertical distance between the water level at point B and the water level at point A. For convenience, let $\delta = \delta_1 + \delta_2$.

Consider the flowline along the surface. Here, $p_A = p_B = p_{atm}$, the atmospheric pressure, and thus Eq. (2.74) becomes:

$$\tfrac{1}{2}v_A^2 + gz_A = \tfrac{1}{2}v_B^2 + gz_B,$$

where $\phi = gz$ (appropriate near the surface of the earth), and where z is the vertical height relative to the bottom of the arroyo. With $v_A \sim 0$ and $z_A - z_B = \delta_2$, we have:

$$\tfrac{1}{2}v_B^2 = g\delta_2 \quad \Rightarrow \quad v_B = \sqrt{2g\delta_2}.$$

Let the width of the arroyo be W. Then at B, the cross-sectional area of the flow is $\delta_1 W$, the flow rate is $Q = v_B \delta_1 W$ m^3 s^{-1}, and thus the flow rate per unit width of the arroyo is $q = Q/W = v_B \delta_1$. Therefore:

$$q = \delta_1 \sqrt{2g\delta_2} \quad \Rightarrow \quad \delta_2 = \frac{q^2}{2g\delta_1^2}.$$

But,

$$\delta = \delta_1 + \delta_2 = \delta_1 + \frac{q^2}{2g\delta_1^2}, \tag{2.75}$$

and this distance is minimised when point B is directly over the crest of the weir. To find this minimal distance, we set:

$$\frac{d\delta}{d\delta_1} = 1 - \frac{q^2}{g\delta_1^3} = 0 \quad \Rightarrow \quad \delta_1 = \left(\frac{q^2}{g}\right)^{1/3}.$$

Substituting this result into Eq. (2.75) yields:

$$\delta_{\min} = \left(\frac{q^2}{g}\right)^{1/3} + \frac{q^2}{2g}\left(\frac{g}{q^2}\right)^{2/3} = \frac{3}{2}\left(\frac{q^2}{g}\right)^{1/3}$$

$$\Rightarrow \quad q = \sqrt{g\left(\frac{2}{3}\delta_{\min}\right)^3}, \tag{2.76}$$

which is our final result. With Eq. (2.76), one can calibrate a pole upstream of the weir to read the flow rate, q, instead of, or in addition to, the water depth above the crest of the weir, $\delta_{\min}$.

This is a good place to pause and either introduce or remind the reader of a very useful technique known as *dimensional analysis* which can often yield qualitatively and sometimes quantitatively useful formulæ based on nothing more than an analysis of the units of the variables involved. While this method cannot be applied in all situations, in those where it can be applied it serves as a very useful check on what may have been a rather extensive and physically complex derivation.

In the present example of the broad-crested weir, one must first list all variables upon which the flow rate past the weir could possibly depend. Ignoring dissipative forces as we have done, one might list the density of water, ρ, the width of the arroyo, W, in addition to quantities such as g and $\delta_{\min}$, the water depth. Upon reflection, however, one soon realises that the rate at which things fall under the influence of gravity does not depend on mass, and thus ρ cannot be relevant. Further, since q is the flow rate *per unit width* of the arroyo, W is also irrelevant leaving us only with g and $\delta_{\min}$. The question then is, how could a quantity such as q, with units m^{2}s^{-1}, depend upon g, with units m s^{-2}, and $\delta_{\min}$, with units m? This can be easily determined by first supposing that:

$$q \propto g^a \delta_{\min}^b,$$

and then comparing the *units* on the left and right sides of the equation to determine a and b. Thus,

$$\mathrm{m^2 s^{-1}} = (\mathrm{m s^{-2}})^a (\mathrm{m})^b$$

$$\Rightarrow \quad 2 = a + b \quad \text{and} \quad -1 = -2a \quad \Rightarrow \quad a = \tfrac{1}{2} \quad \text{and} \quad b = \tfrac{3}{2}.$$

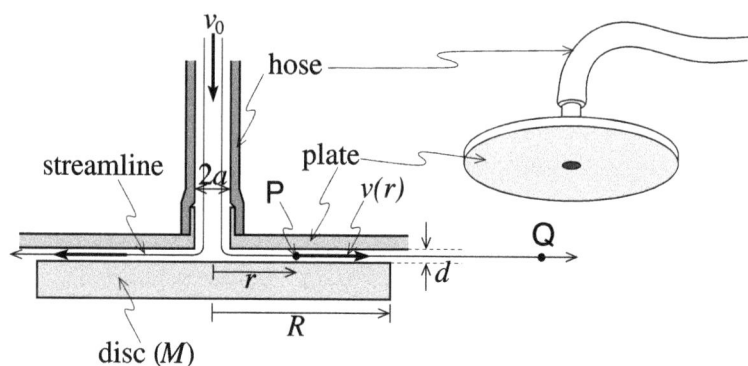

Figure 2.10. A cross-sectional cut through an apparatus designed to demonstrate *Bernoulli levitation*. Air coming down the hose at speed v_0 is deflected radially by the disc of mass M, and forced to flow through the gap of thickness d between the plate and the disc.

And so with *no physics* other than identifying the independent variables, we have determined that:

$$q \propto \sqrt{g\,\delta_{\min}^3},$$

in agreement with Eq. (2.76). Evidently, the physics is needed in this case only to determine the order-unity proportionality constant, namely $(2/3)^{3/2} \sim 0.54$.

Bernoulli levitation

Figure 2.10 depicts a device designed to demonstrate a somewhat counter-intuitive consequence of Bernoulli's theorem. A circular plate with a hole in the middle is fitted with a hose, the other end of which is connected to an air supply. By setting the air pump to draw air in through the hole, a disc of mass M brought up under the plate would, with sufficient suction, be drawn into the plate and supported against its own weight. No surprise there. But what if the airflow were reversed, and air were forced out of the hole? Surely then the disc would be pushed away from the plate and fall.

Bernoulli would beg to differ.

As shown in Fig. 2.10, an air supply forces air at a constant speed v_0 through a hole of radius a at the centre of the circular plate. A circular disc of mass M and radius R is brought up to within a distance d of the circular plate, both held horizontally, so that the centre of the disc is directly under the centre of the hole through which air is flowing. This redirects air from its initially downward direction to a radially outward direction through the narrow gap between the plate and disc.

We shall use the form of Bernoulli's theorem applicable to ideal gases to demonstrate a phenomenon known as *Bernoulli levitation*. Consider two points, P and Q, along the streamline as indicated in Fig. 2.10. P is taken to be a distance r from the centre of the hole, and thus $p_P = p(r)$ and $v_P = v(r)$. Q is taken to be sufficiently far away from the apparatus so that $v_Q = 0$ and $p_Q = p_{\text{atm}}$, the atmospheric pressure.

Since P and Q have the same gravitational potential, Eq. (2.72) requires:

$$p(r) + \tfrac{1}{7}\rho v^2(r) = p_{\text{atm}}, \tag{2.77}$$

where $\gamma = 7/5$, appropriate for a diatomic gas such as air, and where ρ is taken as constant since all velocities are well under the sound speed.[20]

Next, from Eq. (2.70), we conserve mass in the steady state by equating the mass flux through the hose (with cross-sectional area πa^2) to the mass flux flowing past any radius r inside the gap (with cross-sectional area $2\pi r d$). Thus,

$$\rho v_0 \pi a^2 = \rho v(r) 2\pi r d$$

$$\Rightarrow \quad v(r) = \frac{v_0 a^2}{2rd}. \tag{2.78}$$

Substituting Eq. (2.78) into Eq. (2.77), we get:

$$p(r) = p_{\text{atm}} - \frac{\rho v_0^2 a^4}{28\, r^2 d^2}. \tag{2.79}$$

Thus, the pressure inside the gap is *less* than atmospheric pressure, and it is this difference in pressure that holds up the disc when air flows out of the hole.

The net force acting on the disc is the vector sum of the pressure difference, $p_{\text{atm}} - p(r)$, integrated over the surface of the disc acting upwards, F_p, and the thrust of the air flow acting downwards, T (what our intuition might have lead us to conclude that the disc would be pushed away from the plate). For the former, we integrate over annuli concentric with the hole to get:

$$F_p = \int_a^R 2\pi r \bigl(p_{\text{atm}} - p(r)\bigr)\, dr = \int_a^R \frac{\pi \rho v_0^2 a^4}{14\, r d^2}\, dr = \frac{\pi \rho v_0^2 a^4}{14\, d^2} \ln(R/a).$$

For the latter, we consider the time rate of change of momentum, S, as air is redirected by the disc from vertical to horizontal:

$$T = \frac{\delta S}{\delta t} = \frac{\delta m}{\delta t} v_0 = \frac{\rho \pi a^2 v_0 \delta t}{\delta t} v_0 = \rho \pi a^2 v_0^2.$$

To find the thickness of the air gap, d, we set:

$$F_p - T = Mg \quad \Rightarrow \quad \frac{\pi \rho v_0^2 a^4}{14\, d^2} \ln(R/a) - \rho \pi a^2 v_0^2 = Mg$$

$$\Rightarrow \quad \frac{a^2 \ln(R/a)}{14\, d^2} = \frac{Mg}{\rho \pi a^2 v_0^2} + 1,$$

where, had we ignored the thrust acting downwards, the last term (1) on the right-hand side (RHS) would be absent. Solving for d, we get:

$$d = \sqrt{\frac{1}{14} \frac{a^2 \ln(R/a)}{Mg/(\rho \pi a^2 v_0^2) + 1}}. \tag{2.80}$$

For "typical" numbers appropriate for a device constructed to demonstrate Bernoulli

---

[20] For comfortably subsonic speeds, gases are largely incompressible in the sense that by their own actions and motions, ρ will not change appreciably.

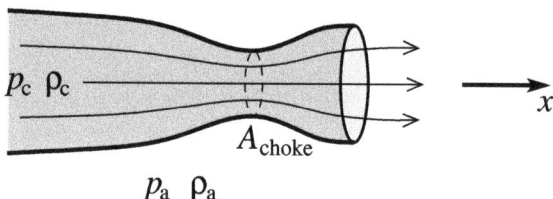

Figure 2.11. Inside the chamber of a de Laval nozzle, gas with pressure p_c and density ρ_c is "choked down" to a cross-sectional area A_{choke} at the *choke point* as it rushes to escape through the nozzle and into the ambient gas of pressure p_a and density ρ_a.

levitation, $M = 0.1$ kg, $R = 0.1$ m, $a = 0.01$ m, $v_0 = 10$ m s^{-1}, and $\rho = 1.2$ kg m^{-3}. Equation (2.80) then yields $d = 0.78$ mm. If one ignores the thrust [*i.e.*, dropped the "+1" term in the denominator of Eq. (2.80)], one finds $d = 0.80$ mm. Thus, the thrust makes $\sim 2.5\%$ difference for these parameters. The fact that d is *smaller* with the thrust taken into account than without may seem surprising, at first. Note, however, that a greater pressure difference is required to balance both the weight and the thrust than the weight alone, and thus d must be correspondingly smaller.

Equation (2.80) tells us that no matter how small v_0 may be, there is a finite gap, d, however small, that allows the disc to be supported against its own weight. But surely this cannot be true for a mass of arbitrary size! In fact, we can see from Eq. (2.79) that if d is too small, $p(r)$ becomes negative. By requiring $p(r) > 0$, it is left to Problem 2.17 to show that, for a given a and R, there is indeed a maximum mass Bernoulli levitation can support. However, for the parameters given above, this mass is almost *five times greater* than the maximum mass supportable by suction. Thus, blowing air out of the hose can actually be more effective at supporting weight than the "intuitively obvious" case of sucking it in!

de Laval nozzle

Gustaf de Laval (1845–1913), a Swedish engineer of French descent, developed the theory and performed the first experiments on what would prove to be the crux of rocket design: a nozzle that can optimally convert the thermal energy of extremely hot gas into kinetic energy of a projectile, such as a rocket.

Figure 2.11 is a schematic diagram of what is now called a *de Laval nozzle*. It turns out that for a given chamber pressure and density (presumably set by the rate of combustion in the rocket), there is a critical cross-sectional area of the throat of the nozzle, A_{choke}, that maximises the thrust. A nice application of Bernoulli's theorem will show that if $A >$ or $< A_{\text{choke}}$, thrust is necessarily reduced.

Start with Bernoulli's theorem for an ideal gas (Eq. 2.72 ignoring ϕ):

$$\frac{v^2}{2} + \frac{\gamma}{(\gamma-1)} \frac{p}{\rho} = \text{constant}.$$

Differentiating this and the steady-state continuity equation (Eq. 2.70) with respect to position along the chamber, x, yields the following two expressions:

$$v\frac{dv}{dx} + \frac{c_s^2}{\rho}\frac{d\rho}{dx} = 0; \tag{2.81}$$

$$\frac{1}{\rho}\frac{d\rho}{dx} + \frac{1}{v}\frac{dv}{dx} + \frac{1}{A}\frac{dA}{dx} = 0, \tag{2.82}$$

where $c_s^2 = dp/d\rho = \gamma p/\rho$ has been explicitly assumed in deriving Eq. (2.81). Dividing Eq. (2.81) by c_s^2 gives:

$$\frac{1}{\rho}\frac{d\rho}{dx} = -\frac{M^2}{v}\frac{dv}{dx},$$

where $M = v/c_s$ is the Mach number. When substituted into Eq. (2.82), we get de Laval's equation:

$$\boxed{(M^2 - 1)\frac{A}{v}\frac{dv}{dx} = \frac{dA}{dx}.} \tag{2.83}$$

Equation (2.83) is surprisingly subtle. For example, for the flow to be transonic ($M^2 = 1$) requires *either* $dA/dx = A' = 0$ (thus locating the transonic point at the choke point where A is an extremum), *or* $dv/dx = v' \to \infty$ (*i.e.*, a shock). As shocks convert kinetic energy into thermal energy, a shock along the flow is undesirable for the purpose of maximising thrust. Thus, we either need to avoid a transonic point altogether, or arrange for it to be located at the choke point of the nozzle.

Second, given that gas is accelerated from rest within the chamber, gas must start off as subsonic flow. Equation (2.83) tells us that when $M^2 < 1$, flow accelerates along the chamber (and thus $v' > 0$) provided $A' < 0$, *i.e.* the cross-sectional area of the chamber decreases in the direction of flow, as is the case on the left side of the choke point, A_choke, in Fig. 2.11. However, by the same token, if the flow is still subsonic to the right of the choke point where $A' > 0$, $v' < 0$ and the flow decelerates, contrary again to the purpose of maximising the thrust.

Now, should the flow become supersonic past the choke point, $v' > 0$ for $A' > 0$. That is, for $M^2 > 1$, the flow speed actually *increases* with increasing cross-sectional area. Thus, subsonic flow before the choke point and supersonic flow after seems to be the best way to maximise thrust, and this is possible only if a transonic point is located at the choke point of the nozzle. This was de Laval's discovery and design, and is why every nozzle on every rocket ever built is called a *de Laval nozzle*. Indeed, as explored in Problem 2.20, de Laval nozzles could arise in nature too.

Of course, this places a requirement on the value of A_choke in Fig. 2.11. If A_choke is too small, subsonic flow will become supersonic before reaching the choke point, triggering a shock and reconverting some of the kinetic energy of the gas back to thermal energy. Conversely, if A_choke is too large, flow will not reach the transonic point by the time it reaches the throat of the chamber, and post-throat flow will be subsonic and decelerate. In practice, the optimal value for A_choke then depends upon the volume of gas forced out of the chamber per second ($\text{m}^3\,\text{s}^{-1}$), the average mass of particles in the gas (kg), and the power liberated by the combustion ($\text{J}\,\text{s}^{-1}$). These ideas are explored further in Problem 2.18.

Problem Set 2

2.1 Show by direct substitution that Eq. (2.14) in the text solves the 3-D wave equation, Eq. (2.9). You may do your proof in Cartesian coordinates, if you like.

2.2 Equation (2.9) is a wave equation in p derived from Eq. (2.7) and (2.8).

a) Derive from these same equations a wave equation for $\vec{v}$, namely,
$$\frac{\partial^2 \vec{v}}{\partial t^2} = p_0' \nabla(\nabla \cdot \vec{v}). \tag{2.84}$$

b) Show that for a plane wave propagating in the x direction (and thus $\vec{v} = v_x \hat{x}$), this reduces to:
$$\frac{\partial^2 v_x}{\partial t^2} = p_0' \frac{\partial^2 v_x}{\partial x^2}.$$

2.3 Equation (2.30) in the text gives the pressure and velocity fluctuations for a propagating 1-D sound wave for all positions and time.

a) By integrating over all space (*e.g.*, $-\infty < x < \infty$), show that the momentum of the gas is conserved.

b) By looking at Fig. 2.1 in the text, can you tell if total energy is being conserved? If you think it is, explain why. If you think it isn't, explain where the energy is either coming from or going to.

2.4 Suppose, in Fig. 2.1 in the text, the initial pressure perturbation in a sound wave at $t = 0$, $\tilde{p}(x)$, is centred at $x = 0$ and has a width $\Delta x = 2$. Suppose further that the height of the pulse on the left (right) is 0.003 (0.001), all in arbitrary units, and that the sloped portion of the perturbation begins and ends at $x = \mp 0.5$. Let $\rho_0 = 1$, $p_0 = 0.6$, and $v_0 = 0$, again in arbitrary units.

a) If the gas is monatomic, what is the sound speed in these arbitrary units (neglecting the pressure and density fluctuations caused by the passage of the sound wave itself)?

b) At what time does the left edge of the pulse reach $x = -4$?

c) What is the pressure and velocity at $x = 0$ and $t = 0.5$, again in these arbitrary units?

d) What is the pressure and velocity at $x = 3$ and $t = 3$?

2.5* Suppose, at $t = 0$, the pressure and velocity perturbation profiles of a sound wave are given by:

$$\tilde{p}(x) = \begin{cases} 0.002 & -0.5 \leq x \leq 0, \\ 0.001 & 0 < x \leq 0.5, \\ 0 & \text{elsewhere}; \end{cases} \qquad \tilde{v}(x) = \begin{cases} 0.003 & -0.5 \leq x \leq 0.5, \\ 0 & \text{elsewhere}. \end{cases}$$

Suppose further that $\rho_0 = \frac{1}{2}$, $p_0 = \frac{5}{14}$, $v_0 = 0$, and the gas is diatomic. All units are arbitrary.

  a) What is the unperturbed (adiabatic) sound speed in these arbitrary units?

  b) Sketch with some quantitative accuracy and in a manner similar to Fig. 2.1 in the text the profiles of the pressure and velocity perturbations at $t = 0, 0.3,$ and 0.6.

2.6*

  a) Similar to what is done in the text for an adiabatic gas, solve the Rankine–Hugoniot jump conditions for an isothermal shock, and include in your solutions an expression for the post-shock Mach number.

  Hint: To start, you'll want to replace Eq. (2.32) in the text with the integral form of Eq. (1.40) in 1-D and steady state, from which you should find $p_1 v_1 = p_2 v_2$. Recall also that the *isothermal* sound speed is given by $c_s^2 = p/\rho$ (Eq. 2.11).

  b) Show that across an isothermal shock, the temperature remains the same and entropy actually *decreases*. Why is the latter result not a violation of the second law of thermodynamics?

  c) Is there such a thing as an isothermal contact discontinuity? Explain.

2.7* A *polytropic* gas is one where $p = \kappa \rho^\gamma$, and where κ is strictly constant. Thus, for the two states depicted in Fig. 2.2 in the text, $\kappa_2 = \kappa_1$ and Eq. (2.53) then requires that entropy be uniform and strictly conserved across a shock. A polytropic gas differs from an adiabatic gas in that the latter admits spatially and temporally varying values of κ, and thus entropy is not strictly conserved.

  a) Similar to what we did in class for an adiabatic gas, solve the Rankine–Hugoniot jump conditions for a polytropic shock, and show that the jump in speed relative to the shock, $\phi = v_2/v_1$, is given by the transcendental equation:

$$\frac{1 - \phi^{-\gamma}}{\phi - 1} = \gamma M_1^2, \qquad (2.85)$$

  where $M_1 = v_1/c_{s_1}$ is the Mach number, and the sound speed is given by Eq.

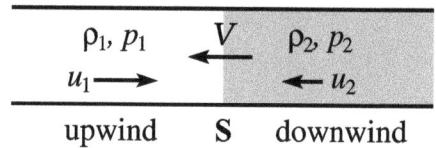

Figure 2.12. (Problem 2.9) A 1-D shock tube viewed from a reference frame relative to which the shocked gas has speed, u_2, and the shock, **S**, has speed $\mathcal{V}$.

(2.10) in the text. Express the jumps in ρ and p in terms of ϕ, as well as the post-shock Mach number in terms of ϕ.

Hint: For this problem, consider only Eq. (2.31) and (2.33) in the text, and replace p directly with $\kappa \rho^\gamma$.

b) Are contact discontinuities even admitted? Explain.

2.8*

a) Verify all three of Eq. (2.58) in the text.

b) Suppose you wanted to excite a 1-D shock inside a long tube with one end closed off and with *shock strength* $\zeta = p_2/p_1 = 5.00 \pm 0.02$, where p_2 (p_1) is the postshock (preshock) pressure. From the frame of reference of the postshock gas (*e.g.*, the "lab frame" as described in §2.2.3), with what speed (with uncertainty) must air be launched into the tube to attain such a shock? (You may assume $\gamma = 7/5$ for air.)

2.9**

a) In terms of the shock strength, $\zeta = p_2/p_1$ where p_2 (p_1) is the postshock (preshock) pressure, show that the density jump, ρ_2/ρ_1, is given by:

$$\frac{\rho_2}{\rho_1} = \frac{1 + \beta\zeta}{\beta + \zeta}, \qquad (2.86)$$

where $\beta = (\gamma + 1)/(\gamma - 1)$.

b) A slightly more general "lab frame" than introduced in §2.2.3 is depicted in Fig. 2.12 where the postshock gas has a non-zero residual speed, u_2, relative to the observer. From this frame of reference, show that in terms of the preshock speed, $u_1 > 0$, and the shock strength, ζ, we have:

$$\mathcal{V} = u_1 - c_{s,1}\sqrt{\alpha(1 + \beta\zeta)} \quad \text{and} \quad u_2 = u_1 - c_{s,1}\frac{\zeta - 1}{\gamma\sqrt{\alpha(1 + \beta\zeta)}}, \qquad (2.87)$$

where $\mathcal{V}$ is the propagation speed of the shock, $\alpha = (\gamma - 1)/2\gamma$, and $c_{s,1} = \sqrt{\gamma p_1/\rho_1}$ is the sound speed in the preshock fluid. Note that if the postshock

Problem Set 2

(downwind) state were on the left in Fig. 2.12, $u_1 < 0$ and the terms proportional to $c_{s,1}$ would be added rather than subtracted.

c) *Relative to the preshock fluid, which has the greatest magnitude: $\mathcal{V}$ or u_2? i.e., which of $|\mathcal{V} - u_1|$ and $|u_2 - u_1|$ is larger?*

2.10 Show that the shock jump conditions in the lab frame, namely Eq. (2.58) in the text, go to the same limiting functions (namely Eq. 2.45, 2.46, and 2.47) as $M_1^2 \to \infty$ as do the shock jump conditions in the shock frame (namely Eq. 2.40, 2.41, and 2.42), as $M_1^2 \to \infty$.

2.11 Suppose the temperature jump across a contact is T_2/T_1. Show that the entropy jump across the contact is given by:

$$\Delta S = \gamma \ln\left(\frac{T_2}{T_1}\right).$$

2.12** *The tides of the Bay of Fundy.*

a) As depicted in Fig. 2.13, a bathtub of length L is filled with water to a depth h. If the water is driven to slosh back and forth in such a way that the water surface remains planar, show that the potential energy of the bathwater, U, is given by:

$$U = \frac{\rho L w g}{6} y^2,$$

where ρ is the density of water, w is the width of the tub, and y is the additional depth of water above h at the left end of the tub ($x = 0$).

b) Show that the kinetic energy, K, of the horizontal motion of the water is given by:

$$K = \frac{\rho w L^3}{60 h}\left(\frac{dy}{dt}\right)^2.$$

Here, we assume the amplitude of oscillation is sufficiently small that we can ignore the contribution of the vertical motion to the kinetic energy.

c) Assuming the oscillation is weakly damped and thus energy dissipation is very small, set $E = U + K = $ constant and show that the water level rises and falls as a simple harmonic oscillator. Thus, find its natural frequency, ω_0, and period of oscillation, T_0.

For small damping, the resonant frequency $\omega_r \approx \omega_0$. Therefore, what is the resonant period of oscillation for a tub with $L = 1.5$ m and $h = 0.3$ m? Does this jibe with your intuition? (Surely *every* kid discovered that sliding back and forth along the bottom of the tub with just the "right" frequency will spill water onto the floor!)

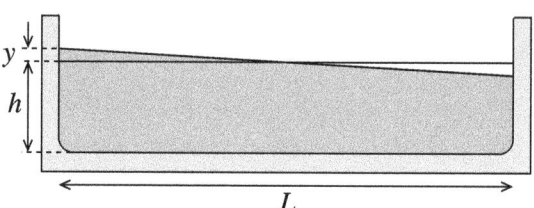

Figure 2.13. (Problem 2.12) A bathtub with an oscillating water level.

d) The Bay of Fundy is about 250 km long. If we model it as a big bathtub sloshing back and forth as it is driven by solar and lunar tidal forces, the bay itself would represent half of the tub with the other half stuck into the Gulf of Maine. Thus, take $L \sim 500$ km and use $h \sim 50$ m as its effective depth. What is the resonant period of oscillation? Compare this with the driving frequency of the moon, which you can work out from the fact that there are two high tides every time the moon returns to the same right ascension (longitude) in the sky.

You should find the driving frequency of the moon to be darn near the resonant frequency of the bay, and we can explain its extraordinarily high tides as a resonance phenomenon. To calculate the actual amplitude ($y \sim 8$ m) is a bit more complicated, as this depends on damping terms such as the viscosity of sea water and the resistance to flow from the bottom and sides of the bay.

2.13* The claim was made on page 45 of the text that one cannot simultaneously conserve mass, momentum, and energy across a bore. This problem explores that assertion.

a) Starting with the energy equation, Eq. (1.23) in the text, where the total energy density, e_T, is given by Eq. (1.20), show that in the steady state, integrating around the control volume in Fig. 2.7 results in a net power per unit width leaving the control volume given by:

$$P_{CV} = \frac{\rho g^{3/2}}{4} \sqrt{\frac{h_2 + h_1}{2 h_2 h_1}} (h_2 - h_1)^3,$$

where h_2 and h_1 are the depths of the water of density ρ downstream and upstream of the bore. Since $P_{CV} \neq 0$ except for $h_2 = h_1$ (no jump), energy is *not* conserved across a bore.

b) In terms of $\xi = h_2/h_1$, what percentage of the *mechanical power* (*i.e.*, excluding thermodynamic terms arising from internal energy and atmospheric pressure) entering the control volume from the upstream side is dissipated by the bore?

c) Using your result from part b), what fraction of mechanical power is dissipated by an undulating bore with $\xi = 1.2$? By a foaming bore with $\xi = 1.5$? (Answers: 0.10% and 1.08% respectively).

2.14 As shown in the inset, a *sluice gate* with pressure head, d, behind it is raised a height $h_1 \ll d$, allowing water to stream out at a constant speed. Because of an *obstacle* downstream, an hydraulic jump (H; stationary in the frame of reference of the gate) is established in the flow.

a) What is the depth, h_2, and speed, v_2, of the water downstream of the jump?

b) If $d = 1$ m and $h_1 = 1$ cm, is the jump (bore) undulating or foaming?

Hint: To find the constant flow speed, v_1, between the gate and the jump, construct a suitable streamline between the reservoir and the region between the gate and the jump, and use Bernoulli's theorem for liquids.

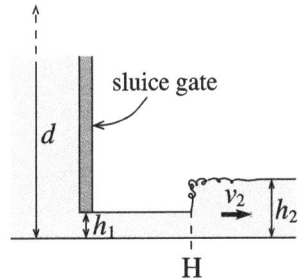

2.15* As shown in the inset, a *sluice gate* with a constant pressure head $d = 5.00$ m behind it is raised a height $h_1 = 5.00 \times 10^{-3}$ m, allowing water to stream out at a constant speed, u_1 (to be determined).

Because of a *barrier* at a distance $L = 3.00$ m downstream of the gate, a bore moving from the barrier towards the gate is established in the flow. From the moment the sluice gate is opened, how long does it take for the bore to reach the gate in seconds? You should assume the problem is "slab symmetric"; that is, derivatives into the page are zero.

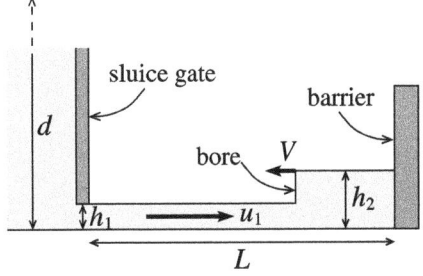

2.16 Show that Bernoulli's theorem is valid across a steady-state, normal shock ('normal' meaning the streamline is parallel to the shock normal). That is to say with $\phi = 0$, show that Eq. (2.72) in the text is consistent with the integrated conservative equations, Eq. (2.31)–(2.32), used to derive the Rankine–Hugoniot jump conditions.

2.17

a) Consider again the example of Bernoulli levitation described in §2.4 and depicted in Fig. 2.10 in the text. Show that the maximum mass that can be supported by this mechanism (ignoring the counter-thrust) is given by:

$$M_{\max} = \frac{2\pi a^2 \ln(R/a)\, p_{\text{atm}}}{g},$$

where all variables are defined in the text.

b) What is the maximum mass supportable by sucking air in through the hose?

c) What is the ratio of the maximum mass supportable by Bernoulli levitation to that supportable by suction?

2.18 Consider again the *de Laval nozzle* discussed in §2.4 in the text. Using no more than simple *dimensional analysis* arguments, determine how the cross sectional area of the choke point, A_{choke}, depends upon the physical quantities that could reasonably determine it (listed at the end of §2.4).

2.19* Extragalactic jets are long ($\sim 10^6$ light years), collimated supersonic "beams" of plasma emitted roughly along the rotation axes and from the "accretion discs" of giant black holes ($\sim 10^9 M_\odot$[21]) residing at the centres of the most massive galaxies.[22] Jets transport a small fraction of the accreting matter back into the *intergalactic medium* (IGM) and with it much of the angular momentum, thus allowing the material left behind in the accretion disc to fall into the black hole (*a.k.a.*, an *active galactic nucleus*; AGN). Figure 2.14 shows a 6-cm radio image of what is often referred to as the "prototypical" extragalactic radio source, Cygnus A.[23] Emission from most every extragalactic radio source is via the *synchrotron mechanism* by which relativistic electrons spiral about magnetic fields.

Figure 2.15 shows the "anatomy" of a jet, as known from numerical simulations such as those in Fig. 2.16. Ultra-hot (10^9 K or more) jet material from the AGN moves at near-light speeds from left to right in the figure, and impacts against the much denser IGM at the right. By terrestrial standards, the IGM is a near-perfect vacuum with about 1–100 particles per cm^3. However, the jet is even lighter at a few tens of particles per cubic *metre*, a factor of a *million* less. As a comparator, water is only about 800 times denser than air. So, even more so than a jet of air blasting into water, the progress of an astrophysical jet is slowed dramatically by the inertia of the IGM, causing a terminal "Mach disc" or "Mach stem" to form at the end of the jet. Here, jet material shocks, has most of its kinetic energy converted to thermal energy, and forms a "hot spot" from which highly pressurised material "squirts" out laterally in all directions inflating a "bubble", more commonly known as a "cocoon" or "lobe".

As much as the jet may be slowed at the hot spot, it still moves through the IGM supersonically (relative to the low sound speed there), exciting a "bow shock"

---

[21] $M_\odot \sim 2 \times 10^{30}$ kg is the mass of the sun.
[22] For a quick review of the subject, see Clarke *et al.*, 2008.
[23] The "A" is used by radio astronomers to indicate this is the brightest radio source (at 1400 MHz) in the constellation Cygnus. The second brightest is thus Cygnus B, and so on.

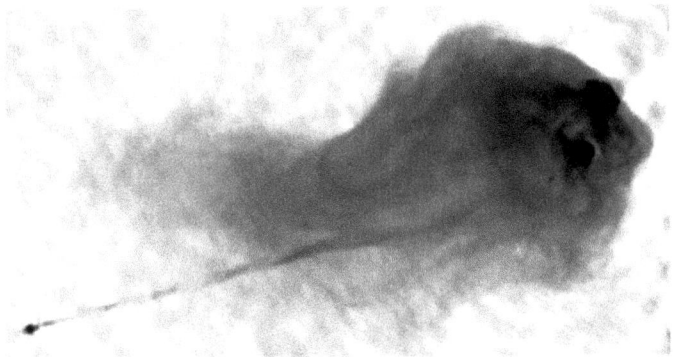

Figure 2.14. (Problem 2.19) A 6-cm radio image of the western half of the extragalactic radio source, *Cygnus A*, observed with the National Radio Astronomy (NRAO) *Very Large Array* (VLA) radio telescope near Socorro NM. A "reversed" grey scale is used, rendering the brightest emission black. The AGN is buried in the bright spot in the lower left from which a thin jet emanates. The supersonic jet impacts against the IGM at the brightest (darkest) "hot spots" in the right of the image, inflating the lobe/cocoon as seen (from Perley *et al.*, 1984, © AAS, reproduced with permission).

that leads the advancing jet. Separating the shocked jet material (cocoon) from the shocked ambient material is a *contact discontinuity* which is Kelvin–Helmholtz (§7.1) and Rayleigh–Taylor (§7.2) unstable, as indicated by the wiggly line representing the contact in Fig. 2.15. It is only the jet material (which does not cross the contact) that contains any appreciable magnetic field (coming from the AGN), and thus it is only the jet and cocoon/lobe that emit synchrotron radiation and are visible in radio images. Thus, the emission in Fig. 2.14 corresponds to everything inside the contact discontinuity only. In particular, then, the bow shock excited in the ambient medium is *not* visible even though it must be there.

A small but significant subset of extragalactic radio sources exhibit "nested lobes" in which a younger, more energetic, and smaller lobe seems to be inflating within an older, less energetic, and larger lobe.[24] A prevailing model to describe such a phenomenon is the "restarting jet" model[25] in which a jet, for whatever reason, ceases to flow for some time, then resumes well after the original jet has disappeared. Simulations show that while the jet is "off", the lobe inflated by the previous jet relaxes and cools only slightly, so that the "restarted jet" propagates through a hot and diffuse *relic lobe* as opposed to the denser and colder IGM through which the original jet propagated. Among other things, this model suggests that the restarting jet should be more ballistic in nature than the original jet.

Figure 2.16 shows stills from a *ZEUS-3D* simulation of a restarting jet, where the first image is shown just before the first jet is turned off, the second after the

---

[24] For example, see Saripalli *et al.* (2003).
[25] For example, see Bridle *et al.* (1986).

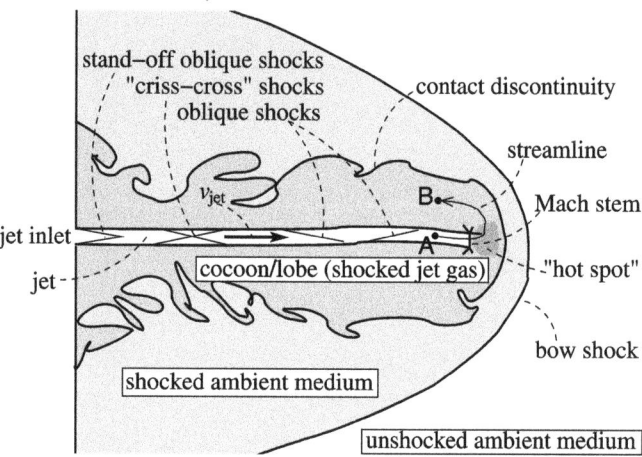

Figure 2.15. (Problem 2.19) A cartoon of a numerical simulation of the jet in Cygnus A (Fig. 2.14). All features discussed in the text are labelled. The streamline to be used for the problem is the line joining points A and B to the right of the diagram.

lobe has relaxed somewhat and just before the jet is resumed, the third soon after the jet is restarted, and the fourth just after the restarted jet has penetrated the relic lobe and resumed pushing its way into the IGM. One of the controversies of the "restarting jet" model has to do with the putative bow shock excited by the restarted jet as it moves through the relic lobe. Such a shock should re-energise the relativistic particles in the old lobe material, and thus be plainly visible. To date, among the many known candidate restarting jets, no bow shock has ever been observed drawing into question the restarting jet model.

So, finally to the problem. Assume that the ratio of jet density to IGM density, $\eta = \rho_j/\rho_x$ ('x' for "external"), and the Mach number of the jet relative to the sound speed in the IGM, M_x, are known and the same for the original and restarted jets. Assuming a monatomic gas, use Bernoulli's theorem to show that *relative to the relic lobe material*, the Mach number of the restarting jet, M_c ('c' for "cocoon"), cannot exceed 2, and that for "typical" values for η ($\sim 10^{-6}$) and M_x (~ 100) in an extragalactic jet, restarting jets actually move *subsonically* through the relic lobe, thus explaining the observed absence of bow shocks.

Here's how to proceed:

- You have M_x and you want M_c. So start by finding M_j, the Mach number of the jet relative to the *jet* sound speed, in terms of M_x and η.

- Make the reasonable assumption that the thermal pressure in the jet and the IGM are roughly equal; if they weren't, one would expand into the other until they were.

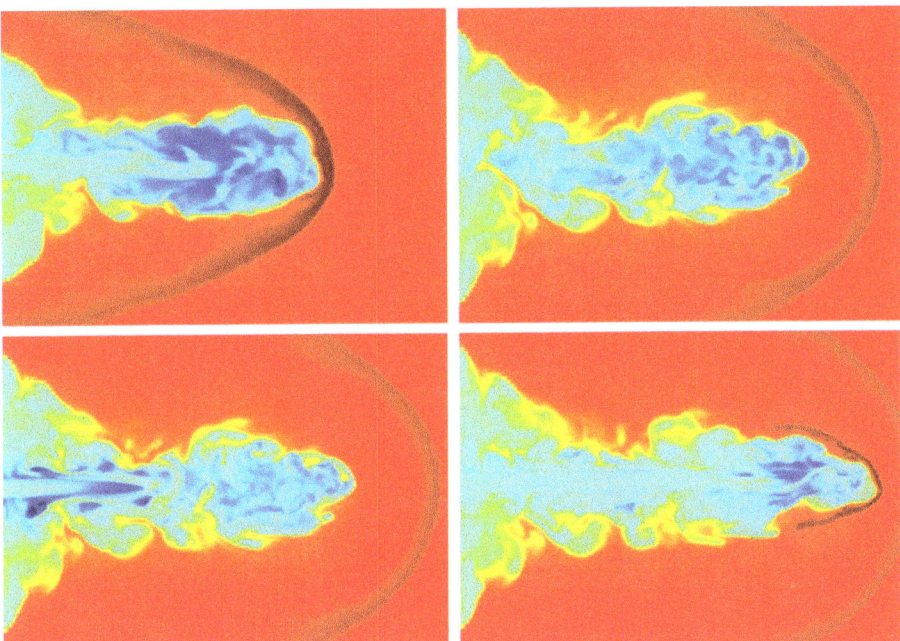

Figure 2.16. (Problem 2.19) Snapshots of a *ZEUS-3D* simulation of a restarting jet in 3-D showing 2-D density slices (red, high; blue, low) through the central plane of the computational domain at four different epochs: *upper left*, just before the jet is turned off; *upper right*, just before the jet is resumed; *lower left*, just after the jet has been resumed; *lower right*, just after the restarted jet resumes penetrating the IGM.

- To get M_c from M_j, you need to relate unshocked jet conditions to shocked jet (cocoon) conditions. You do this by linking a point in the jet (point **A** in Fig. 2.15) with a point in the cocoon (point **B**) by a streamline, and applying Bernoulli's theorem.

- Another fair assumption here is that the cocoon is in a state of *transonic turbulence* (as confirmed by the simulations) and, as such, you can safely take $v_c \sim c_{s,c}$.

- algebra ...

If you make any further assumptions in your calculations, be sure to state them clearly, and give some justification for them as I have done for those I suggest above.

2.20* Figure 2.17 shows the *Wide Angle Tail* radio source, 1919+479 with its 320 kpc western jet that is apparently deflected by almost 90° to the south, before continuing on for another 800 kpc as a much wider "plume".[26] (See the prologue to

---

[26] Burns *et al.* (1986)

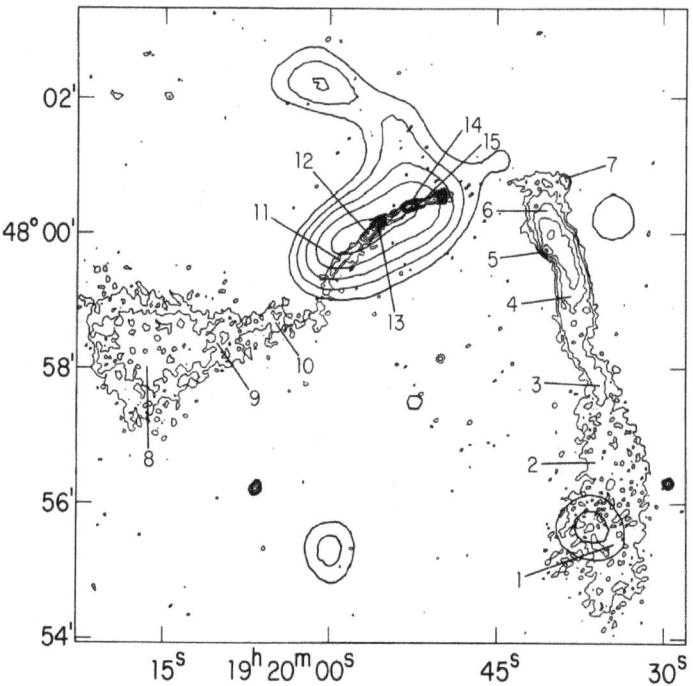

Figure 2.17. (Problem 2.20) A VLA image of the Wide-Angle Tail radio galaxy, 1919+479 showing 20-cm radio emission (4''.5 resolution) in fine contours and x-ray emission (1'.8 resolution) in heavier contours. The core of the radio source (AGN) – an unlabelled feature halfway between features 13 and 14 – launches two jets. The "western jet" whose brightest spots are features 14 and 15, seems to disappear from this image shortly after feature 15, but can be traced in higher sensitivity images through to feature 7 where it does, in fact, end. A long (800 kpc) "plume" emerges from feature 7, continuing on through and beyond feature 1. The "eastern jet" is also traceable from features 13 through and beyond 8, but doesn't exhibit the same abrupt turn as the western jet (from Burns *et al.*, 1986; © AAS, reproduced with permission).

Problem 2.19 for an introduction to jets.) Both the length and the brightness of the plume indicate the flow in the plume is likely supersonic yet, in the laboratory, it is not known how to deflect a supersonic jet much past its *Mach angle* [$\tan^{-1}(1/M)$] without completely disrupting it.

As illustrated in Fig. 2.18, one explanation for the bend in the western jet of 1919+479 is the spontaneous formation of a de Laval nozzle. Suppose a "cloud" (*e.g.*, an HI region) of mass m_c is parked at feature 7 in Fig. 2.17. If a narrow, supersonic jet of cross-sectional area A_j impinges upon the cloud and bores a hole into it, jet material will shock on entry, slow to the co-moving frame of the cloud, and inflate a cavity. As the cavity grows, it will breach the surface of the cloud somewhere else venting shocked jet material back into the IGM. In the case of 1919+479, this vent is presumably at the base of the plume and, if this vent forms a de Laval nozzle, gas will be reaccelerated to supersonic speeds allowing the plume to carry on, as observed.

That a de Laval nozzle would form all on its own is actually quite plausible. As shown in §2.4 in the text, the flow rate through the nozzle is maximum when the flow speed at the throat of the nozzle is transonic. With material inside the cavity "anxious" to escape, it is quite conceivable that the system would attain the right conditions at the vent to achieve maximum flow-through, and indeed, such a scenario has been observed in numerical simulations. The big question is, could a cloud stay put long enough for a plume 800 kpc in length (impressive, even for this class of radio sources) to form? Wouldn't the peculiar motion of the cloud (say its orbit about the nucleus of 1919+479) move it out of the way over the lifetime of the plume? Wouldn't the jet carve the cloud up into shards? Even if both of these questions turn out null, what of the thrust of the plume? Wouldn't the thrust of the venting material force the cloud to move northwards, and thus out of the way of the jet long before the plume could achieve its observed length?

So finally, here is the question. In order for the thrust of the venting plume material not to push the cloud out of the way, what is the lower limit of the mass of the cloud, m_c, in terms of the jet flow variables, ρ_j and v_j, the radius of the jet as it enters the cloud, r_j, the presumed age of the plume, τ_p, and whatever other variable may seem applicable and observationally accessible?

Author's note: This problem was posed to me as a beginning PDF just as I have posed it to you here. I give hints below to lead you to my solution, but I encourage you to try the problem first without using these hints; who knows, my hints might bias you and prevent you from coming up with a better answer!

Hints: Set up the problem with two sets of primitive variables as suggested in Fig. 2.18: (ρ_j, p_j, v_j) just before the jet enters the cloud, and (ρ_n, p_n, v_n) at the throat of the purported de Laval nozzle. Let the cross-sectional areas of the jet as it enters the cloud and the throat of the nozzle be A_j and A_n respectively.

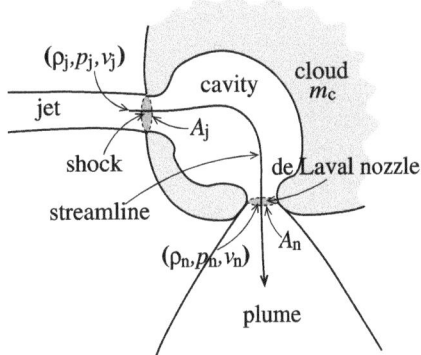

Figure 2.18. A schematic of a de Laval model to explain the apparent ability of the supersonic jet in 1919+479 to bend by almost 90° without disrupting.

- Write down an expression for the reaction force, F_c, on the cloud from the thrust of the venting plasma in terms of ρ_n, v_n, *etc.*

- Write down an expression for the conservation of mass between the entrance and exit of the cloud in terms of jet and nozzle values.

- Connect the jet and nozzle values by a streamline and conserve the Bernoulli function. You might note that the Mach number for an extragalactic jet relative to its own sound speed, M_j, is typically ≥ 10 and sometimes much greater than 10, which could help simplify one of your expressions along the way.

- A bit of kinematics, plus the notion that the cloud shouldn't move more than a jet diameter over the lifetime of the plume (otherwise the jet has to bore a new hole), and you should arrive at:

$$m_c \geq \frac{\pi}{8} \rho_j v_j^2 r_j \tau_p^2 \qquad \text{for } \gamma = 5/3. \tag{2.88}$$

For "typical" values ($\rho_j \sim 10^{-25}$ kg m^{-3}, $v_j \sim 0.1c$, $r_j \sim 0.5$ kpc, $\tau_p \sim 10^8$ yr), Eq. (2.88) requires that $m_c > 3 \times 10^9 \, M_\odot$, and this is a problem since extensive optical observations at the bend of the western jet fail to show any significant Hα emission which would be expected if more than $10^9 M_\odot$ of gas were parked there.[27] I have not since seen a convincing resolution to this problem, though my (and I would have thought Occam's) suggestion is projection effects are exaggerating what is in fact a modest deflection angle at feature 7. However, in a private communication, one of this project's investigators dismissed this suggestion on the grounds that it would then make this already longest jet-plume structure known to be "ridiculously longer".

---

[27]Pinkney *et al.* (1994)

3
The Hydrodynamical Riemann Problem

With every simple act of thinking, something lasting and substantial enters our soul.[†]

Bernhard Riemann (1826–1866)

BERNHARD RIEMANN of Hanover, Germany may not have been as prolific in his short life as many of the mathematical greats such as Euler, Gauss, and Dirichlet by whom he was most influenced, but he is still regarded as one of the most brilliant mathematicians of the 19th century.[1] He is best known for his work in complex analysis (*e.g.*, the *Cauchy–Riemann conditions*) and for what is now known as *Riemannian geometry*, the framework upon which Einstein built his general theory of relativity some 60 years later. But most famous of all is what is known as the *Riemann hypothesis*,[2] which he stated without proof in a paper published in 1859 concerning the number of primes less than x. Also known as the Riemann conjecture, it is the last of the great unsolved problems from the 19th century and still considered by some to be *the* most important unsolved problem in number theory. To the person who finally proves or disproves this conjecture goes a "Millennium Prize" of $1 million, still not collected as of this writing!

While Riemann was not a fluid dynamicist, his pioneering methods in the theory of *hyperbolic* partial differential equations – of which the HD equations are prime examples (see App. C) – has had a profound effect on the way we think of fluids. In particular, the *Riemann problem* is defined as a system of hyperbolic differential equations with two piecewise constant states as initial conditions. At the heart of this problem lies the realisation that the solution to hyperbolic equations can be considered to be the superposition of waves, one for each independent variable.

For fluid dynamics, the initialisation of a typical Riemann problem is illustrated in Fig. 3.1. Two constant "left" and "right" states are brought into contact at position **D** and at time $t = 0$. This is *not* the "shock tube" problem addressed in §2.2; no assumption of steady state is being made here. In particular, the question posed by the Riemann problem is 'What is the spatial dependence of the primitive variables at some later time?'

In this chapter, we shall build up the tools necessary to follow Riemann's

---

[†]From H. Weber, 1876, *Bernhard Riemann's Gesammelte Mathematische Werke*, p. 477, translation by *Google Translate*.
[1] www.wikipedia.org/wiki/Bernhard_Riemann
[2]The real part of every non-trivial zero of the (Riemann) zeta function, $\zeta(z)$, is $\frac{1}{2}$.

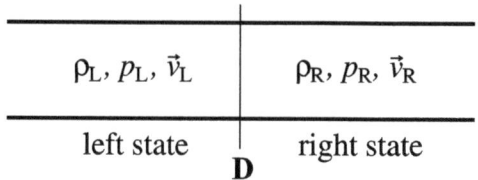

Figure 3.1. Initial set up for the Riemann problem. At $t = 0$, the diaphragm, **D**, is removed, and the two constant states interact with an arbitrary jump in (possibly) all flow variables at **D**.

thinking and to solve the problem that now bears his name. Among other reasons, this problem is important to fluid dynamics because it is at the base of some of the most robust and widely used computational algorithms for solving general problems in (magneto)hydrodynamics.

3.1 Eulerian and Lagrangian frames of reference

By the chain rule, the full time-derivative of a function $f(\vec{r}, t)$ is:

$$\frac{df(\vec{r},t)}{dt} = \frac{\partial f(\vec{r},t)}{\partial t} + \frac{\partial f(\vec{r},t)}{\partial x}\frac{dx}{dt} + \frac{\partial f(\vec{r},t)}{\partial y}\frac{dy}{dt} + \frac{\partial f(\vec{r},t)}{\partial z}\frac{dz}{dt}, \quad (3.1)$$

where, in general, the position of a point in the fluid, $\vec{r} = (x, y, z)$, has its own time-dependence; *i.e.*, it moves. If it doesn't, only the partial time derivative survives.

So what do all the terms in Eq. (3.1) actually mean? Suppose, for the moment, $f = \rho$, the fluid density. Consider an observer comoving with a small parcel of fluid with a fixed mass, ΔM, and a *variable* volume, $\Delta \tau$. As this volume compresses or expands from outside influences, the change in density inside $\Delta \tau$ is measured by the $\partial \rho / \partial t$ term. Let us call this the *intrinsic rate of change* of the density.

Now consider an observer in an inertial "lab frame" watching a *fixed* volume, ΔV, into and out of which fluid is moving at some velocity $\vec{v} = (dx/dt, dy/dt, dz/dt)$. In addition to the intrinsic changes to the density that happen in all frames of reference, matter inside ΔV will leave and other matter of possibly different density will flow in to replace it. These *extrinsic rates of change* will also cause the density inside ΔV to change, and are accounted for by the $\partial \rho/\partial x, \partial \rho/\partial y$, and $\partial \rho/\partial z$ terms.

If we associate the time derivatives of the coordinates $(dx/dt, etc.)$ with the components of the flow velocity, $\vec{v}$ (something we are not necessarily obliged to do), it is customary to write Eq. (3.1) as:

$$\begin{aligned}\frac{D_v f(\vec{r},t)}{Dt} &= \frac{\partial f(\vec{r},t)}{\partial t} + v_x \frac{\partial f(\vec{r},t)}{\partial x} + v_y \frac{\partial f(\vec{r},t)}{\partial y} + v_z \frac{\partial f(\vec{r},t)}{\partial z} \\ &= \frac{\partial f(\vec{r},t)}{\partial t} + \vec{v} \cdot \nabla f(\vec{r},t).\end{aligned} \quad (3.2)$$

Here we have used the upper-case 'D' instead of the lower case 'd' to denote the full time derivative, and the subscript 'v' to indicate which velocity is associated with the time derivatives of the coordinates.

The construct $D_v f/Dt$ is called the *Lagrangian derivative* (also known as the *material derivative*), named after Joseph-Louis Lagrange (1736–1813), the Sardinian (Italian) born but French bred physicist/mathematician responsible for many ideas in classical mechanics and hydrodynamics.[3] The velocity associated with the Lagrangian derivative is called the *Lagrangian* or *characteristic* velocity, and when that velocity is the fluid flow velocity, the subscript is normally dropped from the notation (*e.g.*, Df/Dt).

In fact, we've been using Lagrangian derivatives all along, but never referred to them as such. For example, by rewriting the continuity equation:

$$\frac{\partial \rho}{\partial t} + \nabla \cdot \rho \vec{v} = 0 \quad \Rightarrow \quad \frac{\partial \rho}{\partial t} + \vec{v} \cdot \nabla \rho + \rho \nabla \cdot \vec{v} = 0 \quad \Rightarrow \quad \frac{D\rho}{Dt} = -\rho \nabla \cdot \vec{v},$$

we see that the Lagrangian derivative (with characteristic speed, $\vec{v}$) was always there.

When $\vec{v}$ is the flow velocity relative to the lab frame, we call this the *Eulerian reference frame* in which Eq. Set 3 (page 20) may be written:

Equation Set 4 (Eulerian fluid equations):

$$\frac{D\rho}{Dt} = -\rho \nabla \cdot \vec{v}; \quad \text{continuity}$$

$$\frac{Dp}{Dt} = -\rho c_s^2 \nabla \cdot \vec{v}; \quad \text{pressure equation}$$

$$\frac{D\vec{v}}{Dt} = -\frac{1}{\rho}\nabla p - \nabla \phi. \quad \text{Euler's equation}$$

Alternately, we can examine the fluid from within its co-moving frame, as though we were corks bobbing up and down in a stream. This is known as the *Lagrangian reference frame* where $\vec{v}$ is set to 0 and the Lagrangian derivative becomes:

$$\frac{Df(\vec{r},t)}{Dt} = \frac{\partial f(\vec{r},t)}{\partial t}.$$

Thus, in the Lagrangian frame, Eq. Set 3 becomes:

Equation Set 5 (Lagrangian fluid equations):

$$\frac{\partial \rho}{\partial t} = -\rho \nabla \cdot \vec{v}; \quad \text{continuity}$$

$$\frac{\partial p}{\partial t} = -\rho c_s^2 \nabla \cdot \vec{v}; \quad \text{pressure equation}$$

$$\frac{\partial \vec{v}}{\partial t} = -\frac{1}{\rho}\nabla p - \nabla \phi. \quad \text{Euler's equation}$$

Note that being in the co-moving frame of the fluid means that $\vec{v} = 0$ *locally*,

---

[3] www.wikipedia.org/wiki/Joseph-Louis_Lagrange

and thus $\nabla \cdot \vec{v}$ is not necessarily zero too. Further, since no two elements of the fluid are necessarily in the same frame of reference, examining all the fluid in its co-moving frame means, in principle, having to go to a different frame of reference for every point of the fluid at every point in time. Thus, the concept of a Lagrangian frame of reference has limited applicability for a general fluid. However, for a steady-state, uniformly moving fluid, or even a general 1-D fluid, the concept of a Lagrangian frame of reference can be very useful, as we shall see later in this section. Lagrangian and Eulerian frames of reference will also play a role in our understanding the all-important *magneto-rotational instability* in §7.3.

3.2 The three characteristics of hydrodynamics

We start with the 1-D non-linear form of the pressure and Euler's equations, Eq. (1.41) and (1.36), in the Eulerian frame with $\phi = 0$:

$$\frac{1}{\rho c_s}\frac{\partial p}{\partial t} + \frac{v}{\rho c_s}\frac{\partial p}{\partial x} + c_s \frac{\partial v}{\partial x} = 0; \qquad (3.3)$$

$$\frac{\partial v}{\partial t} + v\frac{\partial v}{\partial x} + \frac{c_s}{\rho c_s}\frac{\partial p}{\partial x} = 0, \qquad (3.4)$$

where Eq. (3.3) was divided through by ρc_s to give it the same units as Eq. (3.4), and where the placement of c_s in both the numerator and denominator of the third term in Eq. (3.4) is deliberate. Note that by starting with the *primitive differential form* of the equations, we are restricting our discussion to *smooth flow* (§1.5) and, in particular, exclusive of the formation of shocks and even contacts. Should we encounter discontinuities in our drive to solve Riemann's problem (and we will!), we shall have to incorporate, somehow, the Rankine–Hugoniot jump conditions (§2.2) into our discussion.

Now, for an adiabatic gas where $c_s^2 = dp/d\rho = \gamma p/\rho$:

$$dp = c_s^2 d\rho = c_s^2 \gamma d\left(\frac{p}{c_s^2}\right) = \gamma dp - \frac{2\gamma p}{c_s}dc_s = \gamma dp - 2\rho c_s dc_s$$

$$\Rightarrow \boxed{\frac{1}{\rho c_s}dp = \frac{2}{\gamma - 1}dc_s,} \qquad (3.5)$$

a rather useful identity. Thus, Eq. (3.3) and (3.4) become:

$$\frac{2}{\gamma - 1}\frac{\partial c_s}{\partial t} + \frac{2v}{\gamma - 1}\frac{\partial c_s}{\partial x} + c_s \frac{\partial v}{\partial x} = 0; \qquad (3.6)$$

$$\frac{\partial v}{\partial t} + v\frac{\partial v}{\partial x} + \frac{2c_s}{\gamma - 1}\frac{\partial c_s}{\partial x} = 0. \qquad (3.7)$$

Adding and subtracting Eq. (3.6) and (3.7) yield:

$$\frac{\partial}{\partial t}\left(v + \frac{2c_s}{\gamma - 1}\right) + (v + c_s)\frac{\partial}{\partial x}\left(v + \frac{2c_s}{\gamma - 1}\right) = 0; \qquad (3.8)$$

$$\frac{\partial}{\partial t}\left(v - \frac{2c_\text{s}}{\gamma - 1}\right) + (v - c_\text{s})\frac{\partial}{\partial x}\left(v - \frac{2c_\text{s}}{\gamma - 1}\right) = 0, \tag{3.9}$$

and suddenly the inherent symmetry in the pressure and Euler's equations is revealed. So strong is this symmetry, that an entirely new way of thinking about the fluid equations emerges.

By defining the *characteristics*,

$$\boxed{\mathcal{J}^+ = v + \frac{2c_\text{s}}{\gamma - 1};} \tag{3.10}$$

$$\boxed{\mathcal{J}^- = v - \frac{2c_\text{s}}{\gamma - 1},} \tag{3.11}$$

with *characteristic speeds*,

$$c^\pm = v \pm c_\text{s}, \tag{3.12}$$

($c^\pm = \pm c_\text{s}$ in the Lagrangian frame), Eq. (3.8) and (3.9) can be written more compactly as:

$$\frac{D_+\mathcal{J}^+}{Dt} = 0; \quad \frac{D_-\mathcal{J}^-}{Dt} = 0, \tag{3.13}$$

using the Lagrangian derivative defined by Eq. (3.2), and where $D_\pm$ is compact notation for $D_{c^\pm}$. Thus, the pressure and Euler's equations have been converted into Lagrangian derivatives of two characteristics, $\mathcal{J}^\pm$, and we interpret Eq. (3.13) as follows: *In the co-moving reference frame with characteristic speed, $c^\pm$, the characteristic $\mathcal{J}^\pm$ is constant.*

Equation Set 3 has three equations and three unknowns, and yet our analysis has only uncovered two characteristics of the flow. Symmetry suggests a third characteristic must exist independent of $\mathcal{J}^\pm$ and whose Lagrangian derivative is also zero, and one need not search long to find such a quantity. For an adiabatic gas, entropy is conserved,

$$\frac{dS}{dt} = \frac{\partial S}{\partial t} + v\frac{\partial S}{\partial x} = 0,$$

while for an isothermal gas, the specific internal energy ($\varepsilon = p/\rho \propto T$) is conserved,

$$\frac{d\varepsilon}{dt} = \frac{\partial \varepsilon}{\partial t} + v\frac{\partial \varepsilon}{\partial x} = 0.$$

Interpreting these conservation statements as Lagrangian derivatives each with characteristic speed,

$$c^0 = v, \tag{3.14}$$

($c^0 = 0$ in the Lagrangian frame) allows us to identify S as the third characteristic for an adiabatic fluid, and ε for an isothermal fluid.

We can combine these two seemingly different variables into a single third characteristic, $\mathcal{S}^0$, as follows:

$$\boxed{\mathcal{S}^0 = \frac{p}{\rho^n},} \tag{3.15}$$

where $n = \gamma$ (1) for an adiabatic (isothermal) gas. From Eq. (2.53), $S \propto \ln(\mathcal{S}^0)$ and

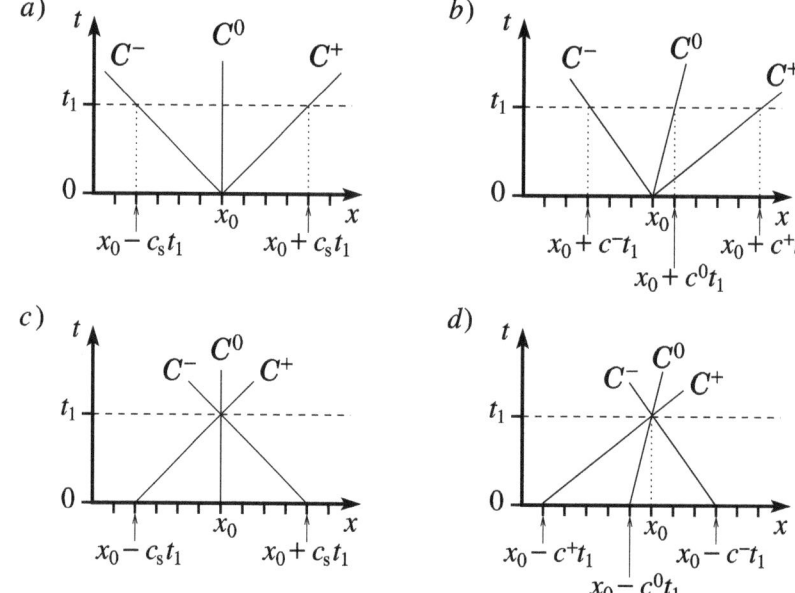

Figure 3.2. Space-time diagrams showing *a)* three characteristic paths emanating from the event $(x_0, 0)$ and arriving at three separate events at $t = t_1$ from a Lagrangian frame ($\mathcal{C}^0$ is vertical $\Rightarrow c^0 = v = 0 \Rightarrow$ Lagrangian), *b)* same as panel *a* but for an Eulerian frame, *c)* three characteristic paths from different events at $t = 0$ converging on the same event (x_0, t_1) from a Lagrangian frame, and *d)* same as panel *c* but for an Eulerian frame. Here, the flow speed relative to the Eulerian frame is evidently $v_0 = c_s/4$ from left to right.

so if S is constant in the co-moving frame of an adiabatic fluid, so is $\mathcal{S}^0$. Similarly, $\varepsilon = e/\rho = p/[\rho(\gamma-1)]$, and so if ε is constant in the co-moving frame of an isothermal fluid, so is $\mathcal{S}^0$.

Together, the three quantities – $\mathcal{J}^\pm$ and $\mathcal{S}^0$ – are known as *Riemann invariants* or, alternately, *characteristics* of the flow and associated with each is a *characteristic speed*; $c^\pm = v \pm c_s$ and $c^0 = v$ respectively. The first two characteristics are essentially *forward-* $(v + c_s)$ and *backward-* $(v - c_s)$ moving sound waves, while the third characteristic in the co-moving frame of the fluid is known as an *entropy wave*.

3.3 Characteristic paths and space-time diagrams

A useful construct to visualise the role characteristics play in fluid dynamics are *space-time diagrams*. Space-time diagrams such as those in Fig. 3.2 plotted with one spatial (position) coordinate along the horizontal axis and time along the vertical axis are referred to as 1+1-D diagrams. A point on a space-time diagram is called an *event*, while a path on a space-time diagram is called a *worldline*. Worldlines

characteristic (Riemann invariant)	characteristic speed	characteristic path
$\mathcal{J}^+ = v + 2c_s/(\gamma - 1)$	$c^+ = v + c_s$	$\mathcal{C}^+$
$\mathcal{S}^0 = p/\rho^{n*}$	$c^0 = v$	$\mathcal{C}^0$
$\mathcal{J}^- = v - 2c_s/(\gamma - 1)$	$c^- = v - c_s$	$\mathcal{C}^-$

*$n = \gamma$ for adiabatic gas, $n = 1$ for isothermal gas.

Table 3.1. Summary of symbols and expressions used to specify the three fluid characteristics. Table entries in the first two columns are for an Eulerian reference frame. For a Lagrangian reference frame, set $v = 0$.

record how a particle's position changes in time, and those parallel to the t-axis represent stationary particles while those with a horizontal component indicate moving particles. Evidently, the local slope of a worldline is inversely proportional to the particle speed, and typically one chooses a critical speed to the dynamics to be represented by worldlines inclined at some prominent angle, say 45°, relative to the t-axis. Finally, worldlines of characteristics are called *characteristic paths*, or just *paths* for short. Table 3.1 summarises the quantities and symbols associated with the three characteristics, their speeds, and paths introduced in this section.

For fluid dynamics, an important critical speed is the sound speed and so here we'll incline worldlines of particles or waves propagating at c_s relative to the origin of a space-time diagram at $\pm 45°$. Figures 3.2a and 3.2c are plotted from a Lagrangian reference frame where x_0 is in the co-moving frame of the fluid. Thus, the worldline of a fluid element at x_0 is vertical (parallel to the t-axis) since, in this frame, its velocity is zero. Suppose further that from event $(x_0, 0)$, all three characteristic waves are launched as shown in Fig. 3.2a. Since the entropy wave is also co-moving with the fluid, its characteristic path, $\mathcal{C}^0$, must be vertical too while the paths of the right- and left-moving sound waves – $\mathcal{C}^+$ and $\mathcal{C}^-$ respectively – are inclined at $\pm 45°$ relative to the t-axis, all as shown.[4]

Exploiting the third dimension, one could plot a 2+1-D space-time diagram with two spatial dimensions along the horizontal (x) and perpendicular (z) axes and time still along the vertical axis, as shown in the inset. Obviously, plotting a fully 3+1-D space-time diagram is problematic, unless one is adept at drawing in 4-D! On the 2+1-D space-time diagram, the volume of revolution one obtains by rotating the paths $\mathcal{C}^\pm$ about the path $\mathcal{C}^0$ (*e.g.*, Fig. 3.2a) is called

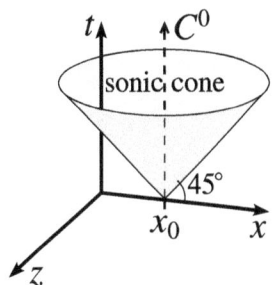

---

[4] As the waves move away from $(x_0, 0)$, both the speed of sound and the local rest frame may change, causing the worldlines to stray from the straight lines used to depict them in Fig. 3.2. However, for the region in the space-time diagram sufficiently local to $(x_0, 0)$, we can safely construct our worldlines as straight lines.

the *sonic cone* and events which lie within the sonic cone can be reached by event $(x_0, 0)$ at subsonic speeds. Meanwhile, events lying above the z–x plane but below the sonic cone can be reached from $(x_0, 0)$ only supersonically. For simplicity, all remaining discussion will be confined to 1+1-D spacetime diagrams.

Figure 3.2b is the space-time diagram for the same waves illustrated in Fig. 3.2a but from an Eulerian frame of reference. In this frame, the flow speed, v_0, is evidently $c_s/4$ from left to right (> 0) since the entropy wave in panel b moves one quarter the distance moved by the sound wave in panel a in the same time, t_1. In general, for flow speeds greater (less) than zero and relative to the t-axis, the path $\mathcal{C}^+$ is inclined at an angle greater (less) than 45°, the path $\mathcal{C}^-$ is inclined at an angle less (greater) than 45°, and the path $\mathcal{C}^0$ leans to the right (left).[5] Note that $\mathcal{C}^0$ will *always* lie between paths $\mathcal{C}^\pm$.

Analysis of a fluid system using its characteristics and characteristic paths is known loosely as the *method of characteristics* (MoC) whose power lies in the fact that at each point along a characteristic path, the corresponding characteristic (Riemann invariant) remains constant. The "ingredients" of a given characteristic, namely ρ, p, and v, may well change, but they will do so in a manner that will preserve the invariance of the affected characteristic. Thus, the value of $\mathcal{J}^+$ at the event $(x_0 + c^+ t_1, t_1)$ in Fig. 3.2b is the same as its value at the event $(x_0, 0)$, and every event along $\mathcal{C}^+$ in between. Similarly, the values of $\mathcal{S}^0$ at event $(x_0 + c^0 t_1, t_1)$ and $\mathcal{J}^-$ at event $(x_0 + c^- t_1, t_1)$ are the same as they were at event $(x_0, 0)$.

Figures 3.2a and 3.2b depict three characteristic paths diverging from a common event at $t = 0$ and intersecting the $t = t_1$ line at three different places. In this way, we know *something* about the flow variables at three different events at $t = t_1$, but not enough to determine all three flow variables at any one of them. For example, knowing the value of $\mathcal{J}^+$ at event $(x_0 + c^+ t_1, t_1)$ is not enough to determine what the values of ρ, v, and p are from Eq. (3.10); two more pieces of information are needed to make a unique determination. Thus, we use the paths as depicted in Fig. 3.2c and 3.2d where $\mathcal{C}^\pm$ and $\mathcal{C}^0$ are launched from different events at $t = 0$ which, by virtue of their characteristic speeds and where their *footprints* are placed on the $t = 0$ axis, converge on the same future event, (x_0, t_1). Thus, if $\mathcal{J}_0^\pm$ and $\mathcal{S}_0^0$ are known at $t = 0$ and since they are invariant along their respective paths,

$$\mathcal{J}^\pm(x_0 - c^\pm t_1, t_0) = \mathcal{J}_0^\pm \quad \text{and} \quad \mathcal{S}^0(x_0 - c^0 t_1, t_0) = \mathcal{S}_0^0,$$

then at event (x_0, t_1), Eq. (3.10), (3.11), and (3.15) yield:

$$\mathcal{J}_0^\pm = v(x_0, t_1) \pm \frac{2}{\gamma - 1}\sqrt{\frac{\gamma p(x_0, t_1)}{\rho(x_0, t_1)}} \quad \text{and} \quad \mathcal{S}_0^0 = \frac{p(x_0, t_1)}{\rho(x_0, t_1)^\gamma}, \qquad (3.16)$$

which can be solved (Problem 3.1) for the primitive variables at event (x_0, t_1):

---

[5]Indeed, for supersonic flow, the designation *left*-moving and *right*-moving may lose their literal meaning; from the Eulerian frame, "left-moving" waves can actually move rightwards for $v_0 > c_s$. Thus, the designation left- and right-moving will be used as though viewed from the co-moving Lagrangian frame of reference.

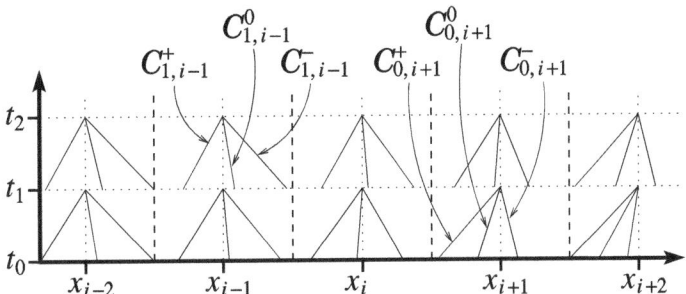

Figure 3.3. Schematic diagram showing how the method of characteristics could be used to design a numerical algorithm for solving the fluid equations in 1-D in an Eulerian reference frame. The first subscript on the characteristic path labels ($\mathcal{C}^{\pm,0}$) indicate from which time the characteristics are evaluated, and the second subscript indicates the nearest grid point to the characteristic footprint.

$$\rho(x_0, t_1) = \left[\frac{1}{\gamma \mathcal{S}_0^0} \left(\frac{\gamma - 1}{4} (\mathcal{J}_0^+ - \mathcal{J}_0^-) \right)^2 \right]^{\frac{1}{\gamma - 1}} ; \qquad (3.17)$$

$$p(x_0, t_1) = \left(\frac{1}{\mathcal{S}_0^0} \right)^{\frac{1}{\gamma - 1}} \left[\frac{1}{\gamma} \left(\frac{\gamma - 1}{4} (\mathcal{J}_0^+ - \mathcal{J}_0^-) \right)^2 \right]^{\frac{\gamma}{\gamma - 1}} ; \qquad (3.18)$$

$$v(x_0, t_1) = \frac{\mathcal{J}_0^+ + \mathcal{J}_0^-}{2} . \qquad (3.19)$$

Herein lies a possible numerical scheme for hydrodynamics. If, for example, at $t = t_0$ one had a 1-D array of locations, x_i, $i = 1, n$, at which each of the flow variables were known, then at each location and for a given time step, δt, one could compute all the characteristics that will converge at all the grid points at the future time, $t_1 = t_0 + \delta t$. This would require interpolating the flow variables to the footprints of each of the characteristic paths as suggested in Fig. 3.3, computing all the characteristics from the interpolated data, then combining them as done in Eq. (3.17)–(3.19) to obtain the flow variables at the advanced time t_1. This algorithm is then repeated for as many time steps as needed to advance the problem forward the desired amount of time. Such schemes have been tried, and are not widely used because they lack the numerical constraints to ensure mass and energy conservation and, as shown in Problem 3.2, small uncertainties in the flow variables at a given time are grossly exaggerated as time progresses. However, characteristics do play a prominent role in a full-scale *upwinded* MHD computer program,[6] where *flux densities* (rather than the flow variables themselves) are computed using data that are "upwind" along each characteristic path (*i.e.*, from $t = t_1$ towards $t = t_0$) converging on the event where the flux density is computed.

Finally, Fig. 3.4 shows four space-time diagrams in the Eulerian frame in which

---

[6] My fluid solver, ZEUS-3D, is a partially upwinded MHD code owing to its use of characteristics and is available open source from www.ap.smu.ca/~dclarke/zeus3d/.

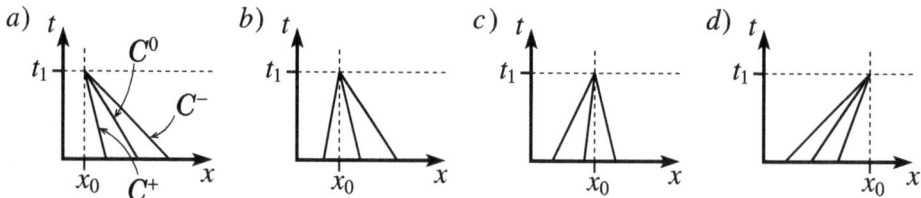

Figure 3.4. Space-time diagrams from an Eulerian reference frame showing how characteristic paths converge onto an event (x_0, t_1) in the case of: *a)* supersonic motion to the left; *b)* subsonic motion to the left; *c)* subsonic motion to the right; and *d)* supersonic motion to the right.

characteristic paths from the present converge on a unique point in the future. Panel *a* shows the diagram for $v < 0$ and supersonic ($|v| > c_s$), panel *b* shows the case for $v < 0$, subsonic; panel *c* shows the case for $v > 0$, subsonic; and panel *d* shows the case for $v > 0$, supersonic. Notice in the supersonic cases, all characteristic paths converge onto the future point from the same side of the $x = x_0$ axis, while for the subsonic cases, characteristic paths converge onto the future point from both sides.

Figure 3.4 sheds some light on why numerical hydrodynamics codes can have difficulty transmitting subsonic fluid flow cleanly across a boundary of the computational domain. Imagine in panels *c* and *d* that x_0 corresponds to the right-hand boundary of the grid. In the case of supersonic flow (panel *d*), all information needed at x_0 to determine the flow variables there at t_1 is provided by the right-leaning characteristic paths whose footprints lie in the computational domain. The footprint values of the characteristics can then be calculated from (interpolations of) the known grid values. However, in the case of subsonic flow, one of the pieces of information needed at x_0 to determine the flow variables there at t_1 must come from a left-leaning characteristic path (in this case, C^-) whose footprint lies outside the computational domain and is therefore unknown. A guess must be made for this value (possibly an extrapolation; always dodgy), and this guess will invariably lead to an incomplete transmission of the subsonic wave; *e.g.*, part of the wave may be reflected back into the grid, introducing errors – possibly large – into the computational data.

3.4 The MoC and the Riemann problem

Clean and useful applications of the method of characteristics described in the previous section exist (*e.g.*, Problems 5.12–5.15 in Chap. 5), but the Riemann problem is *not* such a case. However, the concept of the Riemann invariants and their characteristic paths still play a fundamental role.

Here, we'll examine what happens in the Riemann problem illustrated in Fig. 3.1 for $t > 0$ using the MoC. For illustration, we'll assume $v_L = v_R = 0$, $c_{s,L} > c_{s,R}$, and $p_L > p_R$, although our discussion won't depend unduly on these choices.

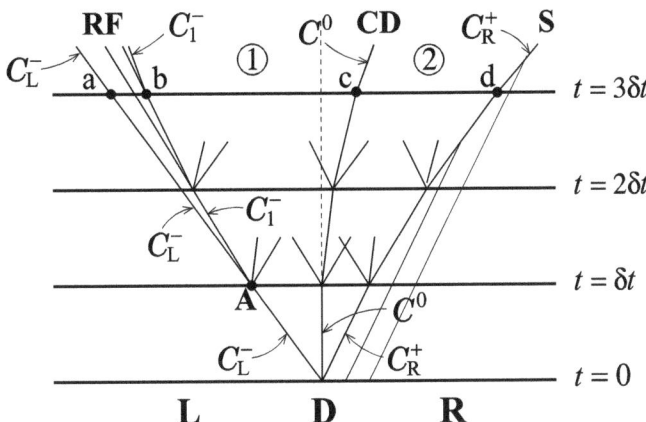

Figure 3.5. An attempt to follow the Riemann problem for three "time steps" using the MoC. From the original discontinuity at **D** develop three waves, namely the rarefaction fan (**RF**) on the left side, a contact discontinuity (**CD**) in the middle, and a shock (**S**) to the right. See text for details.

Figure 3.5 is a space-time diagram showing the development of the solution over three "time steps". At $t = 0$, the only point of interest is **D**, where p and ρ (and in general, v as well) are discontinuous. The $\mathcal{C}^0$ path is launched vertically from **D** at $t = 0$ since there is no fluid motion at this time, while $\mathcal{C}^\pm$ move off into the right (**R**) and left (**L**) states respectively at the local sound speeds. Thus, $\mathcal{C}^\pm$ are given subscripts R and L respectively in the figure. Because $c_{s,L} > c_{s,R}$, $\mathcal{C}_L^-$ is inclined more from the vertical than $\mathcal{C}_R^+$.

Since $p_L > p_R$, state **L** pushes into state **R**, accelerating fluid between $\mathcal{C}_L^-$ and $\mathcal{C}_R^+$ to the right. Thus, all characteristic paths launched at $t = \delta t$ lean slightly more to the right than they did at $t = 0$. During the second time step, fluid between $\mathcal{C}_L^-$ and $\mathcal{C}_R^+$ is further accelerated to the right, and thus characteristic paths launched at $t = 2\,\delta t$ lean even more to the right, and so it goes. During this process, two new states open up: state ① between $\mathcal{C}_L^-$ and $\mathcal{C}^0$ and state ② between $\mathcal{C}^0$ and $\mathcal{C}_R^+$.

As the characteristic paths lean more and more to the right, two interesting phenomena occur. First, the path $\mathcal{C}_L^-$ launched at $t = 0$ continues to propagate into state **L**, demarcating fluid to its left that has yet to be affected by the original discontinuity from fluid to its right which has. However, at point **A** in Fig. 3.5, the new left characteristic path, $\mathcal{C}_1^-$, does not lie on top of the original left characteristic path, $\mathcal{C}_L^-$, because of the new rightward motion of the fluid (v is now > 0) and thus fluid between these paths is aware of the conditions at **D**, but not at **A**. The region between $\mathcal{C}_L^-$ and $\mathcal{C}_1^-$ (whose intersections with $t = 3\,\delta t$ are shown respectively as points **a** and **b** in the figure) is known as a *rarefaction fan* in which the fluid is only partially affected by the original discontinuity. In a space-time diagram, rarefaction fans are characterised by *diverging characteristic paths*, and represent another type of wave supported by the fluid equations. They did not arise in our examination of the Rankine–Hugoniot jump conditions because a rarefaction fan is not a steady-state solution. By its very nature, a rarefaction fan is always widening and thus continually reducing the gradients of the flow variables within it.

A rarefaction fan is consistent with smooth flow, and thus is a solution of both the conservative and primitive fluid equations. A reasonable way to think of a rarefaction fan is like a sandbox divided into two halves by a divider with the sand level on one side of the divider much higher than the other. If the divider is suddenly removed, sand from the high side will spill into the low side, causing what was a discontinuity in the sand level to evolve into a profile with ever-diminishing slope. In the case of a fluid, the sand height is analogous with the fluid pressure.

Second, in the right state $\mathcal{C}_R^+$ demarcates fluid to its right that has not yet been affected by the discontinuity at **D** from fluid to its left that has. However, unlike in state **L** where the rightward leaning of the characteristic paths caused them to diverge, here paths launched from the right state at $t = 0$ converge to the same worldline labelled **S** in Fig. 3.5.

What is the nature of **S**? At an infinitesimal distance to its right, the fluid remains completely in the right state, whereas at an infinitesimal distance to its left, the fluid is completely in the new state ②. Unlike state **L** where a "transition" region – the rarefaction fan – opened up between states **L** and ①, no such transition region opens up between states **R** and ②, and we say that $\mathcal{C}_R^+$ has *steepened into a shock* (whence the designation **S**). The fact that a shock has spontaneously formed from the initial conditions of a Riemann problem indicates its solution cannot be found with the assumption of smooth flow, and the conclusions of the previous section must be augmented with the Rankine–Hugoniot jump conditions (§2.2). Further, we can no longer think of $\mathcal{C}_R^+$ (**S**) as a worldline of constant $\mathcal{J}^+$, embedded in a smoothly flowing medium. Indeed and as we shall see, $\mathcal{C}_R^+$ has developed into a worldline of converged $\mathcal{C}^+$ characteristic paths with multiple values of $\mathcal{J}^+$. Thus, $\mathcal{C}_R^+$ carries with it a *jump* in $\mathcal{J}^+$, consistent with its identification as a shock.

Finally, at the end of §3.5.3 we'll identify the worldline $\mathcal{C}^0$ as a contact discontinuity, carrying with it jumps in density, temperature, and entropy, but not in pressure nor velocity.

A simplified space-time diagram for the Riemann problem is shown in Fig. 3.6. All of Fig. 3.5 can be thought of as being contained within the "dot" at **D** ($t_1 \gg \delta t$), and the slopes of all characteristic paths have attained their asymptotic values. As suggested in §3.3, we can imagine trying to find the values of the flow variables in the intermediate states by tracing certain paths from a typical point in state ②, say, back to the original state at $t = 0$. The problem with this strategy is immediately clear from Fig. 3.6. Both paths $\mathcal{C}^-$ and $\mathcal{C}^0$ must cross a shock while $\mathcal{C}^+$ must cross a contact, and the conclusions that the Riemann invariants must be constant along their respective characteristic paths all the way to the $t = 0$ baseline no longer holds *because those conclusions were based on the assumption of smooth flow*. In a very real sense, the shock and contact become baselines for characteristic paths that strike them, and since the flow variables along lines of discontinuity are not knowable from the method of characteristics, we have no hope of evaluating the Riemann invariants $\mathcal{J}^-$ and $\mathcal{S}^0$ at their bases for use in Eq. (3.17)–(3.19).

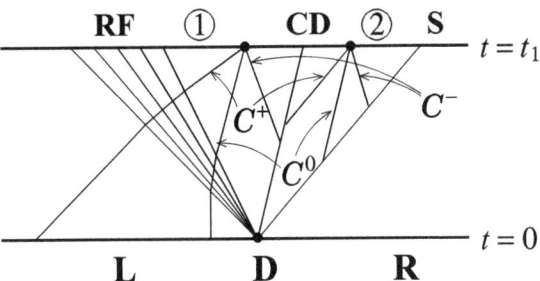

Figure 3.6. In attempting to trace characteristics from the intermediate states, ① and ②, to the initial states, **L** and **R**, at $t = 0$, characteristics encounter discontinuities which are discontinuous in the Riemann "invariants", thereby thwarting the MoC.

A similar problem exists for characteristic paths converging on a point in intermediate state ①. There, while both $\mathcal{C}^0$ and $\mathcal{C}^+$ can be traced back to $t = 0$ crossing only the continuous rarefaction fan, $\mathcal{C}^-$ must cross both the contact and the shock.

Clearly, the Riemann problem cannot be solved using the MoC alone. A different strategy must be developed to account for changes in the primitive variables as various waves created in the Riemann problem are crossed. The Rankine–Hugoniot jump conditions already tell us how the flow variables change across discontinuities such as shocks and contacts. What remains is to determine is how the flow variables change across smooth transitions such as the rarefaction fan.

3.5 Non-linear hydrodynamical waves

3.5.1 Hyperbolic system of equations

Consider now all three primitive equations of 1-D ideal HD (Eq. Set 3 on page 21 with $\phi = 0$), deliberately arranged in a fashion to suggest a matrix formalism:

$$
\begin{aligned}
\frac{\partial \rho}{\partial t} + v\frac{\partial \rho}{\partial x} \quad\quad\quad + \rho\frac{\partial v}{\partial x} &= 0; \\
\frac{\partial p}{\partial t} \quad\quad + v\frac{\partial p}{\partial x} + \gamma p\frac{\partial v}{\partial x} &= 0; \\
\frac{\partial v}{\partial t} \quad\quad + \frac{1}{\rho}\frac{\partial p}{\partial x} + v\frac{\partial v}{\partial x} &= 0,
\end{aligned}
\quad (3.20)
$$

which can be written in matrix notation as,

$$
\frac{\partial}{\partial t}\begin{bmatrix}\rho\\p\\v\end{bmatrix} + \begin{bmatrix}v & 0 & \rho\\0 & v & \gamma p\\0 & 1/\rho & v\end{bmatrix}\frac{\partial}{\partial x}\begin{bmatrix}\rho\\p\\v\end{bmatrix} = 0 \quad\Rightarrow\quad \frac{\partial |q\rangle}{\partial t} + \mathsf{J}\frac{\partial |q\rangle}{\partial x} = 0, \quad (3.21)
$$

where,

$$|q\rangle = \begin{bmatrix} \rho \\ p \\ v \end{bmatrix} \quad \text{and} \quad \mathsf{J} = \begin{bmatrix} v & 0 & \rho \\ 0 & v & \gamma p \\ 0 & 1/\rho & v \end{bmatrix}, \qquad (3.22)$$

are, respectively, the ket[7] of primitive variables and the Jacobian matrix, reminiscent of the formalism introduced in §2.1.2. Unlike our analysis there for sound waves, we have made no attempt to linearise the equations, and thus this discussion is general for smooth flow.

As defined in App. C, a system of equations in the form of Eq. (3.21) is *hyperbolic* if the following conditions are met:

1. The eigenvalues of the Jacobian matrix J are real;

2. The eigenvectors of J are linearly independent.

Hyperbolicity is significant since solutions of such systems of equations can be described in terms of waves with speeds, u_i, given by the eigenvalues of the Jacobian. If all eigenvalues are distinct (as they are for HD), the system of equations is said to be *strictly hyperbolic*. Conversely, if conditions permit degeneracy among the eigenvalues (as is the case in MHD), the system is said to be *not* strictly hyperbolic which, as we shall see in §5.3 and Chap. 6, poses numerous algebraic and computational challenges – difficult but surmountable – in arriving at a solution.

So to proceed, we postulate – as we did in §2.1.2 – a normal mode wave solution to Eq. (3.21) of the form:

$$|q(x,t)\rangle = |\tilde{q}(\xi_i)\rangle, \qquad (3.23)$$

where $\xi_i = x - u_i t$ is a coordinate co-moving with the so-called *i-family of waves* moving at wave speed u_i, and where $|\tilde{q}(x)\rangle$ represents the initial conditions of the ket of variables $|q\rangle$.

On substituting Eq. (3.23) into Eq. (3.21), we get Eq. (2.23) renumbered here for convenience,

$$\mathsf{J}|\tilde{q}'\rangle = u_i|\tilde{q}'\rangle, \qquad (3.24)$$

where the prime (') indicates differentiation with respect to ξ_i, where u_i is an eigenvalue of J (a wave speed), and where $|\tilde{q}'\rangle$ is proportional to the corresponding eigenket, $|r_i\rangle$. As we saw for the linearised hydrodynamical case in §2.1.2, the eigenkets determine the profiles of the primitive variables as a function of position and time. We'll return to this in §3.5.3 for the non-linear case begun here.

3.5.2 Left and right eigenvectors (optional)

For non-linear HD, Eq. (3.21) represents three equations in three unknowns which, as we'll see, give rise to three distinct *wave families*. In 1-D MHD with seven equations in seven unknowns (§5.2), *seven* wave families are admitted. So, before evaluating the eigenvalues and eigenkets specifically for non-linear HD, let's develop the

---

[7] *Rappel*: Dirac's "bra-ket notation" is explained in footnote 5 on page 30.

theory for a general hyperbolic system of n equations in n unknowns, where n is the dimension of $|q\rangle$.

Consider the "eigen-equation",

$$\mathsf{J}|r_i\rangle = u_i|r_i\rangle, \quad i = 1, 2, \ldots, n. \qquad (3.25)$$

Because they appear to the right of J in Eq. (3.25), the eigenkets, $|r_i\rangle$, are frequently referred to as *right eigenvectors*. Now, we can assemble all n of Eq. (3.25) into a single matrix equation by creating a matrix of right eigenvectors, R, whose i^{th} column is $|r_i\rangle$:

$$\mathsf{J}\begin{bmatrix} r_{11} & r_{21} & \cdots & r_{n1} \\ r_{12} & r_{22} & \cdots & r_{n2} \\ \vdots & \vdots & \ddots & \vdots \\ r_{1n} & r_{2n} & \cdots & r_{nn} \end{bmatrix} = \begin{bmatrix} u_1 r_{11} & u_2 r_{21} & \cdots & u_n r_{n1} \\ u_1 r_{12} & u_2 r_{22} & \cdots & u_n r_{n2} \\ \vdots & \vdots & \ddots & \vdots \\ u_1 r_{1n} & u_2 r_{2n} & \cdots & u_n r_{nn} \end{bmatrix}$$

$$= \begin{bmatrix} r_{11} & r_{21} & \cdots & r_{n1} \\ r_{12} & r_{22} & \cdots & r_{n2} \\ \vdots & \vdots & \ddots & \vdots \\ r_{1n} & r_{2n} & \cdots & r_{nn} \end{bmatrix} \begin{bmatrix} u_1 & 0 & \cdots & 0 \\ 0 & u_2 & \cdots & 0 \\ \vdots & \vdots & \ddots & \vdots \\ 0 & 0 & \cdots & u_n \end{bmatrix}$$

$$\Rightarrow \quad \mathsf{JR} = \mathsf{RU}. \qquad (3.26)$$

Note that R_{ji}, the $(j,i)^{\text{th}}$ element of R, is r_{ij} (indices reversed), the jth component of the ith eigenket, and that $\mathsf{U}_{ij} = u_i \delta_{ij}$, where $\delta_{ij} = 1$, $i = j$; 0, $i \neq j$ is the usual Kronecker-delta. Thus, U is a diagonal matrix with the eigenvalues, u_i, strung along its main diagonal. Note further the order of matrix multiplication is important; for example, the right-hand side (RHS) of Eq. (3.26) *cannot* be written as UR. (If this is not obvious to you, try it!)

Invoking once again the hyperbolicity of Eq. (3.21), the eigenkets (and thus the columns of matrix R) are linearly independent, and R^{-1} exists. Thus, we can multiply Eq. (3.26) on the right by R^{-1} to get:

$$\mathsf{JRR}^{-1} = \mathsf{J} = \mathsf{RUR}^{-1},$$

and, in the language of linear algebra, J is *similar* to U. That is, a set of *elementary row reduction operations* (*e.g.*, Gauss–Jordan elimination steps) can be used to row-reduce the Jacobian matrix, J, to the diagonal matrix of eigenvalues, U.

Now multiplying Eq. (3.26) on both sides by R^{-1}, we get:

$$\mathsf{R}^{-1}\mathsf{JRR}^{-1} = \mathsf{R}^{-1}\mathsf{RUR}^{-1} \quad \Rightarrow \quad \mathsf{LJ} = \mathsf{UL}, \qquad (3.27)$$

where $\mathsf{L} \equiv \mathsf{R}^{-1}$. In a similar way that Eq. (3.25) is the "column-decomposition" of Eq. (3.26), we can write down the "row-decomposition" of Eq. (3.27) as:

$$\langle l_i|\mathsf{J} = \langle l_i|u_i, \quad i = 1, 2, \ldots n.$$

This, too, is an eigen-equation, but this time the eigenvectors are row vectors (bras) appearing on the left side of the matrix J. Accordingly, L is the matrix of the *eigenbras* or *left eigenvectors*, $\langle l_i|$. Note that $|r_i\rangle$ is the ith column of the matrix R, while $\langle l_i|$ is the ith row of the inverse matrix L = R^{-1}, and thus will not, in general, be the same as $\langle r_i|$ (the eigenket written as a bra). However, both types of eigenvectors share the same eigenvalues.

Note that if $\langle l_i| = \langle r_i|$, this would require that the rows of R^{-1} be the same as the columns of R; that is to say R^{-1} = R^{T} where the superscript ^{T} indicates a matrix *transpose* (rows made into columns). A matrix whose inverse is its transpose is known as an *orthogonal matrix*, which form an interesting subset of matrices in their own right (*e.g.*, they can be interpreted as rotations of coordinate systems). We shan't divert our attention to the properties of orthogonal matrices here, and instead refer the interested reader to any elementary text in linear algebra.[8]

For examples of how the left- and right-eigenvectors can interact, consider the following two theorems.

Theorem 3.1. *Let $\xi_i = x_i - u_i t$ be the coordinate co-moving with the i-family of waves, and let $|\tilde{q}(x)\rangle$ be the known initial conditions of $|q\rangle$. Then the general solution to Eq. (3.21) is,*

$$\boxed{|q(x,t)\rangle = \sum_{i=1}^{n} \langle l_i|\tilde{q}(\xi_i)\rangle |r_i\rangle.} \qquad (3.28)$$

Proof: As eigenkets of a hyperbolic system of equations, $|r_i\rangle$ form a linearly independent set of n vectors and thus span the n-dimensional vector space. As such, any ket may be written as a linear combination of $|r_i\rangle$, including the solution ket itself:

$$|q(x,t)\rangle = \sum_{i=1}^{n} f_i(\xi_i)|r_i\rangle, \qquad (3.29)$$

where $f_i(\xi_i)$ are to-be-determined functions of ξ_i. Multiplying through by $\langle l_j|$, we get:

$$\langle l_j|q(x,t)\rangle = \langle l_j|\sum_{i=1}^{n} f_i(\xi_i)|r_i\rangle = \sum_{i=1}^{n} f_i(\xi_i)\langle l_j|r_i\rangle = \sum_{i=1}^{n} f_i(\xi_i)\delta_{ij} = f_j(\xi_j),$$

since LR = I (identity matrix), and thus $\langle l_j|r_i\rangle = \delta_{ij}$. Imposing initial conditions at $t = 0$, $|q(x,0)\rangle = |\tilde{q}(x)\rangle$, we can write:

$$f_i(x) = \langle l_i|q(x,0)\rangle = \langle l_i|\tilde{q}(x)\rangle \quad \Rightarrow \quad f_i(\xi_i) = \langle l_i|\tilde{q}(\xi_i)\rangle,$$

after time, t. Substituting this into Eq. (3.29) leads us to Eq. (3.28). □

Theorem 3.1 is little more than a formalisation of the method already used in §2.1.2: find the normal mode solutions ($\propto |r_i\rangle$), write the general solution as a linear

---

[8]For example, Gerald Bradley's *A Primer of Linear Algebra* (1975) is an excellent first book on the subject.

combination of these (Eq. 3.29), find the coefficients, and then apply boundary conditions.

Theorem 3.2. *The characteristics of the flow (Riemann invariants) are given by,*

$$\boxed{\chi_i = \int \langle l_i | dq \rangle,} \tag{3.30}$$

where $|dq\rangle = d|q\rangle$ is the ket of differential flow variables. Note that the construct $\langle a|b\rangle$ in elementary vector notation is just $\vec{a} \cdot \vec{b}$, the inner or dot product.

Proof: Multiplying Eq. (3.21) on the left by L, we get,

$$\mathsf{L}\frac{\partial |q\rangle}{\partial t} + \mathsf{LJ}\frac{\partial |q\rangle}{\partial x} = 0 \;\Rightarrow\; \mathsf{L}\frac{\partial |q\rangle}{\partial t} = -\mathsf{UL}\frac{\partial |q\rangle}{\partial x}, \tag{3.31}$$

using Eq. (3.27). Now, defining $|d\chi\rangle \equiv \mathsf{L}|dq\rangle$, Eq. (3.31) becomes:

$$\frac{\partial |\chi\rangle}{\partial t} = -\mathsf{U}\frac{\partial |\chi\rangle}{\partial x} \;\Rightarrow\; \frac{\partial \chi_i}{\partial t} + u_i \frac{\partial \chi_i}{\partial x} = 0 \;\Rightarrow\; \boxed{\frac{D_{u_i}\chi_i}{Dt} = 0,} \tag{3.32}$$

since U is a diagonal matrix, and where D_{u_i}/Dt is the Lagrangian derivative with respect to the eigen (characteristic) speed, u_i, defined by Eq. (3.2). As shown by Eq. (3.13), a quantity whose Lagrangian derivative is zero is a characteristic of the flow (remains constant in the co-moving frame) and thus χ_i is a characteristic. Therefore, from its definition,

$$|d\chi\rangle = \mathsf{L}|dq\rangle \;\Rightarrow\; d\chi_i = \langle l_i|dq\rangle \;\Rightarrow\; \chi_i = \int \langle l_i|dq\rangle. \;\square$$

Problems 3.5, 3.6, and 3.7 show how Theorems 3.1 and 3.2 may be used in practice.

3.5.3 Hydrodynamical rarefaction fans

Returning now to the specific case of non-linear HD begun in §3.5.1, we evaluate the three eigenvalues of J given by Eq. (3.22). Thus,

$$|\mathsf{J} - u\mathsf{I}| = \begin{vmatrix} v-u & 0 & \rho \\ 0 & v-u & \gamma p \\ 0 & 1/\rho & v-u \end{vmatrix} = -(u-v)\big[(u-v)^2 - c_\mathrm{s}^2\big] = 0,$$

(since $c_\mathrm{s}^2 = \gamma p/\rho$) and the wave speeds (*eigenspeeds*) are:

$$u_1 = v - c_\mathrm{s}; \quad u_2 = v; \quad u_3 = v + c_\mathrm{s}, \tag{3.33}$$

arranged in ascending order. These are none other than the characteristic speeds uncovered in §3.2 (Eq. 3.12 and 3.14), where each speed is associated with one of three families of waves. The *1-family wave* with eigenspeed u_1 is left-moving (relative to the fluid) and can be a shock wave for discontinuous flow, or a rarefaction fan (RF) for continuous flow. The *2-family wave* has eigenspeed $u_2 = v$ and, as such, is co-moving with the fluid. This is an entropy wave, an example of which is the contact discontinuity discussed in §2.2.2. Finally, the *3-family wave* with eigenspeed u_3 is similar to the 1-family wave but right-moving. Affiliated with the 1-, 2-, and 3-wave families are, respectively, the three characteristic paths $\mathcal{C}^-$, $\mathcal{C}^0$, and $\mathcal{C}^+$ and their associated Riemann invariants $\mathcal{J}^-$, $\mathcal{S}^0$, and $\mathcal{J}^+$ described in §3.2.

While the eigenvalues tell us how fast each wave propagates, the eigenkets tell us how the primitive variables vary across each *continuous* wave. To this end, following the usual methods (and as detailed in §2.1.2), it is left to Problem 3.3 to show that the eigenkets of J associated with each of the eigenvalues are:

$$|r_1\rangle = \begin{bmatrix} -\rho \\ -\gamma p \\ c_s \end{bmatrix}; \quad |r_2\rangle = \begin{bmatrix} 1 \\ 0 \\ 0 \end{bmatrix}; \quad |r_3\rangle = \begin{bmatrix} -\rho \\ -\gamma p \\ -c_s \end{bmatrix}, \quad (3.34)$$

with suitable "normalisation".[9] From Eq. (3.24), the derivatives of the solution vectors, $|q'(\xi_i)\rangle$, $i = 1, 2, 3$, must be proportional to the eigenvectors, $|r_i\rangle$, and thus we may write:

$$|q'(\xi_i)\rangle = \begin{bmatrix} \rho'(\xi_i) \\ p'(\xi_i) \\ v'(\xi_i) \end{bmatrix} = w_i(\xi_i)|r_i\rangle, \quad (3.35)$$

where $w_i(\xi_i)$ is an arbitrary scaling factor and a function of the co-moving coordinate, ξ_i, this time – again, for convenience – *not* defined as its derivative (*cf.*, Eq. 2.27). As Problem 3.8 shows, Eq. (3.34) and (3.35) are enough to prove the following theorem on how the Riemann invariants relate to continuous 1- and 3-family waves:

Theorem 3.3.

1. *While $\mathcal{J}^-$ is constant along the $\mathcal{C}^-$ characteristic path, $\mathcal{J}^+$ is constant across the associated 1-family RF.*

2. *While $\mathcal{J}^+$ is constant along the $\mathcal{C}^+$ characteristic path, $\mathcal{J}^-$ is constant across the associated 3-family RF.*

3. *$\mathcal{S}^0$ is constant across* both *1- and 3-family RFs.*

Now, unlike in §2.1.2, there is no need here to evaluate w_i. Rather, we'll use it to define a *generalised coordinate*, $ds_i = w_i d\xi_i$, where s_i varies from 0 on the upwind side of the i-family wave, to some maximum value on the downwind side, $s_{i,\mathrm{d}}$. Thus, for the 1-family wave, Eq. (3.35) becomes:

$$\frac{d}{w_1 d\xi_1}\begin{bmatrix}\rho\\p\\v\end{bmatrix} \equiv \frac{d}{ds_1}\begin{bmatrix}\rho\\p\\v\end{bmatrix} = \begin{bmatrix}-\rho\\-\gamma p\\c_s\end{bmatrix} \Rightarrow \begin{cases}\dfrac{d\rho}{\rho} = -ds_1;\\[4pt] \dfrac{dp}{p} = -\gamma ds_1;\\[4pt] dv = c_s ds_1.\end{cases} \quad (3.36)$$

---

[9] Each component of $|r_3\rangle$ is negative to accommodate a sign convention we shall adopt where the generalised coordinate, s_3, increases from right to left; all will be made clear within the next page or two.

The first two equations integrate trivially to get:

$$\rho(s_1) = \rho_u e^{-s_1} \quad \text{and} \quad p(s_1) = p_u e^{-\gamma s_1}, \qquad (3.37)$$

where ρ_u and p_u are the known upwind values of ρ and p. Next, since the sound speed is given by:

$$c_s(s_1) = \sqrt{\frac{\gamma p(s_1)}{\rho(s_1)}} = c_{s,u} e^{-\frac{\gamma-1}{2} s_1}, \qquad (3.38)$$

where $c_{s,u} = \sqrt{\gamma p_u/\rho_u}$, we can integrate the last of Eq. (3.36) to get:

$$v(s_1) = v_u + \frac{2 c_{s,u}}{\gamma - 1}\left(1 - e^{-\frac{\gamma-1}{2} s_1}\right), \qquad (3.39)$$

where v_u is the known upwind value of v. Together, Eq. (3.37) and (3.39) give the profiles of the primitive variables across the continuous 1-family wave – a 1-rarefaction fan – as a function of the generalised coordinate, s_1. Evidently, as one traverses a rarefaction fan from the upwind to downwind side, the density and pressure decrease exponentially (as a function of s_1) while the flow speed relative to the upwind state increases, all of which are opposite to what happens across a shock (*e.g.*, Zemplén's theorem; Eq. 2.55).

A measure of the *strength* or *width* of a rarefaction fan is how close to zero the pressure drops. As in Problem 2.8 for shocks, we define for the 1-rarefaction,

$$\zeta_1 = \frac{p_d}{p_u}, \qquad (3.40)$$

where p_d is the pressure downwind of the 1-rarefaction. For shocks, $\zeta > 1$ with shock strength increasing as $\zeta \to \infty$, whereas for rarefactions, $\zeta < 1$ with rarefaction width increasing as $\zeta \to 0$. Thus, immediately downwind of the 1-rarefaction, we have from the second of Eq. (3.37):

$$p(s_{1,d}) = p_d = p_u e^{-\gamma s_{1,d}} \;\Rightarrow\; \zeta_1 = e^{-\gamma s_{1,d}} \;\Rightarrow\; e^{-s_{1,d}} = \zeta_1^{1/\gamma}.$$

Substituting this into Eq. (3.37)–(3.39), the transitions for all variables including c_s across a 1-rarefaction from the upwind to downwind states in terms of ζ_1 are:

$$\rho_d = \rho_u \zeta_1^{1/\gamma}; \quad p_d = p_u \zeta_1; \quad c_{s,d} = c_{s,u} \zeta_1^{\frac{\gamma-1}{2\gamma}};$$
$$v_d = v_u + \frac{2 c_{s,u}}{\gamma - 1}\left(1 - \zeta_1^{\frac{\gamma-1}{2\gamma}}\right). \qquad (3.41)$$

These expressions will be useful in solving the HD Riemann problem in §3.6.

A useful physical interpretation of $\nabla \cdot \vec{v}$ is that it distinguishes regions of compression from expansion. Thought of in 1-D, if $\nabla \cdot \vec{v} = dv_x/dx > 0$, fluid is accelerating in the direction of flow (expansion), whereas if $dv_x/dx < 0$, fluid is decelerating in the direction of flow (compression). For shocks, $\nabla \cdot \vec{v} < 0$, and thus, as we found in §2.2.3, shocks waves are compressive. Conversely, from the third of Eq. (3.36),

$$\frac{dv}{ds_1} = c_s > 0,$$

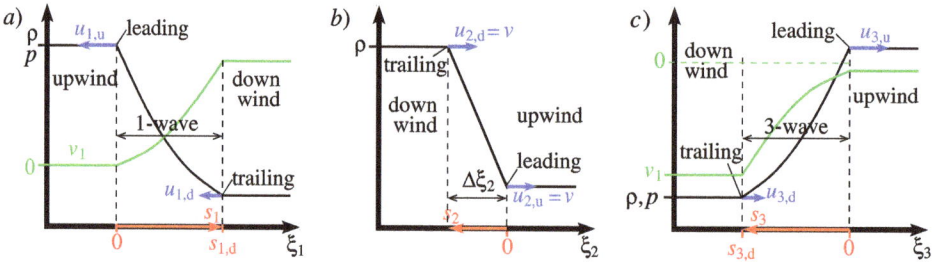

Figure 3.7. Typical profiles across the three non-linear wave families: *a*) 1-family rarefaction fan; *b*) 2-family entropy wave; and *c*) 3-family rarefaction fan. The leading edge of each wave advances into the upwind state while the trailing edge leaves behind a new downwind state. For the 1- and 3-waves, velocity profiles (green, v_1) are given in the reference frame of the upwind state and are such that the wave always widens with time. For the entropy wave (panel *b*), both p and v are continuous and the width of the wave ($\Delta \xi_2$; = 0 for a contact discontinuity) remains constant. The generalised coordinates, s_i, are represented in red (along horizontal axes) which, by definition, exist only within the wave itself beginning at zero at the upwind state, increase monotonically across the fan, and end at some maximum value at the downwind state.

since, in the upwind reference frame, flow is in the $+x$-direction.[10] Thus, rarefaction fans are *expansive* waves whose leading edge moves into upwind fluid at speed $u_{1,u} = v_u - c_{s,u}$ which forever pulls away from its trailing edge moving away from the downwind state at the lesser speed $u_{1,d} = v_d - c_{s,d}$ (Fig. 3.7a).

Turning now to the profiles of ρ, p, c_s, and v across the 1-rarefaction, instead of functions of s_1 as in Eq. (3.37)–(3.39), we prefer to express them as functions of the *self-similar* coordinate, x/t.[11] For this, we set the characteristic coordinate, $\xi_1 = x - u_1 t$ (e.g., Eq. 3.23), to zero to put us in the co-moving frame of the 1-wave, from which we can write,

$$\frac{x}{t} = u_1 = v - c_s, \qquad (3.42)$$

using the first of Eq. (3.33). Thus, for a fixed time t, plotting profiles of ρ, p, and v against u_1 would look just like plotting them against x (position across the wave scaled by t) from the co-moving frame of the wave, which is what we want.

To this end, from Eq. (3.38) and (3.39), Eq. (3.42) becomes,

$$u_1 = v - c_s = v_u + \frac{2c_{s,u}}{\gamma - 1}\left(1 - e^{-\frac{\gamma-1}{2}s_1}\right) - c_{s,u} e^{-\frac{\gamma-1}{2}s_1},$$

$$\Rightarrow e^{-\frac{\gamma-1}{2}s_1} = \frac{2c_{s,u} + (v_u - u_1)(\gamma - 1)}{c_{s,u}(\gamma + 1)}. \qquad (3.43)$$

---

[10] As depicted in Fig. 3.7a, the *wave speed* of the 1-rarefaction fan, $u_1 = v - c_s$, is in the $-x$-direction, but the *flow speed* of the fluid relative to the upwind state is in the $+x$-direction.

[11] A "self-similar system" is one whose only change in time is its scale. For example, other than its diameter (scale), a spherical balloon always looks the same as it's blown up; it's always "similar to itself". When x and t appear exclusively as x/t in an equation, doubling each of x and t does nothing to x/t and thus the equation value, and the system described by this equation remains "similar to itself". Such an equation is said to be "self-similar" and x/t its "self-similar coordinate".

Substituting Eq. (3.43) into each of Eq. (3.37)–(3.39), we find after a little algebra:

$$\rho(u_1) = \rho_u \left(\frac{2c_{s,u} + (v_u - u_1)(\gamma - 1)}{c_{s,u}(\gamma + 1)} \right)^{\frac{2}{\gamma-1}} ; \tag{3.44}$$

$$p(u_1) = p_u \left(\frac{2c_{s,u} + (v_u - u_1)(\gamma - 1)}{c_{s,u}(\gamma + 1)} \right)^{\frac{2\gamma}{\gamma-1}} ; \tag{3.45}$$

$$c_s(u_1) = \frac{2c_{s,u} + (v_u - u_1)(\gamma - 1)}{\gamma + 1} ; \tag{3.46}$$

$$v(u_1) = \frac{2(c_{s,u} + u_1) + v_u(\gamma - 1)}{\gamma + 1} . \tag{3.47}$$

These expressions will also be useful in the next section in finding the full solution to the hydrodynamical Riemann problem.

It is left as an exercise to show that in terms of the pressure drop $\zeta_3 = p_d/p_u$, transitions across the continuous 3-wave (3-rarefaction fan) are given by:

$$\left. \begin{array}{l} \rho_d = \rho_u \zeta_3^{1/\gamma}; \quad p_d = p_u \zeta_3; \quad c_{s,d} = c_{s,u} \zeta_3^{\frac{\gamma-1}{2\gamma}} ; \\[2mm] v_d = v_u - \dfrac{2c_{s,u}}{\gamma - 1}\left(1 - \zeta_3^{\frac{\gamma-1}{2\gamma}}\right), \end{array} \right\} \tag{3.48}$$

identical in form to Eq. (3.41) except for a single sign change in the expression for v_d. It is left to Problem 3.10 to show that in terms of the eigenspeed $u_3 = v + c_s$, profiles across the 3-rarefaction fan (*e.g.*, Fig. 3.7c) are given by:

$$\rho(u_3) = \rho_u \left(\frac{2c_{s,u} + (u_3 - v_u)(\gamma - 1)}{c_{s,u}(\gamma + 1)} \right)^{\frac{2}{\gamma-1}} ; \tag{3.49}$$

$$p(u_3) = p_u \left(\frac{2c_{s,u} + (u_3 - v_u)(\gamma - 1)}{c_{s,u}(\gamma + 1)} \right)^{\frac{2\gamma}{\gamma-1}} ; \tag{3.50}$$

$$c_s(u_3) = \frac{2c_{s,u} + (u_3 - v_u)(\gamma - 1)}{\gamma + 1} ; \tag{3.51}$$

$$v(u_3) = \frac{2(u_3 - c_{s,u}) + v_u(\gamma - 1)}{\gamma + 1} , \tag{3.52}$$

identical to Eq. (3.44)–(3.47) except for differences in sign and subscript number.

Finally, for the continuous 2-wave with eigenspeed $u_2 = v$, we have from Eq. (3.35):

$$\frac{d}{ds_2} \begin{bmatrix} \rho \\ p \\ v \end{bmatrix} = \begin{bmatrix} 1 \\ 0 \\ 0 \end{bmatrix}, \tag{3.53}$$

which means both p and v are constant, and $\Delta \rho = \Delta s_2$, an arbitrary transition across the 2-wave (*e.g.*, Fig. 3.7b). If the change in ρ is abrupt, the 2-wave is a contact discontinuity as identified in §2.2.2. However, for a continuous wave the change in ρ will be smooth, such as – to revisit an example cited in §2.2.2 – any *real* weather front, where heat conductivity and turbulence can render the transitions in density and temperature continuous over many kilometres.

3.6 Solution to the HD Riemann problem

Believe it or not, a strategy for solving the hydrodynamical Riemann problem has emerged! For simplicity, we'll consider all 2-family waves as contacts, while 1- (left-leaning) and 3- (right-leaning) family waves are rarefaction fans for smooth flow or shocks for discontinuous flow. For smooth flow, transitions across rarefactions are given by Eq. (3.41) or (3.48), while for discontinuous flow, jumps across contacts and shocks are given by the Rankine–Hugoniot jump conditions developed in §2.2. To determine whether the 1- and 3-family waves are rarefactions or shocks, we need only look at the pressure which drops downwind across a rarefaction but rises across a shock. Thus, between states **L** (upwind) and ① (downwind) in Fig. 3.6, the left-moving rarefaction fan requires $p_1 < p_L$, while a left-moving shock would require the opposite, $p_1 > p_L$. Of course, p_1 is not known *a priori*, and thus one needs to check both possibilities for each of the 1- and 3-families. This gives rise to the four possible space-time diagrams illustrated in Fig. 3.8, with the correct solution being the one that gives the appropriate transitions in pressure (rise across a shock, drop across a rarefaction).

Referring to any panel of Fig. 3.8, the initial data for Riemann's problem are (ρ_L, p_L, v_L) and (ρ_R, p_R, v_R), acting respectively as the upwind states for wave families 1 and 3. Our task is to find their respective downwind states, (ρ_1, v_1, p_1) and (ρ_2, v_2, p_2). Now, the fact that the 2-wave is *always* a contact discontinuity simplifies the task considerably, since we can set $v_1 = v_2$ and $p_1 = p_2$. Thus, we must find four variables: ρ_1, p_1, v_1, and ρ_2. In the case of a rarefaction fan, it propagates at the local characteristic speed which varies across the fan such that $\nabla \cdot \vec{v} > 0$ (positive divergence). Thus, and as we've already seen, the rarefaction fan broadens as it propagates. Conversely, a shock propagates at the local shock speed with essentially zero width and a negative divergence, as determined by the Rankine–Hugoniot jump conditions.

1-wave:

→ If the 1-wave is a rarefaction, use the first and last of Eq. (3.41) to find ρ_1 and v_1 in terms of $\zeta_1 = p_1/p_L$ (Eq. 3.40), the pressure transition across the 1-rarefaction:

$$\rho_1 = \rho_L \, \zeta_1^{1/\gamma}; \tag{3.54}$$

$$v_1 = v_L + \frac{c_{s,L}}{\alpha \gamma} \left(1 - \zeta_1^\alpha\right), \tag{3.55}$$

where $\alpha = (\gamma - 1)/2\gamma$. Note that for Eq. (3.54) and (3.55) to apply, $\zeta_1 < 1$ and thus $v_1 > v_L$.

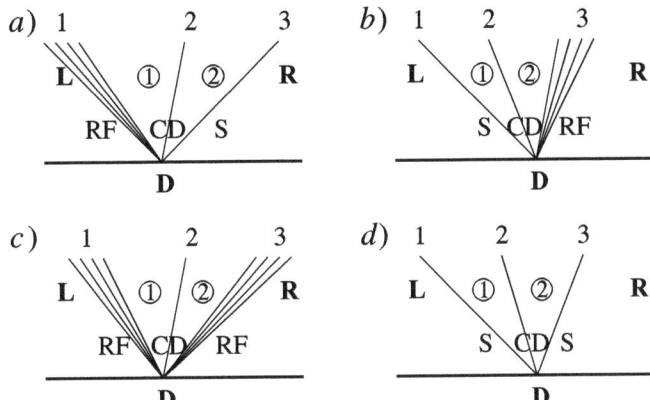

Figure 3.8. Four possible configurations for the solution to the Riemann problem: *a)* $p_L > p_1 = p_2 > p_R$; *b)* $p_L < p_1 = p_2 < p_R$; *c)* $p_L > p_1 = p_2 < p_R$; and *d)* $p_L < p_1 = p_2 > p_R$. Labels indicating family numbers are above the wave characteristic paths, while labels indicating wave type (rarefaction fan, contact discontinuity, shock) are near the base. Contacts necessarily lean in the direction shown in cases *a* and *b*, while in cases *c* and *d* the contact could lean either way.

→ If the 1-wave is a shock, the form of the Rankine–Hugoniot jump conditions suitable for this frame of reference was derived in Problem 2.9 (Eq. 2.86 and 2.87). Thus, in terms of ζ_1, the post-shock density and velocity, ρ_1 and v_1, are given by:

$$\rho_1 = \rho_L \frac{1 + \beta \zeta_1}{\beta + \zeta_1}; \tag{3.56}$$

$$v_1 = v_L - c_{s,L} \frac{\zeta_1 - 1}{\gamma \sqrt{\alpha(1 + \beta \zeta_1)}}, \tag{3.57}$$

where $\beta = (\gamma + 1)/(\gamma - 1)$. The advance speed of the shock into the left state is given by:

$$\mathcal{V}_S = v_L - c_{s,L} \sqrt{\alpha(1 + \beta \zeta_1)}. \tag{3.58}$$

It was shown in the same problem that $|\mathcal{V}_S - v_L| > |v_1 - v_L|$; the shock moves faster relative to the left medium than the speed of the shocked gas following it. Note that for Eq. (3.56) and (3.57) to apply, $\zeta_1 > 1$ and thus $v_1 < v_L$.

3-wave:

→ If the 3-wave is a rarefaction, use the first and last of Eq. (3.48) to find ρ_2 and $v_2 = v_1$ in terms of $\zeta_3 = p_1/p_R$ (Eq. 3.40), the pressure transition across the 3-rarefaction:

$$\rho_2 = \rho_R \zeta_3^{1/\gamma}; \tag{3.59}$$

$$v_1 = v_R - \frac{c_{s,R}}{\alpha \gamma} \left(1 - \zeta_3^\alpha\right). \tag{3.60}$$

Note that for Eq. (3.59) and (3.60) to apply, $\zeta_3 < 1$ and thus $v_1 < v_R$.

→ If the 3-wave is a shock, the post-shock density and velocity, ρ_2 and v_1, are given by:

$$\rho_2 = \rho_R \frac{1 + \beta\zeta_3}{\beta + \zeta_3}; \tag{3.61}$$

$$v_1 = v_R + c_{s,R} \frac{\zeta_3 - 1}{\gamma\sqrt{\alpha(1 + \beta\zeta_3)}}. \tag{3.62}$$

The advance speed of the shock into the right state is given by:

$$V_S = v_R + c_{s,R}\sqrt{\alpha(1 + \beta\zeta_3)}. \tag{3.63}$$

Note that for Eq. (3.62) and (3.62) to apply, $\zeta_3 > 1$ and thus $v_1 > v_R$.

With these expressions, we can construct an explicit algorithm for a Riemann-solver.

1. Assuming two rarefactions, equate Eq. (3.55) to Eq. (3.60) and solve for p_1 (Problem 3.11).

    If $\zeta_1 < 1$ and $\zeta_3 < 1$, then:

    a) compute v_1 using either Eq. (3.55) or (3.60);

    b) compute ρ_1 from Eq. (3.54);

    c) compute ρ_2 from Eq. (3.59).

    Else,

2. Assuming a 1-rarefaction and a 3-shock, equate Eq. (3.55) to Eq. (3.62) and solve for p_1 using a secant root finder (*e.g.*, §D.1 in App. D).

    If $\zeta_1 < 1$ and $\zeta_3 > 1$, then:

    a) compute v_1 using either Eq. (3.55) or (3.62);

    b) compute ρ_1 from Eq. (3.54);

    c) compute ρ_2 from Eq. (3.61).

    Else,

3. Assuming a 1-shock and a 3-rarefaction, equate Eq. (3.57) to Eq. (3.60), and solve for p_1 using a secant root finder.

    If $\zeta_1 > 1$ and $\zeta_3 < 1$, then:

    a) compute v_1 using either Eq. (3.57) or (3.60);

    b) compute ρ_1 from Eq. (3.56);

    c) compute ρ_2 from Eq. (3.59).

    Else,

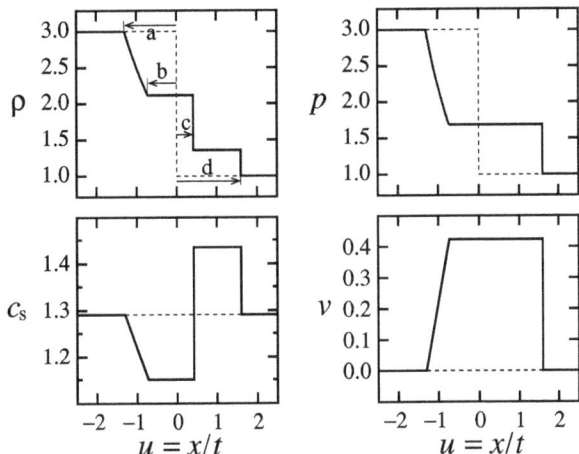

Figure 3.9. The Sod shock-tube solution given by the Riemann solver designed in this section (solid lines) including the internal structure of the rarefaction fan (note the slight curvature in each of the ρ and p plots) with the initial data superposed (dashed lines). Features a–d are explained in the text.

4. Assuming two shocks, equate Eq. (3.57) to Eq. (3.62), and solve for p_1 using a secant root finder.

   If $\zeta_1 > 1$ and $\zeta_3 > 1$, then:

   a) compute v_1 using either Eq. (3.57) or (3.62);

   b) compute ρ_1 from Eq. (3.56);

   c) compute ρ_2 from Eq. (3.61).

This algorithm will find the levels in the two intermediate states ① and ②. To determine how far each wave has travelled in a given time, one uses the characteristic speeds for the rarefaction fans, and the shock speeds given by Eq. (3.58) and (3.63) for shocks.

Finally, profiles across the 1-rarefaction fan as functions of the self-similar coordinate, $u_1 = x/t$ (Eq. 3.42), are given by Eq. (3.44)–(3.47). Similarly, profiles across the 3-rarefaction fan in terms of $u_3 = x/t$ (for $\xi_3 = 0$) are given by Eq. (3.49)–(3.52).

Figure 3.9 depicts the so-called *Sod shock-tube problem*, first introduced by G. A. Sod as a test and discriminator for hydrodynamical computer algorithms (Sod, 1978). The left and right states match Sod's original problem, namely $(\rho_L, p_L, v_L) = (3, 3, 0)$ and $(\rho_R, p_R, v_R) = (1, 1, 0)$ with $\gamma = 5/3$. Shown are the primitive variables and the sound speed after some arbitrary time. Locations of the wave fronts are indicated by the "vectors" a–d, and correspond to the labels in Fig. 3.5. Thus, in units of the self-similar variable $u = x/t$, $|\mathbf{a}| = (v_L - c_{s,L})$, $|\mathbf{b}| = (v_1 - c_{s,1})$, $|\mathbf{c}| = v_1$, and $|\mathbf{d}| = \mathcal{V}_S$, where $\mathcal{V}_S$ is given by Eq. (3.63).

This is an example of a Riemann problem with one rarefaction fan and one shock (*e.g.*, Fig. 3.8*a*, *b*), and is typical of problems in which $v_L = v_R = 0$. Whether the rarefaction fan is on the left or right simply depends on whether $p_L > p_R$ or $p_L < p_R$. To obtain two rarefaction fans (*e.g.*, Fig. 3.8*c*), the left and right sides must initially "pull apart" (thus $v_R - v_L > 0$ and typically $> c_s$) while two shocks (*e.g.*, Fig. 3.8*d*) are obtained when the left and right sides "push together" (thus $v_R - v_L < 0$ and typically $< -c_s$).

Problem Set 3

3.1 Show how Eq. (3.17)–(3.19) in the text follow from Eq. (3.16).

3.2** This problem comes under the category "*You don't really understand something until you can compute it*". So, get ready to calculate...

a) Consider a 1-D compressible monatomic gas, and suppose at $t = 0$ the profiles for the three primitive variables are given by:

$$\rho(x,0) = 1.0 + 0.4\,x; \qquad p(x,0) = 0.6 + 0.6\,x; \qquad v(x,0) = 0.5 - 0.2\,x,$$

where all units are arbitrary. Thus, at event $(x,t) = (0.0, 0.0)$, the primitive variables have values $(\rho, p, v) = (1.0, 0.6, 0.5)$. Use the method of characteristics to estimate the primitive variables at event $(0.0, 0.1)$, and report your results to five decimal places.

b) Suppose we wanted ρ to be estimated within a fractional uncertainty, δ, of 0.01 (1%). If each of the Riemann invariants were known to within a fractional uncertainty of ϵ, find ϵ in terms of δ using whatever (standard) means of uncertainty propagation you like. Specifically, replace ρ with $\rho(1 \pm \delta)$, $\mathcal{J}_0^\pm$ with $\mathcal{J}_0^\pm(1 \pm \epsilon)$, and $\mathcal{S}_0^0$ with $\mathcal{S}_0^0(1 \pm \epsilon)$ in Eq. (3.17), and solve for ϵ. (You should get $\epsilon \sim 0.0022$.)

c) We would now like an estimate of what ϵ might actually be, and see how this compares to the value of ϵ we computed in part b to keep the uncertainty in ρ to within 1%.

One way to proceed is to note that the base of the characteristic path $\mathcal{C}^+$ was set by c^+ evaluated *using primitive variables at $x = 0$*. But one could, with equal justification, use primitive variables at $x = x^+$ to evaluate c^+. The main reason we don't is the obvious "catch-22": We'd need to know where x^+ is to evaluate the primitive variables in order to get c^+, but we need c^+ to know what x^+ is! So, we do the next best thing. Think of c^+ evaluated using the primitive variables at $x = 0$ as your "preliminary guess" of c^+, and think of x^+ based on the preliminary c^+ as your "preliminary guess" of x^+.

Using the primitive variables evaluated at the preliminary x^+ (part a), update your guess of c^+ and thus your guess of x^+; call this $\tilde{x}^+$. Now, compute the primitive variables at $\tilde{x}^+$ and, from these, compute $\mathcal{J}_0^+(\tilde{x}^+)$. *One estimate of your uncertainty in $\mathcal{J}_0^+$ is the fractional difference between $\mathcal{J}_0^+(\tilde{x}^+)$ and $\mathcal{J}_0^+(x^+)$ computed in part a.* How much greater (or smaller) is this quantity than what you found in part b to keep the fractional uncertainty in ρ to within 1%? (Answer: 4–5 times smaller).

3.3 Consider the primitive Jacobian, J, for non-linear HD given by Eq. (3.22) in the text. Given the eigenvalues of J in Eq. (3.33), show that the corresponding eigenkets are given by Eq. (3.34).

3.4***

a) Equation (3.21) in the text was derived using the 1-D primitive equations (and setting $\phi = 0$). Start instead with the 1-D conservative equations, and show that they can be written in the form:

$$\frac{\partial |q_c\rangle}{\partial t} + \frac{\partial |f\rangle}{\partial x} = 0, \qquad (3.64)$$

where $|q_c\rangle = \begin{bmatrix} \rho \\ e_T \\ s \end{bmatrix}$ (ket of *conservative* variables), and $|f\rangle$ is the ket of flux densities.

b) Show that Eq. (3.64) may be written in the same form as Eq. (3.21), where the matrix elements of the conservative Jacobian, J_c, are given by Eq. (2.20). In particular, you should find that:

$$J_c = \begin{bmatrix} 0 & 0 & 1 \\ -\frac{\gamma s e_T}{\rho^2} + (\gamma - 1)\frac{s^3}{\rho^3} & \frac{\gamma s}{\rho} & \frac{\gamma e_T}{\rho} - 3(\gamma-1)\frac{s^2}{2\rho^2} \\ -(3-\gamma)\frac{s^2}{2\rho^2} & \gamma - 1 & (3-\gamma)\frac{s}{\rho} \end{bmatrix}.$$

c) Find the eigenvalues and eigenkets of J_c. For the latter, you should find:

$$|r_1\rangle \propto \begin{bmatrix} 1 \\ h_T - vc_s \\ v - c_s \end{bmatrix}; \quad |r_2\rangle \propto \begin{bmatrix} 1 \\ v^2/2 \\ v \end{bmatrix}; \quad |r_3\rangle \propto \begin{bmatrix} 1 \\ h_T + vc_s \\ v + c_s \end{bmatrix},$$

where $h_T \equiv \frac{c_s^2}{\gamma - 1} + \frac{v^2}{2}$ is the "total specific enthalpy" of the system.

d) Why do you suppose the primitive and conservative sets of equations yield the same eigenvalues (e.g., compare your eigenvalues in part c with Eq. 3.33 in the text), but different eigenvectors (Eq. 3.34)?

3.5*

a) Use Theorem (3.1) in the text to derive Eq. (2.30) describing the propagation of sound waves. Thus, for the linearised hydrodynamic Eq. (2.17), you will need to compute the left and right eigenvectors of the Jacobian, do the appropriate inner products, *etc.*, to finally obtain Eq. (2.30).

b) Use Theorem (3.2) to find the Riemann invariants for sound waves, and use Eq. (3.32) to confirm the functions you found are indeed characteristics of the flow.

3.6** This problem works with the full set of non-linear HD Eq. (*e.g.*, Eq. 3.21 and 3.22 in the text), rather than the linearised pressure and Euler equations of Problem 3.5.

a) Using the eigenkets in Eq. (3.34), find the three eigenbras, $\langle l_1|$, $\langle l_2|$, and $\langle l_3|$.

b) From Theorem (3.1), find the evolution equation for the ket:

$$|q(x,t)\rangle = \begin{bmatrix} \rho(x,t) \\ p(x,t) \\ v(x,t) \end{bmatrix},$$

given the initial variable profiles:

$$|\tilde{q}(x)\rangle = \begin{bmatrix} \tilde{\rho}(x) \\ \tilde{p}(x) \\ \tilde{v}(x) \end{bmatrix}.$$

In addition to recovering Eq. (2.30) for $p(x,t)$ and $v(x,t)$ (but with a non-constant impedance), you will find an evolutionary equation for $\rho(x,t)$.

c) In the context of the discussion in §2.1.2 (and thus, for example, assume the co-moving frame where $\tilde{v} = 0$), how do you interpret your expression for $\rho(x,t)$? In particular, describe what happens to an initial density perturbation,

$$\tilde{\rho}(x) = \begin{cases} \rho_0 + \rho_p, & |x| < a; \\ \rho_0, & |x| \geq a, \end{cases}$$

for $t > 0$, and check that mass is conserved.

3.7* Consider the three primitive equations of 1-D ideal hydrodynamics, as embodied by Eq. (3.20) and (3.21) in the text.

a) Use Theorem 3.2 to find the Riemann invariants, $\mathcal{J}^\pm$ given by Eq. (3.10) and (3.11).

b) Using the same logic, try finding the Riemann invariant associated with the entropy wave, $\mathcal{S}^0$, given by Eq. (3.15). What goes wrong, and why doesn't Theorem 3.2 work for this case?

Enormous hint: If you start blindly with the eigenkets in Eq. (3.34) of the text, the eigenbras will lead you to trouble in doing the required integrals. However, by first multiplying $|r_1\rangle$ and $|r_3\rangle$ by $1/c_s$ (renormalisation), the integrals are much easier. Note that such a move is tantamount to multiplying a first-order ODE through by an "integrating factor" to render it "exact".

3.8* In §3.2, we found that along the worldline of each characteristic, $\mathcal{C}^\pm$ and $\mathcal{C}^0$, its associated Riemann invariant, $\mathcal{J}^\pm$ and $\mathcal{S}^0$, is conserved. This problem shows that these same Riemann invariants are conserved *across* 1- and 3-rarefaction fans, but perhaps not in a way you might expect.

a) Show that across a 1-rarefaction, both $\mathcal{J}^+ = v + 2c_s/(\gamma - 1)$ and p/ρ^γ are conserved.

b) Show that across a 3-rarefaction, both $\mathcal{J}^- = v - 2c_s/(\gamma - 1)$ and p/ρ^γ are conserved.

c) Finally, show that $\mathcal{J}^+ = v + 2c_s/(\gamma-1)$ is *not* conserved across a 3-rarefaction, and that $\mathcal{J}^- = v - 2c_s/(\gamma - 1)$ is *not* conserved across a 1-rarefaction.

d) The 1-characteristic path is a locus of "events" at the same location, but at different times – an historical record of a single point – along which $\mathcal{J}^-$ is conserved but $\mathcal{J}^+$ and $\mathcal{S}^0$ aren't. Conversely, a line across a 1-wave is a locus of events at different locations but at the same time – a "snapshot" of a continuum of points – along which both $\mathcal{J}^+$ and $\mathcal{S}^0$ are conserved, but $\mathcal{J}^-$ isn't. Similarly, *along* a 3-characteristic path, $\mathcal{J}^+$ is conserved but $\mathcal{J}^-$ and $\mathcal{S}^0$ aren't, whereas *across* a 3-wave, $\mathcal{J}^-$ and $\mathcal{S}^0$ are conserved and $\mathcal{J}^+$ isn't. What is the analogous statement for the 2-characteristic path/wave?

3.9* Equations (3.44)–(3.47) in the text give the profiles of the primitive variables and the sound speed across a 1-rarefaction fan and were derived by direct integration of the right 1-eigenvector of the Jacobian (Eq. 3.36).

Rederive these equations by using just the fact that, as shown in Problem 3.8, the Riemann invariants,

$$\mathcal{J}^+ = v + \frac{2c_s}{\gamma - 1} \quad \text{and} \quad \mathcal{S}^0 = \frac{p}{\rho^\gamma},$$

are conserved *across* a 1-rarefaction fan.

3.10* Starting with Eq. (3.35) in the text for the 3-family wave, verify Eq. (3.49)–(3.52) for the profiles across the 3-rarefaction.

3.11 If the solution to a Riemann problem consists of two rarefactions, show that the pressure between the rarefactions is given by:

$$p_1 = \left(\frac{c_{s,L} + c_{s,R} + \gamma\alpha(v_L - v_R)}{c_{s,L}\, p_L^{-\alpha} + c_{s,R}\, p_R^{-\alpha}} \right)^{1/\alpha}, \qquad (3.65)$$

where $\alpha = (\gamma - 1)/(2\gamma)$.

3.12*** Following the algorithm set out in §3.6 in the text, write a computer program to generate the intermediate values, $(\rho_1, v_1, p_1, \rho_2)$ of a Riemann hydrodynamical problem for given (input) left and right states, (ρ_L, v_L, p_L) and (ρ_R, v_R, p_R). If there are any shocks in the solution, your code should also report the shock speed. You do not need to include the internal profiles of rarefaction fans.

Use your program to fill in the following table for a $\gamma = 5/3$ adiabatic gas.

left state			right state			intermediate states				shock speeds	
ρ_L	p_L	v_L	ρ_R	p_R	v_R	ρ_1	p_1	v_1	ρ_2	v_{S_L}	v_{S_R}
3.0	3.0	0.0	1.0	1.0	0.0	—	—	—	—	—	—
1.0	0.6	0.0	0.2	3.0	0.0	—	—	—	—	—	—
1.0	0.6	1.5	0.4	0.8	−2.0	—	—	—	—	—	—
2.0	1.0	−2.0	1.0	0.4	−0.2	—	—	—	—	—	—

For three of the possible configurations, a root finder will be needed to find the intermediate pressure. See App. D in the text for a "cheap-and-cheerful" root finder using the secant method.

Computer project

P3.1 Write a computer program to calculate the intermediate states, including the smooth transitions across rarefaction fans, of the hydrodynamical Riemann problem. If you've already done Problem 3.12, you may use that program as a starting point for this, if you like.

Your program should report on the screen, in a data file, or both all intermediate values $(\rho_1, p_1, v_1,$ and $\rho_2)$ as well as speeds of any shocks that may arise. In addition, you should plot ρ, p, v, and c_s as functions of the self-similar variable, $u = x/t$, similar to Fig. 3.9 on page 93 (without the reference points and dashed lines). Once your program is working, do all four Riemann problems listed in Problem 3.12, giving all numerical values asked for in the table as well as generating a plot as described above for each problem.

4 The Fundamentals of Magnetohydrodynamics

When pondering the heavens, and what it all could be:
the sun, the stars, the northern lights; I'm thinking MHD!

4.1 A brief introduction to MHD

WHEN A GAS is heated enough that random collisions start knocking off electrons from constituent atoms, a fourth state of matter emerges: a *plasma*. For a hydrogen gas, temperatures of a few thousand K will cause some of the H atoms to lose their electrons, and the gas becomes a plasma with a mix of neutral and charged particles. At higher temperatures still ($\sim 10^5$ K for hydrogen), plasmas become fully ionised and all particles are charged. Evidently, unlike the transition between a solid and a liquid or for the most part a liquid to a gas, there is no sudden *phase change* demarcating a transition from gas to plasma.

Still, the behaviour of a significantly ionised plasma is radically different from that of a neutral gas, and at the root of these differences lies the differential motion of the negatively and positively charged ions. Because the mass of an ion is two thousand or more times greater than that of an electron, fluid dynamical and electromagnetic forces within the plasma accelerate the negative and positive ions very differently. And while local charge neutrality can still be assumed, positive and negative currents *within the same fluid element* will differ, and it is this difference that leads to the generation of a net magnetic field.[1] Since the Lorentz force "binds" charged particles to magnetic field (that is, they can spiral along magnetic field lines but cannot cross them), the presence of a magnetic field makes the dynamics of a plasma completely different from that of a neutral gas.

It is often stated that 99.9% of the baryonic matter of the universe is in the plasma state,[2] and thus the role of magnetism in astrophysics is ubiquitous. Sun spots and solar flares are tied to the solar magnetic field and its internal dynamos. The coronal wind from the sun and the interception of these particles by the earth's

---

[1] This is what is known in MHD and plasma physics as a *dynamo*, a mechanism by which magnetic field is generated. A very famous dynamo is the one at work in the earth's core where differentially moving currents within the molten rock generate the earth's magnetic field, protecting us from the solar wind and making life as we know it possible on the surface of the planet. We look at dynamos in some detail in §10.3.4.

[2] See, for example, www.plasma-universe.com/99-999-plasma/.

magnetic field result in the most spectacular terrestrial example of astrophysical magnetism; the aurora borealis/australis (cover image).

However, these phenomena are mere hints of the role magnetism plays in astrophysics. The very existence of the sun and the conditions in which life could evolve would not be possible without magnetism which plays a critical role – as important as gravity – in allowing stars to form from interstellar gas clouds. In the interstellar medium, energy seems to be divided into three roughly equal parts: thermal energy within the gas; kinetic energy stored in cosmic rays;[3] and magnetic energy in the field permeating the galaxy. Indeed, the principle of equipartition[4] is widely believed to spread well beyond the galaxy and apply to the huge extragalactic radio lobes associated with a class of galaxies known as *active galaxies* (*e.g.*, Fig. 2.14), and the intergalactic medium itself; magnetic field is *everywhere*.

Even in systems where the magnetic energy density is orders of magnitude less than the thermal and cosmic ray components, magnetic field still plays a fundamental role. Consider an extragalactic radio jet (*e.g.*, Fig. 2.14 and 2.17), in which the matter number density, n, is as scant as one particle per m^{-3}. So rarefied is such a medium that the mean-free path of a single particle, given by $l = 1/(\sigma n)$ where $\sigma \sim 10^{-20}\,m^2$ is its effective cross-sectional area, is of order 10^{20} m, greater than a typical jet diameter (~ 1 kpc $\sim 3 \times 10^{19}$ m). In such a system, particle collisions simply cannot isotropise the medium, and thus if collisions were the only interaction among the particles, material in extragalactic jets and their associated lobes could not be treated as a fluid.

However, with even a trace magnetic induction (*e.g.*, $B \sim 10^{-12}$ T), the *Larmor radius*, r_L, of a proton moving at an appreciable fraction of the speed of light (*e.g.*, $v \sim 10^8$ m/s), is given by:

$$r_L = \frac{mv}{qB} \sim 10^{12}\,m,$$

seven orders of magnitude *smaller* than the diameter of the jet. Therefore, in this setting it is the trace magnetic induction that provides the isotropisation of individual particle motions and not the particle–particle collisions as has been assumed until now. Since the fluid model doesn't care what causes isotropy, only that the system is isotropic, a trace magnetic induction among the particles saves the fluid model in even the most rarefied of systems in the universe.

Modern astrophysics, therefore, *starts* with magnetohydrodynamics (MHD), with further complications coming from dissipation, radiative processes, relativistic effects, (self-) gravitation, *etc.*, much of which (but not all!) lie beyond the scope of this text.[5]

---

[3]Cosmic rays (CR) consist of a small fraction of gas particles which are promoted by various means to relativistic and ultra-relativistic energies. We'll talk more about CR in §7.4.
[4]*Rappel*: See footnote 3 on page 11.
[5]For a more comprehensive and graduate level text on the physics of astrophysical and terrestrial MHD, see Goedbloed *et al.*, 2019.

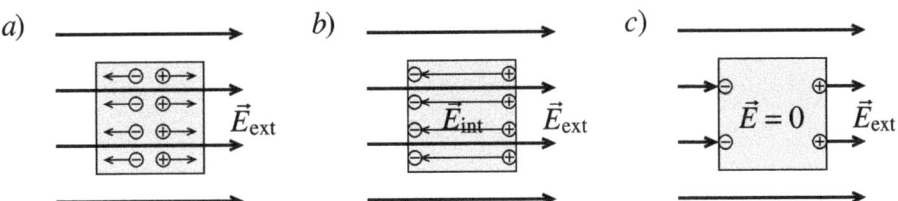

Figure 4.1. *a)* A flat square of metal (grey) is placed within an external electric field, $\vec{E}_{\text{ext}}$, instantly driving opposite charges apart. *b)* Charges move as far as they can go (*e.g.*, to opposite edges of the metal), setting up an internal electric field, $\vec{E}_{\text{int}}$, which in equilibrium (so that no further charges move), has the same magnitude but opposite direction as $\vec{E}_{\text{ext}}$. *c)* $\vec{E}_{\text{int}}$ and $\vec{E}_{\text{ext}}$ cancel, leaving $\vec{E} = 0$ within the metal only.

4.2 The ideal induction equation

When a plasma is only partially ionised, many complicating dissipative effects emerge such as fluid resistivity, the Hall effect, and ambipolar diffusion. We shall discuss these in detail in Chap. 10. In this chapter, we begin our exploration of non-dissipative *ideal MHD* where one assumes the plasma to be fully ionised (and thus a perfect conductor), and where one considers fluid elements of sufficient size that charge neutrality (but not equal positive and negative current densities!) within each element can be assumed.

Now, as any upper-year physics student knows, a perfect conductor cannot support an electric field.[6] What this means in practice is if a metal is placed within a background electric field (*e.g.*, Fig. 4.1*a*), then within a light-crossing time (shortest possible time scale of the system), charges within the metal begin to move in opposite directions; free negative electrons move against $\vec{E}_{\text{ext}}$ whilst positive electron "holes" move with $\vec{E}_{\text{ext}}$. This continues until the internal electric field, $\vec{E}_{\text{int}}$, induced by this charge separation (Fig. 4.1*b*), exactly cancels out $\vec{E}_{\text{ext}}$ and renders the net electric field within the metal zero (Fig. 4.1*c*).

The situation within a plasma is rather more complicated. Here, both negative *and* positive charges are free to move and, as already explained, the difference in how the negative and positive currents are driven results in a net magnetic field. Thus, the generalisation of the "no electric field within a metal" conclusion for a completely ionised fluid is that in a very short time scale, the net *electromagnetic force* – the sum of the electric *and* Lorentz forces, $\vec{F}_{\text{EM}}$ (Eq. B.4 in §B.1) – is what goes to zero by virtue of charge motion and redistribution, and not the electric field. That is to say, in an ideal MHD fluid we have,

$$\vec{F}_{\text{EM}} = q(\vec{E} + \vec{v} \times \vec{B}) = 0, \qquad (4.1)$$

where q is the charge on the particle and $\vec{B}$ is the *magnetic induction* (*a.k.a.*,

---

[6]I'll get back to this piece of "common knowledge" in footnote 8.

*magnetic flux density*⁷) with mks units *Tesla* (T = Wb m⁻²). Thus,

$$\vec{E}_{\text{ind}} = -\vec{v} \times \vec{B}, \tag{4.2}$$

is the *induced* electric field (units V m⁻¹) within the ionised medium.[8]

Combining Eq. (4.2) with the differential form of Faraday's law (third of Eq. B.1), namely,

$$\nabla \times \vec{E} = -\frac{\partial \vec{B}}{\partial t}, \tag{4.3}$$

gives us,

$$\boxed{\frac{\partial \vec{B}}{\partial t} = \nabla \times (\vec{v} \times \vec{B}).} \tag{4.4}$$

This is the so-called *ideal induction equation* that will be part of our focus for the rest of this chapter and indeed Part I. In passing, I note that since so many (astro)physical applications (but by no means all!) assume a non-dissipative medium, the adjective *ideal* is often dropped out of convenience when referring to Eq. (4.4). However, this usage is strictly incorrect and I shall continue to refer to it by its full name, the "ideal induction equation", particularly since *non*-ideal MHD is discussed in such depth in Chap. 10.

Before closing out this section, a quick reminder to the reader of the difference between a *static* and an *induced* electric field is in order. A *static* electric field is one generated by isolated charges, either within or external to the medium. It is conservative ($\nabla \times \vec{E}_{\text{stat}} = 0$) and can be represented with *open* electric field lines whose ends coincide with point charges, such as in the metal square depicted in Fig. 4.1. In non-relativistic, ideal MHD, the fluid is considered to be a conductor of such great efficacy that under the influence of a background static electric field, charges move in times short compared to all other relevant time scales thereby cancelling out the static electric field within the fluid, again just as in Fig. 4.1.

On the other hand, an *induced* electric field, such as that given by equation (4.2), is not conservative (since, by Eq. 4.3, its curl is non-zero), and its field lines are *closed*. Induced electric field lines do not start and end on isolated point charges and, as such, cannot be eliminated by a redistribution of point charges within the fluid. It is a necessary consequence of a time-varying magnetic induction, can drive currents just like a static electric field, and is the reason why the physics of even ideal MHD is so rich.

---

[7] $\vec{B}$ is an example of a *non*advective flux density. See App. B for a review of the essential elements of electrodynamics for MHD.

[8] It is precisely this fact that the physics community writ large failed to recognise in the 1940s and 50s that led to the resistance to Alfvén's theory of magnetohydrodynamics and much of his grief. The criticism of the day was that no conductor could support an electric field, based on the common experience that conductors are metals for which this assertion is true. However, in a system such as a plasma – also a conductor – where negative and positive charges are free to move but in different ways, it is the net *electromagnetic force* (Eq. 4.1) that is zeroed out, not just $\vec{E}$.

4.3 Alfvén's theorem

As defined by the second of Eq. (1.18), the *flux*, Φ, of a vector field, $\vec{\phi}$, passing through an open surface, Σ, is given by,

$$\Phi \equiv \int_\Sigma \vec{\phi} \cdot \hat{n} d\sigma, \qquad (4.5)$$

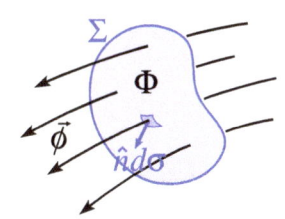

where, as shown in the inset, $\hat{n}$ is a unit vector normal to the surface where a given field line of $\vec{\phi}$ passes through. In this context, $\vec{\phi}$ is also considered as the flux density of Φ and, in the language of Chap. 1, Φ is also the *extensive quantity* (defined over a region) associated with the *intensive vector quantity* $\vec{\phi}$ (defined at a point). However, unlike pure hydrodynamics where the extensive quantity Q is defined as the *volume integral* of its associated intensive quantity, q (Eq. 1.15), here Φ is defined as the *open surface integral* of $\vec{\phi}$.

The Theorem of hydrodynamics (Theorem 1.1) introduced in §1.3 relates the time-evolution equation of an intensive quantity, q, to that of its associated *volume*-integrated extensive quantity Q, the proof of which depended upon Gauss' theorem. It might come as no surprise, then, that a similar theorem can be constructed to relate the time evolution equation of an intensive vector quantity, $\vec{\phi}$, to that of its associated *surface*-integrated extensive quantity Φ whose proof depends upon Stokes' theorem.

Theorem 4.1. The flux theorem. *Let Φ be the flux (extensive quantity) of a vector field, $\vec{\phi}$ (intensive quantity), over an open surface Σ. If its time dependence is,*

$$\frac{d\Phi}{dt} = \Psi, \qquad (4.6)$$

where Ψ is the possibly time-dependent "source term" (reason why Φ might not be "conserved"), then the evolution equation for its associated vector field, $\vec{\phi}(\vec{r},t)$, is given by,

$$\frac{\partial \vec{\phi}}{\partial t} = \nabla \times (\vec{v} \times \vec{\phi}) + \vec{\psi}, \qquad (4.7)$$

where $\vec{v} = d\vec{r}/dt$, $\Psi = \int_\Sigma \vec{\psi} \cdot \hat{n} d\sigma$ is the flux of vector field $\vec{\psi}$, $\hat{n}$ is the unit vector normal to the area differential $d\sigma$, and where $\vec{v} \times \vec{\phi}$ is a differentiable function of the coordinates.

Proof: We start with,[9]

$$\frac{d\Phi}{dt} = \Psi \quad \Rightarrow \quad \frac{d}{dt} \int_\Sigma \vec{\phi} \cdot \hat{n} \, d\sigma = \int_\Sigma \vec{\psi} \cdot \hat{n} \, d\sigma,$$

---

[9] The reader might want to glance back at Theorem 1.1 to convince themself that the following argument involving Stokes' theorem is fundamentally the same.

where $\Sigma = \Sigma(t)$ is also, in principle, time variable. Then, using the standard definition of the derivative,

$$\frac{d\Phi}{dt} = \frac{d}{dt}\int_{\Sigma(t)} \vec{\phi}(\vec{r},t)\cdot\hat{n}\,d\sigma$$

$$= \lim_{\Delta t\to 0}\frac{1}{\Delta t}\left[\int_{\Sigma(t+\Delta t)} \vec{\phi}(\vec{r},t+\Delta t)\cdot\hat{n}\,d\sigma - \int_{\Sigma(t)} \vec{\phi}(\vec{r},t)\cdot\hat{n}\,d\sigma\right]$$

$$= \lim_{\Delta t\to 0}\frac{1}{\Delta t}\left[\int_{\Sigma(t+\Delta t)-\Sigma(t)} \vec{\phi}(\vec{r},t+\Delta t)\cdot\hat{n}\,d\sigma + \int_{\Sigma(t)} \vec{\phi}(\vec{r},t+\Delta t)\cdot\hat{n}\,d\sigma \right.$$
$$\left. - \int_{\Sigma(t)} \vec{\phi}(\vec{r},t)\cdot\hat{n}\,d\sigma\right]$$

$$= \lim_{\Delta t\to 0}\frac{1}{\Delta t}\int_{\Delta\Sigma} \vec{\phi}(\vec{r},t+\Delta t)\cdot\hat{n}\,d\sigma + \lim_{\Delta t\to 0}\int_{\Sigma(t)}\frac{\vec{\phi}(\vec{r},t+\Delta t)-\vec{\phi}(\vec{r},t)}{\Delta t}\cdot\hat{n}\,d\sigma,$$

where, as shown in the inset, integrating over the difference in areas, $\Delta\Sigma = \Sigma(t+\Delta t) - \Sigma(t)$, is the same as integrating around the closed contour, $\partial\Sigma$, with area differential $\hat{n}\,d\sigma = \vec{v}\Delta t \times d\vec{l}$:[10]

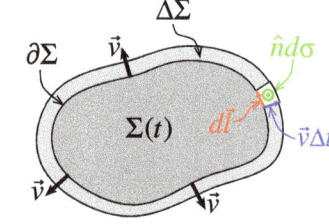

$$\frac{d\Phi}{dt} = \lim_{\Delta t\to 0}\frac{1}{\Delta t}\oint_{\partial\Sigma} \vec{\phi}(\vec{r},t+\Delta t)\cdot(\vec{v}\Delta t\times d\vec{l})$$
$$+ \int_{\Sigma(t)}\frac{\partial\vec{\phi}(\vec{r},t)}{\partial t}\cdot\hat{n}\,d\sigma$$

$$= \oint_{\partial\Sigma}(\vec{\phi}(\vec{r},t)\times\vec{v})\cdot d\vec{l} + \int_{\Sigma(t)}\frac{\partial\vec{\phi}(\vec{r},t)}{\partial t}\cdot\hat{n}\,d\sigma,$$

using vector identity (A.1). We now use the first flavour of Stokes' theorem in §A.2, namely Eq. (A.35), to restore the first integral to a surface integral:

$$\frac{d\Phi}{dt} = \int_{\Sigma(t)}\nabla\times(\vec{\phi}\times\vec{v})\cdot\hat{n}\,d\sigma + \int_{\Sigma(t)}\frac{\partial\vec{\phi}}{\partial t}\cdot\hat{n}\,d\sigma = \int_{\Sigma(t)}\vec{\psi}\cdot\hat{n}\,d\sigma$$

$$\Rightarrow \int_{\Sigma(t)}\left(\frac{\partial\vec{\phi}}{\partial t} - \nabla\times(\vec{v}\times\vec{\phi}) - \vec{\psi}\right)\cdot\hat{n}\,d\sigma = 0,$$

true for any open surface, Σ. Thus, the integrand is zero, proving the theorem. □

Parroting our discussion after Theorem 1.1, $\vec{\phi}$ is not the conserved quantity here, Φ is. However, since Φ is the surface-integral of $\vec{\phi}$, we'll refer to $\vec{\phi}$ as a *surface-conserved quantity*. Further, Eq. (4.6) involves surface integrals Φ and Ψ over a macroscopic open surface Σ, and thus represents a *global* statement (valid over a finite area) on the evolution of the extensive, conserved quantity Φ. On the other hand, Eq. (4.7) is a *differential* equation (differential form of Eq. 4.6) and represents

---

[10]*Rappel*: the area of a parallelogram is the cross product of the two non-parallel sides with order determined by the right-hand rule.

a *local* statement (valid at a point) on the evolution of Φ expressed in terms of its corresponding intensive, surface-conserved quantity $\vec{\phi}$.

The flux theorem is general for any vector field and its associated flux. In fact, the reader who did Problem 1.7 relating the fluid *vorticity*, $\vec{\omega} = \nabla \times \vec{v}$, to its flux, namely the *circulation*, Γ, should have found this discussion very familiar.

Alfvén's theorem applies specifically to the magnetic induction, and gives the conditions for which the magnetic flux,

$$\Phi_B = \int_\Sigma \vec{B} \cdot \hat{n} d\sigma, \tag{4.8}$$

is conserved (units of Φ_B are *Webers*, Wb = T m^2). Given our efforts to derive the ideal induction equation and prove the flux theorem, the proof of Alfvén's theorem is straightforward.

Theorem 4.2. Alfvén's theorem. *In an ideal MHD system, the magnetic flux within any open surface that evolves with the fluid is conserved. That is,*

$$\boxed{\frac{d\Phi_B}{dt} = 0.} \tag{4.9}$$

Proof: Suppose for the moment that the evolution equation for Φ_B is given by,

$$\frac{d\Phi_B}{dt} = \Psi,$$

where Ψ is a source term for Φ_B. Now, from Eq. (4.7), we can immediately write down the evolution equation for its associated intensive vector field, $\vec{B}$:

$$\frac{\partial \vec{B}}{\partial t} = \nabla \times (\vec{v} \times \vec{B}) + \vec{\psi},$$

where $\Psi = \int_\Sigma \vec{\psi} \cdot \hat{n} d\sigma$ and $\vec{\psi}$ serves as a source term for $\vec{B}$. But from Eq. (4.4),

$$\frac{\partial \vec{B}}{\partial t} = \nabla \times (\vec{v} \times \vec{B}) \quad \Rightarrow \quad \vec{\psi} = 0 \quad \Rightarrow \quad \Psi = 0,$$

proving the theorem. □

Physically, Alfvén's theorem asserts that in ideal MHD, the magnetic induction is "frozen in" the fluid, a concept known as *flux-freezing*. Any fluid motion causing a given open surface to distort, shrink, twist, expand, or whatever does so without gaining or losing any magnetic flux through it. One can imagine, therefore, a "pencil" of fluid or, as it is commonly referred to, a *flux tube* (Fig. 4.2) whose sides are made up of lines of induction and, by

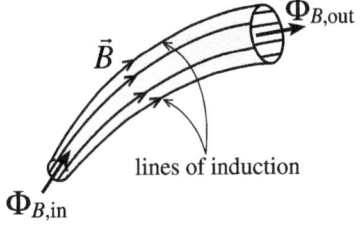

Figure 4.2. A *flux tube* with sides defined by $\vec{B}$, and where at the ends, $\Phi_{B,\text{in}} = \Phi_{B,\text{out}}$.

Alfvén's theorem, whose magnetic flux passing through its ends remains constant regardless of how that flux tube evolves in the fluid.[11] Indeed, since $\nabla \cdot \vec{B} = 0$ (why does this matter?), we also know that the net flux entering the flux tube, $\Phi_{B,\text{in}}$, must equal the net flux leaving, $\Phi_{B,\text{out}}$, since no flux enters or leaves the flux tube along its sides where $\vec{B} \perp \hat{n}$, the surface normal.

Furthermore, if one considers a flux tube of arbitrarily small radius, then the fluid within the flux tube can only follow along lines of induction and never cross them, lest the fluxes through the ends of the tube change. This is not to say that fluid motion is entirely dictated by where lines of magnetic induction lie; far from it. In the give and take of all the MHD forces, either the fluid follows $\vec{B}$, or lines of induction get dragged around and twisted mercilessly by the fluid, or something in between; all cases are consistent with the idea of magnetic flux being frozen within the fluid, *i.e.* flux-freezing. We shall come to quantify this "tug-of-war" between $\vec{B}$ and $\vec{v}$ in Chap. 5 when we define the *MHD-alpha*.

We have therefore added to our august list of three conservation principles for ideal fluid dynamics a fourth, which in their entirety now include:

1. conservation of mass, M (Eq. 1.1);

2. conservation of total energy, E_T (Eq. 1.2);

3. conservation of momentum, $\vec{S}$ (Newton's second law, Eq. 1.3); and now,

4. conservation of magnetic flux, $\vec{\Phi}_B$ (Eq. 4.9).

4.4 Modifications to the momentum equation

As we've seen, in the presence of magnetic induction, a charged particle, q, moving at velocity $\vec{v}$ is subject to the *Lorentz force*,

$$\vec{F}_L = q\vec{v} \times \vec{B}.$$

Therefore, an ensemble of charged particles described by a *charge density* $\rho_q = dq/dV$ moving collectively at an ensemble velocity $\vec{v}$ is subject to a *Lorentz force density*,

$$\vec{f}_L = \rho_q \vec{v} \times \vec{B} = \vec{J} \times \vec{B}, \qquad (4.10)$$

where $\vec{J} = \rho_q \vec{v}$ is the *current density* (SI units: A m^{-2}). Now, from Ampère's law[12] (App. B), we have,

$$\vec{J} = \frac{1}{\mu_0} \nabla \times \vec{B}, \qquad (4.11)$$

---

[11] This should remind the reader of a *streamtube* defined in §2.4 and illustrated in Fig. 2.8.

[12] The *displacement current* in the more general Maxwell–Ampère law, namely $(1/c^2)\,\partial \vec{E}/\partial t$, is ignored on the grounds that the factor $1/c^2$ makes it negligible for non-relativistic fluid motions.

and thus Eq. (4.10) can be written as,

$$\vec{f}_{\mathrm{L}} = \frac{1}{\mu_0}(\nabla \times \vec{B}) \times \vec{B}. \tag{4.12}$$

Adding this force density directly to the RHS of Eq. (1.27) gives us the momentum equation for ideal MHD:

$$\boxed{\frac{\partial \vec{s}}{\partial t} + \nabla \cdot (\vec{s}\vec{v}) = -\nabla p - \rho \nabla \phi + \frac{1}{\mu_0}(\nabla \times \vec{B}) \times \vec{B}.} \tag{4.13}$$

The effect of the Lorentz force density to Euler's Eq. (1.36) is evidently:

$$\boxed{\frac{\partial \vec{v}}{\partial t} + (\vec{v} \cdot \nabla)\vec{v} = -\frac{1}{\rho}\nabla p - \nabla \phi + \frac{1}{\rho\mu_0}(\nabla \times \vec{B}) \times \vec{B}.} \tag{4.14}$$

4.5 The MHD Poynting power density

From Eq. (B.17) in App. B and Eq. (4.2) above, we define the *MHD Poynting vector* in free space as:

$$\vec{S}_{\mathrm{P}} = \frac{1}{\mu_0}\vec{E}_{\mathrm{ind}} \times \vec{B} = \frac{1}{\mu_0}\vec{B} \times (\vec{v} \times \vec{B}). \tag{4.15}$$

Further, from Eq. (B.18), the *MHD Poynting flux* through an open surface Σ is:

$$\Phi_S = \int_\Sigma \vec{S}_{\mathrm{P}} \cdot \hat{n}\, d\sigma = \frac{1}{\mu_0} \int_\Sigma [\vec{B} \times (\vec{v} \times \vec{B})] \cdot \hat{n}\, d\sigma, \tag{4.16}$$

while, according to Eq. (B.19), the *MHD Poynting power density* must be:

$$\boxed{p_S = -\nabla \cdot \vec{S}_{\mathrm{P}} = -\frac{1}{\mu_0}\nabla \cdot [\vec{B} \times (\vec{v} \times \vec{B})].} \tag{4.17}$$

The Poynting flux, whether for ideal MHD or other more general electromagnetic systems, is interpreted as the rate at which energy is transported into or out of a given volume by the electromagnetic fields. In particular, Φ_S, whose units are Watts, is the magnetic power passing through a given surface (a flux), while p_S, whose units are $\mathrm{W\,m^{-3}}$, is the rate of magnetic energy transport per unit volume, and is thus a power density.

The reader unfamiliar with the Poynting flux and its relationship to Maxwell's equations is directed to App. B which contains a brief, but self-contained review of the subject.

4.6 Modifications to the total energy equation

From Eq. (B.13) in App. B, the energy density of the magnetic induction is $B^2/(2\mu_0)$. Thus, we define the total MHD energy density, e_{T}^*, to be:

$$e_T^* = e_T + \frac{B^2}{2\mu_0} = e + \frac{\rho v^2}{2} + \rho\phi + \frac{B^2}{2\mu_0}, \tag{4.18}$$

and then examine the time derivative of e_T^*. Thus,

$$\frac{\partial e_T^*}{\partial t} = \frac{\partial e}{\partial t} + \rho \frac{\partial v^2/2}{\partial t} + \frac{v^2}{2}\frac{\partial \rho}{\partial t} + \phi\frac{\partial \rho}{\partial t} + \frac{1}{\mu_0}\frac{\partial B^2/2}{\partial t}$$

$$= -\nabla\cdot(e\vec{v}) - p\nabla\cdot\vec{v} + \rho\vec{v}\cdot\left(-\vec{v}\cdot\nabla\vec{v} - \frac{1}{\rho}\nabla p - \nabla\phi + \frac{1}{\rho\mu_0}(\nabla\times\vec{B})\times\vec{B}\right)$$

$$- \frac{v^2}{2}\nabla\cdot(\rho\vec{v}) - \phi\nabla\cdot(\rho\vec{v}) + \frac{1}{\mu_0}\vec{B}\cdot[\nabla\times(\vec{v}\times\vec{B})],$$

where the continuity equation (Eq. 1.19), the internal energy equation (Eq. 1.34), the ideal induction equation (Eq. 4.4), and Euler's equation modified for the Lorentz force (Eq. 4.14) have all been used to replace the time derivatives. Note also the use of the identity:

$$\frac{\partial A^2/2}{\partial t} = \vec{A}\cdot\frac{\partial \vec{A}}{\partial t},$$

for any vector $\vec{A}$, and the fact that the gravitational potential, ϕ, has no explicit time dependence (e.g., Eq. 1.28), whence $\partial\phi/\partial t = 0$. Using numerous vector identities from App. A, we can continue to develop the expression for $\partial e_T^*/\partial t$ to get:

$$\frac{\partial e_T^*}{\partial t} = -\nabla\cdot(e\vec{v}) - \nabla\cdot(p\vec{v}) - \nabla\cdot(\rho\phi\vec{v}) - \frac{v^2}{2}\nabla\cdot(\rho\vec{v}) - \rho\vec{v}\cdot(\vec{v}\cdot\nabla\vec{v})$$

$$+ \frac{1}{\mu_0}\left(\vec{v}\cdot(\nabla\times\vec{B})\times\vec{B} + \vec{B}\cdot\nabla\times(\vec{v}\times\vec{B})\right)$$

$$= -\nabla\cdot[(e+p+\rho\phi)\vec{v}] - \nabla\cdot\left(\frac{\rho v^2}{2}\vec{v}\right) + \rho\vec{v}\cdot\left(\nabla\frac{v^2}{2} - \vec{v}\cdot\nabla\vec{v}\right)$$

$$- \frac{1}{\mu_0}\left(\nabla\times\vec{B}\cdot(\vec{v}\times\vec{B}) - \vec{B}\cdot\nabla\times(\vec{v}\times\vec{B})\right)$$

$$= -\nabla\cdot\left[\left(e+\frac{\rho v^2}{2}+\rho\phi+p\right)\vec{v}\right] - \frac{1}{\mu_0}\nabla\cdot\left(\vec{B}\times(\vec{v}\times\vec{B})\right)$$

$$+ \rho\vec{v}\cdot\left(\vec{v}\times(\nabla\times\vec{v})\right)^{\!\!0},$$

where the last term is zero since $\vec{v}\perp\vec{v}\times\vec{A}$ for any vector $\vec{A}$. Note that e_T (and not e_T^*!) has appeared in the first term on the RHS, and the second term is just the MHD Poynting power density, $p_S = -\nabla\cdot\vec{S}_P$, from Eq. (4.17). Thus, the total energy equation for MHD can be written more compactly as:

$$\boxed{\frac{\partial e_T^*}{\partial t} + \nabla\cdot[(e_T+p)\vec{v} + \vec{S}_P] = 0.} \tag{4.19}$$

The fact that the Poynting power density has shown up reflects its interpretation as the rate at which work is done on (by) the magnetic field by (on) the fluid. Like the "pdV" term that contributed to Eq. (1.34), the Poynting term is an "applied power density" that prevents total energy from being conserved in the co-moving frame of the fluid.

4.7 The equations of ideal MHD

We are now in a position to write down the complete set of ideal MHD equations.

Equation Set 6:

$$\frac{\partial \rho}{\partial t} + \nabla \cdot (\rho \vec{v}) = 0; \qquad (4.20)$$

$$\frac{\partial e_T^*}{\partial t} + \nabla \cdot (e_T \vec{v}) = -\nabla \cdot \left(p\vec{v} + \frac{1}{\mu_0} \vec{B} \times (\vec{v} \times \vec{B}) \right); \qquad (4.21)$$

$$\frac{\partial \vec{s}}{\partial t} + \nabla \cdot (\vec{s}\vec{v}) = -\nabla p - \rho \nabla \phi + \frac{1}{\mu_0} (\nabla \times \vec{B}) \times \vec{B}; \qquad (4.22)$$

$$\frac{\partial \vec{B}}{\partial t} - \nabla \times (\vec{v} \times \vec{B}) = 0, \qquad (4.23)$$

where:

$$e_T^* = e_T + \frac{B^2}{2\mu_0}; \qquad p = (\gamma - 1)\left(e_T - \tfrac{1}{2}\rho v^2 - \rho \phi\right); \qquad \vec{s} = \rho \vec{v},$$

are the constitutive equations. As usual and where convenient, the total energy and first constitutive equations may be replaced with either the internal energy equation (Eq. 1.34) or the pressure equation (Eq. 1.41) where use of the latter also eliminates the need for the second of the constitutive equations. Note that neither Eq. (1.34) nor (1.41) are modified by the addition of the magnetic induction. Further, the momentum and third constitutive equations may be replaced with the MHD Euler equation (Eq. 4.14).

The ideal induction equation is fundamentally different from the other hydrodynamical equations. The second term on the left-hand side (LHS) of Eq. (4.20)–(4.22) are perfect divergences, whereas the second term on the LHS of Eq. (4.23) is a perfect curl. As stressed previously, this stems from the fact that the hydrodynamical variables ρ, e_T, and $\vec{s}$ are volume-conserved quantities whereas $\vec{B}$ is a surface-conserved quantity. At this point, some further discussion on the significance of this distinction is in order.

In the absence of source terms, the hydrodynamical equations have the form:

$$\frac{\partial q}{\partial t} + \nabla \cdot (q\vec{v}) = 0,$$

where q is any of ρ, e_T, and $\vec{s}$. Integrated over a volume, V, with surface, S, this yields:

$$\int_V \frac{\partial q}{\partial t} dV = \frac{dQ}{dt} = -\int_V \nabla \cdot (q\vec{v}) dV = -\oint_S q\vec{v} \cdot d\vec{\sigma},$$

upon invoking Gauss' theorem, and where $Q = \int_V q \, dV$. This has the physical interpretation that the quantity Q inside a fixed volume V changes only if there is a net (advective) flux density of Q, $q\vec{v}$, passing across the closed surface, S. This is the nature of a volume-conserved quantity.

On the other hand by integrating the ideal induction equation over an arbitrary open surface, Σ, we get:

$$\int_\Sigma \frac{\partial \vec{B}}{\partial t} \cdot \hat{n}\, d\sigma = \frac{d\Phi_B}{dt} = \int_\Sigma \nabla \times (\vec{v} \times \vec{B}) \cdot \hat{n}\, d\sigma = -\oint_C \vec{E}_{\text{ind}} \cdot d\vec{l},$$

using Eq. (4.2) and Stokes' theorem, where C is the circumference of the open surface Σ. Now, $\oint_C \vec{E}_{\text{ind}} \cdot d\vec{l} = \mathcal{E}$ is the net *emf* induced around C (Eq. B.8). Thus, the magnetic flux, Φ_B, passing through C changes only if a non-zero $\mathcal{E}$ drives a current around C, inducing additional magnetic flux (Biot and Savart). This is the nature of the surface-conserved vector quantity, $\vec{B}$. That the magnetic induction and momentum density obey different types of conservation laws and, related to this, that $\vec{B}$ is subject to the solenoidal condition ($\nabla \cdot \vec{B} = 0$) whereas $\vec{s}$ is not, point to fundamental mathematical differences between the two vector fields. This has profound effects both on their physical influence on the fluid and, for the computationalist, the design of the numerical algorithms required to solve the equations of MHD accurately (*e.g.*, Clarke, 1996). On that subject, see also Problem 4.8.

There are alternate forms for the total energy, momentum, and ideal induction equations that are often useful. For the former, identity (A.2) allows us to write:

$$\vec{B} \times (\vec{v} \times \vec{B}) = \vec{v}B^2 - (\vec{v} \cdot \vec{B})\vec{B},$$

and thus Eq. (4.21) becomes:

$$\frac{\partial (e_{\text{T}}^*)}{\partial t} + \nabla \cdot (e_{\text{T}} \vec{v}) = -\nabla \cdot \left(p\vec{v} + \frac{B^2}{\mu_0}\vec{v} - \frac{1}{\mu_0}(\vec{v} \cdot \vec{B})\vec{B} \right)$$

$$\Rightarrow \quad \frac{\partial (e_{\text{T}}^*)}{\partial t} + \nabla \cdot \left[\left(e_{\text{T}}^* + p + \frac{B^2}{2\mu_0} \right) \vec{v} - \frac{1}{\mu_0}(\vec{v} \cdot \vec{B})\vec{B} \right] = 0. \qquad (4.24)$$

For the momentum equation, identities (A.15) and (A.23) allow us to write:

$$(\nabla \times \vec{B}) \times \vec{B} = (\vec{B} \cdot \nabla)\vec{B} - \frac{1}{2}\nabla B^2 = \nabla \cdot (\vec{B}\vec{B}) - \frac{1}{2}\nabla B^2,$$

with the latter being aided by the solenoidal condition on $\vec{B}$. Both forms conveniently divide the Lorentz force into two terms. The first term [$(\vec{B} \cdot \nabla)\vec{B}$ or $\nabla \cdot (\vec{B}\vec{B})$] contains "shear" derivatives (*e.g.*, $B_y\, \partial B_z/\partial y$ that involve *different* components of the magnetic induction) and thus can be thought of as describing "transverse Lorentz forces". Meanwhile, the second term ($\frac{1}{2}\nabla B^2$) contains nothing but "compressive" derivatives (*e.g.*, $\partial B_y^2/\partial x$, each involving only *one* component of the magnetic induction) and thus can be thought of as describing "longitudinal Lorentz forces". The latter term may be combined with the pressure gradient in the momentum equation to give an alternate form for the MHD momentum equation:

$$\frac{\partial \vec{s}}{\partial t} + \nabla \cdot (\vec{s}\vec{v}) = -\nabla \left(p + \frac{B^2}{2\mu_0} \right) - \rho\nabla\phi + \frac{1}{\mu_0}\nabla \cdot (\vec{B}\vec{B}). \qquad (4.25)$$

Note that in both places where the pressure, p, appears in Eq. (4.24) and (4.25), it is accompanied by $B^2/(2\mu_0)$. Thus, we define the *MHD pressure* to be:

$$p^* = p + \frac{B^2}{2\mu_0} \equiv p + p_M, \tag{4.26}$$

where the *magnetic pressure*, p_M, exists by virtue of the magnetic induction. Physically, p_M acts just like a thermal pressure; it is isotropic and its gradients can push fluid around. Note also that the magnetic pressure is identical in form to the magnetic energy density (Eq. B.13 in App. B) and thus, while $e/p = \gamma - 1$ for a gas, $e_M/p_M = 1$. It turns out for many applications, the MHD equations are easier to handle if both thermal and magnetic ratios of energy density to pressure are the same, and for this reason in a few applications, some investigators have set $\gamma = 2$, even though this doesn't correspond to any real gas (*e.g.*, Brio & Wu, 1988).

Finally and notwithstanding the discussion distinguishing the *volume*-conservative hydrodynamic equations from the *surface*-conservative ideal induction equation, we can "force" the latter into what *looks like* volume-conservative form by invoking vector identity (A.22):

$$\nabla \times (\vec{v} \times \vec{B}) = \nabla \cdot (\vec{B}\vec{v} - \vec{v}\vec{B}),$$

noting that in general, the *dyadic product* of two vectors (see Eq. A.16) doesn't commute (with the notable exception of two parallel vectors, like $\vec{s}$ and $\vec{v}$). With this identity, we can write down an alternate form of the ideal induction equation:

$$\frac{\partial \vec{B}}{\partial t} + \nabla \cdot (\vec{v}\vec{B} - \vec{B}\vec{v}) = 0,$$

which has both theoretic and practical importance and is the form we shall use in Chap. 6 where we solve the MHD Riemann problem.

We therefore have the following alternate set of ideal MHD equations:

Equation Set 7:

$$\frac{\partial \rho}{\partial t} + \nabla \cdot (\rho \vec{v}) = 0; \tag{4.27}$$

$$\frac{\partial e_T^*}{\partial t} + \nabla \cdot \left[(e_T^* + p^*)\vec{v} - \frac{1}{\mu_0}(\vec{v} \cdot \vec{B})\vec{B} \right] = 0; \tag{4.28}$$

$$\frac{\partial \vec{s}}{\partial t} + \nabla \cdot \left(\vec{s}\vec{v} + p^*\mathsf{I} - \frac{1}{\mu_0}\vec{B}\vec{B} \right) = -\rho \nabla \phi; \tag{4.29}$$

$$\frac{\partial \vec{B}}{\partial t} + \nabla \cdot (\vec{v}\vec{B} - \vec{B}\vec{v}) = 0, \tag{4.30}$$

---

where:

$$p^* = (\gamma - 1)(e_T^* - \frac{1}{2}\rho v^2 - \rho\phi) + (2 - \gamma)\frac{B^2}{2\mu_0}; \quad \vec{s} = \rho\vec{v},$$

are the constitutive equations. Note also the use of the vector identity $\nabla p^* = \nabla \cdot p^* \mathsf{I}$, introduced in footnote 12 on page 22, where I is the identity tensor (matrix).

With the gravity term ignored, we declared Eq. Set 1 on page 18 (describing hydrodynamics in 3-D) to be in *conservative form* because, when integrated over volume, all spatial derivatives disappeared via applications of various flavours of Gauss' theorem. The same is true for Eq. Set 7 (again, when $\phi = 0$) with Stokes replacing Gauss for Eq. (4.30), and thus these are the ideal MHD equations in conservative form. It is left to Problem 4.9 to gather together the ideal MHD equations in primitive form.

Finally, by defining kets of conservative variables, $|q\rangle$, and flux densities, $|\vec{f}\,\rangle$,

$$|q\rangle \equiv \begin{bmatrix} \rho \\ e_T^* \\ \vec{s} \\ \vec{B} \end{bmatrix} \quad \text{and} \quad |\vec{f}\,\rangle \equiv \begin{bmatrix} \rho\vec{v} \\ (e_T^* + p^*)\vec{v} - \dfrac{1}{\mu_0}(\vec{v}\cdot\vec{B})\vec{B} \\ \vec{s}\vec{v} + p^*\mathsf{I} - \dfrac{1}{\mu_0}\vec{B}\vec{B} \\ \vec{v}\vec{B} - \vec{B}\vec{v} \end{bmatrix},$$

and dropping ϕ, Eq. Set 7 can be written in its most compact form:

$$\boxed{\partial_t|q\rangle + \nabla\cdot|\vec{f}\,\rangle = 0,} \tag{4.31}$$

where $\partial_t \equiv \partial/\partial t$. Advanced texts and journal articles on MHD will often start off their discussion by writing the equations in this or a similar form, what I think of as "the $F = ma$ of MHD".

4.8 Vector potential and magnetic helicity (optional)

Problem 4.5 asks the reader to prove what may already be taken for granted: any solenoidal vector field, $\vec{B}$, may be written as the curl of another vector, $\vec{A}$. That is,

$$\nabla\cdot\vec{B} = 0 \quad \Rightarrow \quad \vec{B} = \nabla\times\vec{A}, \tag{4.32}$$

where, if $\vec{B}$ is the magnetic induction, $\vec{A}$ is known as the *vector potential*. The proof is actually rather subtle, and for the reader who has yet to go through it, I strongly recommend Problem 4.5. It provides a mathematical appreciation for what it means to perform an *anti-curl*, and yields very practical expressions for anyone with aspirations to work in *numerical* MHD. Consider it a right-of-passage!

Given Eq. (4.32), the ideal induction equation (Eq. 4.4) can be written as,

$$\frac{\partial}{\partial t}\nabla\times\vec{A} = \nabla\times\frac{\partial\vec{A}}{\partial t} = \nabla\times(\vec{v}\times\vec{B})$$

$$\Rightarrow \quad \nabla\times\left(\frac{\partial\vec{A}}{\partial t} - \vec{v}\times\vec{B}\right) = 0 \quad \Rightarrow \quad \frac{\partial\vec{A}}{\partial t} - \vec{v}\times\vec{B} = \nabla\mathcal{V}, \tag{4.33}$$

where $\mathcal{V}$ is a scalar function of the coordinates with units of electric potential

($J\,C^{-1}$). In electrodynamics, how one selects $\mathcal{V}$ is referred to as a *choice of gauge* and, for the particular choice $\nabla \mathcal{V} = 0$, Eq. (4.33) reduces to,

$$\boxed{\frac{\partial \vec{A}}{\partial t} = \vec{v} \times \vec{B} = -\vec{E}_{\text{ind}},} \qquad (4.34)$$

using Eq. (4.2). This gives a very convenient evolution equation for $\vec{A}$.

An important quantity in MHD is the so-called *magnetic helicity*. In general, the *helicity density* of a vector field, $\vec{\phi}$, is defined as,

$$h_\phi = \vec{\phi} \cdot (\nabla \times \vec{\phi}), \qquad (4.35)$$

an intensive quantity, whereas the *helicity* of the same vector field is,

$$H_\phi = \int_V h_\phi \, dV = \int_V \vec{\phi} \cdot (\nabla \times \vec{\phi}) \, dV, \qquad (4.36)$$

an extensive quantity. As we shall see, the helicity is a measure of the complexity or *topology* of the vector field. For example, if $\vec{\phi}$ is completely planar and thus no "streamlines" cross or loop each other, H_ϕ will turn out to be zero. However, should loops of $\vec{\phi}$ link other loops of $\vec{\phi}$ (think "chainlinks"), $H_\phi \neq 0$ with the sign of H_ϕ depending on whether loops of like-parity or counter-parity are linked.

For $\vec{\phi} = \vec{A}$, the vector potential, we get the *magnetic helicity density*,

$$h_A \equiv \vec{A} \cdot (\nabla \times \vec{A}) = \vec{A} \cdot \vec{B}. \qquad (4.37)$$

Examining first its time-variation, we have,

$$\frac{\partial h_A}{\partial t} = \vec{A} \cdot \frac{\partial \vec{B}}{\partial t} + \vec{B} \cdot \frac{\partial \vec{A}}{\partial t} = \vec{A} \cdot \left(\nabla \times (\vec{v} \times \vec{B})\right) + \vec{B} \cdot \left(\vec{v} \times \vec{B} + \nabla \mathcal{V}\right), \qquad (4.38)$$

using Eq. (4.4) and (4.33). Now,

$$\vec{A} \cdot \left(\nabla \times (\vec{v} \times \vec{B})\right) = \nabla \cdot \left[(\vec{v} \times \vec{B}) \times \vec{A}\right] + (\vec{v} \times \vec{B}) \cdot \overbrace{\nabla \times \vec{A}}^{\vec{B}}$$

$$= \nabla \cdot \left[(\vec{A} \cdot \vec{v})\vec{B} - \underbrace{(\vec{A} \cdot \vec{B})}_{h_A}\vec{v}\right] + \cancelto{0}{(\vec{v} \times \vec{B}) \cdot \vec{B}},$$

using vector Identity (A.2). Further,

$$\vec{B} \cdot (\vec{v} \times \vec{B} + \nabla \mathcal{V}) = \cancelto{0}{\vec{B} \cdot (\vec{v} \times \vec{B})} + \nabla \cdot (\mathcal{V}\vec{B}),$$

since $\nabla \cdot \vec{B} = 0$. Substituting these two vector identities into Eq. (4.38), we get,

$$\boxed{\frac{\partial h_A}{\partial t} + \nabla \cdot (h_A \vec{v}) = \nabla \cdot \left[(\vec{A} \cdot \vec{v} + \mathcal{V})\vec{B}\right],} \qquad (4.39)$$

an evolution equation for the magnetic helicity density. This should remind the reader very much of the time evolution equation for total energy density (*e.g.*, Eq. 1.23 for ideal HD, Eq. 4.24 for ideal MHD) which, like Eq. (4.39), have two terms on the LHS of the form $\partial q/\partial t + \nabla \cdot (q\vec{v})$ which express the conservative nature of

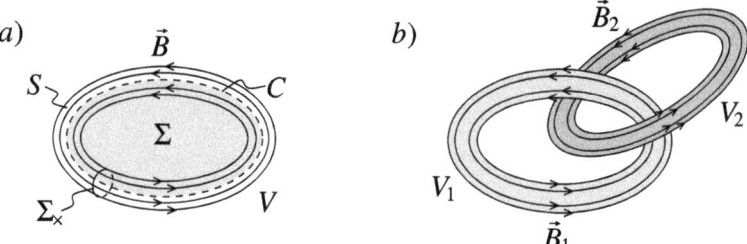

Figure 4.3. *a*) A single flux loop is shown with volume V, surface area S enclosing V, circumference C looping S (dashed line), interior surface Σ bounded by C (grey disc), and cross-sectional area $\Sigma_\times$. *b*) Two flux loops of volumes V_1 and V_2 with magnetic inductions $\vec{B}_1$ and $\vec{B}_2$ linking each other like two links of a chain.

q. Further, the "source terms" on the RHS of Eq. (4.39) form a perfect divergence, again just like the total energy density equations.

Writing Eq. (4.39) as,

$$\frac{\partial h_A}{\partial t} + \nabla \cdot \left[h_A \vec{v} - (\vec{A} \cdot \vec{v} + \mathcal{V}) \vec{B} \right] = 0,$$

so that it looks even more like Eq. (1.23) and (4.24), we identify the *helicity flux density*,

$$\vec{\mathcal{F}}_h \equiv h_A \vec{v} - (\vec{A} \cdot \vec{v} + \mathcal{V}) \vec{B} \quad \Rightarrow \quad \frac{\partial h_A}{\partial t} + \nabla \cdot \vec{\mathcal{F}}_h = 0, \qquad (4.40)$$

casting the evolution equation for the magnetic helicity density very much like a volume-conserved *hydrodynamical* quantity.

Finally, integrating Eq. (4.39) over a volume, V, we have,

$$\int_V \left[\frac{\partial h_A}{\partial t} + \nabla \cdot (h_A \vec{v}) \right] dV = \int_V \nabla \cdot \left[(\vec{A} \cdot \vec{v} + \mathcal{V}) \vec{B} \right] dV$$

$$\Rightarrow \quad \boxed{\frac{dH_A}{dt} = \oint_S (\vec{A} \cdot \vec{v} + \mathcal{V}) \vec{B} \cdot \hat{n}\, d\sigma,} \qquad (4.41)$$

where S is the surface enclosing V, and where the theorems of hydrodynamics (Theorem 1.1) and Gauss (Eq. A.30) were used. Eq. (4.41) is an evolution equation for the magnetic helicity. In particular, if the volume, V, encloses the entire system, then no flux density crosses S and the surface integral on the RHS of Eq. (4.41) vanishes. Thus, like the total mass, total energy, and magnetic flux, the magnetic helicity, H_A, is a conserved quantity of ideal MHD.

4.8.1 Magnetic topology

A good application of magnetic helicity is its role in quantifying *magnetic topology*. Consider the single *flux loop* depicted in Fig. 4.3*a*, where a flux loop is a flux tube wrapped into a toroid. Like a flux tube, the sides of a flux loop are defined by the *same* lines of induction as the fluid evolves in time. Thus, by definition, $\vec{B}$ along the surface, S, is everywhere perpendicular to the surface normal and the surface integral in Eq. (4.41) is zero. That is, *magnetic helicity is conserved within a flux loop*. What's more, we can evaluate the conserved value of the magnetic helicity within the isolated flux loop.

In the inset, a portion of the toroid in Fig. 4.3a is shown with a typical infinitesimal volume, $dV = dl\, d\sigma_\times$, indicated where $\vec{dl}$ lies along the circumference contour C and thus parallel to $\vec{B}$, and $d\sigma_\times$ is an infinitesimal of the cross-sectional area, $\Sigma_\times$. And so,

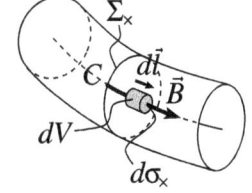

$$H_A = \int_V \vec{A} \cdot \vec{B}\, dV = \int_{\Sigma_\times} \oint_C \vec{A} \cdot \vec{B}\, dl\, d\sigma_\times$$
$$= \int_{\Sigma_\times} \oint_C \vec{A} \cdot \vec{dl}\, B\, d\sigma_\times,$$

using Eq. (4.36) and since $\vec{B} \parallel \vec{dl}$. Now, since magnetic flux is conserved, the magnetic flux through the cross section $\Sigma_\times$ must be the same at each point around C, and we may pull the flux integral outside the contour integral over $\vec{dl}$ to get,[13]

$$H_A = \underbrace{\int_{\Sigma_\times} B\, d\sigma_\times}_{\Phi_{B_\times}\ (\text{Eq. 4.8})} \oint_C \vec{A} \cdot \vec{dl} = \Phi_{B_\times} \int_\Sigma \underbrace{(\nabla \times \vec{A})}_{\vec{B}} \cdot \hat{n}\, d\sigma = \Phi_{B_\times} \Phi_B,$$

using the first flavour of Stoke's theorem (Eq. A.35). To be clear, $d\sigma_\times$ is an infinitesimal of area on $\Sigma_\times$ while $d\sigma$ is an infinitesimal of area on Σ (Fig. 4.3a). And since $\vec{B}$ either lies along Σ (and thus perpendicular to $\hat{n}$) within V or else is zero within the "hole" of the torus, $\Phi_B = 0$ and so, for the single flux loop, $H_A = 0$.

Well, that's pretty neat. So what if we complicate things a bit by linking a flux loop of volume V_1 with magnetic induction $\vec{B}_1$ by another flux loop of volume V_2 with magnetic induction $\vec{B}_2$, as shown in Fig. 4.3b? This is an example of a *non-co-planar* topology and, in a qualitative sense, is more complex than the single co-planar loop in Fig. 4.3a whose magnetic helicity we just found to be zero.

The good news is the mathematics is identical except for one easy difference. The magnetic flux passing through the hole of V_1 is now the magnetic flux carried within V_2:

$$\Phi_{B_2} = \int_{\Sigma_{\times,2}} B_2\, d\sigma_{\times,2} \neq 0,$$

and so the magnetic helicity of flux loop V_1 is,

$$H_{A,1} = \Phi_{B_1} \Phi_{B_2} \neq 0.$$

By symmetry, the magnetic helicity of flux loop V_2 is,

$$H_{A,2} = \Phi_{B_2} \Phi_{B_1},$$

---

[13]This mathematical manoeuvre is actually a little more subtle than presented, for one must also justify that the contour integral of $\vec{A}$ over $\vec{dl}$ can be "pulled out" of the flux integral. This is justified *a posteriori* by the fact that $\oint_C \vec{A} \cdot \vec{dl}$ ends up being the magnetic flux passing through the "hole" of the flux loop (area Σ in Fig. 4.3a), which does not depend upon which cross-section, $\Sigma_\times$, one chooses to integrate over.

Figure 4.4. Coronal loops on the sun's surface, many of which intertwine other loops (photo credit: Solar Dynamics Observatory, NASA).

and so the total magnetic helicity of the combined volume $V_1 + V_2$ is,

$$H_{A,1+2} = 2\Phi_{B_1}\Phi_{B_2} = 2\int_{\Sigma_{\times,1}} B_1 d\sigma_{\times,1} \int_{\Sigma_{\times,2}} B_2 d\sigma_{\times,2}.$$

More generally, should loops V_1 and V_2 actually be spirals and loop each other N_1 and N_2 times respectively, the total magnetic helicity becomes,

$$H_{A,1+2} = (N_2 + N_1)\Phi_{B_1}\Phi_{B_2}.$$

The take-away from this analysis is that magnetic helicity can be interpreted as a quantitative measure of the topological complexity of an MHD system. A coplanar field configuration – being the simplest – has zero-helicity. However, should magnetic field loops link each other, the configuration is fully 3-D and it can no longer be described as co-planar. In this case, helicity is non-zero with the sign of H_A depending on whether the normals of the interlinking flux loops (as determined from the right-hand rule for $\vec{B}$) are inclined by an angle $0 \leq \theta \leq \pi/2$ ($H_{A,1+2} > 0$) or $\pi/2 \leq \theta \leq \pi$ ($H_{A,1+2} < 0$).

A spectacular example of flux-linking in nature are coronal loops on the surface of the sun. Figure 4.4, from NASA's Solar Dynamics Observatory (SDO), shows numerous loops of plasma following lines of magnetic induction many of which are intertwined in a complex fashion giving rise to numerous solar phenomena including sunspots, solar flares, and solar prominences. A NASA video showing the dynamics of coronal loops is mesmerising.[14]

---

[14]www.youtube.com/watch?v=HFT7ATLQQx8

Problem Set 4

4.1 The statement was made near the end of §4.1 that "even a trace magnetic induction" can reduce the Larmour radius to much less than the diameter of the jet, and thus preserve the fluid model even for extragalactic outflow. However, the adjective *trace* was not justified. Sure, 10^{-12} T (10^{-8} G) seems "small", but what is small in an environment where the density is only one particle per m^3?

We'll introduce quantities such as the "plasma-beta" and "MHD-alpha" in §5.2 to quantify what we mean by a "weak" or "strong" magnetic induction. For now, let's just use the ratio of kinetic to magnetic energy densities,

$$f_{\text{KM}} = \frac{\frac{1}{2}\rho v^2}{B^2/2\mu_0}, \tag{4.42}$$

as a measure of magnetic strength (which we'll come to know as the square of the *Alfvén number*).

a) For the numbers given in §4.1 and assuming the jet is made up of ionised hydrogen, find a numerical value for f_{KM}. Therefore, is the energy of jet flow magnetically or kinetically dominated?

b) For a monatomic gas ($\gamma = 5/3$), what is the maximum Mach number of the jet flow before thermal energy density falls below magnetic energy density?

4.2 Gauss' Law for magnetic induction (one of Maxwell's four equations of electromagnetism) states that $\nabla \cdot \vec{B} = 0$, and is often characterised as "the redundant of Maxwell's equations". Show that if one can assume that $\nabla \cdot \vec{B} = 0$ at $t = 0$ (presumably corresponding to conditions during the Big Bang), then the ideal induction equation, Eq. (4.4) in the text which never invoked either of Gauss' laws for its derivation, requires that $\nabla \cdot \vec{B} = 0$ for all time.

4.3 This is a semi-qualitative problem to aid in the students' understanding of the two main theorems in this text: Theorems 1.1 (Theorem of hydrodynamics) and 4.1 (The flux theorem).

a) List two similarities and two differences between these two theorems.

b) Consider the *circulation*, Γ, defined in Problem 1.7 as,

$$\Gamma = \oint_C \vec{v} \cdot d\vec{l},$$

where C is the closed loop defining an open surface Σ within an ideal fluid. Show that Γ can also be thought of as a *vorticity flux*, where the vorticity is given by $\vec{\omega} = \nabla \times \vec{v}$ (Problem 1.6). Then, assuming boundary conditions such that *Kelvin's circulation theorem*,

$$\frac{d\Gamma}{dt} = 0,$$

applies (Problem 1.7), use Theorem 4.1 to write down an evolution equation for $\vec{\omega}$ and thus $\vec{v}$.

c) The integral form of Gauss' Law for an electric field in free space is (first of Eq. B.2),

$$\Phi_E = \oint_S \vec{E} \cdot d\vec{\sigma} = \frac{q_{\text{enc}}}{\epsilon_0},$$

where q_{enc} is the free charge enclosed within a closed surface S. If q_{enc} is constant and thus $d\Phi_E/dt = 0$, why doesn't Theorem 4.1 apply, and thus why can't we immediately write,

$$\vec{E} = \nabla \times (\vec{v} \times \vec{E}) ?$$

(Hint: One sentence is sufficient to answer this part.)

4.4

a) Show that $\nabla \cdot \vec{\omega} = 0$, where $\vec{\omega}$ is the vorticity as defined in Problem 1.6 of Problem Set 1. (This is a one-liner.)

b) Compare the time-evolution equation for the vorticity, as given in Problem 1.6, with the ideal induction equation, Eq. (4.4) in the text. This, along with the fact that both $\vec{B}$ and $\vec{\omega}$ are solenoidal (their divergences are zero) make for a very tempting argument that vorticity and magnetic vector fields must be everywhere and for all time proportional to each other, but they're not. What is it that makes vorticity and magnetism so fundamentally different? From the full set of MHD equations, identify at least two important deviations from the otherwise perfect symmetry implied by the preamble to this problem.

4.5* Show that $\nabla \cdot \vec{B} = 0 \Leftrightarrow \exists \vec{A} \mid \vec{B} = \nabla \times \vec{A}$, where $\vec{A}$ is the *magnetic vector potential*. The $\Leftarrow$ part is trivial, but the $\Rightarrow$ part may take some thinking.

Hint: One way to approach the $\Rightarrow$ problem is to come up with a functional form for $\vec{A}$ that works using a process that is sometimes referred to as taking the *anti-curl*. Note that whatever you find, your solution will not be unique, since $\nabla \times \vec{A} = \nabla \times (\vec{A} + \nabla \psi)$ and $\vec{A}$ is therefore ambiguous to within a term $\nabla \psi$, where ψ is an arbitrary scalar function. In electrodynamics, this ambiguity is known as *gauge freedom*. Thus, you might start by considering a particular and possibly simpler form for $\vec{A}$ such as $\vec{A} = (A_x, A_y, 0)$, and see where that leads you. Whatever you do get for $\vec{A}$, you should verify that taking its curl does indeed give you back $\vec{B}$.

4.6 Show that the MHD total energy equation (either Eq. 4.21 or Eq. 4.28 in the text) is equivalent to the hydrodynamical total energy equation, Eq. (1.23), with an explicit magnetic source term on the right-hand side. Thus, show that:

$$\frac{\partial e_T}{\partial t} + \nabla \cdot \vec{v}(e_T + p) = \vec{E}_{\text{ind}} \cdot \vec{J},$$

where e_T is given by Eq. (1.20), $\vec{E}_{\text{ind}}$ is given by Eq. (4.2), and $\vec{J}$ is given by Eq. (4.11), all in the text.

4.7 If $\vec{B}$ is the magnetic induction, prove the vector identities:

$$(\nabla \times \vec{B}) \times \vec{B} + \nabla \frac{B^2}{2} = (\vec{B} \cdot \nabla)\vec{B} = \nabla \cdot (\vec{B}\vec{B}),$$

required to derive the alternate form of the MHD momentum equation, that is Eq. (4.25) in the text.

4.8* This problem reads long because it enters an important realm that isn't discussed in the text: computational methods. Have no fear, the solution is not as long as the problem description!

In computational MHD, one sets up a *discrete grid* made up of a number of *zones* where, within each zone, the fluid variables are considered constant (Fig. 4.5a). Differences in fluid variables from one zone to the next give rise to differences in pressure, inertia (momentum), and magnetic stresses which, when reconciled by the *differenced* MHD equations, determine updated fluid variables after the given time step, the duration of which is set by criteria for numerical accuracy and stability.

Again for reasons of numerical accuracy and stability, where one places the variables within each zone can be critical. As shown in Fig. 4.5b, numerical codes such as *ZEUS-3D* are based on a *staggered mesh* in which scalars such as ρ are *zone-centred*, components of primary vectors such as $\vec{v}$ and $\vec{B}$ are *face-centred*, and components of secondary vectors such as $\vec{A}$ are *edge-centred*. See the caption of Fig. 4.5b for further explanation. As this problem is designed to show, this strategy is particularly important in ensuring that the *numerical* divergence of $\vec{B}$ is both initialised and remains at zero, which is not necessarily guaranteed even if $\nabla \cdot \vec{B} = 0$ analytically.

a) For illustration, consider the magnetic induction described analytically as,

$$\vec{B} = \left(2\sqrt{x}, -\frac{y}{\sqrt{x}}, 0\right), \tag{4.43}$$

where all quantities are unitless and where, for the purpose of this exercise, B_z and all z-derivatives are assumed to be zero. Show that analytically, $\nabla \cdot \vec{B} = 0$. (This should be a one-liner.)

b) Let's now evaluate the magnetic divergence *numerically*, which we do using

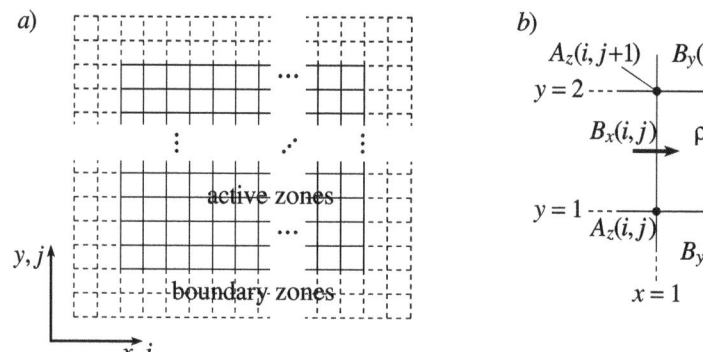

Figure 4.5. (Problem 4.8) *a*) A 2-D grid suitable for a numerical MHD calculation is depicted. Zones in the x- and y-directions are indexed with i and j respectively. "Active zones" (upon which the full MHD equations are applied at each time step) are marked with solid lines, "boundary zones" (upon which certain boundary conditions are specified) with dashed lines. *b*) The $(i,j)^{\text{th}}$ zone is shown with the locations of certain flow variables indicated. On a staggered mesh such as ZEUS-3D, scalars such as density are considered to be "zone-centred", and thus $\rho(i,j)$ takes on the value of the density at the centre of the zone. Primary vector components such as B_x and B_y are "face-centred", and thus $B_x(i,j)$ and $B_y(i,j)$ take on the values of B_x and B_y at the centres of the zone faces to which they are normal. Shown also are $B_x(i+1,j)$ and $B_y(i,j+1)$. Secondary vector components such as A_z (vector potential) are typically "edge-centred", and thus $A_z(i,j)$ takes on the value of A_z at the indicated zone-edge.

the *differenced* form of $\nabla \cdot \vec{B}$. Thus, on our discrete grid and at the $(i,j)^{\text{th}}$ zone (Fig. 4.5*b*),

$$\nabla \cdot \vec{B} \approx \frac{\delta B_x}{\delta x} + \frac{\delta B_y}{\delta y}$$
$$= \frac{B_x(i+1,j) - B_x(i,j)}{\delta x} + \frac{B_y(i,j+1) - B_y(i,j)}{\delta y}, \quad (4.44)$$

where, in the case depicted in Fig. 4.5*b*, $\delta x = \delta y = 1$.

Using Eq. (4.43), evaluate each of $B_x(i,j)$, $B_x(i+1,j)$, $B_y(i,j)$, and $B_y(i,j+1)$ given the values of x and y at the relevant faces and zone-centres in Fig. 4.5*b*. Then, using Eq. (4.44), evaluate the *numerical divergence*. (Answer: 0.0119)

Commentary: Such a non-zero numerical divergence gives rise to "numerical monopoles" which can have devastating first-order effects on a numerical solution causing it to depart qualitatively from an analytical solution, should one exist. We absolutely must ensure that $\nabla \cdot \vec{B} = 0$, even numerically!! But how?

c) By integrating the components of Eq. (4.43), find a suitable vector potential of the form $\vec{A} = A_z \hat{z}$. Don't overthink this; the integrations are pretty easy.

Evaluate your expression for A_z at each of the four edges depicted in Fig. 4.5*b*.

Then, using the *differenced form* of the curl (where z-derivatives are zero),

$$\left.\begin{aligned} B_x(i,j) &= \frac{A_z(i,j+1) - A_z(i,j)}{\delta y}; \\ B_y(i,j) &= -\frac{A_z(i+1,j) - A_z(i,j)}{\delta x}, \end{aligned}\right\} \quad (4.45)$$

evaluate again $B_x(i,j)$, $B_x(i+1,j)$, $B_y(i,j)$, and $B_y(i,j+1)$. From *these* values, evaluate the numerical divergence using Eq. (4.44) (you should get dead zero).

d) Generalise the previous result by substituting Eq. (4.45) directly into Eq. (4.44) without assuming values for δx and δy. In a numerical application, this should convince you of the importance of:

- a staggered mesh; and

- using the vector potential to initialise the magnetic induction.

e) Having taken the trouble to use the vector potential to initialise a numerically divergence-free magnetic induction, we'd like to evolve $\vec{B}$ numerically in such a way to keep it that way. There are a variety ways to do this,[15] but a particularly simple one is to update the vector potential, A_z, with Eq. (4.34) from the text which, for the z-component and in differenced form, is given by,

$$\frac{\delta A_z(i,j)}{\delta t} = -E_{\text{ind}}(i,j) \quad \Rightarrow \quad \delta A_z(i,j) = -E_{\text{ind}}(i,j)\,\delta t,$$

where δt is the time step, and where the edge-centred induced electric field,

$$E_{\text{ind}}(i,j) = \tilde{v}_y(i,j)\tilde{B}_x(i,j) - \tilde{v}_x(i,j)\tilde{B}_y(i,j),$$

is evaluated from suitable interpolations (indicated by the tilde ˜) of the face-centred velocity and magnetic induction components at the present time step. $A_z(i,j)$ is then replaced with $A_z(i,j) + \delta A_z(i,j)$, and these updated vector potentials are used in Eq. (4.45) to update the components of the magnetic induction which, as part d) shows, remains divergence-free.

Now suppose after n time steps we have an active 2-D grid of updated, divergence-free magnetic induction, B_x and B_y (heavy arrows in the inset), and we now wish to set values for B_x and B_y in the *boundary zones* (dashed lines in the inset). Typically, boundary values are not set using an evolution equation like the ideal induction equation, but rather on the physical conditions prevailing at that boundary. Thus, a boundary could be reflecting (*e.g.*, a non-conducting wall), periodic, or "flow out", meaning all waves of material and field are passed across the boundary with minimal (ideally zero) reflection back into the active zones.

---

[15] One of the first such algorithms is *Constrained Transport*, developed by Evans & Hawley (1988).

And so now finally the last question. Consider a 2-D grid with two layers of boundary zones as depicted in Fig. 4.5a and the inset. Suppose that the boundary condition is applied only to the *perpendicular* (to the boundary normal) components of the magnetic induction (light arrows in the inset). Show that the parallel components can then be set by using the numerical divergence (Eq. 4.44) and imposing the solenoidal condition.

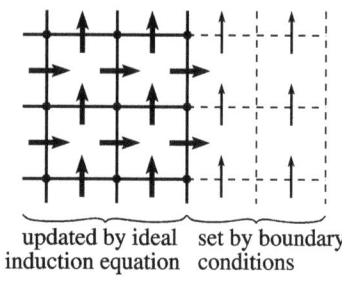

Additional commentary: All these ideas can be applied straight-forwardly (but cautiously!) to a 3-D grid where, in particular to part e), there are now two components of the magnetic induction perpendicular to each surface normal.

4.9 Write down the ideal MHD equations in *primitive form*. Thus, you seek a set of evolution equations, most if not all are already derived and/or written down in the text, for the primitive variables ρ, p, $\vec{v}$, and $\vec{B}$ for which no constitutive equations are required for closure.

4.10* The *cross helicity density*, $h_\times$, is defined as,

$$h_\times = \vec{v} \cdot \vec{B},$$

where $\vec{v}$ and $\vec{B}$ are the usual velocity and magnetic induction vector fields. It is a measure of the linkages between the *vorticity* ($\vec{\omega} = \nabla \times \vec{v}$; see Problems 4.3 and 1.6) and the magnetic induction.

a) Show that for ideal MHD, the time evolution for $h_\times$ is given by,

$$\frac{\partial h_\times}{\partial t} + \nabla \cdot (h_\times \vec{v}) = \nabla \cdot \left(\frac{v^2}{2} - \gamma\varepsilon\right)\vec{B}, \quad (4.46)$$

where ε is the specific internal energy defined by Eq. (1.31) in §1.4.

b) From Eq. (4.46) and the Theorem of hydrodynamics (Theorem 1.1), under what circumstances is the *cross helicity*, $H_\times = \int_V h_\times dV$, a conserved quantity?

5 MHD Waves and Discontinuities

We have to learn again that science without contact with experiments is an enterprise which is likely to go completely astray into imaginary conjectures.[†]

Hannes Alfvén (1908–1995)
father of MHD

WE OPEN this chapter with two changes of note. First, from this point forward we shall use the "abbreviated Leibniz notation" for partial derivatives, namely,

$$\partial_\xi \equiv \frac{\partial}{\partial \xi}; \quad \partial_{\xi\eta} \equiv \frac{\partial^2}{\partial\eta\partial\xi}; \quad etc.$$

Second, our discussion in this and Chap. 6 shall be limited to $1\frac{1}{2}$-D, first defined at the beginning of §2.2. Thus, while quantities may vary only in the x-direction, say (and thus $\partial_y = \partial_z = 0$; that's the '1-D' part), transverse vector components (*e.g.*, v_y, B_z, *etc.*) can, in general, be non-zero (that's the extra '$\frac{1}{2}$-D' bit).

With that, let us proceed...

In one dimension, the ideal MHD equations simplify sufficiently for us to make some significant analytic progress. In particular, we'll be able to identify the nature of the waves governing the physics of an MHD system and use this insight to obtain quantitative solutions. To begin with, we see that in 1-D we have only two components of the magnetic induction to worry about. For if in Cartesian coordinates, quantities vary in the x-direction only, then the x-component of the ideal induction equation becomes:

$$\partial_t B_x = \partial_y(v_x B_y - v_y B_x) - \partial_z(v_z B_x - v_x B_z) = 0,$$

and B_x is constant in time. Further, the solenoidal condition requires:

$$\nabla \cdot \vec{B} = \partial_x B_x = 0,$$

and B_x is also constant in space. Thus, B_x is constant everywhere for all time and we need only concern ourselves with the y- and z-components of the ideal induction equation. In our discussions, the fact that $B_x =$ constant will be used frequently as B_x is passed back and forth across the differential operators ∂_t and ∂_x as needed.

[†]From *Evolution of the Solar System* with G. Arrhenius, 1976, p. 257 (NASA SP; 345).

5.1 Primitive and conservative equations of MHD

As can be readily derived from either Eq. Set 6 or 7 or indeed from the set of 3-D primitive equations in Problem 4.9, the 1-D ($\partial_y = \partial_z = 0$) primitive equations of MHD in Cartesian geometry with $\phi = 0$ are:

Equation Set 8: Primitive form

$$\partial_t \rho + v_x \partial_x \rho + \rho \partial_x v_x = 0; \tag{5.1}$$

$$\partial_t p + v_x \partial_x p + \gamma p \partial_x v_x = 0; \tag{5.2}$$

$$\partial_t v_x + v_x \partial_x v_x + \frac{1}{\rho} \partial_x p + \frac{B_y}{\mu_0 \rho} \partial_x B_y + \frac{B_z}{\mu_0 \rho} \partial_x B_z = 0; \tag{5.3}$$

$$\partial_t v_y + v_x \partial_x v_y - \frac{B_x}{\mu_0 \rho} \partial_x B_y = 0; \tag{5.4}$$

$$\partial_t v_z + v_x \partial_x v_z - \frac{B_x}{\mu_0 \rho} \partial_x B_z = 0; \tag{5.5}$$

$$\partial_t B_y + v_x \partial_x B_y + B_y \partial_x v_x - B_x \partial_x v_y = 0; \tag{5.6}$$

$$\partial_t B_z + v_x \partial_x B_z + B_z \partial_x v_x - B_x \partial_x v_z = 0, \tag{5.7}$$

where no constitutive equations are required to close the system (Problem 5.1). These seven equations may be written more compactly as:

$$\partial_t |q_\mathrm{p}\rangle + \mathsf{J}_\mathrm{p} \partial_x |q_\mathrm{p}\rangle = 0, \tag{5.8}$$

where the ket of primitive variables, $|q_\mathrm{p}\rangle$, and the Jacobian matrix for the primitive variables, J_p, are given by:

$$|q_\mathrm{p}\rangle = \begin{bmatrix} \rho \\ p \\ v_x \\ v_y \\ v_z \\ B_y \\ B_z \end{bmatrix}; \quad \mathsf{J}_\mathrm{p} = \begin{bmatrix} v_x & 0 & \rho & 0 & 0 & 0 & 0 \\ 0 & v_x & \gamma p & 0 & 0 & 0 & 0 \\ 0 & 1/\rho & v_x & 0 & 0 & B_y/\mu_0\rho & B_z/\mu_0\rho \\ 0 & 0 & 0 & v_x & 0 & -B_x/\mu_0\rho & 0 \\ 0 & 0 & 0 & 0 & v_x & 0 & -B_x/\mu_0\rho \\ 0 & 0 & B_y & -B_x & 0 & v_x & 0 \\ 0 & 0 & B_z & 0 & -B_x & 0 & v_x \end{bmatrix}. \tag{5.9}$$

Similarly, from Eq. Set 7, one can write down the 1-D "conservative" equations of MHD in Cartesian geometry:

> **Equation Set 9: Conservative form**
>
> $$\partial_t \rho + \partial_x(\rho v_x) = 0; \tag{5.10}$$
>
> $$\partial_t e_T^* + \partial_x \left(v_x(e_T^* + p^*) - \frac{B_x}{\mu_0} \vec{v} \cdot \vec{B} \right) = 0; \tag{5.11}$$
>
> $$\partial_t s_x + \partial_x(\rho v_x^2 + p^*) = 0; \tag{5.12}$$
>
> $$\partial_t s_y + \partial_x \left(\rho v_x v_y - \frac{B_x B_y}{\mu_0} \right) = 0; \tag{5.13}$$
>
> $$\partial_t s_z + \partial_x \left(\rho v_x v_z - \frac{B_x B_z}{\mu_0} \right) = 0; \tag{5.14}$$
>
> $$\partial_t B_y + \partial_x(v_x B_y - v_y B_x) = 0; \tag{5.15}$$
>
> $$\partial_t B_z + \partial_x(v_x B_z - v_z B_x) = 0, \tag{5.16}$$

where the constitutive equations:

$$e_T^* = \frac{1}{2}\rho v^2 + \frac{p}{\gamma - 1} + \frac{B^2}{2\mu_0}; \quad p^* = p + \frac{B^2}{2\mu_0}; \quad s_i = \rho v_i; \quad B^2 = \sum_i B_i^2,$$

$i = x, y, z$, are required to close the equation set. Note that in the conservative equations, all pressures are p^*, the sum of thermal and magnetic pressures (Eq. 4.26), as opposed to p (the thermal pressure alone) for the primitive equations.

The conservative equations can be written in their most compact form as (*cf.*, Eq. 4.31):

$$\partial_t |q_c\rangle + \partial_x |f\rangle = 0, \tag{5.17}$$

where $|q_c\rangle$ and $|f\rangle$ are the kets of conservative variables and flux densities respectively:

$$|q_c\rangle = \begin{bmatrix} \rho \\ e_T^* \\ s_x \\ s_y \\ s_z \\ B_y \\ B_z \end{bmatrix}; \quad |f\rangle = \begin{bmatrix} \rho v_x \\ v_x(e_T^* + p^*) - B_x(\vec{v} \cdot \vec{B})/\mu_0 \\ \rho v_x^2 + p^* \\ \rho v_x v_y - B_x B_y/\mu_0 \\ \rho v_x v_z - B_x B_z/\mu_0 \\ v_x B_y - v_y B_x \\ v_x B_z - v_z B_x \end{bmatrix}. \tag{5.18}$$

In terms of the Jacobian matrix for the conservative variables, J_c, Eq. (5.17) is written:

$$\partial_t |q_c\rangle + \mathsf{J}_c \, \partial_x |q_c\rangle = 0, \tag{5.19}$$

where the $(i,j)^{\text{th}}$ element of J_c is given by Eq. (2.20), and repeated here for convenience:

$$J_{ij} = \partial_{q_j} f_i. \tag{5.20}$$

In order to evaluate these elements, it is first necessary to express all flux densities in terms of the conserved variables, using the constitutive equations above. Performing the straight-forward, though tedious manipulations, we get (Problem 5.2):

$$
\mathsf{J}_c = \begin{bmatrix}
0 & 0 & 1 & 0 & 0 & 0 & 0 \\
\partial_\rho f_2 & \gamma\dfrac{s_x}{\rho} & \partial_{s_x} f_2 & \partial_{s_y} f_2 & \partial_{s_z} f_2 & \partial_{B_y} f_2 & \partial_{B_z} f_2 \\
\partial_\rho f_3 & \gamma-1 & (3-\gamma)\dfrac{s_x}{\rho} & (1-\gamma)\dfrac{s_y}{\rho} & (1-\gamma)\dfrac{s_z}{\rho} & \partial_{B_y} f_3 & \partial_{B_z} f_3 \\
-s_x s_y/\rho^2 & 0 & s_y/\rho & s_x/\rho & 0 & -B_x/\mu_0 & 0 \\
-s_x s_z/\rho^2 & 0 & s_z/\rho & 0 & s_x/\rho & 0 & -B_x/\mu_0 \\
\partial_\rho f_6 & 0 & B_y/\rho & -B_x/\rho & 0 & s_x/\rho & 0 \\
\partial_\rho f_7 & 0 & B_z/\rho & 0 & -B_x/\rho & 0 & s_x/\rho
\end{bmatrix} \quad (5.21)
$$

where:

$$\partial_\rho f_2 = -\gamma \frac{s_x e_T^*}{\rho^2} + (\gamma-1)\frac{s_x s^2}{\rho^3} - (2-\gamma)\frac{s_x B^2}{2\mu_0 \rho^2} + \frac{B_x}{\mu_0 \rho^2}\vec{s}\cdot\vec{B};$$

$$\partial_{s_x} f_2 = \gamma \frac{e_T^*}{\rho} - (\gamma-1)\frac{2s_x^2 + s^2}{2\rho^2} + (2-\gamma)\frac{B^2}{2\mu_0 \rho} - \frac{B_x^2}{\mu_0 \rho};$$

$$\partial_{s_y} f_2 = -(\gamma-1)\frac{s_x s_y}{\rho^2} - \frac{B_x B_y}{\mu_0 \rho};$$

$$\partial_{s_z} f_2 = -(\gamma-1)\frac{s_x s_z}{\rho^2} - \frac{B_x B_z}{\mu_0 \rho};$$

$$\partial_{B_y} f_2 = (2-\gamma)\frac{s_x B_y}{\rho} - \frac{s_y B_x}{\rho};$$

$$\partial_{B_z} f_2 = (2-\gamma)\frac{s_x B_z}{\rho} - \frac{s_z B_x}{\rho};$$

$$\partial_\rho f_3 = -\frac{s_x^2}{\rho^2} + \frac{\gamma-1}{2}\frac{v^2}{\rho^2};$$

$$\partial_{B_y} f_3 = (2-\gamma)\frac{B_y}{\mu_0};$$

$$\partial_{B_z} f_3 = (2-\gamma)\frac{B_z}{\mu_0};$$

$$\partial_\rho f_6 = -\frac{1}{\rho^2}(s_x B_y - s_y B_x);$$

$$\partial_\rho f_7 = -\frac{1}{\rho^2}(s_x B_z - s_z B_x).$$

Evidently, the Jacobian for the conservative equation (Eq. 5.21) is rather more complicated than that for the primitive equation (Eq. 5.9) and thus used only when absolutely needed.

5.2 MHD wave families

In §3.5.3, we introduced the ideas of *wave families*. As we saw, the three independent equations contained in Eq. (3.21) led to three wave families for 1-D HD. Thus, with the seven independent equations contained in Eq. (5.8), we might expect *seven* wave families to emerge for 1-D MHD, and this is exactly what happens. Our first task, then, is to identify and describe these wave families. For now, the waves we seek will be *linear*, and thus use of the simpler primitive equations – valid only for smooth-flow – is justified.

As we did in §3.5.3, start by finding the eigenvalues of the Jacobian, J_p:

$$\begin{vmatrix} v_x - u & 0 & \rho & 0 & 0 & 0 & 0 \\ 0 & v_x - u & \gamma p & 0 & 0 & 0 & 0 \\ 0 & 1/\rho & v_x - u & 0 & 0 & B_y/\mu_0\rho & B_z/\mu_0\rho \\ 0 & 0 & 0 & v_x - u & 0 & -B_x/\mu_0\rho & 0 \\ 0 & 0 & 0 & 0 & v_x - u & 0 & -B_x/\mu_0\rho \\ 0 & 0 & B_y & -B_x & 0 & v_x - u & 0 \\ 0 & 0 & B_z & 0 & -B_x & 0 & v_x - u \end{vmatrix} =$$

$$-(u - v_x)\big((u - v_x)^2 - a_s^2\big)\big((u - v_x)^2 - a_x^2\big)\big((u - v_x)^2 - a_f^2\big) = 0, \quad (5.22)$$

where:

$$a_s^2 = \frac{1}{2}\left(a^2 + c_s^2 - \sqrt{(a^2 + c_s^2)^2 - 4a_x^2 c_s^2}\right); \quad (5.23)$$

$$a_x^2 = \frac{B_x^2}{\mu_0 \rho}; \quad (5.24)$$

$$a_f^2 = \frac{1}{2}\left(a^2 + c_s^2 + \sqrt{(a^2 + c_s^2)^2 - 4a_x^2 c_s^2}\right), \quad (5.25)$$

and where:

$$a^2 = \frac{B^2}{\mu_0 \rho} = \frac{B_x^2 + B_y^2 + B_z^2}{\mu_0 \rho}; \quad c_s^2 = \frac{\gamma p}{\rho}, \quad (5.26)$$

with details left to Problem 5.3. Here, $a_{f,s}$ are the *fast* and *slow magnetosonic speed*, a is the *Alfvén speed* with a_x being the x-component, and c_s is, as usual, the ordinary sound speed.

The *Alfvén speed* is named for the Swedish physicist, Hannes Alfvén (1908–1995; page ii)[1] who, among other accomplishments, was the first (with N. Herlofson) in 1950 to detect and identify non-thermal (synchrotron) radiation from the cosmos, was the first in 1963 to predict the large-scale filamentary nature of the universe, and is universally regarded as the inventor and father of MHD for which he was awarded the 1970 Nobel Prize in physics.

---

[1] See www.ap.smu.ca/~dclarke/AfciMHD/HA_bio_Peratt.pdf for Anthony Peratt's short biography of Alfvén and www.ap.smu.ca/~dclarke/AfciMHD/HA_centennial.pdf for Bibhas De's pictorial prepared on the occasion of Alfvén's centenary.

When his theory predicting the existence of what are now known as *Alfvén waves* was first published (Alfvén, 1942), it was received with considerable criticism and ridicule. This was the first time anyone ever suggested that electromagnetic waves of any sort could be supported by a conducting medium – in this case a plasma – and exactly opposite to what had been presumed "obvious" since Maxwell, namely that all conductors attenuate electromagnetic waves within a *skin depth*. Comments such as *'If such a thing were possible, Maxwell himself would have discovered it!'* were often how Alfvén's ideas were dismissed. The turning point is said to have occurred during one of Alfvén's lectures at the University of Chicago in the late 1950s where none other than Enrico Fermi was in attendance. Fermi is said to have uttered audibly at the end of Alfvén's presentation: *'Of course!'*, after which – so continues the story – everyone was saying *'Of course!'* Very soon thereafter, Alfvén waves were indisputably created in the lab, and MHD quickly gained wide acceptance. Relatively speaking, then, MHD is a rather new branch of physics.

The seven eigenvalues of the 1-D MHD equations – the characteristic speeds – are given by the roots of Eq. (5.22):[2]

$$u_1 = v_x - a_f; \quad u_2 = v_x - a_x; \quad u_3 = v_x - a_s; \quad u_4 = v_x;$$
$$u_5 = v_x + a_s; \quad u_6 = v_x + a_x; \quad u_7 = v_x + a_f, \quad (5.27)$$

and since these are real (and, as we'll see in §6.2.1, the eigenkets are linearly independent from each other), the system of 1-D MHD equations is *hyperbolic* (App. C). However, unlike the HD equations (§3.5), the eigenvalues of the 1-D MHD equations can be degenerate (*e.g.*, in certain cases, it's possible for $a = a_x = c_s$ or $a_f = a_x = a_s$) and, as such, the 1-D MHD equations are *not strictly hyperbolic*. As we'll see, this has profound consequences for both MHD shocks and rarefaction fans and will introduce a myriad of algebraic headaches along the way. More on this in due course.

Because the equations are hyperbolic (albeit not strictly), the seven eigenvalues represent seven wave speeds each identifying an MHD *wave family* with an associated *characteristic path*. Following the convention introduced in §3.5.3, the seven wave families in MHD are referred to as the *i*-family, $i = 1, \ldots, 7$, with associated characteristic paths $\mathcal{C}_f^-$, $\mathcal{C}_x^-$, $\mathcal{C}_s^-$, $\mathcal{C}^0$, $\mathcal{C}_s^+$, $\mathcal{C}_x^+$, and $\mathcal{C}_f^+$ illustrated in the space-time diagram of Fig. 5.1.

It is left to Problem 5.5 to show that:

$$\boxed{\begin{aligned} a_s &\leq c_s \leq a_f; \\ a_s &\leq a \leq a_f; \\ a_s &\leq a_x \leq a_f, \end{aligned}} \quad (5.28)$$

and thus, the wave speeds listed in Eq. (5.27) are in ascending order from the most negative to the most positive, justifying the order in which the characteristic paths are arranged in Fig. 5.1. Note the $\leq$ signs (rather than $<$) in Ineq. (5.28). These portend the possible degeneracy of the eigenvalues, and the "not-strictly hyperbolic"

---

[2] We shall postpone listing the eigenkets until §6.2 when we actually need them.

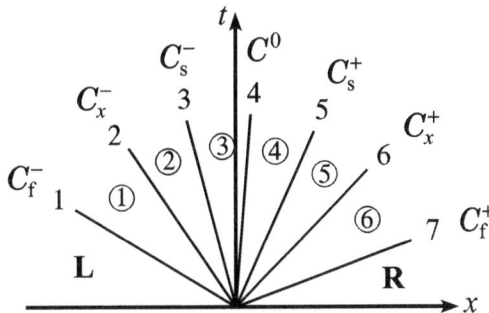

Figure 5.1. A space-time diagram illustrating the seven MHD wave families labelled as described in the text. From arbitrary left, **L**, and right, **R**, states, six piecewise constant intermediate states, ①–⑥, open up between neighbouring characteristic paths. The subscripts f, x, and s indicate the fast, Alfvén, and slow paths respectively, superscripts $-$, $+$ indicate left-leaning and right-leaning paths respectively, while $\mathcal{C}^0$ indicates the entropy characteristic path.

nature indicated above. Among other things, degenerate eigenvalues will mean one or more of the characteristic paths actually lie on top of each other on a space-time diagram, and careful consideration of these cases will have to be made when it comes time to solve the 1-D MHD equations in §5.3.5 and 6.2.2.

Referring again to Fig. 5.1, the seven MHD wave families come in four distinct types. Families 1 and 7 are *fast magnetosonic waves*, families 2 and 6 are *Alfvén waves*, families 3 and 5 are *slow magnetosonic waves*, and family 4 is an *entropy wave*, the one wave type shared by the HD and MHD cases. Families 1, 2, and 3 are left-moving waves (with left-leaning, $-$, characteristic paths), while families 5, 6, and 7 are right-moving waves (with right-leaning, $+$, characteristic paths). Recall that the designations "left" and "right" are with respect to the co-moving (Lagrangian) frame of the fluid. Family 4 is the entropy wave co-moving with the fluid, and thus neither left- nor right-moving.

At times, we will find it useful to consider how the MHD equations reduce to the ordinary HD case as $\vec{B} \to 0$, and for this we need a convenient comparator. The most common in use is the so-called *plasma-beta*, defined as the ratio of thermal to magnetic pressures:

$$\beta \equiv \frac{2\mu_0 p}{B^2}. \tag{5.29}$$

Magnetically dominated MHD is characterised by $\beta < 1$ while thermally dominated MHD is characterised by $\beta > 1$ and, as $\beta \to \infty$, the MHD system becomes hydrodynamical. Interestingly, many systems exhibit a qualitative transition in their properties across a relatively narrow range of β, typically within a factor of a few on either side of unity, while outside this transition region, changing the relative importance of the magnetic induction – even by orders of magnitude – can have quantitatively minor effects only. The numerical simulation of astrophysical jets (*e.g.*, Fig. 2.16) is a prime example. Thermally dominated jets with $\beta > 5$, say, are qualitatively indistinguishable even for $\beta \gg 1$, while magnetically dominated jets with $\beta < \frac{1}{5}$, say, are qualitatively indistinguishable even for $\beta \ll 1$. Yet, with all other parameters the same (density, flow speed, *etc.*), thermally and magnetically dominated jets look nothing like each other.[3]

In our discussions, however, it will be more useful to have a comparator based on the ratio of Alfvén (and later on, magnetosonic) to sound speeds that goes to 1 as the speeds become degenerate, and goes to 0 as $B \to 0$. For this, we define the *MHD-alpha*,

$$\boxed{\alpha \equiv \frac{B^2}{\mu_0 \gamma p}.} \tag{5.30}$$

Evidently, $\frac{6}{5} \leq \alpha\beta = 2/\gamma \leq \frac{3}{2}$, and the two measures are each other's reciprocal to within a factor of order unity.

In the context of wave families, as $\alpha \to 0$, we'll see that the fast wave in MHD becomes an ordinary sound wave in HD, while both the slow and Alfvén waves become degenerate with the entropy wave. Alfvén waves have no analogue in ordinary HD. Again as we'll see, they are characterised by jumps in the transverse components of $\vec{v}$ and $\vec{B}$, and are thus often referred to as *shear* Alfvén waves. On the other hand, fast and slow waves (wave families 1, 7 and 3, 5 in MHD) bear some resemblance to wave families 1 and 3 in HD. While they have both a longitudinal and transverse character, they can develop into either discontinuous shock waves or smooth-flow rarefaction fans. Thus, one speaks of fast and slow shocks and fast and slow rarefaction fans, all studied in depth in §5.3 and 6.2.

Before that, however, we need more insight into the nature of the simple (linear) waves in MHD. In the next two subsections, we first examine Alfvén waves and then fast and slow magnetosonic waves by isolating and studying the "kernels of physics" within the full set of 1-D MHD equations responsible for each wave.

5.2.1 (Shear) Alfvén waves

Mathematically, shear Alfvén waves pose the simplest wave problem in all of fluid dynamics, simpler even than that of ordinary sound waves studied in §2.1. This affords us an opportunity to review the methodology used so far to analyse the wave properties of (M)HD with a minimum of algebraic distraction, while at the same time revealing much about this important type of wave. As mentioned, it was the discovery of Alfvén waves in the lab that brought MHD into the mainstream of physics.

In §2.1.1, we identified and studied sound waves by examining a subset of the full hydrodynamical equations (Euler and pressure equations) and, in that case, the linearised versions. This allowed us to derive a wave equation, determine the wave propagation speed, deduce which variables are oscillating (v_x and p), and conclude that the waves are longitudinal. In §2.1.2, we re-examined the Euler and pressure equations in terms of their eigenvalues and eigenkets, which in turn allowed us to write down very specific expressions for how wave forms are transmitted in a 1-D

---

[3] Compare Clarke et al. (1986, magnetised jet) with Clarke et al. (1989, thermal jet) which differ only in their plasma-beta.

system. Finally, in §3.2, we exploited the inherent symmetry of the pressure and Euler's equations and identified both the Riemann invariants and their characteristic paths for 1-D HD in general, and sound waves in particular. We shall revisit each of these methodologies here for Alfvén waves.

Wave equation approach

Starting with the conservative equations[4] (5.13) and (5.15), we assume that each of ρ, v_x, and B_x are constant. Dividing Eq. (5.13) through by ρ, Eq. (5.15) through by $\sqrt{\rho}$, and setting $a_x = B_x/\sqrt{\mu_0\rho}$ (Eq. 5.24), $a_y = B_y/\sqrt{\mu_0\rho}$, they become:

$$\partial_t v_y + v_x \partial_x v_y - a_x \partial_x a_y = 0; \tag{5.31}$$

$$\partial_t a_y + v_x \partial_x a_y - a_x \partial_x v_y = 0, \tag{5.32}$$

which, in the co-moving frame of the fluid ($v_x = 0$), give:

$$\partial_t v_y = a_x \partial_x a_y \quad \text{and} \quad \partial_t a_y = a_x \partial_x v_y.$$

These combine to yield a wave equation for each of v_y and a_y:

$$\partial_t^2 v_y = a_x^2 \partial_x^2 v_y \quad \text{and} \quad \partial_t^2 a_y = a_x^2 \partial_x^2 a_y. \tag{5.33}$$

These are Alfvén waves causing oscillations in both v_y (transporting kinetic energy density $\frac{1}{2}\rho v_y^2$) and a_y (transporting potential energy density $\frac{1}{2}\rho a_y^2 = B_y^2/2\mu_0$). Since the oscillating quantities are the transverse components of the velocity and the magnetic induction, Alfvén waves are *transverse* (whence the designation *shear*), as opposed to sound waves which are *longitudinal* (since their oscillating quantities are v_x and p; §2.1.1). Note further that in deriving Eq. (5.33), there was no need to linearise Eq. (5.31) and (5.32) as was necessary for ordinary sound waves (Eq. 2.9). Thus, Alfvén waves are linear and can have arbitrary amplitude without developing into non-linear waves such as shocks and rarefactions.

Shear Alfvén waves have no analogue in a non-magnetised fluid. As the magnetic induction is reduced to zero, these waves simply disappear; they don't, for example, "morph" into sound waves. As transverse waves propagating along the prevailing lines of magnetic induction, they are analogous to vibrations propagating along a wire under tension; no wire, no waves. Like a tension wire, lines of induction resist being bent, and this affords the fluid a restorative capacity to transverse perturbations that does not exist in a non-magnetised medium.

Eigenvalue (normal mode) approach

Writing Eq. (5.31) and (5.32) in matrix form, we have,

$$\partial_t |q(x,t)\rangle + \mathsf{J}\,\partial_x |q(x,t)\rangle = 0, \tag{5.34}$$

---

[4]Unlike sound waves, Alfvén waves result in discontinuous jumps in the affected flow variables (transverse components of $\vec{v}$ and $\vec{B}$), and thus we start with the conservative equations which admit such solutions. However, if you follow the math carefully, you will see that in the end it does not matter in this case whether we start with the conservative or primitive form of the equations.

where,
$$|q(x,t)\rangle = \begin{bmatrix} v_y \\ a_y \end{bmatrix} \quad \text{and} \quad \mathsf{J} = \begin{bmatrix} v_x & -a_x \\ -a_x & v_x \end{bmatrix},$$

and where again, J is the Jacobian matrix. As we did in §2.1.2, we look for normal mode solutions by setting $|q(x,t)\rangle = |\tilde{q}(\xi)\rangle$ with $\xi = x - ut$, and then note that,

$$\partial_t|\tilde{q}(\xi)\rangle = (\partial_t\xi)|\tilde{q}'\rangle = -u|\tilde{q}'\rangle \quad \text{and} \quad \partial_x|\tilde{q}(\xi)\rangle = (\partial_x\xi)|\tilde{q}'\rangle = |\tilde{q}'\rangle,$$

where $'$ indicates differentiation with respect to ξ. Thus, Eq. (5.34) becomes:

$$\mathsf{J}|\tilde{q}'\rangle = u|\tilde{q}'\rangle, \tag{5.35}$$

the same eigen-equation as Eq. (2.23), whose eigenvalues are given by:

$$\begin{vmatrix} v_x - u & -a_x \\ -a_x & v_x - u \end{vmatrix} = 0 \quad \Rightarrow \quad (v_x - u) - a_x^2 = 0 \quad \Rightarrow \quad u^\pm = v_x \pm a_x,$$

the same Alfvén characteristic speeds found in Eq. (5.27) and identified (in the co-moving frame) by Eq. (5.33). These correspond to left- and right-moving Alfvén waves moving at speed a_x within a fluid moving at speed v_x relative to a "lab frame".

The eigenkets tell us how the affected variables vary across the wave. Starting with the left-moving wave, we find $|r^-\rangle$ by solving:

$$(\mathsf{J} - u^-\mathsf{I})|r^-\rangle = 0 \quad \Rightarrow \quad \begin{bmatrix} a_x & -a_x \\ -a_x & a_x \end{bmatrix}\begin{bmatrix} r_1^- \\ r_2^- \end{bmatrix} = 0$$

$$\Rightarrow \quad a_x r_1^- - a_x r_2^- = 0 \quad \Rightarrow \quad r_2^- = r_1^- \quad \Rightarrow \quad |r^-\rangle = r_1^- \begin{bmatrix} 1 \\ 1 \end{bmatrix},$$

where r_1^- is a normalisation constant chosen for convenience (e.g., $r_1^- = 1/\sqrt{2}$). Similarly, for the right-moving wave,

$$|r^+\rangle = r_2^+ \begin{bmatrix} -1 \\ 1 \end{bmatrix},$$

where r_2^+ is the normalisation constant.

Since $|\tilde{q}'\rangle$ are proportional to the eigenkets (Eq. 5.35), we have for the left- and right-moving waves:

$$|\tilde{q}^{-'}\rangle = w^{-'}(\xi^-)|r^-\rangle \quad \text{and} \quad |\tilde{q}^{+'}\rangle = w^{+'}(\xi^+)|r^+\rangle,$$

where $w^{-'}$ and $w^{+'}$ are the proportionality functions (expressed as derivatives for convenience; cf. §2.1.2), which are functions of $\xi^- = x - u^-t$ and $\xi^+ = x - u^+t$ respectively. These integrate trivially (because $|r^-\rangle$ and $|r^+\rangle$ are independent of $\xi^\pm$) to:

$$|\tilde{q}^-\rangle = w^-(\xi^-)|r^-\rangle \quad \text{and} \quad |\tilde{q}^+\rangle = w^+(\xi^+)|r^+\rangle,$$

and the complete solution is just their sum:

$$|q(x,t)\rangle = |\tilde{q}^-\rangle + |\tilde{q}^+\rangle = w^-(\xi^-)|r^-\rangle + w^+(\xi^+)|r^+\rangle.$$

To find the functions w^- and w^+, we apply initial conditions, namely $|q(x,0)\rangle = |\tilde{q}(x)\rangle$ at $t = 0$. Thus,

$$|\tilde{q}(x)\rangle = \begin{bmatrix} \tilde{v}_y(x) \\ \tilde{a}_y(x) \end{bmatrix} = w^-(x)\begin{bmatrix} 1 \\ 1 \end{bmatrix} + w^+(x)\begin{bmatrix} -1 \\ 1 \end{bmatrix}$$

$$\Rightarrow \quad w^-(x) = \frac{1}{2}(\tilde{a}_y(x) + \tilde{v}_y(x)) \quad \text{and} \quad w^+(x) = \frac{1}{2}(\tilde{a}_y(x) - \tilde{v}_y(x))$$

$$\Rightarrow \quad \boxed{|q(x,t)\rangle = \begin{bmatrix} v_y(x,t) \\ a_y(x,t) \end{bmatrix} = \frac{1}{2}\begin{bmatrix} \tilde{v}_y(\xi^-) + \tilde{v}_y(\xi^+) + \tilde{a}_y(\xi^-) - \tilde{a}_y(\xi^+) \\ \tilde{a}_y(\xi^-) + \tilde{a}_y(\xi^+) + \tilde{v}_y(\xi^-) - \tilde{v}_y(\xi^+) \end{bmatrix}.} \quad (5.36)$$

This solution is very similar to Eq. (2.30) for linearised HD (sound) waves and, other than variable name changes, is essentially a repeat of its derivation. As shown in Problem 5.11, Eq. (5.36) can also be obtained using the formalism developed in §3.5.2 and, in particular, from Eq. (3.28).

To give a concrete example, consider a uniform $\vec{B}$ in the x-direction such that $B_x/\sqrt{\mu_0} = 1$ in a uniform, quiescent medium where $\rho = 1$ (all units arbitrary). Thus, the x-component of the Alfvén speed is $a_x = B_x/\sqrt{\mu_0 \rho} = 1$. Let the system be disturbed by a "wind shear" pulse of magnitude $v_y = 1$ and width $\Delta x = 1$ centred at the origin. For $B_y = 0$ at $t = 0$, the initial conditions are thus:

$$\tilde{v}_y(x) = \begin{cases} 1, & -0.5 \leq x \leq 0.5; \\ 0, & \text{elsewhere,} \end{cases} \quad \text{and} \quad \tilde{a}_y(x) = \frac{\tilde{B}_y}{\sqrt{\mu_0 \rho}} = 0, \quad (5.37)$$

as depicted in the top panels of Fig. 5.2. Note that this is not a "perturbation", but rather a "wollop" (since $v_y = a_x$) which is fine for Alfvén waves which remain linear regardless of amplitude.

To find profiles for v_y and a_y at $t = 0.3$, say, evaluate Eq. (5.36) at $\xi^- = x + 0.3$ ($v_x = 0$) and $\xi^+ = x - 0.3$ using the functional form for $\tilde{v}_y$ and $\tilde{a}_y$ in Eq. (5.37):

$$\tilde{v}_y(x+0.3) = \begin{cases} 1, & -0.5 \leq x+0.3 \leq 0.5; \\ 0, & \text{elsewhere} \end{cases} = \begin{cases} 1, & -0.8 \leq x \leq 0.2; \\ 0, & \text{elsewhere,} \end{cases}$$

$$\tilde{v}_y(x-0.3) = \begin{cases} 1, & -0.5 \leq x-0.3 \leq 0.5; \\ 0, & \text{elsewhere} \end{cases} = \begin{cases} 1, & -0.2 \leq x \leq 0.8; \\ 0, & \text{elsewhere.} \end{cases}$$

Substituting these along with $\tilde{a}_y(\xi^-) = \tilde{a}_y(\xi^+) = 0$ into Eq. (5.36), we get:

$$v_y(x, 0.3) = \frac{1}{2}\left[\begin{cases} 1, & -0.8 \leq x \leq 0.2; \\ 0, & \text{elsewhere} \end{cases} + \begin{cases} 1, & -0.2 \leq x \leq 0.8; \\ 0, & \text{elsewhere} \end{cases}\right]$$

$$= \begin{cases} \frac{1}{2}, & -0.8 \leq x < -0.2; \\ 1, & -0.2 \leq x < 0.2; \\ \frac{1}{2}, & 0.2 \leq x \leq 0.8; \\ 0, & \text{elsewhere,} \end{cases} ;$$

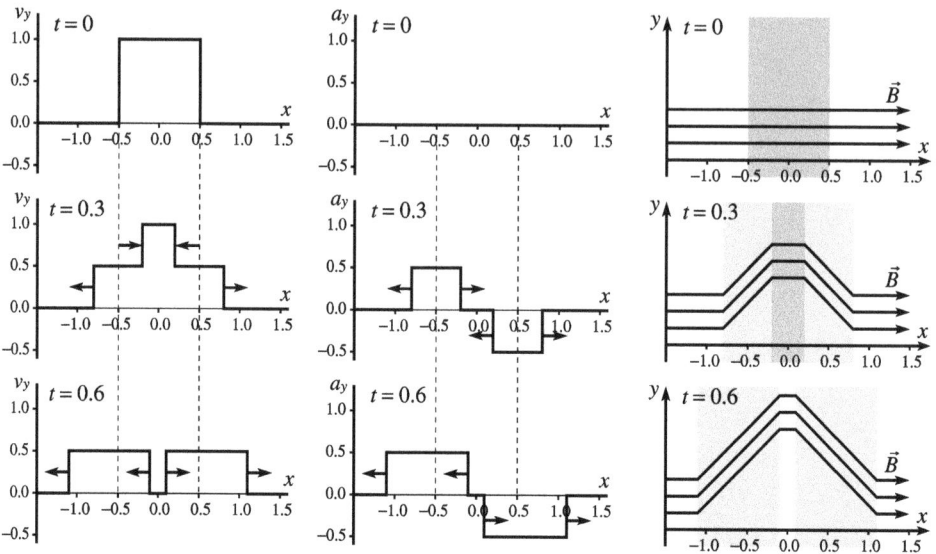

Figure 5.2. Left two columns: Profiles of $v_y(x,t)$ and $a_y(x,t)$ at $t = 0$ (top), $t = 0.3$ (middle), and $t = 0.6$ (bottom) after a uniform medium with $\vec{B}/\sqrt{\mu_0} = \hat{x}$, $\rho = 1$ (arbitrary units) is disturbed by a velocity shear given by Eq. (5.37). Arrows indicate direction of motion of the discontinuities. Right column: Distortion of lines of magnetic induction by the passage of the Alfvén wave. Grey bands indicate where $v_y = 1$ (dark), $v_y = 0.5$ (light) and $v_y = 0$ (white).

$$a_y(x, 0.3) = \frac{1}{2}\left[\begin{cases}1, & -0.8 \leq x \leq 0.2; \\ 0, & \text{elsewhere}\end{cases} - \begin{cases}1, & -0.2 \leq x \leq 0.8; \\ 0, & \text{elsewhere}\end{cases}\right]$$

$$= \begin{cases}\frac{1}{2}, & -0.8 \leq x < -0.2; \\ 0, & -0.2 \leq x < 0.2; \\ -\frac{1}{2}, & 0.2 \leq x \leq 0.8; \\ 0, & \text{elsewhere.}\end{cases}$$

These profiles are shown in the second row of Fig. 5.2. It is left as an exercise to verify the profiles at $t = 0.6$ in the bottom row of the figure.

The right column of Fig. 5.2 provides an intuition boost of why oppositely directed pulses of $a_y = B_y/\sqrt{\mu_0 \rho}$ should have opposite signs. At first glance, this is a necessary consequence of magnetic flux conservation, but this still begs the question: 'Which direction gets the negative pulse?' As velocity shear bends the axial magnetic induction, a y-component in $\vec{B}$ develops. For the line of induction to remain contiguous, B_y must be positive and point upward across the left-propagating wave, and negative and point downward across the right-propagating wave. Note that the solution after $t = 0.6$ is simply the two left- and right-moving pulses separating ever further apart, with the region in between once again quiescent ($v_y = 0$) with a purely axial magnetic induction, $\vec{B}/\sqrt{\mu_0} = \hat{x}$. The effect of the Alfvén wave's passage is simply to have shifted all lines of induction upwards by a distance $\Delta x\, v_y/a_x$.

Characteristic approach (optional)

Finally, let us examine shear Alfvén waves in terms of their *characteristics* and, as described in §3.3, determine their evolution using the *Method of Characteristics* (MoC). As this is one of only a few problems where the MoC can be used to obtain an exact solution, it is a good problem to examine. Here, we shall relax the requirement that ρ be constant (but still require v_x = constant), and start again with Eq. (5.4) and (5.6), this time in the form:[5]

$$\partial_t v_y + v_x \partial_x v_y - \frac{1}{\sqrt{\mu_0 \rho}} a_x \partial_x B_y = 0; \qquad (5.38)$$

$$\frac{1}{\sqrt{\mu_0 \rho}} \partial_t B_y + v_x \frac{1}{\sqrt{\mu_0 \rho}} \partial_x B_y - a_x \partial_x v_y = 0, \qquad (5.39)$$

where, as before, $a_x = B_x/\sqrt{\mu_0 \rho}$ and Eq. (5.39) has been divided by $\sqrt{\mu_0 \rho}$ to give it the same units as Eq. (5.38). Adding and subtracting Eq. (5.38) and (5.39) then yields:

$$\partial_t v_y + (v_x - a_x) \partial_x v_y + \frac{1}{\sqrt{\mu_0 \rho}} \left[\partial_t B_y + (v_x - a_x) \partial_x B_y \right] = 0;$$

$$\partial_t v_y + (v_x + a_x) \partial_x v_y - \frac{1}{\sqrt{\mu_0 \rho}} \left[\partial_t B_y + (v_x + a_x) \partial_x B_y \right] = 0,$$

which can be written more compactly as:

$$D_t^\pm v_y \mp \frac{1}{\sqrt{\mu_0 \rho}} D_t^\pm B_y = 0, \qquad (5.40)$$

where,

$$D_t^\pm = \partial_t + (v_x \pm a_x) \partial_x,$$

is the Lagrangian derivative, first introduced in §3.1. Equations (5.40) aren't quite in the same form as previous characteristic equations (*e.g.*, Eq. 3.13). Having relaxed the requirement that ρ be constant, the LHS of Eq. (5.40) is not a perfect derivative, and thus these equations are not in the form:

$$D_t^\pm \mathcal{A}^\pm = 0,$$

which would clearly identify the Riemann invariants, $\mathcal{A}^\pm$, as explicit functions of v_y, B_y, and possibly ρ. Nevertheless, Eq. (5.40) still identify two characteristic speeds, namely $c^\pm = v_x \pm a_x$ corresponding to oppositely directed Alfvén waves, and we can still proceed with the MoC (§3.3) without (yet) having Riemann invariants explicitly identified.

The Alfvén characteristic paths, $\mathcal{C}^\pm$, are shown in Fig. 5.3*a*. This two-characteristic system is considerably simpler than the three-characteristic hydrodynamical system described in §3.3 not only because there is one less characteristic, but also because the problem remains linear regardless of the amplitude of the waves.

For small enough time steps, or for problems where the profiles for v_y and B_y are piecewise constant and thus where the length of the time step does not

---
[5] Note that an entirely analogous development can be followed for v_z and B_z. Indeed, Alfvén waves can operate independently in the y- and z-directions.

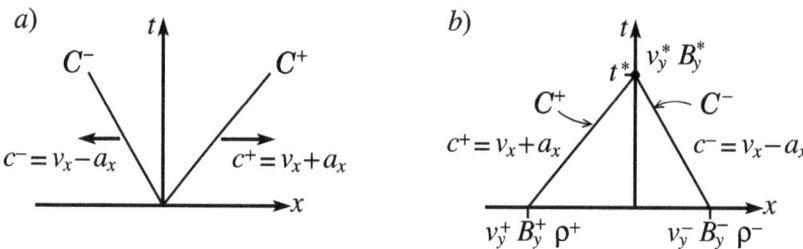

Figure 5.3. *a*) Two Alfvén waves are launched from a common point and propagate in opposite directions at characteristic speeds $c^{\pm}$ tracing out characteristic paths $\mathcal{C}^{\pm}$ in a space-time diagram. As drawn, $0 < v_x < a_x$. *b*) Two Alfvén characteristic paths from different points at $t = 0$ intersect at a common event at $t = t^*$. Using the MoC, the advanced values v_y^* and B_y^* can be determined from the values of v_y, B_y, and ρ at the path footprints.

matter, we can replace the *differential* Eq. (5.40) with the corresponding *forward-differenced* equations between the advanced-time values of v_y and B_y (v_y^* and B_y^*), and the present values of v_y and B_y at the bases, or *footprints* of the characteristic paths, $\mathcal{C}^{\pm}$ ($v_y^{\pm}$ and $B_y^{\pm}$; see Fig. 5.3*b*). This yields two algebraic equations:

$$v_y^* - v_y^+ - \frac{1}{\sqrt{\mu_0 \rho^+}}(B_y^* - B_y^+) = 0; \tag{5.41}$$

$$v_y^* - v_y^- + \frac{1}{\sqrt{\mu_0 \rho^-}}(B_y^* - B_y^-) = 0. \tag{5.42}$$

It is left to Problem 5.14 to show that the appropriate values for ρ to use in these differenced equations are indeed the values at the path footprints, namely $\rho^{\pm}$.

Upon rearranging Eq. (5.41) and (5.42), we find:

$$v_y^* + \frac{B_y^*}{\mu_0 \sqrt{\rho^-}} = v_y^- + \frac{B_y^-}{\sqrt{\mu_0 \rho^-}} \quad \text{and} \quad v_y^* - \frac{B_y^*}{\sqrt{\mu_0 \rho^+}} = v_y^+ - \frac{B_y^+}{\sqrt{\mu_0 \rho^+}},$$

and thus, we make the small leap of logic to assert that:

$$\mathcal{A}^- = v_y + \frac{B_y}{\sqrt{\mu_0 \rho^-}} \quad \text{and} \quad \mathcal{A}^+ = v_y - \frac{B_y}{\sqrt{\mu_0 \rho^+}}, \tag{5.43}$$

are constant along $\mathcal{C}^-$ and $\mathcal{C}^+$ respectively; that is, $\mathcal{A}^{\pm}$ are the Riemann invariants for Alfvén waves.

Next, solving Eq. (5.41) and (5.42) directly for v_y^* and B_y^* (the advanced values) yields:

$$v_y^* = \frac{1}{\sqrt{\rho^-} + \sqrt{\rho^+}}\left(\sqrt{\rho^-}\, v_y^- + \sqrt{\rho^+}\, v_y^+ + \frac{B_y^- - B_y^+}{\sqrt{\mu_0}}\right); \tag{5.44}$$

$$B_y^* = \frac{\sqrt{\rho^- \rho^+}}{\sqrt{\rho^-} + \sqrt{\rho^+}}\left[\frac{B_y^-}{\sqrt{\rho^-}} + \frac{B_y^+}{\sqrt{\rho^+}} + \sqrt{\mu_0}(v_y^- - v_y^+)\right]. \tag{5.45}$$

In the case where $v_y^- = v_y^+ = v_0$ and $B_y^- = B_y^+ = B_0$ (*i.e.*, footprints for both

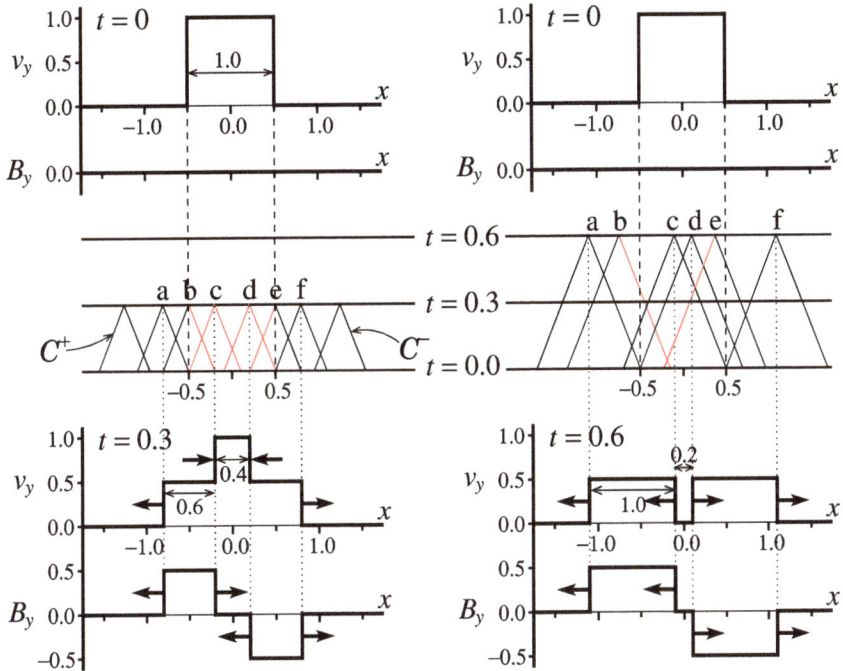

Figure 5.4. The MoC as applied to the same shear Alfvén wave problem as depicted in Fig. 5.2. The left panels show the MoC solution of the left- and right-moving Alfvén waves from $t = 0$ to $t = 0.3$ after they have moved a distance 0.3 in arbitrary units, whereas the right panels show the MoC solution from $t = 0$ to $t = 0.6$ after moving a distance 0.6. Heavy arrows on the bottom panels show the direction of motion of each jump. See text for further explanation.

characteristic paths are in the same state), these equations reduce to:

$$v_y^* = v_0; \qquad B_y^* = B_0, \tag{5.46}$$

and neither v_y nor B_y changes, as expected. When $\rho^- = \rho^+ = \rho$, Eq. (5.44) and (5.45) reduce to:

$$v_y^* = \frac{1}{2}\left[v_y^- + v_y^+ + \frac{1}{\sqrt{\mu_0 \rho}}(B_y^- - B_y^+)\right]; \tag{5.47}$$

$$B_y^* = \frac{1}{2}\left[B_y^- + B_y^+ + \sqrt{\mu_0 \rho}(v_y^- - v_y^+)\right], \tag{5.48}$$

as one would expect from Eq. (5.36).

To exemplify the MoC, we revisit the problem portrayed in Fig. 5.2 in units where $\mu_0 = 1$. Both top panels of Fig. 5.4 echo the initial conditions of the problem, while the middle panels are space-time diagrams illustrating characteristic paths originating at $t = 0$ constructed so that pairs of $\mathcal{C}^\pm$ meet at two designated times in the future. The left half of Fig. 5.4 corresponds to $t = 0.3$ after an Alfvén wave has propagated a distance of 0.3 (all in arbitrary units), while the right half corresponds to $t = 0.6$. Note that all paths have the same slope (to within a minus sign), since

$a_x = 1$ everywhere. As described below, the bottom panels of the figure illustrate the profiles of the variables after the indicated times as determined by the MoC.

In the middle panels of Fig. 5.4, the footprints of red characteristic paths lie within the original pulse in v_y ($-0.5 < x < 0.5$) and carry with them information that at $t = 0$, $v_y = 1$ and $B_y = 0$. The footprints of black characteristic paths lie outside this domain ($|x| > 0.5$) and carry with them information that both v_y and B_y are zero. By $t = 0.3$ (bottom left panel of Fig. 5.4), the distance between footprints of each pair, $\mathcal{C}^\pm$, arriving at the same point is 0.6 (less than the width of the original pulse in v_y) and there exist points at $t = 0.3$ – those between points 'c' and 'd' in the left-middle panel – for which *both* path footprints lie within $-0.5 < x < 0.5$. Since characteristics propagate along their paths with the Alfvén speed (Eq. 5.26), point 'c' is located at $a_x t = 0.3$ right of the left side of the original pulse (*i.e.*, at -0.2), and point 'd' is at 0.2. Thus, characteristic paths intersecting at points within $-0.2 < x < 0.2$ all transmit information that $v_y = 1, B_y = 0$ and, from Eq. (5.46),

$$v_y(|x| < 0.2, t = 0.3) = 1 \quad \text{and} \quad B_y(|x| < 0.2, t = 0.3) = 0,$$

a region of width 0.4 at $t = 0.3$ in which the profiles of the original pulse are preserved (bottom left panel of Fig. 5.4). As time progresses, this domain narrows until by $t = 0.5$ (time for an Alfvén wave to move half the original width of the pulse in v_y), it is "squeezed out" of existence, after which no points retain the initial profiles.

Returning to the left-middle panel, both characteristic paths of points left of point 'a' and right of point 'f' have their footprints outside the region $-0.5 < x < 0.5$, and both carry information that $v_y = 0$ and $B_y = 0$. At $t = 0.3$, point 'a' is located at $x = -0.5 - a_x(0.3) = -0.8$ and point 'f' is located at $x = 0.8$ and thus, again by Eq. (5.46), we have,

$$v_y(|x| > 0.8, t = 0.3) = 0 \quad \text{and} \quad B_y(|x| > 0.8, t = 0.3) = 0.$$

For points between points 'a' and 'c' (*e.g.*, point 'b'), the footprint of $\mathcal{C}^+$ lies outside $-0.5 < x < 0.5$ (black) while the footprint of $\mathcal{C}^-$ lies within (red). Thus, the information these points receive is $v_y^+ = 0$, $v_y^- = 1$, and $B_y^\pm = 0$, and Eq. (5.47) and (5.48) yield:

$$v_y(-0.8 < x < -0.2, t = 0.3) = \frac{1}{2} \quad \text{and} \quad B_y(-0.8 < x < -0.2, t = 0.3) = \frac{1}{2}.$$

At $t = 0.3$, this region has a width of 0.6 and represents a new state that manifests as a "shelf" of height $\frac{1}{2}$ on the left side of the original v_y profile, and a pulse in B_y of height $\frac{1}{2}$ (bottom panels of Fig. 5.4). As time progresses, this shelf widens (heavy arrows in the bottom panels indicate direction of motion of each jump) until $t = 0.5$ when its width reaches its steady-state value of 1, the same as the original pulse.

Similarly, for points between 'd' and 'f' (*e.g.*, point 'e'), the footprint of $\mathcal{C}^+$ lies inside $-0.5 < x < 0.5$ while the footprint of $\mathcal{C}^-$ lies without. Thus, the information these points receive is $v_y^+ = 1$, $v_y^- = 0$, and $B_y^\pm = 0$, and Eq. (5.47)

and (5.48) yield:
$$v_y(0.2 < x < 0.8, t = 0.3) = \frac{1}{2} \quad \text{and} \quad B_y(0.2 < x < 0.8, t = 0.3) = -\frac{1}{2}.$$

At $t = 0.3$, this region also has a width of 0.6 and manifests as a "shelf" of height $\frac{1}{2}$ on the right side of the original v_y profile, and a pulse in B_y of depth $-\frac{1}{2}$.

At $t = 0.5$ when characteristic paths from each side of the initial pulse in v_y have met in the middle, the left-shelf meets up with the right-shelf and the original v_y profile has been completely consumed. At this time, the new states have reached their maximum width of 1, and move off in opposite directions as pulses (Alfvén waves) with speed ± 1.

By $t = 0.6$ (right side of Fig. 5.4), the distance between footprints of each pair, $\mathcal{C}^\pm$, arriving at the same point is 1.2, too wide to fit inside the original pulse. Thus, no points at $t = 0.6$ receive information from both characteristic paths about the original pulse, and nowhere is the original profile of v_y preserved (bottom right panel). Indeed, in the middle right panel, paths arriving at points between points 'c' and 'd' *straddle* the region $-0.5 < x < 0.5$ and both footprints lie outside the original pulse. Thus, these points receive information from both characteristic paths that $v_y = 0$ and $B_y = 0$ creating a "gap" in the profiles between the left- and right-moving Alfvén waves which, at $t = 0.6$, has a width of 0.2.

As it must, the MoC yields the same solution as normal mode analysis (eigenalgebra), and which of the two methods one uses is a matter of taste. While the former may seem more dependent upon visuals and less formulaic than the latter, it is no less rigorous and being able to think in terms of characteristics can be of great benefit to the intuition when thinking of more complicated systems. While the shear Alfvén wave problem consists of just two characteristics and can be handled with very simple algebra, the full MHD problem has *seven* characteristics, each carrying certain information (a Riemann invariant) about the state of the variables at the footprint of their respective characteristic path to future times. Each point in the future, therefore, must be consistent with each of the seven pieces of information arriving there. Indeed, many of the better known MHD research codes incorporate the transmission of information along characteristic paths into their design not so much for "accuracy", but for existential reasons. It turns out, without the use of characteristics, numerical (M)HD codes can be manifestly unstable, and incapable of many applications.

5.2.2 Fast and slow magnetosonic waves

To gain insight into the nature of waves travelling at speeds a_f and a_s (Eq. 5.25 and 5.23), start by linearising the primitive equations for 1-D MHD. In the spirit of §2.1.1 where two of the three hydrodynamical equations were linearised for sound waves, consider a 1-D MHD fluid in its unperturbed rest frame ($\vec{v}_0 = 0$) with a constant axial magnetic induction, B_x, and write:

$$\rho = \rho_0 + \epsilon \rho_p; \qquad p = p_0 + \epsilon p_p; \qquad B_y = B_{y,0} + \epsilon B_{y,p};$$
$$v_x = \epsilon v_{x,p}; \qquad v_y = \epsilon v_{y,p}; \qquad v_z = \epsilon v_{z,p},$$

where here, the subscript "p" stands for "perturbation", not "primitive". Thus, $\rho_p \ll \rho_0$, etc., and $\epsilon = 1$ is used as a "smallness label", just as in §2.1.1. Without loss of generality, assume that the magnetic induction lies entirely in the x–y plane. Then, if θ is the angle between $\vec{B}$ and $\hat{x}$, $\tan\theta = B_{y,0}/B_x$.

Substituting the above expressions for the flow variables in Eq. (5.1)–(5.7), then dropping all terms of order ϵ^2 yields the following linearised equations:

$$\partial_t p_p + \gamma p_0 \, \partial_x v_{x,p} = 0; \tag{5.49}$$

$$\partial_t v_{x,p} + \frac{1}{\rho_0} \partial_x p_p + \frac{B_{y,0}}{\mu_0 \rho_0} \partial_x B_{y,p} = 0; \tag{5.50}$$

$$\partial_t v_{y,p} - \frac{B_x}{\mu_0 \rho_0} \partial_x B_{y,p} = 0; \tag{5.51}$$

$$\partial_t B_{y,p} + B_{y,0} \partial_x v_{x,p} - B_x \partial_x v_{y,p} = 0. \tag{5.52}$$

The linearised equation for v_z reduces to $\partial_t v_{z,p} = 0$, and thus we take v_z to be constant for the remainder of this discussion. B_z is taken as zero and, as in the hydrodynamical case, the continuity equation is isolated by the linearisation process. Thus, there remain a system of four equations and four unknowns to solve.

As usual, Eq. (5.49)–(5.52) can be written in the more compact form,

$$\partial_t |q(x,t)\rangle + \mathsf{J}\, \partial_x |q(x,t)\rangle, \tag{5.53}$$

where,

$$|q(x,t)\rangle = \begin{bmatrix} p_p \\ v_{x,p} \\ v_{y,p} \\ B_{y,p} \end{bmatrix} \quad \text{and} \quad \mathsf{J} = \begin{bmatrix} 0 & \gamma p_0 & 0 & 0 \\ 1/\rho_0 & 0 & 0 & B_{y,0}/\mu_0\rho_0 \\ 0 & 0 & 0 & -B_x/\mu_0\rho_0 \\ 0 & B_{y,0} & -B_x & 0 \end{bmatrix}. \tag{5.54}$$

Here, $|q(x,t)\rangle$ is the ket of perturbations to the variables, and J is the usual Jacobian matrix.

Next, assume a normal mode solution of the form $|q(x,t)\rangle = |\tilde{q}(\xi)\rangle$, where $\xi = x - ut$ and u is the wave speed. Substituting this into Eq. (5.53) gives us:

$$\mathsf{J}|\tilde{q}'\rangle = u|\tilde{q}'\rangle, \tag{5.55}$$

where $'$ indicates differentiation with respect to ξ, an eigen-equation we've seen numerous times already (e.g., Eq. 2.23 in §2.1.2, Eq. 3.24 in §3.5.1, and Eq. 5.35 in §5.2.1). Thus, the eigenvalues, u, of J are found by solving the secular equation,

$$\det(\mathsf{J} - u\mathsf{I}) = 0,$$

which after a little algebra (Problem 5.7) yields the following quadratic in u^2:

$$u^4 - u^2(a_0^2 + c_{s,0}^2) + c_{s,0}^2 a_x^2 = 0, \tag{5.56}$$

where:

$$c_{s,0}^2 = \frac{\gamma p_0}{\rho_0}; \qquad a_0^2 = a_x^2 + a_{y,0}^2; \qquad a_x^2 = \frac{B_x^2}{\mu_0 \rho_0}; \qquad a_{y,0}^2 = \frac{B_{y,0}^2}{\mu_0 \rho_0}.$$

The two roots of Eq. (5.56) are therefore given by:

$$a_s^2 = \frac{1}{2}\left(a_0^2 + c_{s,0}^2 - \sqrt{(a_0^2 + c_{s,0}^2)^2 - 4a_x^2 c_{s,0}^2}\right); \tag{5.57}$$

$$a_f^2 = \frac{1}{2}\left(a_0^2 + c_{s,0}^2 + \sqrt{(a_0^2 + c_{s,0}^2)^2 - 4a_x^2 c_{s,0}^2}\right), \tag{5.58}$$

the fast and slow magnetosonic speeds given by Eq. (5.23) and (5.25) but here expressed in terms of the unperturbed variables. Since Eq. (5.56) is in fact a quartic in u, there must be *four* eigenvalues of J which, in ascending order, are:

$$u_f^- = -a_f; \qquad u_s^- = -a_s; \qquad u_s^+ = a_s; \qquad u_f^+ = a_f. \tag{5.59}$$

Note that here, since the fluid is considered in its unperturbed rest frame, $v_{x,0} = 0$.

Having recovered the fast and slow waves from Eq. (5.49)–(5.52), we identify these as the "kernel of physics" from the more general set of 1-D Eq. (5.1)–(5.6) that capture the essence of fast and slow magnetosonic waves. All we need to know about linear magnetosonic waves can be gleaned from Eq. (5.49)–(5.52) or, equivalently, Eq. (5.55). These equations describe the simultaneous oscillation of the longitudinal (v_x) and transverse (v_y) velocity components ("carriers" of kinetic energy density), along with the thermal pressure (p) and the transverse magnetic (B_y) component ("carriers" of potential energy density). Since both v_x and v_y oscillate, these waves are simultaneously transverse and longitudinal. They are neither sound waves nor Alfvén waves, although in certain limits, their nature reduces to one of these types of waves (see discussion near the end of this section, as well as §5.2.3).

The eigenkets tell us about how the flow variables – in this case their perturbations – behave across the waves. To find them, we solve,

$$(\mathsf{J} - u_m^\pm \mathsf{I})|r_m^\pm\rangle = 0,$$

where $m = $ f or s. After a little algebra and with suitable "normalisation", the eigenkets are found to be,

$$|r_m^\pm\rangle = \begin{bmatrix} \pm c_{s,0}^2 \rho_0 (a_x^2 - a_m^2)/a_m \\ a_x^2 - a_m^2 \\ a_x a_{y,0} \\ \mp a_{y,0} a_m \sqrt{\mu_0 \rho_0} \end{bmatrix}. \tag{5.60}$$

Like the eigenvalues, the eigenkets are completely made up of the undisturbed variable values.

From Eq. (5.55), $|\tilde{q}'\rangle$ are evidently proportional to the eigenkets and, as we've done before (*e.g.*, Eq. 2.27 in §2.1.2), we can write:

$$|\tilde{q}'\rangle = w'(\xi)|r_m^\pm\rangle \quad \Rightarrow \quad |\tilde{q}\rangle = w(\xi)|r_m^\pm\rangle,$$

since $|r_m^\pm\rangle$ are constant and independent of $\xi = x - ut$, and where $w'(\xi)$ is an arbitrary integrable function of ξ. Thus,

$$|\tilde{q}\rangle \propto |r_m^\pm\rangle \quad \Rightarrow \quad \begin{bmatrix} \tilde{p} \\ \tilde{v}_x \\ \tilde{v}_y \\ \tilde{B}_y \end{bmatrix} \propto \begin{bmatrix} \pm c_{s,0}^2 \rho_0 (a_x^2 - a_m^2)/a_m \\ a_x^2 - a_m^2 \\ a_x a_{y,0} \\ \mp a_{y,0} a_m \sqrt{\mu_0 \rho_0} \end{bmatrix}.$$

From this, we see immediately that:

$$\frac{\tilde{v}_x}{\tilde{v}_y} = \frac{a_x^2 - a_m^2}{a_x a_{y,0}}$$

$$\Rightarrow \quad \frac{\tilde{v}_x/\tilde{v}_y}{B_x/B_{y,0}} = \frac{a_x^2 - a_m^2}{a_x^2} \begin{cases} < 0 & \text{for } a_m = a_f; \\ > 0 & \text{for } a_m = a_s, \end{cases} \tag{5.61}$$

where branching into two cases is a direct consequence of the third of Ineq. (5.28).

Herein lies our first piece of real insight into the difference between fast and slow waves. Evidently, the ratio of the x- and y-components of the velocity perturbation has a *different* sign than the ratio of the x- and y-components of the unperturbed magnetic induction for fast waves, and the *same* sign for slow waves. As shown in Fig. 5.5, this means that the compression and shearing of the magnetic induction by the MHD forces work together to maximise the variation of B_y for fast waves, but work against each other to minimise the variation of B_y for slow waves. With a greater amplitude of magnetic induction variation comes a greater restoring magnetic force and thus a higher wave frequency. For a given wavelength, a higher frequency means a higher propagation speed ($u = f\lambda$); whence $a_\text{f} > a_\text{s}$.

Another way to see this effect is to examine the ratio of the fourth and third components of $|r_m^\pm\rangle$, namely,

$$\frac{\tilde{B}_y}{\tilde{v}_y} = \mp \frac{a_{y,0} a_m \sqrt{\mu_0 \rho_0}}{a_x a_{y,0}} \quad \Rightarrow \quad \frac{\tilde{a}_y}{\tilde{v}_y} = \mp \frac{a_m}{a_x}, \tag{5.62}$$

whose magnitude is greater than unity for fast waves ($a_\text{f} > a_x$) and less than unity for slow waves ($a_\text{s} < a_x$; see again Ineq. 5.28). Thus, for a given perturbation in v_y, fast waves evoke a greater perturbation in a_y (and thus B_y), while slow waves evoke a smaller perturbation in a_y (and thus B_y); the same effect shown in Fig. 5.5.

Still another way to understand the difference between fast and slow waves is to examine the ratio of magnetic pressure perturbations to thermal pressure perturbations caused by the passage of the two waves. Magnetic pressure is given by $p_\text{M} = \frac{1}{2}(B_x^2 + B_y^2)$, and since $B_x = $ constant in 1-D (x-direction), perturbations to the magnetic pressure are given by,

$$\delta p_\text{M} = \tilde{p}_\text{M} = \delta\left(\frac{B_y^2}{2\mu_0}\right) = \frac{B_y}{\mu_0}\delta B_y = \frac{B_{y,0}}{\mu_0}\tilde{B}_y, \tag{5.63}$$

keeping only first-order terms. Therefore,

$$\frac{\tilde{p}_\text{M}}{\tilde{p}} = \frac{B_{y,0}}{\mu_0}\frac{\tilde{B}_y}{\tilde{p}} = -\frac{B_{y,0}}{\mu_0}\frac{a_{y,0} a_m^2 \sqrt{\mu_0 \rho_0}}{c_{s,0}^2 \rho_0 (a_x^2 - a_m^2)} = \frac{a_{y,0}^2}{c_{s,0}^2}\frac{a_m^2}{a_m^2 - a_x^2}, \tag{5.64}$$

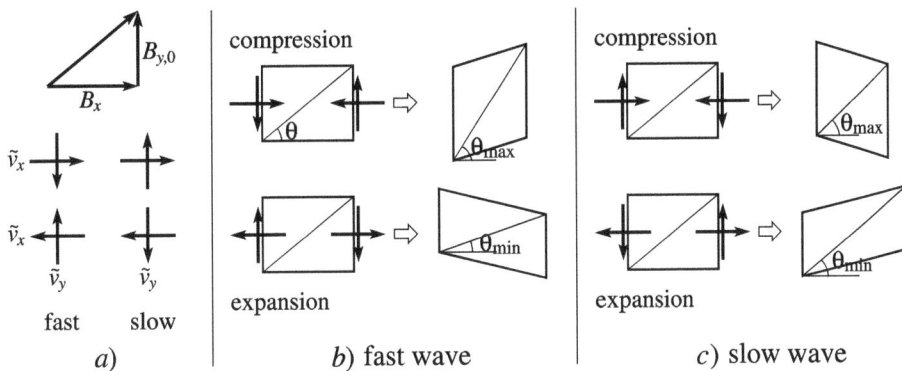

Figure 5.5. For an unperturbed magnetic induction with $B_x/B_{y,0} > 0$ as shown in panel a, velocity perturbations for the fast and slow waves have two possible orientations as shown, and as determined by Eq. (5.61). Consider a rectangular parcel of fluid whose main diagonal gives the orientation (but not the magnitude) of $\vec{B}$. In particular, $B_y = B_x \tan\theta$, where B_x is constant and where θ is the angle between the main diagonal and the x-axis. In panel b, the passage of the fast magnetosonic wave causes a compression of the fluid parcel (horizontal velocity vectors pointed inwards) followed by an expansion (horizontal velocity vectors pointed outwards). Accompanying the horizontal velocity vectors are the appropriately directed vertical velocity vectors as indicated in panel a, and together the horizontal and vertical velocity vectors distort the fluid parcel as shown in panels b and c. In the case of the fast wave, both the compression and the shear of the fluid parcel tend to increase θ (and thus B_y), while both the expansion and its accompanying shear tend to decrease it half a period later. In panel c, the passage of the slow magnetosonic wave has a different effect. While the compression of the fluid parcel works to increase θ (B_y), the shear works to decrease it. Then, as the fluid parcel expands, the expansion tends to decrease θ (B_y) while its accompanying shear tends to increase it. Thus, the amplitude of variation for B_y (indicated by the difference $\theta_{\max} - \theta_{\min}$) is greater for the fast wave than for the slow wave.

using the ratio of the fourth and first components of $|r_m^\pm\rangle$. Thus, for $a_m = a_f > a_x$, the magnetic and thermal pressure perturbations have the same sign, and thus work together in driving the wave. On the other hand, for $a_m = a_s < a_x$, the magnetic and thermal pressure perturbations have opposite sign, and work against each other thereby reducing the restoring force and as a consequence the speed of the wave.

Figure 5.6 shows how these ideas distinguish the two types of magnetosonic waves. According to Eq. (5.64), a local maximum of thermal pressure must correspond to a local maximum of magnetic pressure for a fast wave, and a local minimum of magnetic pressure for a slow wave. In either case, the resulting x-variation of B_y generates a current density,

$$\vec{J} = \frac{1}{\mu_0}\nabla \times \vec{B} = \frac{1}{\mu_0}(0,\, 0,\, \partial_x B_y),$$

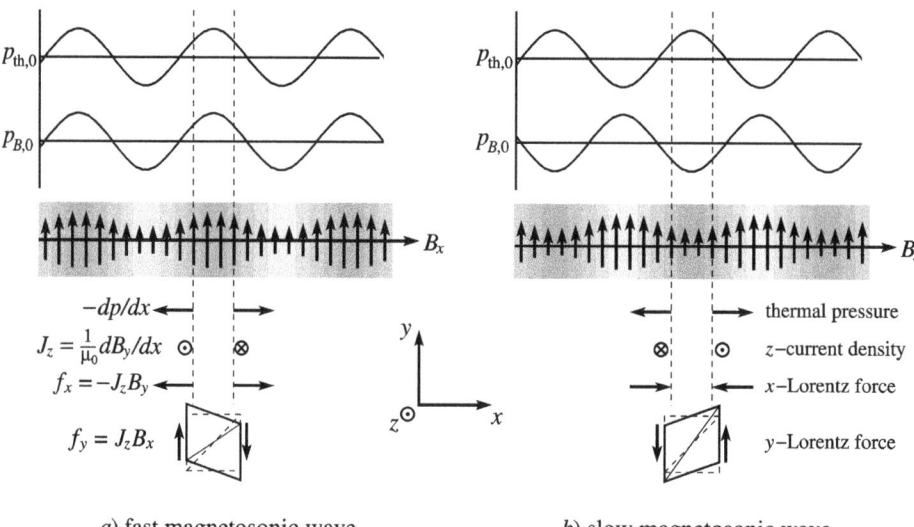

Figure 5.6. A succession of compressions and rarefactions in a magnetosonic wave results in an x-variation of the thermal pressure (grey scale) and transverse magnetic induction (arrows), the latter establishing Lorentz forces that contribute to the net restoring force available to the waves. For fast waves where the thermal and magnetic pressure variations are in phase (Eq. 5.64), the thermal pressure and magnetic restoring forces ($-dp/dx$ and f_x respectively) are additive (panel a), while for slow waves where the two variations are out of phase, the restoring forces partially cancel (panel b). These drive the longitudinal aspect of the waves. The transverse aspect of the waves is driven by the Lorentz shear force, f_y. For both wave types, when B_y is at a peak, the Lorentz shear force distorts the original fluid cell (dashed rectangle) to a parallelogram so that the slope of the main diagonal (a measure of B_y; see caption of Fig. 5.5) is reduced. Conversely, when B_y is at a low, the Lorentz shear forces distort the fluid cell so that the slope of the main diagonal (and thus B_y) increases.

which combines with the magnetic induction to create a Lorentz force density,

$$\vec{f}_L = \vec{J} \times \vec{B} = (-J_z B_y, J_z B_x, 0) = \frac{1}{\mu_0}(-B_{y,0}, B_x, 0)\partial_x \tilde{B}_y, \qquad (5.65)$$

where only the first-order terms have been retained. The x-component of the Lorentz force density is a magnetic pressure term, $-\partial_x p_M$ (e.g., Eq. 5.63), which contributes to the thermal pressure force density, $-\partial_x \tilde{p}$. As shown in Fig. 5.6, these pressure forces are additive for fast waves and partially (perhaps completely) cancel for slow waves. The y-component of the Lorentz force density is solely responsible for the transverse nature of magnetosonic waves and, as shown in Fig. 5.6, is always restorative (tries to bring the perturbation back into equilibrium), regardless of whether the wave is slow or fast. For the fast wave, the added restorative force is not required to assure the propagation of the wave, since the two pressure forces are additive. However, for the slow wave where the two pressure forces work against each other and may actually cancel, the transverse restorative force is critical to its propagation; without it, the slow wave could be stifled.

In the $\theta \to \pi/2$ ($B_x \to 0$) limit, it is easy to show that $a_s \to 0$, and thus the slow waves are extinguished. So are shear Alfvén waves since they require a longitudinal magnetic induction to carry the transverse waves and thus, only the fast magnetosonic waves remain. In this limit, the fast waves propagate with a speed $a_M^2 = a_f^2(\theta = \pi/2) = a^2 + c_{s,0}^2$ and, because there is no longitudinal component of the magnetic induction, the transverse component of the Lorentz force density is zero (Eq. 5.65). Thus, such waves are purely compressional (longitudinal) in nature and are often referred to as *magneto-acoustic* waves.[6] In the further limit as $B_y \to 0$, magneto-acoustic waves become ordinary sound waves moving at the sound speed, while in the opposite limit of a cold (*i.e.*, pressureless) gas, magneto-acoustic waves move at the Alfvén speed and are known as *compressional Alfvén waves*, as opposed to the shear Alfvén waves discussed in §5.2.1.

In the $\theta \to 0$ ($B_{y,0} \to 0$) limit, all three types of waves exist, but with some degeneracy. From Eq. (5.65), we see the x-component of the restoring Lorentz force density disappears (but not the y-component, as $\tilde{B}_y$ can still oscillate to first order about zero), and thus the magnetic pressure term no longer contributes to the longitudinal restoring force. From Eq. (5.58) and (5.57), we see that for $\theta = 0$ (and thus $a_0 = a_x$), the fast and slow speeds become:[7]

$$\left. \begin{aligned} a_f^2(\theta = 0) &= \tfrac{1}{2}\left(c_{s,0}^2 + a_0^2 + \sqrt{(c_{s,0}^2 + a_0^2)^2 - 4c_{s,0}^2 a_0^2}\right) \\ &= \tfrac{1}{2}\left(c_{s,0}^2 + a_0^2 + \sqrt{(c_{s,0}^2 - a_0^2)^2}\right) = \max(c_{s,0}^2, a_0^2); \\ a_s^2(\theta = 0) &= \tfrac{1}{2}\left(c_{s,0}^2 + a_0^2 - \sqrt{(c_{s,0}^2 - a_0^2)^2}\right) = \min(c_{s,0}^2, a_0^2). \end{aligned} \right\} \quad (5.66)$$

Thus, disturbances launched directly along lines of magnetic induction are either sound or Alfvén waves or, at the very least, propagate with one of these speeds.

In fact, fast and slow waves *do* reduce to ordinary shear Alfvén and sound waves when launched along induction lines; the faster of the two being the vestigial fast wave, the slower the slow wave. Setting $B_{y,0} = 0$, Eq. (5.49)–(5.52) reduce to:

$$\partial_t p_p + \gamma p_0 \, \partial_x v_{x,p} = 0; \tag{5.67}$$

$$\partial_t v_{x,p} + \frac{1}{\rho_0} \partial_x p_p = 0; \tag{5.68}$$

$$\partial_t v_{y,p} - \frac{B_x}{\mu_0 \rho_0} \partial_x B_{y,p} = 0; \tag{5.69}$$

$$\partial_t B_{y,p} - B_x \, \partial_x v_{y,p} = 0. \tag{5.70}$$

---

[6]There is no agreement within the literature on how the terms "magnetosonic" and "magneto-acoustic" are to be used. Some authors use them as I do, some use them interchangeably, while others use magneto-acoustic as the general case (of which there is the slow and fast variety) and magnetosonic for the special case being discussed here. Sorry, it's just the way it is!

[7]The appearance of "max-min" functions is most easily verified by noting that $\sqrt{q^2} = |q|$. As we'll see, much of the curious algebra arising in 1-D MHD can be traced to the presence of these max-min functions.

	$\alpha \ll 1$	$\alpha = 1$	$\alpha \gg 1$
a_f	$c_\mathrm{s}\sqrt{1 + \alpha \sin^2\theta}$	$c_\mathrm{s}\sqrt{1 + \sin\theta}$	$a\sqrt{1 + \dfrac{\sin^2\theta}{\alpha}}$
a_s	$a\cos\theta\sqrt{1 - \alpha \sin^2\theta}$	$c_\mathrm{s}\sqrt{1 - \sin\theta}$	$c_\mathrm{s}\cos\theta\sqrt{1 - \dfrac{\sin^2\theta}{\alpha}}$

Table 5.1. Magnetosonic speeds for three different limits of the MHD-alpha, $\alpha = a^2/c_\mathrm{s}^2$, and where $\cos\theta = a_x/a$.

Equations (5.67) and (5.68) are simply the 1-D version of Eq. (2.7) and (2.8) which describe an ordinary sound wave, while Eq. (5.69) and (5.70) are equivalent to Eq. (5.31) and (5.32) in the rest frame of the fluid ($v_{x,0} = 0$) which describe a shear Alfvén wave. No variable in one pair of equations appears in the other, and we conclude that without a transverse component of the magnetic induction, the magnetic and thermal effects decouple completely into sound and Alfvén waves. In magnetically dominated systems ($\alpha > 1$; see Eq. 5.30), the fast wave becomes the transverse shear Alfvén wave and the slow wave becomes the longitudinal sound wave, while in thermally dominated systems ($\alpha < 1$), the reverse is true. For $\alpha = 1$, the waves are degenerate; both the sound wave and shear Alfvén wave move with the same speed and it becomes academic whether one associates the vestigial fast or slow wave with the sound or shear Alfvén wave.

5.2.3 Summary of MHD waves

If we, as a species, had evolved within an MHD (ionised) atmosphere instead of the hydrodynamical (neutral) one we live in,[8] surely our senses would have evolved to use the additional wave information available to us. A thunder clap within an MHD atmosphere, such as that surrounding the sun, would launch all types of waves, and our array of senses would first detect the fast waves (which carry pressure fluctuations just like sound waves, and therefore would be audible) as relatively high-pitched sound, followed sometime later by the shear Alfvén wave which, while not carrying pressure fluctuations and thus inaudible, would result in a sudden gust of wind transverse to the direction between the observer and the disturbance, followed still later by the audible slow waves at a relatively low pitch. The statements on the relative pitches of the fast and slow waves stem from the requirement that $v = f\lambda$. For a given wavelength (determined to first order by the scale length of the disturbance causing the thunder clap), the fast wave must have a higher frequency than the slow wave in order to have a higher propagation speed.

Further, various observers would have differing impressions of the same thunder clap depending on their position relative to where the clap occurred. Those connected to the disturbance by a prevailing line of induction (and according to whom the transverse component of $\vec{B}$ is zero) would feel first a wind shear, then an

---

[8]Let's not worry about how life could have evolved in temperatures exceeding 10^5 K!

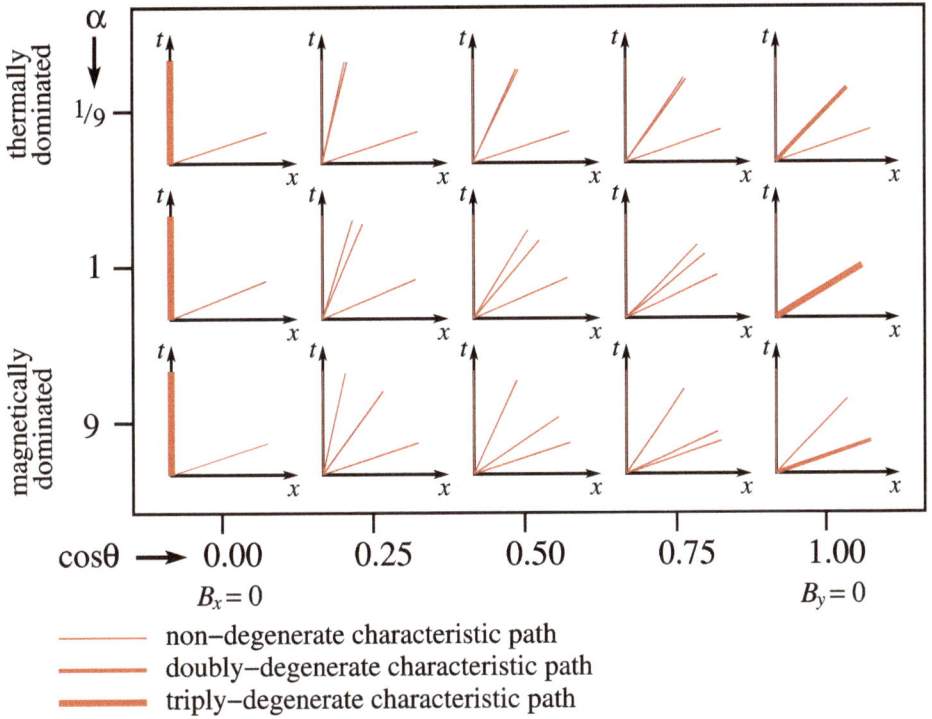

Figure 5.7. Space-time diagrams showing the plus-characteristic paths for various points in α-$\cos\theta$ space, where the top (bottom) row is indicative of a thermally (magnetically) dominated gas. The slope of each characteristic path is proportional to the reciprocal of the characteristic speed as measured from the co-moving frame of the fluid ($v_x = 0$). Thus, from left to right starting at the vertical, the first characteristic path encountered is always the entropy path (coincides with the t-axis in the co-moving frame of the fluid), followed by the slow, Alfvén, and fast characteristic paths.

audible rumble for $\alpha > 1$ (magnetically dominated) or, for $\alpha < 1$, the rumble first, then the wind shear. Meanwhile, an observer whose line of sight to the disturbance is perpendicular to the prevailing lines of induction would hear only the fast wave since, in the co-moving frame of the fluid, neither the Alfvén wave nor the slow wave can propagate in this direction. Further, since the fast speed is a maximum in this orientation, this observer would detect the clap as a higher pitched sound than any other observer, and before any other observer at the same distance from the clap.

It is a useful exercise to understand how the MHD waves become the hydrodynamical waves in the limit as $a \to 0$ and, in the other extreme, what happens to MHD waves in the limit of a cold (*e.g.*, collisionless or pressureless) gas where $c_s \to 0$. Table 5.1 lists the fast and slow speeds for various regimes of the MHD-alpha whose derivation is left as an exercise.

In the thermally dominated regime ($\alpha \ll 1$), it's clear from Table 5.1 that the slow wave speed becomes degenerate with the Alfvén speed ($a_x = a\cos\theta$). This

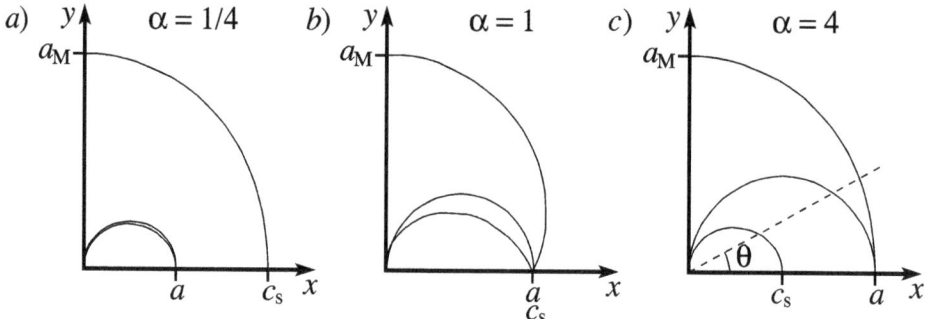

Figure 5.8. Polar diagrams representing the characteristic speeds of the slow (inner curve), Alfvén (middle curve), and fast (outer curve) waves for $\alpha = \frac{1}{4}$ (panel a), $\alpha = 1$ (panel b), and $\alpha = 4$ (panel c). In each case, the wave speed is proportional to the distance between the origin and where a ray inclined at angle θ (angle between $\vec{B}$ and the x-axis) intersects a given curve, as exemplified by the dashed line in panel c.

is apparent in the top row of Fig. 5.7, where even when the sound speed is only three times that of the Alfvén speed, the slow and Alfvén characteristics are barely discernible. It is shown in Problem 5.10 (Eq. 5.117) that compressional forces in slow waves cancel as $\alpha \to 0$, leaving only the transverse magnetic forces to drive the wave. Thus, in this limit, the slow and Alfvén waves are, in fact, the same wave. On the other hand, Problem 5.10 also shows that the magnetic driving force for the fast wave falls off as α, leaving only the thermal pressure gradient. Thus, in the limit $\alpha \to 0$, fast waves becomes ordinary sound waves.

Regardless of α, the slow and Alfvén waves are degenerate with the entropy wave when $B_x = 0$ (left side of Fig. 5.7), and become non-degenerate for $\cos\theta > 0$ ($\theta < \pi/2$). The slow wave path reaches a maximum angle relative to the vertical at $\cos\theta = 1$ becoming degenerate with the fast wave path only when $\alpha = 1$. Conversely, at both the $\alpha \to 0$ and $\alpha \to \infty$ extremes, the slow and entropy waves remain degenerate to first order, and thus their paths are vertical on a space-time diagram. As for the Alfvén characteristic path, for $\alpha < 1$ it too reaches a maximum inclination relative to the vertical, where it once again becomes degenerate with the slow path (top row of Fig. 5.7). However, for $\alpha > 1$, the Alfvén path continues to separate from the slow path as $\theta \to 0$, leaning further and further to the right until it becomes degenerate with the fast characteristic path at $\cos\theta = 1$ (bottom row). The $\alpha = 1$ case is interesting in that both the slow and Alfvén paths become degenerate with the fast path at $\cos\theta = 1$. This is the only point in α-$\cos\theta$ parameter space where this occurs, and we will come to know this as the *triple umbilic* (§6.2.1).

Finally, Fig. 5.8 is a polar plot of the speeds of the three MHD waves where the characteristic speed is proportional to the distance between the origin and where a ray at angle θ intersects a given curve (as exemplified by the dashed line in panel c). Plots for $\alpha = \frac{1}{4}$, 1, and 4 are given. Note that the fast "curve" always contains the Alfvén "curve" which always contains the slow "curve", a direct consequence

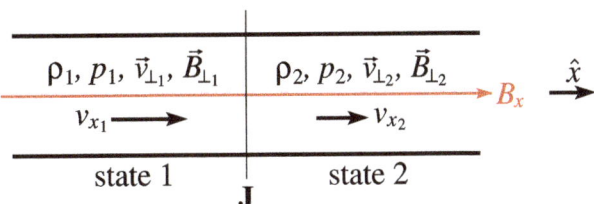

Figure 5.9. $1\frac{1}{2}$-D MHD fluid flow, as viewed from the reference frame **J**.

of the last of Ineq. (5.28). There are, however, places where the curves touch, and these indicate where two and even all three waves become degenerate. For thermally dominated gas (Fig. 5.8a), the fast wave is always distinct from the slow and Alfvén waves whose curves shrink into the origin and disappear altogether in the limit $\alpha \to 0$ ($B \to 0$). In this limit, only the fast wave remains and has become an ordinary sound wave. For magnetically dominated gas (Fig. 5.8c), the Alfvén wave is degenerate with the slow (and entropy) wave at $\cos\theta = 0$, and with the fast wave at $\cos\theta = 1$. Lastly, regardless of α, only the fast wave has a non-zero propagation speed at $\theta = \pi/2$, and it is here that the fast wave is referred to as the magneto-acoustic wave. As discussed in §5.2.2, the magneto-acoustic wave becomes a sound wave as $\alpha \to 0$, and a compressional Alfvén wave as $\alpha \to \infty$.

As much as the degenerate behaviour of the eigenvalues affect the nature of the linear waves, they pose even greater challenges in the non-linear waves, as we shall see in the next section for shocks and in §6.2 for rarefaction fans.

5.3 The MHD Rankine–Hugoniot jump conditions

In §2.2, we considered the $1\frac{1}{2}$-D steady-state *shock tube* problem for ideal hydrodynamics. Here, we revisit this subject for $1\frac{1}{2}$-D ideal MHD. The principles and procedures are identical as in §2.2; only the algebra is trickier and the results correspondingly richer. To wit, for the MHD problem we must solve five equations in three scalars and two 2-D vectors.

Figure 5.9 depicts an MHD shock tube with an upwind (1) and downwind (2) state delineated by and in the reference frame of **J**. Without loss of generality, let the flow be from left to right and identify states 1 and 2 by their primitive variables $(\rho_1, p_1, v_{x_1}, \vec{v}_{\perp_1}, \vec{B}_{\perp_1})$ and $(\rho_2, p_2, v_{x_2}, \vec{v}_{\perp_2}, \vec{B}_{\perp_2})$ respectively. As was shown at the beginning of this chapter, B_x is constant in time and space in 1-D (represented by the long, red arrow in Fig. 5.9), and we need not include it among the downwind variables to be determined. However, B_x cannot be ignored, as the downwind variables do depend upon it in sometimes unexpected ways.

Now, if **J** corresponds to nothing physical in the flow, the upwind and downwind states would be identical; only if **J** were to coincide with a discontinuity would the two states differ (recognise this discussion from §2.2?). Therefore, the question posed here is: Given values for the primitive variables in the upwind state, what steady-state values could these same variables assume in the downwind state?

To this end,[9] we start with the conservative form of the $1\frac{1}{2}$-D MHD equations, Eq. (5.10)–(5.16), which, as has been emphasised before, is the form of the equations valid across discontinuous flow. In 1-D steady state where $\partial_t \to 0$ and $\partial_y = \partial_z = 0 \Rightarrow \partial_x \to d/dx$, these equations reduce to:

$$\frac{d}{dx}(\rho v_x) = 0; \tag{5.71}$$

$$\frac{d}{dx}\left[v_x\left(e_T + p + \frac{B^2}{\mu_0}\right) - \frac{B_x}{\mu_0}\vec{B}\cdot\vec{v}\right] = 0; \tag{5.72}$$

$$\frac{d}{dx}\left(\rho v_x^2 + p + \frac{B_\perp^2}{2\mu_0}\right) = 0; \tag{5.73}$$

$$\frac{d}{dx}\left(\rho v_x \vec{v}_\perp - \frac{B_x \vec{B}_\perp}{\mu_0}\right) = 0; \tag{5.74}$$

$$\frac{d}{dx}\left(v_x \vec{B}_\perp - B_x \vec{v}_\perp\right) = 0, \tag{5.75}$$

where:

1. $p^* = p + B^2/2\mu_0$ and $e_T^* = e_T + B^2/2\mu_0$ have been substituted everywhere;

2. in Eq. (5.72), $B^2 = B_x^2 + B_\perp^2$ with the B_x^2 term retained because, while $d(B_x^2)/dx$ is zero, $d(v_x B_x^2)/dx$ may not be and will instead cancel out part of the $B_x(\vec{B}\cdot\vec{v})$ term;

3. y- and z-components of momentum and induction equations have been combined into two 2-D vector equations, (5.74) and (5.75).

As was true in §2.2, these ODEs are trivial to integrate and, when applied to the conditions of Fig. 5.9, lead to the *MHD Rankine–Hugoniot jump conditions*:

$$\Delta(\rho v_x) = 0; \tag{5.76}$$

$$\Delta\left[v_x\left(e_T + p + \frac{B^2}{\mu_0}\right) - \frac{B_x}{\mu_0}\vec{B}\cdot\vec{v}\right] = 0; \tag{5.77}$$

$$\Delta\left(\rho v_x^2 + p + \frac{B_\perp^2}{2\mu_0}\right) = 0; \tag{5.78}$$

$$\Delta\left(\rho v_x \vec{v}_\perp - \frac{B_x \vec{B}_\perp}{\mu_0}\right) = 0; \tag{5.79}$$

$$\Delta\left(v_x \vec{B}_\perp - B_x \vec{v}_\perp\right) = 0, \tag{5.80}$$

where $\Delta q \equiv q_2 - q_1$ indicates the difference, if any, in the quantity q across **J**.

---

[9] Much of the development presented in this section follows Takahashi & Yamada (2014), hereafter referred to as TY14.

Before proceeding, we need to establish a few theorems from *difference theory* most of which the reader will surely recognise as analogous to well-known theorems of differential calculus. In what follows, f and g are two independent functions of x[10] and a is a constant.

Theorem 5.1. (Addition Rule) $\Delta(f+g) = \Delta f + \Delta g$.

Proof: This is really quite self-evident. Developing the LHS, we get:
$$\Delta(f+g) = (f_2 + g_2) - (f_1 + g_1) = (f_2 - f_1) + (g_2 - g_1) = \Delta f + \Delta g. \quad \square$$

Theorem 5.2. (Product Rule) $\Delta(fg) = \langle f \rangle \Delta g + \langle g \rangle \Delta f$, *where* $\langle q \rangle \equiv \tfrac{1}{2}(q_2 + q_1)$.

Proof: Developing the RHS, we get:
$$\begin{aligned}
\langle f \rangle \Delta g + \langle g \rangle \Delta f &= \frac{f_2 + f_1}{2}(g_2 - g_1) + \frac{g_2 + g_1}{2}(f_2 - f_1) \\
&= \frac{f_2 g_2 - \cancel{f_2 g_1} + \cancel{f_1 g_2} - f_1 g_1 + g_2 f_2 - \cancel{g_2 f_1} + \cancel{g_1 f_2} - g_1 f_1}{2} \\
&= f_2 g_2 - f_1 g_1 = \Delta(fg). \quad \square
\end{aligned}$$

Corollary 5.1. (Scalar Rule) $\Delta(ag) = a\Delta g$.

Proof: This, too, is self-evident. Formally, it is a special case of Theorem 5.2 in which $f = a$ is a constant and thus $\langle f \rangle = a$, $\Delta f = 0$. $\quad \square$

Theorem 5.3. (Average Theorem) $\langle fg \rangle = \langle f \rangle \langle g \rangle + \tfrac{1}{4} \Delta f \Delta g$.

Proof: Starting with the RHS:
$$\begin{aligned}
\langle f \rangle \langle g \rangle + \frac{1}{4}\Delta f \Delta g &= \frac{f_2 + f_1}{2} \frac{g_2 + g_1}{2} + \frac{1}{4}(f_2 - f_1)(g_2 - g_1) \\
&= \frac{f_2 g_2 + \cancel{f_2 g_1} + \cancel{f_1 g_2} + f_1 g_1 + f_2 g_2 - \cancel{f_2 g_1} - \cancel{f_1 g_2} + f_1 g_1}{4} \\
&= \frac{f_2 g_2 + f_1 g_1}{2} = \langle fg \rangle. \quad \square
\end{aligned}$$

And now to the task at hand. By applying Theorem 5.1 and Corollary 5.1 to Eq. (5.78)–(5.80), one can easily show that:

$$\mathcal{M}\Delta v_x + \Delta p + \frac{1}{2\mu_0}\Delta B_\perp^2 = 0; \tag{5.81}$$

---
[10]The functions f and g could be scalar or vector, and if vector the multiplicative operator is understood to be the dot product.

$$\mathcal{M}\Delta\vec{v}_\perp - \frac{B_x}{\mu_0}\Delta\vec{B}_\perp = 0; \tag{5.82}$$

$$\Delta(v_x\vec{B}_\perp) - B_x\Delta\vec{v}_\perp = 0, \tag{5.83}$$

where $\mathcal{M} \equiv \rho_1 v_{x_1} = \rho_2 v_{x_2}$ is the constant mass flux density (Eq. 5.76). Further, Eq. (5.82) and (5.83) can be used to eliminate $\Delta\vec{v}_\perp$:

$$\mathcal{M}\Delta(v_x\vec{B}_\perp) - \frac{B_x^2}{\mu_0}\Delta\vec{B}_\perp = 0. \tag{5.84}$$

Turning now to the differenced total energy equation, Eq. (5.77), start by setting $e_T = e + \frac{1}{2}\rho v^2$ (Eq. 1.20) and then expand the dot products to get:

$$\Delta\left[v_x\left(e + \frac{\rho v_x^2}{2} + \frac{\rho v_\perp^2}{2} + p + \frac{B_x^2}{\mu_0} + \frac{B_\perp^2}{\mu_0}\right) - \frac{B_x}{\mu_0}\left(B_x v_x + \vec{B}_\perp \cdot \vec{v}_\perp\right)\right] = 0$$

$$\Rightarrow \frac{\Delta(pv_x)}{\gamma - 1} + \frac{\mathcal{M}}{2}(\Delta v_x^2 + \Delta v_\perp^2) + \Delta(pv_x) + \frac{1}{\mu_0}\underbrace{\Delta(v_x B_\perp^2 - B_x\vec{B}_\perp \cdot \vec{v}_\perp)}_{\mathcal{Q}} = 0, \tag{5.85}$$

using Theorem 5.1 and Corollary 5.1 repeatedly and the ideal gas law (Eq. 1.11) to replace e with $p/(\gamma - 1)$. While it may be tempting to combine the two terms proportional to $\Delta(pv_x)$, we won't do that just yet (see, however, Problem 5.17). Instead, proceed by examining the quantity $\mathcal{Q}$:

$$\mathcal{Q} = \Delta(v_x B_\perp^2 - B_x\vec{B}_\perp \cdot \vec{v}_\perp) = \Delta[\vec{B}_\perp \cdot (v_x\vec{B}_\perp - B_x\vec{v}_\perp)]$$

$$= \langle\vec{B}_\perp\rangle \cdot \underbrace{\Delta(v_x\vec{B}_\perp - B_x\vec{v}_\perp)}_{= 0;\ \text{Eq. 5.83}} + \Delta\vec{B}_\perp \cdot \langle v_x\vec{B}_\perp - B_x\vec{v}_\perp\rangle \quad \text{(Theorem 5.2)}$$

$$= \Delta\vec{B}_\perp \cdot \langle v_x\vec{B}_\perp\rangle - \underbrace{B_x\Delta\vec{B}_\perp}_{\mu_0\mathcal{M}\Delta\vec{v}_\perp;\ \text{Eq. 5.82}} \cdot \langle\vec{v}_\perp\rangle$$

$$= \Delta\vec{B}_\perp \cdot \langle v_x\rangle\langle\vec{B}_\perp\rangle + \Delta\vec{B}_\perp \cdot \frac{1}{4}\Delta v_x \Delta\vec{B}_\perp - \mu_0\mathcal{M}\Delta\vec{v}_\perp \cdot \langle\vec{v}_\perp\rangle \quad \text{(Theorem 5.3)}$$

$$= \frac{\langle v_x\rangle}{2}\Delta B_\perp^2 + \frac{\Delta v_x}{4}(\Delta B_\perp)^2 - \frac{\mu_0\mathcal{M}}{2}\Delta v_\perp^2 \quad \text{(Theorem 5.2 used twice)}.$$

Substituting this result into Eq. (5.85), we get:

$$0 = \frac{\Delta(pv_x)}{\gamma - 1} + \frac{\mathcal{M}}{2}(\Delta v_x^2 + \cancel{\Delta v_\perp^2}) + \Delta(pv_x) + \frac{\langle v_x\rangle}{2\mu_0}\Delta B_\perp^2 + \frac{\Delta v_x}{4\mu_0}(\Delta B_\perp)^2 - \cancel{\frac{\mathcal{M}}{2}\Delta v_\perp^2}$$

$$= \frac{\Delta(pv_x)}{\gamma - 1} + \mathcal{M}\Delta v_x\langle v_x\rangle + \Delta p\langle v_x\rangle + \langle p\rangle\Delta v_x + \frac{\langle v_x\rangle}{2\mu_0}\Delta B_\perp^2 + \frac{\Delta v_x}{4\mu_0}(\Delta B_\perp)^2$$

$$\text{(Theorem 5.2 used twice)}$$

$$= \frac{\Delta(pv_x)}{\gamma - 1} + \langle p\rangle\Delta v_x + \frac{\Delta v_x}{4\mu_0}(\Delta B_\perp)^2 + \langle v_x\rangle\underbrace{\left(\mathcal{M}\Delta v_x + \Delta p + \frac{\Delta B_\perp^2}{2\mu_0}\right)}_{= 0;\ \text{Eq. 5.81}},$$

The MHD Rankine–Hugoniot jump conditions

and we arrive at the simplified form of the differenced energy equation we'll use to examine discontinuities at **J**:

$$\frac{\Delta(pv_x)}{\gamma - 1} + \langle p \rangle \Delta v_x + \frac{\Delta v_x}{4\mu_0}(\Delta B_\perp)^2 = 0. \tag{5.86}$$

Aside: I pause to comment on the curious nature of Eq. (5.86). The 1-D internal energy equation, Eq. (1.34), is the same whether $\vec{B} = 0$ or not. Thus, if one were to go about differencing it directly assuming steady state, one might write:

$$\cancel{\partial_t e}^0 + \partial_x(ev_x) = -p\partial_x v_x \quad \Rightarrow \quad \frac{\Delta(pv_x)}{\gamma - 1} + \tilde{p}\Delta v_x = 0,$$

since $e = p/(\gamma - 1)$. Granted, one would never attempt to do this in the first place given the non-conservative nature of the $p\partial_x v_x$ term which leads to the conundrum of what $\tilde{p}$ is. Still, if one were to "guess" that $\tilde{p}$ should be $\langle p \rangle$, one would then recover all of Eq. (5.86) except the magnetic term. How could a "bad guess" for $\tilde{p}$ squelch a magnetic term?

I don't really have a satisfying answer to this question. I'm not at all suggesting that something is suspicious with Eq. (5.86); indeed the fault is in trying to treat the non-conservative internal energy equation in a conservative fashion.

Still, I do find this curious...

To recap our progress so far, we have reduced the MHD Rankine–Hugoniot jump conditions to a system of three equations in three unknowns:

$$\mathcal{M}\Delta v_x + \Delta p + \frac{1}{2\mu_0}\Delta B_\perp^2 = 0; \quad \text{Eq. (5.81)}$$

$$\mathcal{M}\Delta(v_x \vec{B}_\perp) - \frac{B_x^2}{\mu_0}\Delta \vec{B}_\perp = 0; \quad \text{Eq. (5.84)}$$

$$\frac{\Delta(pv_x)}{\gamma - 1} + \langle p \rangle \Delta v_x + \frac{\Delta v_x}{4\mu_0}\Delta(B_\perp)^2 = 0, \quad \text{Eq. (5.86)}$$

with the unknowns being Δv_x, Δp, and $\Delta \vec{B}_\perp$. Once these variables have been found, the remaining two can be easily determined. From Eq. (5.76), it is left to Problem 5.18 to show:

$$\Delta \rho = -\frac{\rho_1 \Delta v_x}{v_{x_1} + \Delta v_x}, \tag{5.87}$$

while the jump in the transverse velocity, $\Delta \vec{v}_\perp$, can be found using either Eq. (5.82) or (5.83):

$$\Delta \vec{v}_\perp = \frac{B_x}{\mu_0 \mathcal{M}}\Delta \vec{B}_\perp = \frac{1}{B_x}\Delta(v_x \vec{B}_\perp). \tag{5.88}$$

5.3.1 Case 1: Trivial solution

Obviously, $\Delta\rho = \Delta v_x = \Delta\vec{v}_\perp = \Delta\vec{B}_\perp = \Delta p = 0$ solves all jump conditions, and such a solution is the possibility already suggested, namely **J** corresponds to nothing significant in the flow. As in §2.2.1, this is known as the *trivial solution*, and is consistent with a uniform gas flowing at a constant velocity, $v_{x_1} = v_{x_2}$.

5.3.2 Case 2: Contact discontinuity

If $v_x = 0$ (and thus $\mathcal{M} = 0$) and $B_x \neq 0$, then Eq. (5.84) $\Rightarrow \Delta\vec{B}_\perp = 0$, from which it follows that Eq. (5.81) $\Rightarrow \Delta p = 0$ (recall from Theorem 5.2 that $\frac{1}{2}\Delta B_\perp^2 = \Delta\vec{B}_\perp \cdot \langle\vec{B}_\perp\rangle$). Further, Eq. (5.83) $\Rightarrow \Delta\vec{v}_\perp = 0$ and right away we know that the thermal pressure, p, and transverse vectors, $\vec{v}_\perp$ and $\vec{B}_\perp$, are all constant. Since $v_x = 0$, none of the equations can tell us anything about the density jump, and thus $\Delta\rho$ is arbitrary.

This is identical to the *contact discontinuity* (contact) identified in §2.2.2. As noted there, because the pressure is constant, the density jump must be accompanied by an inverse jump in temperature as well as a jump in entropy and, just as for HD, contacts are sometimes called *entropy waves*.

5.3.3 Case 3: Tangential discontinuity

If $v_x = 0$ and $B_x = 0$ (but $\vec{B}_\perp \neq 0$), a *tangential discontinuity* (TD) is the result. In this case, all Rankine–Hugoniot jump conditions, Eq. (5.76)–(5.80) are trivialised ($0 = 0$) except for Eq. (5.78), which reduces to,

$$\Delta\left(p + \frac{B_\perp^2}{2\mu_0}\right) = \Delta p^* = 0, \tag{5.89}$$

and thus the *total pressure* – thermal plus magnetic – must be continuous across **J**, but not necessarily p and $\vec{B}_\perp$ individually. Thus, across a TD, $v_x = 0$, $B_x = 0$, $\Delta\rho$ and $\Delta\vec{v}_\perp$ are arbitrary, as are Δp and $\Delta\vec{B}_\perp$ so long as $\Delta(p + B_\perp^2/2\mu_0) = 0$.

Note that if $\vec{B}_\perp = 0$ as well, we recover the hydrodynamical limit where the distinction between a TD and a contact depends only on whether a previously existing wind shear – a jump in $\vec{v}_\perp$ – exists across **J** (§2.2.2). In pure HD, there is no way in 1-D to generate a shear and thus it would need to be present as part of the initial conditions. On the other hand, for MHD even in 1-D, shear Lorentz forces can cause a wind shear to be generated spontaneously, and the distinction between a contact and a TD is much more certain depending, as it does, on the presence of B_x.

And so here is another curiosity. How close to zero must $B_x/B_\perp$ actually be before a contact yields to a TD? Does it have to be dead zero, for example, or could it be 10^{-6}? 10^{-12}? 10^{-100}? And what power does B_x hold over the other variables that not until its last "epsilon" of value is extinguished can $\Delta\vec{v}_\perp$ – which is supposed to be zero across a contact – suddenly take on any arbitrary value across a TD? Or when Δp and $\Delta\vec{B}_\perp$ are supposed to be zero individually for a contact, suddenly need only obey the constraint $\Delta(p + B_\perp^2/2\mu_0) = 0$ for a TD?

The answer has to do with the imposed symmetries and the time scales over which steady state is achieved. Still, such conundrums give glimmers of the touchy properties posed by the algebra of MHD which we'll encounter head-on in §5.3.5.

5.3.4 Case 4: Rotational discontinuity

Having exhausted all possibilities where $v_x = 0$ (remember, v_x is the speed *relative to* **J**), we now consider the case where $v_x \neq 0$. However and for the moment, let's continue to suppose $\Delta v_x = 0 \Rightarrow \Delta\rho = 0$, and density is constant across **J**. Further suppose that $B_x \neq 0$. Then, Eq. (5.86) requires $\Delta(\rho v_x) = v_x \Delta\rho = 0$ (Corollary 5.1) $\Rightarrow \Delta p = 0$, and pressure is constant across **J**. Therefore, Eq. (5.81) $\Rightarrow \Delta B_\perp^2 = 0$ too, although this doesn't necessarily mean $\Delta\vec{B}_\perp = 0$; $\vec{B}_\perp$ could still rotate!

Since $\Delta\vec{B}_\perp \neq 0$, $v_x =$ constant, and $\mathcal{M} = \rho v_x$, Eq. (5.84) requires that,

$$\left(\rho v_x^2 - \frac{B_x^2}{\mu_0}\right)\Delta\vec{B}_\perp = 0 \quad\Rightarrow\quad v_x^2 = \frac{B_x^2}{\mu_0\rho} \quad\Rightarrow\quad |v_x| = \frac{|B_x|}{\sqrt{\mu_0\rho}} = a_x, \quad (5.90)$$

and, relative to **J**, the fluid moves at the Alfvén speed, a condition described as *trans-Alfvénic*. Of course, from the fluid's frame of reference, it is **J** that propagates at the Alfvén speed and therefore **J** must represent a type of Alfvén wave.

Finally, substituting Eq. (5.90) into Eq. (5.88), we find,

$$\Delta\vec{v}_\perp = \text{sgn}(v_x B_x)\frac{\Delta\vec{B}_\perp}{\sqrt{\mu_0\rho}} \equiv \text{sgn}(v_x B_x)\Delta\vec{a}_\perp, \quad (5.91)$$

where $\text{sgn}(x) \equiv x/|x|$. Thus, across **J**, both direction *and* magnitude of $\vec{v}_\perp$ jump by an amount, modulo a sign, equal to the change in the transverse Alfvén velocity.

Therefore, for $B_x \neq 0$, $v_x \neq 0$, and $\Delta v_x = 0$, ρ, p, and $B_\perp$ are all constant across **J**, $\vec{B}_\perp$ can rotate, and $\vec{v}_\perp$ can jump both in magnitude and direction. This phenomenon is known as a *rotational discontinuity* (RD), and is unique to MHD; there is no analogue in pure hydrodynamics. Propagating along the fluid at the Alfvén speed, an RD is nothing more than a shear Alfvén wave that rotates $\vec{B}_\perp$ without changing its magnitude.

5.3.5 Case 5: Fast, slow, and intermediate shocks

And now the real fun begins.

We now relax all assumptions about v_x, presuming neither v_x nor Δv_x is zero. Right away, Eq. (5.76) tells us there will be a density jump which, from our experience in §2.2.3, means we expect to form *shock waves*. Further and unlike the previous cases, no terms in Eq. (5.81), (5.84), and (5.86) drop out and, in order to determine the jumps across **J**, we must now solve for the downwind values, those with subscript '2' in Fig. 5.9, in terms of the upwind values with subscript '1'.

To this end, start by spelling out the differences in Eq. (5.88) to find,

$$\frac{1}{B_x}(v_{x_2}\vec{B}_{\perp_2} - v_{x_1}\vec{B}_{\perp_1}) - \frac{B_x}{\rho_1 v_{x_1}\mu_0}(\vec{B}_{\perp_2} - \vec{B}_{\perp_1}) = 0.$$

Multiplying through by B_x/v_{x_1} we get:

$$\frac{v_{x_2}}{v_{x_1}}\vec{B}_{\perp_2} - \vec{B}_{\perp_1} - \frac{a_{x_1}^2}{v_{x_1}^2}(\vec{B}_{\perp_2} - \vec{B}_{\perp_1}) = 0, \qquad (5.92)$$

where $a_{x_1} = |B_x|/\sqrt{\mu_0 \rho_1}$ is the upwind longitudinal Alfvén speed.

Now, let $\phi_x \equiv v_{x_2}/v_{x_1}$. In §2.2.3, we established the *entropy condition* (a.k.a., Zemplén's theorem) in which only *compressive* shocks ($\eta \equiv \rho_2/\rho_1 > 1$) obey the second law of thermodynamics whereby entropy is *increased*. Since ρv_x is a conserved quantity (Eq. 5.76), $\phi_x = 1/\eta$ and we must have $\phi_x < 1$ for a physical shock.[11]

Continuing with Eq. (5.92), we have:

$$\phi_x \vec{B}_{\perp_2} - \vec{B}_{\perp_1} - \frac{1}{A_{x_1}^2}(\vec{B}_{\perp_2} - \vec{B}_{\perp_1}) = 0, \qquad (5.93)$$

where $A_{x_1} = v_{x_1}/a_{x_1}$ (a known quantity) is the so-called *Alfvén number* in the upwind state relative to B_x. The Alfvén number plays a similar role as the Mach number (defined after Eq. 2.38), where the divisor is now the Alfvén speed, a_x (or more generally, a) rather than the sound speed, c_s. Then, solving for the downwind value, $\vec{B}_{\perp_2}$, in Eq. (5.93), we get:

$$\vec{B}_{\perp_2} = \frac{A_{x_1}^2 - 1}{A_{x_1}^2 \phi_x - 1}\vec{B}_{\perp_1}. \qquad (5.94)$$

This equation tells us immediately that $\vec{B}_{\perp_2}$ is either parallel or antiparallel to $\vec{B}_{\perp_1}$, and we write,

$$\vec{B}_{\perp_1} = B_{\perp_1}\hat{e}_\perp \quad \text{and} \quad \vec{B}_{\perp_2} = B_{\perp_2}\hat{e}_\perp,$$

where $\hat{e}_\perp$ is a unit vector giving the direction of both $\vec{B}_{\perp_1}$ and $\vec{B}_{\perp_2}$ if we allow the magnitude, $B_{\perp_2}$, to take on negative values. (Without loss of generality, we may assume $B_{\perp_1} \geq 0$.) Note that $\hat{e}_\perp$ could be in the y-direction, z-direction, or somewhere in between. Regardless, $\hat{e}_\perp$ remains unchanged by the shock and Eq. (5.94) may be reduced to the scalar equation,

$$\boxed{b = \frac{A_{x_1}^2 - 1}{A_{x_1}^2 \phi_x - 1},} \qquad (5.95)$$

where $b \equiv B_{\perp_2}/B_{\perp_1}$. Note that as $A_{x_1}^2 \phi_x \to 1$, $b \to \infty$, and the jump in $B_\perp$ is *infinite* across **J**. The only way this can happen is if $B_{\perp_1} = 0$ and $B_{\perp_2} \neq 0$; this is the so-called "switch-on shock" discussed further starting on page 164. Note also that as $B_x \to 0$ (and $A_{x_1} \to \infty$), $b \to 1/\phi_x = \eta$, and the jump in transverse magnetic induction is the same as the jump in density. This is a common understanding of the term "flux-freezing" though this view is inaccurate. Flux-freezing – introduced

---

[11]Should the reader wish to consult TY14 (see footnote 9), note that they use $\hat{v}$ instead of ϕ_x and state incorrectly twice that $\hat{v} > 1$ for a physical shock. These clearly must be typos, since their subsequent algebra all seems correct.

in §4.3 in the context of Alfvén's theorem – means that in ideal MHD, magnetic flux through an open surface transported by the fluid is conserved; it says nothing about whether the density and contained magnetic induction must track each other. For example, across an MHD shock, fluid can be compressed along the longitudinal magnetic induction without necessarily affecting the transverse component, $B_\perp$. As we'll see, a prime example of this is across a slow shock where the transverse magnetic induction actually *decreases* even as the fluid is compressed.

Next, we turn to Eq. (5.81). With the differences spelled out and using machinations similar to those in deriving Eq. (5.95), we get:

$$\boxed{\zeta = 1 + \gamma M_1^2(1 - \phi_x) - \frac{\gamma \alpha_{\perp_1}}{2}(b^2 - 1).} \tag{5.96}$$

Here, $\zeta \equiv p_2/p_1$ is the ratio of pressures (defined as the *shock strength* in Problem 2.8), $M_1 = v_{x_1}/c_{s_1}$ is the upwind Mach number, $c_{s_1} = \sqrt{\gamma p_1/\rho_1}$ is the upwind adiabatic sound speed, $\alpha_{\perp_1} = a_{\perp_1}^2/c_{s_1}^2 = A_{\perp_1}/M_1^2$ is the upwind MHD-alpha (first defined on page 130) relative to $B_\perp$, $a_{\perp_1} = |B_{\perp_1}|/\sqrt{\mu_0 \rho_1}$ is the upwind transverse Alfvén speed, and $A_{\perp_1} = v_1/a_{\perp_1}$ is the upwind Alfvén number relative to $B_\perp$. All quantities with a subscript '1' are known.

Finally, it is left to Problem 5.18 to show that with some straight-forward algebra, Eq. (5.86) can be developed to yield,

$$\boxed{\frac{(\zeta - 1)(1 + \phi_x)}{\gamma} - (\zeta + 1)(1 - \phi_x) - \frac{(\gamma - 1)\alpha_{\perp_1}}{2}(1 - \phi_x)(b - 1)^2 = 0,} \tag{5.97}$$

with no new quantities defined. Equations (5.95)–(5.97) are partial solutions (no variable has yet been isolated) to the Rankine–Hugoniot jump conditions giving the jumps in v_x, $B_\perp$, and p (variables ϕ_x, b, and ζ respectively) in terms of each other across a shock. Solving these equations for ϕ_x, say (as we'll do starting with Eq. 5.100 on page 162), the jump in density is then given by:

$$\boxed{\eta \equiv \frac{\rho_2}{\rho_1} = \frac{1}{\phi_x},} \tag{5.98}$$

while the jump in $\vec{v}_\perp$ can be found from Eq. (5.79):

$$\boxed{\vec{v}_{\perp_2} - \vec{v}_{\perp_1} = \frac{B_x}{\rho_1 v_{x_1} \mu_0}(b - 1)\vec{B}_{\perp_1}.} \tag{5.99}$$

Note that in general, $\vec{v}_\perp$ is *not* proportional to $\hat{e}_\perp$; the latter indicates the direction of $\vec{B}_\perp$, not $\vec{v}_\perp$.[12] Thus, Eq. (5.99) tells us that only the component of $\vec{v}_\perp$ parallel to $\vec{B}_\perp$ is modified by the shock; the component of $\vec{v}_\perp$ perpendicular to $\vec{B}_\perp$ remains unchanged. Note further, with a suitable Galilean transformation (*e.g.*, one in which $\vec{v}_{\perp_1} \to 0$), $\vec{v}_{\perp_2}$ can be made proportional to $\hat{e}_\perp$; this observation will be relevant when it comes time to design our MHD Riemann solver in Chap. 6.

---

[12]The $\perp$ subscript indicates only that $\vec{v}_\perp$ and $\vec{B}_\perp$ are perpendicular to the x-axis. The actual direction of each vector may be different.

Qualitative description of MHD shocks

Before we take the bull by the horns and actually *solve* Eq. (5.95)–(5.99), there is much we can learn from a qualitative assessment of Eq. (5.95).

First, note that,

$$A_{x_1}^2 \phi_x = \frac{v_{x_1}^2 \mu_0 \rho_1}{B_x^2} \frac{v_{x_2}}{v_{x_1}} = \frac{v_{x_2} \mu_0 \rho_2}{B_x^2} v_{x_2} = A_{x_2}^2,$$

using Eq. (5.76) for the second equality and where A_{x_2} is the downwind Alfvén number relative to B_x. Thus, Eq. (5.95) can be written as:

$$b = \frac{A_{x_1}^2 - 1}{A_{x_1}^2 \phi_x - 1} = \frac{A_{x_1}^2 - 1}{A_{x_2}^2 - 1},$$

and four possible cases present themselves:

1. If both upwind and downwind flows are super-Alfvénic ($A_{x_1}^2 - 1 > 0$, $A_{x_2}^2 - 1 > 0$), $b > 0 \Rightarrow \vec{B}_{\perp_2} \parallel \vec{B}_{\perp_1}$. Further, since $\phi_x < 1$ for a physical shock, the numerator, $A_{x_1}^2 - 1$, is greater (more positive) than the denominator, $A_{x_1}^2 \phi_x - 1$, and $b > 1$. Thus, $B_\perp$ is *enhanced* and $B_{\perp_2} > B_{\perp_1} \geq 0$.

2. If both upwind and downwind flows are sub-Alfvénic ($A_{x_1}^2 - 1 < 0$, $A_{x_2}^2 - 1 < 0$), $b > 0$ (ratio of two negative numbers) and $\vec{B}_{\perp_2}$ is still parallel to $\vec{B}_{\perp_1}$. Further, since $\phi_x < 1$, the numerator is still greater than the denominator, but this time because the former is less negative (and therefore smaller magnitude) than the latter. Thus, in this case, $b < 1 \Rightarrow B_\perp$ is *diminished* and $0 \leq B_{\perp_2} < B_{\perp_1}$.

3. If the upwind flow is super-Alfvénic ($A_{x_1}^2 - 1 > 0$) and the downwind flow is sub-Alfvénic ($A_{x_2}^2 - 1 < 0$), $b < 0$ and $\vec{B}_\perp$ changes sign across **J** (*i.e.*, $\vec{B}_{\perp_2}$ is *anti*parallel to $\vec{B}_{\perp_1}$; $B_{\perp_2} < 0$ if $B_{\perp_1} > 0$). Since the numerator is greater than the denominator,

$$A_{x_1}^2 - \cancel{1} > A_{x_1}^2 \phi_x - \cancel{1} \Rightarrow \cancel{A_{x_1}^2} > \cancel{A_{x_1}^2} \phi_x \Rightarrow 1 > \phi_x,$$

satisfying the entropy condition. Whether this possibility enhances or diminishes $|B_\perp|$ cannot be determined from these qualitative arguments.

4. Finally, if the upwind flow is sub-Alfvénic ($A_{x_1}^2 - 1 < 0$) and the downwind flow is super-Alfvénic ($A_{x_2}^2 - 1 > 0$), the numerator is less than the denominator,

$$A_{x_1}^2 - 1 < A_{x_1}^2 \phi_x - 1 \Rightarrow 1 < \phi_x,$$

violating the entropy condition. This possibility is therefore discarded as unphysical.

Since the flow speeds for possibility 1 are faster than for possibility 2, the former is designated a *fast shock* while the latter a *slow shock*. For a fast shock, both the upwind and downwind speeds are super-Alfvénic and $B_\perp$ is enhanced, while for a slow shock, both the upwind and downwind speeds are sub-Alfvénic and $B_\perp$ is diminished. For both cases, the sign of $\vec{B}_\perp$ is conserved. Thus, possibility 3 seems

discontinuity	v_x	Δv_x	B_x	$\Delta\rho$	$\Delta\vec{v}_\perp$	$\Delta B_\perp$	Δp
contact	$=0$	$=0$	$\neq 0$	arbitrary	$=0$	$=0$	$=0$
tangential	$=0$	$=0$	$=0$	arbitrary	arbitrary	$\neq 0^{(1)}$	$\neq 0^{(1)}$
rotational	$=a_x$	$=0$	$\neq 0$	$=0$	$\Delta \vec{B}_\perp/\sqrt{\mu_0 \rho}$	$=0^{(2)}$	$=0$
fast shock	$>a_x$	<0	arbitrary	>0	$\neq 0$	$>0^{(3)}$	>0
slow shock	$<a_x$	<0	arbitrary	>0	$\neq 0$	$<0^{(3)}$	>0
intermediate	$\gtrless a_x^{(4)}$	<0	arbitrary	>0	$\neq 0$	$\neq 0^{(5)}$	>0

$^{(1)}$so that $\Delta(p + B_\perp^2/2\mu_0) = 0$. $\qquad ^{(2)}$direction can change.
$^{(3)}$direction remains constant. $\qquad ^{(4)} v_{x_1} > a_{x_1}$ but $v_{x_2} < a_{x_2}$.
$^{(5)}$direction flips by $180°$.

Table 5.2. Summary of properties of the six types of discontinuities allowed by the MHD Rankine–Hugoniot jump conditions.

like an intermediate case, where the upwind speed is super-Alfvénic, the downwind speed is sub-Alfvénic, $B_\perp$ may be enhanced or diminished, and the sign of $\vec{B}_\perp$ is switched. Accordingly, possibility 3 is referred to as an *intermediate shock*. Table 5.2 gives a summary of what we have learned qualitatively so far about the six discontinuities allowed by the MHD Rankine–Hugoniot jump conditions.

A designation scheme for MHD shocks based on the upwind and downwind flow speeds introduced by Liberman & Velikovich (1986) is depicted in Fig. 5.10. Here, the fast magnetosonic (a_f), Alfvén (a_x), and slow magnetosonic (a_s) speeds introduced in §5.2 (Eq. 5.25–5.23) are used to divide "velocity space" into four regions: region 1 is super-fast; region 2 sub-fast, super-Alfvénic; region 3 sub-Alfvénic, super-slow; and region 4 sub-slow. In Problem 5.26, you will show that upwind of a fast shock, the flow speed is super-fast (region 1), while downwind the speed is sub-fast/super-Alfvénic (region 2). Accordingly, fast shocks are designated as $1 \to 2$ shocks. Similarly, slow shocks are designated as $3 \to 4$ shocks, while the remaining four possibilities ($2 \to 4$, $1 \to 4$, $2 \to 3$, $1 \to 3$) are all intermediate shocks since each switches the sign of $B_\perp$ (case 3 above). These four flavours of intermediate shocks are designated respectively as types Ia, Ib, IIa, and IIb (Fig. 5.10).

There is nothing controversial about fast and slow shocks. Both are well studied, observed extensively in the laboratory, known to be stable (*e.g.*, survive unchanged when perturbed slightly), and, at least for ideal MHD, perfectly understood mathematically.

Intermediate shocks, on the other hand, are a different matter. Even though they satisfy the entropy condition, many have argued they should be rejected as unphysical on the grounds they violate something called the *evolutionary condition*. Simply put, a physical state is *evolutionary* if it remains intact when perturbed. In classical mechanics, a pencil standing on its sharpened tip is certainly consistent with Newton's laws of mechanics and yet, because even the slightest vibration in

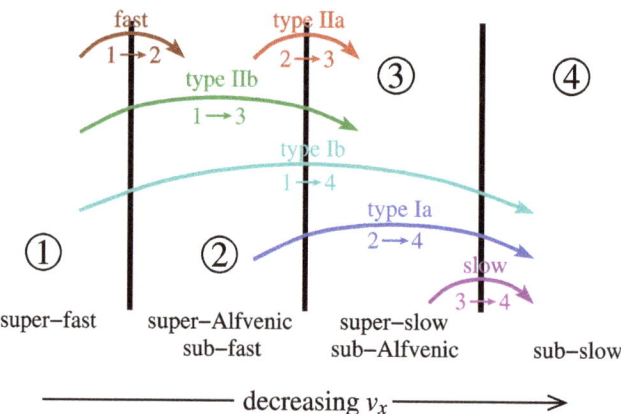

Figure 5.10. Liberman–Velikovich designation scheme for MHD shocks based on the upwind and downwind velocity quartiles spanned by the shock.

the table or gust of air would send it toppling, it is not regarded as a "physically interesting solution". It would never occur spontaneously in nature and is considered *non-evolutionary*. An example in general relativity is the worm hole. While formally a solution to Einstein's field equations, a worm hole collapses to a black hole should a single photon enter its midst. While technically physical, non-evolutionary states do not necessarily represent physically interesting solutions.

By many accounts, intermediate shocks are non-evolutionary. Of course, the motivation to declare them as such is great because with them, the solution to the MHD jump conditions (and therefore the Riemann problem; next chapter) is not always unique. It would require additional physics – which at the time of this writing forms much of the controversy – to determine whether an intermediate shock is to be inserted as part of a solution, or whether an "ordinary" fast or slow shock should be used instead. This makes life more complicated. On the other hand, if intermediate shocks can be rejected as non-evolutionary, the need to search for this additional physics disappears, and life is made simple again.

There are two "gotchas" to this line of thinking. First, certain numerical simulations, such as the famous Brio & Wu (1988) Riemann problem depicted in Fig. 5.11, seem to insist on inserting an intermediate shock plus a rarefaction fan where a slow shock and a rotational discontinuity would do. The analytical solution (lines) and the ZEUS-3D[13] numerical solution (bubbles) agree everywhere except one place where they disagree *qualitatively*. In the portion of the domain $1.76 < x_1 < 1.91$, the "evolutionary" analytical solution inserts an RD and a slow shock, whereas the numerical solution inserts a type Ia intermediate shock attached to a slow rarefaction; a so-called *slow compound wave*. What's more, it's not just ZEUS that favours the intermediate shock. Virtually all numerical codes do the same and, at the time of this writing, it is still an open question why.

---

[13] *Rappel*: For those interested in examining ZEUS-3D, see footnote 6 on page 77.

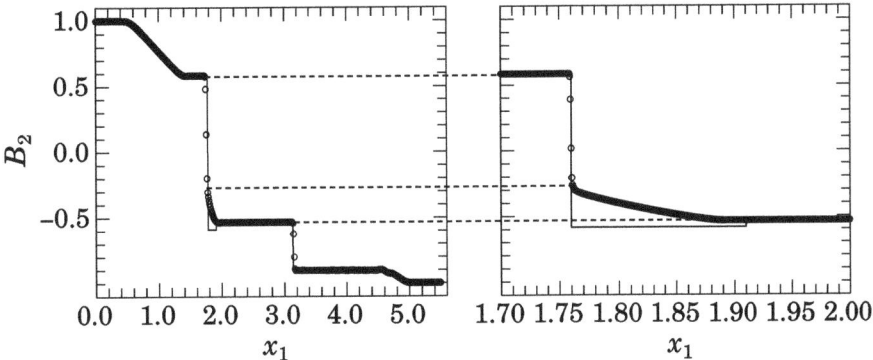

Figure 5.11. Profile of the transverse magnetic induction, $B_2 = B_\perp$, in the "Brio & Wu problem" showing the *ZEUS-3D* solution (open circles; one per zone) and the analytical solution (solid lines) with dashed lines for reference. The left panel shows the entire computational domain ($0 \leq x_1 \leq 5.5$) resolved with 550 zones showing, from left to right, a left-moving fast rarefaction, a left-moving intermediate shock with an attached slow rarefaction (*ZEUS-3D*) or a left-moving RD followed closely by but not attached to a slow shock (analytical), a right-moving slow shock, and finally a right-moving fast rarefaction. The right panel shows the region where the numerical and analytical solutions disagree in a *ZEUS-3D* simulation with 20 times the resolution in the left panel. Regardless of resolution (and thus numerical dissipation), the numerical solution puts a type Ia intermediate shock at $x_1 \sim 1.76$ (which changes the sign and magnitude of B_2 and takes a super-Alfvénic flow to sub-slow) and attaches to it a slow rarefaction, forming a compound wave. Conversely, the analytical solution puts an RD at $x_1 \sim 1.76$ (which changes the sign but not the magnitude of B_2) and a weak slow shock (which changes the magnitude but not the sign of B_2) at $x_1 \sim 1.91$. Both pairs of features are consistent with the rest of the solution and thus, with intermediate shocks allowed, the solution is not unique.

Hada (1994), Markovskii (1998), and Inoue & Inutsuka (2007) all suggest that *dissipation* may be key. All numerical codes are slightly dissipative – some more than others – and these authors suggest that even the slightest amount of dissipation could render some intermediate shocks evolutionary, allowing them to appear spontaneously in a physical solution. What's more – and this is the second "gotcha" – Chao (1995), Feng & Wang (2008), and Feng et al. (2009) claim to have *observational evidence of intermediate shocks in the earth's magnetosphere.*

Therefore, it may end up being that in 100% ideal MHD, intermediate shocks do not appear spontaneously, whereas with even the slightest amount of dissipation, they may become physically relevant. Further, dissipation could be that extra bit of physics to make a solution unique. This is another curio of MHD, by the way, in which some quantity not being exactly zero – dissipation in this case – can have a qualitative effect on the solution.

It's a fascinating subject, and one that will engage people for years. We'll certainly not come to any conclusions about intermediate shocks in this text and after this chapter, we'll continue without examining them any further.

Quantitative solution to R-H equations

And now the algebra. Start by using Eq. (5.95) to eliminate b in Eq. (5.96) and (5.97):

$$\zeta = 1 + \gamma M_1^2(1 - \phi_x) - \frac{\gamma \alpha_{\perp_1}}{2}\left[\left(\frac{A_{x_1}^2 - 1}{A_{x_1}^2 \phi_x - 1}\right)^2 - 1\right], \quad \text{and} \tag{5.100}$$

$$\frac{(\zeta - 1)(1 + \phi_x)}{\gamma} - (\zeta + 1)(1 - \phi_x)$$
$$- \frac{(\gamma - 1)\alpha_{\perp_1}}{2}\left(\frac{A_{x_1}^2 - 1}{A_{x_1}^2 \phi_x - 1} - 1\right)^2 (1 - \phi_x) = 0. \tag{5.101}$$

Next, substitute Eq. (5.100) into Eq. (5.101):

$$M_1^2(1 - \phi_x^2) - \frac{\alpha_{\perp_1}}{2}\left[\left(\frac{A_{x_1}^2 - 1}{A_{x_1}^2 \phi_x - 1}\right)^2 - 1\right](1 + \phi_x)$$
$$- 2(1 - \phi_x) - \gamma M_1^2(1 - \phi_x)^2 + \frac{\gamma \alpha_{\perp_1}}{2}\left[\left(\frac{A_{x_1}^2 - 1}{A_{x_1}^2 \phi_x - 1}\right)^2 - 1\right](1 - \phi_x) \tag{5.102}$$
$$- \frac{(\gamma - 1)\alpha_{\perp_1}}{2}\left(\frac{A_{x_1}^2 - 1}{A_{x_1}^2 \phi_x - 1} - 1\right)^2 (1 - \phi_x) = 0,$$

and we get our first expression in just one unknown, ϕ_x. The first thing to note is:

$$\left(\frac{A_{x_1}^2 - 1}{A_{x_1}^2 \phi_x - 1}\right)^2 - 1 = \frac{A_{x_1}^4 - 2A_{x_1}^2 + \cancel{1} - A_{x_1}^4 \phi_x^2 + 2A_{x_1}^2 \phi_x - \cancel{1}}{(A_{x_1}^2 \phi_x - 1)^2}$$
$$= \frac{A_{x_1}^4(1 - \phi_x^2) - 2A_{x_1}^2(1 - \phi_x)}{(A_{x_1}^2 \phi_x - 1)^2}$$
$$= \frac{M_1^2(1 - \phi_x)}{(M_1^2 \phi_x - \alpha_{x_1})^2}\left(M_1^2(1 + \phi_x) - 2\alpha_{x_1}\right),$$

where $\alpha_{x_1} = M_1^2/A_{x_1}^2 = a_{x_1}^2/c_{s_1}^2$ is the upwind MHD-alpha for B_x. Similarly,

$$\frac{A_{x_1}^2 - 1}{A_{x_1}^2 \phi_x - 1} - 1 = \frac{A_{x_1}^2 - \cancel{1} - A_{x_1}^2 \phi_x + \cancel{1}}{A_{x_1}^2 \phi_x - 1} = \frac{M_1^2(1 - \phi_x)}{M_1^2 \phi_x - \alpha_{x_1}}.$$

Substituting these into Eq. (5.102) and multiplying through by: $2\frac{(M_1^2 \phi_x - \alpha_{x_1})^2}{M_1^2(1 - \phi_x)}$, (allowed since $1 - \phi_x > 0$), we get:

$$2(M_1^2 \phi_x - \alpha_{x_1})^2\left(1 + \phi_x - \frac{2}{M_1^2} - \gamma(1 - \phi_x)\right) - (\gamma - 1)M_1^2 \alpha_{\perp_1}(1 - \phi_x)^2$$
$$- \alpha_{\perp_1}\left(M_1^2(1 + \phi_x) - 2\alpha_{x_1}\right)(1 + \phi_x) \tag{5.103}$$
$$+ \gamma \alpha_{\perp_1}\left(M_1^2(1 + \phi_x) - 2\alpha_{x_1}\right)(1 - \phi_x) = 0.$$

This is evidently a cubic in ϕ_x which, for convenience, I rewrite gathering all like powers in ϕ_x together with the coefficients simplified as much as practical:

$$
\begin{aligned}
\phi_x^3 - \phi_x^2 &\left(\mathcal{H} + 2\frac{\alpha_{x_1}}{M_1^2} + \frac{\gamma}{\gamma+1}\frac{\alpha_{\perp_1}}{M_1^2} \right) \\
&+ \frac{\phi_x}{M_1^2} \left(2\mathcal{H}\alpha_{x_1} + \frac{\alpha_{x_1}\alpha_1}{M_1^2} - \frac{2-\gamma}{\gamma+1}\alpha_{\perp_1} \right) \\
&- \frac{\alpha_{x_1}}{M_1^4} \left(\mathcal{H}\alpha_{x_1} + \frac{\gamma-1}{\gamma+1}\alpha_{\perp_1} \right) = 0,
\end{aligned}
\quad (5.104)
$$

where $\alpha_1 = \alpha_{x_1} + \alpha_{\perp_1}$ is the upwind MHD-alpha (Eq. 5.30), and where:

$$\mathcal{H} \equiv \frac{1}{\gamma+1}\left(\frac{2}{M_1^2} + \gamma - 1\right).$$

This is the "hydrodynamical term", which is all that remains in the limit as $\vec{B} \to 0$, and thus $\alpha_{x_1} \to \alpha_{\perp_1} \to 0$. That is, in this limit Eq. (5.104) reduces to:

$$\phi_x = \mathcal{H}, \quad (5.105)$$

which is Eq. (2.39). The pure HD case seems so simple, now, doesn't it!

Equation (5.104) gives ϕ_x, the jump in longitudinal speed relative to the shock and the inverse of η, the density jump, completely in terms of the thermodynamics, γ, the upwind Mach number, M_1, and the two upwind MHD-alphas, α_{x_1} and $\alpha_{\perp_1}$. There is something very satisfying about this result since it expresses ϕ_x so cleanly and explicitly in terms of the union of all the physics that could possibly affect it.

Being a cubic, Eq. (5.104) admits as many as three independent solutions. Using a suitable root-finder (*e.g.*, App. D) or the Cardano–Tartaglia formula for a cubic (App. E), the physical root is the one that is real, and satisfies the entropy and possibly evolutionary conditions. Imposition of the latter (if we choose to exclude intermediate shocks) is most conveniently done by requiring $b \geq 0$ (Eq. 5.95). With ϕ_x selected, Eq. (5.95) and (5.98) determine b and η respectively. Once b is known, Eq. (5.96) is used to find ζ while Eq. (5.99) is used to find the jump in $\vec{v}_\perp$. This, then, determines the jumps in all variables, and the shock is specified.

Limits of $\vec{B}$ and the "switch-on shock"

An interesting limit of Eq. (5.104) is $B_x \to 0$ ($\alpha_{x_1} \to 0$), where the magnetic induction is completely transverse to the propagation direction. In this case, we arrive at the quadratic:

$$\phi_x^2 - \left(\mathcal{H} + \frac{\gamma}{\gamma+1}\frac{\alpha_{\perp_1}}{M_1^2}\right)\phi_x - \frac{2-\gamma}{\gamma+1}\frac{\alpha_{\perp_1}}{M_1^2} = 0, \quad (5.106)$$

where only the + root gives $\phi_x > 0$. Thus, Eq. (5.106) admits a unique solution for ϕ_x, and this will correspond to a fast shock. Recall in §5.2.2 it was determined that in the direction perpendicular to the prevailing magnetic induction, only the fast wave propagates since in that limit, both the slow and Allfvén speeds go to zero.

It follows, then, that in the absence of longitudinal magnetic induction, only a fast shock can be triggered.

An even more interesting limit – because of its hidden surprise – is $B_{\perp_1} \to 0$ ($\alpha_{\perp_1} \to 0$), where the magnetic induction is now completely longitudinal. At first blush, one might think that since in 1-D, B_x is constant $\forall x, t$, B_x would act just like an inert wire running through the fluid. Thus, in this limit, the shock should simply reduce to the pure hydrodynamical case.

So let's investigate this. In the limit where $\alpha_{\perp_1} \to 0$ and setting $M_1^2 = A_{x_1}^2 \alpha_{x_1}$, Eq. (5.104) reduces to:

$$\phi_x^3 - \phi_x^2 \left(\mathcal{H} + \frac{2}{A_{x_1}^2} \right) + \phi_x \left(\frac{2\mathcal{H}}{A_{x_1}^2} + \frac{1}{A_{x_1}^4} \right) - \frac{\mathcal{H}}{A_{x_1}^4}$$
$$= (\phi_x - \mathcal{H}) \left(\phi_x - \frac{1}{A_{x_1}^2} \right)^2 = 0, \quad (5.107)$$

and there are *two* roots (one doubly degenerate) to this cubic. The first root, $\phi_x = \mathcal{H}$, is the HD limit (commonly referred to as an *Euler shock*) which we intuited above. The second root,

$$\phi_x = \frac{1}{A_{x_1}^2}, \quad (5.108)$$

is a purely MHD phenomenon known as a *switch-on shock*, so-called because of its rather surprising property of creating $B_\perp$ seemingly out of nothing! $B_{\perp_1} = 0$ and B_x is constant across the shock, yet $B_{\perp_2} \neq 0$. It is left to Problem 5.21 to show that for a switch-on shock, the strength of the post-shock transverse magnetic induction is given by:

$$\frac{B_{\perp_2}}{B_x} \equiv b_\perp = \sqrt{(A_{x_1}^2 - 1)\left(\gamma + 1 - \frac{2}{\alpha_{x_1}} - (\gamma-1)A_{x_1}^2\right)}, \quad (5.109)$$

and that in terms of $b_\perp$,

$$\zeta = 1 + \gamma M_1^2(1 - \phi_x) - \frac{\gamma \alpha_{x_1}}{2} b_\perp^2, \quad \text{and} \quad (5.110)$$

$$\vec{v}_{\perp_2} - \vec{v}_{\perp_1} = \frac{B_x}{\mu_0 \mathcal{M}} \vec{B}_{\perp_2} = \frac{B_x^2 b_\perp}{\mu_0 \mathcal{M}} \hat{e}_\perp. \quad (5.111)$$

Here, $\hat{e}_\perp$ is the unit vector giving the direction of $\vec{B}_{\perp_2}$ which, as unsatisfying as it may be, is completely unconstrained by the Rankine–Hugoniot jump conditions (Eq. 5.76–5.80). Once again, the extreme limits in 1-D MHD lead to puzzling conclusions.

We can always bunt, and suggest that the transverse direction is restricted to the y-direction, in which case $\hat{e}_\perp = \hat{y}$. But this is a bunt, for there is no escaping the fact there are two independent orthogonal directions, $\hat{y}$ and $\hat{z}$, and where $\hat{e}_\perp$ points in that 2-D subspace is a legitimate question.

As a member of the fast family of waves, a switch-on shock becomes a fast shock should the upwind perpendicular field differ from zero even infinitesimally. And, since a fast shock is incapable of rotating the perpendicular component of the magnetic induction, $\hat{e}_\perp$ should be in the direction of $\vec{B}_{\perp_1}$, however small it may be. But if we insist on taking $B_{\perp_1}$ to *exactly* zero – even if just in our imaginations – then to set $\hat{e}_\perp$, all we can resort to is to suggest its direction be determined by the asymptotic downwind field onto which it must match with rotational discontinuities being the only possibility of altering its direction along the way.

Failing even that, well then, your guess on $\hat{e}_\perp$ is as good as mine.

It is also left to Problem 5.21 to show that switch-on shocks occur only for:

$$\max(1, \alpha_{x_1}) < M_1^2 < \frac{1}{\gamma - 1}(\alpha_{x_1}(\gamma + 1) - 2) \equiv M_{\text{on}}^2, \qquad (5.112)$$

where M_{on} is the maximum Mach number for a "switch-on shock". The right inequality follows from the fact that since $\phi_x < 1$, Eq. (5.108) $\Rightarrow A_{x_1}^2 > 1$ and thus the second factor under the radical sign in Eq. (5.109) must be positive so that $b_\perp \in \mathbb{R}$. Then, since $A_{x_1}^2 > 1$, the "switch-on" shock must be a member of the fast family of shocks and the left inequality follows by demanding that $v_{x_1} > a_{f_1}$, the upwind fast speed which, for a longitudinal field, is given by Eq. (5.66).

It is left to Problem 5.22 to show that no value of M_1^2 can satisfy Ineq. (5.112) unless $\alpha_{x_1} > 1$. Thus, the LHS of Ineq. (5.112) can be replaced with $1 < \alpha_{x_1}$. So, while an Euler shock is always mathematically possible in the limit as $B_{\perp_1} \to 0$, "switch-on" shocks are allowed only for $\alpha_{x_1} > 1$ (thus magnetically dominant), and then only when $M_1 < M_{\text{on}}$. Within this range, both the Euler and "switch-on" shocks are permitted by the physics, and we are faced with a non-unique solution. As we shall see, uniqueness can be restored by applying the "evolutionary condition" to discard one of the possibilities as "physically uninteresting".

Shock loci, "switch-off shocks", and "Euler branches"

We now turn to the general solutions for an adiabatic MHD gas ($\gamma = 5/3$) depicted in Fig. 5.12 ($\alpha_{x_1} = 0.5$) and 5.13 ($\alpha_{x_1} = 5.0$). In each case, the variables ζ, ϕ_x, and $b_\perp$ are plotted against M_1 for the values of $\alpha_{\perp_1}$ indicated in the legends.

Starting with Fig. 5.12 (where $\alpha_{x_1} < 1 \Rightarrow B_x$ is relatively weak), the left panels show loci of points corresponding to fast shocks (solid lines) For these, loci of $b_\perp$ start where $B_{\perp_2} = B_{\perp_1}$, and then rise for increasing M_1, consistent with $B_\perp$ rising across fast shocks. As Problem 5.26 shows, each fast locus starts on a *fast point* (where the *fast magnetosonic number*, $M_{f_1} = v_{x_1}/a_{f_1} = 1$; different for each $\alpha_{\perp_1}$), and continues indefinitely as $M_1 \to \infty$. The combined slow and intermediate shock loci (dashed lines) are huddled on the left side of the left panels, and shown in better detail and for additional values of $\alpha_{\perp_1}$ in the right panels of the figure. As is also shown in Problem 5.26, these begin on a *slow point* (where the *slow magnetosonic number*, $M_{s_1} = v_{x_1}/a_{s_1} = 1$) and end on an *Alfvén point* (where the Alfvén number, $A_{x_1} = v_{x_1}/a_{x_1} = 1$). In the lower right panel, one can see how the slow (solid lines) and intermediate (dashed lines) shock loci can be distinguished. As noted previously, slow shocks *decrease* $B_\perp$ while preserving its sign, while intermediate shocks change the sign of $B_\perp$.

Given what we know about hydrodynamical shocks (*e.g.*, review Fig. 2.3 in

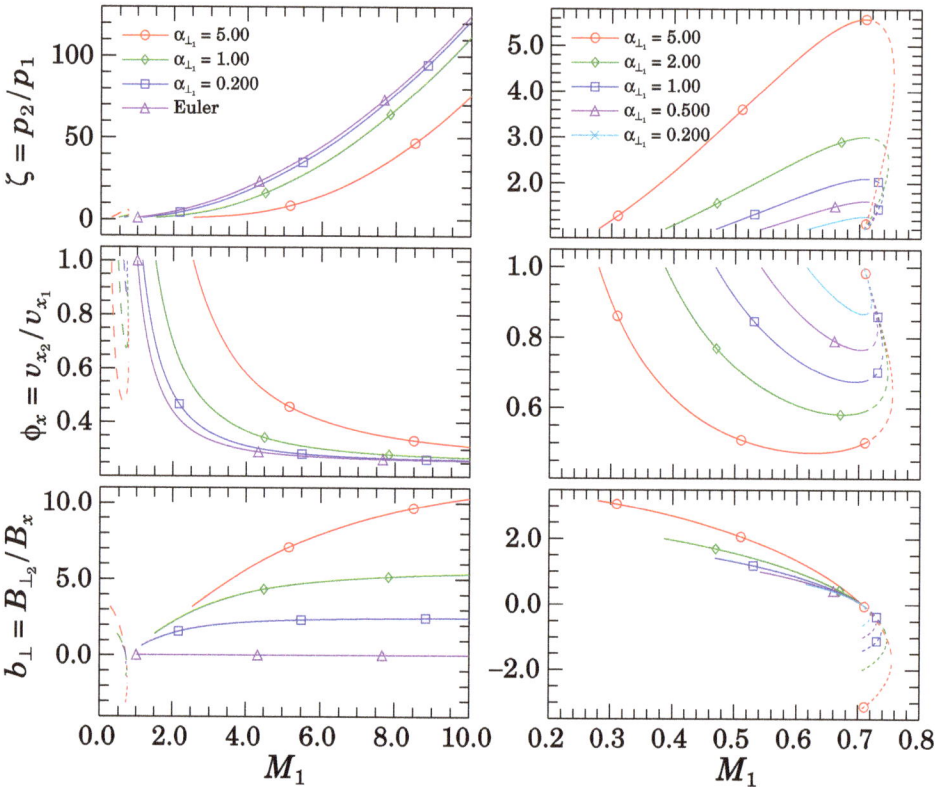

Figure 5.12. Solutions to the MHD Rankine–Hugoniot shock jump conditions for an adiabatic ($\gamma = 5/3$) gas with $\alpha_{x_1} = 0.5$ (weak B_x). Panels from top to bottom show ζ, ϕ_x, and $b_\perp$ (all defined in the text) as a function of upwind sonic Mach number, M_1. The left panels show loci of fast shocks (solid lines) and slow/intermediate shocks (dashed lines) for $\alpha_{\perp_1} = 5, 1, 0.2$, and 0 (Euler). The right panels are essentially a blow-up of the portion of the left panels in $0.2 \le M_1 \le 0.8$ showing the "super-loci" (as defined in the text) of slow shocks (solid lines), intermediate type I shocks (longer dashes), and intermediate type II shocks (shorter dashes) at values of $\alpha_{\perp_1}$ ranging from 5 (strong $B_{\perp_1}$) to 0.2 (weak $B_{\perp_1}$). Note that where (super-)loci begin and end is determined by the "entropy condition" which demands that all physical shocks increase the pressure ($\zeta > 1$) and decrease the speed ($\phi_v < 1$).

§2.2), there is nothing particularly surprising about the fast shock loci in the left panels of Fig. 5.12. These loci are nestled in such a fashion that as $\alpha_{\perp_1} \to \infty$, each locus begins at a higher value of M_1 (faster fast point) and, as $\alpha_{\perp_1} \to 0$, the loci tend towards the "Euler" solution (purple △) which is identically the $\gamma = 5/3$ HD shock in Fig. 2.3. Note that the Euler shock locus starts at $M_1 = 1$, as expected for an HD shock. Since $\alpha_{x_1} < 1$, no switch-on shock appears in the figure.

In the right panels of Fig. 5.12, we see something new. The loci of slow shocks (solid lines) begin at the slow point and continue until the Alfvén point where

$b_\perp = 0$. From this point, the slow shock loci join continuously with the type I intermediate shock loci (longer dashes), and these continue until the *intermediate point*, where M_1 reaches the maximum value it attains along any given shock locus. Here, the type I loci join smoothly with the type II loci (shorter dashes) which curve back towards lower M_1, ending finally at the Alfvén point which, as shown in Problem 5.26, corresponds to a rotational discontinuity where $B_{\perp_2} = -B_{\perp_1}$. Taken collectively, the types I and II loci form double-valued functions of M_1, starting and ending on an Alfvén point and reaching an extremity in M_1 at the intermediate point where the transition between type I shocks ($2, 1 \to 4$ where the post-shock speed is sub-slow like a slow shock) and type II shocks ($1, 2 \to 3$ where the post-shock speed is super-slow but sub-Alfvénic) occurs. Taken individually, the types I and II loci are single-valued functions of M_1.[14]

Evidently, the slow and both types of intermediate shock loci form a single, continuous, smooth "super-locus" of points, and it is for this reason that intermediate shocks are generally considered to be part of the slow family of shocks. Indeed, mathematically there are really just two loci of points per pair of values $(\alpha_{x_1}, \alpha_{\perp_1})$, namely the fast locus and the super-locus just identified.[15] It is only for physical reasons that we bother subdividing the super-locus into three sub-loci (five, when types Ib and IIb show themselves as they do for $\alpha_{x_1} > 1$).

In the top two right panels of Fig. 5.12, note that as $\alpha_{\perp_1} \to 0$, the super-loci shrink upon themselves and collapse to the Alfvén point at $B_{\perp_1} = 0$. Thus, as is true when $B_x \to 0$, there is no slow family of shocks for $B_{\perp_1} = 0$ when $\alpha_{x_1} < 1$. Glancing at the bottom right panel of Fig. 5.12, all super-loci of $b_\perp$ intersect at $b_\perp = 0$ and the Alfvén point, where the slow and intermediate type I loci join. This critical point is known as a *switch-off shock*, since post-shock the transverse magnetic induction, $B_{\perp_2}$ ($b_\perp$), is reduced to zero. This is the opposite effect of a switch-on shock, which manages to produce a non-zero $B_{\perp_2}$ from none.

Turning now to Fig. 5.13 (where $\alpha_{x_1} > 1 \Rightarrow B_x$ is relatively strong), things start to get interesting. Again, the left panels show the fast shock loci (with the slow/intermediate loci suppressed so as not to clutter the plots), and the right panels show the slow/intermediate shock super-loci.

The super-loci in the right panels bear some resemblance to those in Fig. 5.12 with two important exceptions. First, unlike $\alpha_{x_1} < 1$ where the super-loci collapses to the Alfvén point as $B_{\perp_1} \to 0$ (switch-off shock), when $\alpha_{x_1} > 1$ the slow portion of the super-loci converge to a locus of points representing slow shocks across which $B_\perp$ remains zero. This is known as the *slow Euler branch*. Second, notice the domains of the abscissæ in the left and right panels. While the slow portions of the super-loci never overlap (in M_1) with the fast loci, the intermediate loci do. When $\alpha_{x_1} > 1$, portions of the intermediate loci are super-fast, and thus overlap the fast loci providing *three* possible shock solutions for each value of M_1. Note that these super-

---

[14] Technically, the intermediate shocks in the right panels of Fig. 5.12 are type Ia and type IIa; types Ib and IIb don't show up here. See footnote 15 for further discussion.

[15] Note that even the fast locus and slow super-locus join smoothly, but appear separate because portions of these "hyper-loci" have been omitted on grounds they violate the entropy condition (Fig. 5.14).

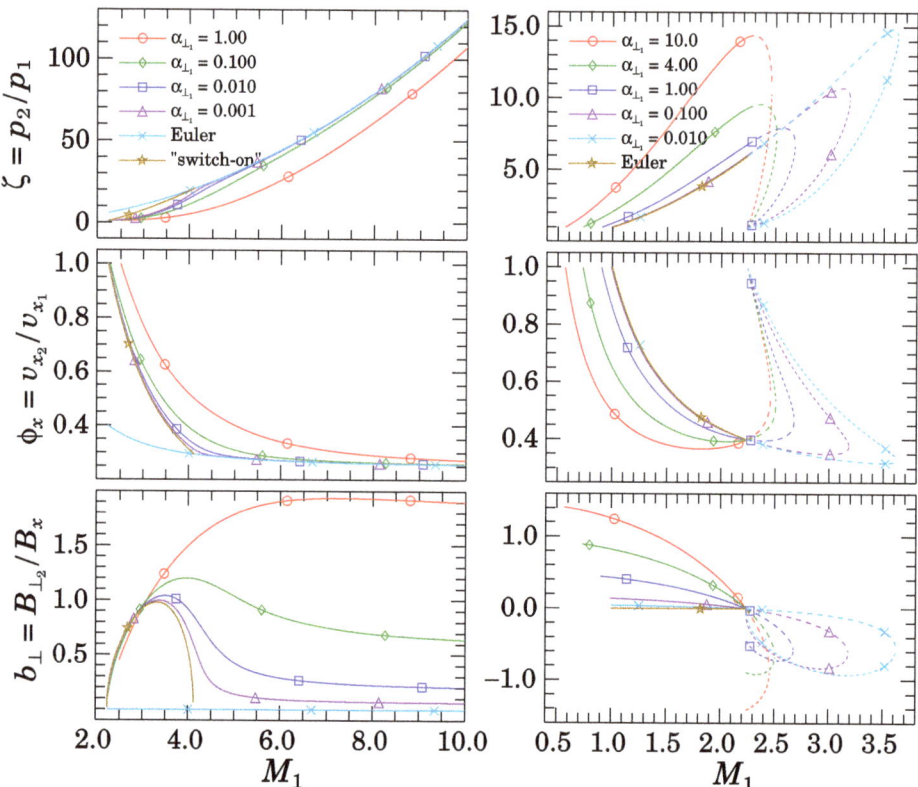

Figure 5.13. Same as Fig. 5.12, except $\alpha_{x_1} = 5$ (stronger B_x). Left panels show loci of fast shocks for $\alpha_{\perp_1}$ ranging from 1 ($c_{s_1} = a_{\perp_1}$, "equipartition") to 0 ($B_{\perp_1} = 0$, Euler) in which the "switch-on" shock (brown ☆) emerges. The right panels show the slow (solid lines) and intermediate (dashed lines) shocks for $\alpha_{\perp_1}$ ranging from 10 (strong $B_{\perp_1}$) to 0. Here, the "Euler branch" (brown ☆) is limited to its "slow half", in which the upwind speed is sub-Alfvénic.

fast portions of the super-loci correspond to type Ib ($1 \to 4$; upwind super-fast, downwind sub-slow) and type IIb ($1 \to 3$; upwind super-fast, downwind super-slow but sub Alfvénic) intermediate shocks. This behaviour is not seen in Fig. 5.12, and it is useful to understand its algebraic origin before delving further into the physics.

To this end, examine ϕ_x for the case of $\alpha_{x_1} = 5$ and $\alpha_{\perp_1} = 0.1$, shown as the green ◇ line in the left panel of Fig. 5.13 and the violet △ line in the right panel where only the portions of the roots declared as *physical* are shown. Conversely, Fig. 5.14 shows both the real (solid lines) and imaginary (dashed lines) parts of *all* three roots (candidate values of ϕ_x) as functions of M_1 without any of the physical constraints applied.

It may take a little staring at the two figures before recognising the portions of the totality of roots shown in Fig. 5.14 which survive the physical constraints to warrant inclusion in the middle panels of Fig. 5.13. For $\alpha_{x_1} = 5$, the Alfvén point (where $A_{x_1} = 1$) is located at $M_1 = A_{x_1}\sqrt{\alpha_{x_1}} = \sqrt{5} \sim 2.24$, precisely where the

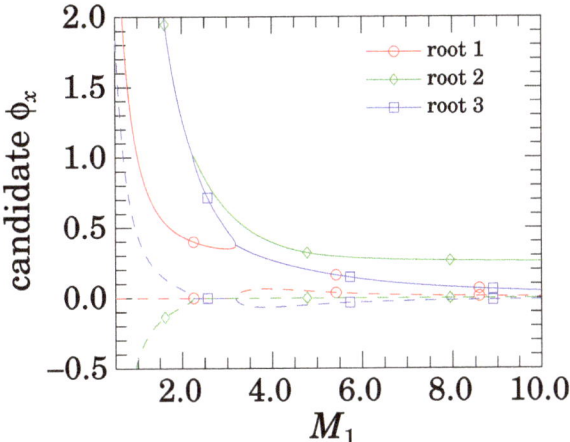

Figure 5.14. Shown are all three roots (root numbering 1, 2, 3 follows convention in App. E) of Eq. (5.104) as a function of M_1 for the case $\alpha_{x_1} = 5$, $\alpha_{\perp_1} = 0.1$ (middle panels of Fig. 5.13; green $\Diamond$ line left, violet $\triangle$ line right), with solid/dashed lines representing the real/imaginary parts. Because of the nature of a cubic, roots are either real or complex conjugates. For the latter, this means any complex roots come in pairs with the imaginary part of one being the negative of the other, and the two real parts degenerate. Thus, for $M_1 \lesssim 2.24$ ($\sqrt{\alpha_{x_1}}$, the Alfvén point), the imaginary parts of roots 2 and 3 are each other's reflection through the $\phi_x = 0$ axis, and only the real part of root 3 can be seen (blue $\square$), since the plotter plotted root 3 after and thus on top of root 2 (green $\Diamond$). Similarly for $M_1 \gtrsim 3.18$ (intermediate point) where the complex conjugate roots reappear for roots 1 and 3.

imaginary parts of roots 2 (green $\Diamond$) and 3 (blue $\square$) converge to the $\phi_x = 0$ axis. Thus, for $M_1 < \sqrt{5}$, roots 2 and 3 are rejected on the grounds they are not real, leaving only root 1 (red $\bigcirc$) corresponding to the slow shock in Fig. 5.13. Note further that the slow point, where $M_{s_1} = v_{x_1}/a_{s_1} = 1$, corresponds to $M_1 \sim 0.98$ (see Eq. 5.125 in Problem 5.23), and that for $M_1 < 0.98$, the real part of root 1 is greater than 1. Thus, based on the entropy condition (second law of thermodynamics requiring $\phi_x < 1$), we reject even root 1 for values of M_1 below the slow point where Fig. 5.13 shows no *physical* shocks of any sort.

For the relatively narrow domain of $2.24 \lesssim M_1 \lesssim 3.18$, all three roots are real. For values of M_1 beyond the Alfvén point, root 1 becomes a type I intermediate shock since, for slow shocks, the pre-shock speed must be sub-Alfvénic. Where the imaginary parts of roots 2 and 3 converge (Alfvén point), the real parts (green $\Diamond$ and blue $\square$ solid lines) begin to diverge. As it happens, their value where they diverge is just over 1 (~ 1.02) and thus both are rejected by the entropy condition. However, both roots fall rapidly with M_1 – root 3 faster than root 2 – and where root 3 first falls below 1.0, it corresponds to the type II intermediate shock in Fig. 5.13. At a slightly greater value of M_1 (the fast point, as a matter of fact, where $M_1 \sim 2.26$; see Eq. 5.125), root 2 falls below 1.0 and emerges as the fast shock locus in Fig. 5.13 which remains real and physical for all greater values of M_1.

Where the real parts of root 1 (representing a type I intermediate shock) and root 3 (a type II intermediate shock) join, $M_1 \to \sim 3.18$. This is evidently the intermediate point (where type I transitions to type II) and thus the downwind slow point. Thereafter, the imaginary parts of roots 1 and 3 emerge, rendering these roots no longer physical. Note that with increasing M_1, the Alfvén point is where the imaginary parts disappear, while the intermediate point is where they re-emerge. Their real parts remain degenerate for all values of M_1 beyond the intermediate point which represents the furthest extent in M_1 for the slow super-loci. Note further that while the Alfvén point is always left of the fast point, the intermediate point can be either to the left or the right of the fast point. In this case, it is to the right which means that for the range of M_1 between the fast and intermediate points, there are *three* physical solutions, as noted above. Incidentally, where the intermediate solutions occur at the same values of M_1 as the fast solution, the preshock speeds are all super-fast, which makes each of the intermediate shocks of subtype b (Fig. 5.10). Where intermediate shocks appear before the fast point, the pre-shock speed is super-Alfvénic and sub-fast, making them subtype a.

To recap, for all values of M_1, the algebra always admits three candidate shock solutions. Physical shocks are selected as those corresponding to real roots of ϕ_x whose values are between 0 and 1 (entropy condition). This will lead to no solution for points slower than the slow point and a unique (slow) solution between the slow and Alfvén points. Then, starting from the Alfvén point, there is at least one (type I), two (types I and II) or even three (types I, II, and fast) solutions available for each value of M_1 until the intermediate point. If the intermediate point is before the fast point, there are two solutions (types I and II) immediately before the intermediate point and none between it and the fast point. If the intermediate point is beyond the fast point, there are three solutions between the two points, then a unique solution (fast) beyond the intermediate point. Lastly, if one were to apply the evolutionary condition thereby eliminating intermediate shocks, one would always have either a unique solution (slow or fast) or, in the regions before the slow point and between the Alfvén and fast points, no shock solution at all.

Returning our attention now to Fig. 5.13, the prominent feature in the left panels is the switch-on shock locus (brown☆), particularly apparent in the lower left panel of $b_\perp$. It has appeared because $\alpha_{x_1} > 1$ and exists within the finite range of M_1 determined by Ineq. (5.112). As mentioned above, it – together with the Euler shock locus (cyan×) – is simply a fast shock in the limit as $B_{\perp_1} \to 0$. However, a little further examination reveals things may not be quite so simple. Notice that over the range of M_1 in which the switch-on shock exists, there also exists part of the Euler branch and, if both solutions are valid at the same time, the shock locus for $B_{\perp_1} = 0$ is double-valued over this part of the domain of M_1. Yet, all fast shock loci, including in the limit as $B_{\perp_1} \to 0$, must be single-valued. So what gives?

Figure 5.15 shows how the complete set of shock loci (in $b_\perp$) in the left panel tend towards the Euler + switch-on shock configuration in the right panel. The left panel shows for a progression of ever-decreasing $\alpha_{\perp_1}$ ($B_{\perp_1} \to 0$) the slow shock loci (long dashes) merging into type I intermediate shock loci (medium dashes) merging

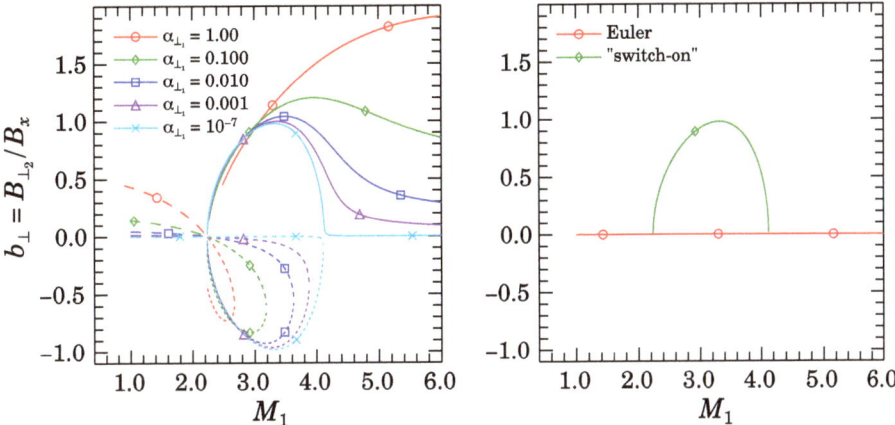

Figure 5.15. Left panel shows the progression of all MHD shock loci for $b_\perp$ from $\alpha_{\perp_1} = 1$ ($c_{s_1} = a_{\perp_1}$, red○), to $\alpha_{\perp_1} = 10^{-7}$ (cyan×) for a strong longitudinal magnetic induction ($\alpha_{x_1} = 5$, same as Fig. 5.13). Long dashes represent slow shock loci, medium dashes type I loci, short dashes type II loci, and solid lines fast loci. On comparing this to the right panel showing the Euler (red○) and switch-on (green ◇) shock loci only ($B_{\perp_1} = 0$), it becomes clear that the switch-on branch corresponds entirely to a portion of a fast locus, while the Euler branch corresponds to three separate shocks, namely fast (right of the switch-on arc), type I intermediate (within the switch-on arc) and slow (left of the switch-on arc).

into type II intermediate shock loci (short dashes). There is then a real gap between where the type II loci end and the fast shock loci (solid lines) begin, with the gap narrowing for decreasing $\alpha_{\perp_1}$ ($B_{\perp_1} \to 0$). As we saw above, type II intermediate and fast loci do join, but with points that violate the entropy condition (*i.e.*, where $\phi_{x_1} > 1$).

In addition, as $\alpha_{\perp_1}$ decreases from $1 \to 10^{-7}$ (red○ → cyan× in the left panels of Fig. 5.15), the type I intermediate shock loci (medium dashes) becomes more horizontal and extend further under the fast shock loci until at $\alpha_{\perp_1} = 10^{-7}$ (cyan×), the type I locus as one moves to the right just about touches the fast locus as one moves to the left. Just at the point where they might touch, the type I locus darts downward yielding to the type II locus (short dashes), while the fast locus darts upward and continues in a single-valued fashion to form what, in isolation, might be identified as the switch-on shock locus of points.

In comparing this with the right panel of Fig. 5.15 which shows the Euler (red○) and switch-on (green ◇) branches only, it should be clear, now, what corresponds to what. The Euler locus actually corresponds to *three* shocks in the limit as $B_{\perp_1} \to 0$. To the right of the arched switch-on shock locus of points, the Euler branch is a fast shock; within the arc, a type I intermediate shock; and to the left of the arc, a slow shock. Finally, the switch-on shock itself is evidently a continuation of the fast shock portion of the Euler branch.

In this picture, the portion of the Euler branch within the switch-on shock arc – the portion that renders the combined Euler + switch-on solution double-valued – is as physical or unphysical as a type I intermediate shock. If we are dismissing intermediate shocks on the grounds of the evolutionary condition, we must also dismiss this portion of the Euler branch and, as such, the limit of the fast shock as $B_{\perp_1} \to 0$ is single valued, as expected.

Indeed, the division of the $\alpha_{\perp_1} = 0$ case into Euler and switch-on branches is perhaps a confusing and even inaccurate distinction, but one we are probably stuck with given the historical development of the subject. As we've seen, the fast portion of the Euler branch combined with the switch-on branch is really a single locus of points corresponding to a fast shock in the limit as $\alpha_{\perp_1} \to 0$ (where $M_1 > M_{\text{on}}$ for the Euler part and $\sqrt{\alpha_{x_1}} < M_1 < M_{\text{on}}$ for the switch-on part; Ineq. 5.112), with the remaining portion of the Euler branch ($1 < M_1 < \sqrt{\alpha_{x_1}}$) belonging to the slow family of waves and having nothing do to with fast shocks at all.

And with that, we terminate our discussion of intermediate shocks.

Problem Set 5

5.1

a) Derive Eq. Set 8 (page 124 in the text) from the primitive equations of MHD (*e.g.*, such as those you would have written down for Problem 4.9).

b) Show how Eq. Set 8 may be written in the forms of Eq. (5.8) by determining the primitive Jacobian.

5.2**

a) Derive Eq. Set 9 (page 125 in the text) from the conservative equations of MHD (either Eq. Set 6 on page 109 or Eq. Set 7 on page 111).

b) Show how Eq. Set 9 may be written in the forms of Eq. (5.17) by determining the conservative Jacobian.

5.3** Find the eigenvalues of the primitive Jacobian matrix, J_p, given by Eq. (5.9) in the text. If you think your algebra is pretty good, this problem will test that thesis!

Hint: Finding the eigenvalues of J_p requires finding the determinant of a 7×7 matrix (with lots of zeros, true, but still...), and solving a seventh-order polynomial. Without a sensible strategy, this could be a rather daunting task.

On the latter, do *not* attempt to develop the determinant into a polynomial of the

form:
$$au^7 + bu^6 + \cdots + gu + h = 0,$$
as this will tell you little about the roots. Rather, try to get the expression into the form:
$$(u - u_1)(u - u_2) \cdots (u - u_7) = 0,$$
from which the roots can be picked off. You can give yourself an even better algebraic target if you just think a little about what form the roots will actually have. For the three HD equations, the roots (wave speeds) were $(u - v_x)$, $(u - v_x - c_s)$, and $(u - v_x + c_s)$, where the last two, when multiplied together, have the form $[(u - v_x)^2 - c_s^2]$, a difference of squares (page 85 in the text). Guided by this experience, we might expect the 7×7 determinant to boil down to an expression of the form:
$$(u - v_x)[(u - v_x)^2 - a_1^2][(u - v_x)^2 - a_2^2][(u - v_x)^2 - a_3^2] = 0,$$
where a_1, a_2, and a_3 are three wave speeds whose form should become self-evident as the algebra unfolds.

5.4* From Eq. (5.23)–(5.25) in the text, and defining the *perpendicular Alfvén speed* as,
$$a_\perp^2 = \frac{B_y^2 + B_z^2}{\mu_0 \rho} = a^2 - a_x^2,$$
prove the following identities:

a) $\qquad\qquad\qquad a_f a_s = c_s a_x;$ (5.113)

b) $\qquad\qquad\qquad a_f^2 + a_s^2 = c_s^2 + a_x^2 + a_\perp^2;$ (5.114)

c) $\qquad\qquad (a_f^2 - c_s^2)(a_f^2 - a_x^2) = a_f^2 a_\perp^2;$ (5.115)

d) $\qquad\qquad (c_s^2 - a_s^2)(a_x^2 - a_s^2) = a_s^2 a_\perp^2.$ (5.116)

5.5* Verify inequalities (5.28) in the text, namely;
$$a_s \leq c_s \leq a_f;$$
$$a_s \leq a \leq a_f;$$
$$a_s \leq a_x \leq a_f.$$

5.6* Starting with the primitive 1-D MHD Eq. (5.1)–(5.7) given in the text, derive the linearised version of these equations, namely Eq. (5.49)–(5.52) assuming $B_z = 0$. Why does this process allow us to ignore the linearised continuity equation and the equation for v_z?

5.7* Evaluate the eigenvalues and eigenvectors of the linearised Jacobian matrix in Eq. (5.54) in the text, and thus verify Eq. (5.59) and (5.60).

5.8** Consider an MHD atmosphere composed of completely ionised Hydrogen (75%) and Helium (25%) whose temperature is 10^5 K. A detector picks up a "loud" fast magnetosonic wave with an audio spectrum that peaks at 1 kHz. Thirty seconds later, the detector detects the wind shear associated with the passage of the Alfvén wave, and another 45 seconds pass before the slow wave is detected. The on-board magnetometer and directional indicators determine that the waves are propagating at an angle of 30° relative to the magnetic field orientation, which we will assume is the same where the waves were launched. You may also assume that the atmosphere is "still"; that is, the detector is in the co-moving frame of the fluid.

  a) What is the MHD-alpha ($\alpha = a^2/c_s^2$) of the atmosphere? (*Hint*: You're going to end up with a *quartic* in $\sqrt{\alpha}$! Chin up, nothing the root finder on my vintage 1981 HP calculator can't handle!)

  b) How far away is the disturbance that launched the waves?

  c) What is the peak frequency of the slow wave spectrum?

  d) What would the detector detect (including peak frequency, if appropriate) if the same disturbance were propagating 90° relative to the magnetic field?

5.9* Verify the expressions in Table 5.1 in the text. The $\alpha = 1$ entries are exact, whereas all other entries are derived from the first two non-zero terms of a binomial expansion of Eq. (5.59) in the text.

5.10*

  a) Show that when $\alpha \ll 1$, the ratio of magnetic and thermal pressure perturbations in a slow magnetosonic wave is given by:

$$\frac{\tilde{p}_B}{\tilde{p}} = -1 + \alpha \cos^2\theta + \mathcal{O}(\alpha^2). \tag{5.117}$$

  b) Evidently, as $\alpha \to 0$, $\tilde{p}_B/\tilde{p} \to -1$. Show that this means in this limit, slow waves lose their compressional nature. What remains to drive the slow waves?

If nothing else, this problem will test your ability to work with binomial expansions and to know how many terms one must keep in order to maintain consistent order throughout your expression.

5.11 Solve Eq. (5.34) in the text, namely,

$$\partial_t |q(x,t)\rangle + \mathsf{J}\,\partial_x |q(x,t)\rangle = 0,$$

Problem Set 5

for the shear Alfvén wave using Theorem (3.1), namely,

$$|q(x,t)\rangle = \sum_{i=1}^{n} \langle l_i | \tilde{q}(\xi_i) \rangle | r_i \rangle. \qquad \text{Eq. (3.28)}$$

Thus, you will need to find the left eigenvectors as outlined in §3.5.2 to assemble your solutions.

Discussion: The fact that this route gives you the identical result as Eq. (5.36) – obtained by integrating the right eigenvectors then applying initial conditions – may seem a little puzzling. How can these two apparently disparate methods give mathematically equivalent results?

What's going on is actually quite subtle. In matrix algebra, taking the inverse is analogous to a Green's function approach in solving differential equations. Operationally, in solving for the proportionality functions $w^\pm(x)$ just before Eq. (5.36) in the text, one is actually doing the same algebraic manoeuvres one does in taking a matrix inverse; it's just a little bit disguised.

Regardless, the lesson here is the following: One can solve a hyperbolic system of n equations *either* via the left eigenvectors (and thus one must invert an $n \times n$ matrix) *or* by integrating the right eigenvectors and applying initial conditions (and thus one must solve n equations in n unknowns). Each approach gives the identical results.

5.12** The inset shows initial profiles for v_y and B_y at $t=0$ in a 1-D Cartesian geometry where $\partial_y = \partial_z = 0$. In units where $\mu_0 = 1$, suppose $\rho = 1$, $B_x = 1$, and let $v_x = v_z = B_z = 0$.

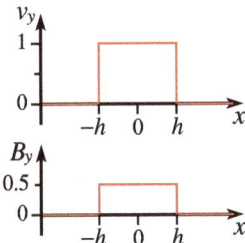

a) Using the eigenvalue approach, determine the profiles of v_y and B_y at time $t = h/2$.

b) Repeat using the Method of Characteristics, being sure to include as part of your solution a diagram analogous to Fig. 5.4 in the text.

5.13* The inset shows initial profiles for v_y and B_y at $t=0$ in a 1-D Cartesian geometry, where $\partial_y = \partial_z = 0$. In units where $\mu_0 = 1$, suppose $\rho = 1$, $B_x = 1$, and let $v_x = v_z = B_z = 0$.

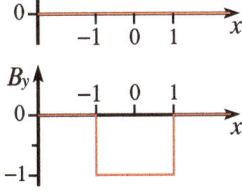

a) Using the eigenvalue approach, determine the profiles of v_y and B_y at time $t = 1.5$.

b) Repeat using the Method of Characteristics, being sure to include as part of your solution a diagram analogous to Fig. 5.4 in the text.

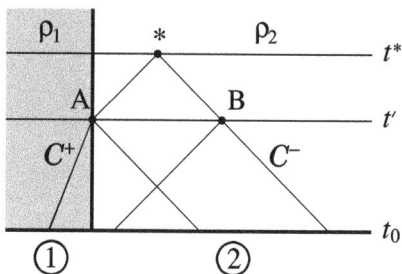

Figure 5.16. (Problem 5.14) While characteristic path C^- passes through a region of constant density, C^+ passes through two regions of different densities.

5.14* Since Eq. (5.41) and (5.42) in the text are the differenced form of Eq. (5.40), replacing $D_t^\pm q_y$ with $q_y^* - q_y^\pm$ (where $q = v$ or B) is obvious. However, replacing ρ – which does not appear inside the differential – with $\rho^\pm$ instead of, say, ρ^* or $\frac{1}{2}(\rho^* + \rho^\pm)$ is not so obvious.

Consider point * in Fig. 5.16 (a D-type point as defined in Fig. 5.4 in the text). Since the characteristic path C^- is entirely embedded in medium 2 (where $\rho = \rho_2$), there is no ambiguity in what value should be assigned to ρ in Eq. (5.42): $\rho^* = \rho^- = \rho_2$, and the use of ρ^- in Eq. (5.42) is justified. However, the path C^+ passes through both medium 1 (where $\rho = \rho_1$) and medium 2 (where $\rho = \rho_2 < \rho_1$ for the sake of argument), and it is not so clear what value of ρ should be used along C^+ when applying Eq. (5.41).

We can, however, justify setting $\rho = \rho^+ = \rho_1$ in Eq. (5.41) by performing the MoC in two steps, as shown in Fig. 5.16. Let t' be an intermediate time between t_0 and t^* that corresponds to when the path C^+ crosses the interface between media 1 and 2. To obtain the values of v^* and B^* at $t = t^*$, one could first evaluate v and B at points A and B at $t = t'$. In this case, all four characteristic paths joining $t = t_0$ to points A and B are completely embedded in *either* medium 1 *or* medium 2, and there is no ambiguity in assigning values for ρ in Eq. (5.41) and (5.42). With v_A, B_A, v_B, and B_B evaluated, we then solve Eq. (5.41) and (5.42) once more to find v^* and B^* from paths originating from points A and B, again with no ambiguity on what values to use for ρ along each of the two characteristic paths originating from $t = t'$.

Perform this two-step process to evaluate v^* and B^*, and show that one gets the same result as would be obtained had one simply done the one step MoC calculation directly from t_0 to t^*, using the densities at the bases of the characteristics paths in Eq. (5.41) and (5.42).

5.15* In §5.2.1, an Alfvén wave propagating in 1-D Cartesian geometry ($\partial_y = \partial_z = 0$) along the x-axis was found to "kink" a magnetic field in its wake by adding a

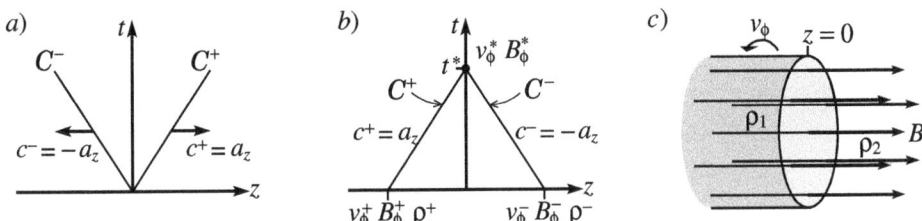

Figure 5.17. (Problem 5.15) *a*) Two characteristics paths, $\mathcal{C}^{\pm}$, emanate from the same location at $t = 0$ in the co-moving frame of the fluid ($v_z = 0$) and propagate in opposite directions at characteristic speeds $\pm a_z$. *b*) Two characteristics paths from different locations at $t = 0$ converge at the same event at $t = t^*$. The fact that the slopes of $\mathcal{C}^{\pm}$ have the same magnitude means that $\rho^+ = \rho^-$ was assumed. *c*) A uniform B_z crosses a density jump at $z = 0$, with $\rho = \rho_1$ for $z < 0$ and $\rho = \rho_2$ for $z > 0$. At $t = 0$, the medium with density ρ_1 begins to rotate about the z-axis.

y (and/or z) component to a pre-existing x component. This was referred to as a *shear* Alfvén wave.

By contrast, in 1-D cylindrical geometry ("axisymmetry"; $\partial_r = \partial_\varphi = 0$), an Alfvén wave propagating along the z-axis "twists" a magnetic field in its wake by adding a φ-component to a pre-existing z-component describing a *torsion* Alfvén wave. In this problem, we'll examine torsion Alfvén waves using the Method of Characteristics (MoC) just as we did for shear Alfvén waves in §5.2.1.

a) Show that in axisymmetric cylindrical coordinates with $v_z = v_r = 0$, the φ components of the MHD Euler and ideal induction equations reduce to:

$$\partial_t v_\varphi = \frac{B_z}{\mu_0 \rho} \partial_z B_\varphi; \qquad \partial_t B_\varphi = B_z \partial_z v_\varphi. \tag{5.118}$$

b) Show how Eq. (5.118) can be written as,

$$D_t^{\pm} v_\varphi \mp \frac{1}{\sqrt{\mu_0 \rho}} D_t^{\pm} B_\varphi = 0, \tag{5.119}$$

where $D_t^{\pm} = \partial_t \pm a_z \partial_z$ are the Lagrangian derivatives.

Discussion: Since Eq. (5.119) are identical in form to Eq. (5.40) in the text, their interpretation in terms of characteristic paths follows suit. Thus, as illustrated in Fig. 5.17a, we can think of each event in the present emanating two characteristic paths into the future, each carrying with it a piece of information (a Riemann invariant) regarding v_φ and B_φ. Alternately, as shown in Fig. 5.17b, we can think of two characteristic paths originating from different events in the present converging on some future event at $t = t^*$. Since a unique piece of information (Riemann invariants, $\mathcal{A}^{\pm}$) is transported along each path, these may be used to solve for v_φ^* and B_φ^* at $t = t^*$ and, in particular, following the reasoning in the text:

$$v_\varphi^* = \frac{1}{\sqrt{\rho^-}+\sqrt{\rho^+}}\left(\sqrt{\rho^-}\,v_\varphi^- + \sqrt{\rho^+}\,v_\varphi^+ + \frac{B_\varphi^- - B_\varphi^+}{\sqrt{\mu_0}}\right); \quad (5.120)$$

$$B_\varphi^* = \frac{\sqrt{\rho^-\rho^+}}{\sqrt{\rho^-}+\sqrt{\rho^+}}\left(\frac{B_\varphi^-}{\sqrt{\rho^-}} + \frac{B_\varphi^+}{\sqrt{\rho^+}} + \sqrt{\mu_0}(v_\varphi^- - v_\varphi^+)\right), \quad (5.121)$$

identical in form to Eq. (5.44) and (5.45). You aren't being asked to derive these; they are being given here for use in the next part.

c) Figure 5.17c depicts a uniform axial magnetic field $B_z = 1$ (in units where $\mu_0 = 1$) in cylindrical geometry where the density is given by,

$$\rho(z) = \begin{cases} \rho_1 = 16, & z < 0; \\ \rho_2 = 1, & z \geq 0. \end{cases}$$

As usual, all units are arbitrary.

Suppose at $t = 0$, the denser fluid ($z < 0$) starts rotating about the z-axis at angular speed $\omega = 1$; i.e., at $r = 1$, $v_\varphi = 1$. Use the MoC to find the profiles of v_φ and B_φ at $t = 1$ and $r = 1$, assuming $B_\varphi = 0$ at $t = 0$.

Your solution should include a diagram similar to Fig. 5.4 in the text showing pairs of characteristic paths, $\mathcal{C}^\pm$, emanating from the $t = 0$ axis converging on various points at $t = 1$, some of which with both footprints in $z < 0$, some with one footprint in $z < 0$ and one in $z > 0$, and some with both in $z > 0$. This diagram should then show explicitly how the invariants transported along these paths result in the profiles of v_φ and B_φ at $t = 1$, and should include specific distances (in the arbitrary units of the problem) travelled by the various wavefronts by that time.

5.16 Consider the rotational discontinuity described in §5.3.4 in the text. What does the RD become if $B_x = 0$?

5.17 As an alternative to Eq. (5.86) in the text, show that the energy equation, Eq. (5.77), can be written as:

$$\frac{\gamma}{\gamma-1}\Delta(pv_x) + \frac{\mathcal{M}}{2}\Delta v_x^2 + \frac{1}{\mu_0}\Delta(v_xB_\perp^2) - \frac{B_x^2}{2\mu_0^2\mathcal{M}}\Delta B_\perp^2 = 0.$$

This is similar to the version of the energy equation used by Torrilhon (2003, J. Plasma Phys., v. 69, p. 253) in his article "Uniqueness conditions for Riemann problems of ideal MHD".

5.18

a) Verify Eq. (5.87) in the text.

b) Show how Eq. (5.97) follows from Eq. (5.86).

5.19 In addition to Eq. 5.104 in the text, a cubic in ϕ_x whose coefficients depend upon the upwind sonic Mach number and the MHD-alphas, show that ϕ_x is also given by the quadratic,

$$\left(\frac{2b}{b-1} - \alpha_x(\gamma+1) - b\alpha_\perp(\gamma+b)\right)\phi_x^2$$
$$+ \left(-2\frac{b+1}{b-1} + 2\alpha_x\gamma + \gamma\alpha_\perp(b+1)\right)\phi_x \qquad (5.122)$$
$$+ \frac{2}{b-1} - (\alpha_x + \alpha_\perp)(\gamma-1) = 0,$$

whose coefficients depend now upon b – the jump in $B_\perp$ – rather than M_1^2. This equation will be useful in building the MHD Riemann solver in Chap. 6.

5.20 Show that a slow shock cannot exist if the upwind perpendicular component of the magnetic field, $B_{\perp_1} \neq 0$ and the downwind Alfvén Mach number, $A_{x_2} = 1$ (*i.e.*, the post-shock flow speed relative to the slow shock is the Alfvén speed).

Hints: First, show that if $A_{x_2} = 1$, then $\phi_x A_{x_1}^2 = 1$. Then use Eq. (5.103) in the text to show that $\phi_x = 1$, and thus there can be no jump.

5.21* For the "switch-on shock", verify from the text:

a) Eq. (5.109), (5.110), (5.111); and

b) Ineq. (5.112),

5.22

a) Show that no value of M_1^2 can satisfy Ineq. (5.112) in the text unless $\alpha_{x_1} > 1$. Thus, to exist, a switch-on shock requires that the longitudinal Alfvén speed be greater than the sound speed.

b) Therefore, show that for a switch-on shock,

$$\frac{\alpha_{x_1}(\gamma-1)}{\alpha_{x_1}(\gamma+1) - 2} < \phi_x < 1. \qquad (5.123)$$

5.23 The MHD-alpha relative to the longitudinal magnetic field,

$$\alpha_x = \frac{B_x^2}{\gamma\mu_0 p} = \frac{a_x^2}{c_s^2} = M_A^2,$$

can also be interpreted as the square of the ordinary Mach number of the flow at the so-called *Alfvén point* (M_A), where the flow speed is equal to the longitudinal

Alfvén speed. In a similar vein, we can define the ordinary Mach number of the fast point (M_+) and the slow point (M_-), where the flow speed is equal to the fast- and slow-speeds respectively. Thus,

$$M_+^2 = \frac{a_f^2}{c_s^2} = \alpha_f \quad \text{and} \quad M_-^2 = \frac{a_s^2}{c_s^2} = \alpha_s, \qquad (5.124)$$

defining the fast and slow MHD-alphas, $\alpha_{f,s}$, in analogy with α_x.

Show that,

$$\begin{aligned} M_\pm^2 = \alpha_{f,s} &= \frac{1}{2}\left(\alpha + 1 \pm \sqrt{(\alpha+1)^2 - 4\alpha_x}\right) \\ &= \frac{1}{2}\left(\alpha + 1 \pm \sqrt{(\alpha-1)^2 + 4\alpha_\perp}\right), \end{aligned} \qquad (5.125)$$

where $\alpha = a^2/c_s^2 = B^2/\gamma\mu_0 p$, as originally defined on page 130 of the text.

5.24 In all the discussion surrounding Fig. 5.15 in the text of what portions of the Euler and switch-on branches correspond to the fast, type I, and slow loci, did it occur to you there was no mention of the type II locus? In the left panel of Fig. 5.15 and for $\alpha_{\perp_1} = 10^{-7}$ (cyan), the type II locus looks to be a reflection through the $b_\perp = 0$ axis of the portion of the fast shock locus that corresponds to the switch-on shock in the limit of $\alpha_{\perp_1} \to 0$. But the right panel shows no analogue of the type II shock locus. Where did it go?

I'm looking for a semi-quantitative, or even qualitative answer here. Your discussion might touch on things such as the double degeneracy of the switch-on shock root (Eq. 5.108) in Eq. (5.107), and what it means for $B_\perp$ to change sign across a shock when the pre-shock value is *zero*.

5.25* When writing the program necessary to generate Fig. 5.12, 5.13, and 5.15 in the text, I found that Eq. (5.95),

$$b = \frac{A_{x_1}^2 - 1}{A_{x_1}^2 \phi_x - 1},$$

was problematic in the limit as $\alpha_\perp \to 0$. For the asymptotic switch-on solutions where $\phi_x \to A_{x_1}^{-2}$, the denominator becomes dominated by machine round off error and starts to yield intolerably inaccurate values for b. When such things occur, a good computational scientist will search for alternate forms for the expression that is giving numerical round-off problems, even if that expression ends up being more cumbersome.

a) Starting with Eq. (5.96) and (5.97) in the text, show that:

$$b = (\gamma-1)\xi \pm \sqrt{\left[(\gamma-1)\xi - 1\right]^2 + 2\xi A_{\perp_1}^2(\gamma+1)(\phi_x - \mathcal{H})}, \qquad (5.126)$$

where $\xi = \dfrac{1-\phi_x}{2\phi_x}$, and the remaining symbols are as they are used in the text.

b) For what values of ϕ_x might you expect machine round-off errors to be problematic for Eq. (5.126)? Thus, explain how one can use Eq. (5.95) and (5.126) in tandem to yield reliable values for b for all values of ϕ_x.

c) How might one determine which of the two roots should be used in Eq. (5.126)?

5.26*

a) Using the entropy condition (i.e., $\phi_x \leq 1$), show that for a fast shock, the upwind speed must be greater than the fast speed, whereas for a slow shock, the upwind speed must be greater than the slow speed.

Hint: You might start with Eq. (5.104) and set $\phi_x = 1$ to find the critical Mach numbers which should correspond to the Mach numbers at the Alfvén, fast, and slow points given in Problem 5.23.

b) Prove that type II intermediate shock loci (e.g., the short-dashed portions of the super-loci on the right panels of Fig. 5.12 and 5.13 in the text) end on a rotational discontinuity.

Hint: Consider using Eq. (5.126) from Problem 5.25.

5.27* Consider the shock loci in $b_\perp$ for $\alpha_{x_1} = 5$ and $\alpha_{\perp_1} = 1$ (and thus $\alpha_1 = 6$) in the left panel of Fig. 5.15 in the text (red lines). State the criterion needed to determine and then calculate to four significant figures the values of M_1 at which:

a) the slow shock locus begins;

b) the slow shock locus ends and the type Ia intermediate shock locus begins;

c) the type Ia intermediate shock locus ends and the type Ib intermediate shock locus begins;

d) the type Ib intermediate shock locus ends and the type IIa intermediate shock locus begins (for this part, just state the criterion and then estimate the value of M_1 from Fig. 5.15 in the text; I have yet to find a way to calculate M_1 for the intermediate point!);

e) the type IIa intermediate shock locus ends and the type IIb intermediate shock locus begins;

f) the type IIb intermediate shock locus ends.

g) the fast shock locus begins.

h) What solutions, if any, exist between where the type IIa shock locus ends and the fast shock locus begins? If there are no solutions, why not?

i) On a $b_\perp$ vs. M_1 graph, draw by hand with some attention to accuracy the slow-family super-locus and the fast locus indicating clearly where each of the six shock loci begin and end.

You may use any of the results in the text as you like, including from other problems in this problem set.

5.28

a) Show that $b_\perp = \sqrt{\alpha_{\perp_1}/\alpha_{x_1}} \equiv b_{\perp_0}$ at the beginning of both the slow and fast shock loci (*i.e.*, at the slow and fast points respectively where $\phi_x \to 1$).

b) At the end of the slow super-locus (*i.e.*, at the end of the type IIa intermediate shock locus) where $A_{x_1} \to 1$ and $\phi_x \to 1$, show that $b_\perp = -b_{\perp_0}$.

Hint: For part a), you might want to start by considering Eq. (5.95) in the text, whereas for part b), Eq. (5.126) from Problem 5.25 might be more suitable.

6 The MHD Riemann Problem

You don't really understand something until you can compute it.

Michael L. Norman
computational astrophysicist

6.1 Overview

HANDS DOWN, the trickiest software I have ever written in my forty years of scientific programming is my program to solve the MHD Riemann problem. It's a venture with zero-divides and near-zero divides around every corner, including the usual and relatively simple-to-solve problems in scalar equations where the denominator gets too close to zero, as well as the much more vexing matrix-vector equations where rows of the Jacobian become zero or near-zero, rendering the matrix equation insoluble or nearly insoluble (*i.e.*, dominated by round-off error). All of these challenges present themselves to those who dare tread forward!

On the plus side, nothing has sealed my own understanding of MHD as has the experience of writing an exact MHD Riemann solver. *Anyone* with serious aspirations of understanding the 1-D MHD problem needs to go through this exercise.

And so let's start with an intuition booster. The precept of the Riemann problem is simple enough. As we did in Chap. 3 and as shown in Fig. 6.1, we start with a left and right state where one state is set completely independently of the other. Before $t = 0$, the two states are separated by an impenetrable diaphragm, **D**, with one state knowing nothing of the other. Then, at $t = 0$ the diaphragm is removed, and suddenly the two states can interact. One state forces its way into the other and yet somehow, at any given time t, one must still be able to get from the left state to the right via a unique set of allowed MHD transitions. The question is, how do we determine these transitions?

Let's approach this by considering a "building block" example, as depicted in Fig. 6.2. I'm thinking of the wooden Brio^{TM}[1] train sets my kids and I used to play with when they were little. As shown in panel *a*, suppose there is a vertical gap between A and B that needs to be spanned, and we may do so only with the pieces that come in our set. We'll allow ourselves the latitude of positioning A and B

---

[1] It should not be lost on the reader that this train set analogy pays homage to one of the seminal papers applying the Riemann problem to computational MHD, namely Brio & Wu (1988).

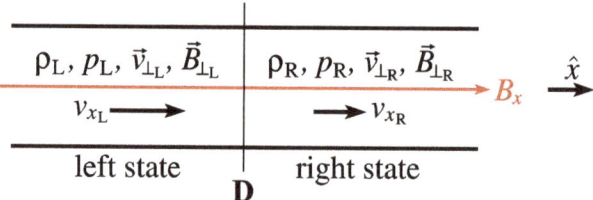

Figure 6.1. Initial set up for the 1-D MHD Riemann problem. At $t = 0$, the diaphragm, **D**, is removed, and the two left- and right-states interact with an arbitrary jump in (possibly) all flow variables at **D** as initial conditions.

horizontally as needed, but their vertical displacement must remain fixed. Suppose further that our Brio^{TM} pieces – to carry on with the analogy – come in a variety of connectors and colours. Standard Brio^{TM} pieces are made of maple or birch, and thus light brown with round male–female track connectors at each end. Let's suppose our set comes in four different colours (with just one side of each piece painted), and that all red pieces have a round male and triangular female connector (panel b), all green pieces have a triangular male and rectangular female connector (panel c), and all blue pieces have a rectangular male and pentagonal female connector (panel d). Meanwhile, there is only one black piece, with two male pentagonal connectors (panel e). Further, the red and blue pieces come as flats, ramps, and jumps with the ramps and jumps coming in an assortment of heights, the green pieces come in flats and jumps, and the one black piece is flat. As an added wrinkle, all red ramps go down from the male connector while all red jumps go up, opposite to the blue

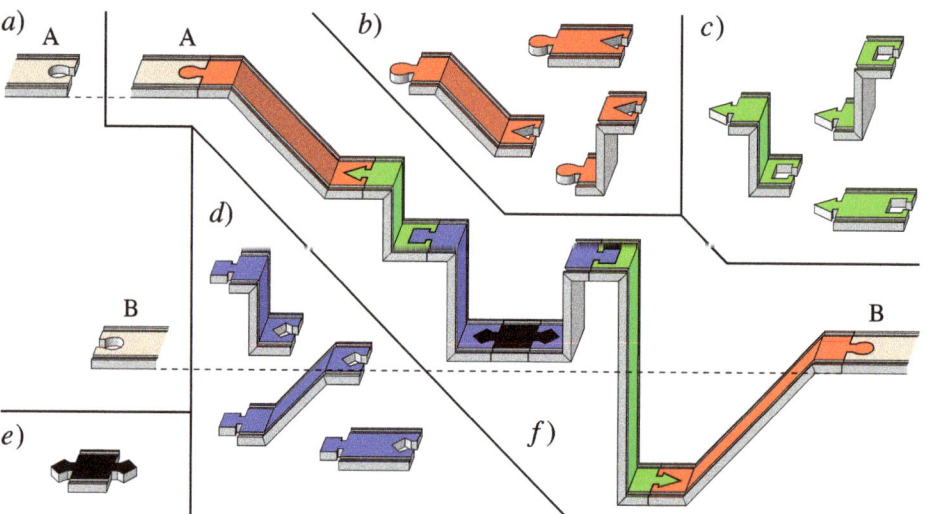

Figure 6.2. The "Brio^{TM} train track Riemann problem"; see text for description.

pieces where the ramps go up and the jumps go down from the male connector. Green jumps go up or down. Finally, in whatever configuration we come up with to connect A and B, all coloured sides must be face up.

OK, that's what we have and those are the rules. Now let's play!

The first thing to notice is that owing to the shapes of the connectors on each piece, to get from A to B we'll need to arrange the pieces in a specific order. Starting from A, we'll need a red piece, then a green, blue, black, blue, green, and finally a red piece to attach to B. So a first step would be to separate the pieces by colour.

Now comes the harder part where we have to – presumably by trial and error – start fitting pieces to see what combination gets us contiguously from A to B with all connectors joining flatly. It may be that the manufacturer was clever enough to make it so that for any given height difference between A and B, one and only one set of pieces will, in aggregate, span the jump (*e.g.*, panel *f* in Fig. 6.2). Or perhaps there are numerous solutions, or maybe even none.

The Brio^{TM} game just described is almost a perfect analogy to the 1-D MHD Riemann problem considering just one of the components of $B_\perp$, B_y say. If the vertical distance represents the value of B_y (positive or negative), then the red pieces are the fast waves coming as either rarefaction fans (ramps) or shocks (jumps), the green pieces are the rotational discontinuities, the blue pieces are the slow waves also coming as rarefaction fans or shocks, and the black piece is the contact across which B_y is constant. The real problem, of course, is much more complicated than this since we have not just one gap to fit, but seven – one for each of the variables ρ, p, v_x, v_y, v_z, B_y, and B_z – where each piece chosen for B_y, say, dictates which piece must be used for each of the other variables. Thus, finding one set of pieces to span the B_y gap doesn't necessarily mean the accompanying pieces for ρ, p, *etc.*, will span their gaps. This is starting to look a bit like a 7-D Rubik's cube^{TM}!

And so, with that bit of discouragement, let's get started!

6.2 Non-linear MHD waves

From the discussion in §5.3, we're familiar with the MHD discontinuities (contacts, RDs, fast and slow shocks) mentioned in the Brio^{TM} example. All that remains to work out before tackling the MHD Riemann problem directly are the profiles across the fast and slow rarefaction fans (RF). Now, from our discussion in §3.5.3 on the hydrodynamical Riemann problem, we know – at least in principle – how to determine these. Starting from Eq. (3.24), namely,[2]

$$\mathsf{J}_\mathrm{p}|q'_\mathrm{p}\rangle = u_i|q'_\mathrm{p}\rangle,$$

where $|q'_\mathrm{p}\rangle$ is the derivative of the ket of primitive variables with respect to its argument ($\xi_i = x - u_i t$) and J_p is the primitive Jacobian matrix (both defined in Eq.

---

[2] *Rappel*: This is the *fifth* time we've seen and used this equation, the last time being Eq. (5.55) in §5.2.2.

5.9), we found the seven eigenvalues (characteristic speeds) of J_p to be $u_i = v_x \pm a_f$, $v_x \pm a_x$, $v_x \pm a_s$, and v_x, where a_f, a_x, and a_s are the fast, Alfvén, and slow speeds given by Eq. (5.23)–(5.25). Thus, across each wave, $|q'_p\rangle$ is proportional to one of the right eigenvectors (eigenkets) of J_p, namely $|r_i\rangle$, and, as given in Eq. (3.35),

$$|q'_p(\xi_i)\rangle = w_i(\xi_i)|r_i\rangle,$$

where w_i is an arbitrary *proportionality* or *scaling* function of the co-moving coordinate, ξ_i. As we did in §3.5.3, define $ds_i = w_i(\xi_i)d\xi_i$ as a differential of a *generalised coordinate*, s_i, that varies from 0 on the upwind side of the i-wave to $s_{i,d}$ on the downwind side which can be thought of as the "width" or "strength" of the rarefaction fan. Then,

$$\boxed{\frac{d|q_p(s_i)\rangle}{ds_i} = |r_i\rangle,} \quad (6.1)$$

gives us a set of seven coupled, first-order ODEs which we integrate through to its width, $s_{i,d}$, to find the profiles (in terms of s_i) of each primitive variable across the i-wave.

And thus we can delay no longer finding the eigenkets of J_p!

6.2.1 Fast and slow eigenkets

The eigenkets of interest here are those associated with eigenvalues $v_x \pm a_f$ and $v_x \pm a_s$ respectively, as these describe the fast and slow rarefaction fans. The Alfvén and entropy eigenkets (those associated with eigenvalues $v_x \pm a_x$ and v_x) correspond to the Alfvén and entropy waves, both typically discontinuous in some of the flow variables and thus better handled by the conservative equations (*e.g.*, §5.3.2–5.3.4). Further discussion of these is relegated to Problem 6.1.

Starting with the left-moving fast wave with wave speed $u_1 = v_x - a_f$, the associated eigenket, $|r_1\rangle = |r_f^-\rangle$, is found by solving the matrix equation (again, see Eq. 5.9 for J_p),

$$(J_p - u_1 I)|r_1\rangle = \begin{bmatrix} a_f & 0 & \rho & 0 & 0 & 0 & 0 \\ 0 & a_f & \gamma p & 0 & 0 & 0 & 0 \\ 0 & 1/\rho & a_f & 0 & 0 & B_y/\mu_0\rho & B_z/\mu_0\rho \\ 0 & 0 & 0 & a_f & 0 & -B_x/\mu_0\rho & 0 \\ 0 & 0 & 0 & 0 & a_f & 0 & -B_x/\mu_0\rho \\ 0 & 0 & B_y & -B_x & 0 & a_f & 0 \\ 0 & 0 & B_z & 0 & -B_x & 0 & a_f \end{bmatrix} \begin{bmatrix} r_{11} \\ r_{12} \\ r_{13} \\ r_{14} \\ r_{15} \\ r_{16} \\ r_{17} \end{bmatrix} = 0,$$

which yields seven linear equations, one of which is redundant. So let's try ignoring the third one (if for no other reason, because it has the most number of terms), and write:

$$a_f r_{11} + \rho r_{13} = 0; \quad (6.2)$$

$$a_f r_{12} + \gamma p r_{13} = 0; \quad (6.3)$$

$$a_f r_{14} - \frac{B_x}{\mu_0 \rho} r_{16} = 0; \tag{6.4}$$

$$a_f r_{15} - \frac{B_x}{\mu_0 \rho} r_{17} = 0; \tag{6.5}$$

$$B_y r_{13} - B_x r_{14} + a_f r_{16} = 0; \tag{6.6}$$

$$B_z r_{13} - B_x r_{15} + a_f r_{17} = 0. \tag{6.7}$$

As r_{13} appears more often than any other component, let's use that as the pivot (scaling factor). Then, Eq. (6.2) and (6.3) give:

$$r_{11} = -\frac{\rho}{a_f} r_{13}; \qquad r_{12} = -\frac{\gamma p}{a_f} r_{13},$$

and multiplying Eq. (6.4) by B_x/a_f and adding Eq. (6.6) gives:

$$-\frac{a_x^2}{a_f} r_{16} + B_y r_{13} + a_f r_{16} = 0 \quad \Rightarrow \quad r_{16} = -\frac{a_f B_y}{a_f^2 - a_x^2} r_{13}. \tag{6.8}$$

Similarly, Eq. (6.5) and (6.7) yield:

$$r_{17} = -\frac{a_f B_z}{a_f^2 - a_x^2} r_{13}. \tag{6.9}$$

Finally, substituting Eq. (6.8) and (6.9) into Eq. (6.4) and (6.5) respectively gives us:

$$r_{14} = -\frac{B_x}{\mu_0 \rho} \frac{B_y}{a_f^2 - a_x^2} r_{13} \quad \text{and} \quad r_{15} = -\frac{B_x}{\mu_0 \rho} \frac{B_z}{a_f^2 - a_x^2} r_{13}.$$

Bringing these results together, the "minus fast eigenket" is,

$$|r_f^-\rangle = \psi_f \begin{bmatrix} -\rho \\ -\gamma p \\ a_f \\ -\dfrac{B_x}{\mu_0 \rho} \dfrac{a_f}{a_f^2 - a_x^2} B_y \\ -\dfrac{B_x}{\mu_0 \rho} \dfrac{a_f}{a_f^2 - a_x^2} B_z \\ -\dfrac{a_f^2}{a_f^2 - a_x^2} B_y \\ -\dfrac{a_f^2}{a_f^2 - a_x^2} B_z \end{bmatrix}, \tag{6.10}$$

where $\psi_f \equiv r_{13}/a_f$ is a "scaling factor" which we'll choose for convenience.

Evidently, the "plus fast eigenket" must be identical to Eq. (6.10) with $-a_f \to +a_f$, and the slow eigenkets are just the fast eigenkets with f $\to$ s. Thus, we have for the four fast and slow eigenkets:

$$|r_{f,s}^{\pm}\rangle = \psi_{f,s} \begin{bmatrix} -\rho \\ -\gamma p \\ \mp a_{f,s} \\ \pm \dfrac{B_x}{\mu_0 \rho} \dfrac{a_{f,s}}{a_{f,s}^2 - a_x^2} \vec{B}_\perp \\ -\dfrac{a_{f,s}^2}{a_{f,s}^2 - a_x^2} \vec{B}_\perp \end{bmatrix} = \psi_{f,s} \begin{bmatrix} -\rho \\ -\gamma p \\ \mp a_{f,s} \\ \pm \mathrm{sgn}(B_x) \dfrac{a_x \, a_{f,s} \, a_\perp}{a_{f,s}^2 - a_x^2} \hat{e}_\perp \\ -\sqrt{\mu_0 \rho} \dfrac{a_{f,s}^2 \, a_\perp}{a_{f,s}^2 - a_x^2} \hat{e}_\perp \end{bmatrix}, \qquad (6.11)$$

where the last four components have been combined into two 2-D vectors $\propto \vec{B}_\perp = (0, B_y, B_z)$, and where $a_x = |B_x|/\sqrt{\mu_0 \rho}$, $\mathrm{sgn}(B_x) = B_x/|B_x|$, $a_\perp = |\vec{B}_\perp|/\sqrt{\mu_0 \rho}$, and $\hat{e}_\perp = \vec{B}_\perp/|\vec{B}_\perp|$.

While that may have seemed straight-forward enough, the kicker is the denominator in the components proportional to $\vec{B}_\perp$ ($\hat{e}_\perp$). Because the 1-D MHD equations are *not strictly* hyperbolic, their eigenvalues can, at times, be degenerate. In particular, we've already seen (*e.g.*, Eq. 5.66) that if $\vec{B}_\perp = 0$ and $a_x > c_s$, $a_f = a_x$ and the components $\propto \hat{e}_\perp$ for the fast eigenkets blow up. Similarly, for $\vec{B}_\perp = 0$ and $a_x < c_s$, $a_s = a_x$ and the components $\propto \hat{e}_\perp$ for the slow eigenkets blow up.

Oops.

Our salvation are the scaling factors $\psi_{f,s}$, which we choose not so much to normalise $|r_{f,s}^{\pm}\rangle$ (which we can't anyway since the components have different units), but to render all singularities removable. It turns out our choices are rather limited since, in addition, $\psi_{f,s}$ must be chosen such that no eigenket is zeroed out. By what has to be described as a stroke of genius and what we'll spend the next few pages justifying, those introduced by Roe & Balsara (1996) and used by Takahashi & Yamada (2014) are:

$$\psi_f = \sqrt{\dfrac{c_s^2 - a_s^2}{a_f^2 - a_s^2}} \quad \text{and} \quad \psi_s = \sqrt{\dfrac{a_f^2 - c_s^2}{a_f^2 - a_s^2}}. \qquad (6.12)$$

The reader might wonder why the more "obvious" choice of, say, $\psi_{f,s} = a_{f,s}^2 - a_x^2$ might not be preferred. However, it doesn't take too long to realise that this would result in $|r_{f,s}^{\pm}\rangle = 0$ for $B_\perp = 0$, which would mean no wave at all.

The mathematics of MHD rarefaction fans is littered with landmines (*i.e.*, zero-divides or, when it comes time to do the programming, *near* zero-divides) which can confound even the most seasoned algebraist. In my experience, the cleanest approach is to express everything in terms of the MHD-alphas, and then to use the various identities among them to eliminate all differences in the denominators where singularities can occur.

So to start, let's recast the four identities among the speeds a_f, a_s, a_x, $a_\perp$, and c_s as listed in Problem 5.4 (Eq. 5.113–5.116), in terms of the MHD-alphas:

$$a_f a_s = c_s a_x \quad \Rightarrow \quad \dfrac{a_f^2}{c_s^2} \dfrac{a_s^2}{c_s^2} = \dfrac{a_x^2}{c_s^2} \quad \Rightarrow \quad \boxed{\alpha_f \alpha_s = \alpha_x;} \qquad (6.13)$$

$$a_f^2 + a_s^2 = c_s^2 + a_x^2 + a_\perp^2 \quad \Rightarrow \quad \boxed{\alpha_f + \alpha_s = 1 + \alpha_x + \alpha_\perp;} \qquad (6.14)$$

$$(a_{\rm f}^2 - c_{\rm s}^2)(a_{\rm f}^2 - a_x^2) = a_{\rm f}^2 a_\perp^2 \quad \Rightarrow \quad \boxed{(\alpha_{\rm f} - 1)(\alpha_{\rm f} - \alpha_x) = \alpha_{\rm f}\alpha_\perp;} \quad (6.15)$$

$$(c_{\rm s}^2 - a_{\rm s}^2)(a_x^2 - a_{\rm s}^2) = a_{\rm s}^2 a_\perp^2 \quad \Rightarrow \quad \boxed{(1 - \alpha_{\rm s})(\alpha_x - \alpha_{\rm s}) = \alpha_{\rm s}\alpha_\perp.} \quad (6.16)$$

Here, $\alpha_{\rm f,s}$ are the fast and slow alphas, first introduced in Problem 5.23 (Eq. 5.124). Next, examine the fast eigenkets which, from Eq. (6.11), we can write as:

$$|r_{\rm f}^\pm\rangle = \begin{bmatrix} -\psi_{\rm f}\rho \\ -\psi_{\rm f}\gamma p \\ \mp\psi_{\rm f}a_{\rm f} \\ \pm{\rm sgn}(B_x)\,\chi_{\rm f}\,c_{\rm s}\dfrac{a_x}{a_{\rm f}}\,\hat{e}_\perp \\ -\chi_{\rm f}\,c_{\rm s}\sqrt{\mu_0\rho}\,\hat{e}_\perp \end{bmatrix}, \quad (6.17)$$

where, in terms of the MHD-alphas, $\psi_{\rm f}$ (Eq. 6.12) and $\chi_{\rm f}$ are given by:

$$\psi_{\rm f} = \sqrt{\dfrac{1-\alpha_{\rm s}}{\alpha_{\rm f}-\alpha_{\rm s}}}; \quad \chi_{\rm f} = \psi_{\rm f}\dfrac{a_{\rm f}^2}{a_{\rm f}^2-a_x^2}\dfrac{a_\perp}{c_{\rm s}} = \sqrt{\dfrac{1-\alpha_{\rm s}}{\alpha_{\rm f}-\alpha_{\rm s}}}\dfrac{\alpha_{\rm f}}{\alpha_{\rm f}-\alpha_x}\sqrt{\alpha_\perp}. \quad (6.18)$$

Then, since $c_{\rm s}a_x = a_{\rm f}a_{\rm s}$ (Identity 6.13), and $c_{\rm s}\sqrt{\rho} = \sqrt{\gamma p}$, we may write Eq. (6.17) in its most compact form:

$$|r_{\rm f}^\pm\rangle = \begin{bmatrix} -\psi_{\rm f}\rho \\ -\psi_{\rm f}\gamma p \\ \mp\psi_{\rm f}a_{\rm f} \\ \pm{\rm sgn}(B_x)\,\chi_{\rm f}\,a_{\rm s}\,\hat{e}_\perp \\ -\chi_{\rm f}\sqrt{\mu_0\gamma p}\,\hat{e}_\perp \end{bmatrix}. \quad (6.19)$$

Now, $\psi_{\rm f}$ and $\chi_{\rm f}$ are much more tightly coupled than Eq. (6.18) appears to suggest. Squaring $\chi_{\rm f}$ and using identity (6.15), we get:

$$\chi_{\rm f}^2 = \dfrac{1-\alpha_{\rm s}}{\alpha_{\rm f}-\alpha_{\rm s}}\dfrac{\alpha_{\rm f}^2}{(\alpha_{\rm f}-\alpha_x)^2}\alpha_\perp = \dfrac{1-\alpha_{\rm s}}{\alpha_{\rm f}-\alpha_{\rm s}}\dfrac{\cancel{\alpha_{\rm f}^2}(\alpha_{\rm f}-1)^2}{\cancel{\alpha_{\rm f}^2}\cancel{\alpha_\perp^2}}\cancel{\alpha_\perp}$$

$$= \dfrac{\alpha_{\rm f}-1}{\alpha_{\rm f}-\alpha_{\rm s}}\dfrac{(1-\alpha_{\rm s})(\alpha_{\rm f}-1)}{\alpha_\perp} = \dfrac{\alpha_{\rm f}-1}{\alpha_{\rm f}-\alpha_{\rm s}}(\underbrace{\alpha_{\rm f}-1-\underbrace{\alpha_{\rm s}\alpha_{\rm f}}_{\alpha_x}+\alpha_{\rm s}}_{\cancel{\alpha_\perp}})\dfrac{1}{\cancel{\alpha_\perp}},$$

using identities (6.13) and (6.14). Thus,

$$\chi_{\rm f}^2 = \dfrac{\alpha_{\rm f}-1}{\alpha_{\rm f}-\alpha_{\rm s}} = \psi_{\rm s}^2 \quad (\text{Eq. 6.12})$$

$$= \dfrac{\alpha_{\rm f}-\alpha_{\rm s}+\alpha_{\rm s}-1}{\alpha_{\rm f}-\alpha_{\rm s}} = 1 - \dfrac{1-\alpha_{\rm s}}{\alpha_{\rm f}-\alpha_{\rm s}} = 1 - \psi_{\rm f}^2.$$

Not only is $\chi_{\rm f} = \psi_{\rm s}$, $\chi_{\rm f}$ and $\psi_{\rm f}$ are related to each other in the same way as sine and cosine! Problem 6.2 completes the symmetry of these factors by showing that,

$$\chi_{\rm s} \equiv \psi_{\rm s}\dfrac{a_{\rm s}^2}{a_x^2-a_{\rm s}^2}\dfrac{a_\perp}{c_{\rm s}} = \psi_{\rm f}, \quad (6.20)$$

and thus $\psi_s^2 + \chi_s^2 = 1$ as well.

So, let's use these relationships to simplify the notation further by setting $\mu = \psi_f = \chi_s$, $\nu = \chi_f = \psi_s$ (and thus $\mu^2 + \nu^2 = 1$)[3] and bring the fast and slow eigenkets to our final form:

$$|r_f^\pm\rangle = \begin{bmatrix} -\mu\rho \\ -\mu\gamma p \\ \mp\mu a_f \\ \pm\mathrm{sgn}(B_x)\,\nu\,a_s\,\hat{e}_\perp \\ -\nu\sqrt{\mu_0\gamma p}\,\hat{e}_\perp \end{bmatrix}; \quad |r_s^\pm\rangle = \begin{bmatrix} -\nu\rho \\ -\nu\gamma p \\ \mp\nu a_s \\ \mp\mathrm{sgn}(B_x)\,\mu\,a_f\,\hat{e}_\perp \\ \mu\sqrt{\mu_0\gamma p}\,\hat{e}_\perp \end{bmatrix}. \quad (6.21)$$

It remains, then, to settle on a final form for the scaling factors μ and ν and to demonstrate that their apparent singularities are, in fact, removable. As given by Eq. (6.18), μ (ψ_f) has a singularity (but, as we'll see, removable) at $\alpha_f = \alpha_s$ which happens only at the so-called *triple umbilic* where $\alpha_\perp = 0$ and $\alpha_x = 1$ (and thus $\alpha_f = \alpha_x = \alpha_s = 1$; that is, the fast, Alfvén, and slow speeds are *triply* degenerate). If, for convenience, we let,

$$\delta_x = \alpha_x - 1, \quad (6.22)$$

(and thus $\delta_x \gtreqless 0$ for $\alpha_x \gtreqless 1$), then the singularity in μ occurs when both $\alpha_\perp$ and δ_x are zero.

By necessity, the MHD Riemann solver we'll design will be semi-analytic. Thus, its reliance on a computer means that even a removable singularity will trigger a zero-divide at the triple umbilic, and we must therefore remove it manually. To do this, it is most convenient to express μ in terms of $\alpha_\perp$ and δ_x, those quantities which when simultaneously zero cause the singular behaviour.

Starting with the definition of the slow speed (Eq. 5.23),

$$a_s^2 = \frac{1}{2}(c_s^2 + a^2 - D) = \frac{1}{2}(c_s^2 + a_x^2 + a_\perp^2 - D), \quad (6.23)$$

where the discriminant, D, is given by,

$$D = \sqrt{(c_s^2 + a_x^2 + a_\perp^2)^2 - 4c_s^2 a_x^2},$$

we have,

$$\alpha_s = \frac{a_s^2}{c_s^2} = \frac{1}{2}(1 + \alpha_x + \alpha_\perp - d), \quad (6.24)$$

where,

$$d \equiv \frac{D}{c_s^2} = \sqrt{(1 + \alpha_x + \alpha_\perp)^2 - 4\alpha_x} = \sqrt{(\alpha_\perp + \delta_x)^2 + 4\alpha_\perp}, \quad (6.25)$$

after a little algebra. This, incidentally, is the most robust form for d one can use for a computer application, since there are no subtractions under the radical. Computers, being of finite precision, can often subtract what ought to be equal values and end up with a small residual of "round-off noise" which can be negative as often as

---

[3] Don't confuse μ with μ_0!!

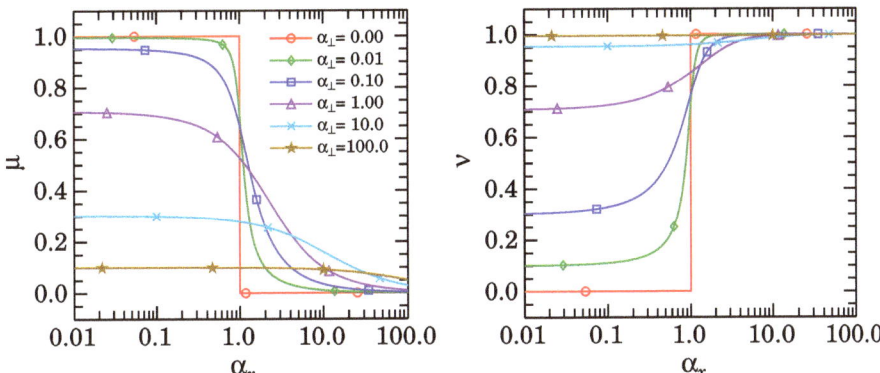

Figure 6.3. Profiles of the fast and slow eigenket "scaling factors", μ (left; Eq. 6.26) and ν (right; Eq. 6.27) as functions of $\alpha_x = \delta_x + 1$, shown for various values of $\alpha_\perp$, including the limiting case of $\alpha_\perp = 0$ (red$\bigcirc$) where μ and ν are step functions with the discontinuity at $\alpha_x = 1$ ($\delta_x = 0$), the triple umbilic.

not. Under a radical sign, this would trigger a "floating point exception" or worse, the dreaded "`NaN`" ("not a number") error messages that cause the program to crash, often without any indication of where the first `NaN` occurred! Further, even if the difference is not purely round-off noise, subtracting two nearly equal numbers can yield results significantly less precise than the stated precision of the machine. Conversely, adding positive quantities suffers no such loss of machine accuracy.

Continuing from Eq. (6.24), we have,

$$\alpha_s = \frac{1}{2}(2 + \delta_x + \alpha_\perp - d) = 1 + \frac{1}{2}(\delta_x + \alpha_\perp - d)$$

$$\Rightarrow \quad 1 - \alpha_s = \frac{1}{2}(d - \delta_x - \alpha_\perp).$$

Noting that $\alpha_f - \alpha_s = d$, we arrive at our final forms for μ and ν:

$$\boxed{\mu^2 = 1 - \nu^2 = \frac{1 - \alpha_s}{\alpha_f - \alpha_s} = \frac{1}{2}\left(1 - \frac{\delta_x + \alpha_\perp}{d}\right);} \quad (6.26)$$

$$\boxed{\nu^2 = 1 - \mu^2 = \frac{\alpha_f - 1}{\alpha_f - \alpha_s} = \frac{1}{2}\left(1 + \frac{\delta_x + \alpha_\perp}{d}\right),} \quad (6.27)$$

with d given in terms of δ_x and $\alpha_\perp$ by Eq. (6.25).

Figure 6.3 shows μ and ν as functions of $\alpha_x = \delta_x + 1$ for various values of $\alpha_\perp$, including the limiting case (red$\bigcirc$) where $\alpha_\perp = 0$ ($B_\perp = 0$). The fact that μ^2 (and thus μ) should be a step function when $\alpha_\perp = 0$ is easy to see from Eq. (6.25) and (6.26). Setting $\alpha_\perp = 0$,

$$d = \sqrt{(\delta_x)^2} = |\delta_x|$$

$$\Rightarrow \quad \mu^2 = \frac{1}{2}\left(1 - \frac{\delta_x}{|\delta_x|}\right) = \begin{cases} 1, & \delta_x < 0, \ (\alpha_x < 1); \\ 0, & \delta_x > 0, \ (\alpha_x > 1), \end{cases} \quad (6.28)$$

which is the red○ profile in the left panel of Fig. 6.3. Evidently, the step function is reversed for $\nu^2 = 1 - \mu^2$ (0 for $\alpha_x < 1$, 1 for $\alpha_x > 1$), as the red○ profile in the right panel shows.

The fact that μ and ν remain finite everywhere means that the apparent singularity at the triple umbilic (when $\alpha_f = \alpha_s$ in Eq. 6.26) is removable; that is, μ^2 remains finite as $\alpha_f \to \alpha_s$ (when both $\alpha_\perp$ and δ_x are zero). To find that limiting value, we first set $\delta_x = 0$ in Eq. (6.25) and (6.26) to get:

$$\lim_{\delta_x \to 0} \mu^2 = \frac{1}{2}\left(1 - \frac{\alpha_\perp}{\sqrt{\alpha_\perp^2 + 4\alpha_\perp}}\right) = \frac{1}{2}\left(1 - \sqrt{\frac{\alpha_\perp}{\alpha_\perp + 4}}\right) \to \frac{1}{2}, \quad (6.29)$$

as $\alpha_\perp \to 0$. Similarly, $\nu^2 \to \frac{1}{2}$ as $\delta_x, \alpha_\perp \to 0$.

This completes the justification of the scaling factors, $\psi_{f,s}$, chosen in Eq. (6.12).

6.2.2 Fast and slow rarefaction fans

Part of the job of the Riemann solver is to integrate Eq. (6.1) using the kets in Eq. (6.21) to find the primitive variable profiles across any rarefaction fan that may be part of the solution. However, without doing the actual integrations, we can determine qualitative properties of the fast and slow fans just by examining the kets as differential changes in the variables.

Assuming for now that $B_x \neq 0$ ($\alpha_x > 0$), let's start by examining the first three components of each ket. Other than the factors μ and ν, these are identical to the purely hydrodynamical kets in Eq. (3.34), with c_s replaced with the appropriate wave speed. So, for the moment, let's set $B_\perp$ (and thus $\alpha_\perp$) to zero so that for $\alpha_x < 1$, $\mu = 1$ and $\nu = 0$ (Fig. 6.3). This makes the fast kets *identical* to the hydrodynamical kets for which, in §3.5.3, we concluded that density and pressure *decrease* from the upwind to downwind side of the fan, while the flow speed relative to the upwind state *increases*. A similar comparison between the slow and hydrodynamical fans may be made for $\alpha_x > 1$ where $\nu = 1$ and $\mu = 0$.

Noting from Fig. 6.3 that the primary effect of increasing $\alpha_\perp$ from zero is to round off the discontinuities in μ and ν without changing their monotonic dependence on α_x, we conclude that ρ and p should decrease across *any* RF while the flow speed relative to the upwind state increases. Thus, at least qualitatively, the hydrodynamical variables ρ, p, and v_x behave the same way across a fast and slow fan as they do across a hydrodynamical fan.

A consequence of p dropping across an MHD fan is that $\alpha_x = a_x^2/c_s^2 = B_x^2/(\mu_0 \gamma p)$ rises. This is an important observation that holds for *any* MHD RF. Note that α_x rises only because p falls; in a 1-D problem such as this, B_x is strictly constant. Thus, one can gain a qualitative feel for how μ and ν vary across a fan – quantities critical to determining the variable profiles from Eq. (6.1) – by scanning across Fig. 6.3 from left to right.

The last component in each ket in Eq. (6.21) governs the profiles of $B_\perp$, and it is here where the properties of the fast and slow fans diverge. For the fast fan, the last component, $-\nu\sqrt{\mu_0 \gamma p}$, is negative – just like the first two components governing the density and pressure profile – and $B_\perp$ *falls* across a fast RF when $\nu \neq 0$. Conversely, the last component in the slow ket, $\mu\sqrt{\mu_0 \gamma p}$, is positive, and $B_\perp$ *rises* across a slow fan so long as $\mu \neq 0$.

Fast fans

As the *magnitude* of a vector, $B_\perp$ is positive definite and cannot become negative. This means that a fast fan can *saturate*, and its "width" in terms of its generalised coordinate, s_i, is limited to where $B_\perp$ falls to zero ($s_{i,\max}$). A fast fan need not be wide enough to allow $B_\perp$ to fall to zero but, when it does, it is referred to as a – you guessed it – *switch-off* fan.

If $B_\perp = 0$ on the upwind side, there are two scenarios for a fast fan. First, for $\alpha_x \geq 1$ upwind of the fan, $\nu = 1$ ($\frac{1}{2}$ for $\alpha_x = 1$) and, since α_x increases across the fan, ν remains pegged at 1 (Fig. 6.3). This obliges $B_\perp$ to decrease across the fan which it can't do starting as it does from zero on the upwind side. We conclude, therefore, that a fast RF cannot be launched from an upwind state in which $B_\perp = 0$ and $\alpha \geq 1$. This, by the way, includes the triple umbilic ($B_\perp = 0$, $\alpha_x = 1$).

Second, for $\alpha_x < 1$, $\nu = 0$ and both $B_\perp$ and $v_\perp$ remain constant. Thus, $B_\perp$ can stay at zero (not being obliged to fall in this case) and, with $\mu = 1$ and $a_f = c_s$, the hydrodynamical variables ρ, p, and v_x vary just as they do across a pure hydrodynamical fan. Such a fan is referred to as a *fast Euler fan*, where the moniker "Euler" is used just as it is for shocks when $B_\perp = 0$ (page 164). Note that a fast Euler fan can also saturate since α_x – which starts < 1 and then rises across the fan – cannot rise beyond unity lest ν go from 0 to 1, obliging $B_\perp$ – already at 0 – to start decreasing. Thus, *a fast Euler fan cannot pass through the triple umbilic*.

Let us summarise what we now know about fast fans. Across any fan, fast or slow, ρ and p fall while v_x and α_x rise. Across a fast fan, $B_\perp$ falls and the fan can only be as wide as it takes for $B_\perp$ to get to zero. If $B_\perp > 0$ upwind of a fast fan and reaches zero on the downwind side, it is referred to as a "switch-off fan". If the upwind $B_\perp$ is already zero, then no fast fan can exist for upwind $\alpha_x \geq 1$ (since this would require $B_\perp$ to fall *below* zero), while for upwind $\alpha_x < 1$, a fast Euler fan (where $B_\perp$ starts and remains at zero) can exist, saturating should α_x reach unity (the triple umbilic).

Slow fans

For slow fans, the roles of μ and ν are reversed. Changes in the hydrodynamical variables are governed by ν (μ for fast fans), and changes in the magnetic variables, $v_\perp$ and $B_\perp$, are governed by μ (ν for fast fans). Further and as already noted, $B_\perp$ *rises* across a slow fan when $\mu > 0$, which leads to very different conclusions from those of the fast fan.

Slow fans don't saturate in the same way fast fans do; rather, they *asymptote* to the point where ρ and p approach zero without ever actually reaching it. Since $B_\perp$ rises from its upwind value across a slow fan, $B_\perp$ approaches an asymptotic limit as $p \to 0$, a value which can be determined only by integrating Eq. (6.1).

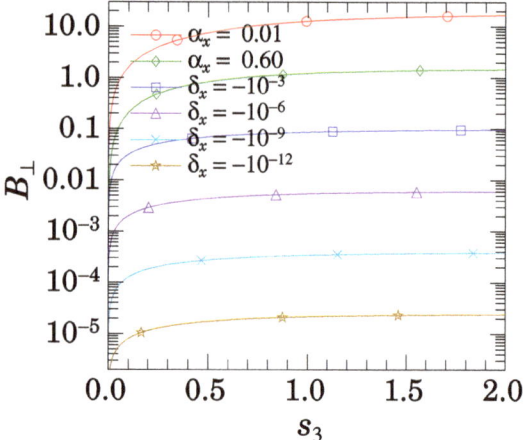

Figure 6.4. Profiles of $B_\perp$ (in units where $\mu_0 = 1$) as functions of the generalised coordinate for the 3-family wave, s_3, across switch-on (slow) fans for $0 < \alpha_x < 1$. Profiles for α_x just under 1 are labelled with $\delta_x = \alpha_x - 1 < 0$. As $\alpha_x \to 1$ ($\delta_x \to 0$), the asymptotic value of $B_\perp$ is progressively smaller and, in the limit where $\alpha_x = 1$ (triple umbilic), $B_\perp$ asymptotes to *zero*. Thus, there is no practical distinction between a switch-on RF launched from the triple umbilic and a slow Euler fan.

For $B_\perp = 0$ on the upwind side, there are again two cases. For $\alpha_x > 1$, $\mu = 0$ and thus $B_\perp$ remains at zero. Further, $\nu = 1$, $a_s = c_s$, and the hydrodynamical variables vary just as they do across a pure hydrodynamical fan. Such a fan is called a *slow Euler fan*, analogous to a fast Euler fan but with different upwind values of α_x: < 1 for a fast Euler fan; > 1 for a slow Euler, the realm from which no fast fan of any sort can be launched.

Staying with a zero $B_\perp$ on the upwind side, for $\alpha_x < 1$, μ starts at 1 and $B_\perp$ immediately starts to rise. Such slow fans are referred to as – you guessed it again – *switch-on* fans. However, since $B_\perp$ does not stay at zero, μ does not remain at unity and instead decreases monotonically across the fan. Referring to the left panel of Fig. 6.3, since both $\alpha_\perp = B_\perp^2/(\mu_0 \gamma p)$ and $\alpha_x = B_x^2/(\mu_0 \gamma p)$ increase as one progresses across the fan, the evolution of μ can be understood by starting at the red ○ $\alpha_\perp = 0$ profile where $\mu = 1$, and then dropping to profiles of progressively higher values of $\alpha_\perp$ (*e.g.*, green ◇, then blue □, then magenta △, *etc.*) as one also scans from left to right, accounting for the increasing value of α_x. Unlike the fast fan, the slow fan with an upwind $B_\perp = 0$ can pass through the point $\alpha_x = 1$, which for it is not the triple umbilic since within the fan, $B_\perp > 0$.

It is the relative rate at which each of these factors, namely an increasing $B_\perp$ (increasing $\alpha_\perp$) and a decreasing p (increasing α_x), cause μ to fall from unity that determines the asymptotic value of $B_\perp$ for a switch-on fan. If the upwind value of $\alpha_x \ll 1$ where the profiles of μ are relatively flat (left panel of Fig. 6.3), then increasing α_x does not affect μ appreciably, and μ drops relatively gradually as $B_\perp$ rises. By the time α_x increases past unity, the growth of $B_\perp$ has been governed by relatively high values of μ, and thus the asymptotic value to which it rises is relatively high (*e.g.*, the red$\bigcirc$ profile in Fig. 6.4).

Conversely, if the upwind value of α_x is just under unity, then as seen in the middle of the left panel of Fig. 6.3, the profiles of μ dive down towards zero rapidly for a small increase in α_x, and $B_\perp$ has little room to grow asymptotically as $p \to 0$, as shown by the sequence of profiles in Fig. 6.4. In the limiting case as the upwind value of $\alpha_x \to 1$ ($\delta_x \to 0$; *i.e.*, a slow fan launched from the triple umbilic), $B_\perp$ has *zero* room to grow (an infinitesimal increase in α_x takes μ to zero), and such a fan is of no practical difference from an Euler fan. Therefore, one can safely include the $\alpha_x = 1$ case with $\alpha_x > 1$, and refer to it as a slow Euler fan as well.

To summarise what we know about slow fans for $\alpha_x > 0$, hydrodynamical variables fall while both α_x and $B_\perp$ rise, with all values approaching their asymptotic limits as $p \to 0$. If the upwind $B_\perp = 0$ and $0 < \alpha_x < 1$ ($a_s = a_x$), $B_\perp$ grows across what is called a "switch-on fan" whose width can take it past the $\alpha_x = 1$ point. If the upwind $B_\perp = 0$ and $\alpha_x \geq 1$ ($a_s = c_s$), $B_\perp$ remains zero across what is called a "slow Euler fan".

Zero axial field

So far, we've tacitly been assuming $B_x \neq 0$. When $B_x \to 0$ (and thus $\alpha_x \to 0$), there are no singularities in Eq. (6.21), (6.26), and (6.27) to concern us and accounting for this case is simple. With a little algebra, Eq. (6.26) and (6.27) become in the limit $\alpha_x \to 0$ ($\delta_x \to -1$),

$$\mu^2 \to \frac{1}{\alpha_\perp + 1} \quad \text{and} \quad \nu^2 \to \frac{\alpha_\perp}{\alpha_\perp + 1}. \quad (6.30)$$

Further, the slow alpha (Eq. 6.24) becomes:

$$\alpha_s = \frac{1}{2}\left(1 + \alpha_\perp - \sqrt{(\alpha_\perp - 1)^2 + 4\alpha_\perp}\right) = 1 + \alpha_\perp - \sqrt{(\alpha_\perp + 1)^2} = 0,$$

since $\alpha_\perp + 1 > 0$. Thus, α_s (and a_s) is zero, and the slow wave neither develops nor propagates. By contrast, the fast alpha is given by:

$$\alpha_f = \frac{1}{2}\left(1 + \alpha_\perp + \sqrt{(\alpha_\perp - 1)^2 + 4\alpha_\perp}\right) = 1 + \alpha_\perp$$

$$\Rightarrow a_f^2 = c_s^2 + a_\perp^2 = a_M^2,$$

where a_M is the magneto-acoustic speed introduced in §5.2.2. Thus, the fast RF survives in the $B_x \to 0$ limit as a *magneto-acoustic fan*. Indeed, for $\alpha_\perp \to 0$ (and thus we are truly at the hydrodynamical limit with *both* components of $\vec{B}$ zero), $\mu \to 1$, $\nu \to 0$, $a_f \to c_s$, and the fast ket in Eq. (6.21) is identical in every way to the hydrodynamical ket in Eq. (3.34). It is therefore the fast MHD fan which becomes the ordinary HD fan in the hydrodynamical limit, while the slow and Alfvén waves merge with the contact forming a tangential discontinuity (§5.3.3).

Examples of MHD rarefaction fans

Figures 6.5–6.7 on pages 197 and 198 show profiles (left → right is upwind → downwind) for the primitive variables ρ, p, v_x, $v_\perp$, and $B_\perp$,[4] and either $\alpha_\perp$ or α_x for the three kinds of fast rarefaction fans discussed. Figures 6.8–6.10 on pages 198 and 199 show the same variables for the three types of slow fans. These profiles were determined using a sixth-order Runge–Kutta integrator (App. F) to solve Eq. (6.1) using the kets in Eq. (6.21) with μ and ν given by Eq. (6.26) and (6.27). The computational details are postponed to §6.4.3 with the algorithm for the MHD Riemann solver. Here, I just wish to give the reader a feel for what various types of MHD fans actually look like and some of the differences among them.

The first thing to note on Fig. 6.5–6.10 is the use of the self-similar independent variable, $x/t = u_i$[5] rather than the generalised coordinate, s_i, used in Fig. 6.4 over which the integrations are actually performed. This is similar to what was done in §3.5.3 for the hydrodynamical fans except there, it was possible to find an expression linking the generalised and self-similar coordinates (Eq. 3.43), which was then used to find profiles for the primitive variables in closed form (Eq. 3.44–3.47). Here, no such closed-form expressions exist, and we must use Runge–Kutta to find the variable profiles in terms of s_i [$e.g.$, $\rho(s_i)$], and then use these solutions to generate the self-similar coordinate, x/t, from the generalised one. And so, starting from the co-moving coordinate, $\xi_i = x - u_i t$, we set ξ_i to zero to put us in the co-moving frame of the fan from which we can write,

$$\frac{x}{t} = u_i = v_x(s_i) \pm a_{f,s}(s_i), \qquad (6.31)$$

where $a_{f,s}$ are given by Eq. (5.23)–(5.25). Once generated, it is then simply a matter of plotting the flow variables against x/t as the independent variable rather than s_i.

The "regular" fan in Fig. 6.5 and the "switch-off" fan in Fig. 6.6 were launched from the same upwind state ($\rho = 3$, $p = 3$, $v_x = v_\perp = 0$, $B_x = 2$, and $B_\perp = 1 \Rightarrow \alpha_x = 0.8$, $\alpha_\perp = 0.2$, all in arbitrary units). The only difference between them is the "switch-off" fan is saturated; that is, integrated until $B_\perp$ reaches zero which, as the reader will note, occurs while p is still well above zero. For these fans, it does not matter whether the upwind values $\alpha_x \gtrless 1$ or $\alpha_\perp \gtrless 1$; qualitatively they look identical. It should also be noted that choosing non-zero upwind values for v_x and $v_\perp$ is tantamount to making a Galilean transformation, leaving profiles for the Galilean invariants (ρ, p, $B_\perp$, and $\alpha_{x,\perp}$) unchanged. Only the velocity profiles are affected, and then just shifted by the choice of upwind $v_{x,\perp}$.

The fast Euler fan in Fig. 6.7 was launched from the upwind state $\rho = 3$, $p = 3$, $v_x = v_\perp = 0$, $B_x = 2$, and $B_\perp = 0.0$, differing from the first two fast fans only in the

---

[4]For convenience, all values for B_x and $B_\perp$ given from here until the end of §6.2 including Fig. 6.5–6.10 are in units where $\mu_0 = 1$. To restore units, multiply any given value of B_x or $B_\perp$ by $\sqrt{\mu_0}$.

[5]*Rappel*: $u_i = v \pm a_f$ are the eigenspeeds for the right- (+) and left- (−) moving fast waves, while $u_i = v \pm a_s$ are the eigenspeeds for the right- and left-moving slow waves. Forgotten what "self-similar" means? See footnote 11 on page 88.

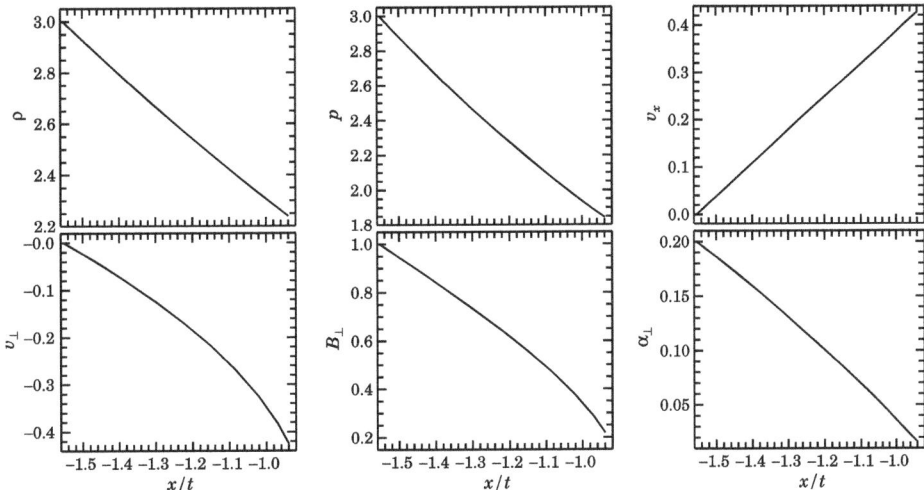

Figure 6.5. Upwind (left) to downwind (right) profiles of the primitive variables and $\alpha_\perp$ across a regular (non-saturated) fast RF. As is typical of fast fans, $B_\perp$ and $\alpha_\perp$ fall monotonically. In this and the following five figures, $B_\perp$ is given in units where $\mu_0 = 1$.

choice for $B_\perp$. Notably, $\alpha_x = 0.8 < 1$ upwind of the fan. As this fan was integrated until $\alpha_x = 1$, this is a saturated fast Euler fan. A non-saturated fast Euler fan would look just like Fig. 6.7, with the profiles ending somewhere to the left of where they do in the figure with α_x, in particular, never reaching 1. As discussed above, fast

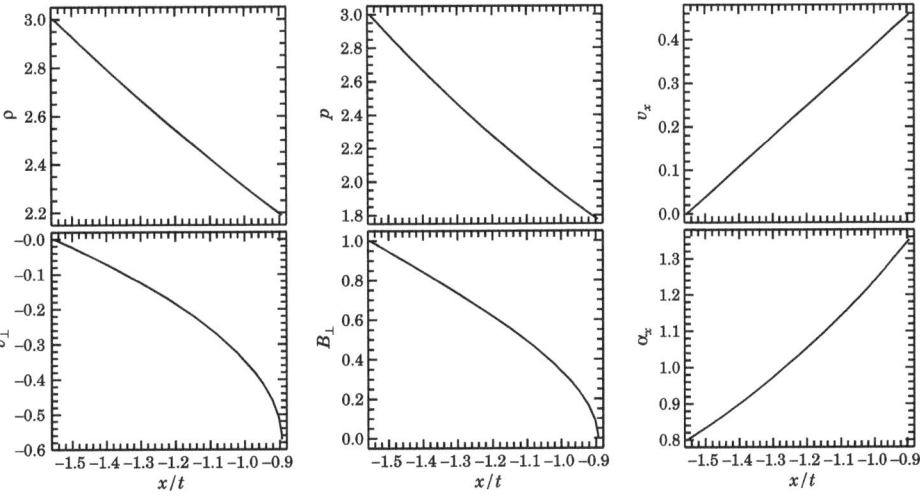

Figure 6.6. Profiles of the primitive variables and α_x across a "switch-off fan", a saturated fast RF where $B_\perp \to 0$. Since $B_\perp > 0$ everywhere except the most immediately downwind point, α_x can pass through 1.

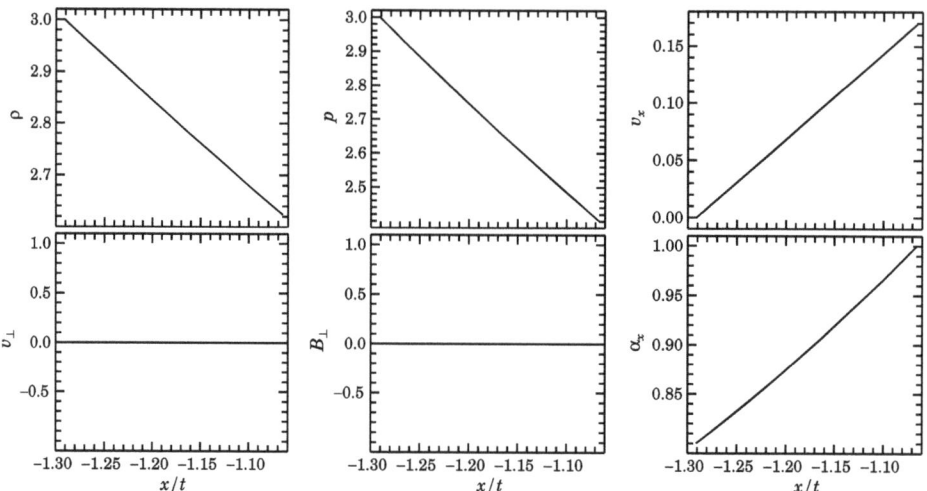

Figure 6.7. Profiles of the primitive variables and α_x across a fast Euler RF. Note that this fan is saturated, since $\alpha_x \to 1$.

fans with $B_\perp = 0$ cannot pass through the $\alpha_x = 1$ point (triple umbilic); indeed, no fast RF can be launched from an upwind state where $B_\perp = 0$ and $\alpha_x \geq 1$.

Turning now to the slow fans, the "regular" slow fan in Fig. 6.8 and the "switch-on" fan in Fig. 6.9 were launched from the upwind state $\rho = 3$, $p = 3$, $v_x = v_\perp = 0$, and $B_x = 2 \Rightarrow \alpha_x = 0.8$, with $B_\perp$ being the only difference: 1 for the regular fan; 0 for the switch-on fan. Other than the steepness at which $B_\perp$ and $v_\perp$ rise from the

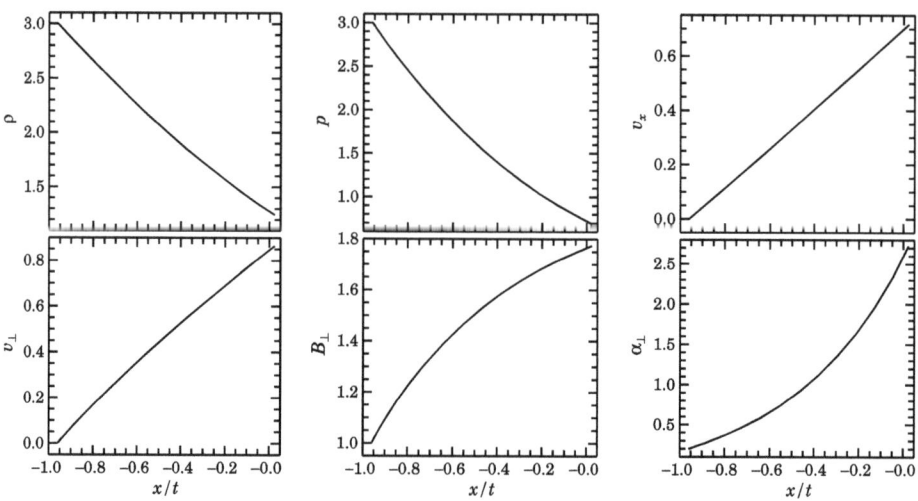

Figure 6.8. Profiles of the primitive variables and $\alpha_\perp$ across a regular slow RF. As is typical of slow fans, $B_\perp$ and $\alpha_\perp$ rise monotonically.

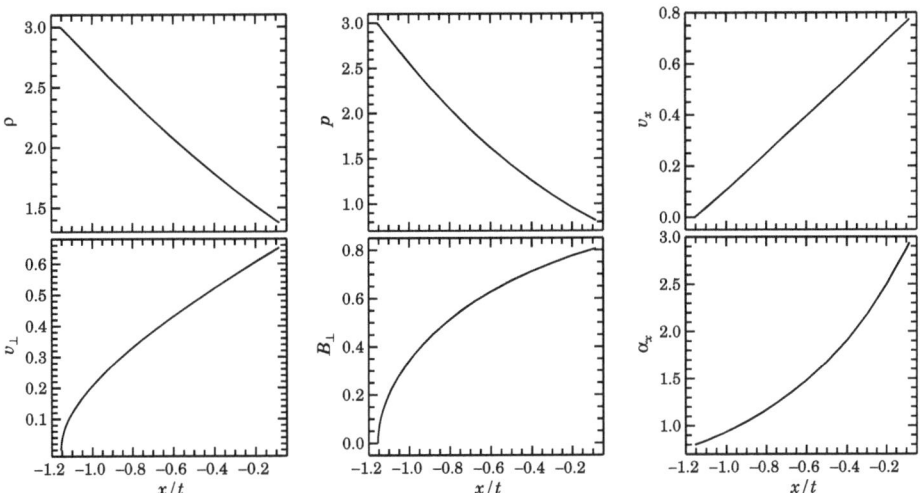

Figure 6.9. Profiles of the primitive variables and α_x across a "switch-on fan", a slow RF where upwind, $B_\perp = 0$ and $\alpha_x < 1$. Because $B_\perp > 0$ everywhere except the most immediately upwind point, α_x can pass through 1.

upwind state in the switch-on fan, there is little to distinguish the profiles in the two figures. Note that the switch-on fan is launched from an upwind state in which $\alpha_x < 1$. Once launched, $B_\perp$ begins rising immediately, and the fan behaves just like a regular slow fan where α_x can rise as far as the fan is integrated (in this case to a final value of $\alpha_x \sim 3$).

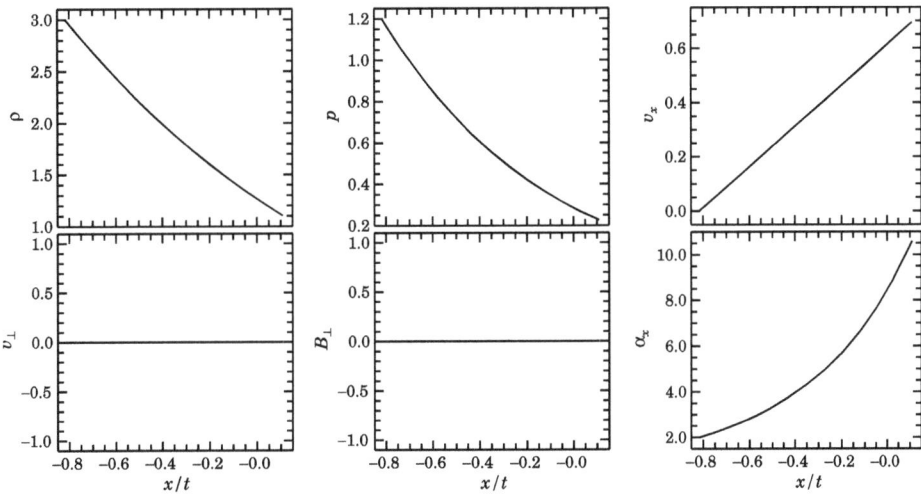

Figure 6.10. Profiles of the primitive variables and α_x across a slow Euler RF, where upwind $B_\perp = 0$ and $\alpha_x \geq 1$. If this fan were launched from the triple umbilic, α_x would have started at 1, and risen from there.

Slow fans don't saturate as fast fans do, but rather asymptote towards $p \to 0$. A fast fan ends at a finite value of its generalised coordinate, s_i, once $B_\perp$ reaches zero, whereas for a slow fan, $s_i \to \infty$ as $p \to 0$. However, with respect to the self-similar coordinate, x/t, as p asymptotes to zero, $x/t = u_i$ reaches a corresponding asymptotic value, and the width of the fan in real space is finite. It is a curious thing, then, that in generalised space the width of a slow fan is not bounded while in real space it is and, as further discussed in §6.4.3, this can pose computational challenges if one attempts to integrate a slow fan out too far.

Finally, Fig. 6.10 shows the profiles for a slow Euler fan launched from the upwind state $\rho = 3$, $p = 1.2$, $v_x = v_\perp = 0$, $B_x = 2$, and $B_\perp = 0$, differing from that for the switch-on fan (Fig. 6.9) only in the value for p. Notably, here the upwind $\alpha_x = 2 > 1$ as required for slow Euler fans and, since α_x rises across all MHD fans, α_x remains above unity and $B_\perp$ remains at zero. Interestingly, there is little to distinguish fast and slow Euler fans (Fig. 6.7 and 6.10) other than α_x which, for a fast fan is ≤ 1 and for a slow fan is ≥ 1. Indeed, a fast saturated Euler fan ending on the triple umbilic ($\alpha_x = 1$, $a_f = a_s$) could join smoothly with a slow Euler fan launched from the triple umbilic forming a compound wave which, in every way, would be described as a single, ordinary, hydrodynamical RF threaded by an axial component of the magnetic induction where $\alpha_x < 1$ on the upwind side and > 1 on the downwind side. This case is revisited in §6.3.

Summary

The phenomenology of MHD fans is complex – at least as complex as MHD shocks – and there is much to consider. Figure 6.11 is therefore offered to give the reader a pictorial summary of the various types of MHD fans described, and to help put them into context with each other.

As we've seen, α_x and/or $\alpha_\perp$ can vary across a given MHD fan. Thus, plotting $\alpha_\perp$ as a function of α_x on an α_x–$\alpha_\perp$ graph such as Fig. 6.11 yields a unique locus of points for each RF. Loci a and b represent typical fast fans while c and d represent slow fans. Arrows on the loci indicate progression across the fan from the upwind to downwind side, reflecting the fact that α_x increases across all MHD fans. Fast fans whose loci end on the $\alpha_x > 1$ portion of

Figure 6.11. Loci for fast (a, b) and slow (c, d) RF described in the text. The triple umbilic ($\alpha_\perp = 0$, $\alpha_x = 1$) is located where the fast and slow Euler regions meet.

the α_x-axis (highlighted in magenta) are the switch-off fans (*e.g.*, b), while slow fans whose loci start on the $\alpha_x < 1$ portion (highlighted in red) are the switch-on fans (*e.g.*, d). Not all fast fans are switch-off (*e.g.*, a). However, by extending locus a downwind (dashed line) to intersect the α_x-axis where $\alpha_\perp = 0$, one can see where it would become a switch-off fan were it integrated to saturation. Similarly, not all slow fans are switch-on (*e.g.*, c). However, by extending locus c upwind to intersect the α_x-axis (dashed line) shows how the fan could have been switch-on for a suitably modified upwind state. Loci of fast fans (or their downwind extensions) end on the $\alpha_x > 1$ portion of the α_x-axis, while loci of slow fans (or their upwind extensions) start on the $\alpha_x < 1$ portion of the α_x-axis.

So what do these loci represent physically? Recall from Theorem 3.3 in §3.5.3 that across the minus (left) hydrodynamical RF, the Riemann invariant $\mathcal{J}^+$ is constant, and similarly across the plus (right) RF, $\mathcal{J}^-$ is constant. Extrapolating this idea to MHD fans, across the minus fast and slow fans, the Riemann invariants $\mathcal{F}^+$ and $\mathcal{S}^+$ are constant, respectively. Thus, were one to express $\mathcal{F}^+$, say, as a function of α_x and $\alpha_\perp$, contours of $\mathcal{F}^+(\alpha_x, \alpha_\perp)$ on an α_x-$\alpha_\perp$ graph would be lines of constant $\mathcal{F}^+$ and thus correspond to loci of minus fast fans such as loci a and b in Fig. 6.11. Similarly, contours of $\mathcal{S}^+(\alpha_x, \alpha_\perp)$ would correspond to loci of minus slow fans, such as c and d in the figure. Note that we didn't need to specify the functional forms for $\mathcal{F}^+(\alpha_x, \alpha_\perp)$ and $\mathcal{S}^+(\alpha_x, \alpha_\perp)$[6] in order to draw this conclusion.

The fact that loci representing RF on an α_x-$\alpha_\perp$ graph correspond to contours of a continuous function such as its associated Riemann invariant means that fast (slow) loci never cross other fast (slow) loci, and the two sets of loci are nested as shown in Fig. 6.11. Further, a fast locus can never start from the magenta region ($\alpha_\perp = 0$, $\alpha_x > 1$), since this is where the fast loci (or their extensions) end, while a slow locus can never end in the red region ($\alpha_\perp = 0$, $\alpha_x < 1$), since this is where all slow loci (or their extensions) begin.

On the other hand, fast loci can start from (and remain within) the red region ($\alpha_\perp = 0$, $\alpha_x < 1$) and slow loci can start from (and remain within) the magenta region ($\alpha_\perp = 0$, $\alpha_x > 1$); these represent the Euler fans. Note that the locus of a fast Euler fan can extend no further than the triple umbilic ($\alpha_\perp = 0$, $\alpha_x = 1$), lest it intersect the end point of a non-Euler fast locus.

Fast loci all point downward since $\alpha_\perp$ drops along them. Those intersecting the α_x-axis at relatively high values of α_x do so with greater slope than those intersecting the axis at $\alpha_x \gtrsim 1$ where, in the extreme, a fast Euler locus intersects the magenta region ($\alpha_\perp = 0$, $\alpha_x > 1$) at $\alpha_x = 1$ with zero slope. Loci for "nearly Euler" fast fans (starting with $\alpha_x \ll 1$) and with a large upwind value of $B_\perp$ (*e.g.*, $\alpha_\perp > 1$) hug the $\alpha_\perp$ axis as $B_\perp$ falls, turn sharply near the origin, then continue by hugging the α_x axis until they intersect it at α_x very slightly greater than 1. The innermost unlabelled fast locus in Fig. 6.11 exemplifies this.

Fast fans launched from within the green region ($\alpha_x = 0$, $\alpha_\perp > 0$) are the magneto-acoustic fans across which B_x remains zero. Thus, loci for such fans are confined to this region, and drop along the $\alpha_\perp$ axis from the upwind to downwind state. Fast fans launched from the blue region ($\alpha_x = \alpha_\perp = 0$) are pure hydrody-

---

[6] In fact, finding $\mathcal{F}^\pm$ and $\mathcal{S}^\pm$ as functions of α_x and $\alpha_\perp$ is a problem to which I do not know the solution, and thus I have neither included it in the text nor relegated it to the problem set! I'd be happy to credit any reader who submits such a solution in any future edition, with a wink to Cambridge University Press!

namical fans. Loci for such fans have zero length (*i.e.*, remain at the origin) on an α_x vs. $\alpha_\perp$ plot, since $\vec{B}$ remains zero across their width. This, of course, doesn't mean they don't exist – hydrodynamical fans most certainly do! – they just don't show up on a magnetic plot such as Fig. 6.11.

Loci for slow fans are straight lines[7] and all point upward reflecting the fact that $B_\perp$ increases across a slow fan. All slow loci (or their upwind extensions) start from within the $\alpha_x < 1$ (red) region of the α_x-axis, and the greater the value of α_x from which they start, the lesser their slope. At the extremes, for $\alpha_x = 0$ the slope of the locus is infinite (vertical), and such slow fans do not propagate (since $a_s = 0$ there). Thus, slow fans cannot be launched from the $\alpha_\perp$-axis (blue and green regions) of Fig. 6.11. At the other extreme where $\alpha_x \to 1$ (triple umbilic), the slope is zero and a slow Euler fan is launched (where $\alpha_\perp$ stays zero).

6.3 Space-time diagrams

In §3.6, we saw that for the *strictly* hyperbolic equations governing ordinary HD, the solution to the Riemann problem *always* consists of three distinct wave families: family 1 (3) is a left (right) moving shock or RF, and family 2 is an entropy wave (contact discontinuity). This simplified the solution to the hydrodynamical Riemann problem a lot more than we may have appreciated at the time.

For the *not strictly* hyperbolic equations governing MHD, the possibility that the eigenspeeds can be partially or fully degenerate means that no single space-time diagram typifies all solutions to the 1-D MHD Riemann problem. Figure 6.12 catalogues the possible diagrams that can describe the left side of an MHD Riemann solution, where the full MHD Riemann solution will be, in general, any two panels (with the right side reflected) spliced together at the family 4 wave.

Panel *a* shows the "normal case", where $B_x \neq 0$ and $B_\perp \neq 0$, the latter even after changes caused by the passage of the fast and slow waves. Here, the left-most characteristic path represents a 1-wave (a fast shock or fan) which can change the magnitude of all primitive variables but not the directions of $\vec{B}_\perp$ and $\vec{v}_\perp$, followed by the path of a 2-wave (rotational discontinuity or Alfvén wave) which can change only the direction of $\vec{B}_\perp$ and the direction and magnitude of $\vec{v}_\perp$, followed by the path of a 3-wave (a slow shock or fan) which also can only affect the magnitudes of the flow variables, followed finally by the path of a 4-wave (contact discontinuity). The contact, the reader will recall (§5.3.2), is an entropy wave in the co-moving frame of the fluid (and thus $v_x = 0$ on both sides) across which p, $\vec{B}_\perp$, and $\vec{v}_\perp$ are constant, while ρ can jump arbitrarily.

For $B_x = 0$ ($\alpha_x = 0$), $a_x = a_s = 0$ and neither the Alfvén nor slow wave exists. This is a non-linear version of a magneto-acoustic wave (§5.2.2), and leads to the

---

[7]I have yet to prove this analytically, and my only evidence of this are the numerous computer-generated plots of $\alpha_\perp$ vs. α_x I created to motivate Fig. 6.11 and the surrounding discussion. Proof lies in finding an analytical function for the slow Riemann invariants $\mathcal{S}^\pm(\alpha_x, \alpha_\perp)$.

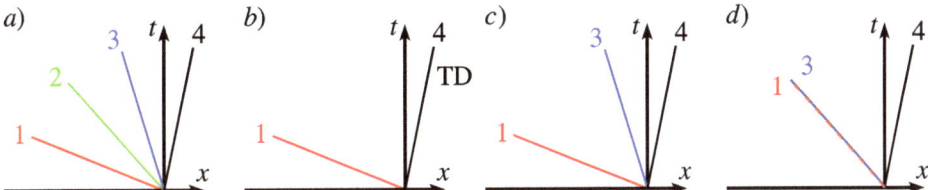

Figure 6.12. Left sides of the four space-time diagrams that can represent solutions to the 1-D MHD Riemann problem. In keeping with Fig. 6.2, red, green, blue, and black characteristic paths represent wave families 1, 2, 3 and 4. For argument's sake, paths are drawn assuming $0 < v_x < a_s$. Each panel corresponds to one of the cases described in the text: a) "normal" ($B_x \neq 0$, $B_\perp \neq 0$); b) "magneto-acoustic" ($B_x = 0$) where the family 4 wave becomes a tangential discontinuity; c) "partial degeneracy" ($B_\perp = 0$, $\alpha_x \gtrless 1$) between families 1 and 3; and d) full degeneracy ($B_\perp = 0$, $\alpha_x = 1$; triple umbilic) between families 1 and 3 whose waves are attached and propagate together as a compound wave.

characteristic paths in Fig. 6.12b. Here, the path of a fast 1-wave precedes that of the 4-wave, resembling the left side of the hydrodynamical space-time diagram in Fig. 3.8. For $B_\perp > 0$ upwind, the fast wave can be an ordinary shock or an ordinary RF, but not, as Problem 6.4 shows, a "switch-off" fan. With $\alpha_x = 0$, $B_\perp$ drops across the fan in lockstep with p and, for a wide enough fan, $B_\perp$ reaches zero asymptotically at the same rate as p, rather than abruptly and before p reaches zero as required of a switch-off fan. Meanwhile, the 4-wave becomes a tangential discontinuity (§5.3.3) where $p + B_\perp^2/2\mu_0$ is continuous rather than p. If $B_\perp = 0$ upwind, the fast wave is pure hydrodynamical (switch-on shock being precluded since this requires $\alpha_x > 1$), and the 4-wave remains an ordinary contact discontinuity. Indeed, with both B_x and $B_\perp$ zero, the MHD Riemann problem reduces to the hydrodynamical Riemann problem of §3.6.

A fast wave can generate $B_\perp = 0$ on its downwind side in one of three ways: as an Euler shock; an Euler fan; or a switch-off fan. In each case, with no $B_\perp$ to rotate, there can be no rotational discontinuity and the characteristic pattern resembles Fig. 6.12c where the characteristic path of the slow 3-wave follows that of the 1-wave with no 2-wave in between. If the fast and slow speeds are degenerate (as happens should the triple umbilic fall between the two waves), they are attached and propagate as one *compound wave* (Fig. 6.12d).

An Euler shock or fan occurs if the upwind $B_\perp$ is already zero and the upwind $\alpha_x < 1$ ($a_f = c_s$; partial degeneracy). Since α_x falls across a shock, an Euler shock can be of arbitrary strength delivering an even lower value of α_x to its downwind side. Conversely, the width of an Euler fan is limited, since α_x rises across a fan but cannot rise beyond 1 (*i.e.*, the triple umbilic where $a_f = a_x = c_s = a_s$; full degeneracy). If the upwind $\alpha_x > 1$, a fast shock would be a switch-on shock and thus would not deliver $B_\perp = 0$, while a fast RF cannot be launched from such an upwind state.

For a non-zero upwind $B_\perp$, only a switch-off fan can deliver a zero $B_\perp$ on the downwind side; a 1-shock can't since $B_\perp$ rises across a fast non-Euler shock. A switch-off fan can be launched from any $\alpha_x > 0$, and can deliver any downwind value of α_x greater than its upwind value. Thus, should the upwind α_x be < 1, it is possible for the switch-off fan to end at $\alpha_x = 1$, *i.e.*, the triple umbilic.

Thus, there are two cases in which the triple umbilic can fall between the fast and slow wave: a saturated fast Euler fan; and a switch-off fan that delivers $\alpha_x = 1$ to its downwind side. In either of these (unusual) cases, the downwind speed of the fast wave is degenerate with the upwind speed of the following slow wave, and the two waves are joined at the triple umbilic, propagating as a single compound wave.

Now, the only slow wave that can be launched from the triple umbilic is a slow Euler fan. A slow non-Euler shock cannot be launched from the triple umbilic because it would require $B_\perp$ to drop below zero, while a slow Euler shock strictly requires $\alpha_x < 1$, not the triple umbilic's value of 1. Finally, starting from $B_\perp = 0$, a non-Euler fan would be a switch-on fan, again requiring $\alpha_x < 1$. Thus, a compound wave crossing the triple umbilic can be either a switch-off fan attached to a slow Euler fan (for $B_\perp \neq 0$ upwind of the fast wave), or two Euler fans (for $B_\perp = 0$ upwind of the fast wave). Both cases are depicted in Fig. 6.12*d*.

6.4 An MHD Riemann solver

Fundamental to the design of our Riemann solver will be our ability to make *very* educated guesses. Starting at the left state (state **L** in Fig. 5.1), we'll "guess our way" to state ③ immediately to the left of the CD by guessing answers to:

1. Is the 1-family wave a shock or an RF and how strong or wide is it?

2. If $\vec{B}_\perp \neq 0$ downwind of the 1-family wave, by what angle does the RD rotate it?

3. Is the 3-family wave a shock or an RF and how strong or wide is it?

Similarly, starting at the right state (state **R** in Fig. 5.1), we'll "guess our way" to state ④ immediately to the right of the CD where we'll compare notes. If $p_3 - p_4$, $v_{\perp,3} - v_{\perp,4}$, $\vec{v}_{\perp,3} - \vec{v}_{\perp,4}$, and $\vec{B}_{\perp,3} - \vec{B}_{\perp,4}$ as must all be true across a CD (remember, the jump in ρ across a CD is arbitrary), then based on the assumption that the solution to the 1-D MHD Riemann problem is unique,[8] we're done!

Of course, odds are slim to none we'll guess all the right answers on the first attempt, and without some quantitative strategy to guide our subsequent guesses, finding a satisfactory solution could take forever!

Fortunately, such a strategy exists, and one so efficient that a solution good to

---

[8]And there's the rub, as it is well known that the full solution to the 1-D MHD Riemann problem is *not* unique (*e.g.*, Takahashi & Yamada, 2014)! However, uniqueness is restored if one assumes the "evolutionary condition" introduced in §5.3.5 in which *intermediate shocks* are not allowed. This is our strategy here.

within ten digits of accuracy can usually be found with less than ten guesses, all within a millisecond on a laptop! Hard to believe? Well then, read on...

6.4.1 Problem parameters

The main engine of the Riemann solver is a secant root finder. Now, a simple *univariate* secant root finder – based on finding the value of *one* variable that satisfies *one* constraint – was used in our algorithm for the HD Riemann problem in §3.6. There, we had two waves, namely a left- and right-moving shock or RF each of unknown strength, and two constraints, namely both pressure and velocity had to be constant across the contact discontinuity (CD). Because we could write down closed expressions for the downwind velocities in terms of the unknown wave strengths (*i.e.*, ζ; see Eq. 3.55, 3.57, 3.60, and 3.62), setting them equal across the CD was tantamount to applying the constraint on velocity, leaving a single equation in a single unknown, namely p. This could be solved analytically when both waves were rarefaction fans (*e.g.*, Eq. 3.65), and with a univariate secant root finder (§D.1) in the three cases when at least one wave was a shock.

Now, had analytical expressions for the downwind velocities not been available and downwind values of the primitive variables had to be determined numerically, we would have been left with two unknown variables – the left- and right-moving wave strengths – with two constraints to apply at the CD. For this, a *bivariate* root finder – one that finds values for two independent variables so that two independent constraints are satisfied simultaneously – would have been needed.

This is more like the case for the MHD Riemann problem. Here, we do not have the luxury of analytically closed expressions for the velocities that link the left and right states all the way to their side of the CD. Worse, we have *six* waves of unknown strength and *six* constraints at the CD (each of p, v_x, v_y, v_z, B_y and B_z constant) rather than two and two for the HD case. Now, as we'll soon see, the MHD problem can be whittled down to five unknown wave strengths (which we'll call *problem parameters*, or just *parameters* for short) and five independent constraints. However, as everything is evaluated numerically, what we'll end up with are ten numbers – five on each side of the CD – and our task will be to find values for all five parameters such that the differences in all five pairs of numbers across the CD are simultaneously zero.

What is needed here is a *multivariate* secant root finder – something I myself had never come across until looking to solve this problem – which is derived for the interested reader in §D.2 in App. D. While this root finder plays a key role in solving the MHD Riemann problem, knowledge of its derivation is not prerequisite to understanding and even implementing the algorithm. Thus, the reader may safely postpone reading §D.2 until a later time.

The first thing to identify are the six parameters describing the current guesses for the strengths of the six waves, namely the left-moving 1-, 2-, and 3-waves [fast, rotational discontinuity (RD), and slow respectively], and the right-moving 5-, 6,- and 7-waves (slow, RD, and fast respectively). The 4-wave, of course, is the CD – or the tangential discontinuity (TD) when $B_x = 0$ – across which the six constraints (two for the TD) are applied.

The strengths of each of the fast and slow waves and whether they are shocks or RFs will be embodied by four parameters, ψ_1, ψ_2, ψ_3, and ψ_4 for the left fast, right fast, left slow, and right slow waves respectively. The reason for this specific order shall become apparent in due course. The parameters ψ_5 and ψ_6 indicate the strengths of the left- and right-moving RDs respectively.

Beginning with the RD, its "strength" is nothing other than the angle through which $\vec{B}_\perp$ is rotated, and so we'll interpret ψ_5 and ψ_6 as these rotation angles. Herein lies the "whittling-down" of independent parameters alluded to above. Suppose the orientation of $\vec{B}_\perp$ in the left and right states is χ_L and χ_R respectively. Then, since neither the slow nor the entropy waves are capable of rotating $\vec{B}_\perp$, both RDs must rotate $\vec{B}_\perp$ to the same orientation, and we have,

$$\chi_L + \psi_5 = \chi_R + \psi_6 \quad \Rightarrow \quad \psi_6 = \psi_5 + \chi_R - \chi_L, \tag{6.32}$$

and ψ_6 is determined by ψ_5. This means we can set our guess for the orientation of $\vec{B}_\perp$ downwind of *both* RDs to $\chi = \chi_L + \psi_5$ (with no further need for ψ_6) where the components of $\vec{B}_\perp$ are evidently given by:

$$B_{y,d} = B_\perp \cos\chi; \quad \text{and} \quad B_{z,d} = B_\perp \sin\chi. \tag{6.33}$$

Here, the upwind value of the magnitude $B_\perp$ can be used since it is constant across an RD. Then, given Eq. (6.33), the downwind perpendicular components of the velocity can be computed from Eq. (5.91), which we rewrite here as:

$$\vec{v}_{\perp,d}^{\pm} = \vec{v}_{\perp,u} \mp \frac{\text{sgn}(B_x)}{\sqrt{\mu_0 \rho}} (\vec{B}_{\perp,d} - \vec{B}_{\perp,u}), \tag{6.34}$$

where the subscripts 'd' and 'u' refer to downwind and upwind values respectively. The superscript $\pm$ on the LHS refers to the right-moving 6-family (+), and left-moving 2-family (−) RD, while the corresponding '$\mp$' on the RHS reflects the fact that *in the frame of reference of the RD*,

$$v_x = \begin{cases} -a_x = -\dfrac{|B_x|}{\sqrt{\mu_0 \rho}}, & \text{for the right-moving RD; and} \\ +a_x, & \text{for the left-moving RD.} \end{cases}$$

With now just five independent parameters, ψ_j, $j = 1:5$, we'll need five independent constraints across the CD, $f_i(\psi_j)$, $i = 1:5$, to evaluate them. For these, we'll take:

$$\left.\begin{aligned} f_1(\psi_j) &= p_4 - p_3 = 0; & f_2(\psi_j) &= v_{x_4} - v_{x_3} = 0; \\ f_3(\psi_j) &= v_{y_4} - v_{y_3} = 0; & f_4(\psi_j) &= v_{z_4} - v_{z_3} = 0; \\ f_5(\psi_j) &= B_{\perp_4} - B_{\perp_3} = 0, & & \end{aligned}\right\} \tag{6.35}$$

where $B_\perp^2 = B_y^2 + B_z^2$, and where subscripts 3 and 4 refer to states ③ and ④ on the left and right sides of the CD respectively.

Turning now to shocks and RFs, in many respects one can think of these two phenomena as yin and yang; two halves of one continuum. Quantities such as $-\nabla \cdot \vec{v}$, $\log \eta$ (density jump) and $\log \zeta$ (pressure jump) are all positive across a shock, negative across an RF, and zero when each becomes infinitesimally weak. Further, a shock of infinitesimal strength generates jumps in the primitive variables infinitesimally different from differences in the primitive variables across an RF of infinitesimal width. Thus, we can reasonably expect a single continuous parameter, ψ_j, to describe the strength of the shock when $\psi_j > 0$, say, and the width of the RF when $\psi_j < 0$, with $\psi_j = 0$ corresponding to either being infinitesimally weak. So the question becomes, what physical parameter of a fast or slow shock that uniquely identifies it shall we attribute to $\psi_j > 0$ and, likewise, what physical parameter of the fast or slow RF shall we attribute to $\psi_j < 0$?

Let's start with the left-moving fast wave separating states **L** and ①, and thus the parameter ψ_1. From the top left panels of Fig. 5.12 and 5.13,[9] we see that the strength of a fast shock, $\zeta = p_1/p_L$, is a monotonic function of M_L, where M_L (upwind Mach number relative to the shock) begins at an easy-to-identify point (namely, the fast point where $M_L = a_{f_L}/c_{s_L} = M_+$; Problem 5.23), and has no upper bound. Further, we know how to compute $\phi_x = v_{x_1}/v_{x_L}$ from M_L (Eq. 5.104) from which jumps in all other variables follow: $b = B_{\perp_1}/B_{\perp_L}$ (Eq. 5.95); $\eta = \rho_1/\rho_L$ (Eq. 5.98); ζ (Eq. 5.96); and $\vec{v}_{\perp_1}$ (Eq. 5.99). Thus, M_L provides a convenient physical parameter that uniquely identifies a fast shock, and we can take for $0 < \psi_1 < \infty$:

$$M_L = \frac{a_{f_L}}{c_{s_L}} + \psi_1.$$

As for the fast RF, the most convenient parameter to characterise it is its width, $s_{1,d}$, the upper limit to which Eq. (6.1) are integrated. Now, as we've seen, fast fans can saturate (page 193); that is, there is a maximum width, $s_{1,\max}$, to which a fast fan can be integrated lest $B_\perp$ be forced below zero or, in the case of a fast Euler fan, lest it be forced through the triple umbilic. Thus, for $-\infty < \psi_1 < 0$, the width of the left-moving fast RF will be taken to be:

$$s_{1,d} = \min(-\psi_1, s_{1,\max}),$$

where details of how $s_{1,\max}$ is computed are left to the algorithm described in §6.4.3 (subroutine RF).

Thus, for a left-moving fast wave – be it a shock or RF – a guess for ψ_1 can be used to set its strength or width:

$$\boxed{\begin{aligned} M_L &= \frac{a_{f_L}}{c_{s_L}} + \psi_1, & \psi_1 &\geq 0; \\ s_{1,d} &= \min(-\psi_1, s_{1,\max}), & \psi_1 &< 0. \end{aligned}} \quad (6.36)$$

A similar expression may be written for ψ_2, the right-moving fast wave parameter.

Referring now to the top right panels of Fig. 5.12 and 5.13, the slow portions of the ζ super-loci (solid lines) are monotonic and single-valued functions of the upwind Mach number, M, and thus, in principle, M could be used as the parameter

---
[9]Note that upwind and downwind states 1 and 2 in Fig. 5.12 and 5.13 correspond to states **L** and ① respectively here.

to identify slow shocks as well. However, for each slow locus, M begins at the slow point ($M_{\text{min}} = a_{\text{s}}/c_{\text{s}} = M_{-}$) and ends at the Alfvén point ($M_{\text{max}} = a_x/c_{\text{s}} = M_{\text{A}}$; Problem 5.23), and it is the latter limit that can be problematic numerically. For example, one gets into round-off error problems near the switch-off limit where even with 64-bit arithmetic, M_{A} may not be known with sufficient precision.

It turns out that for a slow shock, the jump in $B_\perp$, b, is a better quantity to characterise its strength than M. The limits of b are *perfectly* well known – exactly 1 for an asymptotically weak shock, and exactly 0 for a switch-off shock – and, in the case where the upwind $B_\perp$ is zero, the only type of slow shock permitted is an Euler shock which can be treated as a hydrodynamical shock as an easy special case. Further, Eq. (5.122) from Problem 5.19 allows us to compute ϕ_x from b, from which all other flow variable jumps follow. Thus, to identify uniquely a left-moving slow shock, we take for $0 < \psi_3 < \infty$:

$$b = \max(0, 1 - \psi_3).$$

As for slow fans, unlike their fast counterparts, they do not saturate. Rather, both ρ and p asymptotically approach 0 as the width to which the left-moving slow fan is integrated, $s_{3,\text{d}}$, approaches ∞, and this can cause numerical headaches. Should an extreme fan width make ρ and p *too* close to zero, numerical noise could render $c_{\text{s}} = \sqrt{\gamma p/\rho}$ incalculable. In §6.4.3, we shall describe traps to prevent $s_{3,\text{d}}$ from getting *too* wide, but in the meantime we take for $-\infty < \psi_3 < 0$:

$$s_{3,\text{d}} = -\psi_3,$$

as the current guess of the width of the slow fan. Thus, for the left-moving slow wave, we take for the strength of a shock or the width of an RF,

$$\boxed{\begin{array}{ll} b = \max(0, 1 - \psi_3), & \psi_3 \geq 0; \\ s_{3,\text{d}} = -\psi_3, & \psi_3 < 0. \end{array}} \qquad (6.37)$$

A similar expression may be written for ψ_4, the parameter for the right-moving slow wave.

6.4.2 Strategy for the Riemann solver

Now that we have numerical parameters to embody our guesses (however inaccurate) for the strengths/widths of all waves, we use them to compute our way to states ③ and ④ on either side of the CD (Fig. 5.1). Thus, starting from the known left state, **L**, we compute the downwind state for the fast 1-wave whose strength or width is determined by ψ_1, giving us the flow variables in state ①. Taking this guessed state ① as upwind to the RD, we rotate $\vec{B}_\perp$ to orientation angle $\chi_{\text{L}} + \psi_5$ and modify $\vec{v}_\perp$ accordingly to give us state ②. Taking this "doubly guessed" state ② as upwind to the slow 3-wave whose strength/width is determined by ψ_3, we compute its downwind state and thus the flow variables in state ③ just to the left of the CD. This process is repeated starting from the known right state **R** and, using parameters ψ_2, ψ_5, and ψ_4, we determine the flow variables in state ④ just to the right of the CD.

With states ③ and ④ now "triply guessed" each, we evaluate $f_i(\psi_j)$ as defined in Eq. (6.35), none of which – except by an astonishing stroke of blind luck – will be zero. Thus, we need a systematic way to determine by what amounts we should "improve" our guesses of ψ_j – call these $\delta\psi_j$ – so that our next evaluations of $f_i(\psi_j)$ will be closer to zero. This is where the multivariate secant root finder comes in.

As shown in §D.2, we assemble the five parameters and five constraints into 5-kets $|\psi\rangle$ and $|f(\psi_j)\rangle$, and find the ket of "improvements", $|\delta\psi\rangle$, by solving the matrix equation, Eq. (D.6),

$$\mathsf{J}|d\psi\rangle = -|f(\psi_j)\rangle, \tag{6.38}$$

where elements of the *Jacobian matrix*, J, are given by Eq. (D.7),

$$J_{ij} = \frac{\partial f_i}{\partial \psi_j} \approx \frac{f_i(\psi_1,\ldots,\psi_5) - f_i(\psi_1,\ldots,\psi_j-\delta\psi_j,\ldots,\psi_5)}{\delta\psi_j}. \tag{6.39}$$

Note there are actually *six* separate evaluations of states ③ and ④ required for Eq. (6.39). That is, f_i must be evaluated for six sets of parameters: the unperturbed set; for ψ_1 perturbed but not ψ_2–ψ_5; for ψ_2 perturbed but not ψ_1, ψ_3–ψ_5; *etc.* Further, for there to be a solution to Eq. (6.38), J must be invertible (non-singular); that is, $\det \mathsf{J} \neq 0$. As it turns out, this is the most computationally delicate problem to be faced in designing a robust Riemann solver for it is not enough for $\det \mathsf{J} \neq 0$, it also cannot be *nearly* zero. By "nearly", I mean within an order of magnitude or two times machine accuracy. Anything closer, and the solution is vulnerable to undue influence by machine round-off error and completely unreliable.

The possibility that J can be singular stems from the fact that the 1-D MHD equations are *not strictly* hyperbolic; that is, the eigenvalues (wave speeds) can be degenerate. As we've seen, at the triple umbilic (where $B_\perp = 0$ and $\alpha_x = 1$), $a_\mathrm{f} = a_x = a_\mathrm{s}$ and the eigenvalues are *triply* degenerate. As a worst-case scenario, it is one our algorithm *must* be able to handle.

To illustrate, should $B_\perp = 0$ in state ① (downwind of the 1-fast wave and upwind of the 2-RD), there is nothing for the RD to rotate and thus the RD should not exist. How is this manifest in the Jacobian? With nothing to rotate, rotation angles ψ_5 and $\psi_5 + \delta\psi_5$ should give us *exactly* the same values for $f_i(\psi_j)$, and we expect the fifth column of the Jacobian, $\partial f_i/\partial \psi_5$, to be zero. A matrix with a column (or row) of zeros has determinant 0 which means it has no inverse and cannot be used to solve Eq. (6.38).

Further, that the system is independent of ψ_5 means there are only four parameters and thus, in principle, one of the five functions, $f_i(\psi_j)$, should be redundant. If these functions were expressed analytically, we'd say that one of the functions was a linear combination of the other four, and eliminate it from the system of equations to be solved. Numerically, if a column (or row) of the Jacobian is zero, we'll eliminate it and apply some criterion to select a row (or column) to eliminate as well to restore the Jacobian to a square matrix of lower dimension which, without any further rows or columns of zeros, should have an inverse. In the algorithm presented in §6.4.3, you will see how this is put into practice (subroutine J_{ij}).

Last thing before we get to the true nitty-gritty is to address the case where $B_x = 0$, for it is here that fundamental changes to our approach are needed.

When $B_x = 0$, the slow and Alfvén (RD) waves don't exist and the space-time diagram consists of just three characteristic paths; two paths for the left- and right-fast waves and one for a *tangential* discontinuity (TD) in between (*e.g.*, Fig. 6.12*b*). Superficially, this resembles the hydrodynamical case, exemplified by Fig. 3.8. Thus, we have just two parameters to control the fast waves, ψ_1 and ψ_2, along with two constraints across the TD which, in §5.3.3, were determined to be:

$$f_1(\psi_1, \psi_2) = v_{x_4} - v_{x_3} = 0; \quad f_2(\psi_1, \psi_2) = p_4 + \frac{B_{\perp_4}^2}{2} - \left(p_3 + \frac{B_{\perp_3}^2}{2}\right) = 0, \quad (6.40)$$

the last coming from Eq. 5.89. This is why we listed the ψs with the two fast parameters first. In this order, all one has to do to switch from the five-parameter, $B_x \neq 0$ problem to the two-parameter, $B_x = 0$ problem is change the do-loop limit from 5 to 2.

6.4.3 Algorithm for an exact MHD Riemann solver

And so finally, dear reader, we arrive at the reckoning. Do we, as Michael Norman so aptly asks, understand 1-D MHD enough to compute it? All the bits are in place and we need only assemble them. As the saying goes, the devil is in the details and, to be sure, there are *many* details.

This subsection is structured as follows. The main loop – a multivariate secant root finder – is described first, with items requiring further clarification enclosed in a rectangular box. You can think of these as *subroutines* for which detailed algorithms follow the main loop sequentially. Targets are enclosed in boxes with rounded corners; these you can think of as the old *FORTRAN* "go to" statements (gotta love 'em!).

Main loop

1. Set number of parameters and constraints: $n = \begin{cases} 2, & B_x = 0; \\ 5, & B_x \neq 0. \end{cases}$

2. Set error tolerance for convergence ($\varepsilon_{\max} = 10^{-8}$), initial step size for Runge–Kutta integration ($\delta s_0 = 10^{-3}$), and trigger for use of binomial expansions ($\epsilon_{\min} = 10^{-4}$).

3. Initialise $\delta\psi_j$ and ψ_j: $\delta\psi_j = 10^{-6}$ and $\langle\psi_j| = [0.1, 0.1, 0.1, 0.1, \frac{1}{2}(\chi_R - \chi_L)]$ are as good as any (χ_L, χ_R defined on page 206).

A.

4. Compute $\boxed{f_i(\psi_j)}$, and calculate error: $\varepsilon = \left(\sum_{j=1}^{n} f_i^2(\psi_j) \right)^{1/2}$.

   a) If $\varepsilon < \varepsilon_{\max}$, solution has converged; $\boxed{\text{go to B}}$.

5. Compute n sets of "diminished" values for $\boxed{f_i(\psi_j)}$, one for each set of parameters:

   $(\psi_1 - \delta\psi_1, \psi_2, \psi_3, \psi_4, \psi_5)$; $(\psi_1, \psi_2 - \delta\psi_2, \psi_3, \psi_4, \psi_5)$; $(\psi_1, \psi_2, \psi_3 - \delta\psi_3, \psi_4, \psi_5)$;

   $(\psi_1, \psi_2, \psi_3, \psi_4 - \delta\psi_4, \psi_5)$; $(\psi_1, \psi_2, \psi_3, \psi_4, \psi_5 - \delta\psi_5)$,

   for $n = 5$, and just the first two for $n = 2$.

6. Compute n_r^2 Jacobian elements, $\boxed{J_{ij}}$, using data from steps 4 and 5, where $n - n_r =$ number of rows (columns) of zeros or near-zeros removed from J by $\boxed{J_{ij}}$.

7. Using $\boxed{\text{LU}}$ decomposition, solve Eq. (6.38) for n_r values of $\delta\psi_j$.

   a) If no solution is found, $\boxed{\text{LU}}$ re-initialises ψ_j, $\delta\psi_j$. $\boxed{\text{Go to A}}$ to restart convergence.

   b) If solution is found:

   i) align $\delta\psi_j$, $j = 1 : n_r$, with ψ_j, $j = 1 : n$, by inserting rows as needed, setting inserted values of $\delta\psi_j$ to initial value (step 3);

   ii) set new guesses: $\psi_j + \delta\psi_j \rightarrow \psi_j$ (with ψ_5 modulo 2π);

   iii) $\boxed{\text{go to A}}$.

$\boxed{\text{B}}$.

8. Perform $\boxed{\text{output}}$; stop.

In addition, the main loop needs a few niceties, such as counters to prevent too many iterations from being performed for convergence and/or too many reinitialisations of the parameters, ψ_j. In other words, at some point one has to admit defeat! Some screen output would be useful, such as the parameter guesses at the end of each cycle, values of the Jacobian elements particularly as the code is being debugged, and final values of the primitive variables in the intermediate states at convergence.

And now for the subroutines...

Subroutine $\boxed{f_i(\psi_j)}$

Here, we compute our way from the known left and right states to the left and right sides of the CD (TD) using our guesses for ψ_j.

9. Starting from the left state $|q_L\rangle = [\rho_L \; p_L \; v_{x_L} \; v_{y_L} \; v_{z_L} \; B_{y_L} \; B_{z_L}]^T$:

   a) if $\psi_1 > 0$, set M_L from Eq. (6.36), find $|q_1\rangle$ downwind a $\boxed{\text{fast shock}}$;

   b) if $\psi_1 < 0$, set $s_{1,d}$ from Eq. (6.36), find profiles across and $|q_1\rangle$ downwind a fast $\boxed{\text{RF}}$.

If $B_x = 0$ ($n = 2$), skip steps 10 and 11; set $|q_2\rangle$ and $|q_3\rangle$ to $|q_1\rangle$.

10. Using $|q_1\rangle$ as upwind state: set $\chi = \chi_L + \psi_5$, find downwind state $|q_2\rangle$ from Eq. (6.33) and (6.34).

11. Using $|q_2\rangle$ as upwind state:

    a) if $\psi_3 > 0$, set b from Eq. (6.37), find $|q_3\rangle$ downwind a $\boxed{\text{slow shock}}$;

    b) if $\psi_3 < 0$, set $s_{3,d}$ from Eq. (6.37), find profiles across and $|q_3\rangle$ downwind a slow $\boxed{\text{RF}}$.

12. Starting from the right state, $|q_R\rangle = [\rho_R \; p_R \; v_{x_R} \; v_{y_R} \; v_{z_R} \; B_{y_R} \; B_{z_R}]^T$, repeat steps 9–11 to find states $|q_6\rangle$, $|q_5\rangle$, and $|q_4\rangle$.

13. a) For $B_x \neq 0$, evaluate $f_i(\psi_j)$, $i,j = 1,\ldots,5$ from Eq. (6.35).

    b) For $B_x = 0$, evaluate $f_i(\psi_j)$, $i,j = 1,2$ from Eq. (6.40).

Subroutine $\boxed{J_{ij}}$

14. Calculate n^2 elements of **J** from Eq. 6.39 using input $f_i(\psi_j)$ and diminished $f_i(\psi_j)$.

    To protect against "near-zeros", let `zero=1.0d-13` (*i.e.*, within an order of magnitude of machine round-off error for double precision arithmetic) and set to 0 any numerator and any element J_{ij} in Eq. (6.39) whose absolute value is less than `zero`.

As discussed on page 209, the kth column of the Jacobian will be zero if the kth parameter converges before the rest (in which case $\partial_k f_i = 0$), and the kth row will be zero if the kth function is satisfied before the rest (in which case $f_k(\psi_j) = 0$ and $\partial_j f_k = 0$). We eliminate such columns (rows) by "squeezing them out" and then "squeezing out" an equal number of rows (columns) to restore the Jacobian as a square matrix that can be used in Eq. (6.38) to find $\delta\psi_j$ for the parameters that have yet to converge. This is *the* critical step in designing a robust Riemann solver.

First, if $f_i(\psi_j)$ are independent of ψ_k, $\partial_k f_i = 0$ and column k is zero.

15. Flag columns of **J** for which $c_j = \sum_{i=1}^{n} |J_{ij}| = 0$.

16. For each column flagged, flag row with smallest value of $r_i = \sum_{j=1}^{n} |J_{ij}|$.

17. Squeeze out flagged rows and columns.

    For example, if third column is zero, overwrite third column with fourth, and fourth with fifth; doing the same for rows should leave a square matrix.

Second, if $f_k(\psi_j) = 0$, f_k has converged, $\partial_j f_k = 0$, and row k is zero.

18. If $f_i(\psi_j) = 0$, flag row i of J.

19. For each row flagged, flag column with smallest value of $c_j = \sum_{i=1}^{n} |J_{ij}|$.

20. Squeeze out flagged rows and columns.

    Should leave a square matrix with dimension $n_r \leq n$, where $n - n_r = $ total number of rows/columns eliminated in this step and step 17.

Subroutine $\boxed{\text{LU}}$

I'm going to bunt on this one, as there are all kinds of robust linear algebra routines out there, many based on so-called *LU decomposition* (*e.g.* and where else, *Numerical Recipes*, Press *et al.*, 1992).

To give the uninitiated reader an idea of what is being discussed, the premise of LU decomposition is as follows. Any matrix, A, can be expressed as a product of a lower- and upper-triangular matrix,[10]

$$A = LU,$$

so that a system of linear equations,

$$A|x\rangle = |b\rangle,$$

where $|x\rangle$ is the ket of unknowns being solved for, can be written as,

$$LU|x\rangle = |b\rangle,$$

which can be expressed as two separate matrix equations,

$$L|z\rangle = |b\rangle \quad \text{and} \quad U|x\rangle = |z\rangle.$$

Thus, knowing L, U, and $|b\rangle$, one solves the first for $|z\rangle$, then the second for $|x\rangle$. The point is, a system of linear equations with a lower triangular matrix can be solved simply by forward-substitution, while a system with an upper triangular matrix can be solved by backward substitution; no need for Gauss–Jordan elimination. Even including the decomposition, this method is a factor of three times less computationally intensive than inverting a general $n \times n$ matrix (Press *et al.*, 1992).

Thus, what one is looking for is a routine that when given A = J and $|b\rangle =$

---

[10] A lower- (upper-) triangular matrix is one with zeros above (below) the main diagonal.

$-|f(\psi_j)\rangle$, it returns $|x\rangle = |\delta\psi\rangle$, the changes to the parameters you're looking for to make the next – and we hope – better guess.

A feature you'll want to build into such a routine is a trap should the Jacobian turn out to be singular. Despite the effort made in $\boxed{J_{ij}}$ to eliminate rows and columns of zeros, this still doesn't guarantee J will be invertible. One of the tasks of LU decomposition is to shuffle the rows so that no zero lands on the main diagonal. If $\det J \neq 0$, this is always possible; otherwise, it isn't and there's no solution.

Since the Jacobian should *not* be singular, if it is this is a sign the secant root finder has gone astray and may be stuck in a local minimum in the sometimes-complicated solution space with no chance of finding its way out. What I do in this stead is reset the parameters ψ_j randomly to something between ± 0.5, reset the parameter differentials to their initial value (step 3), and instruct the calling routine to restart the convergence process. I find for most cases where my original starting point $[0.1, 0.1, 0.1, 0.1, \frac{1}{2}(\chi_R - \chi_L)]$ leads the secant root finder down a rabbit hole, reissuing random values as described gets it back on track.

Subroutine $\boxed{\text{output}}$

I'm going to bunt on this one as well, as the details will depend very heavily on the nature of the graphics packages available to the reader and the programmer's preference of output. Instead, I just list the output files I have found useful, along with some tips in how these may be assembled.

- I create a ten-column ascii file suitable for use by widely available graphics packages such as GNUPLOT^{TM} able to read and plot column-separated datafiles. My columns include the self-similar position variable $x/t = u$ (where u is the shock speed for shocks, $v_x \pm a_{f,s}$ for RFs, a_x for RDs, and v_x for the CD/TD), followed by nine variables $(\rho, p, e_T, v_x, v_y, v_z, B_y, B_z, \chi)$, where e_T is the total energy density and χ is the orientation angle of $\vec{B}_\perp$.

  The overall width of the solution space is determined by the difference between the speeds of the leading edges of the 1- and 7-waves, which I widen by about 20% so that I can include some of the original left and right states in my solution profiles (*e.g.*, Fig. 6.13–6.30).

  For a discontinuity located at u_i, say, one needs only two points to describe it: the variable values just "epsilon" upwind of u_i and the variable values just "epsilon" downwind of u_i. Continuous waves (RFs) are resolved by whatever steps the Runge–Kutta scheme uses to compute their profiles, and I gather all these data in ascending order of u_i in this single ascii file.

- I create a post-script plot file (using Kevin Kohler's PSPLOT package) based on data in the ascii file described above to create plots such as those in Fig. 6.13–6.30.

- Finally, I create a second ascii file with the convergent values of the parameters, ψ_j, along with the values of the primitive variables in each of the eight levels across the solution space. These data were used to create Tables 6.1 and 6.2 respectively in §6.4.4.

Subroutine $\boxed{\text{fast shock}}$

States **L** and ① are, respectively, the upwind and downwind states for the 1-family (left-moving) fast shock (*e.g.*, Fig. 5.1), while states **R** and ⑥ are those for the 7-family (right-moving) fast shock. In what follows, I shall use subscript 1 for the upwind state and 2 for the downwind state, with the understanding that all discussion is general for both families.

In finding the downwind state of a fast shock, the delicate part is the downwind value for $B_\perp$. For the most part, Eq. (5.95) can be used to find b from which we set $B_{\perp_2} = bB_{\perp_1}$. However, for $B_{\perp_1} = 0$, we must evaluate $b_\perp$ instead (Eq. 5.109 for switch-on shocks, 0 for fast Euler shocks) and set $B_{\perp_2} = b_\perp B_x$.

Still, other numerical issues lurk. It turns out that Eq. (5.95) is rather fragile, and can be dominated by roundoff error when $\phi_x A_{x_1}^2 - 1$ falls below $\sim 10^{-6}$ *even in double-precision arithmetic*. Thus, for $\phi_x A_{x_1}^2 - 1 \lesssim 10^{-3}$, I switch to Eq. (5.126) to evaluate b, which can be used safely so long as one is comfortably away from the Euler branch ($\phi_x \sim \mathcal{H}$; Eq. 5.105 and discussion following Eq. 5.107). This constraint poses no problem, since in the vicinity of an Euler shock, Eq. (5.95) is perfectly reliable (Problem 5.25).

As a consequence of dodging these "numerical landmines" the following algorithm may seem rather riddled with conditionals.

21. Evaluate upwind MHD-alphas, $\alpha_{x_1} = \dfrac{B_x^2}{\mu_0 \gamma p_1}$ and $\alpha_{\perp_1} = \dfrac{B_{\perp_1}^2}{\mu_0 \gamma p_1}$.

22. Find ϕ_x from Eq. (5.104) using a cubic root finder (*e.g.*, App. E).

    a) Root with maximum real part should be pure real; take this as ϕ_x. Other roots are intermediate shocks.

    b) If $B_\perp = 0$, the discriminant Δ (Eq. E.1) is also zero. Since Δ is the difference between two numbers, numerical noise may render $\Delta < 0$ which is a problem for $\sqrt{\Delta}$ needed to evaluate the roots. Thus, set $\Delta = 0$ directly if $B_\perp < \texttt{zero}$.

23. Evaluate $\epsilon = \left| 1 - \dfrac{\alpha_{x_1}}{\phi_x M_1^2} \right|$.

    a) If $\epsilon > \texttt{zero}$, then:

        i) if $\epsilon > 0.001$, evaluate b from Eq. (5.95);

        ii) if $\epsilon \leq 0.001$, evaluate b from Eq. (5.126).

    b) If $\epsilon \leq \texttt{zero}$, evaluate $M_{\text{on}} = \sqrt{\dfrac{\alpha_{x_1}(\gamma+1) - 2}{\gamma - 1}}$ (Eq. 5.112). Then:

i) if $M_1 \geq M_{\text{on}}$, $b_\perp = 0$ (Euler branch);

ii) if $M_1 < M_{\text{on}}$, evaluate $b_\perp$ from Eq. (5.109).

The quantity $\phi_x = v_{x_2}/v_{x_1}$ gives the jump in flow speed *relative to the shock*. To find speeds in the x-direction relative to the "lab frame" (u_{x_1}, u_{x_2}) in which the shock speed is $\mathcal{V}$ (< 0 for 1-family, > 0 for 7-family), we use the Galilean transformations:

$$u_{x_1} = v_{x_1} + \mathcal{V} \qquad \Rightarrow \qquad \boxed{\mathcal{V} = u_{x_1} - v_{x_1};} \qquad (6.41)$$

$$u_{x_2} = v_{x_2} + \mathcal{V} = \phi_x v_{x_1} + u_{x_1} - v_{x_1} \qquad \Rightarrow \qquad \boxed{u_{x_2} = u_{x_1} + v_{x_1}(\phi_x - 1).} \qquad (6.42)$$

where u_{x_1} is the known upwind flow speed in the lab frame, and,[11]

$$v_{x_1} = \begin{cases} M_1 c_{s_1}, & \text{for 1-family waves;} \\ -M_1 c_{s_1}, & \text{for 7-family waves.} \end{cases} \qquad (6.43)$$

With this, we can now evaluate the downwind state.

24. Evaluate $\rho_2 = \dfrac{\rho_1}{\phi_x}$ (Eq. 5.98) and downwind flow speed, u_{x_2} (Eq. 6.42).

25. a) If b were evaluated in step 23 ($\epsilon >$ zero), evaluate:

    i) ζ from Eq. (5.96), and then $p_2 = \zeta p_1$;

    ii) $\vec{v}_{\perp_2}$ from Eq. (5.99), where $\mathcal{M} = \rho_1 v_{x_1}$ and v_{x_1} from Eq. (6.43);

    iii) $\vec{B}_{\perp_2} = b \vec{B}_{\perp_1}$.

    b) If $b_\perp$ were evaluated in step 23 ($\epsilon \leq$ zero), evaluate:

    i) ζ from Eq. (5.110), and then $p_2 = \zeta p_1$;

    ii) $\vec{v}_{\perp_2}$ from Eq. (5.111);

    iii) $\vec{B}_{\perp_2} = b_\perp B_x \hat{e}_\perp$, where $\hat{e}_\perp = (0, \cos\chi, \sin\chi)$.

    As suggested after Eq. (5.111), χ can be set to the asymptotic downwind value. If, for example, the 1-shock is switch-on, then $\vec{B}_{\perp_L} = 0$ and we take $\chi = \chi_R$. If $\vec{B}_{\perp_R} = 0$ as well, then taking $\chi = 0$ is as good a guess as any.

Subroutine slow shock

For a slow shock, the main thing to be aware of is there are four cases that need separate treatment. For all the attention in the literature about "non-regular switch-off waves" and the special consideration they warrant, a switch-off shock is not one

---

[11]Relative to the left-moving shock, upwind fluid moves in the $+x$-direction, whereas relative to the right-moving shock, upwind fluid moves in the $-x$-direction.

of these cases. Rather, it is handled easily as a limit to one of the four identified here.

Relative to a slow shock, the upwind speed is super-slow, sub-Alfvénic. Thus,

$$\frac{a_{s_1}^2}{c_{s_1}^2} \leq M_1^2 \leq \frac{a_{x_1}^2}{c_{s_1}^2} = \alpha_{x_1}, \tag{6.44}$$

where again, subscripts 1 and 2 refer to upwind/downwind states respectively. Further, it's easy to show from Eq. (5.95) that,

$$M_1^2 = \alpha_{x_1} \frac{b-1}{\phi_x b - 1}. \tag{6.45}$$

With these reminders, the downwind state of a slow shock is set as follows.

26. Case 1: $B_x = 0 \Rightarrow$ no slow shock; set downwind state = upwind state; return. Else,

27. Evaluate upwind MHD-alpha, $\alpha_{x_1} = \frac{B_x^2}{\mu_0 \gamma p_1}$.

28. Case 2: $B_{\perp_1} = 0$, $\alpha_{x_1} < 1 \Rightarrow$ no slow shock; set downwind state = upwind state and $\psi_j = 0$; return. Else,

29. Case 3: $B_{\perp_1} = 0$, $\alpha_{x_1} \geq 1 \Rightarrow$ slow Euler shock.

    a) $a_{s_1}^2 = c_{s_1}^2$ (Eq. 5.66), and Eq. (6.44) $\Rightarrow 1 \leq M_1^2 \leq \alpha_{x_1}$.

    b) Since b is irrelevant for an Euler shock, set $M_1^2 = 1 + (\alpha_{x_1} - 1)\tanh \psi$ instead.

    Here, tanh is a convenient function to map $\psi \in [0, \infty) \to M_1^2 \in [1, \alpha_{x_1})$.

    c) Set ϕ_x from Eq. (5.105) and $b = 1$.

30. Case 4: $B_{\perp_1} \neq 0 \Rightarrow$ ordinary slow shock.

    a) If $\psi > 1$, set $\psi = 1$ before setting $b = 1 - \psi$. All $\psi \in [1, \infty)$ is treated as $\psi = 1$, and limiting ψ prevents secant finder from letting $\psi \to \infty$ and getting lost.

    Note that $\psi = 1$ ($b = 0$) is the switch-off shock limit.

    b) Multiply Eq. (5.122) by $(b-1)$ to make it "NaN-proof"; find ϕ_x.

    Take lesser of two roots for ϕ_x; the other is > 1, violating the entropy condition.

    c) Evaluate M_1^2 from Eq. (6.45).

    If $b = 1$, Eq. (6.45) is singular. Thus, set $M_1^2 = a_{s_1}^2/c_{s_1}^2$ (slow point) instead.

31. For cases 3 and 4, we now have b, ϕ_x, and M_1^2 from which we evaluate:

a) $\rho_2 = \dfrac{\rho_1}{\phi_x}$ and downwind flow speed, u_{x_2}, from Eq. (6.42);

b) ζ from Eq. (5.96), and then $p_2 = \zeta p_1$;

c) $\vec{v}_{\perp_2}$ from Eq. (5.99), where $\mathcal{M} = \rho_1 v_{x_1}$ and v_{x_1} is given by Eq. (6.43);

d) $\vec{B}_{\perp_2} = b\vec{B}_{\perp_1}$.

Subroutine RF

We now arrive at *the* most ticklish part of the algorithm. I strongly advise the reader intent on developing an MHD Riemann solver to first write an "RF program", one whose only task is to compute (and plot) the primitive variable profiles across a rarefaction fan – fast or slow – for a given upwind state (Project P6.1). Fig. 6.5–6.10 show examples of such plots determined from my own RF program. Once this has been debugged and understood, it can be incorporated as a module into the Riemann solver itself.

Unlike shocks, I find it more efficient to discuss fast and slow RFs together, as their differences computationally turn out to be rather minor. This is perhaps the only way in which rarefaction fans are "easier" to handle than shocks!

To find the variable profiles across an MHD RF, we solve Eq. (6.1) in terms of the generalised coordinate, s:

$$\frac{d|q(s)\rangle}{ds} = |r_{f,s}^{\pm}\rangle,$$

seven coupled first-order ODEs in seven unknowns for which a sixth-order Runge–Kutta (RK) algorithm is particularly well-suited. There are numerous RK algorithms "out there" (*e.g.*, and as always, *Numerical Recipes*), and the one I use is described in App. F. The reader unfamiliar with RK methods is encouraged to read this appendix before continuing.

At first cut, the generic RK algorithm described in App. F can be pasted directly here. To render it specific to MHD RFs, I set the error tolerance and initial step size (**errmax** and **h** in step 1 of the RK algorithm) to $\varepsilon_{\max}$ and δs_0 set in step 2 above, and the function yprime needs Eq. (6.21) for the function derivatives (eigenkets), including Eq. (6.26) and (6.27) for μ and ν. These are the easy bits. More subtle are the changes needed to accommodate idiosyncrasies of fans with certain upwind states, and to avoid excessive round-off error should differences of very nearly equal values be required. All such issues that I've encountered are enumerated below.

Before entering the RK loop (A on page 473),

32. If upwind $B_\perp = 0$ and $\alpha_x \geq 1$, there is no fast RF (page 193).

    Set downwind state to upwind state and ψ_j to 0; skip RK loop (B on page 474).

33. If $B_x = 0$, there is no slow RF (page 196).

    Set downwind state to upwind state; skip RK loop ($\boxed{B}$ on page 474).

    For $B_x = 0$, there is no slow parameter ψ to reset.

From within the RK loop,

34. Fast fans can saturate (page 193), and their maximum width, $s_{i,\max}$, is finite.

    a) To detect saturation:

        i) compute $\varepsilon_s = \delta s/s$, where δs is current step size (h in App. F);

        ii) if $\varepsilon_s < \varepsilon_{\max}$ (error tolerance), fan is saturated. Set $\psi = -s$, downwind $\vec{B}_\perp = 0$, and exit RK loop ($\boxed{B}$ on page 474).

    b) For upwind $B_\perp > 0$, a saturated fast fan is a switch-off fan which can be no deeper in s than that which makes $B_\perp = 0$. Should either component of $\vec{B}_\perp$ change sign,[12] the step size, δs (h in App. F), is too large:

        i) set $\delta s \to \dfrac{\delta s}{2}$;

        ii) return to top of adaptive RK loop ($\boxed{A}$ on page 473) to redo step.

    c) For upwind $B_\perp = 0$ and $\alpha_x < 1$, a saturated fast fan is a fast Euler fan reaching the triple umbilic ($\alpha_x = 1$; Fig. 6.11). No special considerations are needed since, as it turns out, $\varepsilon_s < \varepsilon_{\max}$ before round-off error pushes s through the triple umbilic.

35. Slow and HD ($\vec{B} = 0$) fans asymptote to their maximum "strength" as $p \to 0$ and $s_i \to \infty$ (page 193). If s gets too large (and p too small), underflow errors can occur (e.g., $c_s = \sqrt{\gamma p/\rho}$) and/or the secant root finder can get lost as it reaches deeper and ever deeper into a fan in a futile search of a solution that may not exist.

    To detect if the asymptotic limit has been reached:

    a) compute $\varepsilon_{v_x} = \delta v_x / v_x$, where δv_x is the difference in v_x over the current step;

    b) if $\varepsilon_{v_x} < \texttt{zero}$[13] (round-off error limit defined in step 14), fan has asymptoted. Set $\psi = -s$ and exit RK loop ($\boxed{B}$ on page 474).

At my count, there are *four* opportunities for numerical round off error to dominate

---

[12] This would be unphysical as the only waves capable of changing the direction of $\vec{B}_\perp$ (a sign change is tantamount to a rotation of π radians) are intermediate shocks and rotational discontinuities.

[13] There is no point in checking the fractional change in p or ρ, since these are approaching zero and their fractional changes may well remain of order unity.

the solution of an RF, all caused by differences of two nearly equal quantities when calculating $|r_{f,s}^{\pm}\rangle$ (Eq. 6.21); these need to be evaded. Quantities dependent upon differences include the discriminant d (Eq. 6.25), the slow speed a_s (Eq. 6.23), the scaling factors μ and ν (Eq. 6.26, 6.27), and $\delta_x = \alpha_x - 1$ (Eq. 6.22).

36. The right-most equality in Eq. (6.25) renders d "difference-free":

$$d = \frac{D}{c_s^2} = \sqrt{(\delta_x + \alpha_\perp)^2 + 4\alpha_\perp}, \qquad (6.46)$$

and this is the form that should be used.

37. Assuming a robust value for a_f (i.e., using Eq. 6.46 with no differences), identity (6.13) allows us to bypass the difference in a_s (Eq. 6.23) altogether:

$$a_s = \frac{c_s a_x}{a_f}. \qquad (6.47)$$

38. There are two numerical considerations for μ and ν.

   a) Problem 6.7 shows that for $B_\perp > 0$ but $\epsilon \ll 1$ (Eq. 6.57), μ^2 (Eq. 6.26) for $\delta_x + \alpha_\perp > 0$, and ν^2 (Eq. 6.27) for $\delta_x + \alpha_\perp < 0$, are dominated by round-off error. Thus, for $\epsilon < \epsilon_{\min}$ (step 2), use power-series expansions in ϵ (Eq. 6.58) for μ^2 or ν^2.

   b) There is a numerical issue in launching a slow fan from the triple umbilic (fast fans cannot be launched from upwind states where $B_\perp = 0$ and $\delta_x \geq 0$, so for them this concern is moot.) While Eq. (6.56) is analytically correct, numerically it is imperative to use instead,

   $$\mu^2 = \begin{cases} 1, & \delta_x < 0; \\ 0, & \delta_x \geq 0, \end{cases} \quad \text{and} \quad \nu^2 = \begin{cases} 0, & \delta_x < 0; \\ 1, & \delta_x \geq 0, \end{cases} \qquad (6.48)$$

   when $\alpha_\perp = 0$. That is, the triple umbilic, where $\delta_x = 0$ and $\mu^2 = \nu^2 = \frac{1}{2}$, should be absorbed into the $\delta_x > 0$ case where $\mu^2 = 0$ and $\nu^2 = 1$.

   Why? Slow fans launched from the triple umbilic are slow Euler fans across which $B_\perp$ remains zero (page 195). Further, as we saw in §6.2, δ_x increases across any MIID RF. Thinking analytically, starting at the triple umbilic with $\mu^2 = \frac{1}{2}$, infinitesimally into the fan both $B_\perp$ and δ_x are infinitesimally greater than zero. However, even as an infinitesimal, $\delta_x > 0$ where $\mu^2 = 0$ and $B_\perp$ can grow no further. It remains at its infinitesimal seed value, and thus effectively zero.

   Numerically, infinitesimals are replaced with small differences and starting with $\mu^2 = \frac{1}{2}$ means that after the first integration step into the fan, $B_\perp$ will be non-zero (second of Eq. 6.21) by a small but non-infinitesimal amount. Thus, integration across the rest of the fan will grow $B_\perp$ as though it were a switch-on fan, ignoring the fact this "seed field" is really a result of *discretisation* error.

Lesson: When launching a slow fan numerically, treat the triple umbilic as though $\delta_x > 0$, and thus use Eq. (6.48) to compute μ^2 and ν^2.

39. Finally, the most awkward case to program is when $\alpha_x \to 1$, and $\delta_x = \alpha_x - 1$ is dominated by round-off error.

    Problem 6.6 shows that for $\varpi \ll 1$ (Eq. 6.52), δ_x should be set by Eq. (6.53) instead of $\alpha_x - 1$. However, it leaves open the question of how to evaluate the numerator, $\Delta p = p - p_{\mathrm{cr}}$ where $p_{\mathrm{cr}} = B_x^2/\gamma$, given that this too is a difference dominated by round-off error when $\alpha_x \to 1$.

    In an RF program (page 218), Δp in the upwind state can be specified as a number free from round-off error. In launching an RF as part of a trial solution, an exact MHD Riemann solver would still set the upwind Δp to the difference between the upwind pressure and critical pressure. While this "seed value" may be "noisy", its noise will be drowned out by the signal of integration across the rest of the RF so long as each integration step adds no more noise to Δp.

    So how do we accomplish this? To the upwind value of Δp, one adds the change in total pressure across each successive RK interval. Since pressure falls monotonically across an RF, p will soon – perhaps only after a few intervals – be significantly different from p_{cr} (*i.e.*, once $\varpi > \epsilon_{\min}$), at which point δ_x can be safely computed from $\alpha_x - 1$. However, within these first few critical steps, each addition to Δp will have full precision and thus, modulo the seed value, so will δ_x set by the power series in ϖ (Eq. 6.53).

    The trick is to ensure that each time a pressure is needed, $p_{\mathrm{cr}} + \Delta p$ agrees with p to within machine round-off error. Thus, the current value of Δp will have to be known to subroutines `RunKut` and `Yprime` (App. F), and Δp will have to be taken through the same Richardson Extrapolation steps as the primitive variables. Further, allowances will have to be made to "undo" the last contribution to Δp should a step need to be retaken (*e.g.*, step size found to be too large). For my own RF program and Riemann solver, I found keeping accurate track of Δp to be rather invasive to the programming. Still, it was *the* critical step that allowed my RF program to integrate *any* MHD RF, including a slow fan with upwind $\alpha_\perp = 0$ and $\delta_x = 10^{-95}$ (thus, just barely a switch-on fan) with full precision without confusing it for a slow Euler fan. (The only reason I could not specify $0 < \delta_x < 10^{-95}$ is such a fan asymptotes with $B_\perp < 10^{-38}$ which my single-precision graphics routines cannot handle without generating underflows.)

6.4.4 Sample problems

In this final section, I give solutions to eighteen Riemann problems in both graphical and tabular form for the reader looking to calibrate their own Riemann solver

(Project P6.2).[14] If eighteen seems excessive, be aware that this is a case where solving one problem perfectly does not guarantee the next problem won't crash your program or worse, yield a wrong but seemingly plausible answer! Indeed, even now that my program can solve *all* these problems satisfactorily, it is still vulnerable to becoming lost in local minima of the solution space, and generating the occasional glitch as noted in the footnotes and figure captions in the following pages.

So, for the reader at the end of their rope trying to get their own solver working, or for the reader who just wants a working (however imperfect) Riemann solver without having to go through the exercise of programming it for themselves (but beware, Mike may not think you really understand it!), my own code along with my RF program – warts and all – is available open source (in *FORTRAN 77*, of course!) from the "software bar" on my website given at the end of the Preface.

Figures 6.13–6.30 (pages 224–232) show my solution to each Riemann problem. Those taken from the literature are cited in the Figure captions, where S76 refers to Sod (1976), RJ95 refers to Ryu & Jones (1995), F02 refers to Falle (2002), and TY13 refers to Takahashi & Yamada (2013). Each figure shows profiles of the primitive variables $(\rho, p, v_x, v_y, v_z, B_y, B_z)$,[15] along with total energy density, e_T, and magnetic orientation angle, $\chi = \tan^{-1}(B_z/B_y)$, the latter being the best indicator of RDs.

Table 6.1 lists the convergent values of the five parameters for each problem referred to by their corresponding figure number. These values may be useful only if my algorithm is followed carefully. It is quite possible to come up with another perfectly good algorithm in which the parameters are scaled or used differently that, while giving essentially identical solutions to each Riemann problem, may do so with different convergent values of ψ_j.

Finally, values of the primitive variables in states **L**, ①–⑥, and **R** (Fig. 5.1) for each problem are given in Table 6.2 following the figures (pages 233–235). Missing states are identical to the state listed immediately before where it would have been inserted in the table. Thus, for the first problem listed in Table 6.2 (the Sod shock-tube in Fig. 6.13), states ② and ③ (not listed) are identical to state ① (listed), and states ⑤ and ⑥ (not listed) are identical to state ④ (listed).

And there you have it! At this point, it should seem in retrospect that the algorithm for the IID Riemann problem described in §3.6 and assigned as project P3.1 was utter child's play!

---

[14] As done for rarefaction fans at the end of §6.2, all values for B_x, B_y and B_z given from here until the chapter's end including all plots and tables are in units where $\mu_0 = 1$. To restore units, multiply any given value of B_x, B_y, or B_z by $\sqrt{\mu_0}$.

[15] I've not been entirely consistent with my use of v_x. In subroutine `fast shock` of the algorithm just described, v_x indicates flow speed relative to the shock while u_x indicates flow speed relative to the lab (*e.g.*, Eq. 6.42). For subroutine `RF`, v_x indicates flow speed relative to the lab. In Fig. 6.13–6.30 and Table 6.2, I use v_x as I do for rarefaction fans: flow speed relative to the lab.

Fig.	ψ_1	ψ_2	ψ_3	ψ_4	ψ_5
6.13	−0.3478274	0.2426354	—	—	—
6.14	−0.2099834	0.0398941	−0.2313041	0.1786934	0.5987685
6.15	1.532796	9.580935	−0.0129561	0.2559795	0.0
6.16	1.132760	−0.1427370	0.7044264	−0.1025533	0.0
6.17	0.3855998	0.3207335	0.1012232	0.1227875	−0.2066066
6.18	1.022167	−0.1255906	0.6094286	−0.2758066	0.2025473
6.19	11.77350	16.95098	—	—	—
6.20	−0.8678615	−0.8678615	—	—	—
6.21	−0.8638215	0.1391506	−0.0917579	0.2682388	0.0
6.22	0.0	−1.037517	0.0	0.0	0.0
6.23	0.0144678	0.0246186	0.9999959	0.0	0.0
6.24	−0.0618703	−0.0202970	−0.7930968	0.2390264	0.5015874
6.25	−0.5904023	−0.2991020	0.0212305	0.3926217	3.141593
6.26	−0.6643923	−0.7833611	0.3208446	0.4637182	3.141593
6.27	0.0710332	5.088679	−0.0326858	0.4084126	0.0
6.28	−1.165195	0.0	0.0	0.0	0.0
6.29	−1.326030	−0.8692138	0.9999569	0.9102257	−3.141580
6.30	0.0963136	0.1096111	−0.7925678	−0.8419474	−1.263669

Table 6.1. Convergent values (with fractional tolerance 10^{-8}) of the five parameters, ψ_j, quantifying the strength/width of each wave in the solutions to the Riemann problems represented in the indicated figures, where $j = 1, \ldots, 5$ correspond to the left and right fast waves, left and right slow waves, and the left RD respectively. For ψ_{1-4}, positive (negative) values indicate shocks (RF). Switch-off shocks are indicated by $\psi_{3,4} \sim 1$; switch-off RFs and switch-on waves don't have such definite values. The units of ψ_5 are radians. For $B_x = 0$, $\psi_{3,4,5}$ are not used, and their values are entered as em-dashes (—). For problems with $B_x \neq 0$, values of ψ_j set to 0.0 indicate that particular wave is absent in the solution, and the corresponding upwind and downwind states are identical. While zeros for ψ_5 were dead zero, those for ψ_{1-4} were, in fact, of order 10^{-6}. Convergence of the functions, $f_i(\psi_j)$, to within tolerance does not necessarily mean the parameters converge to the same tolerance.

The MHD Riemann Problem

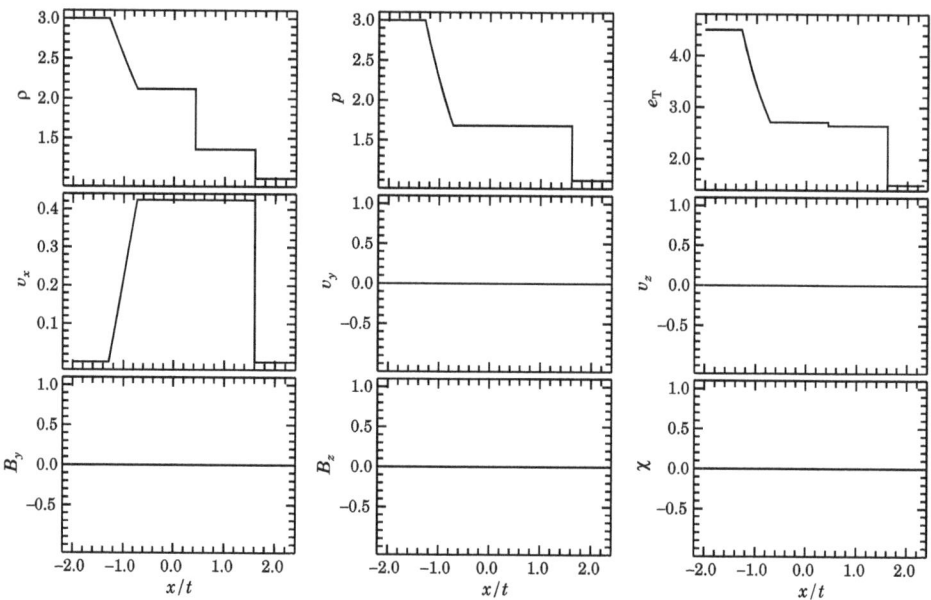

Figure 6.13. Original Sod shocktube from S76; $B = 0$ exactly (hydrodynamical). From left to right: (fast) RF; CD; (fast) shock.

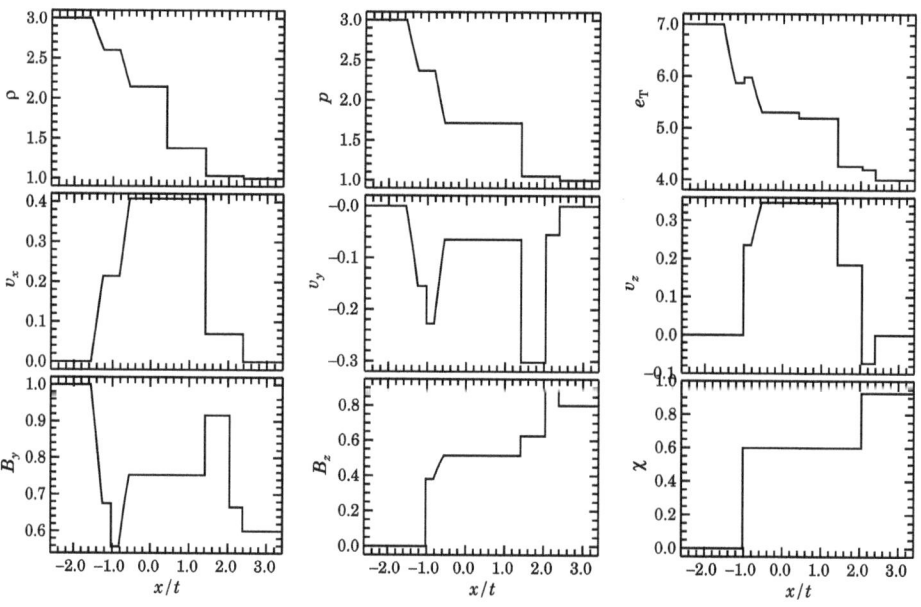

Figure 6.14. Modified Sod shocktube for MHD with $B_x = 2$, $B_{\perp_{L,R}} = 1$. From left to right: fast RF; RD; slow RF; CD; slow shock; RD; fast shock. In this and all remaining figures in this chapter, B_x, B_y, and B_z are given in units where $\mu_0 = 1$.

An MHD Riemann solver

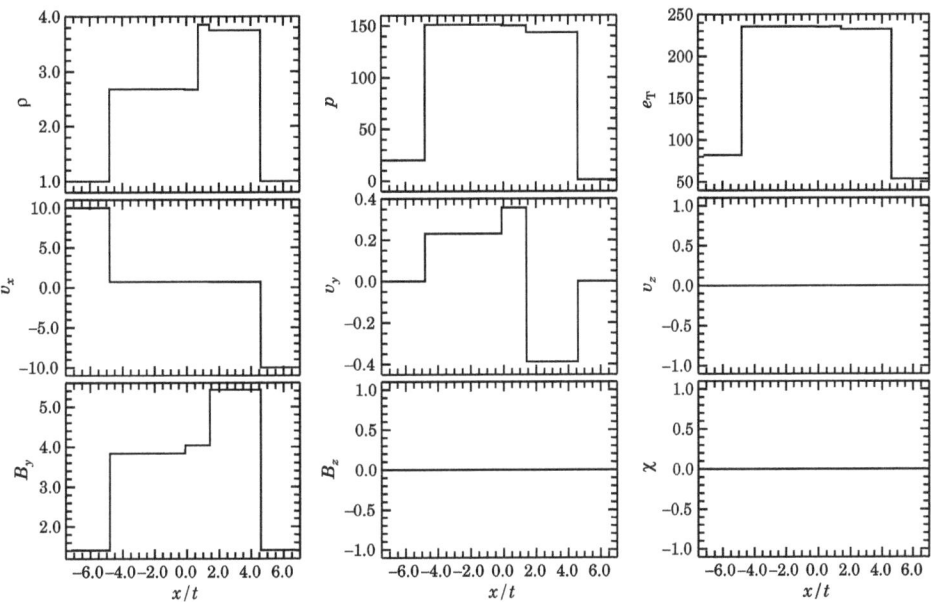

Figure 6.15. Figure 1a from RJ95 (2-D field; $B_z = 0$). From left to right: fast shock; slow RF; CD; slow shock; fast shock. With $B_z = 0$ and $B_{y_{L,R}}$ with the same sign, there are no RDs.

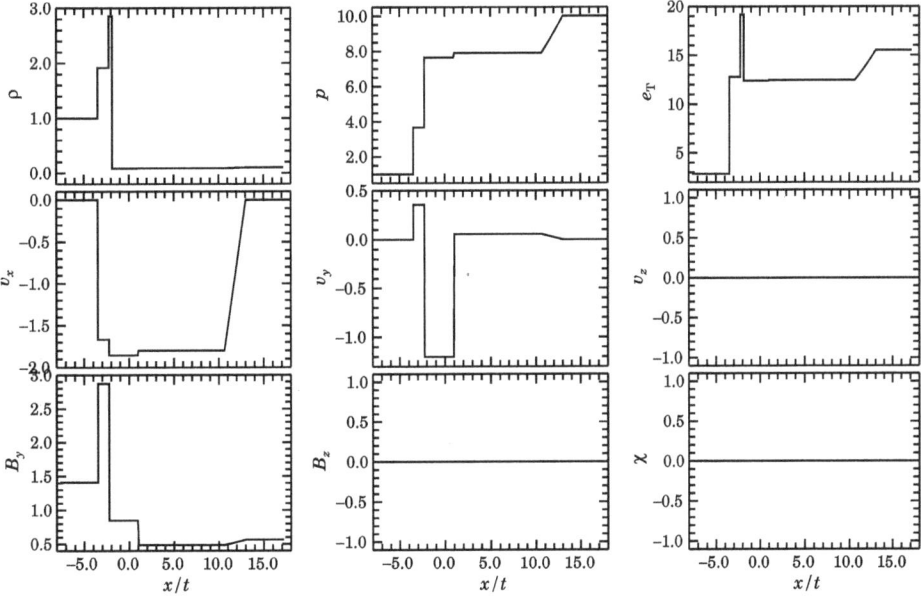

Figure 6.16. Figure 1b from RJ95 (2-D field; $B_z = 0$). From left to right: fast shock; slow shock; CD; slow RF; fast RF.

The MHD Riemann Problem

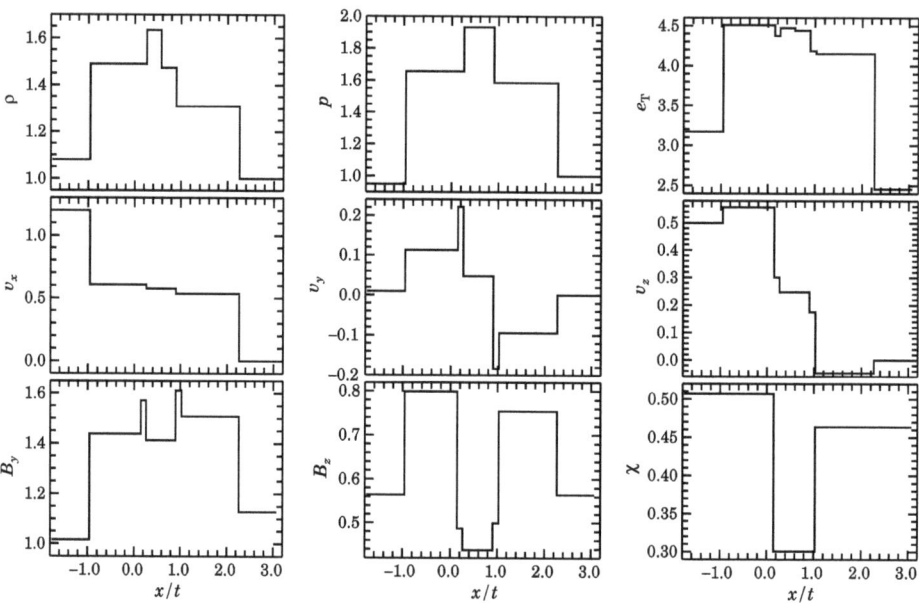

Figure 6.17. Figure 2a from RJ95 (3-D field; $B_z \neq 0$). From left to right: fast shock; RD; slow shock; CD; slow shock; RD; fast shock.

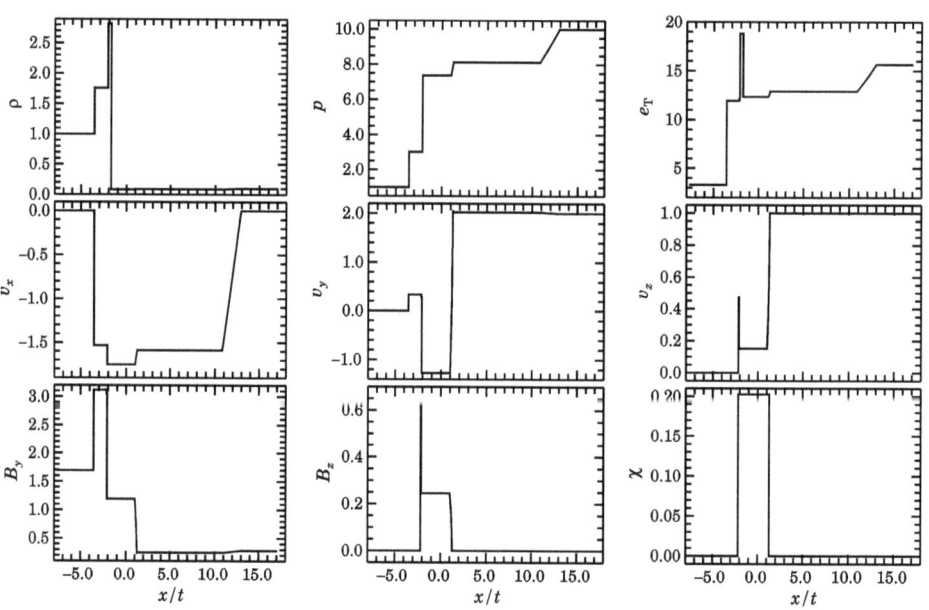

Figure 6.18. Figure 2b from RJ95 (3-D field; $B_z \neq 0$). From left to right: fast shock; RD; slow shock; CD; slow RF; RD; fast RF. The "spikes" in v_z and B_z are real (near degeneracy of slow and Alfvén speeds).

An MHD Riemann solver

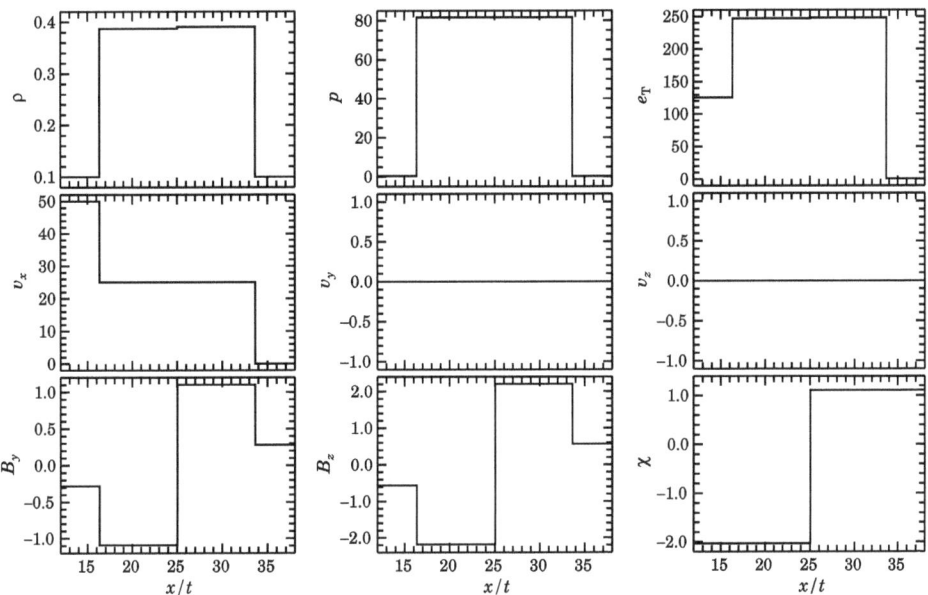

Figure 6.19. Figure 3a from RJ95 (tangential field; $B_x = 0$). From left to right: fast shock; TD; fast shock. Slow waves and RDs do not exist for $B_x = 0$.

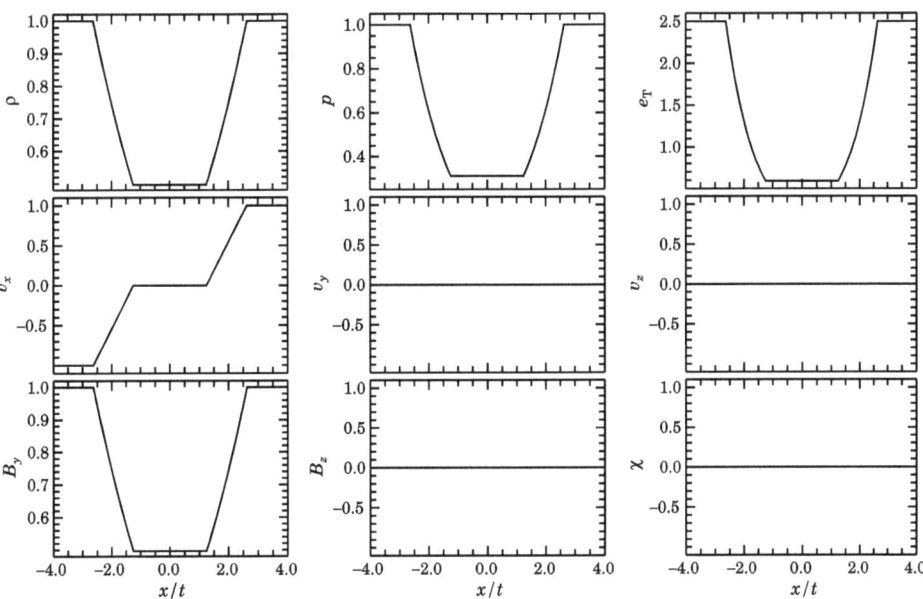

Figure 6.20. Figure 3b from RJ95 (tangential field; $B_x = 0$). From left to right: fast RF; fast RF. The symmetry of the initial conditions preclude a TD.

The MHD Riemann Problem

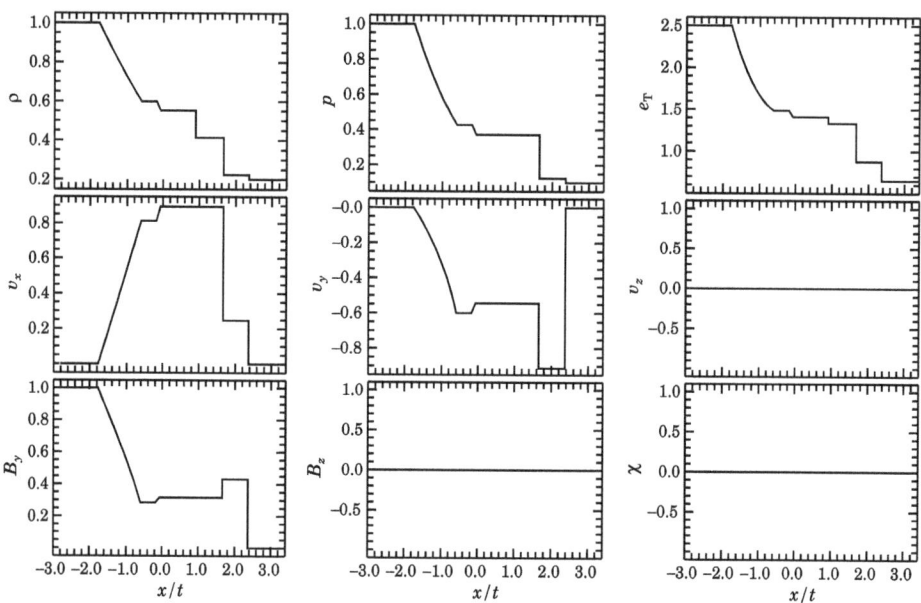

Figure 6.21. Figure 4a from RJ95 (switch-on shock). From left to right: fast RF; slow RF; CD; slow shock; switch-on (fast) shock.

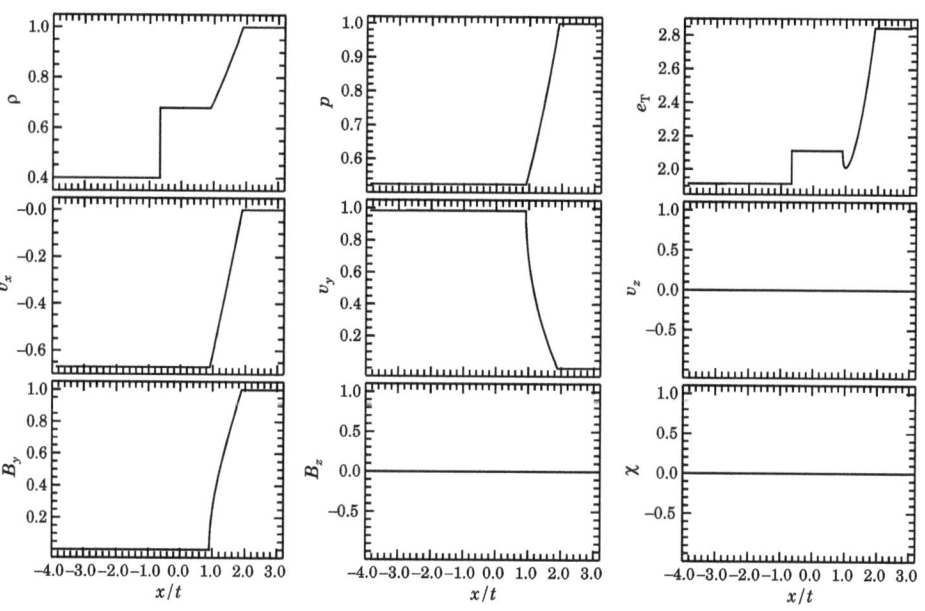

Figure 6.22. Figure 4b from RJ95 (switch-off RF). From left to right: CD; switch-off (fast) RF.

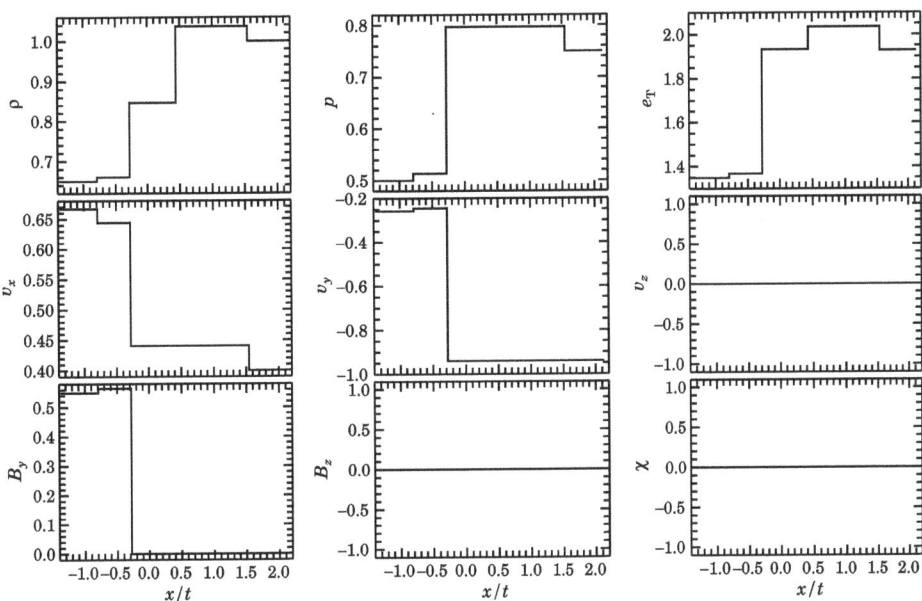

Figure 6.23. Figure 4c from RJ95 (switch-off shock). From left to right: fast shock; switch-off (slow) shock; CD; fast Euler shock.

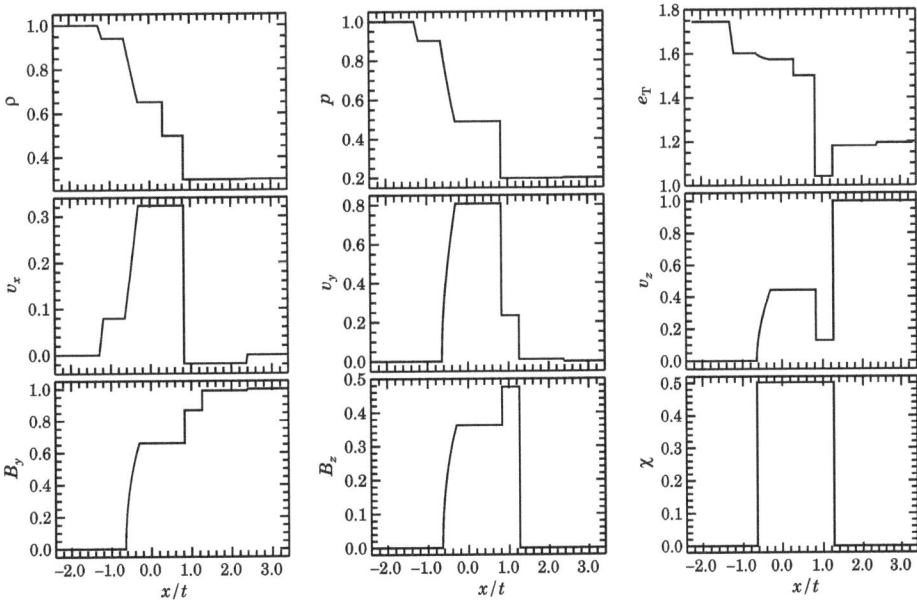

Figure 6.24. Figure 4d from RJ95 (switch-on RF). From left to right: fast Euler RF; switch-on (slow) RF; CD; slow shock; RD; fast RF.

230 The MHD Riemann Problem

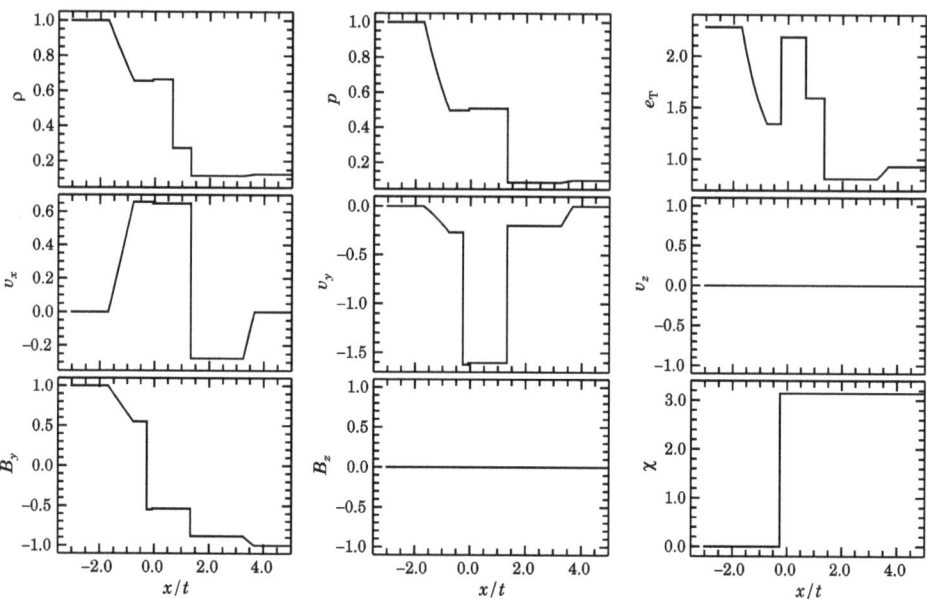

Figure 6.25. Figure 5a from RJ95. From left to right: fast RF; RD; slow shock; CD; slow shock; fast RF. All finite-difference solutions replace the RD + slow shock with a slow compound wave (intermediate shock + slow RF); see Fig. 5.11.

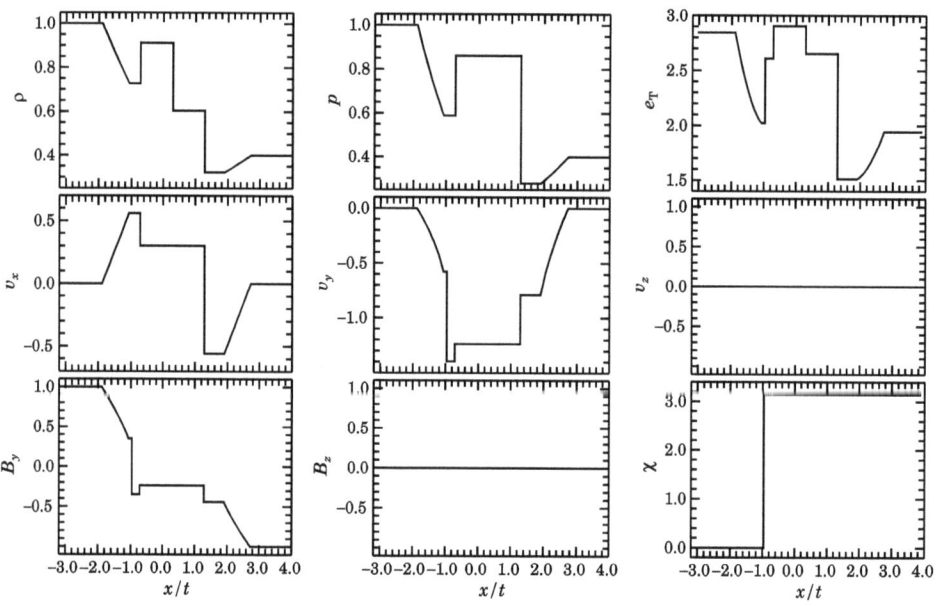

Figure 6.26. Figure 5b from RJ95 (compound wave). From left to right: fast RF; RD; slow shock; CD; slow shock; fast RF. All finite-difference solutions replace the fast RF + RD with a fast compound wave (fast RF + intermediate shock).

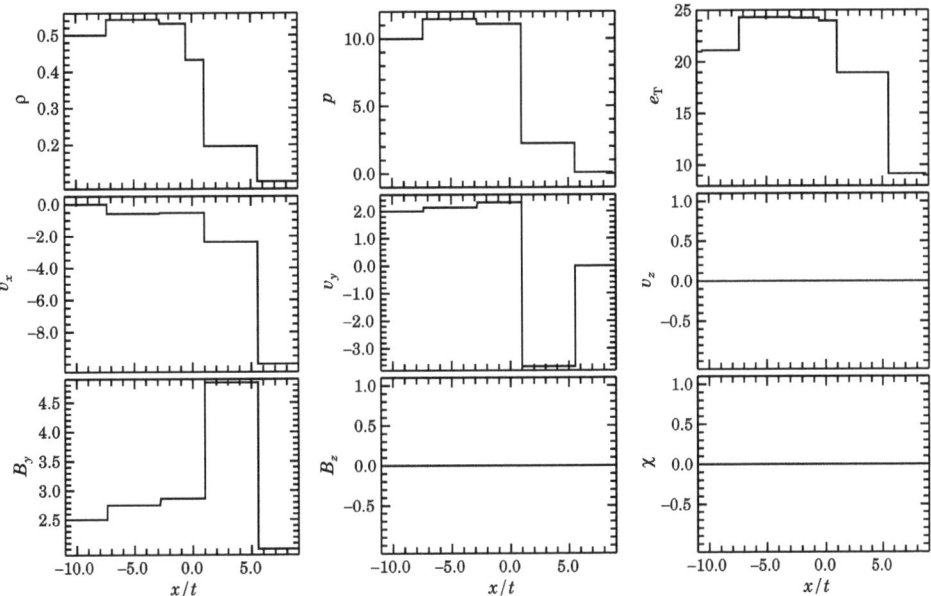

Figure 6.27. Figure 6 from F02. From left to right: fast shock; slow RF; CD; slow shock; fast shock (no RDs).

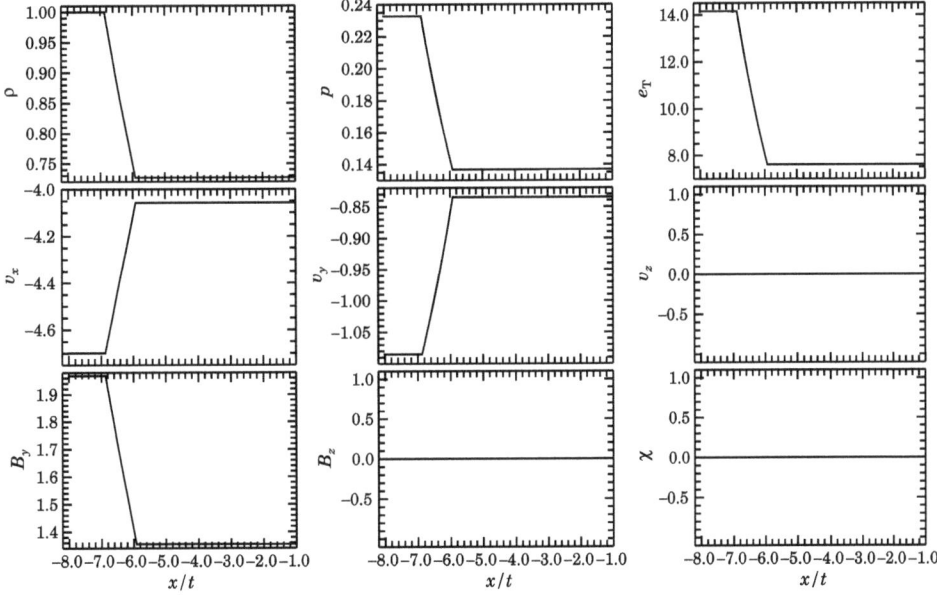

Figure 6.28. Figure 2 from F02 showing a single fast RF.

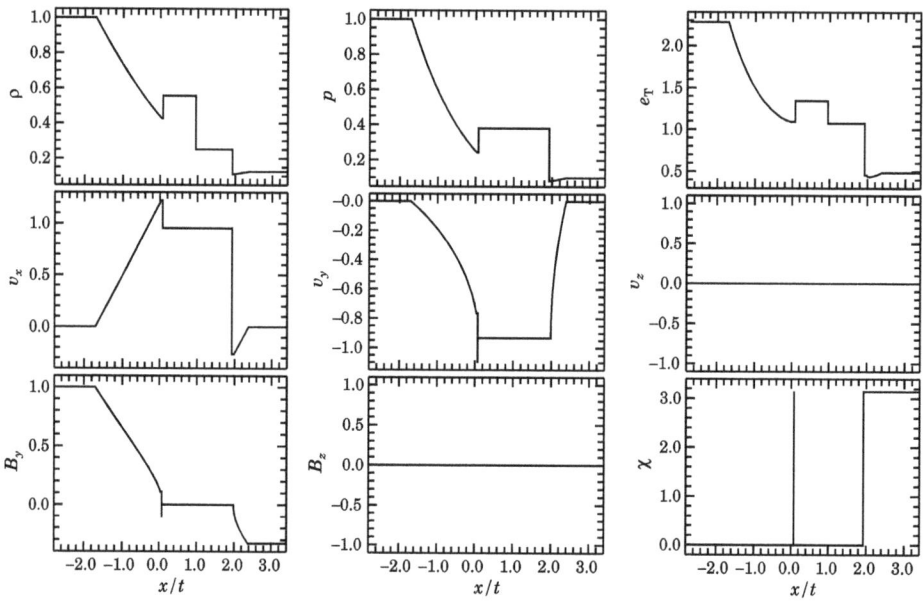

Figure 6.29. Figure 10 from TY13. From left to right: fast RF; switch-off (slow) shock; CD; slow Euler shock, switch-off (fast) RF. The two RDs (absent in TY13) rotating B_y by π radians (including when $B_y = 0$!) point to a bug in my program.

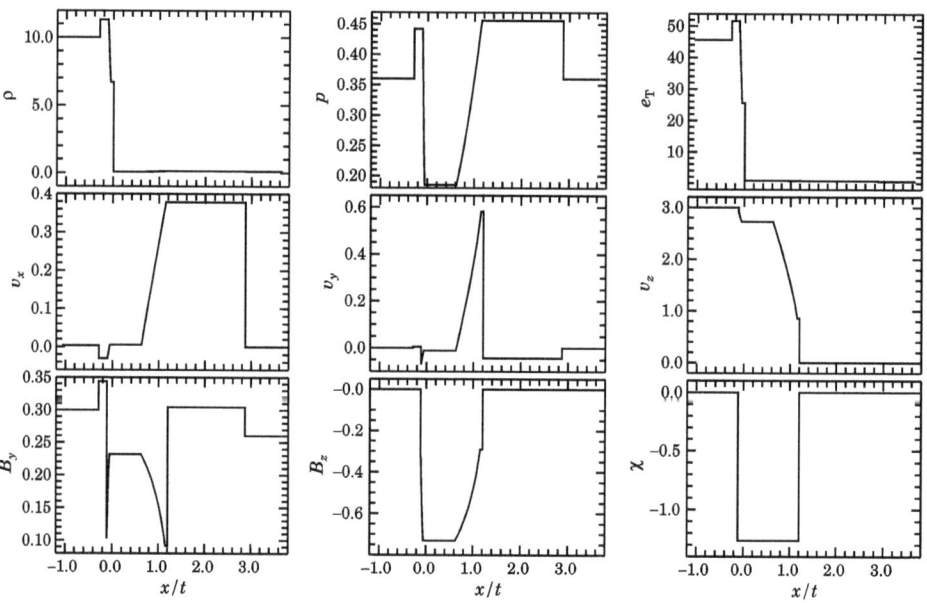

Figure 6.30. A tricky problem. From left to right: fast shock; RD; slow RF; CD; slow RF; RD; fast shock. My program had to restart the problem twice (reshuffling initial guesses each time) before finding this solution.

Fig.	s	ρ	p	v_x	v_y	v_z	B_y	B_z	B_x
6.13	L	3.0	3.0	0.0	0.0	0.0	0.0	0.0	0.0
	1	2.1187	1.6802	0.42399	0.0	0.0	0.0	0.0	
	4	1.3592	1.6802	0.42399	0.0	0.0	0.0	0.0	
	R	1.0	1.0	0.0	0.0	0.0	0.0	0.0	
6.14	L	3.0	3.0	0.0	0.0	0.0	1.0	0.0	2.0
	1	2.6024	2.3671	0.21433	−0.15479	0.0	0.67497	0.0	
	2	2.6024	2.3671	0.21433	−0.22758	0.23582	0.55754	0.38043	
	3	2.1472	1.7181	0.40771	−0.06504	0.34673	0.75259	0.51351	
	4	1.3772	1.7181	0.40771	−0.06504	0.34673	0.75259	0.51351	
	5	1.0301	1.0507	0.06962	−0.30207	0.18500	0.91633	0.62524	
	6	1.0301	1.0507	0.06962	−0.05502	−0.07336	0.66559	0.88746	
	R	1.0	1.0	0.0	0.0	0.0	0.6	0.8	
6.15	L	1.0	20.0	10.0	0.0	0.0	1.4105	0.0	1.4105
	1	2.6797	150.98	0.72113	0.23138	0.0	3.8388	0.0	
	3	2.6713	150.19	0.72376	0.35683	0.0	4.0379	0.0	
	4	3.8508	150.19	0.72376	0.35683	0.0	4.0379	0.0	
	5	3.7481	143.57	0.70505	−0.38803	0.0	5.4271	0.0	
	R	1.0	1.0	−10.0	0.0	0.0	1.4105	0.0	
6.16	L	1.0	1.0	0.0	0.0	0.0	1.4105	0.0	0.84628
	1	1.9185	3.6772	−1.6671	0.35507	0.0	2.8715	0.0	
	3	2.8547	7.6463	−1.8550	−1.2026	0.0	0.84874	0.0	
	4	0.08513	7.6463	−1.8550	−1.2026	0.0	0.84874	0.0	
	5	0.08682	7.9017	−1.7992	0.05545	0.0	0.48640	0.0	
	R	0.1	10.0	0.0	0.0	0.0	0.56419	0.0	
6.17	L	1.08	0.95	1.2	0.01	0.5	1.0155	0.56419	0.56419
	1	1.4903	1.6558	0.60588	0.11235	0.55686	1.4383	0.79907	
	2	1.4903	1.6558	0.60588	0.22157	0.30125	1.5716	0.48702	
	3	1.6343	1.9317	0.57538	0.04760	0.24734	1.4126	0.43772	
	4	1.4735	1.9317	0.57538	0.04760	0.24734	1.4126	0.43772	
	5	1.3090	1.5844	0.53432	−0.18411	0.17554	1.6103	0.49899	
	6	1.3090	1.5844	0.53432	−0.09457	−0.04729	1.5078	0.75392	
	R	1.0	1.0	0.0	0.0	0.0	1.1284	0.56419	
6.18	L	1.0	1.0	0.0	0.0	0.0	1.6926	0.0	0.84628
	1	1.7577	3.0321	−1.5336	0.33873	0.0	3.1166	0.0	
	2	1.7577	3.0321	−1.5336	0.29068	0.47290	3.0529	0.62695	
	3	2.8307	7.3663	−1.7507	−1.2739	0.15159	1.1924	0.24487	
	4	0.08324	7.3663	−1.7507	−1.2739	0.15159	1.1924	0.24487	
	5	0.08823	8.1157	−1.5878	2.0417	0.83250	0.24227	0.04975	
	6	0.08823	8.1157	−1.5878	2.0247	1.0	0.24732	0.0	
	R	0.1	10.0	0.0	2.0	1.0	0.28209	0.0	
6.19	L	0.1	0.4	50.0	0.0	0.0	−0.28209	−0.56419	0.0
	1	0.38714	81.660	25.033	0.0	0.0	−1.0921	−2.1842	
	4	0.39044	81.609	25.033	0.0	0.0	1.1014	2.2029	
	R	0.1	0.2	0.0	0.0	0.0	0.28209	0.56419	
6.20	L	1.0	1.0	−1.0	0.0	0.0	1.0	0.0	0.0
	1	0.49633	0.31114	0.0	0.0	0.0	0.49633	0.0	
	R	1.0	1.0	1.0	0.0	0.0	1.0	0.0	

Table 6.2. Values of the primitive variables in each unique state (state identifier, s, indicated in column 2) of the problems depicted in Fig. 6.13–6.30, and as described in the text.

Table 6.2, continued...

Fig.	s	ρ	p	v_x	v_y	v_z	B_y	B_z	B_x
6.21	L	1.0	1.0	0.0	0.0	0.0	1.0	0.0	1.0
	1	0.59955	0.42629	0.81237	−0.59961	0.0	0.28431	0.0	
	3	0.55151	0.37090	0.89416	−0.54470	0.0	0.31528	0.0	
	4	0.41272	0.37090	0.89416	−0.54470	0.0	0.31528	0.0	
	5	0.22337	0.12402	0.24722	−0.91164	0.0	0.43086	0.0	
	R	0.2	0.1	0.65	0.0	0.0	0.0	0.0	
6.22	L	0.4	0.52467	−0.66991	0.98263	0.0	0.0	0.0	1.3
	4	0.67910	0.52467	−0.66991	0.98437	0.0	0.00110^{16}	0.0	
	R	1.0	1.0	0.0	0.0	0.0	1.0	0.0	
6.23	L	0.65	0.5	0.667	−0.257	0.0	0.54969	0.0	0.75
	1	0.66041	0.51342	0.64401	−0.24512	0.0	0.56470	0.0	
	3	0.84690	0.79673	0.44079	−0.94	0.0	0.0	0.0	
	4	1.0369	0.79673	0.44079	−0.94	0.0	0.0	0.0	
	R	1.0	0.75	0.4	−0.94	0.0	0.0	0.0	
6.24	L	1.0	1.0	0.0	0.0	0.0	0.0	0.0	0.7
	1	0.94000	0.90202	0.07906	0.0	0.0	0.0	0.0	
	3	0.65158	0.48972	0.32267	0.80740	0.44275	0.66003	0.36194	
	4	0.49734	0.48972	0.32267	0.80740	0.44275	0.66003	0.36194	
	5	0.29768	0.19743	−0.01858	0.23388	0.12825	0.86736	0.47563	
	6	0.29768	0.19743	−0.01858	0.01055	1.0	0.98921	0.0	
	R	0.3	0.2	0.0	0.0	1.0	1.0	0.0	
6.25	L	1.0	1.0	0.0	0.0	0.0	1.0	0.0	0.75
	1	0.65673	0.49618	0.65790	−0.26700	0.0	0.54966	0.0	
	2	0.65673	0.49618	0.65790	−1.6235	0.0	−0.54966	0.0	
	3	0.66535	0.50709	0.64841	−1.6053	0.0	−0.53799	0.0	
	4	0.27400	0.50709	0.64841	−1.6053	0.0	−0.53799	0.0	
	5	0.11571	0.08793	−0.27717	−0.19847	0.0	−0.88576	0.0	
	R	0.125	0.1	0.0	0.0	0.0	−1.0	0.0	
6.26	L	1.0	1.0	0.0	0.0	0.0	1.0	0.0	1.3
	1	0.72751	0.58848	0.55942	−0.57673	0.0	0.34899	0.0	
	2	0.72751	0.58848	0.55942	−1.3951	0.0	−0.34899	0.0	
	3	0.91190	0.86059	0.30152	−1.2382	0.0	−0.23702	0.0	
	4	0.60378	0.86059	0.30152	−1.2382	0.0	−0.23702	0.0	
	5	0.32246	0.27931	−0.55835	−0.79046	0.0	−0.44197	0.0	
	R	0.4	0.4	0.0	0.0	0.0	−1.0	0.0	
6.27	L	0.5	10.0	0.0	2.0	0.0	2.5	0.0	2.0
	1	0.54231	11.452	−0.57430	2.1369	0.0	2.7519	0.0	
	3	0.53172	11.082	−0.53022	2.3251	0.0	2.8649	0.0	
	4	0.43257	11.082	−0.53022	2.3251	0.0	2.8649	0.0	
	5	0.19609	2.2397	−2.3747	−3.6537	0.0	4.8427	0.0	
	R	0.1	0.1	−10.0	0.0	0.0	2.0	0.0	

---

[16]The fact that B_y is not exactly zero in the intermediate states means this is an approximate switch-off rarefaction fan; Ryu & Jones' (1995) solution is also approximate by a similar amount. This problem reveals a weakness in my algorithm as changing any of p, v_x, or v_y in the left state L by even the slightest amount can trigger numerous false starts, NaNs, even failure to converge. At this writing, I don't know the cause of this vulnerability, although the fact that v_y is not exactly constant across the CD may be a clue.

Table 6.2, continued...

Fig.	s	ρ	p	v_x	v_y	v_z	B_y	B_z	B_x
6.28	L	1.0	0.23270	−4.6985	−1.0851	0.0	1.9680	0.0	−0.7
	1^{17}	0.72702	0.13678	−4.0577	−0.83485	0.0	1.3551	0.0	
	R	0.72700	0.13680	−4.0577	−0.83490	0.0	1.3550	0.0	
6.29	L	1.0	1.0	0.0	0.0	0.0	1.0	0.0	0.75
	1	0.42671	0.24185	1.2214	−0.76375	0.0	0.10978	0.0	
	2	0.42671	0.24185	1.2214	−1.0999	0.0	−0.10978	0.0	
	3	0.55674	0.37925	0.95327	−0.93182	0.0	0.0	0.0	
	4	0.25132	0.37925	0.95327	−0.93182	0.0	0.0	0.0	
	5	0.11159	0.08276	−0.26216	−0.93167	0.0	−5.27×10^{-5}	0.0	
	R	0.125	0.1	0.0	0.0	0.0	−0.32988	0.0	
6.30	L	10.0	0.36	0.003	0.0	3.0	0.3	0.0	0.28
	1	11.306	0.44198	−0.03032	0.00426	3.0	0.34386	0.0	
	2	11.306	0.44198	−0.03032	−0.06709	2.9025	0.10396	−0.32777	
	3	6.7098	0.18524	0.00552	−0.01150	2.7273	0.23170	−0.73054	
	4	0.06709	0.18524	0.00552	−0.01150	2.7273	0.23170	−0.73054	
	5	0.11519	0.45606	0.37841	0.58237	0.85481	0.09202	−0.29012	
	6	0.11519	0.45606	0.37841	−0.04329	0.0	0.30436	0.0	
	R	0.1	0.36	0.0	0.0	0.0	0.26	0.0	

Problem Set 6

6.1** Show that with appropriate normalisation, the eigenkets associated with the entropy and Alfvén waves are given by:

$$|r_s\rangle = \begin{bmatrix} 1 \\ 0 \\ 0 \\ \vec{0} \\ \vec{0} \end{bmatrix} \quad \text{and} \quad |r_x^\pm\rangle = \begin{bmatrix} 0 \\ 0 \\ 0 \\ \mp\dfrac{\text{sgn}(B_x)}{\sqrt{\mu_0\rho}}\,\hat{e}_x \times \vec{B}_\perp \\ \hat{e}_x \times \vec{B}_\perp \end{bmatrix}, \qquad (6.49)$$

respectively, where $\hat{e}_x$ is a unit vector in the longitudinal (x) direction. You may assume $B_x \neq 0$.

6.2 Prove Eq. (6.20) in the text, namely,

$$\chi_\text{s} \equiv \psi_\text{s}\,\frac{a_\text{s}^2}{a_x^2 - a_\text{s}^2}\,\frac{a_\perp}{c_\text{s}} = \psi_\text{f},$$

---

[17]This problem sets the conditions for the upwind and downwind states (L and R) for a fast RF, and thus the precision of the input values (five significant figures) will mean the intermediate states (1–6) will agree with the downwind state to within a few parts in 10^4. This is reflected by the reported differences between states 1 and R. Similar differences are found among the remaining intermediate states.

where ψ_f and ψ_s are given by Eq. (6.12).

6.3*** To drive home how the eigenkets in Eq. (6.21) are derived and to gain familiarity with the relationships among the MHD-alphas and the scaling factors μ and ν, this problem takes the reader through finding the *eigenbras* (*left* eigenvectors) of the Jacobian for the 1-D primitive ideal MHD equations, J_p (Eq. 5.9 in §5.1).

As discussed in §3.5.2, one way to find the eigenbras, $\langle l_i|$, $i = 1, \ldots, 7$, is to create the right eigenvector matrix, R, with eigenket $|r_j\rangle$ forming the jth column and find its inverse, L, whose i^th row is $\langle l_i|$ (Eq. 3.27). However, unless your algebra is far better than mine, it's a Herculean task to invert any but the simplest 7×7 matrix. If it doesn't seem like it should be so bad, I invite you to give it a try! It took me an entire weekend to find L, and even then I gave up before routing out the last error or two!

A much better way is to evaluate $\langle l_i|$ directly from the Jacobian by solving the matrix equation,
$$\langle l_i|(\mathsf{J}_\text{p} - u_i \mathsf{I}) = 0, \quad i = 1, \ldots, 7. \tag{6.50}$$

a) Following the steps taken in §6.2.1 to obtain the fast and slow (magnetosonic) eigenkets in Eq. (6.21), start with Eq. (6.50) to show that the fast and slow eigenbras are given by:

$$\boxed{\begin{aligned}\langle l_\text{f}^\pm| &= \frac{1}{2}\left[0 \quad -\frac{\mu}{\gamma p} \quad \mp\frac{\mu a_\text{f}}{c_\text{s}^2} \quad \pm\text{sgn}(B_x)\frac{\nu a_\text{s}}{c_\text{s}^2}\hat{e}_\perp \quad -\frac{\nu}{\sqrt{\mu_0 \gamma p}}\hat{e}_\perp\right]; \\ \langle l_\text{s}^\pm| &= \frac{1}{2}\left[0 \quad -\frac{\nu}{\gamma p} \quad \mp\frac{\nu a_\text{s}}{c_\text{s}^2} \quad \mp\text{sgn}(B_x)\frac{\mu a_\text{f}}{c_\text{s}^2}\hat{e}_\perp \quad \frac{\mu}{\sqrt{\mu_0 \gamma p}}\hat{e}_\perp\right], \end{aligned}} \tag{6.51}$$

where all variables have the same meaning as in the text.

Hint: Normalisation of the eigenbras is achieved by noting that $\langle l_i|r_j\rangle = \delta_{ij}$ for which you will make copious use of the relationships among the MHD-alphas (Eq. 6.13–6.16), and the scaling factors, μ and ν (Eq. 6.26 and 6.27).

b) Complete the set by finding the entropy and Alfvén eigenbras, $\langle l_s|$ and $\langle l_x^\pm|$.

c) Assemble the left and right eigenvector matrices, L and R, using your eigenbras for L and the eigenkets in Eq. (6.21) and (6.49) for R, the latter from Problem 6.1. By doing the matrix multiplication, confirm that $\mathsf{LR} = \mathsf{I}$. (And if you found *that* tedious, finding L by inverting R using elementary row operations is $7^2 = 49$ times worse!)

6.4 Show that there can be no switch-off fan for $B_x = 0$. That is, for $B_x = 0$, show that $B_\perp$ asymptotes to zero in the same way that the thermal pressure does, namely $B_\perp \to 0$ as the generalised coordinate $s_i \to \infty$. (For a switch-off fan, $s_i \to s_{i,\text{d}}$, a finite quantity).

Problem Set 6

6.5*

a) Show that on a graph with $x = \sqrt{\alpha_x} = a_x/c_s > 0$ on the abscissa and $y = \sqrt{\alpha_\perp} = a_\perp/c_s > 0$ on the ordinate, contours of constant α_s form a set of confocal hyperbolæ, while contours of constant α_f form a set of confocal ellipses. Where is the focal point for each set?

b) Sketch, with some attempt at accuracy, a representative sample of each set of contours on the same plot over the domain $x, y \in (0, 3)$.[18]

6.6 A critical quantity in evaluating profiles across an MHD rarefaction fan is,
$$\delta_x = \alpha_x - 1, \tag{Eq. 6.22}$$
as it appears both directly and indirectly via the discriminant d (Eq. 6.25) in the normalisation factors for the eigenkets, μ and ν (Eq. 6.26 and 6.27). In a computer application and in the vicinity of $\alpha_x \to 1$, δ_x becomes dominated by machine round-off error which, so long as it is being added to a much greater quantity, is innocuous. However, as is the case for μ and ν, δ_x is often being added to $\alpha_\perp$, and near the triple umbilic where both $\alpha_x \to 1$ and $\alpha_\perp \to 0$, quantities such as μ and/or ν will inherit the machine round-off noise dominating δ_x, leading to all sorts of computational woes.

We therefore need an alternate expression for δ_x that retains its precision as $\alpha_x \to 1$.

a) Define,
$$\varpi = \frac{p - p_{\mathrm{cr}}}{p_{\mathrm{cr}}}, \tag{6.52}$$
where $p_{\mathrm{cr}} = B_x^2/\mu_0\gamma$ is the "critical pressure" which, when $p = p_{\mathrm{cr}}$, $\alpha_x = 1$. Show that for $\varpi \ll 1$,
$$\delta_x = -\varpi + \varpi^2 - \varpi^3 + \cdots. \tag{6.53}$$

With three terms in the expansion, Eq. (6.53) should give δ_x accurate to 12 significant figures (*i.e.*, nearly double precision) for $\varpi \leq 10^{-4}$, say. For $\varpi > 10^{-4}$, one would then use Eq. (6.22).

b) Note that in evaluating ϖ, one just can't set the numerator to $\Delta p = p - p_{\mathrm{cr}}$, as this would entirely defeat the purpose! For $\alpha_x \to 1$, Δp is the difference of two nearly equal numbers, and thus dominated by machine round-off error!

In a numerical scheme, can you suggest an alternate strategy for keeping track of Δp that would not be dominated by numerical noise in the event $\alpha_x \to 1$? *Hint*: domination by numerical noise happens when two nearly equal quantities are *subtracted*. *Adding* numbers, however, even a lot of them...

---

[18]This problem was inspired by Fig. 1 in Roe & Balsara (1996).

6.7 The most critical quantities in evaluating profiles across an MHD rarefaction fan are the eigenket normalisations, μ and ν (Eq. 6.26 and 6.27), rewritten here as:

$$\mu^2 = \tfrac{1}{2}(1-f); \qquad \nu^2 = \tfrac{1}{2}(1+f), \tag{6.54}$$

where,

$$f = \frac{\delta_x + \alpha_\perp}{\sqrt{(\delta_x + \alpha_\perp)^2 + 4\alpha_\perp}}. \tag{6.55}$$

In a computing application, we must avoid differences of nearly equal quantities as these are dominated by machine round-off error. For μ^2 and ν^2, the concern is therefore when $f \to \pm 1$ which cannot be avoided simply by "upping the machine precision".

From Eq. (6.28) and (6.29) in the text, it was shown that for $\alpha_\perp = 0$,

$$\mu^2 = \begin{cases} 1, & \delta_x < 0; \\ \tfrac{1}{2}, & \delta_x = 0; \\ 0, & \delta_x > 0, \end{cases} \quad \text{and} \quad \nu^2 = 1 - \mu^2 = \begin{cases} 0, & \delta_x < 0; \\ \tfrac{1}{2}, & \delta_x = 0; \\ 1, & \delta_x > 0, \end{cases} \tag{6.56}$$

all perfectly well-behaved values. Thus, we may proceed assuming $\alpha_\perp > 0$.

Further, for $\delta_x + \alpha_\perp = 0$ and $\alpha_\perp > 0$, $f = 0$ and, from Eq. (6.54), $\mu^2 = \nu^2 = \tfrac{1}{2}$ (same as the $\delta_x = \alpha_\perp = 0$ case in Eq. 6.56). Thus, we may proceed assuming $\delta_x + \alpha_\perp \neq 0$.

a) So with $\alpha_\perp > 0$ and $\delta_x + \alpha_\perp \neq 0$, define,

$$\epsilon = \frac{\alpha_\perp}{(\delta_x + \alpha_\perp)^2}, \tag{6.57}$$

where $0 < \epsilon < \infty$. Show that: $\lim_{\epsilon \to 0} f = \text{sgn}(\delta_x + \alpha_\perp)$, where $\text{sgn}(x) = \dfrac{x}{|x|}$.

b) Even though $\alpha_\perp > 0$, ϵ can still be arbitrarily close to zero forcing f arbitrarily close to ± 1, which remains problematic for μ^2 or ν^2. Show that for $\epsilon \ll 1$:

$$\left.\begin{aligned}\mu^2 &= \epsilon - 3\epsilon^2 + 10\epsilon^3 - \cdots \quad \text{for } \delta_x + \alpha_\perp > 0; \\ \nu^2 &= \epsilon - 3\epsilon^2 + 10\epsilon^3 - \cdots \quad \text{for } \delta_x + \alpha_\perp < 0,\end{aligned}\right\} \tag{6.58}$$

give values for μ^2 and ν^2 in the domains indicated accurate to at least 10–11 significant figures for $\epsilon < 10^{-4}$ using the three terms shown.

Computer projects

P6.1: Write a computer program to find the profile across any MHD rarefaction fan. In addition to the variable values in the upwind state, your input specifications will need to include whether the fan is fast or slow, and how "wide" the fan should be (upper limit of integration). Thus, you will be solving the seven first-order ODEs in Eq. (6.1) in the text, using Eq. (6.21) for the function derivatives (eigenkets), and Eq. (6.26) and (6.27) for μ and ν, the eigenket "normalisations". See §6.4.3 (subroutine RF on page 218) for guidance.

For starters, your program should be able to reproduce Fig. 6.5–6.10 whose upwind states are described in the figure captions.

In general, your program should be able to:

- detect when a fast or slow rarefaction fan is not possible from the given upwind state;

- detect when a fast fan has saturated, and integrate to this limiting value;

- detect if a fast saturated fan is Euler or switch-off;

- distinguish between a slow Euler fan and a switch-on fan (both requiring $B_\perp = 0$ in the upwind state) even for $|\delta_x| < 10^{-32}$;

- launch a slow fan from the triple umbilic.

Suggestion: You might, like I did, consider writing your program in units where $\mu_0 = 1$. It simplifies what values have to be entered for the magnetic induction components, and it simplifies how the plots are done.

P6.2: Following the algorithm in §6.4.3, write a computer program to give the "exact" and unique solution to any MHD Riemann problem, adhering to both the entropy and evolutionary conditions (*i.e.*, no intermediate shocks, but all switch-on and switch-off waves accounted for).

Confirm the integrity of your program by reproducing Fig. 6.15–6.30 in the text, whose left and right states are all listed in Table 6.2.

Suggestion: See suggestion for project P6.1!

PART II

ADDITIONAL TOPICS IN (M)HD

7. Fluid Instabilities

with thanks to Joel Tanner, B.Sc. (SMU), 2005; M.Sc. (SMU), 2007.[†]

> *I am an old man now, and when I die and go to heaven, there are two matters on which I hope for enlightenment. One is quantum electrodynamics and the other is the turbulent motion of fluids. About the former, I am really rather optimistic.*
>
> Sir Horace Lamb (1849–1934)
> addressing the British Society for the Advancement of Science, 1932

ALL PHYSICAL SYSTEMS have points of equilibrium, where the net force exerted on all or part of the system is zero. In such cases, the question is whether a slight departure from equilibrium results in the system returning to its equilibrium point – a condition referred to as *stable equilibrium* – or whether that slight departure causes the system to move further away from equilibrium, a condition known as *unstable equilibrium*.

The usual analogy made at this point in the discussion is of a ball caught in a dip in the ground *vs.* one perched on top of a mound. As shown in Fig. 7.1 a, a nudge to the ball in the valley causes the ball to roll up the hill a short distance, stop, return to the bottom of the dip with a non-zero speed which carries it through the equilibrium point and up the other side, stop again, and so it goes. Allowing for dissipative forces, the ball eventually settles back to its equilibrium position. This exemplifies a stable equilibrium.

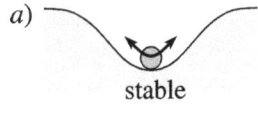

Figure 7.1. A ball in a dip (*a*) and atop a mound (*b*) illustrating stable and unstable equilibria.

Alternatively, to nudge the ball on top of the mound in Fig. 7.1 b causes the ball to roll away from its perch and accelerate, never to return (unaided) to the top of the mound. This exemplifies an unstable equilibrium, or *instability*.

Fluids can also exhibit instabilities and, as we'll see, all lead to a state of chaotic (but not random!) motion in the fluid known as *turbulence* (Fig. 7.2), a subject that has confounded physicists before and since Sir Horace. And while a full treatment

---

[†]Ph. D. (Yale), 2014

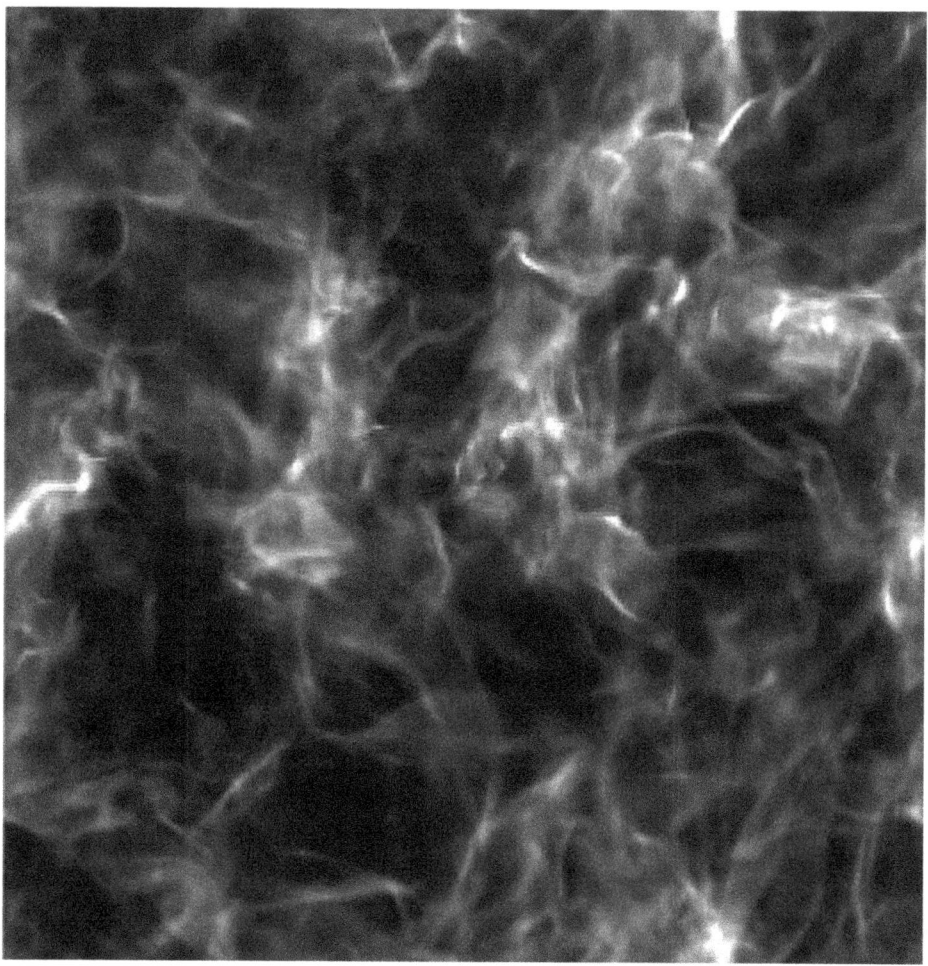

Figure 7.2. An image from a ZEUS-3D simulation of *super-Alfvénic turbulence*, a portion of which is used as chapter banners in this text. Shown is a line-of-sight integration of $B^2/2\mu_0$ through a fluid cube initialised with a weak magnetic field ($\alpha \ll 1$). As the turbulent motion twists and stretches $\vec{B}$, kinetic energy is converted to magnetic energy, and the system asymptotes towards *trans-Alfvénic* turbulence. While turbulence cannot be described as *random* (there is clearly structure), it is *chaotic* producing a filamentary structure reminiscent of the filaments observed in giant radio lobes such as those in Cygnus A (Fig. 2.14). Indeed, super- and trans-Alfvénic turbulence is the state in which the interstellar and inter-galactic media are thought to exist, begun and driven by fluid instabilities.

of turbulence is beyond the scope of this text,[1] a good start is to understand the nature of its precursors, *i.e.*, fluid instabilities.

Like the ball caught in the dip or perched on the mound, certain fluid equilibria

---

[1] See Galtier (2016) for a senior undergraduate level text that treats (M)HD turbulence in considerable depth.

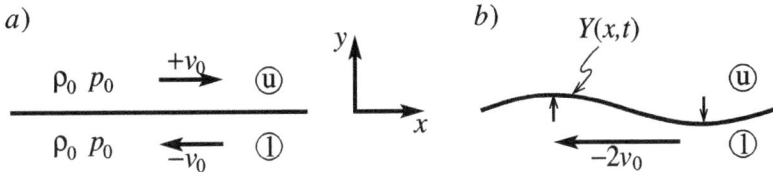

Figure 7.3. *a*) A shear layer in an otherwise uniform fluid in the frame of reference of the lab. *b*) The same shear layer from the frame of reference of the upper fluid (region ⓤ) with a sinusoidal perturbation applied to the shear layer.

can be shown to be stable or unstable. In this chapter, we'll examine four such cases with a similar strategy for each. First, we'll identify one or more variables (*e.g.*, pressure, velocity, *etc.*) with the potential of "running away" from equilibrium. We then derive an equation of motion for that variable from the linearised fluid equations and look to their normal mode solutions[2] and their frequencies, ω, to distinguish between stable ($\omega \in \mathbb{R}$) and unstable equilibria ($\omega \in \mathbb{I}$).[3] Our linearised analysis will only be able to determine what starts the instability and how it grows initially. In two cases, *ZEUS-3D* simulations are used to supplement the discussion to show how the instability progresses from the linear to the *non-linear* regime, and then ultimately towards the onset of turbulence.

7.1 Kelvin–Helmholtz instability

A *shear layer* is a narrow width of fluid (a layer) across which the flow speed along the layer changes significantly. Figure 7.3*a* shows a simplified shear layer from the "lab frame" in which the flow speed changes from $+v_0$ for $y > 0$ (the upper region ⓤ) to $-v_0$ for $y < 0$ (the lower region ⓛ) across an arbitrarily thin layer at $y = 0$. Left alone, an inviscid fluid would continue to flow like this in perpetuity, and thus a shear layer represents an equilibrium state in ideal hydrodynamics.

Now, what if a shear layer is perturbed in the y-direction (as depicted in Fig. 7.3*b*)? Will the disturbance grow in time and thus exhibit an instability, or will it decay in time and thus indicate a stable equilibrium? If the disturbance grows, at what rate does it grow and what determines that rate?

As we'll see, under the right circumstances a shear layer *is* unstable to transverse perturbations and our mathematical approach – which can be described in a number of ways including *linear analysis*, *perturbation theory* or, as we'll call it, *normal mode analysis* – will lead us to what I would argue is the most fundamental fluid instability of ideal hydrodynamics: the *Kelvin–Helmholtz instability* (KHI).

---

[2] *Rappel*: "Normal mode solutions" are defined after Eq. 2.21 on page 30.
[3] See Problem 7.1 on how this method can be applied even to the ball in Fig. 7.1.

7.1.1 Normal mode analysis of the KHI

We wish to consider the time-evolution of a small-amplitude perturbation applied to a shear layer, where $y = Y(x,t)$ describes the shape of the shear layer as a function of x and t. Since the distortions are small, we can use the "linearised" fluid equations from §2.1.1 where all speeds are measured from within the rest frame of the unperturbed fluid. Thus, in the rest frame of the upper fluid here (region ⓤ in Fig. 7.3b), the pressure is governed by the same wave equation developed in §2.1.1, namely:

$$\frac{\partial^2 p}{\partial t^2} = \partial_{tt} p = c_s^2 \nabla^2 p, \qquad \text{Eq. (2.9)}$$

where $c_s^2 = \gamma p_0/\rho_0$ is the square of the sound speed for an adiabatic gas, and where p_0 and ρ_0 are the unperturbed pressure and density respectively. Note that both layers must have the same pressure, p_0, in order for the system to start in equilibrium, but they need not have the same density. Here, we have chosen to set the density of both layers to the same value, ρ_0, to simplify the calculations.

To proceed, we seek normal mode solutions to Eq. (2.9) of the form:

$$p(x,y,t) = P(y)\,e^{i(kx-\omega_u t)}, \qquad (7.1)$$

where $P(y)$ is the y-dependent amplitude of the perturbation, ω_u is the frequency of the normal mode as measured from the rest frame of upper region ⓤ, and k is the wave number, a Galilean invariant (independent of frame from which it is measured). Since the Fourier components of any general function are the normal modes, no loss of generality is suffered by assuming such a form for p.

$P(y)$ is determined by substituting Eq. (7.1) into Eq. (2.9). To this end, we have:

$$\partial_{tt} p = -\omega_u^2 P(y)\,e^{i(kx-\omega_u t)};$$

$$\nabla p = \left(ikP(y), \frac{dP(y)}{dy}, 0\right) e^{i(kx-\omega_u t)};$$

$$\nabla \cdot (\nabla p) = \nabla^2 p = \left(-k^2 P(y) + \frac{d^2 P(y)}{dy^2}\right) e^{i(kx-\omega_u t)}.$$

Thus, Eq. (2.9) becomes:

$$-\omega_u^2 P(y)\,e^{i(kx-\omega_u t)} = c_s^2 \left(-k^2 P(y) + \frac{d^2 P(y)}{dy^2}\right) e^{i(kx-\omega_u t)}$$

$$\Rightarrow \quad \frac{d^2 P(y)}{dy^2} = \left(k^2 - \frac{\omega_u^2}{c_s^2}\right) P(y) \equiv \beta_u^2 P(y)$$

$$\Rightarrow \quad P(y) = A e^{-\beta_u y} + B e^{\beta_u y}, \qquad (7.2)$$

where A and B are constants of integration and β_u is defined for convenience. In region ⓤ, $y > 0$ and we must set $B = 0$ in order for $P(y) \to 0$ as $y \to +\infty$. Thus,

$$P(y) = A e^{-\beta_u y},$$

and Eq. (7.1) becomes:
$$p(x,y,t) = A e^{-\beta_u y} e^{i(kx-\omega_u t)}. \tag{7.3}$$

Aside: For the reader unfamiliar with normal mode analysis, let's take a breather and look at what just happened.

In §2.1.1 where the wave equation (Eq. 2.9) was first developed from the linearised Euler and pressure equations, we found that *any* function of $\vec{r}$ and t would solve it so long as $\vec{r}$ and t came in the form $\vec{k} \cdot \vec{r} \pm \omega t$, where $\vec{k} = k\hat{e}_k$ is the *wave vector*, and where k and ω are as defined above. Now certainly, the function $e^{i(kx-\omega t)}$ satisfies this criterion (taking $\hat{e}_k = \hat{\imath}$, for example), but we could still choose *any* function of $kx - \omega t$. Why the exponential?

There are at least three compelling reasons I can think of.

1. A more "general" solution can always be expressed as a linear combination of its Fourier components, each of which is an exponential with an imaginary argument. Thus, anything we learn about what it means for a single Fourier component to solve Eq. (2.9) applies to the "general solution" too.

2. In perturbation analysis, we actually perturb the fluid in the y-direction with a pure cosine (or sine) wave which is just the real (or imaginary) part of our trial solution. So, in this case, $e^{i(kx-\omega t)}$ actually *is* the general solution!

3. Finally, it's the simplest choice mathematically. Grace of the exponential, by the time we get to Eq. 7.2, all occurrences of x and t have disappeared leaving us with an easy-to-solve ODE for the amplitude, $P(y)$.

Now, we're still only part way there. Our goal is to get a single expression relating ω and k; the so-called *dispersion relation*. But we still have the shape of the interface, $Y(x,t)$, to examine along with the facts that both p and Y remain continuous across the interface as the perturbation evolves. This is what's coming up.

Next, the linearised Euler equation for v_y is,
$$\partial_t v_y = -\frac{1}{\rho_0}\partial_y p = \frac{\beta_u}{\rho_0} A e^{-\beta_u y} e^{i(kx-\omega_u t)},$$

and since the x-component of the velocity in the rest frame of region ⓤ, $v_{x,u}$, is zero, we have at $y=0$:
$$v_y = \frac{dY}{dt} = \partial_t Y + v_{x,u}\partial_x Y = \partial_t Y,$$

where, as mentioned at the top of the section, $Y(x,t)$ gives the shape of the boundary layer as it evolves in time (Fig. 7.3b). Therefore:
$$\partial_{tt} Y = \partial_t v_y \Big|_{y=0} = -\frac{1}{\rho_0}\partial_y p \Big|_{y=0} = \frac{\beta_u}{\rho_0} A e^{i(kx-\omega_u t)}, \tag{7.4}$$

which can be integrated twice over time to get:

$$Y(x,t) = -\frac{\beta_u}{\rho_0 \omega_u^2} A e^{i(kx-\omega_u t)}. \tag{7.5}$$

We can transform the expressions for p and Y (Eq. 7.3 and 7.5) to the lab frame by the following simple argument. Perturbations with wave number k and thus wavelength $\lambda = 2\pi/k$ will have the same wavelength regardless of the frame of reference in which they are observed (as already stated, k is a Galilean invariant). This is not true of the frequency. Suppose in the rest frame of region ⓤ, the waves are moving in the $+x$-direction, and suppose further that we observe $f = \omega_u/2\pi$ crests of the wave passing by each second. Then, to shift to the lab frame as depicted in Fig. 7.3a, we must move with a velocity v_0 into the wavetrain and thus the number of crests we observe passing by per second must *increase*. Thus, transforming the above discussion to the lab frame is tantamount to applying a *Doppler shift* where the frequency observed in the lab frame, ω, is given by:

$$\omega = \omega_u + kv_0 \quad \Rightarrow \quad \omega_u = \omega - kv_0. \tag{7.6}$$

Substituting Eq. (7.6) into each of Eq. (7.3) and (7.5) yields:

$$p_>(x,y,t) = A e^{-\beta_u y} e^{i[kx-(\omega-kv_0)t]}; \tag{7.7}$$

$$Y_>(x,t) = -\frac{\beta_u}{\rho_0(\omega-kv_0)^2} A e^{i[kx-(\omega-kv_0)t]}. \tag{7.8}$$

where the subscript $>$ indicates these expressions are valid for $y > 0$. Further, in terms of ω,

$$\beta_u = \sqrt{k^2 - \frac{(\omega-kv_0)^2}{c_s^2}}.$$

Expressions valid for the lower region ⓛ in Fig. 7.3 can be extracted directly from Eq. (7.7) and (7.8) simply by replacing v_0 with $-v_0$, and by noting that:

$$P(y) = B e^{+\beta_l y}; \quad \beta_l = \sqrt{k^2 - \frac{(\omega+kv_0)^2}{c_s^2}},$$

in order for $P(y) \to 0$ as $y \to -\infty$. Thus:

$$p_<(x,y,t) = B e^{\beta_l y} e^{i[kx-(\omega+kv_0)t]}; \tag{7.9}$$

$$Y_<(x,t) = \frac{\beta_l}{\rho_0(\omega+kv_0)^2} B e^{i[kx-(\omega+kv_0)t]}, \tag{7.10}$$

where the subscript $<$ indicates these expressions are valid for $y < 0$. Note that there is no negative sign leading the RHS of Eq. (7.10). This is because from Eq. (7.4), $\partial_{tt} Y_<$ and thus $Y_< \propto \partial_y p_l \propto e^{+\beta_l y}$ and not $e^{-\beta_u y}$ as is the case for $Y_>$.

To obtain the *dispersion relation* (how ω depends upon k), we exploit the singular property of p and Y that they must be continuous across the shear layer. Thus, at $y = 0$, we set $p_> = p_<$ and $Y_> = Y_<$ to get:

$$A e^{ikv_0 t} = B e^{-ikv_0 t}; \tag{7.11}$$

Kelvin–Helmholtz instability

$$-\frac{\beta_{\mathrm{u}}}{(\omega - kv_0)^2} A e^{ikv_0 t} = \frac{\beta_{\mathrm{l}}}{(\omega + kv_0)^2} B e^{-ikv_0 t}. \tag{7.12}$$

Dividing Eq. (7.12) by Eq. (7.11) yields:

$$-\beta_{\mathrm{u}}(\omega + kv_0)^2 = \beta_{\mathrm{l}}(\omega - kv_0)^2$$

$$\Rightarrow \left(k^2 - \frac{(\omega - kv_0)^2}{c_{\mathrm{s}}^2}\right)(\omega + kv_0)^4 = \left(k^2 - \frac{(\omega + kv_0)^2}{c_{\mathrm{s}}^2}\right)(\omega - kv_0)^4$$

$$\Rightarrow \left(\frac{\omega^2 - k^2 v_0^2}{kc_{\mathrm{s}}}\right)^2 = 2\left(\omega^2 + k^2 v_0^2\right),$$

after some straight-forward algebra. Additional algebraic manipulations yield:

$$\boxed{\left(\frac{\omega}{kc_{\mathrm{s}}}\right)^4 - \left(\frac{\omega}{kc_{\mathrm{s}}}\right)^2 2(M^2 + 1) + M^2(M^2 - 2) = 0,} \tag{7.13}$$

where $M = v_0/c_{\mathrm{s}}$ is the Mach number of the flow relative to the lab frame. This is the dispersion relation, whose roots are given by:

$$\left(\frac{\omega}{kc_{\mathrm{s}}}\right)^2 = M^2 + 1 \pm \sqrt{4M^2 + 1},$$

and thus four possible frequencies are admitted:

$$\left.\begin{aligned}\omega_{1,2} &= \pm kc_{\mathrm{s}}\sqrt{M^2 + 1 + \sqrt{4M^2 + 1}}; \\ \omega_{3,4} &= \pm kc_{\mathrm{s}}\sqrt{M^2 + 1 - \sqrt{4M^2 + 1}},\end{aligned}\right\} \tag{7.14}$$

where the perturbation, $Y(x,t)$, develops as a linear combination of these four modes:

$$Y(x,t) = \sum_{j=1}^{4} \mathcal{C}_j(x,t) e^{-i\omega_j t}.$$

The first two frequencies, ω_1 and ω_2, are real and thus produce oscillatory responses. Since they differ only by a sign, they both represent the same normal mode. The second two frequencies, ω_3 and ω_4, are real when:

$$M^2 + 1 > \sqrt{4M^2 + 1} \;\Rightarrow\; M^4 + 2M^2 + 1 > 4M^2 + 1 \;\Rightarrow\; M^4 > 2M^2$$

$$\Rightarrow \boxed{M^2 > 2.}$$

That is, when $M^2 > 2$, ω_3 and ω_4 also produce oscillatory responses and also represent the same normal mode. Thus, the shear layer responds to the perturbation with a linear combination of two normal modes with frequencies ω_1 and ω_3, and the system is stable.

On the other hand, for $0 < M^2 < 2$, ω_3 and ω_4 are imaginary, and we set:

$$\omega_3 = i\chi; \quad \omega_4 = -i\chi,$$

where :

$$\chi = kc_{\mathrm{s}}\sqrt{\sqrt{4M^2 + 1} - M^2 - 1} \in \mathbb{R} \quad \text{for } M^2 < 2. \tag{7.15}$$

Then:
$$e^{-i\omega_3 t} = e^{\chi t} \quad \text{exponentially growing;}$$
$$e^{-i\omega_4 t} = e^{-\chi t} \quad \text{exponentially decaying.}$$

Thus, after some time t, the ω_3 term dominates all other terms and the shear layer is unstable to normal mode perturbations. In this case,
$$Y(x,t) \propto e^{\chi t},$$
which has an *e-folding growth-time*, τ_{KH}, of,
$$\tau_{\text{KH}} = \frac{1}{\chi} \propto \frac{1}{kc_{\text{s}}} = \frac{\lambda}{2\pi c_{\text{s}}}, \tag{7.16}$$
where λ is the wavelength of the normal mode. Evidently, the shorter the wavelength, the shorter the growth-time, and thus the faster the instability grows. It would seem that infinitely small wavelengths would grow infinitely fast, but this is not the case in practice. Viscous damping and/or surface tension along the shear layer preferentially damp out short-wavelength perturbations, and thus the very short wavelengths do not grow at all. But of the wavelengths not seriously affected by dissipative forces, Eq. (7.16) tells us the shorter ones grow most rapidly.

Finally, we've seen that for $M^2 > 2$, the growth rate of the instability is the least (zero). At what Mach number is the growth rate the greatest? From Eq. (7.16), the shortest growth-time for a given wavelength corresponds to the greatest value for χ, and thus of χ^2. Therefore, from Eq. (7.15),
$$\frac{1}{(kc_{\text{s}})^2} \frac{d(\chi^2)}{d(M^2)} = \frac{4}{2\sqrt{4M^2+1}} - 1 = 0$$
$$\Rightarrow \quad \sqrt{4M^2+1} = 2 \quad \Rightarrow \quad 4M^2+1 = 4 \quad \Rightarrow \quad M^2 = 3/4.$$

Thus, the shear layer is most unstable when the Mach number for each of the upper and lower regions relative to the lab frame is $\sqrt{3}/2$, or when their relative Mach number is $\sqrt{3}$.

This is the Kelvin–Helmholtz instability, named for William Thomson (Lord Kelvin)[4] and Hermann von Helmholtz[5] who first studied the effect in the mid-19th century. Our analysis of the KHI has been entirely in the so-called *linear regime*, a reference to the fact that only the *linearised* Euler and pressure equations have been consulted. As we'll see in the next subsection, the sole non-linear term in the HD equations, namely $\vec{v}\cdot\nabla\vec{v}$ in Euler's equation, is a complete game-changer once the amplitude of the perturbation can no longer be considered "small" compared to its wavelength.

---

[4]Thomson (1824–1907; www.wikipedia.org/wiki/Lord_Kelvin) was the first British scientist elevated to the House of Lords, becoming Lord Kelvin. This is the same Kelvin who formulated the combined first and second laws of thermodynamics and for whom the temperature scale is named.

[5]Helmholtz (1821–1894; www.wikipedia.org/wiki/Hermann_von_Helmholtz) may be better known to the reader for his work in electrodynamics and the "Helmholtz equation".

Figure 7.4. Images from a *ZEUS-3D* simulation of a Kelvin–Helmholtz unstable shear layer showing *a)* the instability still in its "linear regime" where the perturbation can still be described by four wavelengths of a sine wave, *b)* the onset of its "non-linear regime", *c)* well into its non-linear regime where the waves start to "break", *d)* further yet into the non-linear regime as the wave crests wrap up into spirals, and *e)* the development of vortices known as "cat's eyes".

7.1.2 The development of the KHI

Figure 7.4 shows five snapshots from a *ZEUS-3D*[6] simulation (800×200 zones) of a shear layer perturbed by four wavelengths (4λ) of a sinusoidally varying transverse velocity with an amplitude $0.001\,c_s$. The top fluid (blue) moves from left to right, while the bottom fluid (white) moves from right to left, each with the same speed, $M = 0.5$ (and thus a relative Mach number of 1). Panel *a* shows the shear layer after several growth times where the amplitude of the perturbation has grown to $\sim 0.01\lambda$, still small enough to be considered in the linear regime and therefore well-described by four wavelengths of a sine wave. In panel *b* about 1.6 growth times later, the perturbation amplitude has grown to more than 5% of λ, and the non-linear effects are just beginning to appear (little "barbs" to the right of each peak). By panel *c*, non-linear effects are well-developed and waves start to "break", forming a sequence of "curls" like those you might find on top of a soft-serve ice cream cone. As the white and blue fluids wrap themselves around each other, the "curls" become "swirls" and four cells of clockwise-spinning fluid form by panel *d*. As the instability progresses, "swirls" become "whirls" as the two fluids wrap themselves up tightly enough to start mixing, leaving four distinct *vortices* commonly known as "cat's

---

[6] *Rappel*: See footnote 6 on page 77.

eyes" (panel *e*). Left alone, these "whirls" of fluid continue to spin in a clockwise sense, driven as they are by the right-moving fluid above them and left-moving fluid below, much like ball-bearings caught between two oppositely moving planks.

The time from initial perturbation until Fig. 7.4*a* where the sinusoidal distortions can just be seen is roughly the same as the time between panels *a* and *d*. Thus, while it takes a while for the non-linear regime to begin, once it does it rapidly takes over. To be clear, all the analysis that concluded an instability exists with a growth rate given by Eq. (7.15) and (7.16) was done in the linear regime and thus applicable only to a little after Fig. 7.4*a*. Everything thereafter is completely unknown to the linear analysis, and it is only through simulations such as this that we can know how the instability unfolds once tripped. Still, the linear analysis is important since it is what tells us whether an equilibrium state is stable or not and whether there is something interesting to pursue numerically.

Now, with all this mathematical and numerical analysis, we still haven't developed a physical model. On what physical grounds does the KHI operate?

The KHI can be explained in terms of Bernoulli's theorem (§2.4) for a gas:

$$\mathcal{B}_{\text{gas}} = \frac{v^2}{2} + \frac{\gamma}{\gamma-1}\frac{p}{\rho} + \phi = \text{constant along streamlines.} \qquad \text{Eq. (2.72)}$$

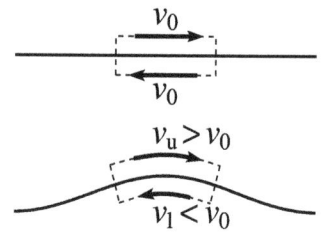

Consider the rectangular region of fluid straddling the shear layer in the top panel of the inset (dashed lines). Flow along the upper boundary of the rectangle (which lies along a streamline) moves from left to right with speed v_0, while along the lower boundary (also a streamline), flow moves from right to left also at speed v_0. As the perturbation bends the shear layer (inset, bottom panel), the rectangular box is also bent, forcing fluid moving along the upper boundary to travel further in the same time than fluid moving along the lower boundary. Thus, as a result of the perturbation, fluid above the shear layer speeds up ($v_u > v_0$) while fluid below slows down ($v_l < v_0$). Assuming the gravitational potential, ϕ, is the same above and below the shear layer, for $\mathcal{B}_{\text{gas}}$ to remain constant, p must decrease above the shear layer and increase below, creating a pressure difference that drives the instability.

7.1.3 The KHI in nature

I give two examples of natural phenomena attributable to the Kelvin–Helmholtz instability, one terrestrial, one extraterrestrial.

Because the KHI is preferentially excited at the subsonic speeds of the upper atmosphere, clouds – which often form at shear layers where two air masses meet – can serve as "tracers" of the KHI as it develops. Figure 7.5 shows a spectacular example of this where clouds "light up" a KHI along an atmospheric shear layer that is well into its non-linear regime and bears a strong resemblance to Fig. 7.4*c*. As

Figure 7.5. Kelvin–Helmholtz instabilities forming a line of "soft-serve ice cream curls" in clouds over Wyoming (photo credit: Brooks Martner, NOAA).

panels *d* and *e* of Fig. 7.4 showed us, the fate of this beautiful sequence of "soft-serve ice cream curls" is to develop into whirling vortices, *i.e.* cat's eyes. Because each vortex spins within non-spinning fluid, its surface is itself a shear layer that, in principle, is Kelvin–Helmholtz unstable and can shed off smaller vortices of its own. As we'll see qualitatively in §8.5, this cascade into smaller and smaller vortices is how the KHI develops into turbulence, something anyone who has ever flown on an airplane will have experienced first hand.

The most spectacular extraterrestrial example of a KHI-driven cat's eye has to be the *Great Red Spot* on Jupiter (Fig. 7.6). Note the examples of smaller amplitude KHI in the surrounding atmosphere and within the red spot itself. The white oval in the lower right of Fig. 7.6 is likely another KHI-driven vortex.

If what we know today is the same feature that was first observed by Robert Hooke in 1664, the Great Red Spot is a very long-lived feature indeed. This means it must essentially be 2-D (planar) in nature because a fully 3-D vortex cascades

Figure 7.6. Jupiter's Great Red Spot thought to be a KHI-driven "cat's eye" (photo credit: Voyager 1, NASA).

into turbulence very quickly (*e.g.*, Ryu *et al.*, 2000). While its origin is still not fully understood, the red spot is widely believed to be a feature confined between two layers of Jupiter's stratified atmosphere, and thus approximately 2-D. Since Jupiter's banded atmosphere rotates differentially with bands nearer the equator rotating

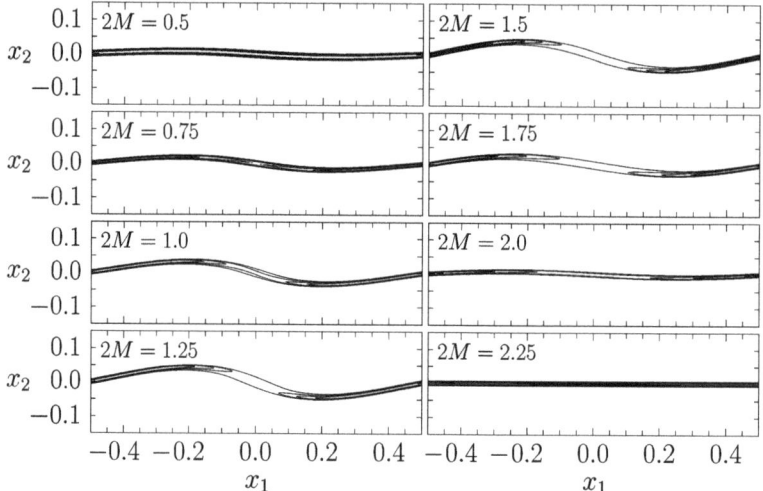

Figure 7.7. Growth of a sinusoidal perturbation along a shear layer as a function of relative Mach number between a right-moving upper and left-moving lower layer ($2M$) as computed by ZEUS-3D on a 500×150 Cartesian grid. Images are taken at $t = 2.7$ (in units of a horizontal sound-crossing time), just at the transition between linear and non-linear behaviour for the $2M = 1.5$ run. Contours show the normal component of vorticity across the shear layer whose thickness is ~ 0.04 in units of x_1.

faster than those nearer the poles, there is shear between the bands which drives the KHI, creating vortices like the Great Red Spot. The spot's colour is thought to result from redder material being dredged up from below by the vortex.

7.1.4 Numerical analysis of the KHI (optional)

Here we examine the linear regime of the Kelvin–Helmholtz instability from a numerical standpoint. Numerous analytical problems (*e.g.*, the Riemann problem) exist that can be used to verify numerical MHD algorithms such as those used in ZEUS-3D, but analytic multi-dimensional solutions are few and far between. Therefore, another good reason to study the linear regime of a multi-dimensional problem such as the KHI is to provide a calibrator/check for an MHD code before setting it loose on the full non-linear problem.

Figure 7.7 shows contours of the vorticity ($\vec{\omega} = \nabla \times \vec{v}$) component normal to the page (an excellent indicator of shear) from a series of ZEUS-3D simulations showing the early development of a KHI across a shear layer. Images differ only in the relative Mach number between the right-moving upper layer and the left-moving lower layer, which label each frame ($2M$). Each image is taken at 2.7 sound-crossing times (the time needed for sound to travel the width of the box, $L = 1$, less than three times) after a sinusoidal perturbation with $\lambda = L$ is applied to the shear layer. It is apparent from these snapshots that the maximum growth rate occurs in $1.25 < 2M < 1.5$, whereas the results we worked out from linear theory predicts maximum growth at

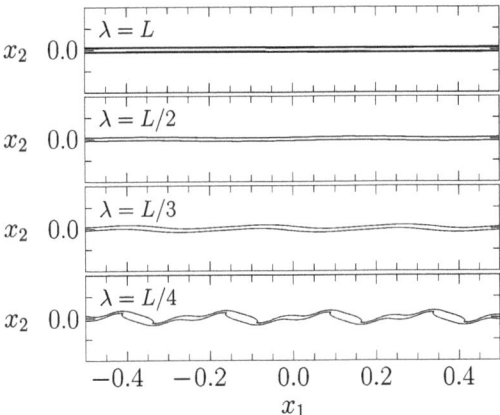

Figure 7.8. Growth of a sinusoidal perturbation along a $2M = 1.5$ shear layer as a function of the wavelength of the perturbation. All images are taken at $t = 0.75$ sound crossing times. The numerical results confirm linear theory's prediction that growth rate varies inversely as the wavelength of the perturbation.

$2M = \sqrt{3} \sim 1.73$. This actually represents reasonable agreement, given that the linear theory is for an infinitely thin shear layer, whereas the simulations require a finite thickness (in this case, 0.04 times the width of the box) to prevent "hair-trigger" instabilities caused by grid discretisation errors from breaking up a too-thin shear layer right from the beginning. Further, some non-linear effects are surely playing a role by $t = 2.7$, particularly for the large-amplitude $2M = 1.5$ simulation.

The simulations also agree with the linear theory prediction that the shear layer is unconditionally stable for $2M \geq 2\sqrt{2} \sim 2.83$. Not shown in Fig. 7.7 are the cases for $2M = 2.5$ and 3.0, the former showing only the slightest perturbation at $t = 3$, and the latter showing little motion even at $t = 12$.

Figure 7.8 shows the development of perturbations of wavelengths L, $L/2$, $L/3$, and $L/4$ each at 0.75 sound crossing times. Shown in each frame are two contours of the normal vorticity at the centre of the shear layer whose thickness, as described above, had to be non-zero for numerical stability. Such a thickness adds a scale length to the problem, which therefore has to be scaled down with the wavelength if it is to have a comparable effect in all runs. With this in mind, the simulations agree with our results from linear theory (Eq. 7.16) that the amplitude of the perturbation varies (roughly) as the inverse of the wavelength. In fact, note that for $\lambda = L/4$, the perturbation has already grown sufficiently for non-linear effects (deviation from a sine-wave) to have become apparent.

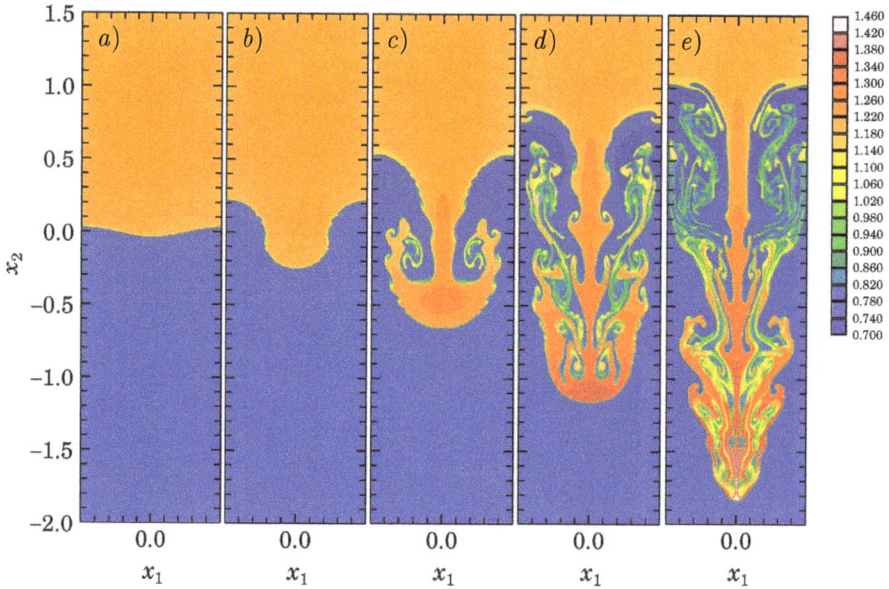

Figure 7.9. Density colour contours of a *ZEUS-3D* 200 × 700 zone simulation of an RTI with $\mathcal{A} = -0.2$ (Atwood number; see text) triggered by a perturbation to the contact discontinuity separating the denser fluid atop the less dense fluid (amplitude 2.5% of box-width, w; panel a). In the initial "linear-regime", growth is exponential as denser material pushes into less dense material in a uniform gravitational field, $g = c_s^2/w$. "Non-linear effects" are already apparent by panel b where minute beginnings of the KHI along what is now a tangential discontinuity (TD) can be seen. By panel c, KHI "wavelets" are clearly seen all along the TD and, through panels d and e, development of the KHI "fractalises" the TD, producing a significant mixing layer as falling heavier material forms a "Rayleigh–Taylor finger".

7.2 Rayleigh–Taylor instability

Named for Lord Rayleigh[7] and Sir Geoffrey Taylor,[8] the Rayliegh–Taylor instability (RTI) is triggered when a lighter (less dense) fluid supports a heavier (denser) fluid in a gravitational field. If the boundary between the fluids is dead flat and still, the heavier fluid can remain on top of the lighter fluid forever. However, any perturbation to the boundary layer will grow in time, resulting in "fingers" of heavier fluid sinking into lighter fluid which is thereby displaced upwards (*e.g.*, Fig. 7.9). This process enables the system to reduce its net potential energy by moving heavier fluid

---

[7]1842–1919, born John William Strutt, Lord Rayleigh made numerous contributions to physics including the discovery of Argon (1904 Nobel Prize) and describing how light scatters off particles smaller than its wavelength (Rayleigh-scattering), explaining for the first time why the sky is blue! (www.wikipedia.org/wiki/John_William_Strutt,_3rd_Baron_Rayleigh)

[8]1886–1975; fluid dynamicist and wave theorist, Taylor is probably best known by students of physics for his first paper as an *undergraduate* on how light produces fringes even when passing through a slit *one photon at a time*. (www.wikipedia.org/wiki/G._I._Taylor)

Figure 7.10. The Crab Nebula (*a.k.a.* M1) in the constellation Taurus is a remnant of a supernova first recorded by Chinese astronomers in 1054, bright enough at the time to be seen in broad daylight. Today where there was once a star, a pulsar resides at the centre of the nebula which modern astronomy interprets as a blast wave from the supernova explosion still accelerating outwards as evidenced by the web of Rayleigh–Taylor fingers giving the nebula its filamentary appearance (photo credit: NASA, STSci).

below the lighter fluid, and is what most people think of as the RTI.

Rayleigh was the one who first described the phenomenon; Taylor was the one to point out that this phenomenon also occurs in an accelerating fluid when less dense fluid pushes denser fluid. Probably the most famous example of this in astrophysics is the Crab Nebula (Fig. 7.10) in which a supernova blast wave is forcing hot, rarefied gas to accelerate into denser shocked material, resulting in the myriad of filaments observed.

7.2.1 Normal mode analysis of the RTI

Unlike the Kelvin–Helmholtz instability (§7.1) which is entirely fluid-driven, the RTI requires an external agent – either a background gravitational field or a driver to accelerate the fluid – to manifest. Therefore, our approach to analyse it will be little different. We will still examine the instability in its "linear regime" using the linearised fluid equations and normal mode solutions to determine the conditions for instability and growth rate. However, not being driven by pressure and velocity fluctuations as the KHI, we shan't analyse the RTI with a wave equation. Instead, we'll examine what happens to the velocity field directly as the heavier fluid begins to displace lighter fluid "beneath it".

Consider two incompressible, inviscid, stagnant fluids with uniform densities $\rho_u \neq \rho_l$ occupying the upper and lower regions of a volume of height $2L$ as depicted in Fig. 7.11. The interface between the two fluids is initially horizontal (dashed line at $y = 0$) and the fluids are embedded in a uniform gravitational field, $\vec{g} = -g\hat{j}$. For simplicity, assume z-symmetry ($\partial_z = 0$), $v_z = 0$, and that at $t = 0$, the fluids are in pressure balance. Thus, the interface can be interpreted as a contact discontinuity.

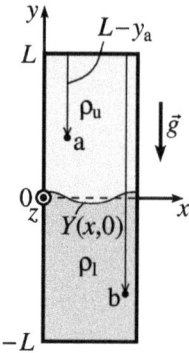

Figure 7.11. Initial conditions for the Rayleigh–Taylor instability.

At $t = 0$, the interface is perturbed by the function $Y(x, 0)$, depicted as a small-amplitude cosine wave in Fig. 7.11. The question we wish to answer is whether the interface returns to its equilibrium position (dashed line), or whether the perturbation grows. In case of the former, what is the relaxation time and/or oscillation frequency and, in case of the latter, what is the growth time?

Our first task, then, is to find an expression for $Y(x, t)$ describing how the initial distortion, $Y(x, 0)$, changes in time. Since $Y(x, t)$ tracks the interface – and thus the contact discontinuity – between the two fluids, its motion in the y-direction must be the fluid motion, namely v_y. Thus, as a start, write (as we did for the KHI),

$$\partial_t Y(x, t) = v_y\big|_{y=Y}, \quad (7.17)$$

anticipating that a time-integration would yield $Y(x, t)$.

To this end, we have seen elsewhere (*e.g.*, Problem 1.6, Eq. 2.59 in §2.3) that for an incompressible fluid, the continuity equation (Eq. 1.19) reduces to $\nabla \cdot \vec{v} = 0$. Thus, like the magnetic induction, the velocity field is solenoidal and may be described by a "vector potential" which, in this context, is called the *stream function*, $\vec{\psi}$:

$$\vec{v} = \nabla \times \vec{\psi} = (\partial_y \psi_z, -\partial_x \psi_z, \partial_x \psi_y - \partial_y \psi_x),$$

since $\partial_z = 0$. Because $v_z = 0$, we are free to set $\psi_x = \psi_y = 0$ and the stream function is adequately described by,

$$\vec{\psi} = (0, 0, \psi) \quad \Rightarrow \quad \vec{v} = (\partial_y \psi, -\partial_x \psi, 0), \quad (7.18)$$

where we have set $\psi_z = \psi$.

Now, by Kelvin's circulation theorem (Problem 1.7), the vorticity flux of an inviscid fluid is conserved and, since the initial conditions in Fig. 7.11 are of a stagnant fluid in which $\vec{\omega} = \nabla \times \vec{v} = 0$ everywhere, the fluid remains irrotational and we have,

$$\nabla \times \vec{v} = \nabla \times (\nabla \times \vec{\psi}) = -\nabla^2 \vec{\psi} + \nabla(\overset{0}{\overbrace{\nabla \cdot \vec{\psi}}}) = 0,$$

using Identity (A.27), and where $\nabla \cdot \vec{\psi} = \partial_z \psi = 0$ by the assumption of symmetry.

Rayleigh–Taylor instability

Thus,
$$\nabla^2 \vec{\psi} = \nabla^2 \psi \hat{z} = 0 \quad \Rightarrow \quad \nabla^2 \psi = 0, \tag{7.19}$$

and the scalar stream function, ψ, solves Laplace's equation. And so, as we did for the pressure in examining the KHI (Eq. 7.1), we seek normal mode solutions of Eq. (7.19) for the stream function, ψ, of the form,

$$\psi(x, y, t) = \Psi(y) e^{i(kx - \omega t + \phi)}, \tag{7.20}$$

where $k = 2\pi/\lambda$ is the wave number of the normal mode of wavelength λ and angular frequency ω, and where ϕ is some arbitrary phase set for convenience.

To find the amplitude function, $\Psi(y)$, substitute Eq. (7.20) into Eq. (7.19):

$$-k^2 \Psi(y) e^{i(kx-\omega t+\phi)} + \frac{d^2 \Psi(y)}{dy^2} e^{i(kx-\omega t+\phi)} = 0 \quad \Rightarrow \quad \frac{d^2 \Psi(y)}{dy^2} = k^2 \Psi(y)$$

$$\Rightarrow \quad \Psi(y) = A e^{-ky} + B e^{ky}.$$

Requiring the amplitude, Ψ, to remain finite (indeed, go to zero) as $y \to \pm\infty$ requires that $B = 0$ for $y > 0$ and $A = 0$ for $y < 0$, and we write:

$$\Psi(y) = \begin{cases} A e^{-ky}, & y > 0; \\ B e^{ky}, & y < 0. \end{cases}$$

Then, since the velocity field is continuous across $y = 0$, so is the stream function,

$$\Psi(0^+) = \Psi(0^-) \quad \Rightarrow \quad A = B,$$

and Eq. (7.20) becomes,

$$\psi(x, y, t) = A e^{-k|y|} e^{i(kx - \omega t + \phi)}. \tag{7.21}$$

Thus, to first order, we have from Eq. (7.17),[9]

$$\partial_t Y = v_y \Big|_{y=0} = -\partial_x \psi \Big|_{y=0} = -ikA e^{i(kx-\omega t+\phi)},$$

and, integrating over time, we get,

$$\boxed{Y(x, t) = \Re\left[\frac{k}{\omega} A e^{i(kx-\omega t+\phi)}\right],} \tag{7.22}$$

where $\Re[\ ]$ indicates the "real part".

To determine whether $Y(x, t)$ is stable to perturbations requires knowledge of the all-important dispersion relation, $\omega(k)$. If $\omega \in \mathbb{R}$, the temporal part of $Y(x, t)$, namely $e^{-i\omega t}$, is oscillatory and the interface is stable. On the other hand, if $\omega \in \mathbb{I}$ (and so let us set $\omega = i\chi$ where $\chi > 0 \in \mathbb{R}$), then $e^{-i\omega t} = e^{\chi t}$ grows exponentially with time and the interface is unstable.

Finding $\omega(k)$ is done most straight-forwardly by considering the pressure perturbations that result as the fluid responds to the evolving distortion. To this end,

---

[9] In fact, Eq. (7.17) would have us evaluate v_y at $y = Y$, where the interface is actually located. Evaluating v_y at $y = 0$ instead introduces "second-order" errors, which can be ignored in the "linear-regime".

consider points 'a' and 'b' in Fig. 7.11. Since $\vec{g}$ is uniform, pressure increases with depth (what we referred to as the "pressure head" in §2.3) and therefore the weight of the overhead fluid (*e.g.*, Eq. 2.64). Thus, at equilibrium (when the fluid interface is at $y = 0$), the pressure at 'a' (with pressure head $L - y_a$) is,

$$p_a = \rho_u g(L - y_a),$$

whereas at 'b', the pressure is,

$$p_b = \rho_u g L - \rho_l g y_b,$$

since the density changes from ρ_u to ρ_l at $y = 0$, and noting that $y_b < 0$. Therefore, at any point in the fluid with vertical coordinate y, we may write at equilibrium,

$$p_0(y) = \begin{cases} \rho_u g L - \rho_u g y, & y > 0; \\ \rho_u g L - \rho_l g y, & y < 0. \end{cases}$$

Perturbing the interface by $Y(x,t)$ introduces an x-dependence to the depth at which ρ_u jumps to ρ_l leading to a pressure perturbation, $p'(x, y, t)$, to the upper and lower regions. Including $p'_{u,l}$, the total pressure at any point in the fluid is now given by,

$$p(x, y, t) = \begin{cases} \rho_u g L - \rho_u g y + p'_u(x, y, t), & y > Y; \\ \rho_u g L - \rho_l g y + p'_l(x, y, t), & y < Y, \end{cases} \quad (7.23)$$

which, because the interface is a contact discontinuity, must be continuous at $y = Y(x,t)$:[10]

$$p(x, Y^+, t) = p(x, Y^-, t)$$
$$\Rightarrow \;\; \cancel{\rho_u g L} - \rho_u g Y + p'_u(x, Y, t) = \cancel{\rho_u g L} - \rho_l g Y + p'_l(x, Y, t)$$
$$\Rightarrow \;\; p'_u(x, Y, t) - p'_l(x, Y, t) = gY(\rho_u - \rho_l). \quad (7.24)$$

To find $p'_{u,l}$, we turn to the x-component of the linearised Euler equation:

$$\partial_t v_x = -\frac{1}{\rho}\partial_x p = \begin{cases} -\dfrac{1}{\rho_u}\partial_x p'_u, & y > Y; \\ -\dfrac{1}{\rho_l}\partial_x p'_l, & y < Y, \end{cases} \quad (7.25)$$

since $p'_{u,l}$ contain the only x-dependence in p (Eq. 7.23). Thus, for $y > Y$ and since $v_x = \partial_y \psi$ (Eq. 7.18),

$$p'_u(x, y, t) = -\rho_u \int \partial_t \partial_y \psi \, dx = -\rho_u \int (-i\omega)(-k) A e^{-k|y|} e^{i(kx - \omega t + \phi)} \, dx$$
$$= -\rho_u \omega A e^{-k|y|} e^{i(kx - \omega t + \phi)},$$

using Eq. (7.21).

---

[10] Here, because the leading-order terms $\rho_u g L$ cancel, we are obliged to carry second-order terms, and thus must make the match at $y = Y$ instead of $y = 0$.

Similarly,
$$p'_1(x,y,t) = \rho_1 \omega A e^{-k|y|} e^{i(kx-\omega t+\phi)},$$

and Eq. (7.24) becomes,

$$-\rho_u \omega A e^{-k|Y|} e^{i(kx-\omega t+\phi)} - \rho_1 \omega A e^{-k|Y|} e^{i(kx-\omega t+\phi)} = g\frac{k}{\omega} A e^{i(kx-\omega t+\phi)}(\rho_u - \rho_1),$$

using Eq. (7.22). Setting $e^{-k|Y|} \sim 1$ and solving for ω, we finally arrive at the dispersion relation,

$$\boxed{\omega(k) = \sqrt{gk\frac{\rho_1 - \rho_u}{\rho_1 + \rho_u}} \equiv \sqrt{gk\mathcal{A}},} \qquad (7.26)$$

where,
$$\mathcal{A} = \frac{\rho_1 - \rho_u}{\rho_1 + \rho_u}, \qquad (7.27)$$

is the so-called *Atwood number*.

Evidently, when the density in the lower fluid is the greater, $\mathcal{A} > 0$, $\omega \in \mathbb{R}$, and the temporal portion of Eq. (7.22) is oscillatory. The interface is stable to perturbations and the system oscillates about its equilibrium point (dashed line in Fig. 7.11) with a period,

$$T = \frac{2\pi}{\omega} = \sqrt{\frac{2\pi\lambda}{g\mathcal{A}}}, \qquad (7.28)$$

since $\lambda = 2\pi/k$. This is entirely analogous to the "kitchen experiment" in which the centre of a level surface of water in a glass is quickly depressed, then released. Buoyancy-driven water waves (*a.k.a.* gravity waves) ensue until damping forces bring the water surface back to rest.

On the other hand, should the density in the upper fluid be the greater and $\mathcal{A} < 0$, $\omega \in \mathbb{I}$ and the temporal part of $Y(x,t)$ is exponential. Letting $\omega = i\chi$ where $\chi > 0 \in \mathbb{R}$, $\chi = \sqrt{gk|\mathcal{A}|} > 0$ and we have from Eq. (7.22),

$$Y(x,t) = Y_0 e^{\chi t} \Re\left[i e^{i(kx+\phi)}\right], \qquad (7.29)$$

where $Y_0 = kA/\chi$. In this case, the perturbation grows exponentially with an e-folding growth time given by,

$$\tau_{\text{RT}} = \frac{1}{\chi} = \sqrt{\frac{\lambda}{2\pi g|\mathcal{A}|}}. \qquad (7.30)$$

Whether oscillatory or exponential, the characteristic times for the RTI are proportional to $\sqrt{\lambda}$. Thus, the period of oscillation, T, or the growth-time, τ_{RT}, are shorter for shorter wavelengths. Recalling Eq. (7.16) from §7.1 in which the KHI growth time $\tau_{\text{KH}} \propto \lambda$, the wavelength dependence of the RTI is evidently weaker than that of the KHI.

As stressed before, this analysis is valid for the linear regime only, based as it is on the *linearised* Euler equation (Eq. 7.25) and invocations of $Y \ll \lambda$. With full non-linear hydrodynamics engaged, the development of the instability departs markedly from the simple conclusion from linear analysis that the amplitude of the initial cosine wave just increases in time. As seen in Fig. 7.9, the boundary layer between "fingers" of heavier fluid and the lighter fluid into which it falls is a tangential discontinuity (shear layer plus density jump) and subject to Kelvin–Helmholtz instabilities. KHI-induced eddies are compounded by those formed as heavy fluid comes into contact with, for example, a lower boundary or obstacle preventing it from falling further and, once again, the fluid cascades to turbulence. While the linear model still predicts that the final state is one in which the two fluids switch places, the ensuing turbulence and, should the fluids be miscible, mixing yields qualitatively different outcomes between the linear and non-linear regimes.

7.2.2 Numerical analysis of the RTI (optional)

Just for fun, I tested the predictions from linear theory with results from a series of *ZEUS-3D* simulations. For a fully non-linear, compressible MHD code such as *ZEUS-3D*, one has to be careful not to trigger the non-linear and compressible aspects of the solution for which it is designed. Further, one must avoid boundary effects and minimise numerical truncation errors that lead to an effective viscosity since the linear theory is insensitive to both. These requirements point to well-resolved runs with distant boundaries above and below the fluid interface, periodic boundaries to each side, and small-amplitude perturbations. Each run can be executed for one or two oscillation periods for stable perturbations ($\mathcal{A} > 0$), or just a few growth times for unstable perturbations ($\mathcal{A} < 0$) since, in the latter case, even after $t = 3\tau_{\text{RT}}$ the perturbation amplitude grows by a factor of $e^3 \sim 20$ which, if the initial amplitude is too large, could mean the simulation is already in the non-linear regime.

Testing first the prediction that $\mathcal{A} > 0$ leads to stable oscillations with a period given by Eq. (7.28), I set up a 2-D (x–y) box of fluid with width $-0.5 \leq x \leq 0.5$ (units arbitrary), height $-10.0 \leq y \leq 10.0$, with density,

$$\rho(y) = \begin{cases} 0.8, & y > 0; \\ 1.2, & y < 0, \end{cases}$$

($\mathcal{A} = 0.2$) in a gravitational field $\vec{g} = -1\hat{j}$, and initialised this in hydrostatic equilibrium (pressure gradient exactly balancing the gravitational field). I then initialised four modest-resolution simulations,[11] perturbing the interface at $y = 0$ with a cosine wave of amplitude $Y_0 = 0.04$ and four different wavelengths: $\lambda = 1, \frac{1}{2}, \frac{1}{3}$, and $\frac{1}{4}$ in units of the box width, w.

Aside: Numerical considerations

As a practical matter, I set $c_s = 1$ at the top of the box which, assuming a diatomic gas, means a pressure $p_{\text{top}} = 4/7$ in these arbitrary units. Thus, at the fluid interface

---

[11]To reproduce these results, I used 250 horizontal zones in $-0.5 \leq x \leq 0.5$ for a zone width of 0.004, 100 vertical zones in $-0.1 \leq y \leq 0.1$ for a zone height of 0.002, and 100 geometrically increasing vertical zones in each of $-10 \leq y \leq -0.1$ and $0.1 \leq y \leq 10$ to remove the boundaries well away from the interface.

where $y = 0$, the pressure has increased by the "weight" of the fluid "overhead" to,

$$p_0 = p_{\text{top}} + \rho g L = 8\tfrac{4}{7},$$

where $L = 10.0$ is the pressure head.

So what do all these "unitless numbers" actually mean? As we'll see in §8.5, the ideal fluid equations are *scale-free*. That is, one can solve the ideal equations of hydrodynamics without attention to the units of the flow variables; everything simply scales to however the flow variables were set at $t = 0$ without any inherent length scale developing in the solution.[12]

Conversely, including gravity introduces an inherent scaling to Euler's equation specified by the *Froude number*, $\mathcal{F}$ (e.g., Problem 8.6). Having already set scaling for length (R) and the sound speed (V), the scaling for $\vec{g}$ (G) cannot be set independently, and whatever value we choose for G then determines $\mathcal{F}$,

$$\mathcal{F} = \frac{V}{\sqrt{GR}},$$

(Eq. 8.63) whose value in the simulation must be the same as its "real-world" value after scaling is restored. For example, where the two fluids meet at $y = 0$ and where the perturbation is imposed,

$$c_s^2 = \frac{\gamma p_0}{\rho_u} = \frac{(7/5)(60/7)}{4/5} = 15,$$

which is our V^2. This with $R = 1$ and $G = 1$ means $\mathcal{F} = \sqrt{15}$ and thus any "real-world" attribution we wish to make to this simulation must have the same Froude number. And so, if we wanted to relate this to an experiment done on Earth with $g = 9.81\,\mathrm{m\,s^{-2}}$ and $c_s = 331\,\mathrm{m\,s^{-1}}$ (273 K),

$$\mathcal{F} = \frac{c_s}{\sqrt{gw}} = \sqrt{15} \quad \Rightarrow \quad w \sim 750\,\mathrm{m},$$

making our box $\sim 15\,\mathrm{km}$ tall; hardly what one would call a practical experiment! Alternately, if one wanted a more "reasonable" box size say of 1 m wide and 20 m tall in Earth's gravitational field,

$$c_s = \sqrt{15\,gw} \sim 12\,\mathrm{m\,s^{-1}},$$

corresponding to a temperature $< 0.4\,\mathrm{K}$! Again, hardly practical. I'll leave it to the reader to work out what g would have to be to have a 1 m wide 20 m high box with a sound speed of $331\,\mathrm{m\,s^{-1}}$ (it's big!).

So why not choose a scaling for gravity, say, that would make this numerical experiment correspond to one done on Earth? For $g = 9.81$, $w = 1$, and $c_s = 331$, $\mathcal{F} \sim 100$, which means that for the numerical simulation, we must choose,

$$g = \frac{15}{100^2\,1} = 0.0015,$$

---

[12] For a more complete description of what it means for an equation to be *scale-free*, the reader might skip ahead to read the first page and a half of §8.5, none of which depends upon anything previous in Chap. 8.

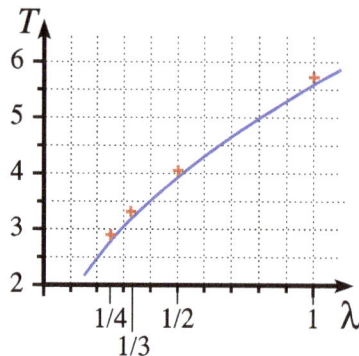 

Figure 7.12. On the left is a plot of Eq. (7.28) (blue line) showing the oscillation period, T, as a function of perturbation wavelength, λ (in units where the box width, $w = 1$), for a fluid with an Atwood number $\mathcal{A} = 0.2$ (see text) in a gravitational field $g = 1$ (units arbitrary). The red crosses show periods measured from ZEUS-3D simulations where the initial amplitude $Y_0 = 0.04\lambda$ is resolved with 20 zones. The numerical simulations show the correct dependence on λ but with slightly longer periods consistent with the small amount of numerical and artificial viscosities not known to the linear theory.

On the right is a sequence of images from the $\lambda = 1$ simulation (full box width, w, resolved with 250 zones) showing from top to bottom one full period of oscillation with each image spaced $\sim T/8$ apart. Heavy fluid is shown in pink, lighter fluid blue.

a low number. Why low? Another numerical issue is numerical stability, which requires one to operate with time steps so that no signal – be it moving fluid or a sound wave – can move more than one zone width during a single time step. To numericists, this is known as the *Courant–Friedrichs–Lewy* (CFL) *condition*,[13] and violating it triggers numerical oscillations so severe that the entire solution can be consumed by infinities within twenty time steps! As Rayleigh–Taylor flow is comfortably subsonic, the time step is governed by the sound speed, $\delta t \propto 1/c_s$ and, with such a low value for g, the small time step means it can take tens of thousands of time steps just to reach one growth time, making the simulation impractical. By "boosting" gravity by almost three orders of magnitude, the results are qualitatively identical, but can be had at a fraction of the compute time.

Figure 7.12 summarises the results from these simulations. The left side shows what Eq. (7.28) predicts for the oscillation period (blue line) as a function of normal mode wavelength, assuming $\mathcal{A} = 0.2$, $w = 1$, and $g = 1$. The results of the four

---

[13]Courant, *et al.* (1928). Well before the invention of the electronic computer, the CFL condition was discovered by fluid simulations done in a room full of desks arranged in a 2-D grid with graduate students seated in each madly doing slide-rule calculations in their "zone", passing their results to the students ahead, behind, and to their sides after each "time step"!

ZEUS-3D simulations at the different perturbation wavelengths are shown with red crosses. The agreement is satisfactory ($\lesssim 2.5\%$ discrepancy for each), with the numerical simulations consistently a little greater than the linear theory. I attribute this to the small amounts of artificial viscosity imposed for numerical stability and numerical viscosity which comes with the territory on a discretised grid.

The right side of Fig. 7.12 shows a sequence of density images (pink $\rho = 1.2$, blue $\rho = 0.8$) roughly $T/8$ apart from the $\lambda = 1$ simulation. Because of the inherent artificial and numerical viscosities, the perturbation evolves like a damped harmonic oscillator for which the amplitude of oscillation is given by,

$$A(t) = A_0 e^{-\gamma t},$$

where γ is the *damping coefficient*. Using data from the simulation [initial amplitude, A_0, and amplitude after one period, $A(T)$], I found that,

$$\gamma = \frac{1}{T} \ln \frac{A_0}{A(T)} \sim 0.01,$$

which, when compared to the natural oscillation frequency $\omega = 2\pi/T \sim 1.1$ (for $T \sim 5.76$) means $\gamma \lesssim 0.01\omega$, corresponding to a rather weakly damped oscillator.

As for the prediction of instability for $\mathcal{A} < 0$ with a growth rate given by Eq. (7.30), the ZEUS-3D simulations can confirm this only qualitatively. Using the same box as the previous test ($-0.5 \leq x \leq 0.5$, $-10.0 \leq y \leq 10.0$) but with the heavier fluid on top ($\mathcal{A} = -0.2$, $g = 1$), I ran three simulations with differing resolutions to follow a single wavelength perturbation, $\lambda = 1$, of initial amplitude 0.0025 (in units of box width w) for 3.4 growth times (where, for the values used, $\tau_{\text{RT}} \sim 0.892$; Eq. 7.30). During this time, the linear theory predicts the amplitude should grow to about 0.075, very near the edge of the linear regime but where the perturbation should still look like a cosine wave.

The solid line in the left side of Fig. 7.13 shows the growth of the amplitude in time as predicted by the linear theory (Eq. 7.29) along with amplitudes measured from the three simulations shown in coloured crossed (blue, green, and red for low,[14] medium, and high resolution). In each case, the expected simple exponential growth rate is not observed. All simulations vastly underestimate the growth rate of the instability at the beginning, and only achieve the growth rate predicted by linear theory by $t \sim 2\tau_{\text{RT}}$. The fact that this holds true regardless of resolution leads me to believe this is not a numerical artefact, but may be pointing to something real. And on that, I'm going to go out on a bit of limb here.

It may just be that there is no true linear regime for the Rayleigh–Taylor instability. The right side of Fig. 7.13 shows the density image of the high-resolution run (2,000 zones horizontally – ten times that of Fig. 7.9 – for a zone width of 0.0005, and 800 zones vertically in $-0.1 \leq y \leq 0.1$ for a zone height of 0.00025 to resolve the initial amplitude with ten zones) at $t = 1.5\tau_{\text{RT}}$ (panel a) and $t = 3.4\tau_{\text{RT}}$ (panel b). A quick glance at panel a seems to confirm what linear theory predicts; a cosine

---

[14] The "low" resolution run actually had 500 zones across w, 2.5 times more than the simulation shown in Fig. 7.9. Therefore, the terms "low", "medium", and "high" should be regarded as *relative* comparators.

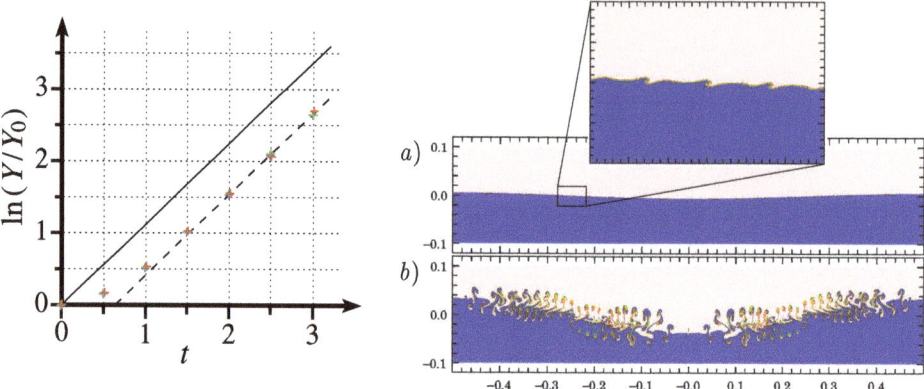

Figure 7.13. On the left is a plot of Eq. (7.29) (with $x = 0$, $\phi = 0$; solid line) illustrating the linear theory prediction that the amplitude of a normal mode perturbation grows with e-folding time τ_{RT} given by Eq. (7.30). For $\mathcal{A} = -0.2$, $g = 1$, and $\lambda = 1$, $\tau_{\text{RT}} \sim 0.892$, the reciprocal of the slope shown. Results from ZEUS-3D runs (low, medium, and high resolutions shown as blue, green, and red crosses; see text) indicate a much slower beginning to the instability than linear theory predicts. While the numerical results reach the expected growth rate by $t \sim 2\tau_{\text{RT}}$ (dashed line), that it takes two growth times to reach this agreement *regardless of resolution* is remarkable.

The right side reveals what may be at the root of the disagreement. Panel a shows the perturbed interface ($\rho_u = 1.2$, pink; $\rho_l = 0.8$, blue) for the high resolution (2,000 zones in $-0.5 \leq x \leq 0.5$, 800 zones in $-0.1 \leq y \leq 0.1$) run at $t = 1.5\tau_{\text{RT}}$. While the interface in the main panel looks "normal" enough, the inset clearly shows the beginning of a Kelvin–Helmholtz instability. The wavelength of the "mini-KHI" is ~ 0.016 which is resolved by 32 zones, ruling out the possibility these may be numerical in origin. By $t = 3.4\tau_{\text{RT}}$ (panel b), the mini-KHI have developed into numerous "mini-RTI" eating into the cosine perturbation and rendering it almost unrecognisable.

wave with growing amplitude. Measurement of that amplitude, however, shows it to be only about 55% of the predicted amplitude and further, one can see along the interface that minute structures have formed which, as the inset shows, are actually the early onset of the Kelvin–Helmholtz instability. As the heavier fluid begins to sink, the contact discontinuity develops a shear (and thus becomes a tangential discontinuity) which makes it vulnerable to the KHI. Each KHI "wavelet" acts as a perturbation in its own right with a wavelength much smaller than the original perturbation (in this case, $\lambda_{\text{KH}} \sim 0.016$). Since the e-folding time for the RTI is $\tau_{\text{RT}} \propto \sqrt{\lambda}$ (Eq. 7.30), perturbations caused by the KHI grow $\sqrt{1.0/0.016} \sim 8$ times faster than the initial $\lambda = 1$ perturbation. Thus, by $t = 3.4\tau_{\text{RT}}$ (panel b), numerous "mini Rayleigh–Taylor instabilities" have developed all along the interface consuming the original $\lambda = 1$ cosine-wave in a myriad of fine "RT-fingers" taking it clearly outside the realm of the "linear regime". It is my suspicion, then, that the development of the KHI and subsequent RTI so soon into the run taps into the potential energy of the falling fluid and slows its progress – at least initially – pre-empting the linear regime almost before it can even get started.

7.2.3 Kruskal–Schwarzchild instability

When a magnetic induction, $\vec{B} = B_z\hat{z}$, is added to the mix, the RTI becomes the *Kruskal–Schwarzchild instability* (KSI). The mathematical principles are the same – normal mode analysis on a perturbed boundary layer – but its execution is rather more complicated than the RTI, and I'm not convinced going through the process again provides much "value-added". Instead, we'll save the heavier-lifting mathematics for the two uniquely MHD instabilities (those with no HD counterpart) in §7.3 (magneto-rotational instability; MRI), and §7.4 (Parker instability) and just quote the final result for the KSI with a bit of physical interpretation.

By adding $\vec{B} = B\hat{z}$ to our initial conditions, Eq. (7.26) becomes,[15]

$$\omega^2(k) = gk\frac{\rho_l - \rho_u}{\rho_l + \rho_u} + \frac{2}{\mu_0}\frac{(\vec{B} \cdot \vec{k})^2}{\rho_l + \rho_u}, \qquad (7.31)$$

where the first term is identical to the RTI criterion, and where $\vec{k} = k\hat{e}_k$ is the *wave vector* of the perturbation (as opposed to the simpler *wave number*, k, introduced for the RTI). In the pure HD case, the direction in which the perturbation is applied within the z–x plane does not matter, and only the scalar wave number was needed. In the MHD case, because $\vec{B}$ sets a preferred direction – in this case $\hat{z}$ – the direction of the applied perturbation *does* matter and the full wave *vector*, $\vec{k}$, is required.

And so we see that provided $\vec{k} \not\perp \vec{B}$, the magnetic term introduces *stability* to the RTI. That is, for $\rho_l > \rho_u$, nothing changes; from Eq. (7.31), $\omega^2(k) > 0$ and the response to the perturbation remains oscillatory. However, for $\rho_l < \rho_u$ which triggers instability in the pure HD case, we see from Eq. (7.31) that provided,

$$\frac{2}{\mu_0}(\vec{B} \cdot \vec{k})^2 > gk(\rho_l - \rho_u),$$

$\omega^2(k)$ remains positive and the RTI is stabilised even if $\rho_u > \rho_l$.

The physical interpretation is simple enough. Think of $\vec{B}$ as providing "floor joists" for the upper, heavier layer. So long as those joists are strong enough and in the "right direction", the heavier fluid is supported against gravity from falling into the lighter fluid below, and the RTI is suppressed. The "right direction" is when perturbing oscillations are along lines of induction, where each line of induction resists being bent and therefore the perturbation. When $\vec{k} \cdot \vec{B} = 0$, perturbing oscillations are not resisted by $\vec{B}$ since there is no cohesion amongst neighbouring lines of induction, and matter simply slips through the "joists".

---

[15] The interested reader is referred to Galtier (2016) for an approachable derivation.

7.3 Magneto-rotational instability

While the first accounts of the *magneto-rotational instability* (MRI) were reported by Velikhov (1959) and Chandrasekhar (1960) in the context of Couette flow in the laboratory (§8.6.5), it was not known to astrophysicists until rather later. Friske (1969) did apply the MRI to the problem of stellar interiors, but it wasn't until the early 1990s that the full impact of the MRI on astrophysics was appreciated.

And this impact is profound.

Since the 1970s, astrophysicists have known that stars form from collapsing gas clouds in the interstellar medium and that on the road to becoming a star, gas collects in what is known as an *accretion disc*.[16] This "gravitational weighstation" is a consequence of angular momentum conservation. While the rotation of a gas cloud at the onset of collapse can be minute, it is magnified many orders of magnitude as the cloud collapses from something of order a parsec in diameter to one comparable to Jupiter's orbit about the sun. Just as the rotation of a figure skater increases as their arms are brought in, the spin of a collapsing gas cloud increases to the point where the centrifugal force prevents further collapse, and gas collects instead in a flat disc rotating at such a rate that the central portion cannot collapse any further to form a star as we know it.

And yet stars form. *Somehow*, sufficient angular momentum is transferred to the outer disc to allow the inner portion to collapse enough to trigger nuclear fusion. But what could that mechanism be? This question so perplexed astrophysicists that for two decades, this unknown property was referred to as "anomalous viscosity". In the 1970s, the only known physical property of a disc that could transport angular momentum outwards was viscosity (Chap. 8), and yet no one had any idea how such a diffuse medium could be sufficiently viscid to allow such transport to occur. Think of "anomalous viscosity" as the astrophysical equivalent to the 17th century cartographers' warning "there be dragons here!".

As the story goes, numerous papers in the late 1980s and early 90s were being published with all kinds of speculation of what this anomalous viscosity could be, and the very pragmatic John Hawley[17] could take no more. He proposed to his colleague, Steven Balbus, that a proper analysis of an accretion disc needed to be done, this time accounting for magnetic field. Their "rediscovery" of the MRI along with their definitive identification of "anomalous viscosity" as a consequence of the MRI was the result, and their 1991 paper won them the 2013 Shaw Prize for astrophysics.

It cannot be overstated how important the MRI is to our understanding of stellar and planetary formation. As is now widely known, the MRI is triggered by the presence of a magnetic field in a differentially rotating flow, exactly what you find in an accretion disc around a compact object such as a protostar or a giant

---

[16] Real and artistic images of an accretion disc forming around a protostar are included at the beginning of Chap. 10, if the reader cares to glance ahead.

[17] 1958–2021; www.wikipedia.org/wiki/John_F._Hawley.

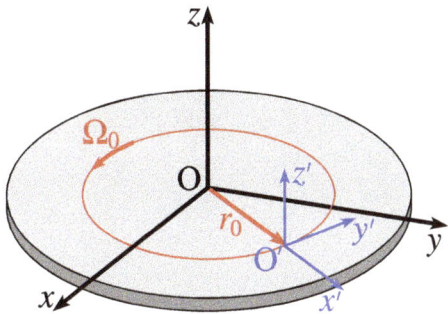

Figure 7.14. A schematic of a differentially rotating disc (such as an astrophysical accretion disc) showing the two coordinate systems introduced in the text: the inertial, non-rotating coordinate system O at the centre of the disc (black), and O′ at a distance r_0 from O (blue) and rotating about O at the local angular speed of the disc, Ω_0.

(10^8–10^9 solar mass) black hole at the centre of a large galaxy. In this section, I take the reader through this road of discovery.

7.3.1 Mathematical model of the MRI

Consider the rotating disc shown in Fig. 7.14.[18] It turns out that the MRI is a *local* instability, meaning one does not need to consider the entire disc to understand its development. Rather, perturbations in an isolated region of the disc are enough to trigger the MRI based entirely on local properties such as the orientation of the magnetic field and the radial gradient of the angular velocity.

And so we consider a small region of the disc displaced by $\vec{r}_0$ from the centre (O) with dimensions $(\delta r, r_0 \delta\varphi)$, small compared to r_0 (Fig. 7.15), and rotating about O at an angular speed Ω_0. Because the region is in a rotating and thus non-inertial frame of reference, we shall take the approach of most sophomore mechanics texts, and examine the problem from the rotating coordinate system[19] indicated in Fig. 7.14 and 7.15 as O′ (blue) with Cartesian unit vectors $\hat{\imath}'$, $\hat{\jmath}'$, and $\hat{k}'$ in the x'-, y'-, and z'-directions, where $\hat{\imath}'$ always points away from O. Thus, x' and y' can be considered a *radial* and *transverse* (azimuthal) coordinate respectively though, to an observer at O′, x' and y' are perfectly Cartesian. Fur-

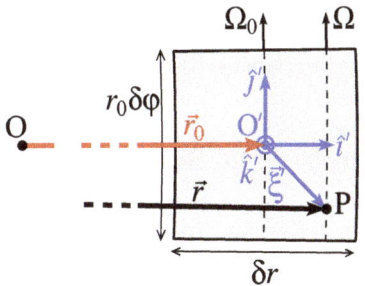

Figure 7.15. A region of the disc in the rotating reference frame O′ displaced from disc centre, O, by $\vec{r}_0$ (red). Point P is displaced from O′ by $\vec{\xi}'$ (blue) and from O by $\vec{r}$ (black).

---

[18] With some exceptions, my development of the MRI follows Balbus (2009).
[19] See App. G for a quick refresher on the *Coriolis theorem*.

ther, at some nearby point P displaced from O' by $\vec{\xi}'$ in the $\hat{i}'$-$\hat{j}'$ plane (Fig. 7.15), we must include the *inertial force densities* – Coriolis, centrifugal, and translational – as observed by O' to the momentum equation. Equivalently, we include the corresponding *inertial accelerations* to Euler's equation (Eq. 4.14) and, for the x'- and y'-components, we write:

$$\partial_t v'_x + \vec{v}' \cdot \nabla' v'_x = -\frac{1}{\rho}\partial'_x p^* - \partial'_x \phi + \frac{1}{\mu_0 \rho}(\vec{B}' \cdot \nabla')B'_x \\ + 2\Omega_0 v'_y + \Omega_0^2 \xi'_x + \Omega_0^2 r_0; \quad (7.32)$$

$$\partial_t v'_y + \vec{v}' \cdot \nabla' v'_y = -\frac{1}{\rho}\partial'_y p^* - \partial'_y \phi + \frac{1}{\mu_0 \rho}(\vec{B}' \cdot \nabla')B'_y - 2\Omega_0 v'_x, \quad (7.33)$$

where some explanation is warranted.

1. All components of vectors and spatial derivatives are with respect to the rotating coordinate system O' and, accordingly, are labelled with a prime (').

2. In Eq. (4.14), the Lorentz force density, $\vec{f}_L$, is proportional to $(\nabla \times \vec{B}) \times \vec{B}$, which has been modified here using Eq. (A.15) to:

$$(\nabla \times \vec{B}) \times \vec{B} = (\vec{B} \cdot \nabla)\vec{B} - \tfrac{1}{2}\nabla B^2.$$

The magnetic pressure, $p_M = B^2/2\mu_0$, is combined with the thermal pressure, p, to give the MHD pressure, $p^* = p + p_M$, as defined by Eq. (4.26), and the remaining magnetic term is broken into its x'- and y'-components.

3. In a reference frame O' rotating about an inertial origin, O, at constant angular velocity, $\vec{\Omega}_0 = \Omega_0 \hat{k}'$,[20] three distinct inertial accelerations must be included.

    a) The *Coriolis acceleration* is given by,

    $$\vec{a}_{\text{Cor}} = -2\vec{\Omega}_0 \times \vec{v}' = 2\Omega_0 v'_y \hat{i}' - 2\Omega_0 v'_x \hat{j}', \quad (7.34)$$

    where $\vec{v}' = v'_x \hat{i}' + v'_y \hat{j}'$ is the fluid velocity at point P relative to O'.

    b) The *centrifugal acceleration* is given by,

    $$\vec{a}_{\text{cen}} = -\vec{\Omega}_0 \times (\vec{\Omega}_0 \times \vec{\xi}') = \Omega_0^2 \xi'_x \hat{i}',$$

    where $\xi'_x \hat{i}'$ is the *radial* displacement of point P relative to O'.

    c) The *translational acceleration* is minus the acceleration of O' relative to O which, in this case, is *centripetal*[21] (since O' orbits O),

    $$\vec{a}_{\text{tr}} = -\vec{a}_{O'} = -[\Omega_0^2 r_0(-\hat{i}')] = \Omega_0^2 r_0 \hat{i}'.$$

These are the three inertial terms appearing explicitly in Eq. (7.32) and (7.33).

---

[20] Since I've been using lower-case ω to represent normal mode frequencies in this chapter, I'll continue that practice and use upper-case Ω for the disc rotation speed. Note also that $\hat{k}$ represents the unit vector in the z-direction while k (no hat, ˆ) will continue to be used for the wavenumber.

[21] Never to be confused with *centrifugal*! The reader for whom this difference is not obvious really needs to review App. G!

For a disc whose differential rotation is in response to a central gravitational field, the real acceleration caused by gravity $(-\nabla\phi)$ at any point, P, in the disc is centripetal (not centrifugal). Since x' is a radial coordinate while y' is azimuthal, we therefore have,
$$-\partial'_x \phi = -r\Omega^2(r) \quad \text{and} \quad -\partial'_y \phi = 0,$$
where $\Omega(r)$ is the orbital angular speed of the disc at P whose distance from O is $r = r_0 + \xi'_x$, and where $-r\Omega^2$ is the centripetal acceleration of P about O. Thus, Eq. (7.32) and (7.33) become:

$$\partial_t v'_x + \vec{v}' \cdot \nabla' v'_x = -\frac{1}{\rho}\partial'_x p^* + \frac{1}{\mu_0 \rho}(\vec{B}' \cdot \nabla')B'_x + 2\Omega_0 v'_y + r(\Omega_0^2 - \Omega^2); \quad (7.35)$$

$$\partial_t v'_y + \vec{v}' \cdot \nabla' v'_y = -\frac{1}{\rho}\partial'_y p^* + \frac{1}{\mu_0 \rho}(\vec{B}' \cdot \nabla')B'_y - 2\Omega_0 v'_x. \quad (7.36)$$

To be clear, Ω_0 is the angular speed of the origin O' about the inertial origin O, and enters into the equations by virtue of the inertial accelerations arising from examining the dynamics from the rotating reference frame, O'. On the other hand, Ω is the angular speed about O of some point P displaced by $\vec{\xi}'$ from O', and enters into the equations by virtue of the gravitational field driving circular orbits. In general, $\Omega_0 \neq \Omega$ in a differentially rotating disc which, as we shall see, is key to triggering the MRI.

Without specifying (for now) the functional dependence of $\Omega(r)$, we can do a Taylor expansion on $\Omega^2(r)$ from O' to point P in Fig. 7.15 whose radial displacement from O' is ξ'_x. Thus, to first order,

$$\Omega^2(r) = \Omega_0^2 + \xi'_x \left.\frac{d\Omega^2}{dr}\right|_{r_0} + \cdots$$

$$\Rightarrow \quad r(\Omega_0^2 - \Omega^2) \approx -r\xi'_x \left.\frac{d\Omega^2}{dr}\right|_{r_0} = -\xi'_x \left.\frac{d\Omega^2}{d\ln r}\right|_{r_0} \equiv -2\kappa\xi'_x, \quad (7.37)$$

where, for convenience, 2κ is defined as the $\ln r$-derivative of Ω^2 evaluated at $r = r_0$.

Next, evidently $\dot{\vec{\xi}}' = \vec{v}'$, and the LHS of Eq. (7.35) and (7.36) can be written as:

$$\left.\begin{aligned}\partial_t v'_x + \vec{v}' \cdot \nabla' v'_x &= \frac{dv'_x}{dt} = \dot{v}'_x = \ddot{\xi}'_x; \\ \partial_t v'_y + \vec{v}' \cdot \nabla' v'_y &= \frac{dv'_y}{dt} = \dot{v}'_y = \ddot{\xi}'_y. \end{aligned}\right\} \quad (7.38)$$

Substituting both Eq. (7.37) and (7.38) into Eq. (7.35) and (7.36), we get:

$$\ddot{\xi}'_x = -\frac{1}{\rho}\partial'_x p^* + \frac{1}{\mu_0 \rho}(\vec{B}' \cdot \nabla')B'_x + 2\Omega_0 \dot{\xi}'_y - 2\kappa\xi'_x; \quad (7.39)$$

$$\ddot{\xi}'_y = -\frac{1}{\rho}\partial'_y p^* + \frac{1}{\mu_0 \rho}(\vec{B}' \cdot \nabla')B'_y - 2\Omega_0 \dot{\xi}'_x. \quad (7.40)$$

We have now gone as far as we can without specifying the nature of the magnetic induction and the dependence of the fluid variables on the coordinates. For utter simplicity, let's suppose before any perturbation that the disc is threaded with a uniform, vertical magnetic induction, $\vec{B}' = B_z \hat{k}'$. Further, let the MHD pressure, p^*, be constant throughout the disc[22] and thus, $\partial'_x p^* = \partial'_y p^* = 0$. Finally, with every point in the disc in equilibrium, let us displace a fluid element from the origin O' to P (Fig. 7.15) according to the real part of,

$$\vec{\xi}_0 = (\xi'_{x,0}\hat{i}' + \xi'_{y,0}\hat{j}')e^{ikz'},$$

(noting that $z = z'$) with no x'- or y'-dependence and no z'-component, and where k is the vertical wavenumber of the perturbation (and not to be confused with the unit vector $\hat{k}'$).

As we shall be looking for a normal mode response to the perturbation, let us suppose the perturbation develops in time as the real part of,

$$\vec{\xi}' = (\xi'_x \hat{i}' + \xi'_y \hat{j}') = \vec{\xi}'_0 e^{-i\omega t} = (\xi'_{x,0}\hat{i}' + \xi'_{y,0}\hat{j}')e^{i(kz-\omega t)}, \quad (7.41)$$

where ω is the angular frequency of the normal mode. As usual, we shall determine the stability or instability of the perturbation based on whether ω is real or imaginary.

Since the magnetic induction is tied to the fluid, any displacement of P from O' distorts the initially uniform $B_z \hat{k}'$ at P to $\vec{B}' = B_z \hat{k}' + \delta \vec{B}'$, where the perturbation, $\delta \vec{B}'$, can be found using the ideal induction equation:

$$\delta \vec{B}' \approx \partial_t \vec{B}' \,\delta t = \nabla' \times (\vec{v}' \times \vec{B}')\, \delta t = \nabla' \times (\vec{v}'\delta t \times \vec{B}') = \nabla' \times (\vec{\xi}' \times \vec{B}').$$

As a perturbation (at least initially), $\vec{\xi}'$ is already first-order small. Thus, in retaining only first-order terms,

$$\vec{\xi}' \times \vec{B}' \approx (\xi'_x \hat{i}' + \xi'_y \hat{j}') \times B_z \hat{k} = B_z(-\xi'_x \hat{j}' + \xi'_y \hat{i}')$$

$$\Rightarrow \quad \delta \vec{B}' \approx \nabla' \times (\vec{\xi}' \times \vec{B}') = ikB_z(\xi'_x \hat{i}' + \xi'_y \hat{j}') = ikB_z \vec{\xi}', \quad (7.42)$$

since B_z is constant and $\vec{\xi}'$ depends only upon z and t so that the curl extracts only a z-derivative: $\partial_z \vec{\xi}' = ik\vec{\xi}'$. Thus, the magnetic terms in Eq. (7.39) and (7.40) become,

$$\frac{1}{\mu_0 \rho}(\vec{B}' \cdot \nabla')\vec{B}' = \frac{1}{\mu_0 \rho}(B_z \partial_z)\delta \vec{B}' = -\frac{k^2 B_z^2}{\mu_0 \rho} \vec{\xi}' \equiv -\omega_A^2 \vec{\xi}', \quad (7.43)$$

retaining only first order terms. Here,

$$\omega_A = \frac{kB_z}{\sqrt{\mu_0 \rho}} = ka_z,$$

is defined as the *Alfvén frequency*, where a_z is the z-component of the Alfvén velocity. Physically, ω_A is the angular frequency of Alfvén waves propagating along otherwise uniform lines of magnetic induction, $B_z \hat{k}'$. Note that the RHS of Eq. (7.43) has the form of a Hooke's law acceleration, since it is proportional to minus the square of a frequency times a displacement. Thus, acting by itself in Eq. (7.39)

---

[22]This assumption is consistent with our supposition that the centripetal acceleration at any point in the disc is exclusively attributable to the central gravitational field.

and (7.40), the magnetic force is restorative and one might expect an oscillatory response.

Substituting Eq. (7.43) into Eq. (7.39) and (7.40) with $\nabla p^* = 0$, the final form for the equations governing the displacement perturbation in the rotating reference frame, O', are:

$$\ddot{\xi}'_x = -(\omega_A^2 + 2\kappa)\xi'_x + 2\Omega_0 \dot{\xi}'_y; \tag{7.44}$$

$$\ddot{\xi}'_y = -\omega_A^2 \xi'_y - 2\Omega_0 \dot{\xi}'_x, \tag{7.45}$$

which have lost all resemblance to Euler's equation from which we began! Still, these are the dynamical equations that carry the essence of the MRI, and into which we substitute the normal mode solutions, Eq. (7.41), to find the all-important dispersion relation. And so:

$$-\omega^2 \xi'_{x,0} = -(\omega_A^2 + 2\kappa)\xi'_{x,0} - 2i\omega\Omega_0 \xi'_{y,0}; \tag{7.46}$$

$$-\omega^2 \xi'_{y,0} = -\omega_A^2 \xi'_{y,0} + 2i\omega\Omega_0 \xi'_{x,0}, \tag{7.47}$$

where the factor $e^{i(kz - \omega t)}$ has cancelled out from each term. Solving Eq. (7.47) for $\xi'_{y,0}$ and then substituting that into Eq. (7.46) we get,

$$-\omega^2 \xi'_{x,0} = -(\omega_A^2 + 2\kappa)\xi'_{x,0} - 2i\omega\Omega_0 \frac{2i\omega\Omega_0}{\omega_A^2 - \omega^2}\xi'_{x,0}$$

$$\Rightarrow \boxed{\omega^4 - 2(\omega_A^2 + 2\Omega_0^2 + \kappa)\omega^2 + \omega_A^2(\omega_A^2 + 2\kappa) = 0,} \tag{7.48}$$

after a little algebra. Like the KHI, the dispersion relation for the MRI is a quadratic in ω^2 which formally has four roots:

$$\omega_{1,2} = \pm\sqrt{\omega_A^2 + 2\Omega_0^2 + \kappa + \sqrt{(2\omega_A\Omega_0)^2 + (2\Omega_0^2 + \kappa)^2}}; \tag{7.49}$$

$$\omega_{3,4} = \pm\sqrt{\omega_A^2 + 2\Omega_0^2 + \kappa - \sqrt{(2\omega_A\Omega_0)^2 + (2\Omega_0^2 + \kappa)^2}}. \tag{7.50}$$

Evidently, $\omega_{1,2} \in \mathbb{R}$ and these frequencies lead to an oscillatory response in the temporal dependence of $\vec{\xi}'$, namely $e^{-i\omega_{1,2}t}$. Note that since $\omega_2 = -\omega_1$, these are one and the same normal mode, just out of phase by π rad and thus by a factor of -1. Similarly, if $\omega_{3,4} \in \mathbb{R}$, these too are the same normal mode and so, in total, the response to the perturbation should be a linear combination of two sinusoids with frequencies ω_1 and ω_3.

On the other hand, $\omega_{3,4} \in \mathbb{I}$ when,

$$\sqrt{(2\omega_A\Omega_0)^2 + (2\Omega_0^2 + \kappa)^2} > \omega_A^2 + 2\Omega_0^2 + \kappa$$

$$\Rightarrow \boxed{2\kappa = \left.\frac{d\Omega^2}{d\ln r}\right|_{r_0} < -\omega_A^2,} \tag{7.51}$$

after a little algebra and replacing 2κ with its definition after Eq. (7.37). Setting $\omega_3 = i\chi$ and $\omega_4 = -i\chi$, $\chi > 0 \in \mathbb{R}$, the temporal responses are $e^{-i\omega_3 t} = e^{\chi t}$ and

$e^{-i\omega_4 t} = e^{-\chi t}$. The latter falls off exponentially with time and is thus inconsequential, but the former increases exponentially and renders the perturbation unstable. In this case, therefore, we expect the initial response to be sinusoidal ($e^{-i\omega_1 t}$) with an exponentially increasing term ($e^{\chi t}$) dominating soon thereafter. The reader will recognise this as the same mathematical behaviour exhibited by the KHI.

Equation (7.51) tells us that a differentially rotating disc is *magneto-rotationally unstable* (MR-unstable) when the logarithmic gradient of Ω^2 falls off faster than the square of the Alfvén frequency. In particular, for a Keplerian disc where the gravitational potential is dominated by a central mass, M,

$$\Omega^2 = \frac{GM}{r^3} \;\Rightarrow\; 2\kappa = \left.\frac{d\Omega^2}{d\ln r}\right|_{r_0} = \left.r_0 \frac{d\Omega^2}{dr}\right|_{r_0} = -3\frac{GM}{r_0^3} = -3\Omega_0^2, \qquad (7.52)$$

and Eq. (7.51) becomes,

$$-3\Omega_0^2 < -\omega_A^2 \;\Rightarrow\; \boxed{\frac{\omega_A}{\Omega_0} = \frac{2\pi B_z}{\lambda\sqrt{\mu_0\rho}\,\Omega_0} \equiv \mu_\rho < \sqrt{3},} \qquad (7.53)$$

where $\lambda = 2\pi/k$ is the vertical wavelength of the perturbation. Here, I define μ_ρ (Greek letters m r) as the unitless *MR-number* that determines when a Keplerian disc threaded with a uniform vertical magnetic induction is MR-unstable to displacement perturbations in the plane of the disc. Since μ_ρ depends only upon fluid variables B, ρ, and the rotation speed, Ω_0, the MRI is inherent to ideal MHD just as much as the KHI is inherent to ideal hydrodynamics. Both require a shear of some sort, and the rest is a direct consequence of the ideal fluid equations. For the MRI, while the quantity GM provides a means by which the fluid can rotate differentially, it does not figure into the criterion for instability, namely Eq. (7.53). To emphasise this point, when Velikhov and Chandrasekhar first recognised this instability in 1959 and 1960, it was in the context of Couette flow in the laboratory where rotating coaxial cylinders provide the differential rotation and where GM is completely irrelevant.

One can think of Eq. (7.53) in a number of ways. For a given magnetic induction strength, disc density, and rotation speed, the critical value $\mu_{\rho,\mathrm{cr}} = \sqrt{3}$ determines the *minimum* wavelength that can trip the MRI, namely,

$$\lambda_{\min} = \frac{2\pi B_z}{\sqrt{3\mu_0\rho}\,\Omega_0}.$$

A perturbation wavelength less than $\lambda_{\min}$ excites an oscillatory response in the disc, and is therefore stable. If $\lambda > \lambda_{\min}$, the response is exponential rendering the disc unstable. This also places a limit on the disc thickness, h. For the disc to be MR-unstable, $h \geq \lambda \geq \lambda_{\min}$ and, for a given $\lambda_{\min}$, a sufficiently thin disc will be MR-stable.

Alternately, for a given Ω_0, λ, and ρ, Eq. (7.53) gives the *maximum* vertical magnetic induction strength that can trigger the MRI, namely,

$$B_{z,\max} = \frac{\sqrt{3\mu_0\rho}\,\lambda\Omega_0}{2\pi}.$$

Note that the limit is on the *maximum* strength of B_z for MR-instability, and *not* the minimum. Thus, a strong B_z can suppress the MRI (the "spring" can be too "stiff"; an analogy that will make more sense after the next subsection), but there is no limit to how weak B_z can be for the MRI to eventually dominate the disc. This isn't the first example we've seen where the difference between a *zero* and *infinitesimal* magnetic induction can have a *qualitative* difference on the physical nature of a system (*e.g.*, see discussion at top of pages 162 and 165). For $B_z = 0$, a differentially rotating disc – as we'll soon see – is unconditionally stable. Conversely, for even the most minute trace of B_z, the MRI shears B_z *relentlessly* into the plane of the disc creating a planar component of the magnetic induction whose energy density becomes comparable to the thermal, gravitational, and/or kinetic energy densities in surprisingly few rotations.

7.3.2 Physical model of the MRI

So what does all this mathematics really mean? What is the nature of the MRI, and what causes it? For this, it is helpful first to examine two limiting cases: $\Omega_0 = 0$; and $B_z = 0$.

On the former, setting Ω_0 and thus κ to zero reduces Eq. (7.48) to,

$$\omega^4 - 2\omega_A^2 \omega^2 + \omega_A^4 = 0 \quad \Rightarrow \quad \omega = \pm \omega_A \in \mathbb{R},$$

and the perturbation is unconditionally stable. In particular, Eq. (7.44) and (7.45) reduce to,[23]

$$\ddot{\vec{\xi}}' = -\omega_A^2 \vec{\xi}', \tag{7.54}$$

the dynamical equation for a simple harmonic oscillator with oscillation frequency ω_A and what was portended at the top of page 273. These are just Alfvén waves that propagate harmlessly along the vertical lines of induction and ultimately away from the disc.

On the latter, setting B_z and thus ω_A to zero reduces Eq. (7.48) to,

$$\omega^4 - 2(2\Omega_0^2 + \kappa)\omega^2 = 0 \quad \Rightarrow \quad \omega^2 = 4\Omega_0^2 + 2\kappa = \Omega_0^2,$$

for a Keplerian disc (Eq. 7.52). Once again, $\omega \in \mathbb{R}$ and the perturbation is unconditionally stable. This time, Eq. (7.44) and (7.45) reduce to,

$$\ddot{\xi}'_x = -2\kappa \xi'_x + 2\Omega_0 \dot{\xi}'_y \quad \text{and} \quad \ddot{\xi}'_y = -2\Omega_0 \dot{\xi}'_x.$$

The last is readily integrated to $\dot{\xi}'_y = -2\Omega_0 \xi'_x$ which can be substituted into the first to get,

$$\ddot{\xi}'_x = -(2\kappa + 4\Omega_0^2)\xi'_x = -\Omega_0^2 \xi'_x, \tag{7.55}$$

for a Keplerian disc. Once again, a classic simple harmonic oscillator is revealed with oscillation frequency – as already determined – Ω_0.

The fact that the oscillation frequency, Ω_0, is the same as the orbital angular speed is a well-known result from celestial mechanics. A perturbed circular orbit in an inverse-square law central force is stable with an oscillation period equal to

---

[23] *Rappel*: The perturbation displacement is $\vec{\xi}' = \xi'_x \hat{\imath}' + \xi'_y \hat{\jmath}'$.

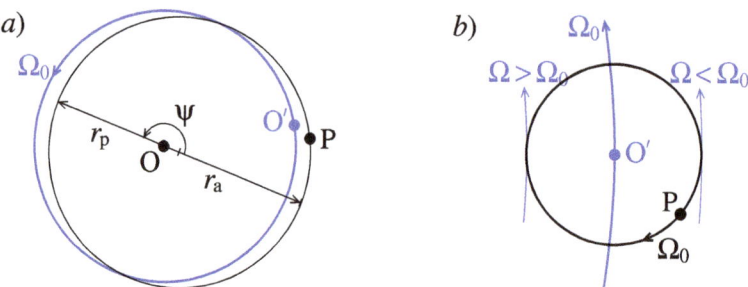

Figure 7.16. *a*) From the inertial reference frame, O, P is perturbed radially from the circular orbit of O′ and oscillates back and forth across the circular orbit with the same period as the orbital period, $2\pi/\Omega_0$, resulting in an apsidal angle, $\psi = \pi$ rad, between the *apicentre* (r_a) and *pericentre* (r_p) of its orbit. *b*) From the rotating reference frame, O′, perturbing P radially causes P to execute *retrograde epicyclic motion* about O′ (black circle) with the same period as the orbital period. See text for clarification.

the orbital period and thus an *apsidal angle* (angle between successive *apsides*, or orbital extrema) of π rad (Fig. 7.16*a*). From the inertial reference frame, O, nudging P further away than O′ puts P into a slightly elliptical orbit such that $\Omega < \Omega_0$ for about half its new orbit, and $\Omega > \Omega_0$ for the other half, causing P to successively lag behind then overtake O′ over the course of a single revolution.

From the rotating reference frame, O′ (Fig. 7.16*b*), the perturbed point P undergoes *retrograde epicyclic motion*, with one complete epicycle executed per orbit of O′ about O. The initial perturbation pushes P further away from O at which point P "goes into orbit about O′" with the same orbital period as O′ has about O. The orbit is maintained by the Coriolis force which accelerates P in a direction always perpendicular to its velocity (Eq. 7.34), much like in an inertial reference frame where the centripetal acceleration of an object in a circular orbit is always perpendicular to its orbital velocity.

Evidently, the MRI is not caused by rotation or magnetism *alone*; it is a consequence of both acting in concert. So once again, let us examine Eq. (7.44) and (7.45) for physical insight, repeated here for convenience:

$$\ddot{\xi}'_x - 2\Omega_0 \dot{\xi}'_y = -(\omega_A^2 + 2\kappa)\xi'_x; \quad\quad \text{Eq. (7.44)}$$

$$\ddot{\xi}'_y + 2\Omega_0 \dot{\xi}'_x = -\omega_A^2 \xi'_y. \quad\quad \text{Eq. (7.45)}$$

The LHS of each equation gives the net acceleration of a particle in the x'- and y'-directions (radial and transverse) including the Coriolis acceleration, and thus describes a particle orbiting a central mass. The RHS describes the acceleration caused by a spring with, evidently, differing "spring constants" in the x'- and y'-directions. Thus a practical model for our system might be two masses, m_1 and m_2, in the same orbit about a central object tethered by a non-isotropic spring. Should m_2 be nudged into a higher orbit with lower Ω, it starts to lag behind m_1 thereby "stretching the spring" and creating a restoring force between the masses.

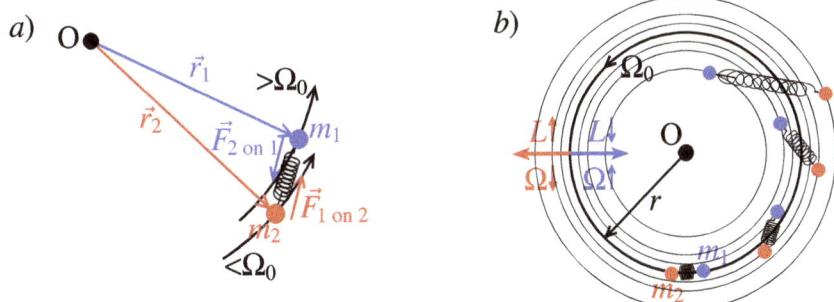

Figure 7.17. *a)* Two masses, m_1 and m_2 in orbit about O are tethered by a "magnetic spring". All motion is counter-clockwise, and thus $\vec{\Omega}$ and $\vec{L}$ are directed out of the page. The torques each mass exerts on the other by virtue of the "spring force" result in angular momentum being transferred from the inner mass, m_1 (blue), to the outer mass, m_2 (red). *b)* As a function of orbit radius, L increases with r while Ω decreases. This sets up the MRI in which the angular momentum transferred from m_1 to m_2 drives the masses ever further apart despite the "restorative" nature of the "spring force". See text for discussion.

As can be seen in Fig. 7.17*a*, relative to the origin O, the torque m_1 exerts on m_2, $\vec{\tau}_{1\,\text{on}\,2} = \vec{r}_2 \times \vec{F}_{1\,\text{on}\,2}$, is directed *out of the page* and points in the same direction as $\vec{L}_2$, the angular momentum of m_2. Conversely, the torque m_2 exerts on m_1 is directed *into the page*, and thus opposite to $\vec{L}_1$. Since $\vec{\tau} = d\vec{L}/dt$, $\vec{L}_2$ increases while $\vec{L}_1$ decreases, and the restoring force of the spring effectively transfers angular momentum from m_1 to m_2.

But here's the thing about Keplerian rotation. While we've already seen that in a Keplerian disc, $\Omega \propto r^{-3/2}$ (Eq. 7.52) and thus the angular speed *decreases* for higher orbits, the angular momentum,

$$\vec{L} = I\vec{\Omega} = mr^2 \vec{\Omega} \propto \sqrt{r},$$

actually *increases* for higher orbits. The fact that the angular momentum of m_2 is increased at the expense of m_1 means that m_2 must seek out a higher orbit just as m_1 is obliged to find a lower one. In this environment, the "restoring force" of the spring is actually *repulsive* and the two masses are driven apart! Of course, this just stretches the "spring" more, increasing the torque each mass exerts on the other, and driving them ever further apart (Fig. 7.17*b*).

Voilà, instability!

And so how aggressive is this instability; how quickly can it grow? For that we return to Eq. (7.50) and the algebra. Since $\omega_3 \in \mathbb{I}$, set $\omega_3 = i\chi$, $\chi \in \mathbb{R}$, to get,

$$\chi^2 = \sqrt{4\omega_A^2 \Omega_0^2 + \frac{\Omega_0^4}{4}} - \omega_A^2 - \frac{\Omega_0^2}{2} = \Omega_0^2 \left(\sqrt{4\mu_\rho^2 + \frac{1}{4}} - \mu_\rho^2 - \frac{1}{2} \right), \qquad (7.56)$$

where I have assumed a Keplerian disc and, accordingly, set $2\kappa = -3\Omega_0^2$. The temporal dependence of the perturbation is now $e^{-i\omega_3 t} = e^{\chi t}$, and χ can be interpreted as the *growth rate* of the instability with an e-folding time $\tau_{\mathrm{MR}} = 1/\chi$.

As for the condition that maximises χ, here I'm going to bunt and leave it to Problem 7.3 to show that the maximum growth rate of the MRI *per period of revolution* occurs when,

$$\mu_\rho = \frac{\sqrt{15}}{4}, \tag{7.57}$$

curiously, just slightly under unity. Thus, for a disc to be MR-unstable, $\mu_\rho < \sqrt{3}$ in which case $\omega_A \lesssim 1.73\Omega_0$ (Eq. 7.53). Peak efficacy of the MRI is reached when μ_ρ is given by Eq. (7.57) and $\omega_A \sim 0.97\Omega_0$. Again, this is not unlike the KHI qualitatively which requires the relative Mach number, $2M$, between the two fluids in the shear layer to be under $2\sqrt{2}$ for instability, and is most efficient when $2M$ is somewhat lower at $\sqrt{3}$ (§7.1).

Substituting Eq. (7.57) back into Eq. (7.56), we find the maximum growth rate to be $\chi_{\max} = 3\Omega_0/4$, in which case the MRI grows a perturbation by a factor,

$$e^{\chi_{\max} T} = \exp\left(\frac{3\Omega_0}{4}\frac{2\pi}{\Omega_0}\right) = e^{3\pi/2} \sim 111,$$

after a *single* revolution. This is the mark of a *very* aggressive instability.

7.3.3 Angular momentum transport

On the *local* scale, a consequence of the "spring model" for the MRI described in the previous subsection is the transfer of angular momentum from a mass inside a given orbit to one outside; m_1 to m_2 in Fig. 7.17a. It is therefore reasonable to expect that on a *global* scale, the MRI could be responsible for the wholesale transport of angular momentum from the interior to the exterior of a disc. Indeed, this demonstration was the triumph of Balbus and Hawley and why they won the 2013 Shaw Prize for astrophysics. Paraphrasing the citation of the Shaw selection committee,

> *The discovery and elucidation of the magneto-rotational instability provides what to this day remains the only viable mechanism for the outward transfer of angular momentum in accretion discs, solving a previously "elusive" problem in astrophysics.*

In this subsection, I take the reader through the calculations that once and for all identified the nature of "anomalous viscosity" in planetary discs. The mathematics is both tricky and beautiful, and I hope my presentation enables the reader to acquire some appreciation for this achievement.

Working now on a global scale, we switch to an inertial (non-rotating) reference frame with the origin, O, at the centre of the disc using cylindrical coordinates (Fig. 7.18). We start with the φ-component of the MHD momentum equation (Eq. 4.13) as given in Eq. Set (A.52) (§A.4), with r times the φ-component of the Lorentz force density inserted (Eq. A.55):

$$\partial_t(rs_\varphi) + \nabla \cdot (rs_\varphi \vec{v}) = -\partial_\varphi p - \frac{1}{2\mu_0}\partial_\varphi B^2 + \frac{1}{\mu_0}\nabla \cdot (rB_\varphi \vec{B}), \quad (7.58)$$

where the gravity term ($\propto \partial_\varphi \phi$) is zero since gravity is a central force. Note that $s_\varphi = \rho v_\varphi$ is the *linear* momentum density while $rs_\varphi \equiv \ell$ is the *angular* momentum density. Thus, Eq. (7.58) is an evolution equation for the quantity whose radial flux we wish to evaluate.

Since $p^* = p + B^2/2\mu_0$ is the MHD pressure (Eq. 4.26), Eq. (7.58) becomes,[24]

$$\partial_t \ell + \nabla \cdot (\ell \vec{v}) = -\partial_\varphi p^* + \frac{1}{\mu_0}\nabla \cdot (rB_\varphi \vec{B}), \quad (7.59)$$

which we rearrange to get our final form for the angular momentum evolution equation:

$$\boxed{\partial_t \ell + \nabla \cdot \vec{f}_L = 0,} \quad (7.60)$$

where,

$$\vec{f}_L \equiv r\rho(v_\varphi \vec{v} - a_\varphi \vec{a}) + rp^*\hat{\varphi},$$

is the angular momentum flux density,[25] and $\vec{a} = \vec{B}/\sqrt{\mu_0 \rho}$ is the usual Alfvén velocity with a_φ being its φ-component.

Our main interest is the *radial* component of the flux density averaged over the disc height, h:

$$\langle f_{L,r}\rangle_z = r\rho\langle v_\varphi v_r - a_\varphi a_r\rangle_z, \quad (7.61)$$

where $\langle\ \rangle_z$ indicates an average over z, *e.g.*,

$$\langle g(z)\rangle_z \equiv \frac{1}{h}\int_{-h/2}^{h/2} g(z)dz, \quad (7.62)$$

and where, for convenience, we set $\lambda = h$. When $\langle f_{L,r}\rangle_z > 0$, angular momentum transport is outward (broad arrows in Fig. 7.18) and when negative, inward. What we

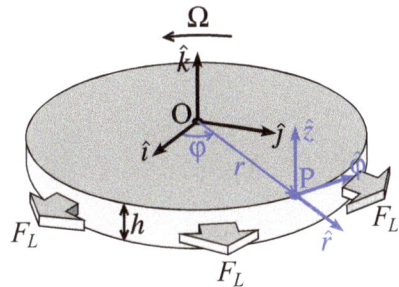

Figure 7.18. Portion of the disc interior to P at a distance r from the disc centre, O, an inertial frame with cylindrical coordinates indicated at P (blue). The disc cross section at r is in light grey, and broad arrows indicate radial transport of angular momentum.

want is the total angular momentum transported radially, *i.e.* the radial angular momentum flux, $\mathcal{F}_L$, which we get by multiplying the radial flux density by the disc cross-sectional area (light grey in Fig. 7.18):

$$\mathcal{F}_L = 2\pi rh\langle f_{L,r}\rangle_z. \quad (7.63)$$

To this end, we first obtain expressions for the velocities, and for that we return to the perturbations in Eq. (7.41), $\vec{\xi}$, that triggered the MRI in the first place. As viewed from the inertial frame of reference O (thus no prime) and in cylindrical coordinates, let the velocity of the particle P located at r_0 before the perturbation

---

[24] See Problem 7.4 for an alternative derivation of Eq. (7.59) using Theorem 1.1 from Chap. 1.
[25] Note the use of the identity $\nabla \cdot (rp^*\hat{\varphi}) = \partial_\varphi p^*$, where $\hat{\varphi}$ is the unit vector in the φ-direction.

be $\vec{v}_0(r_0)$. Since the perturbation moves P from r_0 to r *and* changes its velocity, let the post-perturbation velocity of P be $\vec{v}(r)$, in which case,

$$\Delta \vec{v} = \vec{v}(r) - \vec{v}_0(r_0) = \frac{d\vec{\xi}}{dt} = \partial_t \vec{\xi} + (\vec{v} \cdot \nabla)\vec{\xi}. \tag{7.64}$$

Note that $\Delta \vec{v}$ is a difference of velocities for the *same* particle at two *different* locations (r and r_0); that is, following a particle path. As such, it is a *Lagrangian difference* and many authors would write the total derivative as $D\vec{\xi}/Dt$ (*e.g.*, §3.1).

This, however, is not the difference we need. For if all $\Delta \vec{v}$ did were to change the velocity of P from its natural orbital velocity at r_0 (*e.g.*, $GMr_0^{-1/2}\hat{\varphi}$) to its natural orbital velocity at r, there would be no angular momentum transport, yet there would still be a change in velocity. What we want is the perturbed velocity of P relative to a *cospatial* point in the unperturbed fluid,

$$\delta \vec{v} = \vec{v}(r) - \vec{v}_0(r), \tag{7.65}$$

the so-called *Eulerian difference* where the difference is now between two *different* particles at the *same* location.[26] It is this quantity that ultimately drives angular momentum transport.

The simplest way to relate $\delta \vec{v}$ to $\Delta \vec{v}$ is by linking the unperturbed velocities, $\vec{v}_0$, at r and r_0 which we do by a first-order Taylor expansion. In 3-D, this is given by:

$$\vec{v}_0(r) = \vec{v}_0(r_0) + (\vec{\xi} \cdot \nabla)\vec{v}_0\Big|_{r_0},$$

where we'll now specify that $\vec{v}_0(r) = r\Omega(r)\hat{\varphi}$. Substituting this into Eq. (7.65), we get,

$$\delta \vec{v} = \underbrace{\vec{v}(r) - \vec{v}_0(r_0)}_{\Delta \vec{v} \text{ (Eq. 7.64)}} - (\vec{\xi} \cdot \nabla)\vec{v}_0\Big|_{r_0} = \partial_t \vec{\xi} + (\vec{v} \cdot \nabla)\vec{\xi} - (\vec{\xi} \cdot \nabla)\vec{v}_0\Big|_{r_0},$$

from which we extract the r- and φ-components (last expression in Eq. Set A.40):

$$\delta v_r = \underbrace{\partial_t \xi_r + (\vec{v} \cdot \nabla)\xi_r}_{\dot{\xi}_r} - \cancel{\frac{v_\varphi \xi_\varphi}{r}}^{0} - (\vec{\xi} \cdot \nabla)\cancel{v_{r,0}}\Big|_{r_0} + \cancel{\frac{\xi_\varphi v_{r,0}}{r_0}}^{0} = \dot{\xi}_r;$$

$$\delta v_\varphi = \underbrace{\partial_t \xi_\varphi + (\vec{v} \cdot \nabla)\xi_\varphi}_{\dot{\xi}_\varphi} + \underbrace{\frac{v_\varphi \xi_r}{r}}_{\Omega \xi_r} - \underbrace{(\vec{\xi} \cdot \nabla)v_{\varphi,0}\Big|_{r_0}}_{\xi_r \partial_r(r\Omega)} - \cancel{\frac{\xi_\varphi v_{r,0}}{r_0}}^{0}$$

$$= \dot{\xi}_\varphi + \cancel{\Omega(r)\xi_r} - \cancel{\xi_r \Omega(r_0)} - \xi_r r_0 \frac{d\Omega}{dr}\Big|_{r_0} = \dot{\xi}_\varphi - \xi_r \frac{d\Omega}{d\ln r}\Big|_{r_0},$$

since the radial component of $\vec{v}_0$ is zero, and where all cancellations are good to first order. Thus, from Eq. (7.65), we have:

$$\left.\begin{aligned} v_r &= v_{r,0} + \delta v_r = \dot{\xi}_r; \\ v_\varphi &= v_{\varphi,0} + \delta v_\varphi = r\Omega(r) + \dot{\xi}_\varphi - \xi_r \frac{d\Omega}{d\ln r}\Big|_{r_0}. \end{aligned}\right\} \tag{7.66}$$

---

[26] I thank Steven Balbus (pr. comm.) for this clear articulation!

So what has all this careful exercise in vector calculus bought us? The first of Eq. (7.66) just tells us what surely we already knew: the radial velocity of a particle whose initial velocity is azimuthal, $r\Omega\hat{\varphi}$, is whatever radial velocity the perturbation imparts on it, $\dot{\xi}_r$. The second of Eq. (7.66) might also have been intuited, though here one can easily run into trouble overthinking the problem, and it is here that a careful treatment with vector calculus is warranted. After perturbation, the azimuthal velocity of P is the local unperturbed azimuthal velocity, $r\Omega(r)$, plus the additional azimuthal velocity imparted by the perturbation, $\dot{\xi}_\varphi$, less the difference between the initial orbital angular speeds in the original (at r_0) and final (at r) orbits (the $\ln r$-derivative). Note that this term is zero for solid-body rotation, that is when $\Omega = $ constant and not when $r\Omega = $ constant. In hindsight this may all make sense but, at least for me, I had to go through the calculus in order to convince myself that the form is correct.

Associating ξ_r and ξ_φ, the radial and azimuthal components of the perturbation, with the real parts of ξ'_x and ξ'_y in §7.3.1 ($e^{ikz} \to \cos kz$), we have from Eq. (7.41) and (7.44),

$$\xi_r = \xi_{r,0}\cos(kz)e^{\chi t} \quad \text{and} \tag{7.67}$$

$$\ddot{\xi}_r = \chi^2 \xi_r = -\left(\omega_A^2 + \left.\frac{d\Omega^2}{d\ln r}\right|_{r_0}\right)\xi_r + 2\Omega_0 \dot{\xi}_\varphi = -\omega_A^2 \xi_r + 2\Omega_0\left(\dot{\xi}_\varphi - \xi_r \left.\frac{d\Omega}{d\ln r}\right|_{r_0}\right),$$

where I've set $\omega = i\chi$ for instability and used Eq. (7.37) for 2κ. Thus:

$$\dot{\xi}_r = \chi\xi_r; \qquad \dot{\xi}_\varphi - \xi_r\left.\frac{d\Omega}{d\ln r}\right|_{r_0} = \frac{\chi^2 + \omega_A^2}{2\Omega_0}\xi_r, \tag{7.68}$$

and Eq. (7.66) become,

$$v_r = \chi\xi_r; \quad v_\varphi = r\Omega(r) + \frac{\chi^2+\omega_A^2}{2\Omega_0}\xi_r \quad \Rightarrow \quad v_\varphi v_r = r\Omega\chi\xi_r + \frac{\chi^2+\omega_A^2}{2\Omega_0}\chi\xi_r^2$$

$$\Rightarrow \quad \langle v_\varphi v_r\rangle_z = r\Omega\chi\xi_{r,0}\underbrace{\langle\cos(kz)\rangle_z}_{0}e^{\chi t} + \frac{\chi^2+\omega_A^2}{2\Omega_0}\chi\xi_{r,0}^2\underbrace{\langle\cos^2(kz)\rangle_z}_{1/2}e^{2\chi t}$$

$$= \frac{\chi^2+\omega_A^2}{4\Omega_0}\chi\xi_{r,0}^2 e^{2\chi t}, \tag{7.69}$$

using Eq. (7.67) and where Eq. (7.62) was used for the averages over z (with $k = 2\pi/h$ and $h = \lambda$). Note that the first-order term $\propto \xi_{r,0}$ averages to zero, whereas the second-order term $\propto \xi_{r,0}^2$ does not. One way to describe this is in terms of *correlations*. There is no correlation between v_r and $v_{\varphi,0}$; their product is positive as often as it is negative and therefore integrates to zero. However, there is *high* correlation between v_r and δv_φ; in this simplistic case where the only coordinate dependence is $\cos(kz)$, they always have the same sign and thus their integral is positive-definite.

Turning now to the Alfvén velocities, from Eq. (7.42),

$$\delta B_r = -kB_z\xi_{r,0}\sin(kz)\,e^{\chi t} \quad \text{and} \quad \delta B_\varphi = -kB_z\xi_{\varphi,0}\sin(kz)\,e^{\chi t},$$

once again setting $\omega = i\chi$. As the r- and φ-components of $\vec{B}$ did not exist prior to the perturbation, the perturbed values alone are responsible for the r- and φ-components of the Alfvén velocity, and we can immediately write:

$$a_\varphi a_r = \frac{\delta B_\varphi \delta B_r}{\mu_0 \rho} = \underbrace{\frac{k^2 B_z^2}{\mu_0 \rho}}_{\omega_A^2} \xi_{r,0} \xi_{\varphi,0} \sin^2(kz) e^{2\chi t}. \qquad (7.70)$$

Now, from Eq. (7.47), we have (again, associating $\xi_{\varphi,0}$ with $\xi_{y,0}$ and setting $\omega = i\chi$),

$$\chi^2 \xi_{\varphi,0} = -\omega_A^2 \xi_{\varphi,0} - 2\chi \Omega_0 \xi_{r,0} \quad \Rightarrow \quad \xi_{\varphi,0} = -\frac{2\chi \Omega_0}{\chi^2 + \omega_A^2} \xi_{r,0},$$

where the negative sign is key. Substituting this into Eq. (7.70) and averaging over the disc height, we get,

$$\langle a_\varphi a_r \rangle_z = -\frac{\Omega_0 \omega_A^2}{\chi^2 + \omega_A^2} \chi \xi_{r,0}^2 e^{2\chi t}, \qquad (7.71)$$

since $\langle \sin^2(kz) \rangle_z = \frac{1}{2}$ for $k = 2\pi/h$.

And now, for the home-stretch. Substituting Eq. (7.69) and (7.71) into Eq. (7.61), the average angular momentum flux density becomes,

$$\begin{aligned}\langle f_{L,r} \rangle_z &= r\rho \left(\frac{\chi^2 + \omega_A^2}{4\Omega_0} + \frac{\Omega_0 \omega_A^2}{\chi^2 + \omega_A^2} \right) \chi \xi_{r,0}^2 e^{2\chi t} \\ &= \frac{r\rho}{4\Omega_0} \frac{(\chi^2 + \omega_A^2)^2 + 4\Omega_0^2 \omega_A^2}{\chi^2 + \omega_A^2} \chi \xi_{r,0}^2 e^{2\chi t},\end{aligned} \qquad (7.72)$$

where we first see that the average radial flux density is positive-definite and thus angular momentum is transported *outward*. Further, the exponential factor means the transport is exceedingly efficient; in fact too efficient, as we'll see.

Now, from the dispersion relation (Eq. 7.48) with $\omega = i\chi$,

$$\chi^4 + 2(\omega_A^2 + 2\Omega_0^2 + \kappa)\chi^2 + \omega_A^2(\omega_A^2 + 2\kappa) = 0$$

$$\Rightarrow \quad (\chi^2 + \omega_A^2)^2 = -4\Omega_0^2 \chi^2 - 2\kappa(\chi^2 + \omega_A^2),$$

and Eq. (7.72) becomes,

$$\langle f_{L,r} \rangle_z = \frac{r\rho}{4\Omega_0} \left(\frac{4\Omega_0^2(\omega_A^2 - \chi^2)}{\omega_A^2 + \chi^2} - 2\kappa \right) \chi \xi_{r,0}^2 e^{2\chi t}.$$

Multiplying this by $2\pi r h$, we get from Eq. (7.63) the radial angular momentum flux and thus the rate at which angular momentum is transported radially:

$$\mathcal{F}_L = \frac{m(r)}{2\Omega_0} \left(\frac{4\Omega_0^2(\omega_A^2 - \chi^2)}{\omega_A^2 + \chi^2} - 2\kappa \right) \chi \xi_{r,0}^2 e^{2\chi t}, \qquad (7.73)$$

where $m(r) = \pi r^2 h \rho$ is the mass of the disc interior to r.

Specialising now to a Keplerian disc where $2\kappa = -3\Omega_0^2$ (Eq. 7.52), I leave it to Problem 7.5 to show,

$$\mathcal{F}_L = m(r)\sqrt{4\mu_\rho^2 + \tfrac{1}{4}}\,\Omega_0 \chi \xi_{r,0}^2 e^{2\chi t}. \qquad (7.74)$$

Specialising further to the maximum growth rate where (end of §7.3.2),

$$\mu_\rho^2 = \frac{15}{16} \quad \text{and} \quad \chi = \frac{3}{4}\Omega_0,$$

we get, finally,

$$\boxed{\mathcal{F}_{L,\max} = \frac{3m(r)\Omega_0^2}{2}\xi_{r,0}^2 e^{3\Omega_0 t/2}.}$$

However small the initial perturbation, $\xi_{r,0}$, may be – even squared – the fact that the e-folding time for $\dot{L}$ is,

$$\tau_{\dot{L}} = \frac{1}{2\chi} = \frac{2}{3\Omega_0} = \frac{T}{3\pi},$$

where $T = 2\pi/\Omega_0$ is the rotation period, means that in a *single* revolution, the outward angular momentum transport instigated by the perturbation grows by a factor of $e^{3\pi} > 10^4$! And in the next revolution, another factor of 10^4, and so it goes. After decades where astrophysicists could not explain much of *any* angular momentum transport in protostellar discs, suddenly the spigot was turned on full-blast and the problem quickly became how to mitigate the MRI transport which – left unchecked – would prevent discs as we know them from forming at all! That mitigation comes in the non-linear development of the instability in 3-D to turbulence (Stone *et al.*, 1996) where the MRI still effectively transports angular momentum outward, but allows the disc to exist, albeit in a turbulent state.

7.3.4 Numerical analysis of the MRI (optional)

In an attempt to understand better the various predictions made in our linear analysis, I set up a 2-D azimuthally symmetric grid in cylindrical coordinates representing a local portion of a rotating disc, and performed several numerical experiments using ZEUS-3D. Figure 7.19 shows the computational domain with radial extent $0.9 \leq r \leq 1.1$ and disc thickness $\Delta z = 0.1$ (grey rectangle) containing a point P located at $r = r_0 = 1$ (units could be AU, but consider them arbitrary) from the central gravitating point mass, M. In units where $GM = 1$, the orbital angular speed at P is $\Omega_0 = 1$ and the rotation period is therefore 2π.

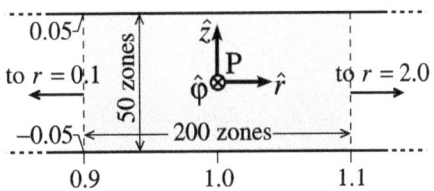

Figure 7.19. The computational domain around point P at $r_0 = 1.0$ with $\Omega_0 = 1.0$ into the page is shown in grey. The complete domain extends to $r = 0.1$ and $r = 2.0$ to move the r-boundaries away from the region of interest. Periodic boundary conditions are imposed at $z = \pm 0.05$.

In §7.3.1, we considered displacement perturbations in both the $\hat{\imath}'$- and $\hat{\jmath}'$-directions, corresponding to $\hat{r}$ and $\hat{\varphi}$ in Fig. 7.19 ($\otimes$ indicates "into the page"). As a matter of practicality, in a fluids code one prescribes a velocity perturbation

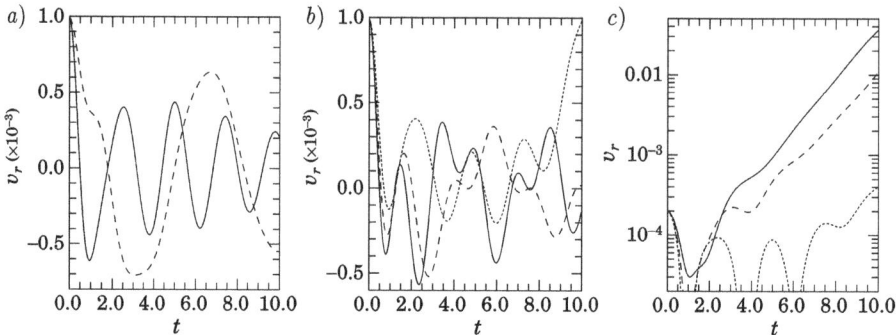

Figure 7.20. *a*) The nearly simple harmonic response to a perturbation of v_r in a disc with either no rotation (solid) or no magnetic induction (dashed). *b*) The bi-modal oscillatory response for super-critical ($\mu_\rho^2 > 3$; see text for definition), and thus MR-stable discs for $\mu_\rho^2 = 7.0$ (solid), 5.0 (dashed), and 3.1 (dotted). *c*) Exponential growth superposed over the oscillatory mode of v_r for sub-critical ($\mu_\rho^2 < 3$), and thus MR-unstable discs for $\mu_\rho^2 = 1.0$ (solid), 2.0 (dashed), and 2.9 (dotted). In all cases, the rotation period is 2π and $t = 10 \sim 1.6$ rotations.

which, after a short time, affects the intended displacement perturbation. At $t = 0$, I specify a small ($< 0.01\, r_0\Omega_0$) radial velocity, $\delta v_r(z, r)$, displacing fluid from P to a slightly higher orbit in the $+\hat{r}$ direction with a smaller orbital speed which pushes fluid in the $-\hat{\varphi}$ direction. The z-dependence of δv_r is one wavelength of a cosine, and thus the perturbation wavelength is $\lambda = 0.1$. The r-dependence is a Gaussian with a peak at r_0 and a standard deviation of $\sigma = 0.03\, r_0$. Thus, δv_r falls to $\sim 0.4\%$ of its peak value at $r = 0.9$ and 1.1 and this radial dependence compresses the "magnetic spring", priming the MRI. Boundary conditions are described in the caption of Fig. 7.19.

While the linear analysis presumes a constant MHD pressure, p^*, this can be difficult to impose numerically. With p^* initially uniform, perturbations launch pressure variations and thus sound waves which do, after some time, have an accumulative effect. To mitigate this, setting a low sound speed ($c_s \sim 0.05$) extends the time it takes for pressure perturbations to reach the boundaries and then get reflected back into the region of interest.

Eight separate experiments were conducted to $t = 10$ (~ 1.6 rotations) when boundary and pressure effects started to be felt. The first two runs tested the prediction that with either Ω_0 or ω_A zero, a displacement perturbation results in a simple harmonic response (Eq. 7.54 and 7.55) with angular frequencies ω_A and Ω_0 respectively. Figure 7.20 *a* shows the amplitude of v_r at point P in Fig. 7.19 over the course of both simulations from which the periods can be measured directly. In the first case where $\Omega_0 = 0$ and $\omega_A = \sqrt{7}$ (solid line), the numerically determined period is $T_{\text{num}} \sim 2.40$ as compared with $T_{\text{lin}} = 2\pi/\omega_A \sim 2.38$ from our linear analysis. Careful measurement shows that the first period (2.86) is somewhat longer than the second two (2.42 and 2.38), indicative of numerical transients.

In the second case where $\Omega_0 = 1$ and $\omega_A = 0$ (dashed line in Fig. 7.20 *a*), the numerical period is $T_{\text{num}} \sim 6.57$, cf. $T_{\text{lin}} = 2\pi/\Omega_0 \sim 6.28$. As only ~ 1.5 full periods are observed, the fact that $T_{\text{num}} > T_{\text{lin}}$ may be another instantiation of the numerical transients supposed above. Regardless, I consider these experiments as evidence that the numerical methods satisfactorily reproduce these results from linear theory, even quantitatively.

Figure 7.20 b shows results from the three super-critical ($\mu_\rho^2 > 3$, MR-stable) runs, with $\mu_\rho^2 = 7, 5$, and 3.1 shown respectively as solid, dashed, and dotted lines. In this scenario, Eq. (7.49) and (7.50) predict a superposition of two normal modes with frequencies,

$$\omega_{1,3} = \Omega_0 \sqrt{\mu_\rho^2 + \tfrac{1}{2} \pm \sqrt{4\mu_\rho^2 + \tfrac{1}{4}}}, \tag{7.75}$$

where I've set $2\kappa = -3\Omega_0^2$ for a Keplerian disc. Examining the solid curve in Fig. 7.20 b where $\mu_\rho^2 = 7$, one can almost "eyeball" the superposition of two normal modes with periods ~ 1.5 and 4.0 corresponding to angular frequencies ~ 4.2 and 1.6 respectively and tantalisingly close to $\omega_1 = 3.58$ and $\omega_3 = 1.48$ as evaluated from Eq. (7.75) using $\Omega_0 = 1$. Reducing μ_ρ^2 to 5 (dashed) then 3.1 (dotted) increases both periods of the bimodal response seen in Fig. 7.20 b, as borne out by Eq. (7.75).

Finally, Fig. 7.20 c shows results from the three sub-critical ($\mu_\rho^2 < 3$, MR-unstable) runs, with $\mu_\rho^2 = 1, 2$, and 2.9 shown respectively as solid, dashed, and dotted lines. In this case, linear theory predicts the response to be a superposition of an oscillatory term with frequency ω_1 given by the positive option in Eq. (7.75), and an exponentially growing term, $e^{\chi t}$ where χ is given by Eq. (7.56). As can be seen in panel c, the growth rate for $\mu_\rho^2 = 2.9$ (dotted) is slow enough for the oscillatory phase to be measured directly from the plot, where I find a period $T_1 \sim 2.5$ and thus $\omega_1 \sim 2.5$, cf. 2.62 from Eq. (7.75). The computational time limit of 10 is too short to capture convincingly the growth portion of $\mu_\rho^2 = 2.9$, but is adequate for $\mu_\rho^2 = 2$ (dashed) and 1 (solid), each showing the tell-tale straight lines for $t > 4$ on the $\log v_r$ vs. t plots in Fig. 7.20 c. For these, I measure directly from the plot $\chi = 0.65$ and 0.75 respectively, cf. $\chi = 0.610$ and 0.749 from Eq. (7.56). While both of these cases show evidence for the oscillatory term when $t < 4$, there isn't enough to make a meaningful estimate of the periods.

I take all of this as strong evidence that these simulations have captured the linear regime of the MRI,[27] confirming the critical value $\mu_{\rho,\text{cr}} = \sqrt{3}$ that differentiates MR-stable ($\mu_\rho > \mu_{\rho,\text{cr}}$) from MR-unstable ($\mu_\rho < \mu_{\rho,\text{cr}}$) Keplerian discs.

As a final comment on the MRI, I note the following. The astute reader may have noticed the similarity between the expressions for the normal mode frequencies and growth rate for the MRI (Eq. 7.75 and 7.56 respectively) and those for the KHI (Eq. 7.14 and 7.15 respectively). Considering the expressions for the growth rate, χ, both have the form,

$$\chi^2 = A^2 \left(\sqrt{4B^2 + C^2} - B^2 - C \right), \quad \text{where}$$

	A	B	C
KHI	kc_s	M	1
MRI	Ω_0	μ_ρ	$\tfrac{1}{2}$

---

[27] *Rappel*: This was *not* the case for the Rayleigh–Taylor instability (§7.2.2) where non-linear effects seem to dominate the progress of the instability right from the get-go.

This plus the fact both instabilities gain their energy from a velocity shear makes one wonder whether the two instabilities aren't somehow related. I am not aware of any literature that explores this possibility, even if to dismiss these similarities as mere coincidence.

7.4 Parker instability

While the Velikhov–Chandrasekhar instability – later known to astrophysicists as the Balbus–Hawley instability and more recently as the MRI (§7.3) – was known several years before the Parker instability was discovered (Parker, 1966), the latter was the first MHD instability motivated by and applied to astrophysics. And while rather easy to describe and explain physically, it poses the most challenging mathematical problem of all four instabilities examined in this chapter, and one of the most challenging in this text. Therefore, the structure of this section differs from the previous three. In the next subsection, I give a qualitative "derivation" of the instability along with a description of Parker's astrophysical motivation and some of the predictions he made. The mathematical details are then delegated to the following subsection which, despite its "optional" designation, I highly recommend the reader go through. The mathematics isn't new, just intense and it represents the most extensive example we do in this text of working with *linearised* equations where all four MHD equations *plus* an equation governing cosmic rays must be accounted for!

7.4.1 A qualitative description

By 1966, it had recently been established that the interstellar medium (ISM) of the Milky Way was permeated by a magnetic induction whose strength, at the time, was thought to be $B_{\text{gal}} \sim 5 \times 10^{-10}$ T (5 μG).[28] To theorists, this posed important fundamental questions including what generates B_{gal} and what ties it to the galaxy? As we know, the presence of a magnetic induction implies a magnetic pressure which cannot be confined by a neutral gas. Nor can it be confined by an external fluid (*e.g.*, an extragalactic medium); ionised or not, in its expansion lines of magnetic induction would simply move around this material.

It was also known that the ISM was made up of roughly equal portions (in energy, not mass) of ordinary thermal matter – primarily neutral and ionised hydrogen and helium – and *cosmic rays* (CR)[29] whose energies are too high for the Milky Way to confine gravitationally. And yet, confined they are. Observations of synchrotron

---

[28] Modern values range from 10^{-10} T in the outer galaxy, to 6×10^{-10} T in the solar neighbourhood, to 4×10^{-9} T towards the galactic core (*e.g.*, Beck, 2007).

[29] While "ordinary thermal matter" and "cosmic rays" are made up of the same "stuff" (*e.g.*, H and He ions, electrons, *etc.*), the "ordinary matter" follows a Maxwellian distribution (like our atmosphere) and can be described by a unique temperature. On the other hand, CR are particles of ordinary matter accelerated by unusual and energetic events such as supernovæ, galactic shocks, and black hole mergers. As such, CR do not follow a Maxwellian distribution and therefore cannot

emission (radiation emitted by charged particles spiralling about a magnetic field) indicate the vast majority of CR remain within the Milky Way, and the only thing that can confine them is the magnetic induction, $\vec{B}_{\text{gal}}$, permeating the ISM. Of course, the high kinetic energies of the CR tugging on $\vec{B}_{\text{gal}}$ gives it all the more reason to expand out of the Milky Way, and still it doesn't.

The explanation formed in the 1960s and which largely prevails today is the bulk of the mass of the ISM is thermal (ionised) matter to which the magnetic induction is also tied and it is the *weight* of this thermal matter within the gravitational field of the Milky Way that ultimately ties $\vec{B}_{\text{gal}}$ and the CR to the galaxy. By analogy, the moon is incapable of holding on to an atmosphere, since individual gas molecules move, on average, too fast for the moon's gravitational field to retain them. However, CO_2 particles bound chemically to the moon's regolith *are* retained since these are effectively held in place by the weight of the regolith.

And so we have a model of the ISM in which neutral gas, ionised gas, and CR co-exist, all permeated by a magnetic induction generated by the turbulent accelerations of charged particles, where the CR continuously vie to escape the confines of the galaxy, and where left to its own devices, $\vec{B}_{\text{gal}}$ would expand out of the galaxy as well. It is then left to the thermal matter to weigh it all down in an uneasy equilibrium that maintains the ISM as we know it.

Now, a property of the ISM known to astronomers in the 1960s was that it is "clumpy". Density contrasts of several between the "in-between" portions of the ISM and what were termed "interstellar clouds" had been reported[30] but these clouds were too small to be formed and held in place by their self-gravity. Eugene Parker[31] was the first to suggest that these clouds were, in fact, an MHD phenomenon and a direct result of a then-unknown instability that now bears his name.

The instability goes like this. As depicted in Fig. 7.21a, the ISM is modelled as a disc of initially uniform gas with an embedded azimuthal magnetic induction, $\vec{B}_{\text{gal}} = B_{\text{gal}}\hat{\varphi}$. In the co-rotating frame of the fluid, the residual vertical component of the acceleration of gravity, $\vec{g}_{\text{gal}} = -g_{\text{gal}}\hat{k}$, points towards the galactic midplane (the radial component being balanced by the centrifugal force). Suppose now that magnetic lines of induction are perturbed sinusoidally, as shown in Fig. 7.21b. Since ionised fluid is confined to move *along* $\vec{B}_{\text{gal}}$, gravity forces matter to "slide down" lines of induction (indicated by δv in Fig. 7.21b), thereby accumulating mass in the troughs of the perturbation at the expense of the crests. The extra "weight" in the troughs weigh lines of induction down more, while the loss of mass at the crests make them more buoyant, dragging lines of induction up. These two actions steepen the perturbation, driving even more ionised matter "down" lines of induction (and also neutral matter; see §10.5) until we achieve the configuration shown in Fig.

---

be described by a unique temperature. Indeed, CR with energies approaching 20 J (*i.e.*, a single α-particle with the energy of a baseball thrown at 60 km/hr!) have been measured, and even the great gravitational well of the entire Milky Way cannot contain them.

[30] Modern values for these density contrasts are well into the hundreds; see Ferriere (2001) and, more accessibly, www.wikipedia.org/wiki/Interstellar_medium.

[31] Eugene Parker (1927–2022; www.wikipedia.org/wiki/Eugene_Parker) was a solar and plasma physicist and, among many contributions, was the first to propose the existence of a solar wind.

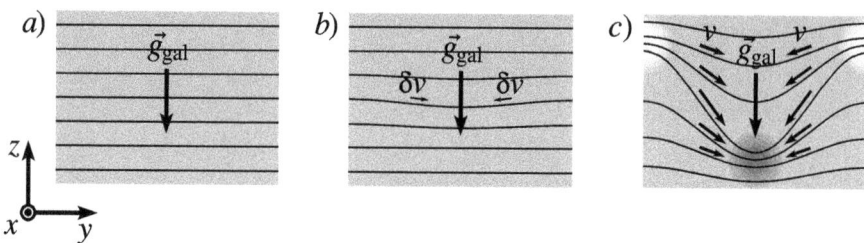

Figure 7.21. *a*) A cross-section of a uniform portion of the ISM, where x points away from the galactic centre, y is the azimuthal coordinate, and z is the altitude above the galactic midplane. Grey indicates a uniform density (in hydrostatic equilibrium), and horizontal lines represent lines of magnetic induction, $\vec{B}_{\text{gal}}$, parallel to the galactic midplane. In a frame of reference co-rotating with the fluid (where the radial gravitational and centrifugal forces balance), the residual gravitational acceleration is towards the galactic midplane, indicated by $\vec{g}_{\text{gal}}$ in the figure. *b*) Perturbing $\vec{B}_{\text{gal}}$ causes ionised matter – which must follow lines of magnetic induction – to "slide down" lines of $\vec{B}_{\text{gal}}$ under the influence of $\vec{g}_{\text{gal}}$. *c*) Matter collecting in the troughs weigh lines of induction down while the paucity of matter at the crests make them buoyant, forcing lines of induction to rise. This is the essence of the Parker instability.

7.21*c*. Here, at the troughs of the instability, the denser clumps form the observed interstellar clouds confined by the magnetic induction of the galaxy rather than by their own self-gravity.

Now, it's not entirely obvious that the scenario just painted should, in fact, lead to an instability. For one thing, bending a line of magnetic induction comes with additional magnetic tension that tends to straighten out the bend. Further, matter piling up in the troughs of the perturbations increases the local thermal pressure which also tends to restore the perturbation. Clearly, there is a balance to be struck. Is the increase in magnetic energy because of bent lines of induction plus the increase in internal energy because of the condensation of matter more or less than the decrease in potential energy afforded by the "falling" gas? The answer to this question is what determines whether the system is unstable, or restores itself to the unperturbed state in a damped oscillatory fashion. Finally, even if there are conditions under which the instability is triggered, does the *wavelength* of the instability correspond to the observed spacings of the interstellar clouds and does the *growth time* of the instability allow them to form within the galaxy's lifetime?

In his treatment, Parker assumed that both the magnetic and CR pressures, p_{M} and p_{CR}, were constant multiples of the thermal pressure, p,[32]

$$\frac{p_{\text{M},0}}{p_0} = \frac{B_0^2/2\mu_0}{p_0} = \sigma_{\text{M}} \quad \text{and} \quad \frac{p_{\text{CR},0}}{p_0} = \sigma_{\text{CR}}, \tag{7.76}$$

where equilibrium quantities are designated with a subscript 0. Further, in equilib-

---

[32]The astute reader will notice that Parker's σ_{M} is just the reciprocal of the plasma-beta (Eq. 5.29) defined in §5.2 and differs from our MHD-alpha (Eq. 5.30) by a factor of $\gamma/2$. Parker's choice, however, does tidy up the algebra enough that we shall retain his definitions here.

rium the thermal gas is assumed to be adiabatic ($p \propto \rho^\gamma$) and governed by the ideal gas law (Eq. 1.12),

$$p_0 = \frac{\rho_0 k_B T_0}{\langle m \rangle} = v_{\text{rms}}^2 \rho_0, \tag{7.77}$$

where k_B is the Boltzmann constant, T_0 is the uniform temperature of the thermal gas in equilibrium, $\langle m \rangle$ is the average particle mass, and $v_{\text{rms}} = (k_B T_0/\langle m \rangle)^{1/2}$ is the rms speed of the thermal gas particles in the z-direction (*i.e.*, with *one* degree of freedom; *cf.* Eq. 1.13 which, for $\gamma = 5/3$, is for *three* degrees of freedom). Note that if the gas were isothermal, v_{rms} would be the *isothermal sound speed*.

Still in equilibrium, Parker assumed all quantities varied only in the vertical (z) direction (except T_0 and g_{gal}, which he assumed constant) with $\vec{B}_{\text{gal}} \propto \hat{y}$ and $\vec{g}_{\text{gal}} \propto -\hat{z}$ (Fig. 7.21a):

$$\rho_0 = \rho_0(z); \quad p_0 = p_0(z); \quad -\nabla\phi = -g_{\text{gal}}\hat{k}; \quad \vec{B}_0 = B_{y,0}(z)\hat{j}. \tag{7.78}$$

With this simplified geometry and dependencies, the Lorentz force becomes,

$$\vec{J}_0 \times \vec{B}_0 = \frac{1}{\mu_0}(\nabla \times \vec{B}_0) \times \vec{B}_0 = -\frac{B_{y,0}}{\mu_0}\frac{dB_{y,0}}{dz}(\hat{i} \times \hat{j}) = -\frac{dp_{M,0}}{dz}\hat{k}, \tag{7.79}$$

and, in equilibrium where $\vec{v} = 0$ and $\partial_t = 0$, the z-component of the momentum equation (Eq. 4.13) in which we include the CR pressure, $p_{\text{CR},0}$, as a contributing *partial pressure*, reduces to:

$$0 = -\frac{d}{dz}(p_0 + p_{\text{CR},0} + p_{M,0}) - \rho_0 g_{\text{gal}} \tag{7.80}$$

$$\Rightarrow \quad \frac{1}{p_0}\frac{dp_0}{dz} = -\frac{g_{\text{gal}}}{v_{\text{rms}}^2(1 + \sigma_M + \sigma_{\text{CR}})} \equiv -\frac{1}{L}, \tag{7.81}$$

using Eq. (7.76), (7.77), and (7.79). The quantity L is referred to as the *scale height* of the galaxy. It is the vertical distance characterising how rapidly, as in the case of Eq. (7.81), the thermal pressure falls. Note that because of Eq. (7.76) and (7.77), we can write,

$$\frac{1}{p_0}\frac{dp_0}{dz} = \frac{1}{p_{M,0}}\frac{dp_{M,0}}{dz} = \frac{2}{B_{y,0}}\frac{dB_{y,0}}{dz} = \frac{1}{p_{\text{CR},0}}\frac{dp_{\text{CR},0}}{dz} = \frac{1}{\rho_0}\frac{d\rho_0}{dz} = -\frac{1}{L}, \tag{7.82}$$

and thus the scale height is also a measure of how rapidly the magnetic pressure, magnetic induction, cosmic ray pressure, and density fall off with scale height.

With all this, Parker found that sinusoidal perturbations (as illustrated in Fig. 7.21b) are unstable provided,

$$\boxed{C(\sigma_M, \sigma_{\text{CR}}, \gamma) > 2\gamma\sigma_M k^2 L^2,} \tag{7.83}$$

where $k = 2\pi/\lambda$ is the wavenumber of the perturbation, λ is its wavelength, and where,

$$C(\sigma_M, \sigma_{\text{CR}}, \gamma) = (1 + \sigma_M + \sigma_{\text{CR}})^2 - \gamma(1 + \tfrac{3}{2}\sigma_M + \sigma_{\text{CR}}). \tag{7.84}$$

Eq. (7.83) tells us that for the longest perturbation wavelength (and thus $k \to 0$),

$C > 0$ putting a constraint on the parameters $\sigma_{\rm M}$, $\sigma_{\rm CR}$, and γ. Further, Parker found the fastest growing perturbation wavelength has an e-folding time,

$$\tau_{\rm P} = \frac{L\sqrt{2(\gamma + 2\sigma_{\rm M})(\gamma\sigma_{\rm M} + C)}}{v_{\rm rms}C}. \tag{7.85}$$

Use of the intermediate quantities L and C bely the complexity with which the criterion for instability and $\tau_{\rm P}$ depend upon the parameters, $\sigma_{\rm M}$, $\sigma_{\rm CR}$, γ, $g_{\rm gal}$, and $v_{\rm rms}$. A quick glance back at the analogous expressions for the Kelvin–Helmholtz, Rayleigh–Taylor, and even the magneto-rotational instabilities and the relative simplicity with which they depend upon their environmental parameters gives one a glimmer of how complicated algebraically the Parker instability actually is. For Parker first to have intuited the existence of this instability, then designed the mathematical strategy, carried out the calculations, and presented his results in such a transparent fashion to prove his intuition correct leaves me, frankly, in awe!

We turn now to some of Parker's quantitative conclusions. First, in the absence of magnetic and CR pressure ($\sigma_{\rm M} = \sigma_{\rm CR} = 0$), Ineq. (7.83) with Eq. (7.84) require:

$$1 - \gamma > 0,$$

which is clearly false for all values of $\gamma \geq 1$, and there is no instability. Thus, without the effects of magnetism and CR in the ISM, the ISM is unconditionally *stable*.[33]

Of course, $\sigma_{\rm M}$ and $\sigma_{\rm CR}$ are not zero in the galaxy, and in 1966 the best values were 0.3 and 0.15 respectively.[34] For an isothermal gas ($\gamma = 1$), Eq. (7.84) gives $C \sim 0.50$. Further, the best value for $v_{\rm rms}$ at the time was $5{,}000\,{\rm m\,s^{-1}}$ (for a mean temperature of $3{,}000\,{\rm K}$)[35] and for $g_{\rm gal}$ was $1.3 \times 10^{-11}\,{\rm m\,s^{-2}}$.[36] Thus, $L \sim 2.8 \times 10^{18}\,{\rm m} \sim 90\,{\rm pc}$[37] and, from Eq. (7.83),

$$\lambda > 2\pi L\sqrt{\frac{2\gamma\sigma_{\rm M}}{C}} \equiv \lambda_{\rm min} \sim 620\,{\rm pc}, \tag{7.86}$$

a value in reasonable agreement with the observed spacings of interstellar "clumps".

Finally, substituting these numerical values into Eq. (7.85) gives the e-folding time for the fastest growing mode:

$$\tau_{\rm P} \sim 5.6 \times 10^7\,{\rm yr},$$

certainly well within the ten-billion plus year age of the Milky Way.

While many of these numbers have been modified significantly over the years (*e.g.*, see footnotes), the theory gives enough latitude to withstand the newer observations. For example, spacings between ISM "clumps" vary from hundreds to

---

[33] For the reader familiar with the *Jeans instability* in which a sufficient density perturbation in a *self-gravitating* interstellar gas cloud can cause the cloud to collapse under its own weight, the Parker instability is insensitive to this since Parker considers *background* gravitational fields only.

[34] Modern values for $\sigma_{\rm M}$ and $\sigma_{\rm CR}$ are both of order unity (*equipartition*; see, *e.g.*, Beck et al., 1996) though these values vary widely across the galaxy and Parker's values are not unreasonable.

[35] Modern values for the ISM temperature range from $10\,{\rm K}$ in the coldest molecular clouds to $10^6\,{\rm K}$ in stellar coronæ, with $v_{\rm rms} \sim 10^4\,{\rm m\,s^{-1}}$; see Ferriere (2001).

[36] Modern value for $g_{\rm gal}$ is $7.44 \times 10^{-11}\,{\rm m\,s^{-2}}$; see Hagen & Helmi (2018).

[37] This is considerably less than the modern value of 300–400 pc; see Carroll & Ostlie (2017).

thousands of pc, and the growth time of 6×10^7 yr is well under the age of the galaxy, even if raised an order of magnitude. While it took some time for Parker's 1966 paper to become widely known among astrophysicists (remember, MHD as a valid branch of physics was still in doubt in 1958, and Alfvén's Nobel Prize was still four years in the future), it has since become regarded as a turning point in our understanding of galactic dynamics.

Parker himself modestly described his instability as "related to the familiar Rayleigh–Taylor instability" (§7.2), where the less-dense gas supporting the denser gas is replaced by the magnetic field. And while the research note by Appenzeller (1971) refers to "Parker's instability" in its title, it seems to have been Frank Shu who coined the currently used moniker in his 1974 paper "The Parker instability in differentially-rotating disks" where he redid Parker's calculations in this more realistic galactic geometry. Shu (1974) found that while the differential rotation could affect, even extend growth times, no amount of shear and rotation could completely stabilise Parker's instability. Perhaps more than any, then, Shu's paper helped seal the role of the Parker instability in astrophysics.

7.4.2 A quantitative description (optional)

We begin by identifying the dynamical equations needed to describe the physics of the Parker problem: a thermal + CR fluid embedded in a uniform magnetic field with a background acceleration of gravity in a slab-symmetric geometry ($\partial_x = 0$) with $v_x = 0$ and $B_x = 0$ at all times. While the need for gravity to trigger the instability may remind us of the RTI (§7.2) or even the KSI (§7.2.3), Parker's instability is much more complicated mathematically because, in part, the condensation of matter into "clouds" precludes the assumption of incompressibility. Thus, $\nabla \cdot \vec{v} \neq 0$ and, unlike the Rayleigh–Taylor analysis, we can't use the "stream function", ψ, to reduce two components of the momentum equation to one scalar equation.

And so for starters, we know we'll need continuity (Eq. 4.20) to track density perturbations, and *two* components of the momentum equation (Eq. 4.22) to track the y- and z-velocity perturbations (see Fig. 7.21a for coordinates). Now, for *some* simplicity, one might be tempted to assume an isothermal equation of state for the thermal gas, thereby obviating the need for an energy equation. However, Parker's analysis used an adiabatic equation of state so that he could examine the effects of the adiabatic index, γ, on his instability. Following his lead then, we'll also need a dynamical equation for the thermal pressure, p, such as Eq. (1.35).

To this point, we've accumulated four dynamical equations to be linearised. For ∂_x, v_x, and B_x all zero, Eq. (4.20), the y- and z-components of Eq. (4.22), and Eq. (1.35) with $dp/d\rho = \gamma p/\rho$ (adiabatic EoS) become:

$$\partial_t \rho + \partial_y(\rho v_y) + \partial_z(\rho v_z) = 0; \tag{7.87}$$

$$\partial_t s_y + \partial_y(s_y v_y) + \partial_z(s_y v_z) = -\partial_y(p + p_{\mathrm{CR}}) + \frac{B_z}{\mu_0}(\partial_z B_y - \partial_y B_z); \tag{7.88}$$

$$\partial_t s_z + \partial_y(s_z v_y) + \partial_z(s_z v_z) = -\partial_z(p + p_{\mathrm{CR}}) - \rho g_{\mathrm{gal}} + \frac{B_y}{\mu_0}(\partial_y B_z - \partial_z B_y); \tag{7.89}$$

$$\partial_t p + v_y \partial_y p + v_z \partial_z p = -\gamma p (\partial_y v_y + \partial_z v_z). \tag{7.90}$$

It is the induction equation where we catch our first and only (small) break. Since $\nabla \cdot \vec{B}$ *is* zero, we have from Eq. (4.32) and (4.33),

$$\vec{B} = \nabla \times \vec{A} \quad \Rightarrow \quad \partial_t \vec{A} = \vec{v} \times \vec{B} + \nabla \mathcal{V}, \qquad \text{Eq. (4.33)}$$

the ideal induction equation for the vector potential, $\vec{A}$, where, for a suitable choice of gauge, $\mathcal{V}$ can be set to zero (see §4.8). Now, for $\partial_x = 0$ and $B_x = 0$, we have,

$$\vec{B} = (0, B_y, B_z) = \nabla \times \vec{A} = (\partial_y A_z - \partial_z A_y, \partial_z A_x, -\partial_y A_x), \tag{7.91}$$

where we are free to set $A_y = A_z = 0$. Thus, the vector potential can be completely described by its x-component, A_x, and Eq. 4.33 reduces to,

$$\partial_t A_x = v_y B_z - v_z B_y, \tag{7.92}$$

giving a single scalar equation describing both magnetic components, B_y and B_z.

It is also worth pointing out that in this 2-D geometry where $A_x = A_x(y, z)$,

$$\vec{B} \cdot \nabla A_x = (B_y \hat{j} + B_z \hat{k}) \cdot (\underbrace{\partial_y A_x}_{-B_z} \hat{j} + \underbrace{\partial_z A_x}_{B_y} \hat{k}) = 0,$$

using Eq. (7.91), and the gradient of A_x is everywhere perpendicular to $\vec{B}$. Thus, *contours* of constant A_x are everywhere *tangential to*, and therefore *are* lines of magnetic induction. This demonstrates graphically how a single *scalar* function, A_x, can be used to describe two components of a magnetic induction *vector*, $\vec{B}$.

As for the CR, Parker argued that their high streaming velocity – making them immune to gravitational effects – and their extremely high effective sound speed – being so much higher than any other wave speed in the system – means that the CR pressure, p_{CR}, is constant along stream lines. Thus, p_{CR} can be described by an *advection* equation,

$$\frac{dp_{\text{CR}}}{dt} = \partial_t p_{\text{CR}} + \vec{v} \cdot \nabla p_{\text{CR}} = 0$$

$$\Rightarrow \quad \partial_t p_{\text{CR}} + v_y \partial_y p_{\text{CR}} + v_z \partial_z p_{\text{CR}} = 0, \tag{7.93}$$

for $\partial_x = 0$.

Our task, then, is to linearise the *six* equations (7.87)–(7.90), (7.92), (7.93), and whittle them all down to a single Helmholtz-like equation[38] in a single quantity – which will turn out to be the perturbation to the vector potential, $A_{x,\text{p}}$ – from which we determine the condition for instability. A tall task indeed!

We start by setting all primitive variables to the sum of their equilibrium values (with subscript '0') and a perturbation (with subscript 'p'):

$$\left. \begin{array}{llll} \rho = \rho_0 + \rho_\text{p}; & v_y = v_{y,\text{p}}; & v_z = v_{z,\text{p}}; & p = p_0 + p_\text{p}; \\ B_y = B_{y,0} + B_{y,\text{p}}; & B_z = B_{z,\text{p}}; & p_{\text{CR}} = p_{\text{CR},0} + p_{\text{CR},\text{p}}, \end{array} \right\} \tag{7.94}$$

---

[38] *Rappel*: The Helmholtz equation has the form $f''(\xi) = -q^2 f(\xi)$, where $q^2 > 0$ gives oscillatory and thus stable solutions, while $q^2 < 0$ gives exponential and thus unstable solutions.

since the equilibrium values of v_y, v_z, and B_z are zero. Recall that equilibrium quantities are functions of z only, whereas in general, perturbations are functions of y, z, and t.

Making substitutions for ρ, v_y, and v_z in Eq. (7.87), we get the linearised continuity equation,

$$\partial_t(\rho_0 + \rho_p) + \partial_y[(\rho_0 + \rho_p)v_{y,p}] + \partial_z[(\rho_0 + \rho_p)v_{z,p}] = 0$$

$$\Rightarrow \partial_t\rho_p = -\rho_0\partial_y v_{y,p} - \rho_0\partial_z v_{z,p} - v_{z,p}\frac{d\rho_0}{dz}, \qquad (7.95)$$

where all "second-order small" terms with more than one perturbation factor have been dropped [$e.g.$, $\partial_y(\rho_p v_{y,p})$],[39] as have terms with $\partial_t\rho_0$ and $\partial_y\rho_0$ since these are zero.

Before linearising the y-component of the momentum equation (Eq. 7.88), we note that the vector potential can also be written as the sum of its equilibrium value and a perturbation,

$$A_x = A_{x,0} + A_{x,p}, \qquad (7.96)$$

and, since $A_{x,0}$ describes the equilibrium magnetic induction ($B_{y,0} = \partial_z A_{x,0}$), $A_{x,p}$ must describe the perturbations to B_y and B_z, namely,

$$B_{y,p} = \partial_z A_{x,p}; \qquad B_{z,p} = -\partial_y A_{x,p}.$$

Evidently, $A_{x,0}$ is a function of z only (since $B_{z,0} = 0$, $A_{x,0}$ cannot have a y-dependence) and, like all perturbed quantities, $A_{x,p}$ is a function of y, z, and t.

With this, we make the appropriate substitutions into Eq. (7.88) to get,

$$\partial_t\big[(\rho_0 + \rho_p)v_{y,p}\big] + \partial_y\big[(\rho_0 + \rho_p)v_{y,p}v_{y,p}\big] + \partial_z\big[(\rho_0 + \rho_p)v_{y,p}v_{z,p}\big]$$
$$= -\partial_y(p_0 + p_p + p_{\mathrm{CR},0} + p_{\mathrm{CR},p}) + \frac{1}{\mu_0}\big[\partial_z(B_{y,0} + B_{y,p}) - \partial_y B_{z,p}\big]B_{z,p},$$

which mercifully reduces to,

$$\rho_0\,\partial_t v_{y,p} = -\partial_y(p_p + p_{\mathrm{CR},p}) - \frac{1}{\mu_0}\frac{dB_{y,0}}{dz}\partial_y A_{x,p}, \qquad (7.97)$$

dropping all second-order terms along with t- and y-derivatives of equilibrium quantities.

Before linearising the z-component of the momentum equation (Eq. 7.89), it is worth examining separately how its Lorentz term is linearised. To wit,

$$\frac{1}{\mu_0}(\partial_y B_z - \partial_z B_y)B_y = \frac{1}{\mu_0}\big[\partial_y B_{z,p} - \partial_z(B_{y,0} + B_{y,p})\big](B_{y,0} + B_{y,p})$$

$$= -\underbrace{\frac{B_{y,0}}{\mu_0}\frac{dB_{y,0}}{dz}}_{dp_{\mathrm{M},0}/dz} + \frac{B_{y,0}}{\mu_0}\underbrace{\big(\partial_y B_{z,p} - \partial_z B_{y,p}\big)}_{-\nabla^2 A_{x,p}} - \frac{1}{\mu_0}\frac{dB_{y,0}}{dz}B_{y,p}$$

---

[39] *Rappel*: In §2.1.1 and 5.2.2 we used ϵ as a label to keep track of "order of smallness" (*e.g.*, $\rho = \rho_0 + \epsilon\rho_p$, where formally, $\epsilon = 1$). Thus, after the algebra, any term with ϵ^2 or higher got dropped. Here, I'm omitting this crutch, relying instead on our ability to recognise second-order small terms on their own merits.

$$= -\frac{dp_{M,0}}{dz} - \frac{B_{y,0}}{\mu_0}\nabla^2 A_{x,p} - \frac{1}{\mu_0}\frac{dB_{y,0}}{dz}\partial_z A_{x,p},$$

dropping second-order terms such as $B_{y,p}\partial_y B_{z,p}/\mu_0$, etc., where $\nabla^2 A_{x,p} = \partial_{yy} A_{x,p} + \partial_{zz} A_{x,p}$, and recalling that the equilibrium magnetic pressure is given by $p_{M,0} = B_{y,0}^2/2\mu_0$ (e.g., Eq. 7.76). Using this and making the appropriate substitutions into Eq. (7.89), we drop all second-order terms and t-derivatives of equilibrium quantities to get,

$$\begin{aligned}\rho_0 \partial_t v_{z,p} = &-\frac{d}{dz}(p_0 + p_{CR,0}) - \partial_z(p_p + p_{CR,p}) - (\rho_0 + \rho_p)g_{gal} \\ &- \frac{dp_{M,0}}{dz} - \frac{B_{y,0}}{\mu_0}\nabla^2 A_{x,p} - \frac{1}{\mu_0}\frac{dB_{y,0}}{dz}\partial_z A_{x,p} \\ = &\underbrace{-\frac{d}{dz}(p_0 + p_{CR,0} + p_{M,0}) - \rho_0 g_{gal}}_{=0 \text{ (Eq. 7.80)}} - \partial_z(p_p + p_{CR,p}) - \rho_p g_{gal} \\ &- \frac{B_{y,0}}{\mu_0}\nabla^2 A_{x,p} - \frac{1}{\mu_0}\frac{dB_{y,0}}{dz}\partial_z A_{x,p},\end{aligned} \quad (7.98)$$

the linearised z-component of the momentum equation.

It is left to Problems 7.7 and 7.8 to show that linearising the equations for thermal pressure (Eq. 7.90), ideal induction (Eq. 7.92), and the CR pressure (Eq. 7.93) lead to:

$$\partial_t p_p = -v_{z,p}\frac{dp_0}{dz} - \gamma p_0(\partial_y v_{y,p} + \partial_z v_{z,p}); \quad (7.99)$$

$$\partial_t A_{x,p} = -B_{y,0}v_{z,p}; \quad (7.100)$$

$$\partial_t p_{CR,p} = -v_{z,p}\frac{dp_{CR,0}}{dz}. \quad (7.101)$$

Equations (7.95) and (7.97)–(7.101) (equivalent to Parker's Eq. III.5, III.3, III.4, III.6, III.2, and III.7 respectively[40]) are the six linearised equations we must combine to deliver a single evolution equation in $A_{x,p}$. And this is where the real fun begins...

The first and most trivial step is to solve Eq. (7.100) for $v_{z,p}$:

$$v_{z,p} = -\frac{1}{B_{y,0}}\partial_t A_{x,p}, \quad (7.102)$$

which will be an important key in unlocking the hidden dependence of all the linearised MHD/CR equations on $A_{x,p}$. For starters, substituting Eq. (7.102) into Eq. (7.101) along with various other machinations (Problem 7.8) yields,

$$p_{CR,p} = \frac{\sigma_{CR} v_{rms}^2}{B_{y,0}}\frac{d\rho_0}{dz}A_{x,p}. \quad (7.103)$$

Looking to the y-component of the momentum equation next, set $\rho_0 = p_0/v_{rms}^2$

---

[40] All equations in this subsection referred to as "Parker's Eq." come from Parker (1966).

(Eq. 7.77) in Eq. (7.97), then differentiate with respect to time to get,

$$\frac{p_0}{v_{\rm rms}^2}\partial_{tt}v_{y,\rm p} = -\partial_{yt}p_{\rm p} - \partial_{yt}p_{\rm CR,p} - \frac{1}{B_{y,0}}\underbrace{\frac{B_{y,0}}{\mu_0}\frac{dB_{y,0}}{dz}}_{dp_{\rm M,0}/dz}\partial_{yt}A_{x,\rm p}$$

$$= \partial_y\left(v_{z,\rm p}\frac{dp_0}{dz} + \gamma p_0\,\partial_y v_{y,\rm p} + \gamma p_0\,\partial_z v_{z,\rm p}\right.$$
$$\left. + v_{z,\rm p}\frac{dp_{\rm CR,0}}{dz} - \frac{1}{B_{y,0}}\frac{dp_{\rm M,0}}{dz}\partial_t A_{x,\rm p}\right),$$

using Eq. (7.99) for $\partial_t p_{\rm p}$ and Eq. (7.101) for $\partial_t p_{\rm CR,p}$.

Next, multiply through by $v_{\rm rms}^2/p_0$, use Eq. (7.102) to eliminate $v_{z,\rm p}$, note that $p_{\rm CR,0} = \sigma_{\rm CR}p_0$ and $p_{\rm M,0} = \sigma_{\rm M}p_0$ (Eq. 7.76), and move the term $\gamma v_{\rm rms}^2\partial_{yy}v_{y,\rm p}$ from the RHS to the LHS to get,

$$\underbrace{\partial_{tt}v_{y,\rm p} - \gamma v_{\rm rms}^2 \partial_{yy}v_{y,\rm p}}_{\equiv \mathcal{L}v_{y,\rm p}} = -\gamma v_{\rm rms}^2 \partial_z\left(\frac{1}{B_{y,0}}\partial_{yt}A_{x,\rm p}\right)$$
$$- \frac{1}{B_{y,0}}\left(\frac{v_{\rm rms}^2}{p_0}\frac{dp_0}{dz} + \frac{v_{\rm rms}^2\sigma_{\rm CR}}{p_0}\frac{dp_0}{dz} + \frac{v_{\rm rms}^2\sigma_{\rm M}}{p_0}\frac{dp_0}{dz}\right)\partial_{yt}A_{x,\rm p}$$

$$\Rightarrow \quad \mathcal{L}v_{y,\rm p} = \gamma\frac{v_{\rm rms}^2}{B_{y,0}^2}\frac{dB_{y,0}}{dz}\partial_{yt}A_{x,\rm p} - \gamma\frac{v_{\rm rms}^2}{B_{y,0}}\partial_z\partial_{yt}A_{x,\rm p}$$
$$- \frac{v_{\rm rms}^2}{B_{y,0}}\frac{1}{p_0}\frac{dp_0}{dz}(1 + \sigma_{\rm CR} + \sigma_{\rm M})\partial_{yt}A_{x,\rm p},$$

where the operator $\mathcal{L}$ is defined for convenience, and where the product rule is applied to the first term on the RHS differentiated with respect to z.

I pause to caution the reader that the abbreviated Leibniz notation, adopted in Chap. 5, can conceal the complexity of the mathematics. For example, never forget that the construct,

$$\partial_z\partial_{yt}A_{x,\rm p} = \frac{\partial}{\partial z}\frac{\partial}{\partial t}\frac{\partial A_{x,\rm p}}{\partial y},$$

is actually a *triple derivative* which could, in principle, be written as $\partial_{ytz}A_{x,\rm p}$ and – wait for it – fourth derivatives are on their way! So while things may seem to be getting complicated, you ain't seen nothin' yet!

Carrying on from the last expression for $\mathcal{L}v_{y,\rm p}$, we write,

$$\mathcal{L}v_{y,\rm p} = \frac{v_{\rm rms}^2}{B_{y,0}}\left(\gamma\underbrace{\frac{1}{B_{y,0}}\frac{dB_{y,0}}{dz}}_{-1/2L} - \underbrace{\frac{1}{p_0}\frac{dp_0}{dz}}_{-1/L}(1 + \sigma_{\rm CR} + \sigma_{\rm M}) - \gamma\partial_z\right)\partial_{yt}A_{x,\rm p},$$

where the underbraces come from Eq. (7.82) defining the scale height, L, and the "term" $\gamma\partial_z$ is treated as an operator. Thus, the time derivative of Eq. (7.97) whittles down to,

$$\boxed{\mathcal{L}v_{y,\rm p} = \frac{v_{\rm rms}^2}{B_{y,0}}(D - \gamma\partial_z)\partial_{yt}A_{x,\rm p},} \qquad (7.104)$$

where,
$$D \equiv \frac{1+\sigma_{\mathrm{M}}+\sigma_{\mathrm{CR}}-\gamma/2}{L}, \qquad (7.105)$$
is defined for convenience. Eq. (7.104) is equivalent to Parker's Eq. (III.9).

Next, I'll let the reader verify in Problem 7.9 that the continuity and pressure equations (Eq. 7.95 and 7.99) can be written as:

$$\boxed{\frac{\mathcal{L}\rho_{\mathrm{p}}}{\rho_0} = -\frac{v_{\mathrm{rms}}^2}{B_{y,0}}(D-\gamma\partial_z)\partial_{yy}A_{x,\mathrm{p}} - \frac{1}{B_{y,0}}\left(\frac{1}{2L}-\partial_z\right)\mathcal{L}A_{x,\mathrm{p}};} \quad \text{and} \quad (7.106)$$

$$\boxed{\frac{\mathcal{L}p_{\mathrm{p}}}{p_0} = -\frac{\gamma v_{\mathrm{rms}}^2}{B_{y,0}}(D-\gamma\partial_z)\partial_{yy}A_{x,\mathrm{p}} - \frac{1}{B_{y,0}}\left(\frac{1-\gamma/2}{L}-\gamma\partial_z\right)\mathcal{L}A_{x,\mathrm{p}},} \quad (7.107)$$

which are equivalent to Parker's Eq. (III.10) and (III.11).

We now tackle the big one, the z-momentum equation. Divide Eq. (7.98) through by ρ_0 (p_0/v_{rms}^2 where convenient), and use Eq. (7.102) to eliminate $v_{z,\mathrm{p}}$ in favour of $A_{x,\mathrm{p}}$ to get,

$$\frac{1}{B_{y,0}}\partial_{tt}A_{x,\mathrm{p}} = \frac{v_{\mathrm{rms}}^2}{p_0}\partial_z p_{\mathrm{p}} + \frac{1}{\rho_0}\partial_z p_{\mathrm{CR,p}}$$
$$+ g_{\mathrm{gal}}\frac{\rho_{\mathrm{p}}}{\rho_0} + \underbrace{\frac{B_{y,0}^2}{2\mu_0}}_{p_{\mathrm{M},0}}\frac{2}{\rho_0 B_{y,0}}\nabla^2 A_{x,\mathrm{p}} + \frac{1}{\rho_0 B_{y,0}}\underbrace{\frac{B_{y,0}}{\mu_0}\frac{dB_{y,0}}{dz}}_{dp_{\mathrm{M},0}/dz}\partial_z A_{x,\mathrm{p}}. \qquad (7.108)$$

Now, from the product rule,
$$\partial_z\left(\frac{p_{\mathrm{p}}}{p_0}\right) = \underbrace{-\frac{1}{p_0^2}\frac{dp_0}{dz}}_{-1/p_0 L}p_{\mathrm{p}} + \frac{1}{p_0}\partial_z p_{\mathrm{p}}$$
$$\Rightarrow \quad \frac{1}{p_0}\partial_z p_{\mathrm{p}} = \partial_z\left(\frac{p_{\mathrm{p}}}{p_0}\right) - \frac{1}{L}\frac{p_{\mathrm{p}}}{p_0} = -\left(\frac{1}{L}-\partial_z\right)\frac{p_{\mathrm{p}}}{p_0}.$$

Substituting this and Eq. (7.103) into Eq. (7.108), and then applying the operator $\mathcal{L}$ to the result, we get,

$$\frac{1}{B_{y,0}}\partial_{tt}\mathcal{L}A_{x,\mathrm{p}} = \underbrace{-v_{\mathrm{rms}}^2\left(\frac{1}{L}-\partial_z\right)\mathcal{L}\frac{p_{\mathrm{p}}}{p_0}}_{\equiv q_1} + \underbrace{\frac{1}{\rho_0}\partial_z\left(\frac{v_{\mathrm{rms}}^2\sigma_{\mathrm{CR}}}{B_{y,0}}\frac{d\rho_0}{dz}\mathcal{L}A_{x,\mathrm{p}}\right)}_{\equiv q_2}$$
$$+ g_{\mathrm{gal}}\mathcal{L}\frac{\rho_{\mathrm{p}}}{\rho_0} + \frac{2p_{\mathrm{M},0}}{\rho_0 B_{y,0}}\nabla^2(\mathcal{L}A_{x,\mathrm{p}}) + \frac{1}{\rho_0 B_{y,0}}\frac{dp_{\mathrm{M},0}}{dz}\partial_z(\mathcal{L}A_{x,\mathrm{p}}). \qquad (7.109)$$

We proceed by examining the terms q_1 and q_2 in Eq. (7.109) separately. Starting with q_1, substitute in Eq. (7.107) to get,

$$q_1 = v_{\mathrm{rms}}^2\left(\frac{1}{L}-\partial_z\right)\left[\frac{\gamma v_{\mathrm{rms}}^2}{B_{y,0}}(D-\gamma\partial_z)\partial_{yy}A_{x,\mathrm{p}} + \frac{1}{B_{y,0}}\left(\frac{1-\gamma/2}{L}-\gamma\partial_z\right)\mathcal{L}A_{x,\mathrm{p}}\right]$$

$$= \frac{\gamma v_{\rm rms}^4}{B_{y,0}} \left(\frac{1}{L}(D - \gamma\partial_z) + \underbrace{\frac{1}{B_{y,0}} \frac{dB_{y,0}}{dz}}_{-1/2L}(D - \gamma\partial_z) - \partial_z(D - \gamma\partial_z) \right) \partial_{yy} A_{x,\rm p}$$

$$+ \frac{v_{\rm rms}^2}{B_{y,0}} \left[\frac{1}{L}\left(\frac{1-\gamma/2}{L} - \gamma\partial_z \right) + \underbrace{\frac{1}{B_{y,0}} \frac{dB_{y,0}}{dz}}_{-1/2L}\left(\frac{1-\gamma/2}{L} - \gamma\partial_z \right) \right.$$
$$\left. - \partial_z\left(\frac{1-\gamma/2}{L} - \gamma\partial_z \right) \right] \mathcal{L} A_{x,\rm p}$$

$$\Rightarrow \quad q_1 = \frac{\gamma v_{\rm rms}^4}{B_{y,0}} \left(\frac{1}{2L} - \partial_z \right)(D - \gamma\partial_z) \partial_{yy} A_{x,\rm p}$$
$$+ \frac{v_{\rm rms}^2}{B_{y,0}} \left(\frac{1}{2L} - \partial_z \right)\left(\frac{1-\gamma/2}{L} - \gamma\partial_z \right) \mathcal{L} A_{x,\rm p}. \tag{7.110}$$

Next, for q_2 we use the product rule for ∂_z to get three terms,

$$q_2 = -\frac{v_{\rm rms}^2 \sigma_{\rm CR}}{B_{y,0}} \underbrace{\frac{1}{B_{y,0}} \frac{dB_{y,0}}{dz}}_{-1/2L} \underbrace{\frac{1}{\rho_0} \frac{d\rho_0}{dz}}_{-1/L} \mathcal{L} A_{x,\rm p} + \frac{v_{\rm rms}^2 \sigma_{\rm CR}}{B_{y,0}} \frac{1}{\rho_0} \partial_z\left(\frac{d\rho_0}{dz} \right) \mathcal{L} A_{x,\rm p}$$
$$+ \frac{v_{\rm rms}^2 \sigma_{\rm CR}}{B_{y,0}} \underbrace{\frac{1}{\rho_0} \frac{d\rho_0}{dz}}_{-1/L} \partial_z(\mathcal{L} A_{x,\rm p}).$$

But,

$$\partial_z \underbrace{\left(\frac{1}{\rho_0} \frac{d\rho_0}{dz} \right)}_{-1/L} = 0 = -\underbrace{\frac{1}{\rho_0^2} \frac{d\rho_0}{dz} \frac{d\rho_0}{dz}}_{1/L^2} + \frac{1}{\rho_0} \partial_z\left(\frac{d\rho_0}{dz} \right) \quad \Rightarrow \quad \frac{1}{\rho_0} \partial_z\left(\frac{d\rho_0}{dz} \right) = \frac{1}{L^2}{}^{41}$$

$$\Rightarrow \quad q_2 = \frac{v_{\rm rms}^2 \sigma_{\rm CR}}{L B_{y,0}} \left(-\frac{1}{2L} + \frac{1}{L} - \partial_z \right) \mathcal{L} A_{x,\rm p} = \frac{v_{\rm rms}^2 \sigma_{\rm CR}}{L B_{y,0}} \left(\frac{1}{2L} - \partial_z \right) \mathcal{L} A_{x,\rm p}. \tag{7.111}$$

Substituting Eq. (7.110), (7.111), and (7.106) into Eq. (7.109) and noting that $p_{\rm M,0}/\rho_0 = v_{\rm rms}^2 \sigma_{\rm M}$ (Eq. 7.76 and 7.77), we get (yikes!),

$$\frac{1}{B_{y,0}} \partial_{tt} \mathcal{L} A_{x,\rm p} = \frac{\gamma v_{\rm rms}^4}{B_{y,0}} \left(\frac{1}{2L} - \partial_z \right)(D - \gamma\partial_z) \partial_{yy} A_{x,\rm p}$$
$$+ \frac{v_{\rm rms}^2}{B_{y,0}} \left(\frac{1}{2L} - \partial_z \right)\left(\frac{1-\gamma/2}{L} - \gamma\partial_z \right) \mathcal{L} A_{x,\rm p}$$
$$+ \frac{v_{\rm rms}^2 \sigma_{\rm CR}}{L B_{y,0}} \left(\frac{1}{2L} - \partial_z \right) \mathcal{L} A_{x,\rm p} - \frac{g_{\rm gal} v_{\rm rms}^2}{B_{y,0}} (D - \gamma\partial_z) \partial_{yy} A_{x,\rm p}$$
$$- \frac{g_{\rm gal}}{B_{y,0}} \left(\frac{1}{2L} - \partial_z \right) \mathcal{L} A_{x,\rm p} + \frac{2 v_{\rm rms}^2 \sigma_{\rm M}}{B_{y,0}} \nabla^2(\mathcal{L} A_{x,\rm p})$$
$$+ \frac{v_{\rm rms}^2 \sigma_{\rm M}}{B_{y,0}} \underbrace{\frac{1}{p_{\rm M,0}} \frac{dp_{\rm M,0}}{dz}}_{-1/L} \partial_z(\mathcal{L} A_{x,\rm p}).$$

---

[41] In hindsight, this result makes sense. If the first derivative of ρ_0 divided by ρ_0 gives the inverse scale height, $1/L$, it seems reasonable that the second derivative divided by ρ_0 should give $1/L^2$.

Multiplying through by $B_{y,0}$ and gathering like terms, we get,

$$\partial_{tt}\mathcal{L}A_{x,\text{p}} = 2v_{\text{rms}}^2\sigma_M\nabla^2(\mathcal{L}A_{x,\text{p}}) + v_{\text{rms}}^4\underbrace{\left(\frac{\gamma}{2L} - \frac{g_{\text{gal}}}{v_{\text{rms}}^2} - \gamma\partial_z\right)}_{-D\ (\text{Eq. 7.81, 7.105})}(D - \gamma\partial_z)\partial_{yy}A_{x,\text{p}}$$

$$+ v_{\text{rms}}^2\underbrace{\left(\frac{\sigma_{\text{CR}}}{L} - \frac{g_{\text{gal}}}{v_{\text{rms}}^2} + \frac{1-\gamma/2}{L} - \gamma\partial_z\right)}_{-(\sigma_M+\gamma/2)/L}\left(\frac{1}{2L} - \partial_z\right)\mathcal{L}A_{x,\text{p}} - \frac{v_{\text{rms}}^2\sigma_M}{L}\partial_z(\mathcal{L}A_{x,\text{p}})$$

$$= 2v_{\text{rms}}^2\sigma_M\nabla^2(\mathcal{L}A_{x,\text{p}}) - v_{\text{rms}}^4(D^2 - \gamma^2\partial_{zz})\partial_{yy}A_{x,\text{p}} - \cancel{\frac{v_{\text{rms}}^2\sigma_M}{L}\partial_z(\mathcal{L}A_{x,\text{p}})}$$

$$- \frac{v_{\text{rms}}^2}{2L^2}\left(\sigma_M + \frac{\gamma}{2} + \cancel{L\gamma\partial_z}\right)\mathcal{L}A_{x,\text{p}} + \frac{v_{\text{rms}}^2}{L}\left(\cancel{\sigma_M} + \cancel{\frac{\gamma}{2}} + L\gamma\partial_z\right)\partial_z(\mathcal{L}A_{x,\text{p}}),$$

and all remaining single z-derivatives magically cancel out! Continuing, we substitute $\mathcal{L} = \partial_{tt} - \gamma v_{\text{rms}}^2\partial_{yy}$ in the last two terms to get,

$$\partial_{tt}\mathcal{L}A_{x,\text{p}} = 2v_{\text{rms}}^2\sigma_M\nabla^2(\mathcal{L}A_{x,\text{p}}) - \frac{v_{\text{rms}}^2}{2L^2}\left(\sigma_M + \frac{\gamma}{2}\right)\partial_{tt}A_{x,\text{p}} + \gamma v_{\text{rms}}^2\partial_{zz}(\partial_{tt}A_{x,\text{p}})$$

$$- v_{\text{rms}}^4(D^2 - \cancel{\gamma^2\partial_{zz}})\partial_{yy}A_{x,\text{p}} + \frac{\gamma v_{\text{rms}}^4}{2L^2}\left(\sigma_M + \frac{\gamma}{2}\right)\partial_{yy}A_{x,\text{p}}$$

$$- \cancel{\gamma^2 v_{\text{rms}}^4\partial_{zz}(\partial_{yy}A_{x,\text{p}})}$$

$$= 2v_{\text{rms}}^2\sigma_M\nabla^2(\mathcal{L}A_{x,\text{p}}) - v_{\text{rms}}^2\left(\frac{1}{4L^2}(\gamma + 2\sigma_M) - \gamma\partial_{zz}\right)\partial_{tt}A_{x,\text{p}}$$

$$- \frac{v_{\text{rms}}^4}{L^2}\underbrace{\left[\left(1 + \sigma_M + \sigma_{\text{CR}} - \frac{\gamma}{2}\right)^2 - \frac{\gamma}{2}\left(\sigma_M + \frac{\gamma}{2}\right)\right]}_{C\ (\text{Eq. 7.84})}\partial_{yy}A_{x,\text{p}},$$

using Eq. (7.105) to replace D^2. Thus, we have finally,

$$\boxed{\partial_{tt}\mathcal{L}A_{x,\text{p}} = 2v_{\text{rms}}^2\sigma_M\nabla^2(\mathcal{L}A_{x,\text{p}}) - v_{\text{rms}}^2\left(\frac{1}{4L^2}(\gamma + 2\sigma_M) - \gamma\partial_{zz}\right)\partial_{tt}A_{x,\text{p}} - \frac{Cv_{\text{rms}}^4}{L^2}\partial_{yy}A_{x,\text{p}}.} \quad (7.112)$$

This is the single dynamical equation in $A_{x,\text{p}}$ we've been seeking, and is equivalent to, but in a different arrangement from, Parker's Eq. (III.12).

Still with me? Good, *because we're still nowhere near finished*!!

Given that Eq. (7.112) is in terms of second (and fourth!!) derivatives in time and space, it is a wave equation (of sorts), and we can try a normal mode solution for $A_{x,\text{p}}$, namely,

$$A_{x,\text{p}} = f(\xi)e^{i(ky-\omega t)} = f(\xi)e^{\chi t}e^{iky}. \quad (7.113)$$

Here, $f(\xi)$ is the amplitude of the perturbation to the vector potential (lines of induction) as a function of the unitless vertical coordinate, $\xi \equiv kz$. The y-dependence of the perturbation is a sinusoid with wavelength λ and wavenumber $k = 2\pi/\lambda$. Stability of the system is dictated by the nature of the angular frequency, ω. If $\omega \in \mathbb{R}$, the temporal response is oscillatory and the system is stable. Conversely, if $\omega = i\chi$ with the growth rate $\chi \in \mathbb{R}$, the temporal response is exponential, and the system is unstable. This is a somewhat different approach than we took for the previous three instabilities, where we used the normal mode solution to find the dispersion relation, $\omega(k)$, and from that the criterion for instability. Here, it's a little more convenient to "build in" the instability by setting $\omega = i\chi$, and then let the mathematics tell us under what conditions $\chi \in \mathbb{R}$.

In preparing to substitute Eq. (7.113) into Eq. (7.112), we first evaluate the derivatives:

$$\left.\begin{aligned}
\partial_{tt} A_{x,\mathrm{p}} &= \chi^2 A_{x,\mathrm{p}}; \\
\partial_{yy} A_{x,\mathrm{p}} &= -k^2 A_{x,\mathrm{p}}; \\
\mathcal{L} A_{x,\mathrm{p}} &= (\partial_{tt} - \gamma v_{\mathrm{rms}}^2 \partial_{yy}) A_{x,\mathrm{p}} = \left(\chi^2 + \gamma v_{\mathrm{rms}}^2 k^2\right) A_{x,\mathrm{p}}; \\
\partial_{tt}(\mathcal{L} A_{x,\mathrm{p}}) &= \chi^2 \left(\chi^2 + \gamma v_{\mathrm{rms}}^2 k^2\right) A_{x,\mathrm{p}}; \\
\partial_{zz}\partial_{tt} A_{x,\mathrm{p}} &= \chi^2 \frac{k^2}{f} f''(\xi) A_{x,\mathrm{p}}; \\
\nabla^2(\mathcal{L} A_{x,\mathrm{p}}) &= \left(-k^2 + \frac{k^2}{f} f''(\xi)\right)\left(\chi^2 + \gamma v_{\mathrm{rms}}^2 k^2\right) A_{x,\mathrm{p}}.
\end{aligned}\right\} \quad (7.114)$$

Then, substituting every one of Eq. (7.114) into Eq. (7.112) and dividing out $A_{x,\mathrm{p}}$, we get,

$$\chi^2\left(\chi^2 + \gamma v_{\mathrm{rms}}^2 k^2\right) = 2v_{\mathrm{rms}}^2 \sigma_{\mathrm{M}}\left(-k^2 + \frac{k^2}{f} f''(\xi)\right)\left(\chi^2 + \gamma v_{\mathrm{rms}}^2 k^2\right)$$
$$- \frac{v_{\mathrm{rms}}^2}{4L^2}(\gamma + 2\sigma_{\mathrm{M}})\chi^2 + \gamma v_{\mathrm{rms}}^2 \chi^2 \frac{k^2}{f} f''(\xi) + \frac{C v_{\mathrm{rms}}^4}{L^2} k^2,$$

where C is given by Eq. (7.84). Next, dividing through by $v_{\mathrm{rms}}^4 k^4$ and then defining the unitless quantity $\zeta = \chi/(v_{\mathrm{rms}} k)$, we get,

$$\zeta^2(\zeta^2 + \gamma) = 2\sigma_{\mathrm{M}}\left(\frac{f''}{f} - 1\right)(\zeta^2 + \gamma) - \frac{\zeta^2}{4k^2 L^2}(\gamma + 2\sigma_{\mathrm{M}}) + \gamma \zeta^2 \frac{f''}{f} + \frac{C}{k^2 L^2}$$

$$\Rightarrow \quad \left(\zeta^2(\gamma + 2\sigma_{\mathrm{M}}) + 2\gamma\sigma_{\mathrm{M}}\right) f'' = \left(\zeta^2(\gamma + 2\sigma_{\mathrm{M}}) + 2\gamma\sigma_{\mathrm{M}}\right. \quad (7.115)$$
$$\left. + \zeta^4 + \frac{\zeta^2}{4k^2 L^2}(\gamma + 2\sigma_{\mathrm{M}}) - \frac{C}{k^2 L^2}\right) f,$$

after a little algebra. Remember, $f = f(\xi)$ is the amplitude of the perturbation applied to $A_{x,\mathrm{p}}$ at $t = 0$ as a function of the vertical coordinate, $\xi = kz$. And so for illustration, if we let,

$$P = \zeta^2(\gamma + 2\sigma_{\mathrm{M}}) + 2\gamma\sigma_{\mathrm{M}} \quad \text{and} \quad Q = \zeta^4 + \frac{\zeta^2}{4k^2 L^2}(\gamma + 2\sigma_{\mathrm{M}}) - \frac{C}{k^2 L^2},$$

then Eq. (7.115) reduces to the deceivingly simple form,

$$f''(\xi) = \frac{P+Q}{P} f(\xi).$$

Evidently, if $(P+Q)/P > 0$, then $f(\xi)$ is an exponential function of ξ whereas if $(P+Q)/P < 0$, $f(\xi)$ is oscillatory. However, an exponential perturbation would mean $f(\xi) \to \infty$ as $\xi \to \infty$ (a rather impractical requirement!), and so we choose to impose a sinusoidal perturbation, thus forcing,

$$\frac{P+Q}{P} \equiv -K^2 < 0 \quad \text{so that} \quad f(\xi) = E \sin(K\xi) + F \cos(K\xi), \tag{7.116}$$

where K is the *vertical* wavenumber (in the z-direction), as opposed to k which is the *horizontal* wavenumber, and where E and F are constants of integration set by boundary conditions. To that point, if we require the perturbation to vanish at the midplane ($\xi = kz = 0$), then $F = 0$ and Eq. (7.113) becomes,

$$A_{x,\mathrm{p}}(y, z, t) = E \sin(K\xi) e^{\chi t} \cos(ky),$$

where E is an arbitrary small amplitude (to keep things linear) and where we've retained only the real part of e^{iky}.

We are finally in a position to uncover Parker's criterion for instability. For an unstable solution, the growth rate $\chi \in \mathbb{R} \Rightarrow \chi^2 > 0$. Thus, $\zeta^2 = \chi^2/(v_{\mathrm{rms}}k)^2 > 0$ which means $P > 0$ and the requirement that $f(\xi)$ be oscillatory (*cf.* Eq. 7.116) can be realised only if,

$$Q < -P \quad \Rightarrow \quad \zeta^4 + \frac{\zeta^2}{4k^2L^2}(\gamma + 2\sigma_{\mathrm{M}}) - \frac{C}{k^2L^2} < -\zeta^2(\gamma + 2\sigma_{\mathrm{M}}) - 2\gamma\sigma_{\mathrm{M}}$$

$$\Rightarrow \quad \zeta^4 + \zeta^2(\gamma + 2\sigma_{\mathrm{M}})\left(1 + \frac{1}{4k^2L^2}\right) + 2\gamma\sigma_{\mathrm{M}} - \frac{C}{k^2L^2} < 0, \tag{7.117}$$

which is Parker's Eq. (III.15). Note that in the case where $K \to 0$ [long vertical wavelength for $f(\xi)$], Ineq. (7.117) becomes an equality; later, we shall impose this limit in finding the growth rate for the fastest growing perturbation.

Aside: Interpreting a quadratic inequality

Inequality (7.117) has the form,

$$y(x) = ax^2 + bx + c < 0,$$

a quadratic inequality in $x = \zeta^2$ where $a > 0$, $b > 0$, and where the sign of c is to be determined.

Such a parabola opens upwards (since $a > 0$) and has one extremum (minimum) at,

$$(x_{\mathrm{ex}}, y_{\mathrm{ex}}) = \left(-\frac{b}{2a}, c - \frac{b^2}{4a}\right),$$

as shown in the inset. Since $b > 0$, the minimum is to the left of the y-axis and is below the x-axis provided,

$$c - \frac{b^2}{4a} < 0 \quad \Rightarrow \quad b^2 > 4ac.$$

This is nothing more than the requirement that the discriminant $b^2 - 4ac$ from the quadratic formula,

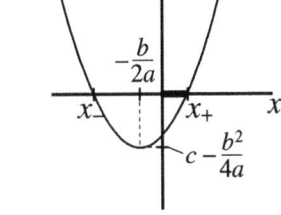

$$x_\pm = \frac{-b \pm \sqrt{b^2 - 4ac}}{2a},$$

be positive for there to be two real roots, $x_\pm$, also indicated in the inset. Thus, for $y(x) < 0$, x must lie between the two roots: $x_- < x < x_+$. Further, if x is positive definite (as is the case for ζ^2 in Ineq. 7.117), then,

$$y(x) < 0 \quad \Rightarrow \quad 0 < x < x_+,$$

(heavy portion of the x-axis in the inset) and, in particular, $x_+ > 0$. That is,

$$x_+ = \frac{-b + \sqrt{b^2 - 4ac}}{2a} > 0 \quad \Rightarrow \quad b^2 - 4ac > b^2 \quad (\text{for } a > 0 \text{ and } b > 0)$$

$$\Rightarrow \quad \boxed{c < 0.} \tag{7.118}$$

Therefore, from Ineq. (7.118), Ineq. (7.117) is consistent with $\zeta^2 > 0$ (which it must be for instability) provided,

$$c = 2\gamma\sigma_\mathrm{M} - \frac{C}{k^2 L^2} < 0 \quad \Rightarrow \quad C > 2\gamma\sigma_\mathrm{M} k^2 L^2$$

$$\Rightarrow \quad \boxed{\lambda = \frac{2\pi}{k} > 2\pi L \sqrt{\frac{2\gamma\sigma_\mathrm{M}}{C}}.}$$

This is Parker's criterion for instability (his Eq. III.16 where his Y is my C) where the RHS represents the *minimum* wavelength for an unstable perturbation.

To find the fastest growing wavelength and its e-folding time, we rearrange Ineq. (7.117) slightly to make explicit its dependence on the horizontal wavenumber, k, or more conveniently, the unitless wavenumber $\kappa = kL$. Defining $n = L\chi/v_\mathrm{rms}$ as the unitless growth rate of the instability, we have $\zeta = \chi/(v_\mathrm{rms} k) = n/\kappa$ and Ineq. (7.117) becomes,

$$h(\kappa^2) \equiv 2\gamma\sigma_\mathrm{M}\kappa^4 + \left(n^2(\gamma + 2\sigma_\mathrm{M}) - C\right)\kappa^2 + \frac{\gamma + 2\sigma_\mathrm{M}}{4}n^2 + n^4 < 0, \tag{7.119}$$

after multiplying through by κ^4 and doing a little algebra. Since $h(\kappa^2)$ is a quadratic in κ^2, its minimum value (maximum instability) occurs at (see the *Aside* above),

$$\kappa_\mathrm{f}^2 = -\frac{b}{2a} = \frac{C - n_\mathrm{f}^2(\gamma + 2\sigma_\mathrm{M})}{4\gamma\sigma_\mathrm{M}} > 0, \tag{7.120}$$

where κ_f is the fastest growing wavenumber, and where n_f is the number of growth times of the fastest growing perturbation within a scale height, L. Further,

$$h(\kappa_\mathrm{f}^2) \equiv h_\mathrm{min} = c - \frac{b^2}{4a} = \frac{\gamma + 2\sigma_\mathrm{M}}{4}n_\mathrm{f}^2 + n_\mathrm{f}^4 - \frac{\left(C - n_\mathrm{f}^2(\gamma + 2\sigma_\mathrm{M})\right)^2}{8\gamma\sigma_\mathrm{M}} < 0$$

$$\Rightarrow (\gamma - 2\sigma_M)^2 n_f^4 - 2(\gamma + 2\sigma_M)(C + \gamma\sigma_M)n_f^2 + C^2 < 0, \quad (7.121)$$

after multiplying through by $8\gamma\sigma_M$ and a little more algebra. Again, we are left with a quadratic inequality, this time in n_f^2. To find an explicit expression for n_f^2, impose somewhat arbitrarily the limit of long vertical wavelength ($K \to 0$) in which case Ineq. (7.121) becomes an equality with two real and positive roots,

$$n_\pm^2 = \frac{(\gamma + 2\sigma_M)(C + \gamma\sigma_M) \pm \sqrt{(\gamma + 2\sigma_M)^2(C + \gamma\sigma_M)^2 - (\gamma - 2\sigma_M)^2 C^2}}{(\gamma - 2\sigma_M)^2}.$$

It's fairly easy to see that the "plus root", n_+^2, is inadmissible. For if $n_f^2 = n_+^2$, then,

$$C - n_f^2(\gamma + 2\sigma_M) = C - \underbrace{\frac{(\gamma + 2\sigma_M)^2}{(\gamma - 2\sigma_M)^2}}_{>1} \underbrace{(C + \gamma\sigma_M)}_{>C} + \underbrace{\frac{(\gamma + 2\sigma_M)}{(\gamma - 2\sigma_M)^2}\sqrt{\sim}}_{>0} < 0,$$

contradicting Eq. (7.120). Thus,

$$n_f^2 = n_-^2 = \frac{-b - \sqrt{b^2 - 4ac}}{2a},$$

where now:

$$a = (\gamma - 2\sigma_M)^2; \quad b = -2(\gamma + 2\sigma_M)(C + \gamma\sigma_M); \quad c = C^2.$$

Now, in the event $b^2 \gg 4ac$ which, using Parker's values ($b^2 \sim 2.56$, $4ac \sim 0.16$) is a decent approximation to make, we have,

$$n_f^2 = \frac{-b - |b|(1 - 4ac/b^2)^{1/2}}{2a} \approx \frac{-b - |b| + |b|2ac/b^2}{2a} = \frac{c}{|b|},$$

since $b < 0$, and retaining just the first two terms of the binomial expansion. Thus, Parker's estimate for the fastest (unitless) growth rate is,

$$\boxed{n_f \approx \sqrt{\frac{c}{|b|}} = \frac{C}{\sqrt{2(\gamma + 2\sigma_M)(C + \gamma\sigma_M)}}.} \quad (7.122)$$

Setting $\tau_P = L/(v_{\rm rms} n_f)$ recovers my Eq. (7.85) for Parker's e-folding time for the fastest growing instability.

Parker points out that since n_f^2 is evaluated for $K \to 0$ (what he refers to as "marginal stability" in the vertical direction), it shouldn't be taken too seriously. Still, it seems this is the best estimate one can make for the e-folding time scale of the Parker instability in the galactic environment and, as pointed out in the previous subsection (§7.4.1), it is by orders of magnitude well within the lifetime of the galaxy.

Last thing to derive is the wavelength of the fastest growing perturbation, λ_f, with e-folding time, τ_P. Substituting Eq. (7.122) into Eq. (7.120) yields,

$$\kappa_f^2 = \frac{1}{4\gamma\sigma_M}\left(C - \frac{C^2}{2(C + \gamma\sigma_M)}\right) = \frac{C}{8\gamma\sigma_M}\frac{C + 2\gamma\sigma_M}{C + \gamma\sigma_M}.$$

Since $\kappa_f = k_f L = 2\pi L/\lambda_f$, we find,

$$\lambda_f = 4\pi L \sqrt{\frac{2\gamma\sigma_M}{C}\frac{C+\gamma\sigma_M}{C+2\gamma\sigma_M}} = 2\lambda_{\min}\sqrt{\frac{C+\gamma\sigma_M}{C+2\gamma\sigma_M}}, \qquad (7.123)$$

using Eq. (7.86) for $\lambda_{\min}$, the minimum wavelength for instability. For Parker's numbers ($C = 0.5$, $\gamma = 1$, $\sigma_M = 0.3$), the fastest growing wavelength is therefore about 1.7 times $\lambda_{\min}$ which we found to be $\sim 620\,\mathrm{pc}$ in §7.4.1. Thus, λ_f is a little over a kpc, corresponding nicely with the then-observed separation of ISM clumps.

In closing out this section, I'll make two quantitatively based comments. First, how well do Parker's equations hold up to "modern-day" observations of the galaxy? This is complicated by the fact that observations in the 21st century are *so* good, that it becomes difficult to specify "typical values". For instance, magnetic induction strengths across the galaxy are known to vary considerably, from $< 10^{-10}\,\mathrm{T}$ in the outer arms of the galaxy to $6 \times 10^{-10}\,\mathrm{T}$, in the solar neighbourhood, to $> 4 \times 10^{-9}\,\mathrm{T}$ closer to the galactic core and much higher still in localised spots. Similarly, values for σ_M and σ_{CR} vary widely across the galaxy and what values one chooses can depend very much on what reference you consult.

So I'll bunt a little. Theorists have good reason to believe the principle of equipartition among the thermal, magnetic, and CR energy densities should prevail over much of the galaxy, and so we'll use rather different values than Parker did: $\sigma_M \sim \sigma_{CR} \sim 1$. Other more modern numbers include (see footnotes and references therein in §7.4.1):

$$\langle B \rangle \sim 2.0 \times 10^{-10}\,\mathrm{T}; \qquad g_{\mathrm{gal}} \sim 7.44 \times 10^{-11}\,\mathrm{m\,s^{-2}}; \qquad v_{\mathrm{rms}} \sim 10^4\,\mathrm{m\,s^{-1}}.$$

The only number we'll use that Parker did is $\gamma = 1$, assuming the thermal component of the ISM is isothermal. Thus, from Eq. (7.81), $L \sim 130\,\mathrm{pc}$ (*cf.* Parker's 90 pc) and from Eq. (7.84), $C \sim 5.5$ (*cf.* Parker's 0.5). Therefore, from Eq. (7.86), (7.85), and (7.123), we find:

$$\lambda_{\min} \sim 500\,\mathrm{pc}\ (620\,\mathrm{pc}); \qquad \tau_P \sim 1.3 \times 10^7\,\mathrm{yr}\ (5.6 \times 10^7\,\mathrm{yr});$$

$$\lambda_f \sim 1.9\,\lambda_{\min} \sim 920\,\mathrm{pc}\ (1{,}060\,\mathrm{pc}),$$

where Parker's values are included parenthetically. And so while modern observations do change the numbers, Parker's conclusion remains unaffected: *The clumpiness of the galactic ISM is driven and maintained by an MHD instability, and not by the self-gravity of the clumps themselves.*

My second comment is whether there could be any terrestrial applications of the Parker instability. Can we set up an experiment in which the minimum wavelength to trigger the instability, $\lambda_{\min}$, can be captured in a plausible laboratory? The short answer is 'no'.

Starting with the obvious, take $\sigma_{CR} = 0$ and suppose σ_M and γ are both 1. Keeping things simple, imagine a lab in which large-ish quantities of 50% ionised hydrogen can be created and contained at atmospheric pressure. The Saha equation (*e.g.*, see Eq. 10.80) tells us that hydrogen gas at number density $N \sim 4.98 \times 10^{23}\,\mathrm{m^{-3}}$ would be 50% ionised at $T \sim 1.45 \times 10^4\,\mathrm{K}$, and the ideal gas law tells us that a gas with those values of N and T would be under $\sim 0.98\,\mathrm{atm}$ of pressure;[42] close enough to 1 for our purposes. So under these conditions, what is λ_{min}?

For 50% ionised hydrogen, the average particle mass is $\frac{2}{3}m_{\mathrm{p}}$, and the rms speed (Eq. 7.77) is,

$$v_{\mathrm{rms}} = \sqrt{\frac{k_{\mathrm{B}} T}{\frac{2}{3} m_{\mathrm{p}}}} \sim 1.34 \times 10^4\,\mathrm{m\,s^{-1}},$$

at $T = 1.45 \times 10^4\,\mathrm{K}$. Thus, the scale height is given by Eq. (7.82),

$$L = \frac{v_{\mathrm{rms}}^2 (1 + \sigma_{\mathrm{M}})}{g} \sim 3.66 \times 10^7\,\mathrm{m},$$

using the earth's acceleration of gravity, $g = 9.81\,\mathrm{m\,s^{-2}}$. Now, from Eq. (7.84),

$$C = (1 + \sigma_{\mathrm{M}})^2 - \gamma(1 + \tfrac{3}{2}\sigma_{\mathrm{M}}) = 1.5,$$

for $\sigma_{\mathrm{M}} = 1$ and $\gamma = 1$,[43] and so the minimum unstable wavelength is given by Eq. (7.86),

$$\lambda_{\mathrm{min}} = 2\pi L \sqrt{\frac{2\gamma\sigma_{\mathrm{M}}}{C}} \sim 2.66 \times 10^8\,\mathrm{m},$$

or about 20 Earth diameters, hardly a practical size for a lab!

So what can we do to make such an experiment more practical? Clearly λ_{min} must come down several orders of magnitude which means decreasing v_{rms}^2 and/or increasing the effective g. The rms speed falls as $\sqrt{T}$, and the Saha equation tells us that T falls with number density. A good vacuum pump in a lab can get to $10^{-14}\,\mathrm{atm}$ or lower, thus reducing the number density by 14 orders of magnitude. Alas, Saha still requires a temperature of about 4,000 K to ionise this lower density at 50%, which only lowers v_{rms}^2 and thus λ_{min} by a factor of ~ 3.6. Now, if we were to break the bank and buy the most powerful centrifuge available and somehow overcome the technical challenge of merging it with our pump, this could increase our effective g by 10^6 buying us a total factor of 3.6×10^6 and reducing λ_{min} to "only" $\sim 75\,\mathrm{m}$. Of course, this still poses a serious problem since ultra-centrifuges typically have volumes of a litre or less! Playing with σ_{M} doesn't buy us much either. Increasing σ_{M} from 1 to 10, for example, decreases $\sqrt{\sigma_{\mathrm{M}}/C}$ from ~ 0.82 to ~ 0.31 giving us another factor of ~ 2.6. We'd still need a vacuum chamber $\sim 30\,\mathrm{m}$ in diameter, and perhaps double that to capture the fastest growing mode.

This is not an Earth-based instability, neither in nature nor in the lab. The Parker instability is uniquely astrophysical and is why the effect of cosmic ray pressure – appearing nowhere else in this text – was included from the outset. As Parker showed analytically in 1966, it provides an elegant explanation for the semi-regular non-uniformity of the galactic ISM. As shown later largely by numerical means (*e.g.*, Shibata *et al.*, 1989), it also helps explain the origin of sun spots,

---

[42] Number density of the earth's atmosphere at STP is $\sim 2.69 \times 10^{25}\,\mathrm{m^{-3}}$.

[43] Note that $C < 0$ for $\sigma_{\mathrm{M}} = 1$ and $\gamma = \tfrac{5}{3}$, so we'll somehow need to maintain isothermality!

prominences, and other phenomena in the solar atmosphere that depend on "magnetic buoyancy", a phrase that has come to describe the physical property at the heart of the Parker instability.

Problem Set 7

7.1 Use the normal mode method developed in the chapter for the fluid instabilities to determine a stability criterion for a ball within a dip or atop a mound, as shown in Fig. 7.1 of the text. Thus, you'll need to identify a variable that has the "potential of running away", an equation of motion governing that variable, and then use a trial normal mode solution to determine the stability criterion. State clearly any assumptions you may need to make consistent with the displacement of the ball from equilibrium being a perturbation.

7.2** Consider the cross section of a "slab jet" with a "half-thickness" a (symmetry in the direction orthogonal to the page) shown in the inset below. Perform a stability analysis on the slab jet, similar to that done in §7.1, but this time equate p and Y across both boundaries between the jet and external medium. You may assume $p_j = p_x$ and $\rho_j/\rho_x = \eta$ which is not necessarily 1. Show that two distinct dispersion relations result, namely:

$$\tan(\alpha_j a) = \eta \frac{\beta_x}{\alpha_j} \left(1 - \frac{kv_j}{\omega}\right)^2 ;$$

$$\cot(\alpha_j a) = -\eta \frac{\beta_x}{\alpha_j} \left(1 - \frac{kv_j}{\omega}\right)^2 ,$$

where:

$$\alpha_j = \sqrt{\frac{(\omega - kv_j)^2}{c_{s,j}^2} - k^2}; \qquad \beta_x = \sqrt{k^2 - \frac{\omega^2}{c_{s,x}^2 \eta}}.$$

Here and in the figure, subscripts 'j' and 'x' refer to quantities in the jet and external medium respectively. Note that these two dispersion relations cannot be true simultaneously and, much like the quantum mechanical solution to the square well problem, they yield two families of solutions (one will be symmetric about the axis, the other antisymmetric), the union of which yield all possible values of the frequency, ω, for a given wavenumber, k.

Hints:

- Treat the external medium as the "lab frame" and analyse the perturbations propagating in the jet in the jet frame. Then, transform these results back to the lab frame before matching p and Y across the interfaces. You can do this much like the single shear layer problem in §7.1 using a Doppler shift.

- Similar to the problem of the single shear layer, you will limit your solution (sum of exponentials) for $P(y)$ for $y > a$ and $y < -a$ on the grounds that the solution must remain finite as $y \to \pm\infty$. However, this argument cannot be made for the solution inside the jet where $-a < y < a$ and where both terms remain finite. Thus, you must keep both terms in the jet solution, which are more conveniently expressed as trig functions rather than exponentials.

- Numerous references for this problem exist in the literature. If you are hunting for old papers on the subject, look for Attilio Ferrari and Phil Hardee in the 1980's. Be warned, these sorts of calculations can consume *reams* of paper.

7.3 Show that the peak growth rate for the MRI occurs when,

$$\omega_A = \frac{\sqrt{15}}{4}\Omega_0,$$

(Eq. 7.57 in the text) where, as defined in the text, $\omega_A = kv_A$ is the angular frequency of Alfvén waves with wave number k propagating at the Alfvén speed, v_A, and where Ω_0 is the local rotation speed of a Keplerian disc.

Hint: If you try to extremise Eq. (7.56) in the text by setting $d\chi/d\Omega_0 = 0$ directly, you'll run into trouble as you'll soon find the condition to be $\omega_A = 0$! You need to think about what it is that's being maximised. To get you started, the text does introduce Eq. (7.57) as "the maximum growth rate of the MRI *per period of rotation*". Hmmm...

7.4 In §7.3.3, we "derived" the evolution equation for angular momentum density, ℓ, (Eq. 7.59) from the φ-component of the MHD momentum equation (Eq. 7.58) already worked out in equation sets (A.52) and (A.55) in §A.4 and §A.5. Here, you'll re-derive Eq. (7.59) using the Theorem of Hydrodynamics (Theorem 1.1).

Consider the flat, circular disc shown in the inset rotating at angular speed $\Omega(r)$. Let the angular momentum of a small fluid element P (blue) relative to O be $\vec{L} = L\hat{z}$ and the torque acting on that same fluid element P be $\vec{\tau}_{\text{ext}} = \tau_{\text{ext}}\hat{z}$. Then, Newton's second law for angular motion can be written in a scalar form as,

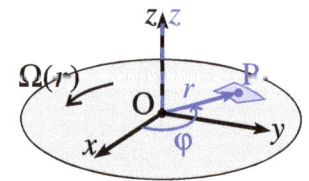

$$\frac{dL}{dt} = \sum \tau_{\text{ext}}. \qquad (7.124)$$

By following Example 3 in §1.3, show how Eq. (7.58) in the text can be derived from Theorem 1.1 accounting for all external *torque densities* acting on P.

Problem Set 7

7.5 For a Keplerian disc, show how Eq. (7.74), namely,

$$\mathcal{F}_L = m(r)\sqrt{4\mu_\rho^2 + \tfrac{1}{4}\Omega_0\chi\xi_{r,0}^2 e^{2\chi t}},$$

follows from Eq. (7.73) in the text.

7.6 In an MR-unstable disc where angular momentum is transported so efficiently, one might expect a scalar quantity such as dust to be diffused efficiently as well. This problem explores that consequence of the MRI.

A scalar quantity, q, that is simply transported with the flow follows the so-called *advection equation*,

$$\partial_t q + (\vec{v} \cdot \nabla)q = 0. \qquad (7.125)$$

Written in terms of the Lagrangian derivative (§3.1), Eq. (7.125) becomes,

$$\frac{Dq}{Dt} = 0,$$

which simply means that q is conserved along a *particle path* (§2.4); one notices neither an increase nor a decrease in q as one moves along with the flow. Put another way, the *Lagrangian difference* of an advected scalar, q, is zero:

$$\Delta q = \delta q + (\vec{\xi} \cdot \nabla)q = 0, \qquad (7.126)$$

where $\vec{\xi} = \vec{v}\Delta t$ is a displacement along the particle path.

a) Assuming $\rho = $ constant as we've been doing, write down Eq. (7.125) in conservative form (*e.g.*, like the continuity equation). This should be at most a two-liner.

b) By following the arguments in §7.3.3 in the text leading to Eq. (7.72), the average radial flux density for angular momentum, find the analogous expression for the average radial flux density for a scalar, q, and show that leads to an effective MRI-induced *diffusion coefficient*,

$$\mathcal{D}_r = \frac{1}{2}\chi\xi_{r,0}^2 e^{2\chi t},$$

where χ and $\xi_{r,0}$ are defined in the text. *Hint*: You might want to consult App. H.

7.7

a) Show how linearising Eq. (7.90) in the text leads to Eq. (7.99).

b) Show how linearising Eq. (7.92) in the text leads to Eq. (7.100).

7.8 Starting with Eq. (7.93) in the text, namely, $\partial_t p_{\text{CR}} + \vec{v} \cdot \nabla p_{\text{CR}} = 0$:

a) show how Eq. (7.101), namely,
$$\partial_t p_{CR,p} = -v_{z,p} \frac{dp_{CR,0}}{dz},$$
is obtained by linearising Eq. (7.93); and

b) show how Eq. (7.103), namely,
$$p_{CR,p} = \frac{\sigma_{CR} v_{rms}^2}{B_{y,0}} \frac{d\rho_0}{dz} A_{x,p},$$
follows from Eq. (7.101).

7.9*

a) Verify Parker's version of the continuity equation, namely Eq. (7.106) in the text.

   Hint: Start by operating on Eq. (7.95) with $\mathcal{L}$, then integrate once over time.

b) Repeat for the thermal pressure equation, Eq. (7.107) in the text.

c) In the isothermal case when $\gamma = 1$, Eq. (7.106) and (7.107) become identical, save for the name of the principle variable. Why do you suppose this is?

8 Viscid Hydrodynamics

with thanks to Patrick Rogers, B.Sc. (SMU), 2006.[†]

> *Big whirls have little whirls that feed on their velocity,*
> *and little whirls have lesser whirls and so on to viscosity.*
>
> Lewis Fry Richardson (1881–1953)
> from *Weather Prediction by Numerical Process*

8.1 Introduction

FOR THE MOST PART, astrophysical fluid dynamics is contemplated as *inviscid*, that is in the absence of *viscosity*. This is not to say, however, that astrophysicists believe the universe to be viscous-free! Indeed, and as discussed in this chapter, even the slightest viscosity can have qualitative effects on the nature of a fluid. For example, with identically zero viscosity, Kelvin's circulation theorem (Problem 1.7, Chap. 1) demands that an initially vorticity-free fluid remain vorticity free, whereas even the slightest viscosity can be exploited by a fluid to develop vortex tubes which play a prominent role in its cascade to *turbulence*. For a given viscosity, however small, there exists a length scale, again however small, at which viscous stresses dominate, and thus viscosity cannot really be ignored in any realistic application if all scale lengths are important.

However, it has been the tradition not to include viscosity for most astrophysical applications (and many physical and mathematical applications, for that matter) largely because of the additional complexity viscosity introduces into the fluid equations and, to some extent, because not all length scales are necessarily critical. As a case in point, examples in this chapter include only the simplest cases in which viscid flow can be described analytically.

On the other hand, the derivation of the celebrated Navier–Stokes equation (essentially Euler's equation with viscosity) and the accompanying modifications required for the energy equation(s) for a so-called Newtonian fluid (applicable to virtually all gases and many liquids) is given in its most general form (*i.e.*, compressible and non-isothermal), suitable for investigation by numerical methods. This is, in fact, where the greatest progress in astrophysical viscid (M)HD is being made today.

[†]Ph. D. (McMaster), 2012

As for the mathematics required for this chapter, most authors present this material assuming the reader has a sound background in *tensor analysis* with which the viscous stresses are most naturally described. Such a background has not been assumed in this text to now, and we shan't start here! Thus, a derivation of the so-called *stress tensor* in the next section is rather more long-winded than most presentations relying, as it does, on ideas in vector analysis (though tensors will still be constructed). This should, therefore, render the discussion accessible to the reader with only a sophomore background in mathematics. The reader who would like a more traditional tensor approach is directed to the classic texts in this field, of which Batchelor (2000) and Landau & Lifshitz (1987) are two of the most cited examples.

As a final introductory comment, this chapter returns us to the zero-field limit of MHD, namely ordinary hydrodynamics. As the mathematical derivation of the Navier–Stokes equation and some of its most approachable applications are most transparently illustrated for $\vec{B} = 0$, this is the approach taken here. We'll return to $\vec{B} \neq 0$ in Chap. 9 and 10.

8.2 The stress tensor

In §1.3 where the fundamental equations of HD are derived, our discussion of the momentum equation starts by applying the Theorem of hydrodynamics (Theorem 1.1) to Newton's second law, which yields:

$$\partial_t \vec{s} + \nabla \cdot (\vec{s}\vec{v}) = \vec{f}_p + \vec{f}_\phi, \qquad (8.1)$$

where $\vec{f}_p = -\nabla p$ is the *force density* (units Nm^{-3}) arising from pressure gradients, and $\vec{f}_\phi = -\rho \nabla \phi$ is the force density arising from gradients in the (self-)gravitational potential. This is essentially Eq. (1.27). Here, we shall retain the gravity term, but replace the pressure term, $\vec{f}_p$, with a term $\vec{f}_T$ that accounts for both compressive and shear stresses[1] not accounted for before, namely those arising from viscosity.

We proceed by imagining that in addition to compressional stresses such as pressure, a fluid is also subject to *shear* stresses. To illustrate this, consider an infinitesimal fluid "cube" (with dimension $\delta x \to 0$) shown in Fig. 8.1a, and the x_1–x_2 cross-section through that fluid infinitesimal shown in Fig. 8.1b, with the numerous components of those stresses in the x_1–x_2 plane labelled. Throughout this discussion, it is important to bear in mind that ultimately, this "cube" will be considered as a single point as δx is taken to be *identically* zero. That is, we are not interested here in the force balance on a macroscopic fluid element; rather we wish to consider the nature of all the stresses acting at a single point.

As there are six faces of the fluid cube and three stress components at each face, there are, in principle, 18 individual stress components to consider. Fortunately, symmetry arguments and the requirement that all accelerations remain finite will

---

[1]Here, a *stress* is defined generically as a force per unit area, of which pressure is one example.

The stress tensor

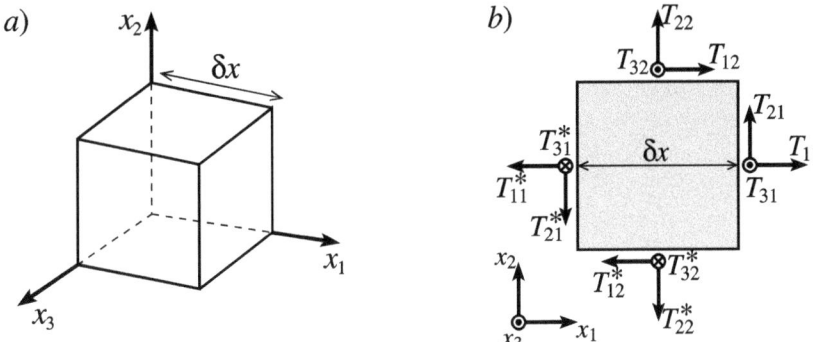

Figure 8.1. *a*) An infinitesimal "cube" of fluid of dimension $\delta x \to 0$ used to illustrate the viscous stresses in Cartesian-like coordinates. *b*) An x_1–x_2 slice through the fluid cube, with the stress components acting on the 1- and 2-faces only. Analogous stresses acting on the 3-faces above and under the page are left out for clarity.

reduce these to six independent components which shall form what we will call the *stress tensor*.

Let T_{ij} be the stress component acting in the $+x_i$-direction on the face whose normal pointing *outside* the cube points in the $+x_j$-direction, and let T_{ij}^* be the stress component acting in the $+x_i$-direction on the face whose normal points in the $-x_j$-direction (Fig. 8.1*b*). Then Newton's second law gives in the 1-direction:

$$(T_{11} - T_{11}^*)\delta x^2 + (T_{12} - T_{12}^*)\delta x^2 + (T_{13} - T_{13}^*)\delta x^2 = \rho \delta x^3 a_1,$$

where a_1 is the acceleration of the fluid cube arising from any imbalance in the 1-forces. Thus,

$$a_1 = \frac{(T_{11} - T_{11}^*) + (T_{12} - T_{12}^*) + (T_{13} - T_{13}^*)}{\rho \delta x}.$$

As $\delta x \to 0$, the numerator must go to zero at least as fast as δx if the acceleration is to remain finite. As we shall see, T_{11}, T_{12}, and T_{13} arise from independent physical processes and are thus independent stresses, unaccountable to each other. Therefore, each term in parentheses ( ) must *individually* go to zero, and we have:

$$T_{ij}^* = T_{ij}.$$

This leaves us with only nine stress components acting on the fluid infinitesimal, which we now arrange conveniently in matrix form as,

$$\mathsf{T} = \begin{bmatrix} T_{11} & T_{12} & T_{13} \\ T_{21} & T_{22} & T_{23} \\ T_{31} & T_{32} & T_{33} \end{bmatrix}, \tag{8.2}$$

giving us the first glimpse of the so-called *stress tensor*. As shown in Fig. 8.1*b*, the diagonal elements, T_{ii}, are all pointing parallel to the surface normal of the fluid cube, while the off-diagonal elements, $T_{i \neq j}$, are pointing in directions relative to T_{ii} consistent with the right-hand rule. Note that these choices are completely arbitrary. Any incorrectly chosen direction will ultimately be detected if the sign of the component magnitude, T_{ij}, ends up being negative. In fact, since the only external fluid stress we know about so far – pressure – is *compressive* in nature, we might already anticipate that T_{ii} – the stress component most plausibly associated with pressure – ought to point inwards rather than outwards as shown in Fig. 8.1*b*. This will indeed be the case, as when we finally get to the point of writing down an expression for T_{ii}, we'll find it to be negative.

Next, we examine the torques generated by the various stress components. Thus, about an axis passing through the centre of the cube and parallel to the x_3-axis (point O in Fig. 8.2), the torque is,

$$\tau_3 = (T_{21} - T_{12} + T_{21} - T_{12})\delta x^2 \frac{\delta x}{2} = (T_{21} - T_{12})\delta x^3 = I_3 \alpha_3,$$

where I_3 is the moment of inertia of the cube and α_3 its angular acceleration, both about O. Now,

$$I_3 = \int_V \rho r^2 dV \propto \delta x^5 \quad \Rightarrow \quad \alpha_3 \propto \frac{T_{21} - T_{12}}{\delta x^2}.$$

Thus, if α_3 is to remain finite as $\delta x \to 0$, $T_{21} - T_{12} \to 0$ at least as fast as δx^2. This yields,

$$T_{ij} = T_{ji},$$

and the stress tensor is symmetric. As a result, there remain just six independent stress components to account for.

8.2.1 The trace of the stress tensor

The *trace* of a tensor (sum of the diagonal elements of its matrix representation) often has a significant meaning, and the stress tensor[2] is no exception. Physically significant quantities are, among other things, independent of the coordinate system so we begin by showing that tr(T) in invariant under a rotation about the x_3-axis.[3]

To this end, consider the x_1–x_2 slice through the fluid cube as shown in Fig. 8.2, and in particular the fluid "prism" ABC. Note that although the coordinate system x'_1–x'_2 is rotated by 45° relative to the original x_1–x_2 coordinate system, the following argument can be made for *any* rotation angle (Problem 8.1). Demanding force balance on ABC in the limit as $\delta x \to 0$, we have in the x'_1-direction,

$$-T'_{11}\sqrt{2}\delta x^2 + 2T_{12}\frac{\delta x^2}{\sqrt{2}} + T_{11}\frac{\delta x^2}{\sqrt{2}} + T_{22}\frac{\delta x^2}{\sqrt{2}} = 0,$$

where the fact that face AC is $\sqrt{2}$ times larger than face AB, that $\cos 45° = 1/\sqrt{2}$, and that T is symmetric have all been taken into account. Thus,

$$2T'_{11} = T_{11} + T_{22} + 2T_{12}. \tag{8.3}$$

---

[2] Arranging nine elements in a 3 × 3 matrix as done in Eq. (8.2) does not a tensor make, and referring to T as a tensor is premature. However, as our analysis is all being done outside the formalism of tensor analysis, no harm can come from this pre-emptive designation.

[3] A well-known theorem of tensor analysis shows that tr(T) is invariant under *any* rotation.

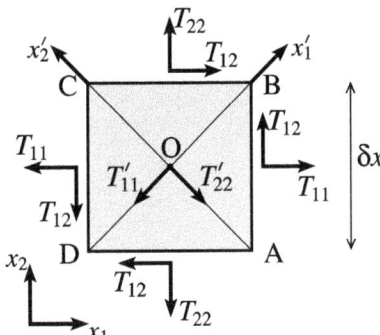

Figure 8.2. The x_1–x_2 slice through the fluid cube of side length $\delta x \to 0$, showing the rotated coordinates x'_1 and x'_2 with the compressive stress components in the rotated coordinates acting on the AC and BD faces.

Similarly for the prism DCB, we have in the x'_2-direction,

$$2T'_{22} = T_{11} + T_{22} - 2T_{12}. \tag{8.4}$$

Subtracting Eq. (8.4) from Eq. (8.3) yields:

$$2T_{12} = T'_{11} - T'_{22}, \tag{8.5}$$

which we shall recall later in §8.2.3. Adding Eq. (8.3) and (8.4) yields:

$$T'_{11} + T'_{22} = T_{11} + T_{22}. \tag{8.6}$$

As a rotation about the x_3-axis, $\hat{x}_3 = \hat{x}'_3$ and the 33-component of T remains unchanged; that is $T_{33} = T'_{33}$. Thus, with Eq. (8.6), this gives us,

$$T'_{11} + T'_{22} + T'_{33} = T_{11} + T_{22} + T_{33} \equiv \text{tr}(\mathsf{T}).$$

While this derivation is restricted to rotations about the x_3-axis, it is nevertheless illustrative of a general result from tensor analysis: *The trace of any tensor T is invariant under an arbitrary coordinate rotation.* Physically, this means that tr(T) is an *isotropic* property of the fluid and can be represented by a *scalar*, independent of the coordinates. For now, let us provisionally define that scalar to be:

$$\boxed{\varpi \equiv -\tfrac{1}{3}\text{tr}(\mathsf{T}).} \tag{8.7}$$

Eventually, we shall come to identify ϖ as the *thermal pressure*. Even still, we will find that the diagonal elements of T are, in general, different and come to interpret this result as the non-isotropic nature of the pressure in a viscid fluid. It is only for an inviscid fluid that the three diagonal elements, T_{ii}, are equal and thus, by Eq. (8.7), each equal to $-\varpi$. Note further that at this stage with ϖ not yet known to be the pressure, the sign in Eq. (8.7) is arbitrary, and we still don't know T_{ii} to be negative.

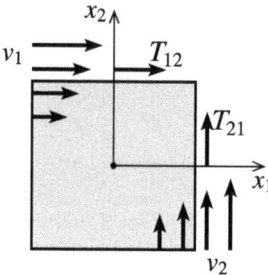

Figure 8.3. The x_1–x_2 slice through the fluid cube, showing a positive 2-gradient in v_1 and a positive 1-gradient in v_2 which, respectively, result in the shear stresses T_{12} and T_{21} acting on the fluid cube in the directions shown. This justifies the directions chosen for the tensor components, T_{ij}, $i \neq j$, in Fig. 8.1.

8.2.2 Viscosity and Newtonian fluids

It was Sir Isaac Newton himself who first devised a mathematical model of a viscid fluid and, rather astonishingly, it is still his description we use today. He postulated that fluids possess a property (we now call viscosity) that causes one "layer" of fluid to drag along a neighbouring layer moving at a different velocity. Thus, the shear stresses experienced by such a fluid would depend upon shear velocity gradients of the form $\partial_j v_i$, $i \neq j$ (e.g., Fig. 8.3), since co-moving layers of fluid would not affect each other's motion.

At first blush, Occam's razor might suggest the following connection between the stresses (T_{ij}) and the strains ($\partial_j v_i$) in the fluid,

$$T_{ij} \propto \partial_j v_i, \qquad i \neq j,$$

but this turns out to be *too* simple. For example, such a prescription requires $T_{12} \propto \partial_2 v_1$ and $T_{21} \propto \partial_1 v_2$, as illustrated in Fig. 8.3. However, the symmetry of the stress tensor requires $T_{12} = T_{21}$, yet there is no reason to expect $\partial_2 v_1 = \partial_1 v_2$! Thus, Newton went to the next level of complexity, and postulated that,

$$\boxed{T_{ij} = \mu(\partial_i v_j + \partial_j v_i), \qquad i \neq j,} \tag{8.8}$$

where the proportionality constant, μ, is called the *shear viscosity*, or often just the *viscosity* of the fluid. The units of T_{ij} are $\mathrm{N\,m^{-2}}$, and those of the viscosity are $\mathrm{N\,m^{-2}\,s} = \mathrm{kg\,m^{-1}\,s^{-1}}$.

If the stress–strain relationship of a fluid is well modelled by Eq. (8.8), it is referred to as a *Newtonian fluid*; otherwise it is a *non-Newtonian fluid*. Experimentally, virtually all gases and most "simple" liquids, such as water, alcohol, and liquid O_2, are excellent examples of Newtonian fluids. Examples of non-Newtonian fluids include liquids made up of long polymers, suspensions, and complex mixtures such as, of all things, cake batter. Indeed, one way to distinguish a clearly Newtonian fluid from a clearly non-Newtonian one is to beat it with an egg beater. Newtonian fluids will form a depression around the beaters, whereas some non-Newtonian fluids (such as cake batter) will climb up the beater shafts. Other common non-Newtonian fluids include blood, ketchup, paint, and shampoo.

Some non-Newtonian fluids – particularly suspensions – are characterised by their ability to switch back and forth between a solid-like state and a liquid-like state. A thick enough suspension of corn starch, for example, will allow a spoon to be immersed into it if the spoon is pushed in slowly. By contrast, an attempt to remove the spoon too quickly will cause the fluid to take on its solid-like state, temporarily "freezing" the spoon in the medium. Quick sand is another example of such a fluid.

8.2.3 Non-isotropic "pressure"

Consider again Fig. 8.2, this time to examine the elements of the velocity gradient in the different coordinate systems.

One way to estimate the x'_1 component of the velocity at point B is to take a first-order Taylor expansion from point O:

$$v'_1(\text{B}) = v'_1(\text{O}) + \frac{\delta x}{\sqrt{2}} \frac{\partial v'_1}{\partial x'_1}. \tag{8.9}$$

Alternatively, $v'_1(\text{B})$ can be expressed in terms of the velocity components in the unprimed coordinates:

$$v'_1(\text{B}) = \frac{v_1(\text{B})}{\sqrt{2}} + \frac{v_2(\text{B})}{\sqrt{2}}, \tag{8.10}$$

which in turn can be estimated by taking two-dimensional Taylor expansions again from point O:

$$v_1(\text{B}) = v_1(\text{O}) + \frac{\delta x}{2} \partial_1 v_1 + \frac{\delta x}{2} \partial_2 v_1; \tag{8.11}$$

$$v_2(\text{B}) = v_2(\text{O}) + \frac{\delta x}{2} \partial_2 v_2 + \frac{\delta x}{2} \partial_1 v_2. \tag{8.12}$$

Substituting Eq. (8.11) and (8.12) into Eq. (8.10) gives,

$$v'_1(\text{B}) = v'_1(\text{O}) + \frac{\delta x}{2\sqrt{2}} (\partial_1 v_1 + \partial_2 v_1 + \partial_1 v_2 + \partial_2 v_2),$$

where $v'_1(\text{O}) = \frac{1}{\sqrt{2}}[v_1(\text{O}) + v_2(\text{O})]$ has been used. Comparing this to Eq. (8.9) and using Eq. (8.8), we find,

$$2\frac{\partial v'_1}{\partial x'_1} = (\partial_1 v_1 + \partial_2 v_2) + \frac{T_{12}}{\mu}. \tag{8.13}$$

Similarly, by examining the velocities at point C in Fig. 8.2 we get,

$$2\frac{\partial v'_2}{\partial x'_2} = (\partial_1 v_1 + \partial_2 v_2) - \frac{T_{12}}{\mu}, \tag{8.14}$$

and subtracting Eq. (8.14) from Eq. (8.13) yields:

$$2T_{12} = 2\mu \left(\frac{\partial v'_1}{\partial x'_1} - \frac{\partial v'_2}{\partial x'_2} \right). \tag{8.15}$$

Now, from Eq. (8.5), we had,
$$2T_{12} = T'_{11} - T'_{22},$$
and comparing this to Eq. (8.15), we arrive at the main result from this analysis:
$$T'_{11} - 2\mu \frac{\partial v'_1}{\partial x'_1} = T'_{22} - 2\mu \frac{\partial v'_2}{\partial x'_2} = T'_{33} - 2\mu \frac{\partial v'_3}{\partial x'_3},$$
where the last equality was included on grounds of symmetry. As there is nothing special about the primed coordinate system, the same relationship must hold in the unprimed coordinate system as well and we have,

$$T_{11} - 2\mu \partial_1 v_1 = T_{22} - 2\mu \partial_2 v_2 = T_{33} - 2\mu \partial_3 v_3$$

$$\Rightarrow \quad T_{11} - 2\mu \partial_1 v_1 = \tfrac{1}{3}(T_{11} - 2\mu \partial_1 v_1 + T_{22} - 2\mu \partial_2 v_2 + T_{33} - 2\mu \partial_3 v_3)$$

$$= \tfrac{1}{3}(T_{11} + T_{22} + T_{33}) - \tfrac{1}{3}(2\mu \nabla \cdot \vec{v})$$

$$\Rightarrow \quad T_{11} = -\varpi + 2\mu \left(\partial_1 v_1 - \tfrac{1}{3}\nabla \cdot \vec{v}\right),$$

using Eq. (8.7). Thus, we have in general,
$$T_{ii} = -\varpi + 2\mu \left(\partial_i v_i - \tfrac{1}{3}\nabla \cdot \vec{v}\right), \quad i = 1, 2, 3, \tag{8.16}$$

and we have succeeded in expressing the diagonal elements of the stress tensor in terms of the isotropic trace (ϖ) and the velocity *compressive* derivatives, rather than the shear derivatives which defined the off-diagonal elements given by Eq. (8.8). Note that the diagonal elements of the stress tensor are not necessarily equal, depending as they do on $\partial_i v_i$. As these elements are compressive, they are to a viscid fluid what pressure is to an inviscid fluid, and thus we see that the compressive stresses in a viscid fluid are not necessarily isotropic, as is the case for an inviscid fluid.

Examining Eq. (8.8) and (8.16), we see they can be combined to give a single expression for all stress elements, namely,
$$\boxed{T_{ij} = -\varpi \, \delta_{ij} + \mu \left(\partial_i v_j + \partial_j v_i - \tfrac{2}{3}\nabla \cdot \vec{v}\,\delta_{ij}\right),} \tag{8.17}$$

where $\delta_{ij} = 1$, $i = j$; 0, $i \neq j$ is the usual Kronecker-delta. Written as matrices (tensors), this becomes:
$$\boxed{\mathsf{T} = -\varpi \, \mathsf{I} + \mu \left(\nabla \vec{v} + (\nabla \vec{v})^{\mathrm{T}} - \tfrac{2}{3}\nabla \cdot \vec{v}\,\mathsf{I}\right),} \tag{8.18}$$

using Eq. (A.19), where I is the identity matrix (tensor), and where – as introduced on page 84 – the superscript ^{T} indicates the matrix transpose (rows become columns).

Let us now define the viscid portion of the stress tensor, S, as,
$$\mathsf{S} \equiv \nabla \vec{v} + (\nabla \vec{v})^{\mathrm{T}} - \tfrac{2}{3}\nabla \cdot \vec{v}\,\mathsf{I} \equiv 2\mathsf{E} - \tfrac{2}{3}\nabla \cdot \vec{v}\,\mathsf{I}, \tag{8.19}$$

where $\mathsf{E} = \tfrac{1}{2}\left[\nabla \vec{v} + (\nabla \vec{v})^{\mathrm{T}}\right]$ is the *symmetrised strain tensor*, first used without being named in Eq. (8.8). Evidently, the $(i,j)^{\text{th}}$ element of S is given by,
$$S_{ij} = \partial_i v_j + \partial_j v_i - \tfrac{2}{3}\nabla \cdot \vec{v}\,\delta_{ij} = 2E_{ij} - \tfrac{2}{3}\nabla \cdot \vec{v}\,\delta_{ij}.$$

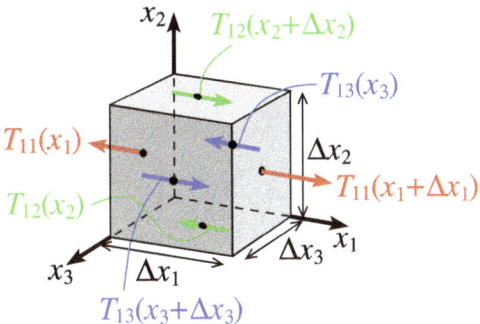

Figure 8.4. A "semi-macroscopic" fluid cube showing the location of the six 1-stresses. Here, $\Delta x_1 = \Delta x_2 = \Delta x_3 = \Delta x$, with the subscripts introduced only for convenience.

Thus, Eq. (8.18) becomes:

$$\boxed{\mathsf{T} = -\varpi\,\mathsf{I} + \mu\,\mathsf{S},} \qquad (8.20)$$

cleanly separating the inviscid and viscid portions of the stress tensor.

8.3 The Navier–Stokes equation

We are now ready to assess the force density arising from the stress tensor, $\vec{f}_\mathsf{T}$. Figure 8.4 depicts a "semi-macroscopic" fluid cube with dimensions Δx_i, $i = 1, 2, 3$, and the 1-stress at each face indicated. Unlike the fluid cubes in Fig. 8.1 and 8.2 where δx was ultimately taken as dead zero, here we are interested in the *limit* as Δx_i tends to zero but never actually reaches it. This is an important mathematical distinction since this means opposite faces of the cube are never actually coincident. Thus, the net 1-force on the cube is,

$$\begin{aligned}F_{1,\text{net}} &= [T_{11}(x_1 + \Delta x_1) - T_{11}(x_1)]\Delta x_2 \Delta x_3 + [T_{12}(x_2 + \Delta x_2) - T_{12}(x_2)]\Delta x_3 \Delta x_1 \\ &\quad + [T_{13}(x_3 + \Delta x_3) - T_{13}(x_3)]\Delta x_1 \Delta x_2 \\ &= \frac{T_{11}(x_1 + \Delta x_1) - T_{11}(x_1)}{\Delta x_1}\Delta V + \frac{T_{12}(x_2 + \Delta x_2) - T_{12}(x_2)}{\Delta x_2}\Delta V \\ &\quad + \frac{T_{13}(x_3 + \Delta x_3) - T_{13}(x_3)}{\Delta x_3}\Delta V,\end{aligned}$$

where $\Delta V = \Delta x_1 \Delta x_2 \Delta x_3$ is the volume of the fluid cube. Then, dividing through by ΔV and taking the *limit* as $\Delta x_i \to 0$, we get the *force density* in the 1-direction,

$$f_{1,\text{net}} \equiv \frac{F_{1,\text{net}}}{\Delta V} = \partial_1 T_{11} + \partial_2 T_{12} + \partial_3 T_{13} = \nabla \cdot \vec{T}_1, \qquad (8.21)$$

where $\vec{T}_1 = (T_{11}, T_{12}, T_{13})$ is, formally, the "vector" formed from the first row of T comprised of the three stress components pointing in the 1-direction. Thus, the full force density arising from viscous stresses is:

$$\vec{f}_\mathsf{T} = \nabla \cdot \mathsf{T}^\mathsf{T},\tag{8.22}$$

where the transpose is used to preserve the rules of ordinary matrix multiplication when performing the dot product.[4] However, since T is symmetric ($\mathsf{T} = \mathsf{T}^\mathsf{T}$), this formality is rather moot and we can write the momentum equation (Eq. 8.1) as,

$$\begin{aligned}\partial_t \vec{s} + \nabla \cdot (\vec{s}\vec{v}) &= \nabla \cdot \mathsf{T} - \rho \nabla \phi \\ &= -\nabla \varpi + \nabla \cdot (\mu \mathsf{S}) - \rho \nabla \phi,\end{aligned}\tag{8.23}$$

using Eq. (8.20) and applying the identity $\nabla \cdot (\varpi \mathsf{I}) = \nabla \varpi$. This is the *viscid momentum equation*, applicable for compressible flow with a variable (in position) shear viscosity, μ. An alternate and slightly expanded form of this equation is:

$$\partial_t \vec{s} + \nabla \cdot \left(\vec{s}\vec{v} + (\varpi + \tfrac{2}{3}\mu \nabla \cdot \vec{v})\mathsf{I} - 2\mu \mathsf{E}\right) = -\rho \nabla \phi,$$

where as many terms have been brought into the divergence operator as possible. This form is more useful for numerical methods, where numerical conservation is maintained by keeping track of flux densities such as the quantity between the large parentheses in the previous equation.

The terms $\nabla \cdot (\vec{s}\vec{v})$ and $\nabla \cdot (\mu \mathsf{S})$ in Eq. (8.23) are known as the *inertial term* and *viscous term* respectively. When applying the viscid fluid equations to a particular problem, one must assess which of these terms dominate the dynamics, and make the appropriate approximation. If neither term dominates, little analytical headway can be made with Eq. (8.23), leaving numerical methods the only viable recourse.

Case 1: $\mu = 0$ *(inviscid flow)*

When $\mu = 0$, Eq. (8.23) becomes,

$$\partial_t \vec{s} + \nabla \cdot (\vec{s}\vec{v}) = -\nabla \varpi - \rho \nabla \phi,$$

recovering the momentum Eq. (1.27), as expected, only if $\varpi = p$, the thermal pressure! Only now, do we know that the trace of the stress tensor is $-\tfrac{1}{3}$ times the thermal pressure, and that $T_{ii} < 0$. Thus, the reader should now be prepared to substitute ϖ for p in all equations since and including Eq. (8.7).

Incidentally, this also proves *Pascal's law* (or *principle*, but not to be confused with Pascal's *theorem*[5]) that states:

> The stresses in an inviscid fluid in mechanical equilibrium are completely described by an isotropic scalar, in this case $\varpi = p$.

Case 2: $\mu = $ constant

In practice, μ can be a sensitive function of position, making its place between the del operator and S in Eq. (8.23) non-negotiable and mathematical analysis difficult.

---

[4]$(\nabla \cdot \mathsf{T}^\mathsf{T})_i = \sum_j \partial_j T^\mathsf{T}_{ji} = \sum_j \partial_j T_{ij}$, as required in Eq. (8.21).
[5]www.wikipedia.org/wiki/Blaise_Pascal

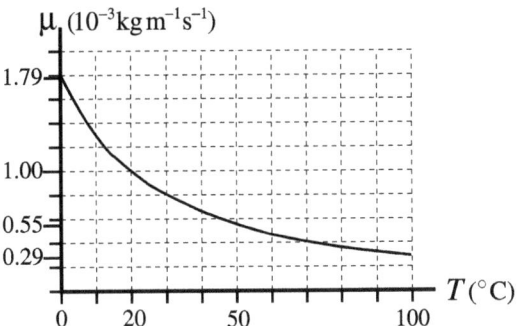

Figure 8.5. The shear viscosity, μ, of fresh water as a function of temperature, with values at $0°C$, $20°C$, $50°C$, and $100°C$ highlighted.

However, experimentation shows that at least for Newtonian fluids, μ is largely a function of temperature (*e.g.*, Fig. 8.5) and independent of density and pressure individually. Thus, for fluids that are at least nearly isothermal, taking μ to be independent of the coordinates is a reasonable approximation to make, allowing it to slip through the *nabla*, ∇.

Following the same procedures used to derive Euler's equation (Eq. 1.36; Problem 1.2) from the momentum equation, we develop from Eq. (8.23):

$$\partial_t \vec{v} + (\vec{v} \cdot \nabla)\vec{v} = -\frac{1}{\rho}\nabla p + \nu \nabla \cdot \mathsf{S} - \nabla \phi, \tag{8.24}$$

where $\nu \equiv \mu/\rho$ is the *kinematic viscosity* of the fluid, so-named because its units are $m^2\,s^{-1}$ and therefore looks like the product of a distance and a velocity. From Eq. (8.19) and using identities (A.27) and (A.28), we get,

$$\nabla \cdot \mathsf{S} = \nabla \cdot (\nabla \vec{v} + (\nabla \vec{v})^{\mathrm{T}} - \tfrac{2}{3}\nabla \cdot \vec{v}\,\mathsf{I}) = \nabla^2 \vec{v} + \tfrac{1}{3}\nabla(\nabla \cdot \vec{v}) = \tfrac{4}{3}\nabla(\nabla \cdot \vec{v}) - \nabla \times \vec{\omega},$$

where $\vec{\omega} = \nabla \times \vec{v}$, the vorticity, was first introduced in Problem 1.6. Thus, Eq. (8.24) becomes:

$$\begin{aligned}\partial_t \vec{v} + (\vec{v} \cdot \nabla)\vec{v} &= -\frac{1}{\rho}\nabla p - \nabla \phi + \nu \left(\tfrac{1}{3}\nabla(\nabla \cdot \vec{v}) + \nabla^2 \vec{v}\right) \\ &= -\frac{1}{\rho}\nabla p - \nabla \phi + \nu \left(\tfrac{4}{3}\nabla(\nabla \cdot \vec{v}) - \nabla \times \vec{\omega}\right).\end{aligned} \tag{8.25}$$

This is the *compressible* form of the so-called *Navier–Stokes* equation, named for Claude-Louis Navier,[6] and George Gabriel Stokes.[7] Navier was the first to develop the theory of elasticity, but his most remembered work is his publication of the first version of what is now known as the Navier–Stokes equation in 1822. Stokes' work in physics and mathematics (he's the same Stokes of *Stokes' theorem*) was done after Navier's death, and is largely responsible for reformulating the fluid equations into the form we recognise today.

---

[6] 1785–1836, www.wikipedia.org/wiki/Claude-Louis_Navier
[7] 1819–1903, www.wikipedia.org/wiki/Sir_George_Stokes,_1st_Baronet

Case 3: $\mu =$ constant, *incompressible*

For incompressible flow, $\rho =$ constant and all derivatives of ρ are zero. Thus, the continuity equation becomes $\nabla \cdot \vec{v} = 0$ ($\vec{v}$ is solenoidal), and Eq. (8.25) becomes:

$$\boxed{\partial_t \vec{v} + (\vec{v} \cdot \nabla)\vec{v} = -\frac{1}{\rho}\nabla p - \nabla \phi + \nu \nabla^2 \vec{v} = -\frac{1}{\rho}\nabla p - \nabla \phi - \nu \nabla \times \vec{\omega}.} \quad (8.26)$$

This is the *incompressible* form of the *Navier–Stokes* equation, and by far the most widely used as it is in liquids where viscosity plays the most obvious role. Note that if an incompressible isothermal fluid is also irrotational (hence $\vec{\omega} = 0$ and, in particular, no shear layers), it will behave as an inviscid fluid regardless of what the viscosity may be.

8.4 The viscid energy equation

Starting with Eq. (1.21), reproduced here for reference,

$$\partial_t e_\mathrm{T} + \nabla \cdot (e_\mathrm{T} \vec{v}) = p_\mathrm{app},$$

we wish to evaluate the applied power density, p_app, of the viscous stress tensor. To this end, the increment of work done by the 1-stresses (as labelled in Fig. 8.4) on the small cube of fluid over a small time step, Δt, is:

$$\begin{aligned}
\Delta W_1 &= T_{11}(x_1 + \Delta x_1)\Delta x_2 \Delta x_3 v_1(x_1 + \Delta x_1)\Delta t - T_{11}(x_1)\Delta x_2 \Delta x_3 v_1(x_1)\Delta t \\
&\quad + T_{12}(x_2 + \Delta x_2)\Delta x_3 \Delta x_1 v_1(x_2 + \Delta x_2)\Delta t - T_{12}(x_1)\Delta x_3 \Delta x_1 v_1(x_2)\Delta t \\
&\quad + T_{13}(x_3 + \Delta x_3)\Delta x_1 \Delta x_2 v_1(x_3 + \Delta x_3)\Delta t - T_{13}(x_1)\Delta x_1 \Delta x_2 v_1(x_3)\Delta t \\
\Rightarrow \quad p_{\mathrm{app},1} &= \frac{1}{\Delta V}\frac{\Delta W_1}{\Delta t} = \partial_1(T_{11}v_1) + \partial_2(T_{12}v_1) + \partial_3(T_{13}v_1) \\
&= \nabla \cdot (\vec{T}_1 v_1),
\end{aligned} \quad (8.27)$$

as $\Delta x_i \to 0$, where $\vec{T}_1$ was first used in Eq. (8.21). Thus, from all three directions the total applied power from the stress tensor is,

$$p_\mathrm{app} = \nabla \cdot (\vec{T}_1 v_1) + \nabla \cdot (\vec{T}_2 v_2) + \nabla \cdot (\vec{T}_3 v_3) = \nabla \cdot (\mathsf{T}^\mathrm{T} \cdot \vec{v}), \quad (8.28)$$

where it is left to Problem 8.2 to verify that the preservation of the rules of matrix multiplication require T^T rather than T. Note that these same rules require $\vec{v}$ to be treated as a column vector (ket) when it is "dotted" from the right.

Once again, since T is symmetric, $\mathsf{T} = \mathsf{T}^\mathrm{T}$ and the *viscid energy equation* can be written as,

$$\boxed{\partial_t e_\mathrm{T} + \nabla \cdot (e_\mathrm{T} \vec{v}) = \nabla \cdot (\mathsf{T} \cdot \vec{v}).} \quad (8.29)$$

It is left to Problem 8.3 to show that we can rewrite Eq. (8.29) as:

$$\partial_t e_T + \nabla \cdot [(e_T + p)\vec{v}] = \nabla \cdot (\mu \mathsf{S} \cdot \vec{v})$$
$$= \nabla \cdot \left[\mu \left((\vec{v} \cdot \nabla)\vec{v} + \tfrac{1}{2}\nabla v^2 - \tfrac{2}{3}\vec{v}\nabla \cdot \vec{v}\right)\right]$$
$$= \nabla \cdot \left[\mu \left(\nabla v^2 - \vec{v} \times \vec{\omega} - \tfrac{2}{3}\vec{v}\nabla \cdot \vec{v}\right)\right]. \tag{8.30}$$

8.4.1 Viscous dissipation

The power density from the stress tensor modifies both the mechanical and internal energy of the fluid cube. We seek here to distinguish between the two.

From the work-kinetic theorem, the kinetic (mechanical) power density is the scalar product of the force density with the velocity field:

$$p_K = \vec{f}_T \cdot \vec{v} = (\nabla \cdot \mathsf{T}) \cdot \vec{v}, \tag{8.31}$$

using Eq. (8.22) and the fact that T is symmetric. Subtracting p_K from the total applied power, p_{app} given by Eq. (8.28), gives us the dissipated power, namely,

$$p_D = \nabla \cdot (\mathsf{T} \cdot \vec{v}) - (\nabla \cdot \mathsf{T}) \cdot \vec{v} = \mathsf{T} : \nabla \vec{v}, \tag{8.32}$$

using Eq. (A.25), which amounts to a "product rule" for the inner product of a tensor with a vector. Note the use of the "colon product" (*a.k.a.* the "double dot product"; essentially a double contraction) between two tensors defined by Eq. (A.18).

On substituting Eq. (8.18) into Eq. (8.32), we get,

$$p_D = -\left(p + \frac{2\mu}{3}\nabla \cdot \vec{v}\right)\mathsf{I} : \nabla \vec{v} + 2\mu \mathsf{E} : \nabla \vec{v}. \tag{8.33}$$

Now, one can show (trivially for Cartesian coordinates) that $\mathsf{I} : \nabla \vec{v} = \nabla \cdot \vec{v}$. Further, since E is symmetric, it's easy to show $\mathsf{E} : \nabla \vec{v} = \mathsf{E} : (\nabla \vec{v})^T$, and Eq. (8.33) becomes:

$$p_D = -p\nabla \cdot \vec{v} - \frac{2\mu}{3}(\nabla \cdot \vec{v})^2 + 2\mu \mathsf{E} : \mathsf{E}. \tag{8.34}$$

The $-p\nabla \cdot \vec{v}$ term in Eq. (8.34) is the same term appearing in the inviscid internal energy equation (Eq. 1.34). It arises from *adiabatic expansion* where it is negative and represents a loss in internal energy, or *adiabatic compression* where it is positive and represents a gain in internal energy. Adiabatic expansion/compression is a *reversible* process. The terms proportional to μ are *viscous dissipation*, one of which is negative-definite, the other positive-definite. Viscous dissipation is an *irreversible* process, and it is left to Problem 8.4 to show that together, the viscous dissipation terms are always positive and result in a gain of thermal energy, never a loss.

As it is the dissipation terms that appear on the RHS of the internal energy equation, we can immediately update Eq. (1.34) for viscous dissipation to get,

$$\boxed{\begin{aligned}\partial_t e + \nabla \cdot (e\vec{v}) &= \mathsf{T} : \nabla \vec{v} = -p\nabla \cdot \vec{v} + \mu \mathsf{S} : \nabla \vec{v} \\ &= -p\nabla \cdot \vec{v} - \frac{2\mu}{3}(\nabla \cdot \vec{v})^2 + 2\mu \mathsf{E} : \mathsf{E},\end{aligned}} \tag{8.35}$$

where the last equality is also the subject of Problem 8.4.

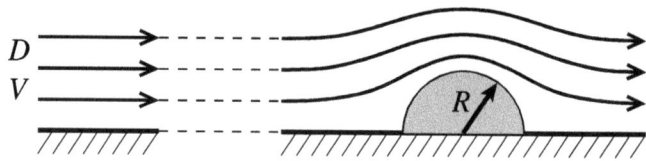

Figure 8.6. Fluid streamlines pass over a cylindrical obstruction, illustrating how the fiducial scaling parameters, D, V, and R, might be chosen to scale the fluid equations.

8.5 The Reynolds number

Consider Euler's equation (Eq. 1.36) reproduced below without the gravity term:

$$\frac{\partial \vec{v}}{\partial t} + (\vec{v} \cdot \nabla)\vec{v} = -\frac{1}{\rho}\nabla p. \tag{8.36}$$

As given, each term of Euler's equation has units of acceleration. However, we can render Euler's equation "unitless" by imposing the following "scaling laws":

$$r = Rr'; \qquad \rho = D\rho'; \qquad \vec{v} = V\vec{v}', \tag{8.37}$$

where R, D, and V are *fiducial* values of length, density, and speed, and carry the same units as r, ρ, and v respectively. Thus, the quantities:

$$r' = \frac{r}{R}; \qquad \rho' = \frac{\rho}{D}; \qquad \vec{v}' = \frac{\vec{v}}{V},$$

are unitless and represent values of the primitive variables in units of the fiducial quantities.

Figure 8.6 gives an illustrative example. Fluid flowing smoothly and uniformly along a solid surface comes across an obstacle such as the half-cylinder shown in cross section. As the streamlines are diverted around the obstacle, the speed, density and pressure of the fluid will vary, and whether the far downstream state returns to the far upstream state will depend upon such things as the Mach number and viscosity of the fluid. In this case, the fiducial length, density, and speed values could be the radius of the obstacle, and the far upstream values of the density and velocity. Typically, each problem will present its own fiducial values.

Note that with only three independent "units" (*e.g.*, length, mass, and time), only three scaling laws may be specified, and Eq. (8.37) fulfil this limit. Thus, all other variables and differential operators must be scaled in terms of Eq. (8.37), whence:

$$t = t'\frac{R}{V}; \qquad \nabla = \nabla'\frac{1}{R}; \qquad c_{\text{s}} = c'_{\text{s}}V; \qquad p = p'DV^2, \tag{8.38}$$

where the scaling law for pressure is most easily seen from the expression for the adiabatic sound speed (Eq. 2.11):

$$p = \frac{\rho c_{\text{s}}^2}{\gamma}.$$

Substituting each of Eq. (8.37) and (8.38) into Eq. (8.36) gives us,

$$\frac{V}{R}\frac{\partial \vec{v}'}{\partial t'}V + \frac{V}{R}(\vec{v}'\cdot\nabla')\vec{v}'V = -\frac{1}{D}\frac{1}{\rho'}\frac{1}{R}\nabla' p' DV^2.$$

All scaling factors cancel out, and we end up with,

$$\frac{\partial \vec{v}'}{\partial t'} + (\vec{v}'\cdot\nabla')\vec{v}' = -\frac{1}{\rho'}\nabla' p'. \qquad (8.39)$$

Equation (8.39) is the "scale-free" version of Euler's equation, and the fact that it is identical in form to the unscaled Eq. (8.36) means that Euler's equation is inherently "scale-free". In fact, the continuity equation (Eq. 1.19) and all flavours of the energy equation (Eq. 1.23, 1.34, and 1.35), excluding gravity, are also scale-free (Problem 8.5) and thus ideal hydrodynamics is said to be scale-free.

So what does this mean? Consider the solution to the Sod shock tube problem, illustrated in Fig. 3.9. The independent coordinate used is the self-similar coordinate, x/t.[8] This means that regardless of the value of x or t individually, the distributions of ρ, v, etc., remain identical. Thus, for example, Fig. 3.9 could be illustrating a shock happening in a five-metre long tube in a laboratory, or it could represent a trans-galactic shock wave excited by the spiral arms as they pass through the interstellar medium on length scales of thousands of parsecs. The relative levels of the primitive variables shown in the figure are the same in both cases.

Let us take this "scale-free" notion to the realm of the engineer. In designing the hull for a new sailboat, for example, the ability of the boat to "slice" through the water faster than the rival's design is paramount. If the hydrodynamics of the water responding to the hull of the boat were as scale-free as the 1-D ideal shock-tube problem would suggest, engineers could do their measurements and tests on a scaled-down toy boat in a bathtub and get identical results to the measurements taken on an actual boat in the sea. *Yet we know this not to be the case!* Design engineers do use scale models, but at quarter and even half scale and do their measurements in multi-million dollar laboratories with enormous water tanks with wave generators, measuring devices, and the like, and not bathtubs!

So what have we left out? Why do the conclusions drawn from our scaling analysis of the ideal fluid equations not apply to the "real world"? The answer is, in part, viscosity.

Let us try, now, to scale the incompressible, isothermal variation of the Navier-Stokes equation without gravity. Scaling Eq. (8.26) with the scaling laws (8.37) and (8.38), we get,

$$\frac{V}{R}\frac{\partial \vec{v}'}{\partial t'}V + \frac{V}{R}(\vec{v}'\cdot\nabla')\vec{v}'V = -\frac{1}{D}\frac{1}{\rho'}\frac{1}{R}\nabla' p' DV^2 + \nu\frac{1}{R^2}\nabla'^2 \vec{v}' V.$$

This time, most but not all of the scaling factors cancel out, and we are left with,

$$\frac{\partial \vec{v}'}{\partial t'} + (\vec{v}'\cdot\nabla')\vec{v}' = -\frac{1}{\rho'}\nabla' p' + \frac{1}{\mathcal{R}}\nabla'^2 \vec{v}',$$

---

[8] *Rappel*: Forgotten what *self-similar* means? See footnote 11 on page 88.

where,

$$\boxed{\mathcal{R} \equiv \frac{VR}{\nu},}\qquad(8.40)$$

is the so-called *Reynolds number*, named for Osborne Reynolds (1842–1912)[9] who was the first to discuss the importance of this number to fluid dynamics in a seminal paper (Reynolds, 1883). He was also one of the first scientists in Great Britain ever appointed as a "Professor of Engineering" which he held at Owen's college in Manchester.

Since V and R are measures of fiducial scaling factors at the *macroscopic* scale of the fluid, whereas the kinematic viscosity, ν, is a measure of the behaviour of fluids at microscopic scales, there is no reason to expect $\mathcal{R}$ to be the same for all applications as it would need to be if we were to declare the Navier–Stokes equation scale-free. Indeed, Reynolds numbers can vary from the very large (10^{20} for some astrophysical fluids) to the very small (10^{-6} for highly viscous, "creeping" flow), and one must specify the Reynolds number – and thus the scale of the problem – before one sets out to do a problem or an experiment in viscid hydrodynamics.

While the Reynolds number compares the relative importance of the inertial and viscous terms, the *Euler number*,[10] $\mathcal{E}$,

$$\mathcal{E} \equiv \frac{p_\mathrm{u} - p_\mathrm{d}}{DV^2},$$

compares the relative importance of the pressure gradient and the inertial terms. Here, $p_\mathrm{u} - p_\mathrm{d}$ is the upstream to downstream pressure drop, and D and V are the fiducial density and velocity already defined. Together, the Reynolds and Euler numbers describe the relative importance of all the terms in the Navier–Stokes equation. When two flows have the same values of $\mathcal{R}$ and $\mathcal{E}$, they are said to be *similar flows* and one can be used to model the other.

This brings us back to the engineer designing the new sailboat. It can be shown that the toy boat in the bathtub and the 10 m sloop slicing through the sea do *not* represent similar flows. In fact, by adding more physical attributes to our fluid description (*e.g.*, surface tension, cavitation[11]), the list of numbers that define a similar flow can get so long that the only truly *similar* flow ends up being an *identical* flow. It is for this reason design engineers are forced to work with scale models as close in size to the real objects as their facilities and budgets can manage.

The Reynolds number and "turbulence"

Loosely speaking, turbulence can be described as the mechanism by which a high-Reynolds number fluid dissipates its kinetic energy into thermal energy.[12] As for a solid, fluids dissipate kinetic energy *via* friction and, for a fluid, friction means viscosity. Thus, a fluid can dissipate its energy when the viscous stresses are compa-

---

[9] www.wikipedia.org/wiki/Osborne_Reynolds.
[10] Not to be confused with *Euler's number*, namely $e = 2.781828\ldots$
[11] Cavitation is the formation of small bubbles of water vapour, for example, owing to the agitation of, say, a propeller, that change the local nature of the fluid.
[12] In MHD turbulence, kinetic energy can also be converted to magnetic energy; see Fig. 7.2.

rable to the inertial term, and this occurs only at scale lengths where the Reynolds number is of order unity.

So how does a high-Reynolds number fluid manage to create a low-Reynolds number environment in which its kinetic energy can be dissipated? Imagine a fluid being agitated by some external mechanism which sets into motion an eddy of some scale length determined by the physical dimension of the agitator. The surface of a large eddy is a shear layer and, as such, subject to the Kelvin–Helmholtz instability (see §7.1). Thus, a high-Reynolds number eddy will immediately start to form smaller eddies (*"big whirls have little whirls..."*), beginning the so-called cascade to smaller and smaller eddies (*"little whirls have lesser whirls..."*) each with a smaller length scale than its parent, and each subject to its own regime of K–H instabilities. As the length scale of an eddy decreases, its Reynolds number decreases until such time as the Reynolds number reaches unity where viscous dissipation begins to dominate the dynamics and where the cascade to smaller and smaller length scales is finally halted. Thus, the range of turbulent motion is dictated, on the large scale, by factors external to the fluid (*e.g.*, the dimension of the agitator) and, at the small scale, by the viscous stresses internal to the fluid.

This then illustrates one qualitative difference viscosity makes to a fluid, however small that viscosity may be. For any non-zero viscosity, there exists a length scale at which the turbulent cascade is halted and at which the kinetic energy of the fluid is dissipated into heat. On the other hand, in a so-called "superfluid" (*e.g.*, "quantum fluids" such as liquid He), the cascade to smaller length scales continues right down to the molecular level where interactions are *elastic* and the fluid viscosity is demonstrably *zero*. In such fluids, motions of eddies, *etc.*, are maintained *forever*, a qualitatively different behaviour from a fluid with even a trace of viscosity.

The adjective "turbulent" is often used subjectively to refer to a fluid simultaneously exhibiting motions at "many" length scales without the word *many* being quantified. A highly viscid fluid is quite likely to appear *laminar* (opposite to "turbulent") as will a relatively inviscid fluid moving slowly enough (*e.g.*, water flowing smoothly down a gently inclined sheet of glass). Such flows exhibit a paucity of length scales and represent flow at one end of a laminar-turbulent spectrum. On the other hand, highly agitated, rapidly moving low-viscosity fluids (and thus high Reynolds number) exhibit a plethora of length scales and clearly "look" turbulent. However, what of the wide range of phenomena in between these extremes? How does one actually *measure* the extent to which a fluid is turbulent? When does turbulence actually begin? How quickly does it end once energy input is ceased? What of magnetic and gravitational effects? These and many other questions are the subject of numerous texts[13] and numerous research programmes in engineering, applied mathematics, and astrophysics, and beyond the scope of this text.

"Yardsticks" for Reynolds numbers

For $\mathcal{R} > 1$, the fluid is said to be *inertially dominated*, for $\mathcal{R} \sim 1$, the viscous and inertial terms are comparable, and for $\mathcal{R} < 1$, the fluid is said to be *dominated*

---

[13] For a senior undergraduate text that discusses (M)HD turbulence in depth, see Galtier (2016).

by viscous stresses. Some practical, "kitchen" examples may be of help here. At room temperature, the kinematic viscosities of molasses, maple syrup, and water are about 0.01, 2×10^{-4}, and 10^{-6} (all in units of $m^2 s^{-1}$) respectively. Imagine each of these fluids pouring out of a measuring cup with a ~ 1 cm wide "spout". For molasses flowing at a speed of 1 cm s^{-1}, $\mathcal{R} \sim 0.01$ (viscous dominated), for maple syrup running a little more quickly at 4 cm s^{-1}, $\mathcal{R} \sim 2$, and for water moving more rapidly still at 10 cm s^{-1}, $\mathcal{R} \sim 1,000$. These all correspond to our "intuitive" notion of viscous fluid flow: molasses is clearly viscous, maple syrup not so much, and water seemingly not at all.

At still-higher Reynolds numbers, the onset of turbulence in a pipe typically occurs at $\mathcal{R} \sim 4,000$, though in some flows confined to a boundary layer, turbulence can be "staved off" until $\mathcal{R} > 10^6$.

Finally, back to the example of the 10 m sloop in the engineer's design lab. As it slices through the water at a crisp 10 m s^{-1} (20 knots), the effective Reynolds number is 10^8, well above the onset of turbulence as can be verified by looking at the boat's wake. One can ask, therefore, at what length scale does the turbulent cascade end? That is, what is the size of the smallest eddy in the water, at which kinetic energy is finally being dissipated into thermal energy? This is easily answered by setting $\mathcal{R} = VR/\nu = 1 \Rightarrow R \sim \nu/V = 10^{-7}$ m. Thus, it is desirable to have the hull of the boat smooth to length scales of a micron or less (*e.g.*, by maintaining the paint on the hull) so that dynamical friction[14] between the boat and the turbulence it generates is kept to a minimum.

8.6 Applications

The Navier–Stokes equation – in any form – is much more complicated than Euler's equation and thus analytical solutions are known for only the simplest of cases. In large part, this added complexity arises because the viscous term is parabolic in nature (*e.g.*, App. C) and, as such, does not lend itself to the wave and eigenvalue analysis appropriate for the hyperbolic equations describing ideal (inviscid) hydrodynamics (§3.5) and even MHD (§5.2).

The applications chosen here make the following assumptions. We seek solutions in which the flow is *laminar* meaning, among other things, there is only one component of the velocity (*e.g.*, $\vec{v} = v\hat{x}_1$). Next, our examples will look for v as a function of the coordinate perpendicular to the flow direction, $v_1(x_2)$ say, with the implicit assumption that these profiles are identical for all values of x_1. Thus, in this case, $\partial_1 v_1 = 0$ (eliminating the inertial term) and $\nabla \cdot \vec{v} = 0$, rendering the fluid effectively incompressible (more specifically, that $d\rho/dt = 0$). Finally, we shall assume *slab symmetry*, and thus all 3-derivatives are zero.

---

[14]Loosely speaking, *dynamical friction* is the momentum transfer caused by collisions between macroscopic elements of two systems in contact.

With all this, Eq. (8.26) becomes:

$$\partial_t v_x = -\frac{1}{\rho}\partial_x p - \partial_x \phi + \nu \partial_y^2 v_x; \qquad (8.41)$$

$$\partial_t v_z = -\frac{1}{\rho}\partial_z p - \partial_z \phi + \frac{\nu}{r}\partial_r(r\partial_r v_z), \qquad (8.42)$$

in Cartesian and cylindrical coordinates respectively. Additional assumptions are made in the various examples as appropriate, including steady state ($\partial_t = 0$), uniform pressure ($\partial_x p = 0$), no gravity ($\phi = 0$), *etc.*

Next, all the examples in this section will involve fluids coming up against solid barriers, so we will need to impose boundary conditions at the fluid-solid interface. It is an experimental fact that if a fluid has even the slightest amount of viscosity (and here again is another qualitative difference between a fluid with exactly zero viscosity, and one with only the smallest amount), then the layer immediately adjacent to a solid surface will be in the co-moving frame of that surface. This is the so-called *no-slip* boundary condition and is represented mathematically by:

$$\vec{v} = \text{constant across boundary} \quad \text{and} \quad \vec{v} \cdot \hat{n} = 0, \qquad (8.43)$$

where $\hat{n}$ is a unit vector normal to the solid surface. The last requirement forces the fluid to have no velocity component perpendicular to the surface (it can't move into the surface, nor can it pull away lest it leave a vacuum behind). The only velocity component the fluid can have at the boundary is tangential to the boundary and, by the no-slip condition, this component must be equal to the velocity of the solid boundary itself.

8.6.1 Plane laminar viscous flow

Consider a viscous fluid between two solid parallel plates separated by a distance D, where the plates move in opposite directions with the same speed, V, as shown in Fig. 8.7. We seek the velocity profile of the fluid, $v_x(y)$, between the plates.

With no pressure or gravity gradients, Eq. (8.41) becomes:

$$\partial_t v_x = \nu \partial_y^2 v_x.$$

This is a *diffusion equation* (Eq. H.3, App. H), a parabolic PDE that describes how a quantity (in this case v_x) is diffused into a medium from the boundaries at a rate governed by the "diffusion coefficient", ν.

We can easily determine the diffusion time scale, τ_ν, by dimensional analysis (*e.g.*, §2.4), noting that it can depend only on two quantities, namely D, the gap between the plates, and ν, the kinematic viscosity. The quantity v_x is what is being diffused and, as such, does not affect the diffusion time scale. Thus, $\tau_\nu \sim D^2/\nu$. The time it takes for the velocity of the boundary layers to be transmitted to the centre of the fluid is inversely proportional to the viscosity, but proportional to the *square* of the gap. Thus, water ($\nu \sim 10^{-6}$) between two plates separated by 1 m

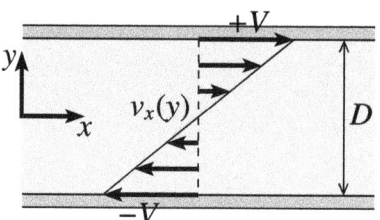

Figure 8.7. Two plates, moving in opposite directions with a viscous fluid in between them, eventually give rise to the fluid velocity profile as shown.

would take about *two weeks* to come into steady state, while a gap of 1 cm would come into steady state in *under two minutes*.

Seeking the steady-state solution, we set $\partial_t v_x = 0$, and are left with a very simple differential equation to solve,

$$\nu \partial_y^2 v_x = 0 \quad \Rightarrow \quad v_x(y) = Ay + B.$$

Applying the no-slip boundary conditions (Eq. 8.43):

$$v(\tfrac{1}{2}D) = V; \qquad v(-\tfrac{1}{2}D) = -V,$$

we solve for A and B and get as our final velocity profile,

$$\boxed{v_x(y) = \frac{2V}{D} y,}$$

which is the profile illustrated in Fig. 8.7. Note that the viscosity itself does not appear in the velocity profile. In this problem, the time it takes for the steady state to be established is dependent on ν, but not the steady-state profile itself, which is the same for all viscid fluids regardless of how small (but not zero!) the viscosity may be.

8.6.2 Forced flow between parallel plates

Consider now the steady-state flow between two stationary plates driven by a pressure gradient as shown in Fig. 8.8. With $\partial_x p = -f_x =$ constant and, once again, no gravity term, Eq. (8.41) becomes:

$$\partial_y^2 v_x = -\frac{f_x}{\mu},$$

where $\mu = \rho \nu$ is the shear viscosity. This too is trivial to integrate,

$$v_x(y) = -\frac{f_x}{2\mu} y^2 + Ay + B. \tag{8.44}$$

Applying no-slip boundary conditions at each plate gives us $v(\tfrac{D}{2}) = v(-\tfrac{D}{2}) = 0$, and thus,

$$\boxed{v_x(y) = \frac{f_x}{2\mu} \left(\frac{D^2}{4} - y^2 \right),} \tag{8.45}$$

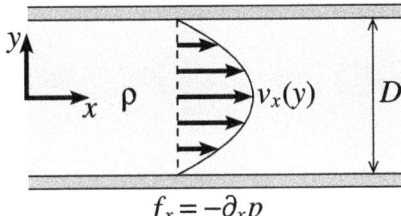

Figure 8.8. Fluid is driven by a pressure gradient between two stationary plates, giving rise to the steady-state fluid velocity profile as shown.

which is the parabolic profile shown in Fig. 8.8. Unlike the previous example, this profile does depend on the fluid viscosity.

An easily measurable and useful experimental quantity is the *discharge rate*, Q (m^3 s^{-1}), or, as we shall calculate in this case, the discharge rate per unit width, q (m^2 s^{-1}):

$$q = \int_{-D/2}^{D/2} v_x(y)\,dy. \tag{8.46}$$

Substituting Eq. (8.45) into Eq. (8.46) gives us,

$$\boxed{q = \frac{f_x D^3}{12\mu},}$$

from which the viscosity of the fluid can be easily measured.

Finally, we can calculate the *average* flow speed from the discharge rate (Eq. 8.46), namely,

$$\langle v_x \rangle D = q \quad \Rightarrow \quad \langle v_x \rangle = \frac{f_x D^2}{12\mu}.$$

8.6.3 Open channel flow

Consider a viscous liquid flowing down a smooth plane (*e.g.*, glass) inclined to the horizontal at an angle θ, as shown in Fig. 8.9a. Suppose that the fluid is *not* being replenished at the top and thus the surface gradually drains of liquid. We wish to find the flow velocity profile, $v_x(y)$, the depth of the fluid, $D(x,t)$ as a function both of position along the plane and time, and the discharge rate, $q(x,t)$. This will be a case of a system in *quasi*-steady state, in which we will still make the assumption of steady state in the mathematical analysis, but check after-the-fact to make sure that whatever time dependence has to be built into the solution is sufficiently slight as to warrant the assumption of steady state – at least to a good approximation – in the first place.

We start by assuming $D \ll L$, the length of the plane, so that all points in the fluid are sufficiently close to the surface that the fluid pressure is essentially the atmospheric pressure and $\partial_x p = 0$. The assumption of laminar flow means $v_y = 0$, though this cannot be exactly true since the decreasing depth of the liquid over

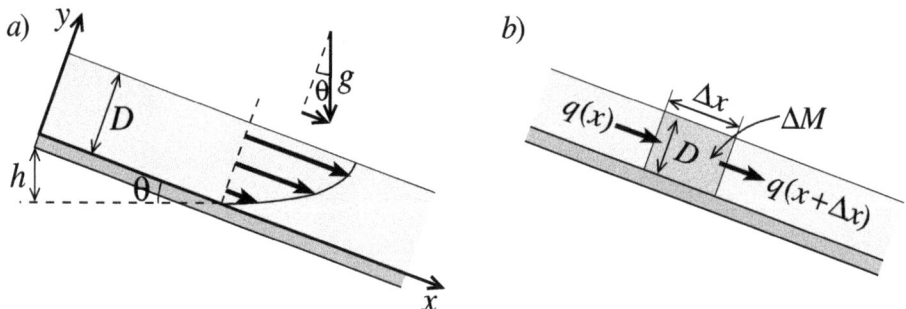

Figure 8.9. *a*) Fluid drains from the surface of a smooth plane inclined at an angle θ to the horizontal. The top "open" surface is free from viscous stresses from above, and thus the strain $(\partial_y v_x)$ is zero there. *b*) The mass, ΔM, of a fluid element of length Δx decreases as the depth, D, decreases which depends upon the difference in discharge rates on the left and right sides.

time means there has to be *some* y-component of the velocity. In as much as we can assume $v_y = 0$, incompressibility ($\nabla \cdot \vec{v} = 0$) means that $\partial_x v_x = 0$ (the problem is "slab-symmetric", and thus all z-derivatives are automatically zero), and v_x does not depend on x. Thus, the inertial term, $v_x \partial_x v_x$, is also (approximately) zero.

Finally, as fluid is flowing downhill, the driving force is gravity where the gravitational potential is given by $\phi = gh = -gx\sin\theta$ (Fig. 8.9*a*). With all these considerations, Eq. (8.41) becomes,

$$\partial_t v_x = g\sin\theta + \nu \partial_y^2 v_x(y) = 0,$$

in the steady state. This equation can be easily integrated twice to get,

$$v_x(y) = -\frac{g\sin\theta}{2\nu}y^2 + Ay + B. \tag{8.47}$$

Boundary conditions for this problem are a little different from the previous two examples. While no-slip conditions (Eq. 8.43) require $v_x(0) = 0 \Rightarrow B = 0$, the upper boundary is a "free boundary", which means it experiences no shear. Specifically, at $y = D$, $T_{xy} = \mu \partial_y v_x = 0 \Rightarrow \partial_y v_x = 0$, and we have:[15]

$$-\frac{g\sin\theta}{\nu}D + A = 0 \quad \Rightarrow \quad A = \frac{g\sin\theta}{\nu}D$$

Thus, Eq. (8.47) becomes,

$$\boxed{v_x(y) = \frac{g\sin\theta}{2\nu}(2yD - y^2),} \tag{8.48}$$

and the discharge rate is,

$$q = \int_{-D/2}^{D/2} v_x(y)dy = \frac{g\sin\theta}{3\nu}D^3. \tag{8.49}$$

---

[15] Boundary conditions in which one is applied to the function, $v_x(y)$, and the other to its first derivative, $\partial_y v_x$, are called *Cauchy boundary conditions*.

Because the fluid is draining from the plane, we expect there to be *some* temporal dependence on the variables, and in particular, D. Figure 8.9b shows a small segment of the flow over which D is very nearly constant. Continuity requires that the time-rate-of-change of the mass per unit width of the fluid element ΔM (indicated in medium grey) must be accounted for by the difference in discharge rates at the left and right sides of the element. Thus,

$$\partial_t \Delta M = \rho \Delta x \partial_t D = \rho[q(x) - q(x + \Delta x)]$$

$$\Rightarrow \quad \partial_t D = -\partial_x q = -\frac{g \sin \theta}{\nu} D^2 \partial_x D. \tag{8.50}$$

This equation can be solved using separation of variables, where we assume:

$$D(x, t) = X(x) T(t). \tag{8.51}$$

Substituting Eq. (8.51) into Eq. (8.50), and separating factors involving x and t on different sides of the equation, we get,

$$\frac{g \sin \theta}{\nu} X \frac{dX}{dx} = -\frac{1}{T^3} \frac{dT}{dt} = c,$$

where c is a constant. Solving each of X and T separately, we get,

$$X^2 = \frac{2\nu c x}{g \sin \theta}, \qquad T^2 = \frac{1}{2ct},$$

and thus,

$$\boxed{D^2(x, t) = X^2 T^2 = \frac{\nu}{g \sin \theta} \frac{x}{t}.} \tag{8.52}$$

D is evidently a self-similar quantity (depending, as it does, on x/t), which we might have guessed from the outset. Thus, the mean velocity, $\langle v_x \rangle$, is given by:

$$\langle v_x \rangle D = q \quad \Rightarrow \quad \langle v_x \rangle = \frac{g \sin \theta}{3\nu} D^2 = \frac{x}{3t},$$

using Eq. (8.49). Substituting Eq. (8.52) into Eq. (8.49), we find the x- and t-dependence of q to be,

$$\boxed{q(x, t) = \frac{1}{3} \sqrt{\frac{\nu}{g \sin \theta}} \left(\frac{x}{t} \right)^3.} \tag{8.53}$$

We are finally in a position to check our assumption of *quasi*-steady state, which is satisfied if,

$$|\partial_t v_x| \ll \nu |\partial_y^2 v_x|.$$

For the LHS, let us set,

$$|\partial_t v_x| \approx |\partial_t \langle v_x \rangle| = \frac{x}{3t^2},$$

and for the RHS, use Eq. (8.48) to find,

$$\nu |\partial_y^2 v_x| = g \sin \theta.$$

Thus, we must have:

$$\frac{x}{3t^2} \ll g\sin\theta \quad \Rightarrow \quad t \gg \sqrt{\frac{x}{3g\sin\theta}}. \tag{8.54}$$

We've also assumed v_y is zero, and thus D cannot be a strong function of x. This is easily checked by differentiating Eq. (8.52) with respect to x, and demanding this be much less than unity. That is,

$$\partial_x D = \frac{\nu}{gt\sin\theta}\frac{1}{2D} = \frac{D}{2x} \ll 1 \quad \Rightarrow \quad D \ll 2x. \tag{8.55}$$

Since $D \ll L$, Ineq. (8.55) is valid everywhere except near $x = 0$ at the top of the incline, and where the uncertain boundary conditions there make that part of the flow poorly understood anyway.

Thus, so long as conditions (8.54) and (8.55) hold, the velocity profile, liquid depth, and discharge rate are given by Eq. (8.48), (8.52), and (8.53) respectively.

A little experiment[16]

Ever rinse out a wine bottle, for example, shake out what you thought was the last of its contents, only to pick it up carelessly an hour or so later and find water pouring out onto the floor? Here's an easy "kitchen experiment" you can do to understand this phenomenon in terms of "open channel flow" described in this subsection.

Fill a glass wine bottle (straight sides) with water, then empty it. Glass works better than plastic as glass reduces surface tension effects which aren't accounted for in our theory. Once water flow is reduced to a drip, hold the bottle upside down and vertical for ten seconds more, then right the bottle, cork it, and set it aside.

After half an hour, say, uncork the bottle and carefully pour the accumulated contents into a graduated cylinder capable of measuring a few ml (or a teaspoon = 5 ml). Typically, you'll find between 1 and 2 ml of liquid pours out of the bottle.

Let us model the inside surface of the bottle – with angle of inclination 90° to the horizontal – as an open channel. For $L = 18\,\text{cm}$ (typical height of the vertical sides of a wine bottle), Ineq. (8.54) requires that,

$$t \gg \sqrt{\frac{L}{3g}} \sim 0.08,$$

which $t = 10\,\text{s}$ surely satisfies and thus the assumption of steady state is upheld. Next, from Eq. (8.52), we find (for $x = 18\,\text{cm}$, $t = 10\,\text{s}$, and $\nu = 10^{-6}\,\text{m}^2\,\text{s}^{-1}$) $D \sim 4.3 \times 10^{-5}\,\text{m}$ which is certainly much less than $2L$, as required by Ineq. (8.55). Thus, the flow may be safely considered laminar and D should be a good estimate of the thickness of the water layer clung to the inside surface of the bottle after $10\,\text{s}$. With an inside bottle radius $r = 3.5\,\text{cm}$, this leads to a volume:

$$V = 2\pi r L D = 1.7 \times 10^{-6}\,\text{m}^3 = 1.7\,\text{ml}.$$

Finally, $D \propto t^{-1/2}$ is why no matter how long you drain the bottle, there always seems to be just a little more water left inside...

---

[16] I first saw this idea in Tom Faber's text *Fluid Dynamics for Physicists*, p. 211 (Faber, 1995).

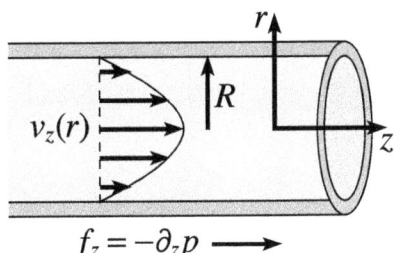

Figure 8.10. Laminar flow of a viscid fluid along the axis of a cylindrical pipe is known as Hagen–Poiseuille flow.

8.6.4 Hagen–Poiseuille flow

In this subsection, we look at the problem of forced flow along a pipe, first studied experimentally in 1838 and then theoretically in 1840 by Jean Louis Poiseuille (1799–1869), and independently by Gotthilf Heinrich Ludwig Hagen (1797–1884) in 1839. While Hagen was the first to publish his results, Poiseuille had actually done his experiments and derived the formula a year earlier.

This problem must be considered in cylindrical coordinates, (z, r, φ),[17] and is the cylindrical analogue of the problem of forced flow between parallel plates considered in §8.6.2. Of course, flow of a viscid fluid through a pipe has all sorts of applications, from anatomy to machine design, and our look here will only brush the surface of this subject.

Consider the laminar flow along the axis of an axisymmetric pipe depicted in Fig. 8.10, where the velocity field is given by $\vec{v} = v_z(r)\hat{z}$. A constant pressure gradient, $\partial_z p = -f_z$, is applied along the pipe, and gravity is ignored for a horizontal configuration. Thus in the steady state, Eq. (8.42) reduces to,

$$\partial_r(r\partial_r v_z) = -\frac{r}{\mu}f_z,$$

which can be easily integrated to yield:

$$r\partial_r v_z(r) = -\frac{f_z}{\mu}\frac{r^2}{2} + A,$$

where $A = 0$ is required at $r = 0$ to satisfy the boundary condition,

$$\partial_r v_z\big|_{r=0} = 0,$$

imposed by axisymmetry. Integrating again yields:

$$v_z(r) = -\frac{f_z}{\mu}\frac{r^2}{4} + B,$$

---

[17]While most authors list the cylindrical coordinates as (r, φ, z), I find it more useful as a computational fluid dynamicist to align the cylindrical and spherical polar coordinate φ in the third slot. The only truly Cartesian-like cylindrical coordinate, z, then goes to the first slot which has additional algorithmic conveniences. I have therefore adopted the somewhat unorthodox ordering (z, r, φ) throughout this text.

and by applying the boundary condition $v(R) = 0$,[18] we get finally:

$$\boxed{v_z(r) = \frac{f_z}{4\mu}(R^2 - r^2).}$$

This is the profile shown in Fig. 8.10.

To find the rate at which fluid flows through the pipe, we integrate the velocity profile over the cross section of the pipe to get:

$$\boxed{Q = \int_0^{2\pi} \int_0^R v_z(r) r\, dr\, d\varphi = \frac{\pi f_z R^4}{8\mu},} \qquad (8.56)$$

which has units $m^3\,s^{-1}$. Equation (8.56) is known as the *Hagen–Poiseuille equation* (or *law*) and describes what is called *Hagen–Poiseuille flow*, namely viscous flow through a pipe. The fact that Q depends on R^4 and not R^2, as one might naïvely expect, explains why a straw need not be *that* much wider to ease the task of drinking a thick milkshake.

8.6.5 Couette flow

Consider now fluid flowing azimuthally between two concentric cylinders, as depicted in Fig. 8.11a. The inner cylinder, with radius a, rotates at some steady angular velocity, ω_a, while the outer cylinder of radius b rotates at ω_b. Under the assumptions of steady-state and laminar flow, we wish to find the radial profile of the *angular* velocity, $\omega(r)$, and, in so doing, the torques transmitted from the inside to outside cylinders. Herein is one of the most oft-used designs for a *viscometer*, a device for measuring the viscosity of a fluid.

To study this problem, we must first revisit our mathematically simplistic interpretation of Newton's model for stresses on a viscid fluid element (Eq. 8.8) discussed in §8.2.2:

$$T_{ij} = \mu(\partial_i v_j + \partial_j v_i); \quad i \neq j, \qquad \text{Eq. (8.8)}$$

which turns out to be inappropriate for a rotating fluid. For example, for a cylinder of fluid rotating at a constant angular speed ω about its axis of symmetry as a "solid body", the velocity of any point in the fluid at a distance r from the axis is,

$$\vec{v} = v_\varphi \hat{\varphi} = r\omega \hat{\varphi}.$$

In this case, according to Eq. (8.8) the viscous stress between adjacent annuli of fluid should be,

$$T_{r\varphi} = \mu(\partial_r v_\varphi + \cancelto{0}{\partial_\varphi v_r}) = \mu\omega \neq 0.$$

Yet, in solid-body rotation where there is no relative motion between one annulus and another, the shear stresses should be zero! So what went wrong?

This is where overlooking the *tensor* nature of T has come to bite us. In particular, in going from Eq. (8.17) to Eq. (8.18), we took the leap of faith that the

---

[18]Again, we have Cauchy boundary conditions.

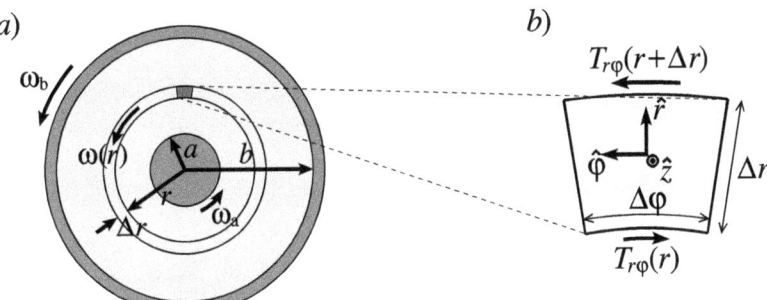

Figure 8.11. *a)* Laminar and azimuthal flow of a viscid fluid between two concentric cylinders of radii a and b is shown. A typical annulus of width Δr rotates at an angular speed of $\omega(r)$ which, if the rotation is not solid-body, will exert viscous stresses on the annuli immediately above and below it. *b)* A z-cross section through a cylindrical fluid element depicting the φ-r elements of the stress tensor as imposed by the differential rotation.

ij^{th} element of $\nabla\vec{v}$ was simply:

$$(\nabla\vec{v})_{ij} = \partial_i v_j, \tag{8.57}$$

which, since $\nabla\vec{v}$ is a tensor, is true only in Cartesian coordinates (*e.g.*, Eq. A.19). For an orthogonal coordinate system where the *length differential* is given by,

$$dl^2 = h_1^2 dx_1^2 + h_2^2 dx_2^2 + h_3^2 dx_3^2,$$

the ij^{th} element of $\nabla\vec{v}$ is, in fact, given by,[19]

$$(\nabla\vec{v})_{ij} = \begin{cases} \partial_i\left(\dfrac{v_i}{h_i}\right) + \dfrac{1}{h_i}\vec{v}\cdot\nabla h_i, & i = j; \\ \dfrac{1}{h_i}\left(\partial_i v_j - \dfrac{v_i}{h_j}\partial_j h_i\right), & i \neq j. \end{cases} \tag{8.58}$$

Here, h_1, h_2, and h_3 are the *coordinate scale factors* related to the so-called *metric* which, for cylindrical coordinates, are given by $(h_z, h_r, h_\varphi) = (1, 1, r)$.[20] Thus, Eq. (8.58) gives:

$$(\nabla\vec{v})_{r\varphi} = \frac{1}{h_r}\left(\partial_r v_\varphi - \frac{v_r}{h_\varphi}\partial_\varphi h_r\right) = \partial_r(r\omega) = \omega + r\partial_r\omega; \quad \text{and}$$

$$(\nabla\vec{v})_{\varphi r} = \frac{1}{h_\varphi}\left(\partial_\varphi v_r - \frac{v_\varphi}{h_r}\partial_r h_\varphi\right) = \frac{1}{r}(\partial_\varphi v_r - r\omega)$$

$$\Rightarrow \quad T_{r\varphi} = \mu\big((\nabla\vec{v})_{r\varphi} + (\nabla\vec{v})_{\varphi r}\big) = \mu\left(r\partial_r\omega + \frac{1}{r}\partial_\varphi v_r\right).$$

For the problem depicted in Fig. 8.11, $v_r = 0$, and we have:

$$T_{r\varphi} = \mu r\partial_r\omega, \tag{8.59}$$

---

[19] For an overview of tensor analysis with most of the basic ideas and theorems derived from first principles, see www.ap.smu.ca/~dclarke/home/documents/byDAC/tprimer.pdf, my *Primer on Tensor Calculus*.

[20] For Cartesian coordinates, $h_1 = h_2 = h_3 = 1$ in which case, Eq. (8.58) reduces to Eq. (8.57).

which gives the expected behaviour for solid-body rotation (*i.e.*, $T_{r\varphi} = 0$ for constant ω).

For $\omega = \omega(r)$, we can use Eq. (8.59) to assess the torque one annulus of fluid exerts on another. Figure 8.11b shows a cylindrical element of fluid in cross section with the r–φ stresses on the top and bottom surfaces indicated. As we are dealing with angular motion, we evaluate the net torque per unit length on the fluid element about the cylinder axis to be:

$$\tau_{\text{net}} = (r + \Delta r)^2 T_{\varphi r}(r + \Delta r)\,\Delta\varphi - r^2 T_{\varphi r}(r)\,\Delta\varphi = \partial_r(r^2 T_{\varphi r})\,\Delta r \Delta\varphi.$$

Dividing through by $r\,\Delta r\,\Delta\varphi$ gives us the "torque density", which we equate to the time rate of change of the "angular momentum density", $\ell = \rho r^2 \omega$, to get,

$$\frac{1}{r}\partial_r(r^2 T_{\varphi r}) = \rho r^2 \partial_t \omega \quad\Rightarrow\quad \partial_t \omega = \frac{\nu}{r^3}\partial_r(r^3 \partial_r \omega),$$

using Eq. (8.59), giving us a diffusion equation of sorts (App. H) for ω. This is the azimuthal Navier–Stokes equation for incompressible, isothermal fluid ignoring the pressure and inertial terms. In the steady state, we set as usual, $\partial_t = 0$, and what remains of the equation is easily integrated twice to get:

$$\omega(r) = -\frac{A}{2r^2} + B,$$

where A and B are set by imposing the boundary conditions $\omega(a) = \omega_a$ and $\omega(b) = \omega_b$. The ensuing algebra gets surprisingly complicated surprisingly quickly, yielding the final and somewhat awkward-looking result:

$$\boxed{\omega(r) = \frac{1}{b^2 - a^2}\left[b^2\omega_b - a^2\omega_a + \frac{a^2 b^2}{r^2}(\omega_a - \omega_b)\right],} \tag{8.60}$$

which simplifies considerably if either ω_a or ω_b is zero.

It is left to Problem 8.9 to show that the torque per unit length of the cylinder on a fluid annulus at radius $a \leq r \leq b$ is then given by,

$$\boxed{\tau(r) = 4\pi\mu\frac{b^2 a^2}{b^2 - a^2}(\omega_b - \omega_a),} \tag{8.61}$$

which is independent of r, as required by Newton's third law. As is also shown in Problem 8.9, this is the basis for an easily constructed and easily used viscometer. Indeed, it was yet another French physicist, Maurice Alfred Couette (1858–1943), who invented and used such a design to make the first accurate measurements of viscosities of various fluids in the late nineteenth century, and it is for this reason that azimuthal flow of a viscid fluid between two concentric cylinders is known as *Couette flow*.

Problem Set 8

8.1* Show that tr(T) is invariant under a coordinate rotation of *any* angle, θ, about the x_3-axis and not just 45° as used to derive Eq. (8.6) in the text.

The invariance of the trace of a tensor is a well-known result, and if you know anything about tensor analysis or even matrix theory, the proof is rather easy. So try instead doing this in the same spirit as was done in the text for a rotation of 45°. Thus, referring to the inset (a modification of Fig. 8.2 in the text), sum the forces on prisms ABCD and GCEF and see where this leads you. Doing it this way emphasises the link between the invariability of the trace of a tensor, and the physical reasons for it in a viscid fluid.

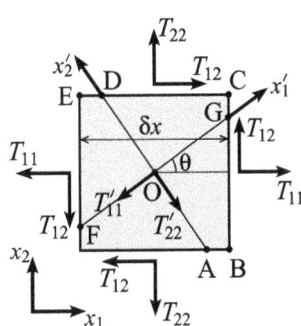

8.2 Starting from Eq. (8.27) in the text, verify Eq. (8.28) making sure that your verification shows how the transpose of T comes about. For this problem, it is sufficient to expand $\vec{T}_i$ into its Cartesian-like components, T_{ij}.

8.3 Show how Eq. (8.30) in the text follows from Eq. (8.29).

8.4*

a) Starting with Eq. (8.18) in the text, use vector identities and the definition of the colon product to verify the last equality in Eq. (8.35), namely:

$$\mathsf{T} : \nabla \vec{v} = -p \nabla \cdot \vec{v} - \frac{2\mu}{3} (\nabla \cdot \vec{v})^2 + 2\mu \, \mathsf{E} : \mathsf{E}. \qquad (8.62)$$

b) Show that together, the last two terms proportional to μ in Eq. (8.62) are never negative. This means that the viscous source terms always increase the internal energy of the fluid (or at best, leave it the same), never decrease it. (You may assume the Cartesian representation of all vector identities.)

8.5 Using the scaling relations in Eq. (8.37) and (8.38) in the text, scale the total energy equation (Eq. 1.23) in the same manner that Euler's equation (Eq. 1.36) was scaled in §8.5 including setting $\phi = 0$. Thus, show that the total energy equation for ideal hydrodynamics is scale-free.

8.6*

a) Scale Euler's equation (Eq. 1.36) as was done in §8.5 in the text, but this time include the gravitational potential. Think carefully about how you are going to scale ϕ. Just because the units of ϕ are m²s⁻², is it reasonable to scale ϕ according to $\phi = \phi' V^2$, with V being the fiducial speed of the fluid?

b) You should find that Euler's equation with $\phi \neq 0$ is *not* scale-free, and has a "left-over" factor:

$$\frac{GDR^2}{V^2} \equiv \frac{1}{\mathcal{F}^2},$$

associated with the $\nabla' \phi'$ term. $\mathcal{F}$ is the so-called *Froude number* and, depending on its value, governs how a fluid behaves under the influence of a gravitational field.

In terms of the Froude number, explain why a portion of the interstellar medium (ISM) might be gravitationally unstable (and thus prone to collapse under its own weight to form stars), whereas a representative portion of the earth's atmosphere is not. For the atmosphere, take as fiducial values $D \sim 1$ and $V \sim c_s \sim 300$ (both in mks), whereas for the ISM, take $D \sim 1.7 \times 10^{-21}$ ($n \sim 10^6$ m⁻³) and $V \sim 5 \times 10^3$ ($T \sim 10^3$ K). Think about what fiducial values for R might be appropriate in both situations.

Discussion: As viscosity leads to the *Reynold's number*, $\mathcal{R}$ (Eq. 8.40), this problem shows how gravity leads to the *Froude number*, named for the English engineer and hydrodynamicist, William Froude (1810–1879). In general, for any external field leading to a local acceleration, g, the Froude number is given by:

$$\mathcal{F} = \frac{V}{\sqrt{gR}}, \qquad (8.63)$$

where V and R are the characteristic speed and scale length of the flow, just as they are for $\mathcal{R}$. Note that for gravity, $g = GM/R^2 = GDR$.

High Froude number flow corresponds to ideal hydrodynamics where all external fields including gravity are negligible. Conversely, when engineers consider ships cutting through water or astrophysicists contemplate stars collapsing from the ISM, the Froude number can be of order unity resulting in fluid behaviour very different from ideal hydrodynamics.

8.7 A viscous fluid is driven between two plane parallel plates by a uniform pressure gradient, $\partial_x p = -f_x$, as shown in Fig. 8.12. In addition, the top plate moves to the right with speed V while the lower plate remains at rest.

a) Find the steady-state velocity profile of the fluid. You may make the same assumptions about the fluid as made in §8.6.2 of the text.

b) Find the discharge rate (per unit width) and thus the average flow velocity across the profile found in part a.

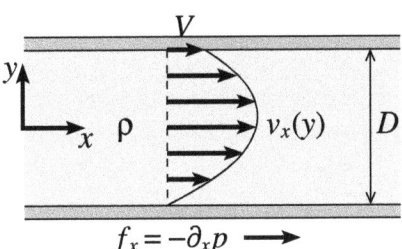

Figure 8.12. (Problem 8.7) Viscous fluid is forced between two plane parallel plates, where the top plate is moving at a velocity V relative and parallel to the bottom plate.

8.8 Suppose the space between the horizontal co-axial cylinders shown in Fig. 8.13 is filled with a viscous fluid. Suppose further the inner cylinder is dragged along the common axis at a speed V while the outer cylinder remains at rest.

a) Find the velocity profile, $v_z(r)$, of the fluid in the steady state. You may assume $v_r = v_\varphi = 0$.

b) Determine the vorticity ($\vec{\omega} = \nabla \times \vec{v}$) of the fluid and show that it has exactly the same form as the azimuthal magnetic field, B_φ, induced by a steady axial current, i. Thus, complete the following statement:

$$B_\varphi \text{ is to } i \text{ what } \omega_\varphi \text{ is to } \underline{\qquad}.$$

Hint: You may want to consult Ampère's Law.

8.9*

a) Confirm Eq. (8.61) in the text.

b) In a Couette viscometer of inner radius a and outer radius b (*e.g.*, Fig. 8.11 in the text), the inner cylinder is rotated at angular velocity ω_a and the outer cylinder is held at rest. What is the torque exerted on a unit length of the outer cylinder by a Newtonian fluid occupying the space between the cylinders?

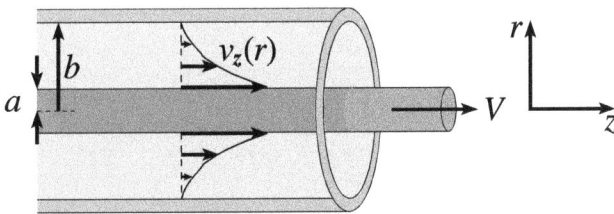

Figure 8.13. (Problem 8.8) The central cylinder is pulled with a velocity V along its axis, dragging forward the viscous fluid between the cylinders.

c) Suppose now that a horizontal spring of spring constant k connects a fixed anchor with the top of the outer cylinder so that when stretched, the spring exerts a tangential retaining force on the cylinder at $r = b$. Suppose further that when the inner cylinder is rotated at constant angular speed ω_a, the spring is stretched by $\delta x \ll b$ from its equilibrium length. What is the shear viscosity, μ, of the fluid in the viscometer in terms of k, a, b, ω_a, δx, and L, the length of the cylinder?

For illustration purposes, set $k = 5.0\,\mathrm{N\,m^{-1}}$, $a = 2.0\,\mathrm{cm}$, $b = 6.0\,\mathrm{cm}$, $L = 10\,\mathrm{cm}$, $\omega_a = 6.0\,\mathrm{rad\,s^{-1}}$, and $\delta x = 0.30\,\mathrm{cm}$, and find a numerical value for μ. What common household substance has this viscosity?

9 Steady-State MHD

with thanks to Jonathan Ramsey, Ph. D. (SMU), 2011 and Nicholas MacDonald, M. Sc. (SMU), 2008[†].

> *The steady-state theory is æsthetically and philosophically pleasing to many people, to whom it is a matter of regret that observations indicate that it is not the correct model.*
>
> Jamal Nazrul Islam (1939–2013)
> from *The Ultimate Fate of the Universe*

"CONSIDER A SPHERICAL COW" is how some of our colleagues in other disciplines of science like to gibe physicists in our approach to problem-solving. And yet, sometimes a spherical cow is all one needs for a suitable answer! While Islam's critique is specifically directed towards the prevailing view of the cosmos prior to the *Big Bang Theory*, it is equally applicable to the assumption of steady state – that nothing changes in time – in any branch of physics. The restriction to steady state can never describe properly any complex physical system writ large, no matter how appealing the simplification of the mathematics may be.

Still and on a more limited basis, the assumption of steady state can often be sufficiently realised in the lab, in simulations, and even in nature that useful insight and measurable results can be achieved. In our case, the equations of MHD are simply too complex to dismiss the opportunity of *some* analytic headway even if it isn't exactly "the correct model".

This chapter is a generalisation of our discussion in §2.4 where Bernoulli's theorem was introduced for ordinary hydrodynamics. Other than the restriction to inviscid flow, the only assumption made there was steady-state flow, where all time derivatives are set to zero ($\partial_t = 0$). As we'll see, MHD requires the additional assumption of azimuthal symmetry ($\partial_\varphi = 0$ in cylindrical or spherical polar coordinates) from which not one, but *four* constants of motion along streamlines (as defined in Fig. 2.8 in §2.4 and which we'll see are the same as lines of induction) can be found. These include three constants of the flow for which there are no hydrodynamical analogues (§9.1), plus a generalisation to the Bernoulli function (§9.2). The rest of the chapter examines how these steady-state constants can be applied to two examples from astrophysics: stellar winds and astrophysical jets.

[†]Ph. D. (Boston), 2016

9.1 The Weber–Davis constants

While it seems that Weber & Davis (1967; hereafter WD67) were the first to derive and apply astrophysically what I'm going to refer to as the three *Weber–Davis constants*, my treatment actually follows Spruit (1996; hereafter S96) whose derivations are more general, and who includes a derivation of the MHD Bernoulli function as well.

Consider an ideal MHD system (*i.e.*, where Alfvén's theorem 4.2 applies and magnetic flux is "frozen in" the fluid) in steady state ($\partial_t = 0$). Further – and in addition to what was required for Bernoulli's theorem in ordinary hydrodynamics (§2.4) – we'll need to assume azimuthal symmetry, a.k.a. axisymmetry, which is most conveniently accomplished in cylindrical coordinates by setting $\partial_\varphi = 0$. As both our astrophysical applications of steady-state MHD are to rotating systems, rotation is included in our derivations.

A fundamental quantity in axisymmetric ideal MHD is the so-called *flux function*, f. Since $\vec{B}$ is solenoidal, we saw in §4.8 that $\vec{B}$ can be written in terms of a *vector potential*, $\vec{A}$, which, in cylindrical coordinates with axisymmetry, appears as (third of Eq. Set A.40),

$$\vec{B} = \nabla \times \vec{A} = \frac{1}{r}\partial_r(rA_\varphi)\hat{z} - \partial_z A_\varphi \hat{r} + \left(\partial_z A_r - \partial_r A_z\right)\hat{\varphi}.$$

In particular, the *poloidal* component of $\vec{B}$ (that part within the z–r plane) is given by,

$$\vec{B}_\mathrm{p} = B_z\hat{z} + B_r\hat{r} = \frac{1}{r}\left(\partial_r f \hat{z} - \partial_z f \hat{r}\right) = \frac{1}{r}\nabla f \times \hat{\varphi}, \qquad (9.1)$$

where $f \equiv rA_\varphi$ is the flux function (units Wb, same as Φ_B in Eq. 4.8). Evidently, $\vec{B}_\mathrm{p} \perp \nabla f$ and thus contours of f are everywhere tangential to $\vec{B}_\mathrm{p}$. Put another way, f is constant along poloidal lines of induction and thus f can be used to *label* each line of induction uniquely.[1] In the discussion that follows, we'll recognise other quantities as constant along poloidal lines of induction (and thus functions only of f) by their gradients being everywhere perpendicular to $\vec{B}_\mathrm{p}$.

At the basis of our discussion are the ideal MHD equations from Chap. 4 (Eq. 4.20, 4.21, 4.14, and 4.23) with the assumption of steady state:

$$\nabla \cdot (\rho \vec{v}) = 0; \qquad \text{(continuity)} \qquad (9.2)$$

$$\nabla \cdot (e_\mathrm{T} \vec{v}) = -\nabla \cdot \left(p\vec{v} + \frac{1}{\mu_0}\vec{B} \times (\vec{v} \times \vec{B})\right); \qquad \text{(total energy)} \qquad (9.3)$$

$$(\vec{v} \cdot \nabla)\vec{v} = -\frac{1}{\rho}\nabla p - \nabla \phi + \frac{1}{\mu_0 \rho}(\nabla \times \vec{B}) \times \vec{B}; \qquad \text{(Euler)} \qquad (9.4)$$

$$\nabla \times (\vec{v} \times \vec{B}) = 0. \qquad \text{(ideal induction)} \qquad (9.5)$$

---

[1] Readers who followed §7.4.2 describing the Parker instability will recognise the flux function f as playing the same role in 2-D cylindrical coordinates as A_x did in 2-D Cartesian coordinates with x-symmetry. That is, contours of each function follow lines of induction in their respective geometry.

Beginning with Eq. (9.5), since $\vec{v} \times \vec{B}$ is irrotational, it may be written as the gradient of a scalar potential, $\mathcal{V}$,[2]

$$\vec{v} \times \vec{B} = \nabla \mathcal{V}. \tag{9.6}$$

Now, consider $\vec{v}$ and $\vec{B}$ broken up into their *poloidal* (in the z–r plane) and *azimuthal* (in the φ-direction) components:

$$\vec{v} = \vec{v}_{\rm p} + v_\varphi \hat{\varphi}; \qquad \vec{B} = \vec{B}_{\rm p} + B_\varphi \hat{\varphi}. \tag{9.7}$$

Equation (9.6) then becomes,

$$\underbrace{\vec{v}_{\rm p} \times \vec{B}_{\rm p}}_{\propto \hat{\varphi}} + \underbrace{B_\varphi \vec{v}_{\rm p} \times \hat{\varphi} + v_\varphi \hat{\varphi} \times \vec{B}_{\rm p}}_{\text{in the poloidal plane}} + v_\varphi B_\varphi \cancelto{0}{\hat{\varphi} \times \hat{\varphi}} = \nabla_{\rm p} \mathcal{V} + \frac{1}{r}\cancelto{0}{\partial_\varphi \mathcal{V}}\hat{\varphi}. \tag{9.8}$$

The fact that $\hat{\varphi} \times \hat{\varphi} = 0$ is self-evident, but the other terms may warrant comment. The first term is a cross product of two vectors within the poloidal plane which therefore points in the azimuthal direction, as indicted by the underbrace. The second underbrace recognises that vectors in the poloidal plane crossed with $\hat{\varphi}$ remain in the poloidal plane; *e.g.*, $\hat{r} \times \hat{\varphi} = \hat{z}$. On the RHS, $\nabla_{\rm p} = \partial_z \hat{z} + \partial_r \hat{r}$ is the poloidal gradient operator, while the azimuthal component of the gradient operator vanishes because of axisymmetry. *It is setting this last term to zero that makes the rest of this chapter possible, and why axisymmetry was imposed in the first place.* For without it, we could not make this critical observation. On equating the φ-components on the LHS and RHS of Eq. (9.8), we get,

$$\vec{v}_{\rm p} \times \vec{B}_{\rm p} = 0 \quad \Rightarrow \quad \vec{v}_{\rm p} \parallel \vec{B}_{\rm p} \quad \Rightarrow \quad \vec{v}_{\rm p} = \lambda(z, r) \vec{B}_{\rm p}, \tag{9.9}$$

where λ is some coordinate-dependent scalar function which could be positive or negative. That is, *in steady-state, axisymmetric, ideal MHD, the poloidal velocity and magnetic induction are everywhere parallel or anti-parallel!* Among other things, this means that anything found to be constant along a line of induction is also constant along a streamline. Quite literally, the rest of our discussion in this chapter depends upon this fact.

Equating the poloidal components on the LHS and RHS of Eq. (9.8) and noting that with axisymmetry, $\nabla_{\rm p} \mathcal{V} = \nabla \mathcal{V}$, we find,

$$B_\varphi \vec{v}_{\rm p} \times \hat{\varphi} + v_\varphi \hat{\varphi} \times \vec{B}_{\rm p} = (v_\varphi - \lambda B_\varphi) \hat{\varphi} \times \vec{B}_{\rm p} = \nabla \mathcal{V}, \tag{9.10}$$

using Eq. (9.9) to replace $\vec{v}_{\rm p}$. Thus, $\vec{B}_{\rm p} \cdot \nabla \mathcal{V} = 0$ and, like the flux function f, the scalar potential $\mathcal{V}$ is constant along poloidal lines of induction. That is, all contours of $\mathcal{V}$ and f line up, and $\mathcal{V}$ can be considered as a function exclusively of f:

$$\mathcal{V} = \mathcal{V}(f) \quad \Rightarrow \quad \nabla \mathcal{V} = \frac{d\mathcal{V}}{df} \nabla f \equiv \Omega_0(f) \nabla f, \tag{9.11}$$

where $\Omega_0 = d\mathcal{V}/df$ is also a function exclusively of f. But, since ∇f lies in the poloidal plane (remember, $\partial_\varphi = 0$),

$$\nabla f = \hat{\varphi} \times (\nabla f \times \hat{\varphi}) = \hat{\varphi} \times (r\vec{B}_{\rm p}),$$

---

[2] This is the same scalar function, $\mathcal{V}$, introduced in Eq. (4.33).

where the first equality is illustrated by the inset, and the second equality is from Eq. (9.1). Substituting this into Eq. (9.11), and then comparing that to Eq. (9.10), we get,

$$(v_\varphi - \lambda B_\varphi) \hat{\varphi} \times \vec{B}_\mathrm{p} = r\Omega_0(f) \hat{\varphi} \times \vec{B}_\mathrm{p}$$

$$\Rightarrow \boxed{\Omega_0(f) = \frac{1}{r}(v_\varphi - \lambda B_\varphi),} \qquad (9.12)$$

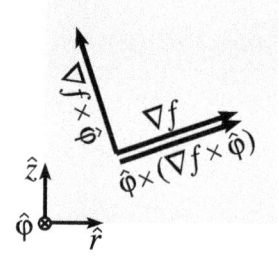

the first of the three Weber–Davis (WD) constants (Eq. 5 in WD67; Eq. 12 in S96) which, evidently, has units of angular frequency whence its label, Ω_0.

Before I give a physical interpretation of Eq. (9.12), I offer Fig. 9.1 for some astrophysical context and a toy model of how an azimuthal component is twisted out of a purely poloidal line of induction (labelled f in the figure) by a rotating, axisymmetric disc. The salient points of this model are described in some detail in the figure caption and so I shan't repeat them here. Most important to the current discussion, along the portion of f inside the so-called *Alfvén point* (labelled A at $r = r_\mathrm{A}$ in Fig. 9.1), the poloidal Alfvén number $A_\mathrm{p} < 1$[3] and the Lorentz force dominates fluid inertia. Thus, the "stiffness" of $\vec{B}$ prevents it from being twisted too much out of the poloidal plane, and the magnitude of the azimuthal component of magnetic induction, B_φ, remains small compared to the poloidal component, B_p. However, beyond A where $A_\mathrm{p} > 1$, fluid inertia dominates magnetic stresses and the rate at which f is bent "into the page" increases significantly (Fig. 9.1b–d) as does the magnitude of B_φ which quickly exceeds B_p.

OK, so back to Eq. (9.12). Since Ω_0 is a function exclusively of f, it must be constant along the line of induction identified by f and so its value can be determined at any point along that line. Thus, at its *footpoint* on the equatorial plane (F in Fig. 9.1) where $r = r_0$, $v_\varphi = v_{\varphi,0}$, and $B_\varphi = 0$, Eq. (9.12) gives us,

$$\Omega_0(f) = \frac{v_{\varphi,0}}{r_0}, \qquad (9.13)$$

and we interpret Ω_0 as the *angular speed* of the footpoint of f as it rotates with the disc. Along f beyond the footpoint, $B_\varphi \neq 0$ and, from Eq. (9.12), the local angular speed is evidently given by,

$$\Omega(f, r) = \frac{v_\varphi}{r} = \Omega_0(f) + \frac{\lambda B_\varphi}{r}. \qquad (9.14)$$

Now, as depicted in Fig. 9.1 and explained in its figure caption, $B_\varphi < 0$ for $z > 0$. Further, both $\vec{B}_\mathrm{p}$ and $\vec{v}_\mathrm{p}$ point away from F ($\vec{B}_\mathrm{p} \parallel \vec{v}_\mathrm{p}$) making $\lambda > 0$. For $z < 0$, $B_\varphi > 0$ while $\vec{B}_\mathrm{p}$ is *anti*parallel to $\vec{v}_\mathrm{p}$ (the former points towards F while the latter still points away), making $\lambda < 0$. Thus, $\lambda B_\varphi < 0$ everywhere and Eq. (9.14) requires $\Omega(f, r) \leq \Omega_0(f)$. That is to say, fluid flowing along f[4] at $r > r_0$ lags *behind* the

---

[3] *Rappel*: The *Alfvén number* is the ratio of the flow and Alfvén speeds. Here, $A_\mathrm{p} = v_\mathrm{p}/a_\mathrm{p}$ where v_p is the flow speed in the poloidal plane and $a_\mathrm{p} = B_\mathrm{p}/\sqrt{\mu_0 \rho}$ is the poloidal Alfvén speed.

[4] We haven't established yet that fluid is driven outwards along lines of induction. We'll assume it for now, with verification awaiting discussion on the MHD Bernoulli function in §9.2.

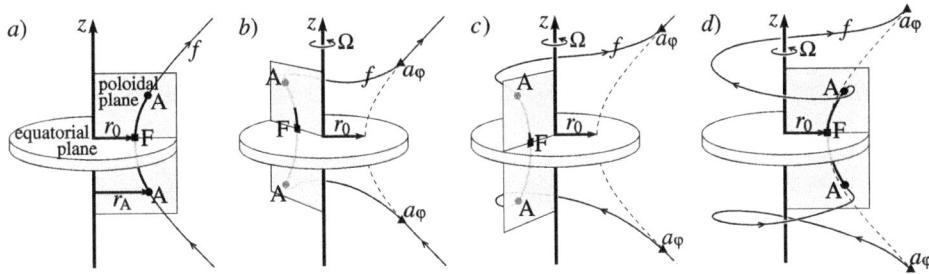

Figure 9.1. *a*) With no rotation, a single line of induction, f, (*e.g.*, of a dipole magnetic field in a stellar accretion disc) confined to a given poloidal plane passes through and is anchored to the equatorial plane (*e.g.*, the accretion disc) at its *footpoint*, F (black squares), a distance r_0 from the symmetry axis, z. With reflection symmetry across the equatorial plane and $\vec{B}$ pointing generally upwards as shown, $\vec{B}$ is axial at F ($B_r = 0$), acquires an outwardly pointing radial component above F ($B_r > 0$), and an inwardly pointing component below ($B_r < 0$). Between F and the *Alfvén point*, A, a distance r_A from the z-axis (black circles) where $A_p = v_p/a_p < 1$, the Lorentz force density, $\vec{f}_L$, dominates fluid inertia and f is drawn with a heavy line. Beyond A where $A_p > 1$, fluid inertia dominates $\vec{f}_L$ and f is drawn with a finer line. *b*) With $\Omega > 0$ and after the disc rotates through 120°, the dominant $\vec{B}$ in $r_0 \leq r \leq r_A$ resists, but does not prevent being twisted out of the poloidal plane and a small component of azimuthal induction, B_φ, develops. Beyond r_A, fluid forces dominate and differential rotation is better able to bend f out of the poloidal plane creating a significant B_φ. Note that counterclockwise rotation with a generally upward $\vec{B}$ twists out a clockwise B_φ (< 0) above the equatorial plane ($z > 0$), and a counterclockwise B_φ (> 0) below. The rotation also launches a torsion Alfvén wave whose progress along f is marked by black triangles and labelled a_φ, the azimuthal Alfvén speed. Thus, the twisted portion of f ends at a_φ beyond which f rejoins the original poloidal line of induction (dashed lines). *c*) As the disc rotates through 240°, B_φ continues to be twisted out of f while the torsion Alfvén wave moves further out along f. *d*) After a full rotation, a complete loop of B_φ has twisted out of f above and below the equatorial plane. As the torsion Alfvén wave moves outwards along f at speed a_φ, it brings more and more of the original poloidal line of induction into what is becoming a tightly wound magnetic helix.

footpoint, F, by an angular speed,

$$\Omega_{\text{lag}} = \frac{\lambda B_\varphi}{r} < 0, \tag{9.15}$$

twisting the azimuthal component, B_φ, out of f. Note that Ω_{lag} both creates B_φ and is proportional to it, giving rise to an exponential growth of B_φ in the vicinity of the Alfvén point. Such growth cannot go on forever and indeed, B_φ saturates as v_φ in Eq. (9.14) asymptotes to zero; remember, $\lambda B_\varphi < 0$. The four panels in Fig. 9.1 illustrate this progression.

Turning now to the steady-state continuity equation (Eq. 9.2), since $\partial_\varphi = 0$, we have,

$$0 = \nabla \cdot (\rho \vec{v}) = \nabla \cdot (\rho \vec{v}_p) = \nabla \cdot (\rho \lambda \vec{B}_p),$$

using Eq. (9.9) for the third equality. Thus,

$$\nabla \cdot (\rho \lambda \vec{B}_\text{p}) = \vec{B}_\text{p} \cdot \nabla(\rho \lambda) + \rho \lambda \underbrace{\nabla \cdot \vec{B}}_{0} = 0, \qquad (9.16)$$

invoking $\partial_\varphi = 0$ once again to complete the divergence of $\vec{B}$. Since $\vec{B}_\text{p} \cdot \nabla(\rho \lambda) = 0$, we can immediately conclude that $\rho \lambda$ is also a function of f only and,

$$\boxed{\eta(f) \equiv \rho \lambda = \frac{\rho v_\text{p}}{B_\text{p}},} \qquad (9.17)$$

is constant along a line of induction (streamline) in steady-state φ-symmetry. This is the second of three WD constants (equivalent to Eq. 8 in WD67; Eq. 13 in S96).

The quantity η, with units $\text{kg}\,\text{m}^{-2}\,\text{s}^{-1}\,\text{T}^{-1}$, is known as the *mass load*, and Eq. (9.17) asserts that the poloidal mass flux density, ρv_p, per unit poloidal magnetic flux density, B_p, is constant along a line of induction. This is the steady-state 2-D manifestation of *flux freezing*, as required by Alfvén's theorem 4.2 for ideal MHD. Note that η inherits the sign of λ. Thus, $\eta > 0$ for $z > 0$ and $\eta < 0$ for $z < 0$.

Unlike Ω_0, we cannot evaluate η at the footpoint, F, anchored as it is to the equatorial (symmetry) plane, for here at $z = 0$, $v_\text{p} = 0$ (but $B_\text{p} = B_z \neq 0$) and thus $\rho \to \infty$ in order for η to remain non-zero. This, of course, is completely unrealistic and one of the vagaries of assuming a steady-state axisymmetric system. So instead, we evaluate η at the Alfvén point (A in Fig. 9.1) where $v_\text{p} = a_\text{p} = B_\text{p}/\sqrt{\rho_\text{A} \mu_0}$, and,

$$\eta = \sqrt{\frac{\rho_\text{A}}{\mu_0}}, \qquad (9.18)$$

where ρ_A is the fluid density at the Alfvén point.

Next, consider the φ-component of the steady-state Euler equation, Eq. (9.4):

$$(\vec{v} \cdot \nabla \vec{v})_\varphi = \frac{1}{\mu_0 \rho} ((\nabla \times \vec{B}) \times \vec{B})_\varphi = \frac{1}{\mu_0 \rho} (\vec{B} \cdot \nabla \vec{B})_\varphi, \qquad (9.19)$$

using Identity (A.19) and setting $\partial_\varphi = 0$. Now, from the fifth of Eq. Set (A.40), the φ-component of $\vec{A} \cdot \nabla \vec{A}$ in cylindrical coordinates for an arbitrary vector, $\vec{A}$, is,

$$(\vec{A} \cdot \nabla \vec{A})_\varphi = \vec{A} \cdot \nabla A_\varphi + \frac{A_\varphi A_r}{r} = \frac{\vec{A}}{r} \cdot \nabla(r A_\varphi).$$

Thus, Eq. (9.19) becomes,

$$\vec{v} \cdot \nabla(r v_\varphi) = \frac{1}{\mu_0 \rho} \vec{B} \cdot \nabla(r B_\varphi)$$

$$\Rightarrow \quad \frac{1}{\lambda} \vec{v}_\text{p} \cdot \nabla(r v_\varphi) = \vec{B}_\text{p} \cdot \nabla(r v_\varphi) = \frac{1}{\rho \lambda \mu_0} \vec{B}_\text{p} \cdot \nabla(r B_\varphi), \qquad (9.20)$$

since, yet again, $\partial_\varphi = 0$ and $\vec{B}_\text{p} = \vec{v}_\text{p}/\lambda$ (Eq. 9.9). Now, since,

$$\vec{B}_\text{p} \cdot \nabla\left(\frac{r B_\varphi}{\rho \lambda}\right) = \frac{1}{\rho \lambda} \vec{B}_\text{p} \cdot \nabla(r B_\varphi) - \frac{r B_\varphi}{(\rho \lambda)^2} \underbrace{\vec{B}_\text{p} \cdot \nabla(\rho \lambda)}_{= 0 \ (\text{Eq. 9.16})},$$

Eq. (9.20) becomes,
$$\vec{B}_p \cdot \nabla \left(r v_\varphi - \frac{r B_\varphi}{\eta \mu_0} \right) = 0,$$

using Eq. (9.17), identifying another scalar whose gradient is perpendicular to $\vec{B}_p$. That is,

$$\boxed{l(f) \equiv r\left(v_\varphi - \frac{B_\varphi}{\eta \mu_0}\right),} \qquad (9.21)$$

is a constant along a line of induction (streamline). Further, since $B_\varphi/\eta < 0$, both terms on the RHS of Eq. (9.21) are positive. This is the third and final WD constant (equivalent to Eq. 9 in WD67; Eq. 19 in S96) which we'll refer to as the *specific angular momentum*.

Writing Eq. (9.21) as,

$$l(f) = \frac{r}{\eta B_p}\left(\eta B_p v_\varphi - \frac{B_p B_\varphi}{\mu_0}\right) = \frac{1}{\eta B_p}\left(v_p r \rho v_\varphi - \frac{r B_p B_\varphi}{\mu_0}\right), \qquad (9.22)$$

then the quantity $v_p r \rho v_\varphi$ on the RHS can be interpreted as the *angular momentum flux density*, i.e., angular momentum per unit area per unit time whose units are $(\mathrm{kg\,m^2\,s^{-1}})\,\mathrm{m^{-2}\,s^{-1}} = \mathrm{kg\,s^{-2}}$. Therefore, the full term, $v_p r \rho v_\varphi/(\eta B_p) = r v_\varphi$, is interpreted as the angular momentum per unit mass transported along a poloidal line of induction.

The quantity $r B_p B_\varphi/\mu_0$ is the *magnetic torque density* (per unit area) with units $\mathrm{m\,T^2/(T\,m\,A^{-1})} = \mathrm{N\,m/m^2} = \mathrm{kg\,s^{-2}}$ that the line of induction exerts on the pencil of flux-frozen fluid flowing along f (a.k.a., a *flux tube*) as it bends in the azimuthal direction. That is, bent lines of induction work to straighten themselves out. The full term, $-r B_p B_\varphi/(\mu_0 \eta B_p) > 0$ with units $\mathrm{N\,m\,s\,kg^{-1}}$ is the *magnetic torque impulse* per unit mass along a poloidal line of induction.

For $r \gtrsim r_0$ (Fig. 9.1), $B_\varphi \sim 0$ and the magnetic torque density is negligible. For $r \gg r_A$, $v_\varphi \sim 0$ and the angular momentum flux density becomes negligible. At points in between, an increasing magnetic torque density comes at the expense of angular momentum flux density so that $l(f)$ remains constant along a line of induction. Thus, magnetic torque works against angular momentum as fluid flows outwards until such time as v_φ asymptotes to zero.

Aside: Magnetic torque and moment

That the second term on the RHS of Eq. (9.22) may be interpreted as a magnetic torque may not be obvious to the reader, whence this aside.

Figure 9.2 illustrates a section of the flux tube tied to f within the poloidal plane. As explained in the Fig. 9.1 caption, $\vec{B}_\varphi$ is in the clockwise $(-\hat{\varphi})$ direction above the equatorial plane, and thus shown pointing out of the page in Fig. 9.2. Meanwhile, $\vec{B}_p$ is everywhere tangent to f and thus shown emerging from the end of the flux tube section.

This is entirely analogous to a current-carrying solenoid (a few coils of which are shown in Fig. 9.2 to reinforce the analogy) supporting a magnetic induction, $\vec{B}_\mathrm{p}$, along its axis given by Ampère's law,[5]

$$\vec{B}_\mathrm{p} = \frac{\mu_0 Ni}{V}\vec{\Sigma}. \qquad (9.23)$$

Here, N is the number of coils in the solenoid, i is the current it carries, and $V = L\Sigma$ is its volume where Σ is its cross-sectional area and L is its length.

If the solenoid is embedded in an *external* magnetic induction, $\vec{B}_\mathrm{ext}$, then by the law of Biot and Savart a net magnetic torque is exerted on the solenoid given by,

$$\vec{\tau} = \vec{\mu} \times \vec{B}_\mathrm{ext},$$

where $\vec{\mu} = Ni\vec{\Sigma}$ is the *magnetic moment* of the solenoid. Evidently, it is the misalignment of $\vec{\mu}$ and $\vec{B}_\mathrm{ext}$ that generates a torque causing one to twist into the other.

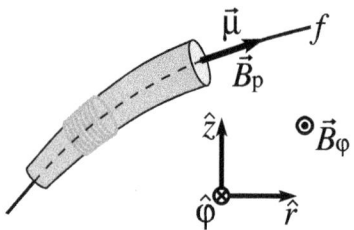

Figure 9.2. A flux tube tied to f with $\vec{B}_\mathrm{p}$ pointing along its axis is analogous to a solenoid (grey coils) with cross-sectional area, Σ, and N coils carrying a current, i, generating the same $\vec{B}_\mathrm{p}$ along its symmetry axis. Its magnetic moment, $\vec{\mu} = Ni\vec{\Sigma}$, interacts with $\vec{B}_\varphi$ generating a torque that acts to straighten out $\vec{B}$. See text for details.

Now, from Eq. (9.23), $Ni\vec{\Sigma} = V\vec{B}_\mathrm{p}/\mu_0$, and we may write,

$$\vec{\tau} = \vec{\mu} \times \vec{B}_\mathrm{ext} = Ni\vec{\Sigma} \times \vec{B}_\mathrm{ext} = \frac{V}{\mu_0}\vec{B}_\mathrm{p} \times \vec{B}_\mathrm{ext},$$

and we see that a magnetic torque can also be interpreted as the misalignment between two magnetic induction vectors or, equivalently, a severely bent single line of induction. Anyone who has played with toy magnets or has used a compass will intuit the phenomenon that misaligned magnetic fields try to align.

For the flux tube in Fig. 9.2, $\vec{B}_\mathrm{ext} = \vec{B}_\varphi$ and the torque per unit area acting on the flux-frozen fluid to realign $\vec{B}_\mathrm{p}$ with $\vec{B}_\varphi$ and reacting to the bend in $\vec{B}$ is,

$$\frac{\vec{\tau}}{\Sigma} = \frac{L}{\mu_0}\vec{B}_\mathrm{p} \times \vec{B}_\varphi = \frac{B_\mathrm{p}B_\varphi}{\mu_0}\vec{L} \times \hat{\varphi},$$

since $\vec{B}_\mathrm{p} \parallel \vec{L} = r\hat{r} + z\hat{z}$. The z-component of this expression is evidently,

$$\boxed{\frac{\tau_z}{\Sigma} = \frac{rB_\mathrm{p}B_\varphi}{\mu_0}}, \qquad (9.24)$$

justifying the second term on the RHS of Eq. (9.22).

Like $\eta(f)$, $l(f)$ as given by Eq. (9.21) is not easily evaluated at the footpoint of f (F in Fig. 9.1). While it is true that $B_\varphi = 0$ at F, so is $\eta = \rho\lambda = \rho v_\mathrm{p}/B_\mathrm{p} = 0$

---

[5]The reader unfamiliar with Ampère's law and the law of Biot and Savart (coming up soon) is referred to the chapters on magnetism in any good first year textbook. My personal favourite is Halliday, Resnick, & Walker (2003).

since, at the equatorial plane, $v_p = v_z = 0$ by symmetry. Instead, we evaluate $l(f)$ at the poloidal Alfvén point (A in Fig. 9.1).

To this end, first solve Eq. (9.12) for B_φ and then substitute that into Eq. (9.21) to get,

$$l = r\left(v_\varphi - \frac{v_\varphi - r\Omega_0}{\lambda\eta\mu_0}\right). \tag{9.25}$$

But,

$$\lambda\eta\mu_0 = \lambda^2\rho\mu_0 = \frac{v_p^2}{B_p^2/\mu_0\rho} = \frac{v_p^2}{a_p^2} = A_p^2, \tag{9.26}$$

using Eq. (9.9) and (9.17) and where A_p is the poloidal Alfvén number (see footnote 3 on page 344). Substituting this into Eq. (9.25) and subtracting $r^2\Omega_0$ from both sides, we get,

$$l - r^2\Omega_0 = r\left(v_\varphi - r\Omega_0 - \frac{v_\varphi - r\Omega_0}{A_p^2}\right)$$

$$\Rightarrow v_\varphi - r\Omega_0 = \frac{l - r^2\Omega_0}{r(1 - 1/A_p^2)}, \tag{9.27}$$

after a little algebra.

Equation (9.27) has a potential pole at the poloidal Alfvén point, A, where $A_p = 1$. Since the LHS of Eq. (9.27) must be finite everywhere including A, the numerator on the RHS must also be zero at $A_p = 1$. That is, at A where $r = r_A$,

$$l = r_A^2 \Omega_0 = \frac{r_A^2}{r_0} v_{\varphi,0}, \tag{9.28}$$

using Eq. (9.13), which is equivalent to Eq. 21 in S96. Thus, the constant value of l may be interpreted as the specific angular momentum at the Alfvén point in a frame co-rotating with the footpoint of f. Problem 9.1 gives an alternate derivation of Eq. (9.28) that avoids the Alfvén point altogether.

I leave it to Problem 9.2 to confirm these alternate versions of Eq. (9.12), (9.17), and (9.21) expressed in terms of A_p:

$$\Omega_0(f) = \frac{1}{r}(v_\varphi - A_p a_\varphi) = \frac{v_{\varphi,0}}{r_0}; \tag{9.29}$$

$$\eta(f) = A_p\sqrt{\frac{\rho}{\mu_0}} = \sqrt{\frac{\rho_A}{\mu_0}}; \tag{9.30}$$

$$l(f) = r\left(v_\varphi - \frac{a_\varphi}{A_p}\right) = \frac{r_A^2}{r_0} v_{\varphi,0}. \tag{9.31}$$

9.2 The MHD Bernoulli function

The fourth and final constant along a line of induction/streamline is the MHD Bernoulli function. For the hydrodynamical case, we approached this in §2.4 by examining the total energy equation in steady state; a similar approach for the

MHD problem is relegated to Problem 9.3. Here, we follow S96 and derive the Bernoulli function from the steady-state Euler equation, Eq. (9.4), in a reference frame, O', with origin at $(z, r) = (0, 0)$ co-rotating with the footpoint, F, at angular speed Ω_0. In this frame, the fluid velocity is given by,

$$\vec{v}' = \vec{v} - r\Omega_0(f)\hat{\varphi} = \underbrace{\vec{v} - v_\varphi \hat{\varphi}}_{\vec{v}_p = \lambda \vec{B}_p} + \lambda B_\varphi \hat{\varphi} = \lambda \vec{B}, \qquad (9.32)$$

using Eq. (9.9) and (9.12). Thus, in frame O' – and in this frame alone – not only is $\vec{v}_p \parallel \vec{B}_p$, but $\vec{v}' \parallel \vec{B}$. Note that O' is different for each line of induction, f, whose footpoints are at different distances, r_0, from the rotation axis and rotate with different angular speeds, Ω_0.

Since O' is an accelerating frame of reference, the appropriate inertial accelerations, namely Coriolis and centrifugal,[6] must be included in Eq. (9.4), whence,

$$(\vec{v}' \cdot \nabla)\vec{v}' = -\frac{1}{\rho}\nabla p - \nabla \phi + \frac{1}{\mu_0 \rho}(\nabla \times \vec{B}) \times \vec{B} + \underbrace{2\vec{v}' \times \Omega_0 \hat{z}}_{\text{Coriolis}} + \underbrace{\Omega_0^2 \vec{r}}_{\text{cent.}}, \qquad (9.33)$$

where $\vec{r} = r\hat{r}$ is the radial displacement of a point along f from the rotation axis.[7]

Let s be the arc length from the equatorial plane along a line of induction, f, and let $\hat{s}$ be a unit vector everywhere tangential to f. Thus, we may write $\vec{B} = B\hat{s}$ and, taking the dot product of $\hat{s}$ with Eq. (9.33), we get,

$$\hat{s} \cdot \left((\vec{v}' \cdot \nabla)\vec{v}'\right) = -\frac{1}{\rho}\hat{s} \cdot \nabla p - \hat{s} \cdot \nabla \phi + \Omega_0^2 \hat{s} \cdot \vec{r}, \qquad (9.34)$$

where both the Lorentz and Coriolis terms drop out (since $\hat{s} \parallel \vec{B} \parallel \vec{v}'$; Eq. 9.32). Now, it might strike the reader as rather curious that of all things, the *Lorentz force* – that which distinguishes MHD from ordinary HD – drops out in this frame of reference! Indeed, even the Coriolis force drops out, and it is left solely to the centrifugal force to distinguish this from the pure hydrodynamical case! As the following discussion shows and exploits, this is a peculiar property of this particular frame of reference; from all other vantage points, Coriolis and/or magnetic terms appear explicitly. Of course, in the end all reference frames must agree on whatever flow is driven by these forces, and that reconciliation comes grace of the WD constants derived in the previous subsection (*e.g.*, Problems 9.4 and 9.5). For now, we remain in the frame of reference, O', co-rotating with the footpoint of f because it is here where the MHD Bernoulli function is most easily derived and interpreted.

For convenience, define the spatial derivative along the line of induction, f, including as it bends into the azimuthal direction as,

$$\frac{\partial}{\partial s} \equiv \partial_s \equiv \hat{s} \cdot \nabla = \frac{1}{B}\vec{B} \cdot \nabla \underset{\substack{\uparrow \\ \because \partial_\varphi = 0}}{=} \frac{1}{B}\vec{B}_p \cdot \nabla = \frac{B_p}{B}\hat{s}_p \cdot \nabla \equiv \frac{B_p}{B}\partial_{s_p}, \qquad (9.35)$$

---

[6] The reader unfamiliar with Coriolis' theorem that derives these terms is directed to App. G.
[7] See Problem 9.4 for how Eq. (9.33) can be derived from an inertial frame of reference with no direct mention of inertial accelerations.

where $\hat{s}_\mathrm{p}$ is the unit vector everywhere tangential to the *poloidal* component of the magnetic induction (exclusive of the azimuthal direction), and where $\partial_{s_\mathrm{p}} = \hat{s}_\mathrm{p} \cdot \nabla$ is the spatial derivative with respect to s_p, the arc length along the *poloidal* component of $\vec{B}$. Evidently from this string of operator equalities, if $\partial_{s_\mathrm{p}} q = 0$ (*e.g.*, $q =$ a WD constant), then $\partial_s q = 0$ as well.

Considering now the LHS of Eq. (9.34), we have from Identity (A.19),

$$\hat{s} \cdot ((\vec{v}' \cdot \nabla) \vec{v}') = \hat{s} \cdot \nabla \frac{v'^2}{2} - \hat{s} \cdot (\vec{v}' \times (\nabla \times \vec{v}'))^{\,0} = \partial_s \frac{v'^2}{2}, \qquad (9.36)$$

since once again, $\hat{s} \parallel \vec{v}'$.

Next, here's an identity I bet most readers aren't familiar with; certainly I wasn't until I prepared this subsection! With I the identity matrix (tensor) and since $\nabla \vec{r} = \mathsf{I}$ (Eq. A.29),

$$\hat{s} \cdot \vec{r} = \vec{r} \cdot (\hat{s} \cdot \mathsf{I}) = \vec{r} \cdot (\hat{s} \cdot \nabla \vec{r}) = \vec{r} \cdot \partial_s \vec{r} = \tfrac{1}{2} \partial_s r^2. \qquad (9.37)$$

This is a general result; the dot product between any unit vector and the displacement vector, $\vec{r}$, is the same as half the spatial derivative in the direction of the unit vector of r^2. Who knew? And, as an immediate corollary, we also have,

$$\hat{s} \cdot \hat{r} = \frac{1}{r} \hat{s} \cdot \vec{r} = \frac{1}{2r} \partial_s r^2 = \partial_s r. \qquad (9.38)$$

Thus, the last term in Eq. (9.34) can be written as,

$$\Omega_0^2 \hat{s} \cdot \vec{r} = \frac{\Omega_0^2}{2} \partial_s r^2 = \frac{1}{2} \partial_s (\Omega_0 r)^2, \qquad (9.39)$$

since $\partial_s \Omega_0 = 0$ (a WD constant).

Finally, for an adiabatic equation of state (first defined on page 28), $p = \kappa \rho^\gamma$ where κ and γ are constants, and we can write,

$$\frac{1}{\rho} \hat{s} \cdot \nabla p = \kappa \gamma \rho^{\gamma - 2} \partial_s \rho = \frac{\gamma}{\gamma - 1} \partial_s \kappa \rho^{\gamma - 1} = \frac{\gamma}{\gamma - 1} \partial_s \left(\frac{p}{\rho}\right) = \partial_s h, \qquad (9.40)$$

where h is the enthalpy of the gas (Eq. 2.73).

Substituting Eq. (9.36), (9.39), and (9.40) into Eq. (9.34) we get,

$$\partial_s \left(\frac{v'^2}{2} + h + \phi - \frac{(\Omega_0 r)^2}{2} \right) = 0$$

$$\Rightarrow \quad \mathcal{B}_\mathrm{M}(f) \equiv \frac{v'^2}{2} + h + \phi - \frac{(\Omega_0 r)^2}{2}, \qquad (9.41)$$

is constant along the line of induction (streamline), and thus a function of the flux function, f, alone. This is the *MHD Bernoulli function* which, with the first part of Eq. (9.32),

$$\vec{v}' = \vec{v} - r \Omega_0 \hat{\varphi} = \vec{v}_\mathrm{p} + (v_\varphi - \Omega_0 r) \hat{\varphi},$$

becomes,

$$\boxed{\mathcal{B}_\mathrm{M}(f) = \frac{v_\mathrm{p}^2}{2} + \frac{1}{2}(v_\varphi - \Omega_0 r)^2 + h + \phi - \frac{(\Omega_0 r)^2}{2},} \qquad (9.42)$$

or, alternately,

$$\mathcal{B}_{\mathrm{M}}(f) = \frac{v^2}{2} + \hbar + \phi - v_\varphi \Omega_0 r, \qquad (9.43)$$

(equivalent to Eq. 30 and 31 in S96). Setting any of Eq. (9.41)–(9.43) to a constant along a line of induction (streamline) is *Bernoulli's theorem for MHD*. Note that the first three terms on the RHS of Eq. (9.43) are identical to those in the hydrodynamical Bernoulli function for gas ($\mathcal{B}_{\mathrm{gas}}$; Eq. 2.72), and it is only the fourth term that distinguishes $\mathcal{B}_{\mathrm{M}}$ from $\mathcal{B}_{\mathrm{gas}}$. Still and curiously enough, this distinguishing term doesn't appear to be magnetic.

From Eq. (9.41), we see that in the rotating frame of reference, $\mathcal{B}_{\mathrm{M}}$ includes terms for the specific kinetic energy, the enthalpy, the gravitational potential, and something loosely referred to as the *specific centrifugal energy*, $\frac{1}{2}(\Omega_0 r)^2$, whose role is much more apparent from Eq. (9.42). If something were to provide the fluid with a certain rigidity, say a strong poloidal magnetic induction (*e.g.*, Fig. 9.1), then at least out to a certain distance, say the Alfvén point, the fluid may more or less undergo solid-body rotation ($v_\varphi \sim \Omega_0 r$), in which case, Eq. (9.42) reduces to,

$$\mathcal{B}_{\mathrm{M}}(f) \approx \frac{v_{\mathrm{p}}^2}{2} + \hbar + \phi - \frac{(\Omega_0 r)^2}{2}.$$

Ignoring for the moment fluctuations in $\hbar$ and ϕ, as r increases $v_{\mathrm{p}}^2/2$ must rise to compensate for an increasingly negative centrifugal term if $\mathcal{B}_{\mathrm{M}}$ is to remain constant along f. Thus, we arrive at the most important physical conclusion from this analysis: *In a rotating, axisymmetric magnetised fluid in the steady state, the poloidal velocity of the fluid must accelerate outward along a line of induction/streamline, f, at least until the Alfvén point.*

Of course, flow cannot accelerate forever and Eq. (9.42) comes with its own built-in limiter. We've already seen that once past the Alfvén point (A in Fig. 9.1), B_φ briefly grows exponentially from $\vec{B}_{\mathrm{p}}$ (discussion after Eq. 9.15) until such time as v_φ asymptotes to zero (Eq. 9.14) and B_φ to its final value. In this stead, Eq. (9.42) reduces to,

$$\mathcal{B}_{\mathrm{M}}(f) \approx \frac{v_{\mathrm{p}}^2}{2} + \hbar + \phi,$$

recovering $\mathcal{B}_{\mathrm{gas}}$ (Eq. 2.72) with no apparent mechanism for flow to continue accelerating outwards.

So what is the role of $\vec{B}$?

To be sure, magnetic effects are fundamentally responsible for the poloidal acceleration. For example, without the presumed rigidity of the fluid afforded by a dominant magnetic induction inside the Alfvén point and to a lesser extent beyond, the mechanism described above could not accelerate flow along f. The analogy that comes to mind is the bead-on-a-rod problem depicted in Fig. 9.3 and presented in detail in Example G.1 (App. G). Here, we ask: 'What is the acceleration of a bead free to slide along a frictionless rod flung out in a horizontal plane?' Intuitively, it should be obvious that the bead is accelerated and launched by such an action. Those familiar with Example G.1 could, in the rotating frame of reference, identify the accelerant as the centrifugal force and, given the angular speed of the fling,

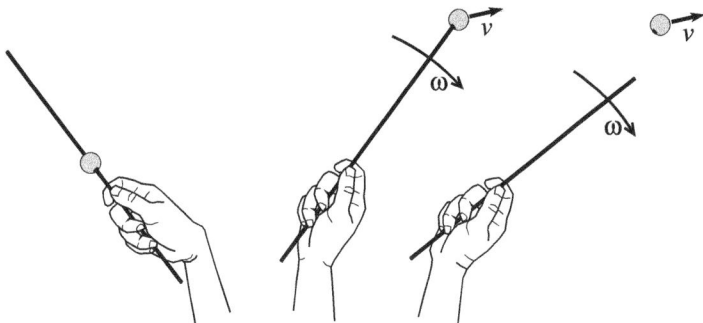

Figure 9.3. A time-sequence of drawings depicting a bead being flung out in the horizontal plane along a frictionless rod. From the inertial frame of reference (*e.g.*, the owner of the hand), the bead is launched by the normal force exerted by the rod on the bead. In the rotating frame of reference of the rod, the bead is flung out by a centrifugal force. While the two frames disagree on the nature of the force, both agree on the bead's acceleration and when the bead passes the tip of the rod.

calculate the speed of the bead along the rod as a function of time, its acceleration, *etc.* Of course, as viewed from an inertial frame of reference, the real force responsible for the acceleration is the *normal force* exerted by the rod on the bead; those for whom this claim is counter-intuitive are particularly directed to Example G.1.

In the current discussion of a rotating ideal MHD fluid, the "rod" is the "stiffness" of the magnetic induction and the "bead" is a parcel of "flux-frozen fluid" flowing along f. Thus, as interpreted by the rotating frame of reference, O', flow is accelerated by the centrifugal terms in Eq. (9.41)–(9.43), whereas the non-rotating inertial frame of reference attributes the poloidal acceleration to, as it turns out, two distinct magnetic effects, only one of which is analogous to the "bead-on-a-rod" illustrated in Fig. 9.3.

To expose these magnetic terms explicitly, set $v^2 = v_\mathrm{p}^2 + v_\varphi^2$ and take the derivative of Eq. (9.43) with respect to s to get,

$$\partial_s \mathcal{B}_\mathrm{M} = \partial_s \left(\frac{v_\mathrm{p}^2}{2} + \frac{v_\varphi^2}{2} + \hbar + \phi - v_\varphi \Omega_0 r \right) = 0. \tag{9.44}$$

I then leave it to Problem 9.5 to show that by examining the problem in an inertial reference frame and invoking all three WD constants, Eq. (9.44) leads to,

$$\partial_s \left(\frac{v_\mathrm{p}^2}{2} + \hbar + \phi \right) = \frac{v_\varphi^2}{r} \hat{s} \cdot \hat{r} - \frac{1}{2\rho\mu_0 r^2} \partial_s (r B_\varphi)^2. \tag{9.45}$$

And so terms dependent upon the magnetic induction (direction of $\vec{B}_\mathrm{p}$, $\hat{s}$, and the azimuthal component, B_φ) were there all along, hiding within the terms dependent upon v_φ. Ignoring for the moment contributions from $\hbar$ and ϕ, Eq. (9.45) tells us that $\partial_s(v_\mathrm{p}^2/2) > 0$ (*i.e.*, flow accelerates along the line of induction) provided $\hat{s} \cdot \hat{r} > 0$ and/or $\partial_s(r B_\varphi^2) < 0$.

The first term on the RHS of Eq. (9.45) is associated with the "bead-on-a-rod" mechanism (BRM) discussed above. This centrifugal-like term accelerates material along a streamline so long as $\hat{s}$ has a significant radial component. While the criterion stated above merely asks that $\hat{s} \cdot \hat{r} > 0$, the discussion below in §9.2.1 shows that by taking proper account of the gravitational potential, ϕ, the BRM is triggered only when $\hat{s} \cdot \hat{r} > 0.5$, that is when $\vec{B}_p$ emerges from the equatorial plane at an angle less than $60°$. Once flow passes the Alfvén point (A in Fig. 9.1), the line of magnetic induction, f, bends dramatically into the azimuthal direction and develops a significant $\hat{\varphi}$-component (*i.e.*, $\hat{s} \cdot \hat{r}$ falls below 0.5), stifling the BRM.

The second term on the RHS of Eq. (9.45) is a gradient of a radially weighted magnetic pressure along f and, so long as this is negative, flow is accelerated outwards. As the disc rotates and the torsional Alfvén wave (marked a_φ in Fig. 9.1) advances, more of the poloidal line of induction is twisted into the magnetic helix, while lines of induction already in the helix get wrapped up more tightly. Thus, the magnetic pressure associated with B_φ^2 is greater just beyond A than further out and, so long as $(rB_\varphi)^2$ decreases with r, the so-called "magnetic tower mechanism" (MTM; Lynden-Bell, 1996) creates an outwardly directed magnetic force. As a result, material can be gently accelerated outwards well beyond the Alfvén point and even beyond the fast point where fluid flow exceeds the fast speed (Eq. 5.25). This is explored a little further in §9.4 while Problem 9.6 explores further the MTM.

9.2.1 Critical launching angle

It seems, then, that in a rotating, strongly magnetised axisymmetric MHD system in steady state, poloidal acceleration of fluid is inevitable by two *magneto-rotational* mechanisms: "bead on a rod" (BRM); and the "magnetic tower" (MTM). Within the Alfvén point, A, where $\vec{B}_p$ dominates the fluid inertia and B_φ is minimal, the BRM is the primary accelerant. Beyond A where fluid inertia dominates $\vec{B}_p$ and $B_\varphi > B_p$, the MTM takes over as the primary mechanism which can continue accelerating fluid even beyond the fast point.

As the dominant mechanism inside A, the BRM must also be responsible for liberating material from the predominant source of matter, namely the accretion disc itself. So let's think about this for a moment. As described above, the BRM is predicated on the rod being swung in the plane into which the bead is flung. That is, the rotation vector of the rod, $\vec{\Omega}$, is perpendicular to the displacement vector along the rod, $\vec{s}$. In fact, $\vec{\Omega}$ and $\vec{s}$ need not be fully orthogonal, but they certainly can't be parallel! That is, a rod spinning about its long axis would never be able to launch a bead centrifugally.

And yet, that's exactly what we have at the footpoint, F, of the line of induction, f, in Fig. 9.1. At the equatorial plane, $\vec{B}$ is pictured with a z-component only and thus parallel to $\vec{\Omega}$. True, further out lines of induction develop a significant radial component along which – as we've demonstrated – material can be accelerated centrifugally. But how does material get *launched* from the disc when there, the "rod" is parallel to the rotation axis?

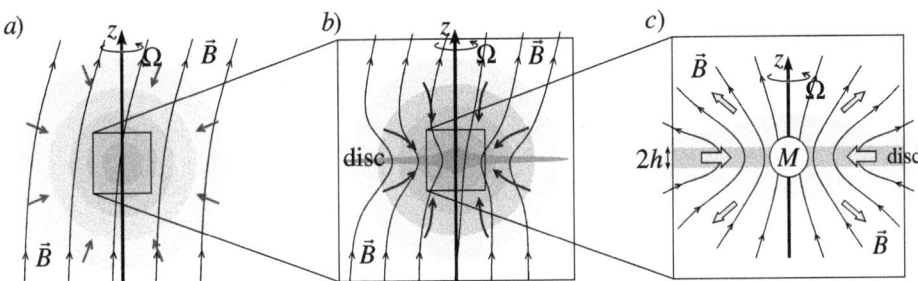

Figure 9.4. *a*) A barely Jeans-stable,[8] slowly rotating interstellar cloud with a weak $\vec{B}$ roughly aligned with its rotation axis is perturbed by the passage of a nearby star or shock wave and begins to collapse under its own weight (grey arrows pointing radially inwards). *b*) As collapse towards the centre progresses, lines of induction are dragged inwards along the equatorial plane forming an "hourglass" configuration. Collapse also increases the rotation speed giving rise to a "centrifugal barrier" where centrifugal and gravitational forces balance, redirecting radial flow towards the equatorial plane forming an accretion disc. *c*) Complex dynamics allow disc material to make its way to the gravitational centre (large inward-pointing arrows) where a "protostar" of mass M forms. Concentration of matter increases both the magnetic induction strength and rotation rate triggering magneto-rotational outflow from the disc surface (smaller outward-pointing arrows). *Key point*: Because of the finite thickness of the disc ($2h$) and the degree to which lines of induction have been dragged inwards from afar, lines of induction can emerge from the disc surface at angles substantially less than $90°$.

At this juncture, the reader would benefit from a quick tutorial on how protostars and their accompanying accretion discs form from an interstellar gas cloud. The theory and current understanding of this truly fundamental question in astrophysics is complicated and involved, and the ambitious reader is referred to the excellent and well-cited *123-page* review article by McKee & Ostriker (2007). Of course, Wikipedia's page on *Star formation* (listed with the McKee & Ostriker reference) gives a more concise review.

The figure caption to Fig. 9.4 gives what I readily acknowledge is a vast oversimplification of the process which I do without apologies so as not to distract unduly from the current discussion. The prescient result here is that once formed, lines of induction actually emerge from the surface of the accretion disc at angles substantially *less* than the $90°$ pictured in Fig. 9.1 (*e.g.*, see Fig. 9.4*c*). In fact, lines of induction can leave the disc surface at such small angles that they practically lie along the disc itself.

Let ϵ be the angle between the line of induction, f, and the surface of the accretion disc at the footpoint, F, as shown in Fig. 9.5*a*. We know that at $\epsilon = 0°$ the BRM is in full force, while at $\epsilon = 90°$, the BRM is completely stifled. Therefore,

---

[8] A "Jeans-stable" interstellar cloud is one whose internal pressure gradients (thermal, magnetic, cosmic rays) are sufficient to prevent collapse of the cloud under its own self-gravity. This so-called *Jeans condition* is named for its discoverer, Sir James Jeans (1877–1946).

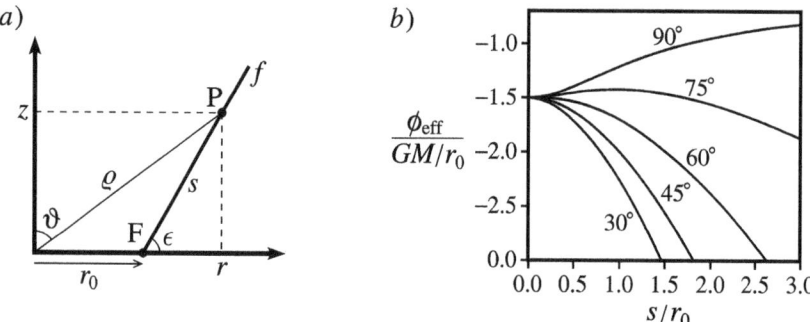

Figure 9.5. *a*) A line of induction, f, exits the surface of an accretion disc at its footpoint, F, with angle ϵ. In axisymmetry, an arbitrary point, P, along f has cylindrical coordinates (z, r), spherical polar coordinates (ϱ, ϑ), and is located a distance s from F which is a distance r_0 from the rotation axis. *b*) The scaled effective potential plotted against the scaled distance along f from the disc surface for five different values of ϵ. Only for $\epsilon = 90°$ does ϕ_{eff} increase monotonically and is thus attractive. For $\epsilon < 90°$, some or all of ϕ_{eff} decreases monotonically where it is repulsive.

there must be some critical angle between these two extremes, ϵ_{crit}, below which the BRM works and above which it does not.

To find ϵ_{crit}, consider the *effective potential* in the co-rotating frame of reference that includes both gravitational and centrifugal terms (last two terms in Eq. 9.42),

$$\phi_{\text{eff}}(s) = -\frac{GM}{\varrho(s)} - \frac{1}{2}\Omega_0^2 r^2(s), \qquad (9.46)$$

where ϱ is the distance between the gravitating mass (protostar), M, and an arbitrary point, P, along f, and r is the distance between P and the rotation axis (Fig. 9.5*a*). For reasons that shall become clear, we choose to parameterise Eq. (9.46) with the distance coordinate along f, s, first defined just before Eq. (9.34). Note that the functional dependence of ϱ and r on s can be gleaned from Fig. 9.5*a*.

There are two things of particular note here. First, I remind the reader we are assuming that this portion of f is co-rotating with its footpoint, F, at angular speed Ω_0. This, of course, presumes $\vec{B}$ is strong enough to enforce co-rotation which, as we've discussed, is a fair assumption only inside the Alfvén point ($s < s_A$) and, in particular, for s not too far from F. Beyond the Alfvén point, of course, this assumption fails completely. Second, we explicitly ignore the enthalpy, h, and thus make the assumption that the disc gas is *cold*.[9]

Figure 9.5*b* shows ϕ_{eff} from Eq. (9.46) in units of the gravitational potential at F, plotted as a function of s in units of r_0, at five different values of ϵ. Only for $\epsilon = 90°$ where $r = r_0$ does ϕ_{eff} increase monotonically with s, whose scaled value

---

[9]In this context, "cold" means both the Alfvén and Keplerian speeds are much greater than the sound speed. It doesn't necessarily mean that the temperature is too low for any of the gas to be ionised.

asymptotes to,
$$\frac{1}{GM/r_0} \lim_{s\to\infty} \phi_{\text{eff}}(s) = -\frac{\Omega_0^2 r_0^3}{2GM} = -\frac{1}{2},$$
since $\varrho(\infty) \to \infty$, and for Keplerian rotation, $\Omega_0^2 = GM/r_0^3$. Thus, at $\epsilon = 90°$, the effective force is always *attractive* and no fluid is accelerated outwards. For $\epsilon = 75°$, say, ϕ_{eff} increases from $s = 0$ until it reaches a maximum at the equilibrium point, $s = s_{\text{eq}}$, then decreases thereafter as the centrifugal term takes over. Evidently, for $s < s_{\text{eq}}$, ϕ_{eff} is attractive while for $s > s_{\text{eq}}$, ϕ_{eff} is repulsive. Thus, if disc gas can somehow rise above the potential barrier in $0 < s < s_{\text{eq}}$ (*e.g.*, thermal agitation if the gas isn't "too cold"), then fluid could be accelerated outwards once it passes s_{eq}. Finally, for $\epsilon = 30°$, say, ϕ_{eff} decreases monotonically right from the disc surface, and material can be directly launched from the disc by the BRM.

And so we see that the property of ϕ_{eff} that distinguishes whether disc gas must first overcome a potential barrier or whether it can be launched immediately from the disc surface is the sign of its first derivative at $s = 0$. For $\partial_s \phi_{\text{eff}}\big|_{s=0} > 0$, ϕ_{eff} is concave upward and the gas is presented with a potential barrier. Conversely, for $\partial_s \phi_{\text{eff}}\big|_{s=0} < 0$, ϕ_{eff} is concave *downward* and repulsive right from the "git-go". The *critical angle*, ϵ_{crit}, that separates these two possibilities is then where ϕ_{eff} is neither concave upward nor downward, that is at the inflection point where its second derivative at $s = 0$ is zero:
$$\frac{d^2 \phi_{\text{eff}}}{ds^2}\bigg|_{s=0} = 0.$$
I leave it to Problem 9.7 to show that this criterion leads to,
$$\boxed{\epsilon_{\text{crit}} = 60°.} \tag{9.47}$$

9.3 Stellar winds

That the sun could drive a spherically symmetric *wind* into what was thought to be the complete vacuum of interplanetary space was a hard sell, but Eugene Parker – the same Parker of the Parker instability in §7.4 – convinced the then editor of the Astrophysical Journal, Chandrasekhar himself, to overrule the referees and allow what would become a seminal work to be published (Parker, 1958). Coincidentally, this paper appeared just as MHD was making its own way into the mainstream of physics and, nine years later, WD67 would bring these two ideas together to explain how mass and angular momentum are transported away from the sun by Parker's *solar wind*.

In this section, we'll use what we've learned about steady-state MHD, namely the Bernoulli MHD function and the three WD constants, to follow how Weber and Davis were able to derive profiles for the density, poloidal velocity, and other quantities as functions of the coordinate, s, as material accelerates outwards along a given line of induction, f_0.

For boundary conditions, we'll assume B_p, Ω_0, and r_0 are known at the footpoint F of f_0, and that $\vec{B}_p$ is strong enough to be *force-free* $(\vec{J} \times \vec{B} = 0)$[10] so that electromagnetic forces that might otherwise alter its initial configuration are zero. This buys us two important simplifications. First, we can take $\vec{B}_p$ to be a given initial and enduring condition *everywhere*. Second, f_0 will more or less co-rotate with the footpoint, F, at least until the Alfvén point.

In this context, consider once again the MHD Bernoulli function, reproduced here for convenience:

$$\mathcal{B}_M(f) = \frac{v_p^2}{2} + \frac{1}{2}(v_\varphi - \Omega_0 r)^2 + \hbar + \phi - \frac{(\Omega_0 r)^2}{2}. \qquad \text{Eq. (9.42)}$$

Now, from Eq. (9.17), we have,

$$v_p = \frac{\eta B_p}{\rho},$$

and, by combining Eq. (9.26)–(9.28), we have,

$$v_\varphi - \Omega_0 r = \frac{\Omega_0(r_A^2 - r^2)}{r(1 - \rho/\eta^2 \mu_0)} = \frac{\Omega_0(r_A^2 - r^2)}{r(1 - \rho/\rho_A)}, \qquad (9.48)$$

using Eq. (9.18) for the last equality. Substitute these and Eq. (2.73) into Eq. (9.42) to get,

$$\mathcal{B}_M(f) = \frac{1}{2}\left(\frac{\eta B_p}{\rho}\right)^2 + \frac{1}{2}\left(\frac{\Omega_0(r_A^2 - r^2)}{r(1 - \rho/\rho_A)}\right)^2 + \frac{\gamma p}{(\gamma - 1)\rho} - \frac{\Omega_0^2 r_0^3}{\varrho} - \frac{(\Omega_0 r)^2}{2}, \qquad (9.49)$$

since $\phi = -GM/\varrho$ and, for Keplerian rotation in the disc, $GM/r_0^2 = \Omega_0^2 r_0$ at the footpoint, F (Fig. 9.5a). Thus, for a barotropic equation of state, $p = p(\rho)$,[11] $\mathcal{B}_M$ in Eq. (9.49) can be considered as a function of ρ, r, and ϱ (with B_p, Ω_0, and r_0 set as boundary conditions) assuming the parameters η and r_A can be determined; more on that later.

However, both r and ϱ can be considered as functions of s. How? For that we go back to the beginning of the chapter and Eq. (9.1), namely,

$$\vec{B}_p = B_z \hat{z} + B_r \hat{r} = \frac{1}{r}(\partial_r f \hat{z} - \partial_z f \hat{r}),$$

which tells us that if we know B_p as a function of (z, r) – which we do as an initial and enduring condition – we also know the flux function, f, as a function of (z, r). Thus, selecting a particular f_0 is tantamount to specifying a particular path in (z, r) space; that is,

$$f(z, r) = f_0 \quad \Rightarrow \quad z = z(r, f_0).$$

Now, an incremental displacement, $d\vec{s}$, along f_0 has magnitude (see inset),

$$ds = \sqrt{dr^2 + dz^2} = dr\sqrt{1 + \left(\frac{dz}{dr}\right)^2},$$

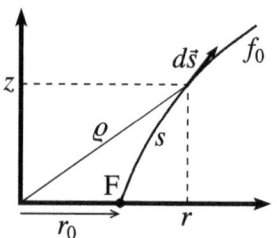

---

[10] See Problem 9.8 for an example of how such a $\vec{B}$ can be set to within a constant.
[11] *Rappel*: Both isothermal ($p = c_{iso}^2 \rho$) and adiabatic ($p = \kappa \rho^\gamma$) equations of state are barotropes.

where dz/dr is also a function of r. Thus,

$$s = \int_{r_0}^{r} ds = \int_{r_0}^{r} \sqrt{1 + \left(\frac{dz}{dr}\right)^2}\, dr \equiv s(r),$$

which, in principle, can be inverted to give $r = r(s)$. Then, since $\varrho = \sqrt{r^2 + z^2}$, we also have,

$$\varrho(s) = \sqrt{r^2(s) + z^2(r(s))},$$

and both r and ϱ are functions of s as claimed. Therefore, given values for η and r_A, $\mathcal{B}_M$ in Eq. (9.49) can be thought of as a function of ρ and s only. Thus, given a value of $\mathcal{B}_M$ for the chosen f_0 (more on this later too), Eq. (9.49) can, in principle, be solved for $\rho(s)$ which is our desired end. For once $\rho(s)$ is known, the functions $v_p(s)$ and others soon follow.

And so it remains to determine values for η (or equivalently, ρ_A), r_A, and $\mathcal{B}_M$ for a given poloidal line of induction, f_0. For these, we'll need an additional constraint independent of Eq. (9.49) which we'll get by examining its so-called *critical points*.

9.3.1 Critical points

The *critical points* of a given function are typically where the first derivative with respect to the independent variables are zero. Since the MHD Bernoulli function, $\mathcal{B}_M(\rho, s)$, is constant along a poloidal line of induction, $\partial_s \mathcal{B}_M = 0$ automatically and so to find the critical points of Eq. (9.49) we need only seek points where $\partial_\rho \mathcal{B}_M = 0$.

To this end, we take derivatives with respect to ρ of Eq. (9.49) term by term. Starting with the first term on the RHS, we have,

$$\frac{1}{2}\partial_\rho \left(\frac{\eta B_p}{\rho}\right)^2 = \frac{\eta B_p}{\rho}\left(-\frac{\eta B_p}{\rho^2}\right) = -\frac{1}{\rho}\left(\frac{\eta B_p}{\rho}\right)^2 = -\frac{v_p^2}{\rho}, \qquad (9.50)$$

using Eq. (9.17). Next,

$$\frac{1}{2}\partial_\rho \left(\frac{\Omega_0(r_A^2 - r^2)}{r(1 - \rho/\rho_A)}\right)^2 = \frac{\Omega_0(r_A^2 - r^2)}{r(1 - \rho/\rho_A)}\, \partial_\rho \left(\frac{\Omega_0(r_A^2 - r^2)}{r(1 - \rho/\rho_A)}\right)$$

$$= \frac{\Omega_0(r_A^2 - r^2)}{r(1 - \rho/\rho_A)}\left(-\frac{\Omega_0(r_A^2 - r^2)}{r(1 - \rho/\rho_A)^2}\right)\left(-\frac{1}{\rho_A}\right)$$

$$= \underbrace{\left(\frac{\Omega_0(r_A^2 - r^2)}{r(1 - \rho/\rho_A)^2}\right)^2}_{v_\varphi - \Omega_0 r = \lambda B_\varphi}\frac{1}{\rho_A - \rho} = \frac{1}{\rho}\frac{(\lambda B_\varphi)^2}{\rho_A/\rho - 1},$$

where the underbrace is justified by Eq. (9.48) and (9.12). But $\lambda = v_p/B_p$ (Eq. 9.9) and $\rho/\rho_A = \eta^2\mu_0/\rho = v_p^2/a_p^2$ (Eq. 9.17, 9.18, and 9.26). Thus,

$$\frac{1}{2}\partial_\rho \left(\frac{\Omega_0(r_A^2 - r^2)}{r(1 - \rho/\rho_A)}\right)^2 = \frac{1}{\rho}\frac{v_p^2}{B_p^2}\frac{B_\varphi^2}{v_p^2/a_p^2 - 1} = \frac{1}{\rho}\frac{v_p^2 a_\varphi^2}{v_p^2 - a_p^2}, \qquad (9.51)$$

since $B_\varphi^2/B_p^2 = a_\varphi^2/a_p^2$.[12] Next, for an adiabatic gas ($p = \kappa \rho^\gamma$),

$$\partial_\rho \hbar = \partial_\rho \left(\frac{\gamma p}{(\gamma-1)\rho} \right) = \frac{\gamma \kappa}{\gamma-1} \partial_\rho \rho^{\gamma-1} = \gamma \kappa \rho^{\gamma-2} = \frac{1}{\rho}\frac{\gamma p}{\rho} = \frac{c_s^2}{\rho}, \qquad (9.52)$$

where c_s is the adiabatic sound speed given by Eq. (2.11) in §2.1.1.[13] Finally, the last two terms in Eq. (9.49) are independent of ρ. Therefore, taking the ρ derivative of Eq. (9.49), we get using Eq. (9.50)–(9.52),

$$\rho \partial_\rho \mathcal{B}_M = -v_p^2 + \frac{v_p^2 a_\varphi^2}{v_p^2 - a_p^2} + c_s^2. \qquad (9.53)$$

I then leave it to Problem 9.9 to show that Eq. (9.53) is equivalent to,

$$\rho \partial_\rho \mathcal{B}_M = -\frac{(v_p^2 - a_s^2)(v_p^2 - a_f^2)}{v_p^2 - a_p^2}, \qquad (9.54)$$

where a_s and a_f are respectively the slow and fast magnetosonic speeds first introduced in §5.2 (Eq. 5.23 and 5.25) and, in this context, are given by:

$$\left. \begin{aligned} a_s^2 &= \frac{1}{2}\left(a^2 + c_s^2 - \sqrt{(a^2+c_s^2)^2 - 4a_p^2 c_s^2}\right); \\ a_f^2 &= \frac{1}{2}\left(a^2 + c_s^2 + \sqrt{(a^2+c_s^2)^2 - 4a_p^2 c_s^2}\right), \end{aligned} \right\} \qquad (9.55)$$

with $a = \sqrt{a_p^2 + a_\varphi^2}$ being the Alfvén speed (Eq. 5.26).

Thus, the critical points of the MHD Bernoulli function are where the ρ-derivative of $\mathcal{B}_M$ is zero which, by Eq. (9.54), are evidently the slow and fast points (where $v_p = a_s$ or a_p) along the selected poloidal line of induction, f_0. As Problem 9.11 shows, the slow point doesn't tell us anything we don't already know about the solution. However, the fast point does provide a useful new constraint, which we exploit in the next subsection.

Finally, the astute reader will note that the *Alfvén* point (where $v_p = a_p$) is a pole of Eq. (9.54) and thus $\rho \partial_\rho \mathcal{B}_M$ is singular there. However, it is evidently an *integrable* singularity[14] since, on physical grounds alone, $\mathcal{B}_M$ must remain finite all along the poloidal line of induction, including the Alfvén point. Mathematically, we can trace the singular nature of Eq. (9.54) to the second term on the RHS of Eq. (9.49) which, by virtue of Eq. (9.48) (where the LHS is evidently finite for all r and thus s[15]), remains finite even at the Alfvén point.

---

[12] *Rappel:* a_φ and a_p are the azimuthal and poloidal Alfvén speeds respectively.

[13] Note that for an isothermal fluid, the result is identical. That is, $\partial_\rho \hbar = c_{iso}^2/\rho$ where c_{iso} is the isothermal sound speed given also by Eq. (2.11).

[14] Much like the function $f(x) = 1/\sqrt{x}$ which is singular at $x = 0$ but integrable over the domain $x \in [0,1]$: $\int_0^1 f(x)dx = 2\sqrt{x}\big|_0^1 = 2$.

[15] The reader is encouraged to review the discussion surrounding Eq. 9.27 and 9.28 if this point is unclear.

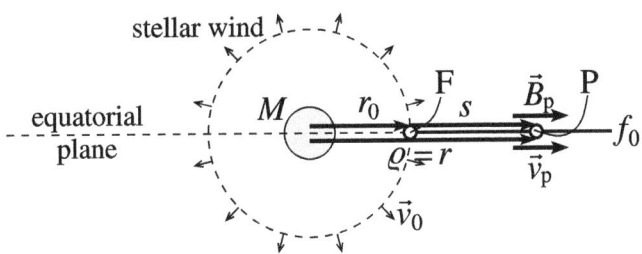

Figure 9.6. The Weber–Davis model for the stellar wind follows the outflow along the equatorial plane ($\vartheta = \pi/2$; $z = 0$) where the spherical and cylindrical radial components are the same ($\varrho = r$). A line of magnetic induction, f_0, is followed from its footpoint, F, a distance r_0 from the origin to essentially ∞ and includes an arbitrary point, P, at a distance ϱ from the origin and s from F. Both the poloidal velocity and magnetic induction are purely radial ($\vec{v}_{\rm p} = v\hat\varrho$, $\vec{B}_{\rm p} = B\hat\varrho$) along f_0.

9.3.2 The Weber–Davis Model

The stellar wind model considered by WD67 includes a few additional assumptions and simplifications beyond those already made. First, using spherical polar coordinates they consider wind only in the equatorial plane; that is flow along a poloidal line of induction, f_0, inclined at an angle $\epsilon = 0$ to the putative disc (Fig. 9.5a) and where its footpoint, F, is a distance r_0 from the centre of a (proto)star with mass M. Thus, $\vartheta = \pi/2$ (the $z=0$ plane), $\vec{v}_{\rm p} = v_{\rm p}(s)\hat\varrho$, and $\vec{B}_{\rm p} = B_{\rm p}(s)\hat\varrho$ with the radial coordinate $\varrho = r = s + r_0$ (see Fig. 9.6).

In addition to the assumption of co-rotation along f_0 (and thus $a_{\rm p} \gg \Omega_0 r_0$), WD67 assume the "disc" at $r = r_0$ to be "cold", and thus,

$$\Omega_0 r_0 \gg c_{s_0} \quad \Rightarrow \quad a_{p_0} \ggg c_{s_0} \quad \Rightarrow \quad a_{s_0} \sim c_{s_0} \to 0,$$

(e.g., first of Eq. 9.55). Since v_0 also vanishes at $r = r_0$ (outflow starts from rest), $v_0 \sim c_{s_0} \sim a_{s_0} \to 0$ identifying the footpoint, F, of f_0 as its slow/sonic point. Further, it is assumed that $r_{\rm A} \gg r_0$ but still finite whereas the fast point, $r_{\rm f} \to \infty$. Recall that the assumption of co-rotation along f_0 is reasonable so long as $r < r_{\rm A}$.

Next, whatever the nature of the outflow may be, both magnetic flux[16] and mass flux[17] must be conserved which, in spherical polar geometry, requires:

$$B_{\rm p} r^2 = \text{constant} = B_0 r_0^2 \quad \text{and} \quad \rho v_{\rm p} r^2 = \text{constant} = \rho_{\rm A} a_{\rm p_A} r_{\rm A}^2. \quad (9.56)$$

Here, the constant magnetic flux is evaluated at r_0 where $B_{\rm p} = B_0$, whereas the constant mass flux is evaluated at the Alfvén point where $\rho = \rho_{\rm A}$ and $v_{\rm p_A} = a_{\rm p_A}$, the poloidal Alfvén speed at $r = r_{\rm A}$. Note that the mass flux was not evaluated at $r = r_0$ since, as noted just before Eq. (9.18), $v_0 \to 0$ and thus $\rho_0 \to \infty$.

With these additional assumptions and considerations, we start by evaluating

---

[16] The first of Eq. (9.56) is easily verified from the definition of Φ_B in Eq. (4.8).
[17] *Rappel*: We've used conservation of mass flux in the steady state before: see Eq. (2.70).

the constant value of $\mathcal{B}_\text{M}$. For if $\mathcal{B}_\text{M}$ is truly constant, it can be evaluated anywhere along f_0, and so we choose to evaluate it at the slow/sonic point (F) where $v_0 \ll \Omega_0 r_0$, $v_\varphi = \Omega_0 r_0$, $\varrho = r = r_0$, and $p \to 0$ by virtue of $c_\text{s} \ll \Omega_0 r_0$. Thus, Eq. (9.42) and equivalently Eq. (9.49) evaluated at F is,

$$\mathcal{B}_\text{M} = -\frac{\Omega_0^2 r_0^3}{\varrho} - \frac{(\Omega_0 r)^2}{2} = -\frac{3}{2}(\Omega_0 r_0)^2, \qquad (9.57)$$

a known quantity since Ω_0 and r_0 are known boundary conditions. Comparing Eq. (9.57) with Eq. (9.49) and then dividing through by $\Omega_0^2 r_0^2$, we get an expression for the *unitless* MHD Bernoulli function,

$$\widetilde{\mathcal{B}}_\text{M} = \frac{1}{2}\frac{\eta^2 B_\text{p}^2}{\rho^2 \Omega_0^2 r_0^2} + \frac{1}{2}\frac{(r_\text{A}^2 - r^2)^2}{r_0^2 r^2 (1 - \rho/\rho_\text{A})^2} - \frac{r_0}{r} - \frac{r^2}{2r_0^2} = -\frac{3}{2}, \qquad (9.58)$$

since $\varrho = r$ and where enthalpy has been dropped by the assumption of a cold disc.

For convenience, we express $\widetilde{\mathcal{B}}_\text{M}$ in terms of unitless quantities by introducing the scaled variables,

$$x \equiv \frac{r}{r_0} \quad \text{and} \quad \tilde{\rho} \equiv \frac{\rho}{\rho_\text{A}}. \qquad (9.59)$$

Further, from the first of Eq. (9.56), $B_\text{p} = B_0/x^2$ and from Eq. (9.18), $\eta^2 = \rho_\text{A}/\mu_0$. With all this, Eq. (9.58) becomes,

$$\widetilde{\mathcal{B}}_\text{M} = \frac{q_0}{\rho_\text{A}}\frac{1}{\tilde{\rho}^2 x^4} + \frac{(x_\text{A}^2 - x^2)^2}{2x^2(1-\tilde{\rho})^2} - \frac{1}{x} - \frac{x^2}{2} = -\frac{3}{2}, \qquad (9.60)$$

where,

$$q_0 \equiv \frac{B_0^2/2\mu_0}{\Omega_0^2 r_0^2}, \qquad (9.61)$$

is the ratio of the magnetic pressure to the square of the rotation speed at F whose units are those of density (and so q_0/ρ_A is unitless).

Eq. (9.60) is still not in a useful form for finding a specific density profile, even for the scaled variables $\tilde{\rho}(x)$. At face value, the ratio q_0/ρ_A is awkward, requiring knowledge of measured quantities such as Ω_0, B_0, and even ρ_A. Fortunately, we haven't yet exhausted all the applicable constraints. We still haven't used the knowledge that whatever the solution may be, it must pass through the critical points. It turns out that passing through the slow point only tells us something about the footpoint we already know (and, if you want to find out what that is, you'll have to do Problem 9.11!). However, forcing the solution through the fast point does provide something new, and gives us the last piece of the puzzle needed to generate a specific stellar wind profile.

For Eq. (9.60), the criterion for critical points used in §9.3.1, namely $\partial_\rho \mathcal{B}_\text{M} = 0$, becomes $\partial_{\tilde{\rho}} \widetilde{\mathcal{B}}_\text{M} = 0$. Applying this to Eq. (9.60), we get,

$$\partial_{\tilde{\rho}} \widetilde{\mathcal{B}}_\text{M} = -\frac{q_0}{\rho_\text{A}}\frac{2}{\tilde{\rho}^3 x^4} + \frac{(x_\text{A}^2 - x^2)^2}{x^2(1-\tilde{\rho})^3} = 0. \qquad (9.62)$$

Taking the fast point to be very much further away than r_0, we evaluate Eq. (9.62) as $x \to \infty$ where the density, $\tilde{\rho}$, tapers off to zero. From the second of Eq. (9.56),

we have,
$$\rho v_p r^2 = \rho_A a_{pA} r_A^2 = \rho_A \frac{B_A r_A^2}{\sqrt{\mu_0 \rho_A}} = \rho_A \frac{B_0 r_0^2}{\sqrt{\mu_0 \rho_A}}, \qquad (9.63)$$

using the first of Eq. (9.56). Dividing through by $\rho_A v_p r_0^2$, we find,

$$\tilde{\rho} x^2 = \frac{B_0}{v_p \sqrt{\mu_0 \rho_A}} \quad \Rightarrow \quad \lim_{x \to \infty}(\tilde{\rho} x^2) = \frac{B_0}{v_\infty \sqrt{\mu_0 \rho_A}} \equiv c, \qquad (9.64)$$

where v_∞ is the flow speed as $x \to \infty$. Thus, so long as $v_\infty \neq 0$, $x \to \infty$ and $\tilde{\rho} \to 0$ in such a way that $\tilde{\rho} x^2$ tends to a finite constant, c.

And so with this insight, we multiply Eq. (9.62) through by $\tilde{\rho}$ and expand out the second term to get,

$$-\frac{q_0}{\rho_A} \frac{2}{\tilde{\rho}^2 x^4} + \frac{\tilde{\rho} x_A^4}{x^2 (1-\tilde{\rho})^3} - \frac{2 \tilde{\rho} x_A^2}{(1-\tilde{\rho})^3} + \frac{\tilde{\rho} x^2}{(1-\tilde{\rho})^3} = 0.$$

Then, taking the limit as $x \to \infty$, $\tilde{\rho} \to 0$, and $\tilde{\rho} x^2 \to c$, the middle two terms disappear leaving us with,

$$-\frac{2 q_0}{\rho_A c^2} + c = 0 \quad \Rightarrow \quad c^3 = \frac{2 q_0}{\rho_A}. \qquad (9.65)$$

Equation (9.65) is half of what requiring the solution to pass through the fast point buys us. The other half comes from evaluating Eq. (9.60) itself at the fast point which first we rearrange a little to get,

$$\tilde{\mathcal{B}}_M = \underbrace{\frac{q_0}{\rho_A} \frac{1}{\tilde{\rho}^2 x^4} + \frac{x_A^4}{2 x^2 (1-\tilde{\rho})^2} - \frac{x_A^2}{(1-\tilde{\rho})^2} - \frac{1}{x}}_{} + \frac{x^2}{2(1-\tilde{\rho})^2} - \frac{x^2}{2}$$

$$= \quad " \quad + \frac{x^2}{2}\left(\frac{1}{(1-\tilde{\rho})^2} - 1\right)$$

$$= \quad " \quad + \frac{\tilde{\rho} x^2}{(1-\tilde{\rho})^2} - \frac{\tilde{\rho}^2 x^2}{2(1-\tilde{\rho})^2} = -\frac{3}{2}.$$

Written in this form we see that by taking the limit as $x \to \infty$, $\tilde{\rho} \to 0$, and $\tilde{\rho} x^2 \to c$, the second, fourth, and sixth terms on the RHS (last line) drop out and we're left with,

$$\frac{q_0}{\rho_A c^2} - x_A^2 + c = -\frac{3}{2}.$$

But, from Eq. (9.65), $q_0/\rho_A = c^3/2$ and we have,

$$\frac{3}{2} c - x_A^2 = -\frac{3}{2} \quad \Rightarrow \quad c = \frac{2}{3} x_A^2 - 1.$$

This is the other half of what the constraint buys us which, when compared with Eq. (9.65), gives us the full constraint we've been seeking:

$$\frac{2}{3} x_A^2 - 1 = \left(\frac{2 q_0}{\rho_A}\right)^{1/3} \quad \Rightarrow \quad \boxed{\frac{q_0}{\rho_A} = \frac{1}{2}\left(\frac{2}{3} x_A^2 - 1\right)^3,} \qquad (9.66)$$

giving us a convenient expression for q_0/ρ_A with no need for "measured values".

Aside: So let's take a breath and review what we've done before trying to generate plots. We've re-examined the MHD Bernoulli equation, Eq. (9.42), in order to determine profiles for ρ and v_p along a "rigid" line of induction, f_0, that lies in the equatorial plane ($\epsilon = 0$) of a cold accretion disc. We further assume $\vec{B}_p$ to be force-free which means it can be set as an initial and enduring condition, and that f_0 is in solid-body rotation with its footpoint, F, at least until the Alfvén point. Boundary conditions include knowledge of B_0, Ω_0, and r_0 at F.

With all this, we first found the constant value of the MHD Bernoulli function, $\mathcal{B}_M$, by evaluating Eq. (9.42) at the slow/sonic point at $r = r_0$; our result was Eq. (9.57). This and the scaling laws in Eq. (9.59) allowed us to write down Eq. (9.60) for the unitless MHD Bernoulli function, $\tilde{\mathcal{B}}_M$, in terms of unitless variables and the ratio q_0/ρ_A. An expression for the latter was found (Eq. 9.66) by forcing the solution to pass through the fast point.

What remains, then, is to use all we've learned to find specific profiles for the scaled variables: $\tilde{\rho}$; $\tilde{v}_p = v_p/\Omega_0 r_0$; and, as we'll find to be very insightful, the "pitch angle" of the magnetic induction, $\tan\psi = B_\varphi/B_p$.

Thus, on substituting Eq. (9.66) into Eq. (9.60), we get our final form for the scaled MHD Bernoulli equation for flow along a "rigid" radial line of induction along the equatorial plane of a rotating disc:

$$\boxed{F(\tilde{\rho}, x, x_A) \equiv \frac{1}{2}\left(\frac{2}{3}x_A^2 - 1\right)^3 \frac{1}{\tilde{\rho}^2 x^4} + \frac{(x_A^2 - x^2)^2}{2x^2(1-\tilde{\rho})^2} - \frac{1}{x} - \frac{x^2}{2} + \frac{3}{2} = 0,} \quad (9.67)$$

where $F = \tilde{\mathcal{B}}_M + \frac{3}{2}$ is defined for convenience. For a given x_A and x, Eq. (9.67) presents a transcendental equation in $\tilde{\rho}$ which can be solved using a suitable root-finder (*e.g.*, the univariate secant method in §D.1 of App. D) to find the value of $\tilde{\rho}$ that makes $F(\tilde{\rho}, x, x_A)$ zero.[18] The resulting function $\tilde{\rho}(x)$ for $x_A = 100$ is shown in Fig. 9.7*a* over a domain of $x \in [2:400]$ (avoiding $x = 1$ where $\tilde{\rho}$ blows up).

For the poloidal velocity, we return to Eq. (9.63) and write,

$$\rho v_p r^2 = \rho_A \frac{B_0 r_0^2}{\sqrt{\mu_0 \rho_A}} \quad \Rightarrow \quad v_p = \frac{1}{\tilde{\rho} x^2} \frac{B_0}{\sqrt{\mu_0 \rho_A}}. \quad (9.68)$$

But, from Eq. (9.61) and (9.66),

$$\frac{q_0}{\rho_A} = \frac{1}{\rho_A} \frac{B_0^2/2\mu_0}{\Omega_0^2 r_0^2} = \frac{1}{2}\left(\frac{2}{3}x_A^2 - 1\right)^3 \quad \Rightarrow \quad \frac{B_0}{\sqrt{\mu_0 \rho_A}} = \Omega_0 r_0 \left(\frac{2}{3}x_A^2 - 1\right)^{3/2},$$

and Eq. (9.68) becomes,

$$\tilde{v}_p \equiv \frac{v_p}{\Omega_0 r_0} = \frac{1}{\tilde{\rho} x^2}\left(\frac{2}{3}x_A^2 - 1\right)^{3/2}, \quad (9.69)$$

---

[18] The ambitious reader who would like to try this themself is directed to *Computer Project* P9.1 in the problem set along with all the numerical cautions explained therein!

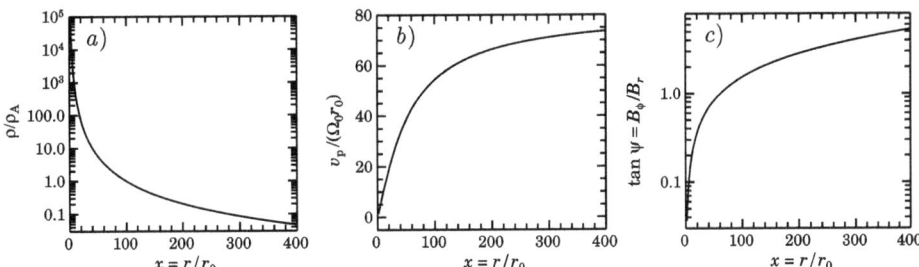

Figure 9.7. Weber–Davis solutions for a) density, b) poloidal velocity, and c) magnetic pitch angle along a radial line of induction, f_0, within the equatorial plane of an accretion disc as functions of radial distance from the origin. The footpoint, F, is located at $x = x_0 = 1$ and the Alfvén point, A, is at $x = x_A = 100$. See Fig. 9.6 for the meaning of some of the quantities and the text for details.

where $\tilde{v}_p$, shown in Fig. 9.7b, is scaled in terms of the rotational speed of the footpoint, F.

I leave it to Problem 9.13 to show that along f_0, the tangent of the *pitch angle* – defined as the ratio of the azimuthal to poloidal components of $\vec{B}$ – is given by,

$$\tan \psi = \frac{B_\varphi}{B_p} = \sqrt{\left(\frac{1}{x} + \frac{x^2}{2} - \frac{3}{2}\right) \frac{\rho_A}{q_0} \tilde{\rho}^2 x^4 - 1}, \qquad (9.70)$$

shown in Fig. 9.7c. It may come as a bit of a surprise that there should be *any* B_φ at all; that is, were the poloidal induction strong enough to enforce solid-body rotation, no part of $\vec{B}_p$ should get twisted into the azimuthal direction and $\tan \psi$ should be zero everywhere.

The fact is, our assumption of a "stiff" line of induction only bought us the ability to prescribe B_p rather than computing it self-consistently with the other flow variables. It does *not* strictly enforce solid-body rotation particularly as the Alfvén point is approached and passed. As seen in Fig. 9.7c, $\tan \psi$ starts off very small (0 at $r = r_0$), and then, as Problem 9.13 shows, reaches almost (but weirdly, not quite) 1 rad at the Alfvén point ($x = x_A = 100$) when $x_A \gg 1$. Beyond x_A, $\tan \psi$ continues to increase and, as $x \to \infty$, $\tan \psi \propto x \to \infty$ (Problem 9.13 again) and $\psi \to \pi/2$. This harkens back to Fig. 9.1 at the beginning of the chapter where it was reasoned that the line of induction should bend sharply out of the plane at the Alfvén point, A, and become almost purely azimuthal at some point thereafter.

The last thing I want to touch on before leaving this section is the asymptotic behaviour of v_p, introduced but not evaluated in Eq. (9.64). Continuing from there, we see that,

$$\lim_{x \to \infty} v_p \equiv v_\infty = \frac{B_0}{c\sqrt{\mu_0 \rho_A}} = \frac{B_0}{(\mu_0 \rho_A)^{1/2}} \left(\frac{\rho_A}{2q_0}\right)^{1/3} = \left(\frac{\mu_0}{\rho_A}\right)^{1/6} \frac{B_0}{\mu_0^{2/3} (2q_0)^{1/3}},$$

using Eq. (9.65) for the second equality. Carrying on, multiplying the RHS top and bottom by $\Omega_0 r_0$ we get,

$$v_\infty = \frac{1}{\eta^{1/3}} \frac{B_0}{(2\mu_0^2 q_0)^{1/3}} \frac{\Omega_0 r_0}{\Omega_0 r_0} = \left(\frac{1}{\eta} \underbrace{\frac{B_0}{\mu_0 \Omega_0 r_0}}_{\eta^*}\right)^{1/3} \Omega_0 r_0,$$

using Eq. (9.61) and after a little algebra for the second equality. The quantity η^* has the same units as η and is composed of nothing but values from the boundary conditions. Thus, defining the *scaled* mass load as $\tilde{\eta} = \eta/\eta^*$, we find for the asymptotic poloidal velocity,

$$\boxed{v_\infty = \frac{1}{\tilde{\eta}^{1/3}} \Omega_0 r_0,} \qquad (9.71)$$

which has a rather simple and profound interpretation. For a mass, m, in Keplerian orbit of radius r_0 about a central mass, M, its escape velocity is $v_{\text{esc}} = \Omega_0 r_0$. Therefore, so long as the scaled mass load, $\tilde{\eta}$, is less than 1 (*i.e.*, a "light" mass load), $v_\infty > \Omega_0 r_0$ and material launched from the accretion disc will escape the gravitational confines of M. Conversely, for a "heavy" mass load ($\tilde{\eta} > 1$), $v_\infty < v_{\text{esc}}$ and launched material eventually falls back into M.

9.4 Astrophysical jets (optional)

We conclude this chapter with an example of steady-state MHD arising spontaneously within a complex numerical simulation. This example comes from work I did with my former graduate student on launching a protostellar jet from an accretion disc (Ramsey & Clarke, 2011; 2019), and thus ties in very nicely with the main result of this chapter: magneto-rotational driven outflow. In this section, I give a minimalist description of what we did, just enough so the connections with the Weber–Davis constants and the MHD Bernoulli function can be made. For those who would like to learn more about these simulations, I would like to think our papers are quite approachable to any reader of this text!

Our simulations were by no means the first to investigate numerically the magneto-rotational acceleration of jets.[19] They were, however, the first done on a computational domain large enough both to resolve the physics necessary to accelerate fluid near the protostar (zone size ~ 0.005 AU) and to follow the outflow to distances where direct comparisons with observations could be made (several thousand AU). To do so directly on a single 2-D grid would require nearly a *trillion* zones which, even at the time of this writing, is completely impractical. Therefore, we used a version of *ZEUS-3D* called AZEuS capable of integrating numerous grids of varying resolution in the same simulation, a technique known as *adaptive mesh refinement* (see references in Ramsey & Clarke, 2019).

Figure 9.8 shows four of the nested grids from the final epoch (~ 100 years[20]) of

---

[19] An extensive bibliography of those preceding us is given in Ramsey & Clarke (2019).

[20] A hundred years doesn't seem like a lot, and indeed our jets got no further than about

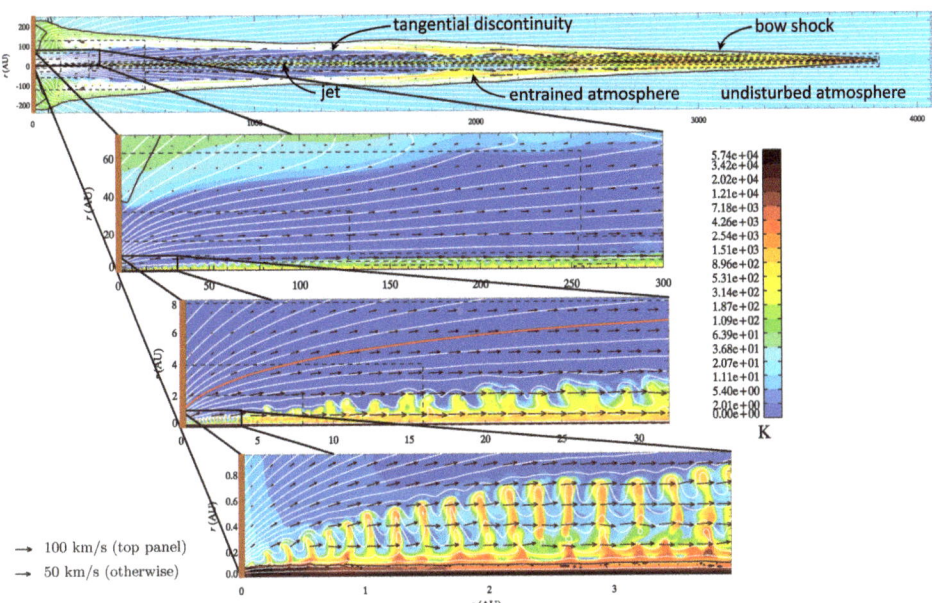

Figure 9.8. Nested images from an AZEuS simulation of a protostellar jet launched from an accretion disc (from Ramsey & Clarke, 2019, © Oxford University Press, reprinted with permission); details in text.

an axisymmetric simulation – remember the axisymmetric part – described in Ramsey & Clarke (2019) in which both magneto-rotational drivers (BRM and MTM) are at work. Flow is from left to right with the location of the accretion disc indicated by the heavy brown lines along the left edges of each panel. Colour contours represent temperature with the scale given in Kelvins, black arrows represent velocity vectors where length of the vector is proportional to magnitude (legend at bottom left in km s^{-1}), white lines represent lines of magnetic induction, and the fine black lines in the top two panels map out the *slow surface*, the locus of slow points where the slow magnetosonic number, $M_s = v_{jet}/a_s = 1$. Dashed black lines indicate nested grid boundaries and the units along the axes are AU.

In the top panel, the slow surface completely surrounds the super-slow[21] (and, for the most part, super-Alfvénic and even super-fast) outflow, separating it from sub-slow material still on the grid. The latter includes the still-static "undisturbed atmosphere" as well as the base of the outflow just "above" the disc (*i.e.*, to the right of the brown line) where material being accelerated by the BRM (and MTM) has yet to reach the slow speed. Since much of the outflow is super-slow and, in fact, super-fast, a "bow shock" is excited in the undisturbed atmosphere (labelled

---

$\sim$4,000 AU, tiny on astrophysical scales. Still, this is enough to reach observational lengths – which was our purpose – and, as it was, this simulation took over two *months* of collective cpu on then state-of-the-art parallel computers.

[21] And of course, by "super-slow" I don't mean *really really* slow! I mean speeds faster than the slow magnetosonic speed, a_s.

in the top panel of Fig. 9.8 and nicely demarcated by the black slow surface) whose narrowness attests to the high magnetosonic numbers ($M_{f,s}$) of the jet.

Still focussed on the top panel of Fig. 9.8, the outflow is actually comprised of two components, separated as indicated by a tangential discontinuity (TD). Inside the TD (deep blue on left, green and orange on right) is the actual *jet*; material magneto-rotationally accelerated from the disc as we've been discussing. The second component trapped between the TD and bow shock is *entrained* atmosphere that has passed through the bow shock and now moves forward with the jet. If you've ever stood too close to the side of a road as a truck speeds by, you'll know what "entrained air" feels like. For a supersonic jet triggering a bow shock, not only is gas entrained, it is condensed and heats up by factors of twenty or more.

Last things to discuss before making the connection with steady-state MHD are the initial conditions. For this, we set up a non-rotating atmosphere in hydrostatic equilibrium ($\nabla p + \rho \nabla \phi = 0$) around a $0.5\,M_\odot$[22] "protostar" located at the origin. Embedded within this atmosphere is a force-free magnetic induction ($\vec{J} \times \vec{B} = 0$) in an "hour-glass configuration" similar to that pictured in Fig. 9.4c. Looking carefully at the white lines of induction in the top panel of Fig. 9.8, you'll see this hour-glass configuration still embedded within the sky-blue "undisturbed atmosphere", the portion of the initial atmosphere yet to be affected of the passage of the jet. To break the equilibrium, an accretion disc specified as a boundary condition (perfectly conducting fluid in Keplerian rotation about the protostar; heavy brown lines on left edges in Fig. 9.8) is "turned on" at $t = 0$, twisting the force-free lines of poloidal induction anchored in the disc into a non-force-free helical configuration. As all things in fluid dynamics, information about changing conditions never transmits instantly. Rather, it propagates throughout the fluid at one or more of the characteristic speeds. In this case, information about the rotating disc propagates along the poloidal lines of induction as a torsional Alfvén wave (*e.g.*, point labelled a_φ in Fig. 9.1b,c,d) leaving in its wake a rotating atmosphere and the conditions necessary for magneto-rotational acceleration.

Because this is an *axisymmetric* simulation, one of the necessary conditions for steady-state MHD is already in place. However, the simulation very definitely includes all ∂_t terms, and if any region within the computational domain exhibits steady-state behaviour, it's only because the physics has determined that $\partial_t = 0$ there, or at least approximately so. Much of the second-from-the-top panel of Fig. 9.8 indicates such a region.

Our first result in §9.1 was an important one. From Eq. (9.9), we learned that in steady state, the poloidal components of velocity and magnetic induction are everywhere parallel or anti-parallel; streamlines lie on top of lines of induction. As seen in the second-from-the-top panel of Fig. 9.8, most poloidal velocity vectors are tangential to the white lines of induction, indicative of a region in steady state, or at least nearly so. The biggest exception is near the top of the panel where poloidal lines of induction bend backwards across the tangential discontinuity while

---

[22] $M_\odot \sim 2 \times 10^{30}$ kg is the mass of the sun.

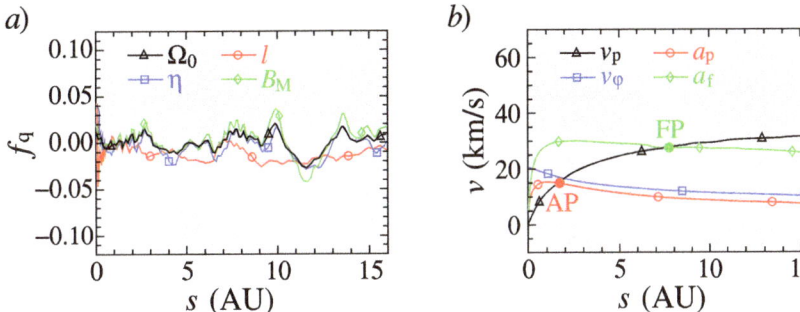

Figure 9.9. *a*) Fractional deviations of the three Weber–Davis constants and the MHD Bernoulli function as a function of distance, s, along the line of induction anchored at 1 AU (red line in the third panel of Fig. 9.8). Where q is one of the four constants, $f_q = (q - q_0)/q_0$ where q_0 is the expected constant value. *b*) Key speeds along the same line of induction where the Alfvén and fast points are indicated by the red (AP) and green (FP) dots respectively. Adapted from panels F of Fig. 12 and 11 respectively from Ramsey & Clarke (2019).

the velocity vectors continue pointing forward. Here, velocity vectors cross lines of poloidal induction indicating non-steady-state behaviour.

Most of the upper portion of the third-from-the-top panel (deep blue; cold) also appears to be in steady state, based again on where $\vec{v}_p \parallel \vec{B}_p$, while the lower portion hugging the symmetry axis (yellow-orange; hot) does not. The bottom panel shows in detail the dynamics right at the innermost portion of the jet and immediately after material is launched. We'll return to discuss this region – which is clearly *not* in steady state – before leaving this section.

First, in the third panel of Fig. 9.8, let's focus on the single red poloidal line of induction anchored in the disc at 1 AU from the origin (protostar) dividing the panel more or less in half. Flow along this line certainly *looks* to be in steady state; let's see what the data say.

Figure 9.9 shows profiles of various quantities as a function of position, s, along the red poloidal line of induction out to 16 AU (half way across the third panel of Fig. 9.8). Plotted on the left (Fig. 9.9*a*) are fractional deviations of the Weber–Davis constants and the Bernoulli function from their expected constant values (Ω_0 and $\mathcal{B}_M$ evaluated at the footpoint, η and l at the Alfvén point). Specifically:

$$f_{\Omega_0} = \frac{\Omega_0 r_0}{v_{\varphi,0}} - 1; \quad f_\eta = \eta\sqrt{\frac{\mu_0}{\rho_A}} - 1; \quad f_l = \frac{lr_0}{v_{\varphi,0}r_A^2} - 1; \quad f_{\mathcal{B}_M} = \frac{2\mathcal{B}_M}{3v_{\varphi,0}^2} - 1,$$

using Eq. (9.13), (9.18), (9.28), and (9.57), and where all quantities ($v_{\varphi,0}$, r_A, *etc.*) are as previously defined. As the profiles show, the WD constants and the Bernoulli function vary by no more than a few percent, indicating the red line of induction is within an approximately steady-state region.

Figure 9.9*b* shows profiles of various speeds along the red line of induction. The profile of the azimuthal speed, v_φ (blue □), shows the behaviour we expect from Eq. (9.14) which states that since $\lambda B_\varphi < 0$, v_φ should drop off steadily and monotonically from its highest value at the footpoint of f.

Conversely, the poloidal velocity, v_p (black profile △), steadily accelerates from zero at the footpoint ($s = 0$) through the Alfvén point (AP, red dot in Fig. 9.9b) where v_p reaches the local poloidal Alfvén speed (a_p, red profile○), and then on through the fast point (FP, green dot) where v_p reaches then exceeds the local fast speed (a_f, green profile ◇). While not evident in Fig. 9.8, a significant azimuthal component of magnetic induction, B_φ, has been twisted out of the original poloidal induction even by the Alfvén point and by the fast point, $B_\varphi \gg B_p$. Thus, for most of the 16 AU shown in Fig. 9.9, we conclude that poloidal outflow is accelerated principally by the magnetic tower mechanism (MTM).

However, within the first AU from the footpoint is the realm of the bead-on-a-rod mechanism (BRM), as the bottom panel of Fig. 9.8 dramatically shows. "Towers" of hot fluid are moving outwards at the local poloidal speed (thus, these are moving "plasmoids" rather than waves) creating an environment near the symmetry axis that is anything but steady state. Notice how lines of poloidal magnetic induction are wrapped around the red-yellow plasmoids whose magnetic tension contains them. In fact, without this magnetic "packaging", these plasmoids would burst open, since the thermal pressure within them is very much greater than in the cooler blue-green material surrounding them.

So where do these plasmoids come from? This, I have to say, was an entirely unexpected result and is why one does simulations!

As seen in the bottom panel of Fig. 9.8, the angles at which lines of induction emerge from the accretion disc (defined as ϵ in Fig. 9.5a) increases as one gets closer to the rotation axis. In this particular simulation, ϵ doesn't rise above $\epsilon_{\text{crit}} = 60°$ – the critical launching angle discussed in §9.2.1 – until fairly close to the axis. Remember, for $\epsilon > \epsilon_{\text{crit}}$, the BRM is stifled; only for $\epsilon < \epsilon_{\text{crit}}$ is outflow by the BRM spontaneous. Thus, over most of the disc where $\epsilon < 60°$, material is gently and continuously launched from the disc and into the outflow, as is evident by the velocity arrows close to the disc in the bottom panel.

However, close to the rotation axis, ϵ rises above $60°$ and, in principle, the BRM should be suppressed. Now, lines of induction are not the ideal rigid rods that our analytical work at times found convenient to assume. For lines of induction emerging from the disc at angles just above $60°$, material that would otherwise be launched accumulates at the footpoint. Owing to the rotation of the disc, this accumulating mass is pressed outwards centrifugally which, at some point, is sufficient to bend the line of induction outward just enough for ϵ to fall below $60°$. The accumulated material is then launched *en masse* as a plasmoid, relieving the strain on the line of induction which then snaps back to its unencumbered inclination ($\epsilon > 60°$). Accumulation of mass resumes, and the cycle repeats. Thus, what we're seeing in the bottom panel of Fig. 9.8 is a simple harmonic oscillator at work, launching material via the BRM as plasmoids rather than the steady wind further from the axis. These plasmoids eventually blend and fuse together to form the hot jet core seen in the three panels above.

In these two examples from astrophysics – stellar winds in §9.3 and jets here – we've seen two good uses for steady-state analysis, not being the "correct model" notwithstanding. In §9.3 we saw, with a bit of effort, how to coax from the equations of MHD analytic profiles for the density, poloidal velocity, and magnetic pitch angle as functions of distance from the star within its equatorial plane (Fig. 9.7). The profiles look simplistic and do not in any detail represent what an actual stellar wind would look like. Still, some predictions from these profiles such as mass, energy, and angular momentum fluxes, and conditions for outflow to reach escape velocity are all more or less borne out by the observations. As a first cut, then, a steady-state MHD model gives a good, first glimpse at how a stellar wind works.

In our second example, the simulations do not rely on steady-state theory at all. MHD codes are designed to work in a time-dependent, fully 3-D environment, and it is only a question of computing resources (and what physics and chemistry may have been neglected) that limits the nature of the solution obtained.

Instead, here steady-state MHD provides a useful method for *in situ* code verification. When developing software such as ZEUS-3D and AZEuS, one relies heavily on problems that can be solved (semi-)analytically (*e.g.*, the Riemann problem in Chap. 6) to check against the code's solution. Certainly, any MHD code worth using will satisfy such tests, but these are *very* limited. It's one thing to reproduce a 1-D result from a shock tube. It's quite another to simulate something such as full-on 3-D *super-Alfvénic turbulence*; the state of 99% of all baryonic matter in the universe with ZEUS-3D's version represented in Fig. 7.2 and the chapter title bars.

It is the appearance of regions of steady state within a full-blown simulation that can provide an important calibrator for the numerical algorithms used to solve the time-dependent equations of MHD. It gives one a great deal of "faith" in the algorithms when, in the middle of a complex calculation, something that can be tested analytically arises spontaneously, and then the code passes those tests. In the simulation depicted in Fig. 9.8, the fact that the WD constants and Bernoulli function are "behaving" in regions identified as approximate steady state gives one faith that maybe – just maybe – the rest of the simulation is OK too.

Never let anyone tell you faith and science don't mix!

Problem Set 9

9.1 One can derive Eq. (9.28) in the text without encountering the Alfvén critical point. By evaluating each of Eq. (9.12) and (9.21) at the Alfvén point, confirm that,

$$l = r_A^2 \Omega_0,$$

where r_A is the radial distance to the Alfvén point (A in Fig. 9.1 in the text), and where l and Ω_0 are two of the Weber–Davis constants defined by Eq. (9.12) and (9.21).

9.2 Show how the alternate forms of the Weber–Davis constants expressed in terms of the poloidal Alfvén number, A_p, namely Eq. (9.29), (9.30), (9.31) in the text can be derived from Eq. (9.12), (9.17), and (9.21).

9.3

a) Starting with the steady-state total energy equation (Eq. 9.3 in the text), show that an alternate MHD Bernoulli function,

$$\mathcal{B}'_M = \frac{v^2}{2} + h + \phi + a_\varphi^2 - \frac{v_\varphi B_\varphi}{\mu_0 \rho \lambda},$$

is a constant along a poloidal line of induction. That is,

$$\vec{B}_p \cdot \nabla \mathcal{B}'_M = 0,$$

and thus, $\mathcal{B}'_M$ is a function of f only. Recall that $a_\varphi = B_\varphi/\sqrt{\mu_0 \rho}$ is the toroidal Alfvén speed.

b) Show that a suitable combination of $\mathcal{B}'_M$ and some or all of the Weber–Davis constants recovers the MHD Bernoulli function, $\mathcal{B}_M$, derived in the text (Eq. 9.43).

9.4 In the text, the derivation of the MHD Bernoulli function begins with Eq. (9.33), the steady-state Euler equation from a frame of reference, O', co-rotating with the footpoint of the line of induction (F in Fig. 9.1). As such, Eq. (9.33) is written down with Coriolis and centrifugal accelerations included on the RHS.

For those for whom the incorporation of such inertial terms does not come naturally, this problem is designed to take some of the mystery out of this approach. Starting with the steady-state Euler equation as written for an *inertial* frame of reference, namely Eq. (9.4) with no Coriolis or centrifugal terms added to the RHS, perform the Galilean transformation,

$$\vec{v} = \vec{v}' + r\Omega_0 \hat{\varphi},$$

to the LHS. Then, by using nothing more than vector calculus identities from App. A, show how Eq. (9.33) can be recovered.

9.5* As presented in the text, the MHD Bernoulli function as given by Eq. (9.43) is derived from the steady-state Euler equation (Eq. 9.4) from a particular rotating reference frame, and differs from the pure hydrodynamical case (Eq. 2.72) by a centrifugal-like term, $-v_\varphi \Omega_0 r$, that has no apparent connection to the magnetic induction. Yet, $\vec{B}$ must *surely* distinguish the two cases! This problem is designed to make that connection by redoing the problem from an inertial reference frame.

a) The first part is a generic vector calculus problem in cylindrical coordinates with φ-symmetry. For an arbitrary vector $\vec{A}$, show that in this geometry,

$$\hat{A}_{\rm p} \cdot [\vec{A} \times (\nabla \times \vec{A})] = \frac{1}{2r^2} \hat{A}_{\rm p} \cdot \nabla (rA_\varphi)^2, \tag{9.72}$$

where $\hat{A}_{\rm p}$ is a unit vector parallel to $\vec{A}_{\rm p} = A_z \hat{z} + A_r \hat{r}$, the poloidal component of $\vec{A}$, and A_φ is the toroidal component.

Hint: Before taking the dot product with $\hat{A}_{\rm p}$, break $\vec{A}$ and $\nabla \times \vec{A}$ up into their poloidal and toroidal components, and evaluate $\vec{A} \times (\nabla \times \vec{A})$. Then take the dot product.

b) To do the physics problem from an *inertial* frame of reference, examine the consequences of Euler's equation along a line of induction, whence the $\hat{s}$ direction. Start by taking the dot product of the steady-state Euler equation (Eq. 9.4 in the text) with $\hat{s}_{\rm p}$ and later, with the help of Eq. (9.35), derive Eq. (9.45), namely,

$$\partial_s \left(\frac{v_{\rm p}^2}{2} + h + \phi \right) = \frac{v_\varphi^2}{r} \hat{s} \cdot \hat{r} - \frac{1}{2\rho\mu_0 r^2} \partial_s (rB_\varphi)^2.$$

Why couldn't you have started off by taking the dot product of Eq. 9.4 with $\hat{s}$ directly?

Note that since this is being done in an inertial frame of reference, there should be no mention of Coriolis and centrifugal accelerations.

Hint: Right off the bat, Identity (A.19) should be of help.

c) To "complete the circuit", we must show that Eq. (9.44) in the text – a direct consequence of the MHD Bernoulli function in Eq. (9.42) and (9.43) derived in a *rotating* reference frame – is equivalent to Eq. (9.45) derived in an *inertial* reference frame.

To this end, show that,

$$\partial_s \left(-\frac{v_\varphi^2}{2} + v_\varphi \Omega_0 r \right) = \frac{v_\varphi^2}{r} \hat{s} \cdot \hat{r} - \frac{1}{2\rho\mu_0 r^2} \partial_s (rB_\varphi)^2. \tag{9.73}$$

With this connection made, we can finally conclude that the terms involving the toroidal speed, v_φ, in the MHD Bernoulli function given by Eq. (9.42) and (9.43) are where the magnetic driving forces stemming from the Lorentz force are "hiding".

Hint: I found it easier to manipulate the RHS to become the LHS and, as a start, you might develop the v_φ^2/r term using Eq. (9.37) in the text. Further, all three Weber–Davis constants are needed to complete the problem.

9.6 Another way of looking at the *magnetic tower mechanism* (MTM) discussed briefly in the text is to consider the poloidal component of the Lorentz force density in an inertial frame of reference.

a) Starting from the Lorentz force density, $\vec{f}_L = \vec{J} \times \vec{B}$, show that the condition for outward acceleration is,

$$\hat{v}_p \cdot \left(\frac{1}{2} \nabla B_\varphi^2 + \frac{B_\varphi^2}{r} \hat{r} \right) < 0, \tag{9.74}$$

where $\hat{v}_p \parallel \hat{s}_p$ is a unit vector in the direction of outward flow along a poloidal line of induction.

Hint: You might begin by considering each of $\vec{J}$ and $\vec{B}$ as a sum of their poloidal and toroidal components.

b) Show how Eq. (9.74) relates to the last term in Eq. (9.45) of the text, thus identifying Eq. (9.74) with the MTM.

Comment: The first term, $\nabla B_\varphi^2/2$, is a magnetic pressure gradient which, like a thermal pressure gradient, exerts an outward force when negative. The second term, B_φ^2/r, is a positive-definite magnetic tension which works against the magnetic pressure gradient. That is, the magnetic pressure gradient must be negative enough to overcome the magnetic tension in order for flow to be accelerated outwards by the MTM.

9.7

a) Referring to Fig. 9.5b in the text, show that at $s = 0$, the effective potential given by Eq. (9.46) is:

$$\left. \frac{\phi_{\text{eff}}}{GM/r_0} \right|_{s=0} = -1.5.$$

This is a two-liner.

b) Show how setting the second derivative of ϕ_{eff} with respect to s to zero at $s = 0$ leads to the critical angle, $\epsilon_{\text{crit}} = 60°$ (Eq. 9.47 in the text).

Hint: You might first consider how to glean the relations $\varrho(s)$ and $r(s)$ from Fig. 9.5a in the text, and then continue by differentiating ϕ_{eff} as given by Eq. (9.46) *implicitly* with respect to s.

9.8 A "force-free" magnetic induction is one where the Lorentz force, $\vec{J} \times \vec{B} = 0$. A general way to achieve this is for $\vec{J} \parallel \vec{B}$, but more simply, one can also have $\vec{J} = \nabla \times \vec{B} = 0$ in which case the magnetic induction can be written as,

$$\vec{B} = \nabla \psi, \tag{9.75}$$

where ψ is a *scalar potential* and a function of the coordinates. A magnetic induction whose curl is zero and thus given by Eq. (9.75) is known as a *potential field*.

a) Show that the scalar potential, ψ, solves Laplace's equation. That is,

$$\nabla^2 \psi = 0. \tag{9.76}$$

This is a two-liner.

b) Solve Eq. (9.76) in cylindrical coordinates (assuming φ-symmetry as we've been doing throughout this chapter) and show that the magnetic induction everywhere is determined by knowing the value of a single free parameter at a boundary (*e.g.*, the disc).

Rappel: A good approach to solving a partial differential equation such as Eq. (9.76) is the technique known as *separation of variables* which most undergraduate curricula include in a "mathematical methods in physics" course. For those in need of a refresher, I recommend the excellent text by Arfken, Weber, & Harris (2013).

9.9 Show how Eq. (9.54) in the text follows from Eq. (9.53).

9.10* Besides mass and momentum, an MHD outflow transports energy. Here, we examine two important energy flux densities,[23] namely Poynting (magnetic) and kinetic:

$$\vec{S}_\text{P} = \frac{1}{\mu_0}\vec{B}\times(\vec{v}\times\vec{B}); \quad \text{and} \quad \vec{\mathcal{K}} = \frac{1}{2}\rho v_\text{p}^2\, \vec{v}_\text{p},$$

where $\vec{S}_\text{P}$, the Poynting vector (*a.k.a.*, Poynting flux density) is first introduced by Eq. (4.15) in §4.5, and derived in App. B (Eq. B.17).

a) Show that the poloidal component of the Poynting flux density is given by,

$$S_\text{P,p} = \vec{S}_\text{P}\cdot\hat{s}_\text{p} = -\Omega_0\frac{T_z}{A},$$

where $\hat{s}_\text{p}$ is defined by Eq. (9.35) in the text, and T_z/A is given by Eq. (9.24). Give a physical interpretation of $S_\text{P,p}$.

Hint: Consider working in the rotating frame of reference, O', where Eq. (9.32) applies.

b) In the limit as $r\to\infty$, show that the ratio of Poynting to kinetic flux densities is given by,

$$q\equiv\frac{|S_\text{P,p}|}{\mathcal{K}}=2\frac{a_\varphi^2}{v_\text{p}^2},$$

where a_φ is the Alfvén speed in the toroidal direction.

c) Thus, at the fast point where $r\to\infty$, show that $q\to 2$ and give a physical interpretation of this result.

9.11 Show that forcing the Weber–Davis solution to pass through the slow critical point only confirms something we already know. What is that something?

Hint: One can approach this problem in a variety of ways, most of which lead nowhere! I recommend starting with Eq. (9.42) in the text and apply to it the

---

[23] *Rappel*: See discussion after Eq. (1.18) for definitions of *flux* and *flux density*.

conditions as $r \to r_0$ (*i.e.*, cold disc, slow/sonic point). Then set $\partial_r \mathcal{B}_\mathrm{M} = 0$ to solve the problem.

9.12

a) Show that as $x \to \infty$, Eq. (9.67) admits *two* asymptotic values for $\tilde{\rho}$.

b) On what basis can we determine which of these two asymptotic values is physical and which is not?

9.13

a) Starting with Eq. (9.42) in the text, show that the tangent of the pitch angle, $\tan\psi$, of the magnetic induction along a line of induction in the Weber–Davis model is given by Eq. (9.70).

b) For $x_\mathrm{A} \gg 1$, show that Eq. (9.70) gives the expected result at $x = 1$ (the footpoint), namely that $\tan\psi = 0$.

c) Again for $x_\mathrm{A} \gg 1$, find a numerical value for $\tan\psi$ at the Alfvén point.

d) Finally, show that for $x \gg 1$ (but not necessarily x_A), $\tan\psi$ grows linearly with x and, in particular,

$$\lim_{x\to\infty} \tan\psi = x\left(\frac{2}{3}x_\mathrm{A}^2 - 1\right)^{-1/2}. \tag{9.77}$$

Thus, as $x \to \infty$ $\psi \to \pi/2$ and the poloidal line of induction is completely twisted into the toroidal direction.

Figure 9.6 in the text illustrates the Weber–Davis model for $x_\mathrm{A} = 100$. Does Eq. (9.77) give the expected result at $x = 400$ in panel *c* of the figure?

Computer project

P9.1 Write a computer program to calculate the scaled density profile, $\tilde{\rho}(x)$, from Eq. (9.67) in the text. Since this equation is transcendental, you will need to use a root finder to solve it. The basic strategy is as follows. Set a fixed value for the distance to the Alfvén point, x_A. For Fig. 9.7, I used $x_\mathrm{A} = 100$, and so you might start with this so that you can aim to reproduce this figure for verification of your program. Next, choose a domain – again for Fig. 9.7, I chose $1 < x \leq 400$ – and then for, say, 500 evenly spaced points within your domain, solve Eq. (9.67) for $\tilde{\rho}$. Then use Eq. (9.69) and Eq. (9.70) to find $\tilde{v}_\mathrm{p}$ and $\tan\psi$ from $\tilde{\rho}$, x, and x_A and plot them up!

Problem Set 9

In principle, one ought to be able to write a simple program using the univariate secant solver in §D.1 as the workhorse. However, two complications arise that may make this routine a little *too* simple; at least I was unable to make it work. The complications I found are:

1. For values of $x_0 < x_6 < x < x_A$ where x_6 is some point between x_0 and x_A, Eq. (9.67) has *six* roots. Now half of them are negative, so these can be eliminated quickly. Still, one must choose among the three positive roots that exist, as only one can be physical. The dumb secant finder in §D.1 will just report the first root it finds, if it can find one at all; see complication 2.

2. Eq. (9.67) has two poles at $\tilde{\rho} = 0$ and 1. Should the root get too close to either, the secant finder in §D.1 may not find a root at all.

Hints: For the second issue, I found I needed a hybrid bisection-secant root finder to handle getting too close to either pole (*e.g.*, Numerical Recipes by Press *et al.* 1992), and for this I needed a routine that could sensibly "bracket" the root before engaging the root finder. By "bracketing the root", what I mean is you need to find two values, $\tilde{\rho}_{\text{left}}$ and $\tilde{\rho}_{\text{right}}$ say, such that the functional value of $F(\tilde{\rho})$ in Eq. (9.67) changes sign between the two all the while ensuring – and this is critical – neither pole lies between $\tilde{\rho}_{\text{left}}$ and $\tilde{\rho}_{\text{right}}$.

Addressing the first issue, for $x < x_A$, $\tilde{\rho} > 1$ and it turns out (at least for $x_A = 100$) that Eq. (9.67) has only one root greater than 1; that's obviously the one you want. On the other hand, for $x > x_A$, $\tilde{\rho} < 1$ and there can be two roots between 0 and 1; you want the greater of the two to force a continuous solution at x_A.

Bon courage!

10 Non-ideal MHD

with thanks to Michael Power, B.Sc. (SMU), 2018,[†] and Christopher Mac-Mackin, B.Sc. (SMU), 2015.[‡]

> *Where did we come from, and how did we get here?*
>
> anybody who ever looked up

10.1 Introducing non-ideal MHD

IN SEARCH for answers to such grand and existential questions, humanity has found its way to the study of stellar and planetary formation. In this adventure, the science of protostars and protoplanets has blossomed into its own subdiscipline within astronomy and astrophysics, as confirmed by the sheer volume of literature generated on the subject over the past four decades.[1] Quite literally, thousands of scientists worldwide spend their entire careers doing little else than delving into the extremely challenging problems, both observational and theoretical, posed by how stars and their planets come into being.

Doing what seemed unfathomable even twenty-five years ago, modern observatories are now peering into the tiniest recesses of our galaxy to image actual planetary discs forming about newly minted protostars (Fig. 10.1). Not to be outdone, astrophysicists are now tackling head-on the daunting complexities of science needed to understand how discs form from the left-overs of

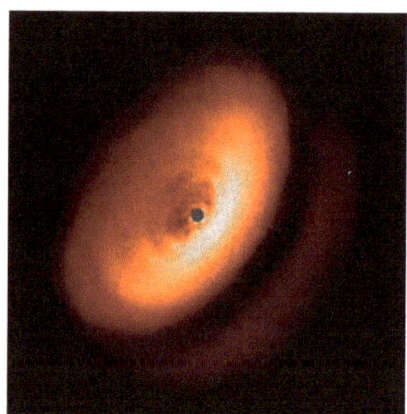

Figure 10.1. Planetary disc of T Tauri star IM Lup, ~ 160 pc from the earth. The disc radius is ~ 335 AU, ~ 10 times Pluto's orbit (credit: ESO, H. Avenhaus *et al.*, E. Sissa *et al.*, DARTT-S, SHINE).

[†]Ph. D. (Toronto), 2028?
[‡]D. Phil. (Oxford), 2019
[1]See, for example, the latest in the series *Protostars and Planets* (Buether *et al.*, 2014).

Figure 10.2. An artist's conception of a planetary disc surrounding a newly formed star (credit: NASA/JPL-Caltech).

stellar collapse and, just as important, how a given disc condenses to form its protostar's future family of planets (*e.g.*, Armitage, 2011).

For a complete understanding of the mechanics and evolution of a planetary disc like the one illustrated in Fig. 10.2, the list of required physics and chemistry is overwhelming: molecular chemistry, dust formation, photo-ionisation, aerosol theory, radiative transport and, of course, non-ideal MHD. In fact, the problem of the planetary disc is among the most illustrative of the complexities of "real-life MHD" because it is here where the three main departures from ideal MHD are revealed and apply: *resistive dissipation*, the *Hall effect* (separation of charge), and *ambipolar diffusion* (coupling of neutrals to ions). Since the temperature within much of a planetary disc is typically below – and even well below – 10^4 K, the fluid is only weakly ionised with most ions coming not from the principle ingredients of H and He, but rather the more easily ionised and relatively plentiful contaminants such as Na and K (5.14 eV and 4.34 eV/atom respectively, *cf.*, 13.6 eV for H). And it is in a weakly ionised medium where the non-ideal effects of MHD are most apparent.

In this, the ultimate chapter of this text, we have left the most challenging aspect of MHD for last. The mathematics is not for the faint-at-heart and it will be quite enough for us to acquire a sense of non-ideal MHD without bringing in the full suite of physics and chemistry that "real-life" astrophysicists grapple with for this problem. Still, no physicist can think of beginning their own journey into answering humanity's most venerable questions of origin without a firm grasp of what underpins it all: non-ideal magnetohydrodynamics.

10.2 The three players

We begin our last journey together by following the excellent review article by Steven Balbus (2009) who outlines, in a most readable form, the basics of non-ideal MHD motivated by the problem of planetary disc formation.

10.2.1 A weakly ionised, isothermal, one-fluid model

Consider a weakly ionised medium consisting primarily of neutral particles (*e.g.*, H, H_2, and He) with a small component of positively charged ions (*e.g.*, Na^+, K^+, and – to quote Balbus directly – other "trace vitamins") and their liberated, negatively charged electrons. Each of these three cospatial components of the fluid is governed by their own set of (magneto)hydrodynamical equations and are coupled together by easily understood forces that each component exerts on another.

For simplicity and illustration, let the fluid be isothermal. By no means is a planetary disc properly considered isothermal! However, the facts that this assumption eliminates the distraction of an energy equation, and the physics of non-ideal MHD – which originate in the momentum equations – end up exclusively in the induction equation, justifies this approach at least for our first look. The more general, non-isothermal case is considered in §10.5.

The momentum equations (Eq. 1.27) for the three components of the fluid including electromagnetic force densities and force densities arising from the interaction between fluid components are given by:

$$\partial_t \vec{s}_n + \nabla \cdot (\vec{s}_n \vec{v}_n) = -\nabla p_n - \rho_n \nabla \phi + \vec{f}^{\,a}_{i,n} + \vec{f}^{\,a}_{e,n}; \tag{10.1}$$

$$\partial_t \vec{s}_i + \nabla \cdot (\vec{s}_i \vec{v}_i) = -\nabla p_i - \rho_i \nabla \phi + Zen_i(\vec{E} + \vec{v}_i \times \vec{B}) + \vec{f}^{\,a}_{e,i} + \vec{f}^{\,a}_{n,i}; \tag{10.2}$$

$$\partial_t \vec{s}_e + \nabla \cdot (\vec{s}_e \vec{v}_e) = -\nabla p_e - \rho_e \nabla \phi - en_e(\vec{E} + \vec{v}_e \times \vec{B}) + \vec{f}^{\,a}_{n,e} + \vec{f}^{\,a}_{i,e}, \tag{10.3}$$

where there are numerous variables to declare, even if most are self-evident:

- $\vec{s}_n$, $\vec{s}_i$, $\vec{s}_e$ are, respectively, the momentum densities ($\vec{s} = \rho\vec{v}$) of neutrals, ions, and electrons ('n', 'i', and 'e');

- $\vec{v}_n$, $\vec{v}_i$, $\vec{v}_e$ are, respectively, the velocities of neutrals, ions, and electrons;

- ρ_n, ρ_i, ρ_e are, respectively, the mass densities of neutrals, ions, and electrons;

- p_n, p_i, p_e are, respectively, the partial pressures of neutrals, ions, and electrons;

- ϕ is, as usual, the gravitational potential which, though not central to fluid dynamics, is included because of the astrophysical context of a planetary disc;

- $\vec{f}^{\,a}_{1,2}$ is the *ambipolar force density* that fluid component '1' exerts on fluid component '2' (1, 2 represent 'n', 'i', or 'e') and, because of Newton's third law, $\vec{f}^{\,a}_{2,1} = -\vec{f}^{\,a}_{1,2}$;

- Ze is the average charge per ion (for low ionisation, $Z \approx 1$);

- $-e$ is the charge on an electron;

- n_i and n_e are, respectively, the number densities of ions and electrons and, because of charge conservation, $n_e = Zn_i$; and

- $\vec{f}_{\rm EM} = \rho_q(\vec{E} + \vec{v} \times \vec{B})$ is the electromagnetic force density on an ionised component of the fluid (ions or electrons), where ρ_q is the charge density, $\vec{E}$ is the background electric field, and $\vec{B}$ is the background magnetic induction. Note that for reasons that will become evident, we are not yet writing this term in the form $\vec{J} \times \vec{B}$.

The nature of $\vec{f}^{\,\rm a}_{1,2}$ warrants some discussion. Two neutral particles approaching each other will have a scattering cross section proportional to their cross-sectional areas (said to be *geometrical*) and, if these particles are essentially point particles, their mutual cross section would be exceedingly small and the likelihood of the particles scattering as a result of their proximity commensurately low.

Conversely, a charged particle, be it a positive ion or a negative electron, will, in the vicinity of a neutral particle, raise poles on the latter much like the moon raises tides on the earth. That is, as a charged particle – let us suppose it to be positive – approaches a neutral particle, electrons within the neutral particle are drawn towards the approaching ion leaving a paucity of negative charge on the opposite side. As a consequence, the now polarised neutral atom and approaching ion feel a residual Coulomb attraction, enhancing their scattering cross section and thus interaction significantly, even in a weakly ionised, rarified medium.

As subpopulations of the fluid, the ions and neutrals will, in general, have different bulk flow velocities, $\vec{v}_1$ and $\vec{v}_2$. However, the enhanced interaction between them encourages the two components to mix, thus acting like a drag force whose effect is to reduce whatever relative velocity may be between them. As a drag force, this *ambipolar force density*[2] is evidently proportional to the density of each component and their relative velocity,

$$\vec{f}^{\,\rm a}_{1,2} = \gamma_{1,2}\rho_1\rho_2(\vec{v}_1 - \vec{v}_2) \equiv \gamma_{1,2}\rho_1\rho_2\vec{v}_{1,2}, \tag{10.4}$$

where $\vec{v}_{1,2}$ is the velocity of fluid component '1' relative to component '2', and where $\gamma_{1,2}$ is the *coupling (drag) coefficient* containing all the physics leading to the enhanced scattering cross section. It is usually expressed as (*e.g.*, Draine *et al.* 1983; Balbus, 2009),

$$\gamma_{1,2} = \frac{\langle \sigma u \rangle_{1,2}}{m_1 + m_2}, \tag{10.5}$$

where m_1 and m_2 are the average masses of particles in components '1' and '2' (*e.g.*, ions and neutrals), and where $\langle \sigma u \rangle_{1,2}$ – the *rate coefficient*, a.k.a. the *collision rate* – is the product of the scattering cross section, σ, and the relative velocity,

---

[2] The etymology of "ambipolar" is "ambi" = both, as in "ambidexterous". The Coulomb residuals from *both poles* contribute to the enhanced scattering cross section between ions and neutrals.

u, for particles in components '1' and '2' averaged over the ensemble assuming a Maxwell–Boltzmann distribution (thermodynamical equilibrium; see Problem 10.1).

In the present discussion, the important rate coefficients are those for electron–neutral and ion–neutral interactions. For a planetary disc that is 84% H_2 and 16% He by number (solar abundances), the average mass per neutral particle is $m_n = 2.32 m_p$, where m_p is the mass of the proton. Assuming the bulk of ions are Na^+ and K^+, then again from solar abundances, $Na:K \sim 5:4$ and the average mass per ion is $m_i \sim 30 m_p$. For such a composition, Draine et al. (1983) find that,

$$\langle \sigma u \rangle_{n,e} = \chi_{n,e} \sqrt{\frac{T}{T_0}} \quad \text{and} \quad \langle \sigma u \rangle_{n,i} = \chi_{n,i}, \tag{10.6}$$

where $\chi_{n,e} \sim 2.62 \times 10^{-14}\,\mathrm{m^3\,s^{-1}}$, $\chi_{n,i} \sim 1.90 \times 10^{-15}\,\mathrm{m^3\,s^{-1}}$, and $T_0 = 10^3\,\mathrm{K}$, the nominal temperature in the inner portion of a planetary disc. These expressions shall be used later when establishing the relative importance of the non-ideal terms.

For our discussions in §10.4 and §10.5, we'll also need the rate coefficient for ion–electron interactions. Problem 10.1 takes the reader through the calculation of this quantity, where you should find,

$$\langle \sigma u \rangle_{i,e} = \chi_{i,e} \left(\frac{T_0}{T}\right)^{3/2}, \tag{10.7}$$

where $\chi_{i,e} \sim 1.84 \times 10^{-9}\,\mathrm{m^3\,s^{-1}}$.

Then, at the risk of confusing matters further, I define yet another quantity – what I refer to as the *ambipolar coefficient* – for convenience and future use:

$$\beta_{1,2} = \frac{1}{\gamma_{1,2}\rho_1\rho_2} = \frac{1}{\langle \sigma u \rangle_{1,2}\mu_{1,2}n_1 n_2}, \tag{10.8}$$

where $\mu_{1,2}$ is the reduced mass of particles in components '1' and '2',

$$\mu_{1,2} = \frac{m_1 m_2}{m_1 + m_2}. \tag{10.9}$$

The important point to bear in mind with the three quantities $\gamma_{1,2}$, $\langle \sigma u \rangle_{1,2}$, and $\beta_{1,2}$ is that each describe the nature of the particle–particle interaction. In a given context, whether one quantity is used or another will depend on the argument being made and/or which makes the algebra look tidiest. The SI units for $\gamma_{1,2}$ are inverse density *per* second ($\mathrm{m^3\,kg^{-1}\,s^{-1}}$), for $\langle \sigma u \rangle_{1,2}$ are area times a speed ($\mathrm{m^3\,s^{-1}}$), for $\beta_{1,2}$ are inverse density *times* second ($\mathrm{m^3\,kg^{-1}\,s}$), and all three are symmetric with respect to the interchange of their subscripts. I note in passing that in the literature, you'll find yet another related quantity,

$$\nu_{1,2} = n_2 \langle \sigma u \rangle_{1,2},$$

which is *not* symmetric in the interchange of '1' and '2' because of the appearance of the number density, n_2. With units $\mathrm{s^{-1}}$, $\nu_{1,2}$ is known to plasma physicists as the *collision frequency* with which a single particle of component '1' collides with *any* particle of component '2'. In an effort to pare down the number of coefficients at least by one, in my presentation I've managed to avoid the specific need for $\nu_{1,2}$, and so I mention it no further!

Finally, from Eq. (10.4), the ambipolar force density exerted by component '2' on component '1' is, by Newton's third law,

$$\vec{f}^{\text{a}}_{2,1} = -\vec{f}^{\text{a}}_{1,2} = \gamma_{1,2}\rho_1\rho_2\vec{v}_{2,1} = \langle\sigma u\rangle_{1,2}\mu_{1,2}n_1n_2\vec{v}_{2,1} = \frac{\vec{v}_{2,1}}{\beta_{1,2}}, \qquad (10.10)$$

since $\rho_1 = m_1 n_1$.

Now, in a weakly ionised medium such as a planetary disc, $\rho_i \ll \rho_n$ and $\rho_e \ll \rho_n$. Equivalently, the partial pressures of the ionised components are negligible compared to that of the neutral particles: $p_i \ll p_n$ and $p_e \ll i_n$. In this limit, the inertial terms in Eq. (10.2) and (10.3) are negligible compared to the electromagnetic and ambipolar force densities which evidently must balance each other. Thus:

$$en_e(\vec{E} + \vec{v}_i \times \vec{B}) + \vec{f}^{\text{a}}_{e,i} + \vec{f}^{\text{a}}_{n,i} = 0; \qquad (10.11)$$

$$-en_e(\vec{E} + \vec{v}_e \times \vec{B}) + \vec{f}^{\text{a}}_{n,e} + \vec{f}^{\text{a}}_{i,e} = 0, \qquad (10.12)$$

since $n_e = Zn_i$. Adding Eq. (10.11) and (10.12) together then yields:

$$\underbrace{en_e(\vec{v}_i - \vec{v}_e)}_{\vec{J}} \times \vec{B} + \vec{f}^{\text{a}}_{n,i} + \vec{f}^{\text{a}}_{n,e} = 0, \qquad (10.13)$$

where $\vec{J}$ is the *current density* in the fluid, proportional to the velocity of the ions relative to the electrons.[3] This is a good time to remind the reader that from the Ampére–Maxwell law (Eq. B.1 in App. B),

$$\vec{J} = en_e\vec{v}_{i,e} = \frac{1}{\mu_0}\nabla \times \vec{B}, \qquad (10.14)$$

when the displacement current can be ignored, as is the case for non-relativistic MHD. Thus, we can rewrite Eq. (10.13) as,

$$\vec{f}^{\text{a}}_{i,n} + \vec{f}^{\text{a}}_{e,n} = -\vec{f}^{\text{a}}_{n,i} - \vec{f}^{\text{a}}_{n,e} = \underbrace{\vec{J} \times \vec{B}}_{\vec{f}_L} = \frac{1}{\mu_0}(\nabla \times \vec{B}) \times \vec{B}, \qquad (10.15)$$

where we finally recognise the Lorentz force density, $\vec{f}_L$, as defined in Eq. (4.10). Thus, Eq. (10.1) becomes,

$$\boxed{\partial_t \vec{s}_n + \nabla \cdot (\vec{s}_n \vec{v}_n) = -\nabla p_n - \rho_n \nabla \phi + \frac{1}{\mu_0}(\nabla \times \vec{B}) \times \vec{B},} \qquad (10.16)$$

which now looks like the standard, ideal momentum equation for MHD (*e.g.*, Eq. 4.13).

Now, if you've been paying attention, something about Eq. (10.16) should strike you as a bit odd. The list of force densities on the RHS of the momentum equation for *neutral* particles includes the Lorentz force to which neutral particles aren't supposed to be affected! So what gives?

As Eq. (10.15) tells us, in the limit of a weakly ionised medium, the sum of the force densities exerted by the charged subpopulations on the neutrals is equal

---

[3] Evidently, if the ions and electrons move in lockstep with $\vec{v}_{i,e} = 0$, there can be no net current.

to the Lorentz force density the magnetic induction exerts on the charges. Thus, $\vec{J} \times \vec{B}$ appears in Eq. (10.16) as a proxy, if you will, for the ambipolar force densities acting on the neutrals, $\vec{f}_{\text{i,n}}^{\text{a}} + \vec{f}_{\text{e,n}}^{\text{a}}$. Now, even though $\vec{f}_{\text{i,n}}^{\text{a}}$ and $\vec{f}_{\text{e,n}}^{\text{a}}$ are proportional to ρ_{i} and ρ_{e} respectively, they are not negligible (evidently not; they sum to $\vec{J} \times \vec{B}$!) because of the enhanced ion–neutral coupling coefficients $\gamma_{\text{e,n}}$ and $\gamma_{\text{i,n}}$. Thus, to emphasise what has already been stated, the ions and electrons have a significant effect on the neutral particles despite their low number densities.

Next, examining the ratio of the ambipolar force densities the electrons and ions each exert on the neutrals, we have,

$$\frac{f_{\text{e,n}}^{\text{a}}}{f_{\text{i,n}}^{\text{a}}} = \frac{\langle \sigma u \rangle_{\text{e,n}} \mu_{\text{e,n}} n_{\text{e}} n_{\text{n}} v_{\text{e,n}}}{\langle \sigma u \rangle_{\text{i,n}} \mu_{\text{i,n}} n_{\text{i}} n_{\text{n}} v_{\text{i,n}}} \sim 3.5 \times 10^{-3} \sqrt{\frac{T}{T_0}}, \qquad (10.17)$$

using Eq. (10.6) and (10.10), where $\mu_{\text{e,n}}$ and $\mu_{\text{i,n}}$ are the reduced masses of the electron–neutral and ion–neutral particles respectively (Eq. 10.9) using solar abundances ($m_{\text{n}} = 2.32 m_{\text{p}}$ and $m_{\text{i}} = 30 m_{\text{p}}$), and presuming $n_{\text{e}} = n_{\text{i}}$ with no systemic difference between the velocities of the electrons and ions relative to the neutral particles. This ratio is even smaller for $T < T_0$ and so, for our purposes, $\vec{f}_{\text{e,n}}^{\text{a}}$ can be safely ignored compared to $\vec{f}_{\text{i,n}}^{\text{a}}$. Thus, Eq. (10.15) becomes,

$$\vec{f}_{\text{i,n}}^{\text{a}} = \vec{J} \times \vec{B} = \vec{f}_{\text{L}} = \frac{\vec{v}_{\text{i,n}}}{\beta_{\text{i,n}}} \quad \Rightarrow \quad \vec{v}_{\text{i,n}} = -\vec{v}_{\text{n,i}} = \beta_{\text{i,n}} \vec{f}_{\text{L}}, \qquad (10.18)$$

using Eq. (10.10). Therefore, in the low-ionisation limit, three prominent velocities of the flow can be attributed unique physical interpretations, namely:

1. $\vec{v}_{\text{n}}$ is the bulk flow velocity of the vast majority of the fluid;

2. $\vec{v}_{\text{i,e}} = \vec{J}/en_{\text{e}}$, proportional to the current density (Eq. 10.14); and

3. $\vec{v}_{\text{i,n}} = \beta_{\text{i,n}} \vec{f}_{\text{L}}$, proportional to the Lorentz force density (Eq. 10.18).

Returning now to Eq. (10.12), $\vec{f}_{\text{i,e}}^{\text{a}} \propto \rho_{\text{e}} \rho_{\text{i}}$ can be safely ignored on the grounds that it is second-order small. Thus, dividing Eq. (10.12) through by $-en_{\text{e}}$, dropping $\vec{f}_{\text{i,e}}^{\text{a}}$, inserting a few gratuitous velocities here and there, and using Eq. (10.4) for $\vec{f}_{\text{n,e}}^{\text{a}}$, we get,

$$\vec{E} + (\vec{v}_{\text{e}} + \vec{v}_{\text{i}} - \vec{v}_{\text{i}} + \vec{v}_{\text{n}} - \vec{v}_{\text{n}}) \times \vec{B} - \frac{\gamma_{\text{n,e}} \rho_{\text{n}} \rho_{\text{e}}}{en_{\text{e}}} (\vec{v}_{\text{n}} - \vec{v}_{\text{e}} + \vec{v}_{\text{i}} - \vec{v}_{\text{i}}) = 0$$

$$\Rightarrow \quad \vec{E} + \vec{v}_{\text{n}} \times \vec{B} - \vec{v}_{\text{i,e}} \times \vec{B} + \vec{v}_{\text{i,n}} \times \vec{B} + \frac{\gamma_{\text{n,e}} \rho_{\text{n}} \rho_{\text{e}}}{en_{\text{e}}} (\vec{v}_{\text{i,n}} - \vec{v}_{\text{i,e}}) = 0.$$

Then, using Eq. (10.14) and (10.18) to replace $\vec{v}_{\text{i,e}}$ and $\vec{v}_{\text{i,n}}$, we get,

$$\vec{E} = -\vec{v}_{\text{n}} \times \vec{B} + \frac{1}{en_{\text{e}}} \underbrace{\vec{J} \times \vec{B}}_{\vec{f}_{\text{L}}} - \beta_{\text{i,n}} \vec{f}_{\text{L}} \times \vec{B} - \frac{\gamma_{\text{n,e}} \rho_{\text{n}} \rho_{\text{e}}}{en_{\text{e}}} \left(\beta_{\text{i,n}} \vec{f}_{\text{L}} - \frac{1}{en_{\text{e}}} \vec{J} \right). \qquad (10.19)$$

The second and fourth terms on the RHS are both proportional to $\vec{f}_L$. Examining the ratio of their scalar coefficients, we find:

$$\frac{\text{term}_4}{\text{term}_2} = \frac{\gamma_{n,e}\rho_n\rho_e}{en_e} \frac{1}{\gamma_{i,n}\rho_i\rho_n} en_e = \frac{\langle\sigma u\rangle_{e,n}\mu_{e,n}n_e n_n}{\langle\sigma u\rangle_{i,n}\mu_{i,n}n_i n_n} \sim 3.5 \times 10^{-3}\sqrt{\frac{T}{T_0}},$$

the same result found in Eq. (10.17). Thus the fourth term can be safely ignored, and Eq. (10.19) can be written as,

$$\boxed{\vec{E} = -\vec{v}_n \times \vec{B} + \eta\vec{J} + \frac{1}{en_e}\vec{f}_L - \beta_{i,n}\vec{f}_L \times \vec{B},} \qquad (10.20)$$

where,

$$\eta \equiv \frac{\gamma_{n,e}\rho_n\rho_e}{(en_e)^2} = \frac{\langle\sigma u\rangle_{n,e}\mu_{n,e}}{e^2}\frac{n_n}{n_e}. \qquad (10.21)$$

Since η has SI units,

$$[\eta] = \frac{\text{m}^3\,\text{s}^{-1}\,\text{kg}\,\text{m}^{-3}}{\text{C}^2}\,\text{m}^{-3}\cdot\text{m}^{-3}\,=\,\frac{\text{V m}}{\text{A}} = \Omega\,\text{m},$$

and is irrespective of the ionised component, it is interpreted as the *resistivity* arising from the frictional interaction between electrons and neutral particles. Thus, $\eta\vec{J}$ is the portion of the electric field, $\vec{E}$, required to drive a current density, $\vec{J}$, in a medium with resistivity, η.

In the literature, Eq. (10.20) is often referred to as the *generalised Ohm's law*, which I suppose stems from the commonly held view that $\vec{E} = \eta\vec{J}$ is a statement of the "ungeneralised" Ohm's law. As any astute first-year physics student will know but many seasoned physicists seem to have forgotten, the expressions $\vec{E} = \eta\vec{J}$ and its cousin from circuit analysis, $V = IR$, are *not* statements of Ohm's law but merely operational definitions of the resistivity, η, and resistance, R. A *correct* statement of "Ohm's law" is for some so-called "Ohmic materials", η (R) is independent of the applied field, $\vec{E}$ (voltage, V); that is, $\vec{E} \propto \vec{J}$ or $V \propto I$.[4] While it is certainly true that η, as given by Eq. (10.21), is independent of $\vec{E}$ (and so the MHD fluid under consideration qualifies as "Ohmic"), I'm still not certain what those who refer to Eq. (10.20) as the "generalised Ohm's law" actually mean. For my part, I shall refer to Eq. (10.20) as the "electric field in a non-ideal MHD fluid" and any effects attributed to the resistivity, η, as "resistive" rather than "Ohmic". I mention all this so the reader is cognizant of why many authors will refer to these as the "generalised Ohm's law" and "Ohmic effects" respectively.

Finally, from Faraday's law of induction (Eq. B.1), we can write from Eq. (10.20),

$$-\nabla \times \vec{E} = \boxed{\partial_t \vec{B} = \nabla \times \left(\vec{v}_n \times \vec{B} - \eta\vec{J} - \frac{1}{en_e}\vec{f}_L + \beta_{i,n}\vec{f}_L \times \vec{B}\right),} \qquad (10.22)$$

giving us the *non-ideal induction equation* with "the three players" of non-ideal MHD fully exposed. Evidently, the leading term behind the curl, $\vec{v}_n \times \vec{B}$, is the *induction term*, which we found by conserving magnetic flux in §4.2 (Eq. 4.4). The

---

[4] See, for example, Halliday, Resnick, & Walker (2003) who make these specific points.

three terms following are what render this version of the induction equation "non-ideal" which, from left to right, are:

1. $-\eta \vec{J}$: *Resistive dissipation*, discussed further in §10.3. As stated, this results from the interaction between the electron and neutral sub-populations via the resistivity, η.

2. $-\vec{f}_\mathrm{L}/en_\mathrm{e}$: the *Hall effect*, discussed further in §10.4. Since the coefficient $1/en_\mathrm{e} = 1/Zen_\mathrm{i}$ is irrespective of the neutral component, this term stems from the interaction between the electron and ionised sub-populations.

3. $\beta_{\mathrm{i,n}}\vec{f}_\mathrm{L} \times \vec{B}$: *ambipolar diffusion*, discussed further in §10.5, so named because of the enhanced diffusion of ions and neutrals afforded by the ambipolar force density as measured by the ambipolar coefficient, $\beta_{\mathrm{i,n}}$.

Including the continuity equation, *Equation Set 10* below embodies our derivation from first principles of a non-ideal MHD model for a weakly ionised three-component isothermal fluid consisting of low number densities each of electrons and ions, mixed with a much higher number density of neutral particles.

Equation Set 10:

$$\partial_t \rho_\mathrm{n} + \nabla \cdot (\rho_\mathrm{n} \vec{v}_\mathrm{n}) = 0; \qquad \text{Eq. (1.19)}$$

$$\partial_t \vec{s}_\mathrm{n} + \nabla \cdot (\vec{s}_\mathrm{n} \vec{v}_\mathrm{n}) = -\nabla p_\mathrm{n} - \rho_\mathrm{n} \nabla \phi + \vec{f}_\mathrm{L}; \qquad \text{Eq. (10.16)}$$

$$\partial_t \vec{B} = \nabla \times \left(\vec{v}_\mathrm{n} \times \vec{B} - \eta \vec{J} - \frac{1}{en_\mathrm{e}} \vec{f}_\mathrm{L} + \beta_{\mathrm{i,n}} \vec{f}_\mathrm{L} \times \vec{B} \right), \qquad \text{Eq. (10.22)}$$

---

where $p_\mathrm{n} = c_\mathrm{iso} \rho_\mathrm{n}$, $c_\mathrm{iso} = \sqrt{k_\mathrm{B} T/m_\mathrm{n}}$ is the isothermal sound speed (Eq. 2.11) in keeping with our assumption of isothermality, $\vec{J} = \nabla \times \vec{B}/\mu_0$ is the current density, $\vec{f}_\mathrm{L} = \vec{J} \times \vec{B}$ is the Lorentz force density, and where η and $\beta_{\mathrm{i,n}}$ are given by Eq. (10.21) and (10.8) respectively.

Other than the non-ideal terms in the induction equation, Eq. Set 10 looks just like Eq. Set 6 on page 109 (less the energy equation, of course), derived for a *single fluid*. The flow variables in Eq. Set 10 (ρ_n, $\vec{v}_\mathrm{n}$, *etc.*) all refer to the dominant neutral component with the only vestiges of the charged components found in the coefficients η and $1/en_\mathrm{e}$ (in which n_e appears), and $\beta_{\mathrm{i,n}}$ (in which n_i is buried). And since $n_\mathrm{e} = Zn_\mathrm{i}$, one only needs input for n_i which, for weakly ionised astrophysical plasmas, can usually be modelled as some function of n_n. For example, Fiedler & Mouschovias (1993) suggest,

$$n_\mathrm{i} = k_1 \left(\frac{n_\mathrm{n}}{n_1}\right)^{1/2} + k_2 \left(\frac{n_2}{n_\mathrm{n}}\right)^2, \qquad (10.23)$$

where $k_1 = 3 \times 10^3 \,\mathrm{m}^{-3}$, $k_2 = 4.64 \times 10^2 \,\mathrm{m}^{-3}$, $n_1 = 10^{11} \,\mathrm{m}^{-3}$, and $n_2 = 10^9 \,\mathrm{m}^{-3}$.

Equation Set 10 is a *one-fluid isothermal model* for non-ideal MHD and forms the basis for theoretical and computational work in planetary discs, as well as any other problem in (astro)physics where the fluid is expected to be weakly ionised and isothermal.

The first to incorporate one-fluid ambipolar diffusion in an isothermal MHD code was Black & Scott (1982). Since then, many works too numerous to list have performed one-fluid isothermal simulations including one and even two non-ideal terms, but precious few have included all three. The first I'm aware of to apply all three terms to a simulation of a planetary disc is Lesur, Kunz, & Fromang (2014). Computational challenges introduced by the non-ideal terms – particularly the Hall term – are discussed by Falle (2003), who also reports on an isothermal algorithm that overcomes these challenges under certain circumstances.

10.2.2 Relative importance of the non-ideal terms

For viscid flow, we defined in Chap. 8 the unitless *Reynolds number*, $\mathcal{R}$ (Eq. 8.40), to distinguish flows in which viscous stresses are negligible compared to the inertial terms ($\mathcal{R} \gg 1$) from those where they dominate ($\mathcal{R} < 1$). In that same spirit, we define the unitless *magnetic Reynolds number*, $\mathcal{R}_\mathrm{M}$, to compare the relative importance of the inductive and resistive terms. By taking $v_\mathrm{n} \to V$ as characteristic of the flow speed and $\nabla \to 1/L$ as a characteristic length (*e.g.*, distance over which variables change appreciably), we can write the ratio of the induction and resistive terms in Eq. (10.22) as,[5]

$$\frac{|\vec{v}_\mathrm{n} \times \vec{B}|}{\eta|\vec{J}|} = \frac{|\vec{v}_\mathrm{n} \times \vec{B}|}{\eta|\nabla \times \vec{B}|/\mu_0} \to \frac{VB}{\eta B/(L\mu_0)} = \frac{\mu_0 V L}{\eta} \equiv \mathcal{R}_\mathrm{M}. \qquad (10.24)$$

When the characteristic speed is taken as the Alfvén speed, $a = B/\sqrt{\mu_0 \rho}$ (Eq. 5.26), the magnetic Reynolds number is called the *Lundquist number*,

$$\mathcal{S} \equiv \frac{\mu_0 a L}{\eta}. \qquad (10.25)$$

Starting with the Lundquist number, we can find a more practical expression (*i.e.*, into which numbers can be plugged) for the ratio of the inductive and resistive terms, which I designate here as R_IR. Substituting Eq. (10.21) into Eq. (10.25), then using the first of Eq. (10.6), we arrive at,

$$\begin{aligned} R_\mathrm{IR} = \mathcal{S} &= \mu_0 a L \frac{e^2}{\langle \sigma u \rangle_\mathrm{n,e} \mu_\mathrm{n,e}} \frac{n_\mathrm{e}}{n_\mathrm{n}} \\ &= \frac{\mu_0 e^2 L_0}{\chi_\mathrm{n,e} \mu_\mathrm{n,e}} \sqrt{\frac{k_\mathrm{B} T_0}{m_\mathrm{n}}} \, a \underbrace{\sqrt{\frac{m_\mathrm{n}}{k_\mathrm{B} T}}}_{1/c_\mathrm{iso}} \frac{n_\mathrm{e}}{n_\mathrm{n}} \frac{L}{L_0} = \chi_\mathrm{IR} \sqrt{\frac{\alpha}{\alpha_0}} \frac{n_\mathrm{e}}{n_\mathrm{n}} \frac{L}{L_0}, \end{aligned} \qquad (10.26)$$

where $\chi_\mathrm{IR} = 3.82 \times 10^{13}$ (unitless), $L_0 = 1\,\mathrm{AU}\,(1.50 \times 10^{11}\,\mathrm{m})$, $\alpha = a^2/c_\mathrm{iso}^2$ is the "MHD-alpha" defined by Eq. (5.30), $\alpha_0 = 0.01$ which corresponds to a "warm"

---

[5] As compared to how we arrived at the Reynolds number, $\mathcal{R}$, in §8.5, this "derivation" of the magnetic Reynolds number is rather "cheap and cheerful". For a more "proper" derivation of $\mathcal{R}_\mathrm{M}$, see Problem 10.4.

planetary disc (where, owing to protostellar heating, the thermal energy density is greater than the magnetic energy density), c_iso is the isothermal sound speed, and $m_\text{n} = 2.32 m_\text{p}$ as set on page 382.

Evidently, for a "warm" planetary disc where $L = 1$ AU is taken to be the scale length, $R_\text{IR} \propto \chi_\text{IR}$ is *ridiculously* larger than 1 (*i.e.*, the fluid can be considered a perfect conductor) unless the ionisation fraction, n_e/n_n, is *exceedingly* low.

As it happens, in much of a planetary disc where $10 \lesssim T \lesssim 10^4$ K, $n_\text{e}/n_\text{n} \lesssim 10^{-10}$ making $R_\text{IR} \lesssim 10^3$ which, while still significantly greater than 1, now makes fluid resistivity arising from electron–neutral coupling a dynamical player.[6] Thus, what seems to be an overwhelmingly large coefficient, χ_IR, in Eq. (10.26) cannot rule out the resistive term in some astrophysical applications nor, for that matter, the Hall and ambipolar diffusion terms.

To compare the resistive and Hall terms, we consider the ratio of the second and third terms on the RHS of Eq. (10.22),

$$R_\text{RH} = \frac{\eta |\vec{J}|}{|\vec{f}_\text{L}|/(en_\text{e})} = \frac{\eta e n_\text{e} |\vec{J}|}{|\vec{J} \times \vec{B}|} \to \frac{\eta e n_\text{e}}{B} = \frac{\langle \sigma u \rangle_\text{n,e} \mu_\text{n,e} n_\text{n}}{Be},$$

using Eq. (10.21). Once again, we insert the first of Eq. (10.6) and factors of unity to get,

$$R_\text{RH} = \frac{\chi_\text{n,e} \mu_\text{n,e}}{e} \frac{1}{\sqrt{k_\text{B} T_0 \mu_0}} \underbrace{\sqrt{\frac{k_\text{B} T}{m_\text{n}}}}_{c_\text{iso}} \underbrace{\frac{\sqrt{\mu_0 m_\text{n} n_\text{n}}}{B}}_{1/a} n_\text{n}^{1/2} = \sqrt{\frac{\alpha_0}{\alpha} \frac{n_\text{n}}{n_\text{RH}}}, \qquad (10.27)$$

where the critical density for R_RH, $n_\text{RH} = 7.83 \times 10^{21}$ m^{-3}, corresponds to a mass density of $\sim 3 \times 10^{-5}$ kg m^{-3}. While small on terrestrial scales (n_air is $\sim$3,500 times greater), n_RH is *enormous* compared to typical ISM densities. For example, in the densest part of a planetary disc (within 1 AU of the protostar; just about as dense as the ISM gets) $n_\text{n} \sim 5 \times 10^{19}$ m^{-3} (Balbus & Terquem, 2001) which is still a factor of $\sim$150 times smaller than n_RH.

Thus, even in the innermost and densest region of a "warm" planetary disc, $R_\text{RH} \sim 0.08$. As this only decreases with distance from the protostar (as n_n falls), the Hall effect generally dominates the resistive term everywhere in a planetary disc.[7]

Finally, to compare the Hall and ambipolar diffusion effects, we consider the ratio of the third and fourth terms on the RHS of Eq. (10.22),

$$R_\text{HA} = \frac{1}{en_\text{e}} \frac{|\vec{f}_\text{L}|}{\beta_\text{i,n} |\vec{f}_\text{L} \times \vec{B}|} \to \frac{\gamma_\text{i,n} \rho_\text{i} \rho_\text{n}}{en_\text{e} B} = \frac{\langle \sigma u \rangle_\text{i,n} \mu_\text{i,n} n_\text{i} n_\text{n}}{en_\text{e} B},$$

using Eq. (10.8), (10.5), and (10.9). To continue, we set $\langle \sigma u \rangle_\text{i,n} = \chi_\text{i,n}$ (second of

---

[6] *Rappel* from Chap. 8: For viscid fluids, a Reynolds number of 10^3 corresponds to water flowing out of a measuring cup (§8.5), an entirely laminar flow (as opposed to turbulent) because of the viscous stresses in water even at such a high Reynolds number.

[7] I would be remiss if I did not mention that this fact was impressed upon the astrophysical community in good part by the efforts of Steven Balbus and his collaborators during the early to mid aughts in a series of very illuminating papers (*e.g.*, see Balbus, 2009, and references therein).

Eq. 10.6) and insert numerous factors of unity to get,

$$R_{\mathrm{HA}} = \frac{\chi_{i,n}\mu_{i,n}}{e\sqrt{k_\mathrm{B}T_0\mu_0}}\frac{n_i}{n_e}\underbrace{\frac{\sqrt{\mu_0 m_n n_n}}{B}}_{1/a}\underbrace{\sqrt{\frac{k_\mathrm{B}T}{m_n}}}_{c_{\mathrm{iso}}}\sqrt{\frac{T_0}{T}}\sqrt{n_n} = \sqrt{\frac{\alpha_0}{\alpha}\frac{T_0}{T}\frac{n_n}{n_{\mathrm{HA}}}}, \qquad (10.28)$$

since $n_e = Z n_i \sim n_i$ for low ionisation, and where $n_{\mathrm{HA}} = 9.51 \times 10^{16}\,\mathrm{m}^{-3}$ – the critical density for R_{HA} – is nearly five orders of magnitude smaller than n_{RH} (Eq. 10.27).

Thus, in the inner AU of a "warm" planetary disc where $n_n \sim 5 \times 10^{19}\,\mathrm{m}^{-3}$, $R_{\mathrm{HA}} \gtrsim 20$ and once again the Hall effect is the dominant non-ideal term. Since n_n falls off more rapidly than T with distance from the protostar, it is only in the outer regions of the disc where ambipolar diffusion overtakes the Hall effect.

Evidently, all other ratios can be determined by multiplying the appropriate ratios from Eq. (10.26), (10.27), and (10.28). Thus, the ratio of the induction to Hall terms is,

$$R_{\mathrm{IH}} = R_{\mathrm{IR}}R_{\mathrm{RH}} = \chi_{\mathrm{IR}}\sqrt{\alpha}\,\frac{n_e}{n_n}\frac{L}{L_0}\sqrt{\frac{1}{\alpha}\frac{n_n}{n_{\mathrm{RH}}}} = \chi_{\mathrm{IR}}\sqrt{\frac{n_n}{n_{\mathrm{RH}}}}\frac{L}{L_0}\frac{n_e}{n_n}, \qquad (10.29)$$

independent of α and thus the relative role of thermal and magnetic energy densities. As with the resistive term, the leading factor χ_{IR} provides a steep hill for the Hall term to climb to be felt next to the induction term. However, the fact that $\sqrt{n_n/n_{\mathrm{RH}}} < 1$ gives the Hall term a "leg up", as it were, on the resistive term particularly as one draws away from the protostar. This is consistent with the conclusion drawn from Eq. (10.27) that the Hall term dominates the resistive term in planetary discs.

Similarly, the ratio of the induction to ambipolar diffusion term is,

$$R_{\mathrm{IA}} = R_{\mathrm{IH}}R_{\mathrm{HA}} = \chi_{\mathrm{IR}}\frac{n_e}{\sqrt{n_{\mathrm{RH}}n_{\mathrm{HA}}}}\frac{L}{L_0}\sqrt{\frac{\alpha_0}{\alpha}\frac{T_0}{T}}, \qquad (10.30)$$

dependent on length scale, energy balance (α), and temperature. This time, χ_{IR} is mitigated by the ratio of n_e to the geometric average of n_{RH} and n_{HA} which, for a planetary disc, is comparable to n_n in the inner regions. Unlike n_n, however, $\sqrt{n_{\mathrm{RH}}n_{\mathrm{HA}}}$ does not fall off with distance, and this leads to the eventual dominance of ambipolar diffusion in the outer disc.

Finally, the ratio of resistive to ambipolar diffusion terms is given by,

$$R_{\mathrm{RA}} = R_{\mathrm{RH}}R_{\mathrm{HA}} = \sqrt{\frac{\alpha_0}{\alpha}\frac{n_n}{n_{\mathrm{RH}}}}\sqrt{\frac{\alpha_0}{\alpha}\frac{T_0}{T}\frac{n_n}{n_{\mathrm{HA}}}} = \frac{\alpha_0}{\alpha}\frac{n_n}{\sqrt{n_{\mathrm{RH}}n_{\mathrm{HA}}}}\sqrt{\frac{T_0}{T}}, \qquad (10.31)$$

which is $\sim$ unity in the inner-most disc where the resistive and ambipolar diffusion terms are comparable but dominated by the Hall term. However, unlike Eq. (10.28), n_n appears outside the radical and, as such, its fall with distance from the protostar has a much greater effect on R_{RA} than the other ratios leading to the dominance of the ambipolar diffusion term over both resistive and Hall terms in the outer disc.

Figure 10.3 illustrates regions of dominance of the "three players" in a portion of n_n-T space consistent with the one-fluid approximation (i.e., $n_e \ll n_n$ and thus

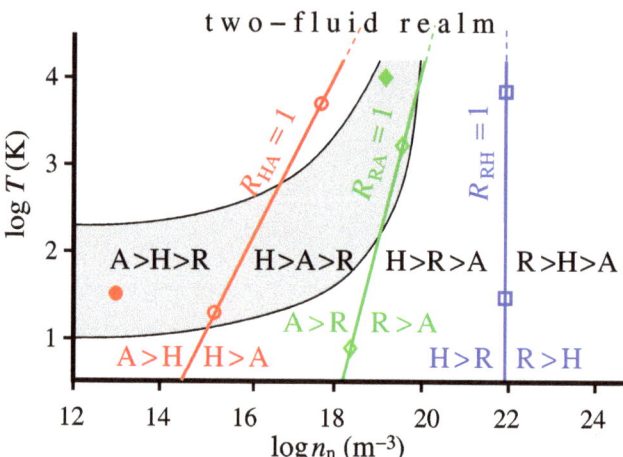

Figure 10.3. Regions in n_n–T space where the resistive (R), Hall (H), and ambipolar diffusion (A) terms are the dominant or second most dominant non-ideal term in Eq. (10.22). The grey swath represents approximately the portion of this space occupied by a typical planetary disc, with the inner disc corresponding to the top, outer disc to the left. See further discussion in the text. (Adapted from Fig. 2 in Balbus, 2009.)

$T < 10^4$ K) and inclusive of the region occupied by a typical proto-planetary disc (shaded swath). Indeed, it is because of the very smallness of n_e/n_n that the non-ideal terms in the form they appear in Eq. (10.22) play any dynamical role at all (Eq. 10.26, 10.29, and 10.30).

The red$\bigcirc$, green$\Diamond$, and blue$\square$ lines are, respectively, where $R_{\rm HA}$, $R_{\rm RA}$, and $R_{\rm RH}$ equal 1 (using $\alpha = 0.01$), while the letters R, H, and A indicate the resistive, Hall, and ambipolar diffusion terms. For example, by setting $R_{\rm RA} = 1$ and for $\alpha = \alpha_0$, we have from Eq. (10.31),

$$T = \frac{T_0}{n_{\rm RH} n_{\rm HA}} n_n^2 \sim (1.34 \times 10^{-36}) n_n^2 \quad \Rightarrow \quad \log T \sim 2 \log n_n - 35.9,$$

which is the equation for the green$\Diamond$ line in Fig. 10.3.

One interprets Fig. 10.3 as follows. To the right (towards higher n_n) of the $R_{\rm IJ} = 1$ line (where I, J = R, H, or A), term I > term J. Conversely, to the left (towards lower n_n) of the $R_{\rm IJ} = 1$ line, term J > term I. Thus and for example, in the outer regions of the planetary disc, the point $(n_n, T) = (10^{13}, 30)$ (filled red circle) is left of all three lines, and thus the ambipolar term is greater than the Hall term (red$\bigcirc$ line) which is greater than the resistive term (blue$\square$ line): A > H > R. By contrast, in the inner disc closest to the protostar, the point $(n_n, T) = (10^{19}, 10^4)$ (filled green diamond) is right of the red$\bigcirc$ line, and thus the Hall term now exceeds the ambipolar term, but still left of the green$\Diamond$ line and thus the ambipolar term remains greater than the resistive term: H > A > R.

For temperatures greater than a few 10^4 K, ionisation of H and He become significant and, as ionisation increases, the one-fluid model no longer applies. As indicated in Fig. 10.3, as $T \to 10^5$ K, the so-called *two-fluid model* – introduced in §10.5 – must be used. Indeed, *none* of the discussion in this section applies to a moderately to highly ionised fluid such as that in the inner-most region of a planetary disc or the hottest regions of stellar jets. In these environments, the dominant non-ideal term is ambipolar diffusion (roughly in the ratio of ion to electron mass) where the ionised and neutral densities are comparable. Carrying on to even higher temperatures such as those found in stellar atmospheres and within a typical extragalactic jet, the fluid is fully ionised and the system can be treated with ideal MHD. That is, for the most part all non-ideal terms at $T \gtrsim 10^6$ K become negligible.

10.3 Resistive dissipation

10.3.1 The resistive induction equation

In a *collisional plasma*,[8] a charge, q, is subject to both the electromagnetic force (Eq. B.4 in §B.1),

$$\vec{F}_{\text{EM}} = q(\vec{E} + \vec{v} \times \vec{B}),$$

arising from background electric and magnetic fields, and a drag force, $\vec{F}_{\text{dr}}$, arising from collisions and/or Coulomb interactions with other particles in the medium. In quasi-equilibrium, the drag and electromagnetic forces balance and charges move at a "terminal velocity", analogous to a mass falling in the earth's atmosphere when its weight is balanced by air drag; this is the approach we'll take to account for $\vec{F}_{\text{dr}}$.

As a charged particle accelerates through a medium, the frequency of interactions with other particles increases linearly with both the number density of the medium, n, and its velocity relative to other particles, $\vec{u}$. Since each interaction contributes to the effective drag, we can write,

$$\vec{F}_{\text{dr}} = -\mathcal{C} n \vec{u},$$

where the $-$ sign indicates $\vec{F}_{\text{dr}}$ points *opposite* to $\vec{u}$, and where $\mathcal{C}$ is a constant of proportionality whose SI units are evidently $\text{kg m}^3 \text{s}^{-1}$. Multiplying the top and bottom of the RHS by $q^2 n_q$ where n_q is the number density of charged particles, and noting that $n_n = (1-f)n$ where n_n is the number density of neutrals and $f = n_q/n$ is the ionisation fraction, we have,

$$\vec{F}_{\text{dr}} = -q\mathcal{C} \frac{n}{q^2 n_q} \underbrace{q n_q \vec{u}}_{\vec{J}} = -q \underbrace{\frac{\mathcal{C}}{1-f} \frac{1}{q^2} \frac{n_n}{n_q}}_{\eta} \vec{J} = -q \eta \vec{J}, \qquad (10.32)$$

where the quantity identified as η has units $\text{kg m}^3 \text{s}^{-1} \text{C}^{-2} = \Omega \, \text{m}$ (units of resistivity), and the quantity identified as $\vec{J}$ has units $\text{C m}^{-3} \text{m s}^{-1} = \text{A m}^{-2}$ (units of

---

[8] A collisional plasma is one with sufficient density that the mean-free-path for a particle between collisions is small compared to other length scales of interest. As shown in Fig. 10.3, the "resistive" regime corresponds to such "high-density" plasmas.

current density). This "derivation" should be taken more as a "plausibility argument", but note that even in its simplicity, the dependencies of η on q, n_n, and n_q revealed in Eq. (10.21) have all been recovered by Eq. (10.32).

Summing the electromagnetic and drag forces to zero for quasi-equilibrium, we get,

$$\vec{F}_\mathrm{EM} + \vec{F}_\mathrm{dr} = q(\vec{E} + \vec{v} \times \vec{B} - \eta \vec{J}) = 0$$

$$\Rightarrow \quad \vec{E} = -\vec{v} \times \vec{B} + \eta \vec{J} = \vec{E}_\mathrm{ind} + \vec{E}_\eta, \tag{10.33}$$

identifying both the induced and resistive terms of the net electric field, $\vec{E}$.

Finally, by combining Eq. (10.33) with Faraday's law (App. B), we get,

$$-\nabla \times \vec{E} = \boxed{\partial_t \vec{B} = \nabla \times (\vec{v} \times \vec{B} - \eta \vec{J}),} \tag{10.34}$$

the *resistive induction equation* where, in this analysis, $\eta \vec{J}$ is the only non-ideal term uncovered. Note that Eq. (10.22) reduces to Eq. (10.34) in the limit of high density and modest temperature where the resistive term dominates over both the Hall and ambipolar diffusion terms (Fig. 10.3). Note further that the high-density limit is critical for the presumption of the drag force, since it is predicated on the predominance of particle–particle interactions which occur only in a collisional plasma. At sufficiently low densities where the plasma can be considered *collisionless* (time between collisions long compared to other time scales of interest), the Hall and ambipolar diffusion terms dominate the resistive term.

10.3.2 Dissipation of magnetic energy

Implicit to an isothermal system is the assumption that any kinetic or magnetic energy converted to thermal energy is whisked away at precisely the rate required to maintain the temperature. Equivalently, any cooling mechanism such as work done by a region of gas expanding into another is compensated by just the right influx of thermal energy. The usual mechanisms cited include heat conduction, Poynting flux, or electromagnetic radiation (bremsstrahlung, synchrotron, *etc.*) all or any of which, under favourable circumstances, could remove or supply thermal energy at the rate required to maintain the temperature.

The opposite extreme to isothermality, of course, is adiabaticity where energy is neither gained nor lost by the system as a whole, but merely redistributed. In this case, one must add an internal (or total) energy equation to Eq. Set 10.

As shown in §B.4, the resistive power density, p_R, delivered to a fluid element of resistivity, η, by an electric field, $\vec{E}_\eta$, is given by Eq. (B.15),

$$p_\mathrm{R} = \vec{J} \cdot \vec{E}_\eta = \eta J^2, \tag{10.35}$$

using Eq. (10.33) in which $\vec{E}_\eta = \eta \vec{J}$ is identified. As a *power density*, it can be added directly to the RHS of the *ideal* internal energy equation, Eq. (1.34), to get,

$$\boxed{\partial_t e + \nabla \cdot (e\vec{v}) = -p \nabla \cdot \vec{v} + \eta J^2,} \tag{10.36}$$

the *non-ideal* internal energy equation with resistive heating. Since the resistive term is positive-definite, it is always a *source* of internal energy, never a sink. It is a one-way conduit that increases internal energy at the expense of magnetic energy. Note too that because of the J^2 dependence, resistive losses can become the dominant source of heating in regions of strong current density corresponding to where $\vec{B}$ varies most rapidly.

It is left to Problem 10.2 to show that the total energy equation for a resistive fluid is,

$$\partial_t e_T^* + \nabla \cdot \left[(e_T + p)\vec{v} + \frac{1}{\mu_0}\vec{B} \times (\vec{v} \times \vec{B}) + \frac{\eta}{\mu_0}\vec{J} \times \vec{B} \right] = 0, \qquad (10.37)$$

where, as usual, $e_T^* = e_T + \dfrac{B^2}{2\mu_0} = e + \tfrac{1}{2}\rho v^2 + \rho\phi + \dfrac{B^2}{2\mu_0}$ is the total MHD energy density of the fluid (Eq. 4.18) and where the quantity in large square brackets is the flux density of the total MHD energy (E_T^*).

10.3.3 Magnetic diffusion and reconnection

Setting $\vec{J} = \dfrac{1}{\mu_0}\nabla \times \vec{B}$, Eq. (10.34) can be written as,

$$\partial_t \vec{B} = \nabla \times (\vec{v} \times \vec{B}) - \frac{\eta}{\mu_0}\nabla \times (\nabla \times \vec{B}), \qquad (10.38)$$

for constant (in space) η. The quantity η/μ_0 is usually defined as the *magnetic diffusivity* ($\mathcal{D}_M$, with SI units m^2 s^{-1}) which is to resistive MHD what the kinematic viscosity, ν, is to viscid hydrodynamics (§8.3; Eq. 8.24). Thus, using Identity (A.27) in App. A, Eq. (10.38) becomes,

$$\partial_t \vec{B} = \nabla \times (\vec{v} \times \vec{B}) + \mathcal{D}_M \nabla^2 \vec{B}, \qquad (10.39)$$

since $\nabla \cdot \vec{B} = 0$, and the resistive term ($\propto \mathcal{D}_M$) can be seen to be *diffusive* (Eq. H.3 in App. H) where $\mathcal{D}_M$ acts as the diffusion coefficient. This should remind the reader who went through Chap. 8 of the viscous diffusion term in the incompressible Navier–Stokes equation (Eq. 8.26 in §8.3), namely $\nu\nabla^2\vec{v}$ where, in the context of viscid hydrodynamics, the kinematic viscosity, ν, acts as the diffusion coefficient.

As shown in App. H, the effect of the diffusion term, $\mathcal{D}_M \nabla^2 \vec{B}$, is to eat away at local gradients of the quantity being diffused, in this case $\vec{B}$. Diffusion is relentless and, in the absence of a mechanism to build up the magnetic induction (*e.g.*, a dynamo; §10.3.4), it doesn't rest until $\vec{B}$ is absolutely uniform everywhere. As worked out for viscid flow in §8.6.1, the magnetic diffusion time scale for a non-uniform magnetic field to relax to uniformity is,

$$\tau_{\text{diff}} \sim \frac{L^2}{\mathcal{D}_M}, \qquad (10.40)$$

where L is a characteristic length (*e.g.*, half the distance between local extrema in $\vec{B}$).

A consequence of this action is a phenomenon known as *magnetic reconnection*. Viewed as a collection of "lines of induction", the left panels of Fig. 10.4 illustrate

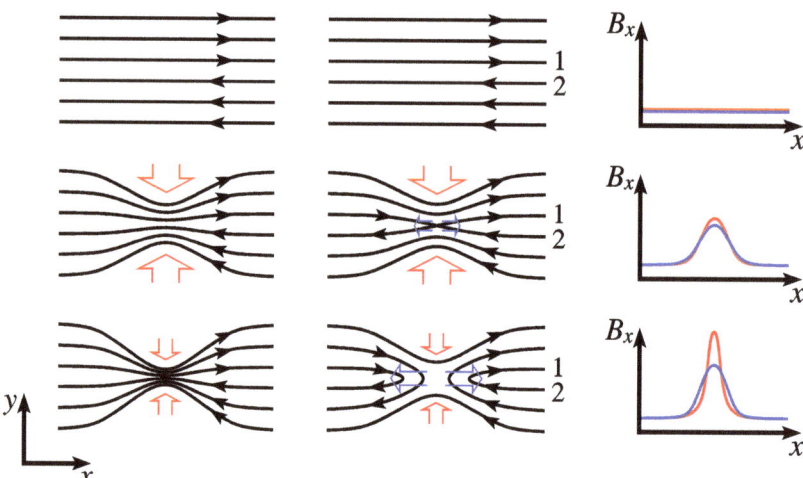

Figure 10.4. The left panels show how, in the absence of diffusion, fluid flow (red open arrows) pinch parallel lines of induction together, thereby increasing magnetic pressure and slowing the flow. In this case, lines of induction can be pushed together as closely as needed to decelerate the flow to zero while leaving lines of induction integral. By contrast, the middle panels show when flow in the y-direction forces lines of induction together beyond their "diffusion limit", diffusion in the x-direction (blue open arrows) reduces the gradient in B_x at the expense of induction line integrity. Thus and for example, lines of induction 1 and 2 in the top middle panel are pushed together in the centre middle panel creating a so-called "X-point" where the path of each line of induction becomes ambiguous. In the bottom middle panel, lines 1 and 2 have broken then "reconnected" to form two loops, one from each of the left and right halves of lines 1 and 2. The right panels show profiles for B_x – red for left panels, blue for middle panels – where the blue profiles are diffused (broader) versions of the red profiles.

how $\vec{B}$ gets pinched together by fluid flow (open red arrows) when governed by the ideal induction equation (Eq. 4.4) and when magnetic flux is strictly conserved (Alfvén's theorem, §4.3). Unencumbered by diffusion, lines of induction can be pinched together as closely as needed to attain sufficient density so that the resulting magnetic pressure resists further pinching. Throughout, all lines of induction continue to traverse the region without being broken or somehow redirected.

Conversely, in a fluid where $\eta \neq 0$, the *resistive* induction equation (Eq. 10.39) governs $\vec{B}$ and Alfvén's theorem no longer applies. In this case, should B vary significantly over a short enough distance, L, so that the diffusion time scale, τ_{diff} (Eq. 10.40), is comparable to or less than the flow time scale, the magnetic gradient is said to be *diffusion-limited*. As illustrated by the middle panels of Fig. 10.4, this necessarily means that lines of induction, whose density is a measure of magnetic strength and local gradient, are forbidden from coming any closer than a certain limit. This results in an X-point from which B_x is diffused left and right (horizontal open blue arrows) effectively "breaking" and "reconnecting" lines of induction to achieve a configuration consistent with both the pinching action of the fluid and the diffusion-limited density of $\vec{B}$. As shown by Eq. (10.36), magnetic energy is converted to internal energy and "loops of magnetic flux" are created from what were once lines of induction passing unbroken across the region (bottom middle panel in Fig. 10.4).

Finally on Fig. 10.4, the right panels show profiles of B_x as a function of position, x, across the pinched region, where the red (blue) profiles correspond to the left (middle) panels (blue profiles generally below and broader than the red ones). Whereas the flow of fluid works against (but does not defeat) diffusion in the y-direction, there is no fluid flow in the x-direction, and diffusion of x-gradients is unchecked as indicated by the broader blue profiles in the right panels.

A primary example of where magnetic reconnection plays an important role is at the base of solar flares, the most energetic phenomena in the solar system. During a "reconnection event", enormous amounts of magnetic energy is released in a short period of time (typically 10 s), with the ensuing explosion resulting in a solar flare whose baseline is typically $L \sim 10^7$ m, roughly the diameter of the earth. If we were to make the astoundingly naïve assumption that such reconnection events occur on the diffusion time scale, then on the solar surface where $\mathcal{D}_M \sim 1\,\mathrm{m^2\,s^{-1}}$,[9] Eq. (10.40) would have us believe that,

$$\tau_{\text{diff}} = \frac{L^2}{\mathcal{D}_M} \sim 10^{14}\,\mathrm{s} \sim 3.0 \times 10^6\,\mathrm{yr}, \tag{10.41}$$

or thirteen orders of magnitude longer than the observed time scale of 10 s!! At the risk of understatement, something else is going on here.

Sweet–Parker model

So let's examine the physics of the region in which reconnection occurs a little more closely. The key aspect of Fig. 10.4 is that two regions of opposite magnetic polarity are brought together and, since $\vec{J} \propto \nabla \times \vec{B}$, this generates a so-called *current sheet* between the regions as illustrated in Fig. 10.5. The Sweet–Parker model[10] examines this region in detail and, using the various conservation laws that apply, shows that magnetic reconnection happens within a pinched current sheet on a time scale much shorter than the diffusion time scale given by Eq. (10.41).

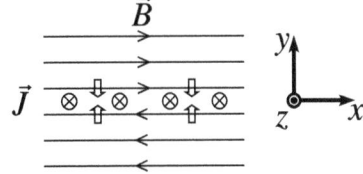

Figure 10.5. A current sheet ($\vec{J}$ marked with $\otimes$) separates two regions of opposing magnetic induction generating a Lorentz force (open arrows) that tends to narrow the current sheet.

As envisioned by Sweet and Parker, the current sheet is contained within a "reconnection zone" of width $2L$ and thickness $2l$ (grey rectangle in Fig. 10.6) inside

---

[9] "First-principles" estimates of the solar magnetic diffusivity comes from knowledge of material properties on the surface of the sun; see, *e.g.*, Solanki *et al.* (2006).

[10] Parker (1957), based on a 1956 conference proceeding of Sweet; and Sweet (1958). This is the same Parker of the Parker instability discussed in §7.4.

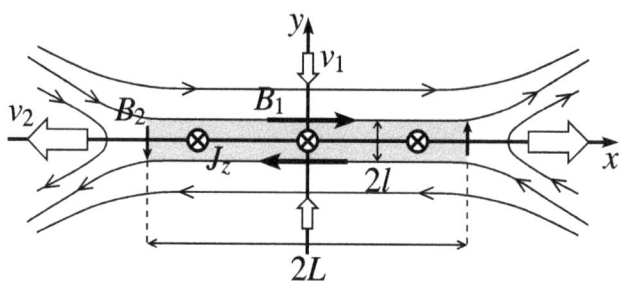

Figure 10.6. A "blow-up" of the current sheet, J_z, in Fig. 10.5 showing the "reconnection zone" (grey rectangle with dimensions $2L \times 2l$) in which magnetic reconnection occurs in the Sweet–Parker model. The Lorentz force density drives fluid along the y-axis transporting flux-frozen magnetic induction, B_1, into the current sheet at speed v_1. Within the current sheet, magnetic diffusivity "unfreezes" the magnetic flux allowing reconnection to occur. Fluid is then expelled along the x-axis at speed $v_2 \gg v_1$ and transporting with it "refrozen" magnetic induction $B_2 \ll B_1$.

which the magnetic induction is reconfigured. Outside this zone, Alfvén's theorem is assumed to apply and magnetic induction remains frozen in the fluid. Inside the zone, however, the model assumes $\vec{B}$ is suddenly "unfrozen" by magnetic diffusion, and the fluid and lines of induction can move more independently.

Within the current sheet, the current density, $\vec{J} = -\partial_y B_x \hat{k}$, is evidently directed into the page ($\otimes$ in Fig. 10.6) and thus the Lorentz force, $\vec{F}_\mathrm{L} = \vec{J} \times \vec{B}$, pinches fluid towards the x-axis, forcing it into the reconnection zone. Entering the zone with speed v_1 and magnetic induction B_1, flux-frozen fluid is suddenly "unfrozen" allowing lines of induction to break, reconnect, and then get pushed back out of the zone along the $\pm x$-axis with the fluid. Exiting the zone, fluid with speed v_2 "refreezes" with the magnetic induction, B_2, reconfigured as shown in Fig. 10.6. As we'll see, the analysis shows that $v_2 \gg v_1$ and $B_2 \ll B_1$ to the same extent as $L \gg l$. This is reflected in the figure by the different sized arrows for B (solid) and v (open).

So in this "what-goes-in-must-come-out" picture, the time-scale for reconnection is evidently the distance fluid must travel before ejection, L, divided by the speed at which fluid is injected into the reconnection zone, v_1, and we have,

$$\tau_\mathrm{SP} = \frac{L}{v_1}. \tag{10.42}$$

It remains to find an expression for v_1 in terms of more accessible properties of the flow.

We'll assume the reconnection zone and its vicinity are incompressible ($\nabla \cdot \vec{v} = 0$), that the dominant non-ideal term is resistive, and that the fluid is in steady state ($\partial_t = 0$). From the latter, we can write $\partial_t \vec{B} = -\nabla \times \vec{E} = 0$ and, for no static field, $\nabla \cdot \vec{E} = 0$. For the configuration depicted in Fig. 10.6, this implies $\vec{E} = $ constant. To wit, since $v_z = B_z = 0$, $\vec{v} \times \vec{B}$ has only a z-component. Further, $\vec{J} = J_z \hat{k}$, and from Eq. (10.33), $\vec{E}$ has only a z-component. Thus, $\nabla \times \vec{E} = 0 \Rightarrow \partial_x E_z = \partial_y E_z = 0$ and further, $\nabla \cdot \vec{E} = 0 \Rightarrow \partial_z E_z = 0$. Therefore, within and around the reconnection zone, E_z is constant.

Now, just outside the reconnection zone in Fig. 10.6 where $\eta = 0$, Eq. (10.33) requires:

$$\vec{E} = -\vec{v} \times \vec{B} = \begin{cases} -(v_1(-\hat{j}) \times B_1\hat{i}) = -v_1 B_1 \hat{k} & \text{above zone;} \\ -(v_2(-\hat{i}) \times B_2(-\hat{j})) = -v_2 B_2 \hat{k} & \text{left of zone;} \\ -(v_1\hat{j} \times B_1(-\hat{i})) = -v_1 B_1 \hat{k} & \text{below zone;} \\ -(v_2\hat{i} \times B_2\hat{j}) = -v_2 B_2 \hat{k} & \text{right of zone.} \end{cases}$$

Meanwhile, inside the zone where $\eta \neq 0$, the electric field is assumed to be dominated by the resistive term, and we have,

$$\vec{E} = \eta \vec{J} = \frac{\eta}{\mu_0} \nabla \times \vec{B} = -\mathcal{D}_\mathrm{M} \partial_y B_x \hat{k} = -\frac{\mathcal{D}_\mathrm{M} B_1}{l} \hat{k}.$$

Since E_z is constant within and around the reconnection zone, all expressions must be the same and we have,

$$v_1 B_1 = v_2 B_2 = \frac{\mathcal{D}_\mathrm{M} B_1}{l}. \tag{10.43}$$

Next, from incompressibility, integrating $\nabla \cdot \vec{v} = 0$ over the reconnection zone yields,

$$\int_V \nabla \cdot \vec{v}\, dV = \oint_S \vec{v} \cdot \hat{n}\, d\sigma = 0,$$

using Gauss' theorem (Eq. A.30). Now, because of uniformity throughout the zone, we can evaluate the surface integral just by taking the dot products of the velocities with the surfaces into which they enter/emerge, and adding the four terms together. Thus, starting from the top surface of the reconnection zone in Fig. 10.6 and moving around it counter-clockwise, we have,

$$-v_1(2L)w + v_2(2l)w - v_1(2L)w + v_2(2l)w = 4(v_2 l - v_1 L)w = 0,$$

where w is some arbitrary width of the zone into the page. Thus,

$$v_2 l = v_1 L, \tag{10.44}$$

and $v_2 \gg v_1$ to the same extent as $L \gg l$.[11] Finally, from the x-component of the magnetic Euler equation (Eq. 4.14) in steady state, we have along the x-axis,

$$\rho v_x \partial_x v_x + \rho v_y \underbrace{\partial_y v_x}_{0} = -\partial_x p + (\vec{J} \times \vec{B})_x = \partial_x p - \frac{1}{\mu_0} B_y \partial_x B_y, \tag{10.45}$$

since, along the x-axis at the location of B_2, $(\vec{J} \times \vec{B})_x = -J_z B_y$ and $J_z = \partial_x B_y/\mu_0$. Further, $B_y = B_2$, $v_x = v_2$, $\rho = $ constant, and so Eq. (10.45) can be written as,

$$\partial_x \left(\frac{1}{2} \rho v_2^2 + p + \frac{1}{2\mu_0} B_2^2 \right) = 0 \quad \Rightarrow \quad \frac{1}{2} \rho v_2^2 + p + \frac{1}{2\mu_0} B_2^2 = C,$$

a constant. Now, at $(x, y) = (0, 0)$ (i.e., the X-point defined in the caption of Fig.

---

[11] And, from Eq. (10.43), $B_1 \gg B_2$ to the same extent as $v_2 \gg v_1$ and thus $L \gg l$, completing the observation made on page 396.

10.4), $v_x = 0$, $B_y = 0$, and so $C = p(0,0)$, the thermal pressure at the origin. Thus, the x-component of the magnetic Euler equation in steady state yields,

$$\frac{1}{2}\rho v_2^2 + p + \frac{1}{2\mu_0}B_2^2 = p(0,0).$$

Similarly, the y-component yields,

$$\frac{1}{2}\rho v_1^2 + p + \frac{1}{2\mu_0}B_1^2 = p(0,0),$$

and, combining these two results, we get,

$$\rho v_2^2 + \frac{B_2^2}{\mu_0} = \rho v_1^2 + \frac{B_1^2}{\mu_0}, \tag{10.46}$$

taking p constant throughout the reconnection zone. But, from Eq. (10.43) we have $B_2 = B_1 v_1/v_2$ and from Eq. (10.44), we have $v_1 = v_2 l/L$. Thus, Eq. (10.46) becomes,

$$\rho v_2^2 + \frac{B_1^2}{\mu_0}\frac{l^2}{L^2} = \rho v_2^2 \frac{l^2}{L^2} + \frac{B_1^2}{\mu_0} \Rightarrow v_2 = \frac{B_1}{\sqrt{\mu_0 \rho}} = a_1, \tag{10.47}$$

after a little algebra. Here, a_1 is the Alfvén speed at the point where v_1 enters the reconnection zone (top and bottom of the grey rectangle in Fig. 10.6).

OK, so we're almost there; we just have to pull it all together. From the left and right sides of Eq. (10.43) we have,

$$v_1 = \frac{\mathcal{D}_M}{l} = \frac{\mathcal{D}_M v_2}{v_1 L} \Rightarrow v_1 = \sqrt{\frac{\mathcal{D}_M a_1}{L}}, \tag{10.48}$$

using Eq. (10.44) and (10.47). Therefore, the reconnection time scale (Eq. 10.42) becomes,

$$\tau_{SP} = \frac{L}{v_1} = \frac{L^{3/2}}{\sqrt{\mathcal{D}_M a_1}} = \underbrace{\frac{L^2}{\mathcal{D}_M}}_{\tau_{\text{diff}}} \underbrace{\sqrt{\frac{\mathcal{D}_M}{L a_1}}}_{S^{-1/2}}, \tag{10.49}$$

where the underbraces are from Eq. (10.40) and (10.25), the latter defining the Lundquist number introduced in §10.2.2. Now, we've already established that on the solar surface at the base of a putative flare, $L \sim 10^7$ m and $\mathcal{D}_M \sim 1 \Rightarrow \tau_{\text{diff}} \sim 10^{14}$ s, *thirteen* orders of magnitude greater than the observed time scale for solar flares to erupt. To find the Sweet–Parker reconnection time scale, we need an estimate of the Lundquist number, S, and thus the Alfvén speed on the solar surface; fortunately, this is reasonably well-known. At the base of a solar flare, $B \sim 0.01$ T (*cf.* 10^{-4} T averaged over solar surface) and $\rho \sim 10^{-12}$ kg m^{-3}. Thus, $a_1 \sim 10^7$ m s^{-1}, $S \sim 10^{14}$, and Eq. (10.49) yields,

$$\tau_{SP} = \tau_{\text{diff}} S^{-1/2} \sim 10^7 \text{ s}, \tag{10.50}$$

still six orders of magnitude greater than the observed time scale of 10 s for solar flares, but a seven-orders-of-magnitude improvement over the diffusion time scale given by Eq. (10.41).

Clearly, we're still not there. Following Sweet–Parker, arguments were made that shocks in the reconnection zone were somehow responsible for an "anomalous resistivity" at the X-point driving the reconnection time scale even lower (*e.g.*, Petschek, 1964). The sceptical reader who read §7.3 may recognise this as another example of "there be dragons here!".

A more promising solution may be through the Hall effect, which takes over when the plasma is more or less collisionless and thus where electron and ion flows are somewhat independent. We'll revisit this idea briefly in §10.4.2.

10.3.4 Dynamo theory (optional)

This subsection depends very weakly on familiarity with the incompressible Navier-Stokes equation, Eq. (8.26).

A *dynamo* is a mechanism by which a magnetic field is enhanced or sustained against dissipation. Dynamos can take on a variety of configurations, some of which can be created in the lab, and others that occur naturally in an astrophysical environment. For life on Earth, the most profound example of a dynamo is that which sustains the planet's magnetic field without which, the solar wind would have stripped the earth of its atmosphere æons ago and life as we know it would never have evolved.

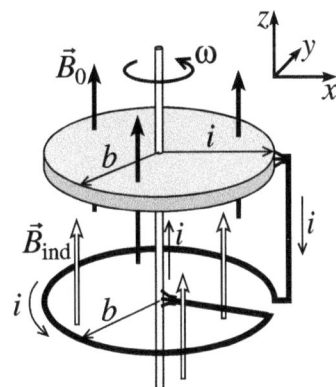

Figure 10.7. A *Bullard dynamo* consists of a rotating disc embedded in a background magnetic induction, $\vec{B}_0$, driving a current, i, that induces an additional magnetic induction, $\vec{B}_{\text{ind}}$, that adds to $\vec{B}_0$.

Figure 10.7 depicts the so-called *Bullard dynamo*,[12] the simplest example of a dynamo that one can create in a lab. A metal disc of radius b rotating at a constant angular velocity $\vec{\omega} = \omega \hat{z}$ (counter-clockwise) is embedded in a background magnetic induction, $\vec{B} = B_0 \hat{z}$ (solid arrows in Fig. 10.7). Since metals have free charge-carriers, a charge carrier $q > 0$[13] at a distance r from the rotation axis and an azimuthal velocity $\vec{v} = r\omega \hat{\varphi}$ ($\hat{\varphi}$ is a unit vector pointing in the counterclockwise direction) experiences a Lorentz force,

$$\vec{F}_{\text{L}} = q\vec{v} \times \vec{B} = qr\omega B_0 \hat{\varphi} \times \hat{z} = q\omega B_0 r\hat{r}.$$

Thus a current, i, is driven radially outwards which, in the configuration shown in

---

[12] Bullard (1955). A student of Ernest Rutherford (1871–1937), Sir Edward Crisp Bullard (1907–1980; www.wikipedia.org/wiki/Edward_Bullard) pioneered the development of the theory of the *geodynamo*. An interesting piece of Canadiana: both Rutherford and Bullard spent some of their careers in Canada, Rutherford at McGill University (1898–1907) and Bullard at the University of Toronto (1948–1950).

[13] Yes, charge carriers in metals are electrons and thus negative. However, since we still abide by Franklin's convention that current is in the direction of *positive* charge flow (see discussion in §10.4), think of these free charges as "electron holes". In the end, it makes no difference to the calculations whether one thinks of electrons moving inwards or electron holes moving outwards.

Fig. 10.7, is captured by a non-rotating circuit (with "frictionless brushes" at each end to allow rotation) and directed around the counterclockwise loop below, also of radius b. By the law of Biot and Savart, such a current induces its own magnetic induction, $\vec{B}_{\text{ind}} = B_{\text{ind}}\hat{z}$ (open arrows in Fig. 10.7) parallel to and thus reinforcing $\vec{B}_0$. That $\vec{B}_0$ is (indirectly) responsible for generating its own reinforcement makes this configuration a *dynamo*. Note that all one has to do to change this to an *anti-dynamo* (where $\vec{B}_0$ is responsible for its own partial annihilation) is reverse the direction of the current loop below the disc, generating $\vec{B}_{\text{ind}} \propto -\hat{z}$.

Conditions for an MHD dynamo

Dynamos can exist in an MHD environment as well, though the nature of a fluid makes it a much more complicated mathematical problem. Analytical headway with the full MHD equations normally requires a number of assumptions on symmetry, steady-state, self-similarity, *etc.*, but these are exactly the assumptions that can prevent a dynamo from forming as the following "anti-dynamo theorem" illustrates:

Theorem 10.1. *If both the velocity and magnetic induction in an incompressible, resistive MHD fluid are axisymmetric ($\partial_\varphi \vec{v} = 0$, $\partial_\varphi \vec{B} = 0$), the magnetic induction necessarily diffuses away.*

Proof: In cylindrical coordinates with $\partial_\varphi = 0$, $\vec{B}$ in terms of the vector potential (§4.8) is,

$$\vec{B} = \nabla \times \vec{A} = \frac{1}{r}\partial_r(rA_\varphi)\hat{z} - \partial_z A_\varphi \hat{r} + B_\varphi \hat{\varphi} = \frac{1}{r}\left(\partial_r f \hat{z} - \partial_z f \hat{r}\right) + B_\varphi \hat{\varphi}, \quad (10.51)$$

revealing two quantities, the flux function, $f = rA_\varphi$ (first defined in §9.1 by Eq. 9.1), and B_φ, the azimuthal component of $\vec{B}$, whose evolution need to be examined to determine whether the magnetic induction increases or diffuses with time.

In terms of the vector potential, the resistive induction equation (Eq. 10.34) with an appropriate choice of gauge (again, see §4.8) is,

$$\partial_t \vec{A} = \vec{v} \times \vec{B} - \eta \vec{J} \quad \Rightarrow \quad \partial_t f = r(v_z B_r - v_r B_z) - \frac{r\eta}{\mu_0}(\partial_z B_r - \partial_r B_z),$$

extracting the φ-component and multiplying through by r. Folding in expressions for B_z and B_r from Eq. (10.51), we get,

$$\partial_t f = -v_z \partial_z f - v_r \partial_r f + r\mathcal{D}_M \left[\partial_z\left(\frac{1}{r}\partial_z f\right) + \partial_r\left(\frac{1}{r}\partial_r f\right)\right]$$

$$\Rightarrow \quad \underbrace{\partial_t f + \vec{v}\cdot\nabla f}_{df/dt} = \underbrace{\mathcal{D}_M\left(\partial_{zz} + \partial_{rr} - \frac{1}{r}\partial_r\right)}_{\widetilde{\nabla}^2}f,$$

where $\widetilde{\nabla}^2$ is the so-called *pseudo-Laplacian operator*, differing from the regular

Laplacian by a sign change in the third term. It arises frequently in cylindrical coordinates and, in particular, in problems of MHD equilibrium. Thus,

$$\Rightarrow \quad \boxed{\frac{df}{dt} = \mathcal{D}_M \tilde{\nabla}^2 f,} \quad (10.52)$$

where, even with the term $-\partial_r/r$ in the definition of $\tilde{\nabla}^2$, the RHS of Eq. (10.52) is completely diffusive, and the quantity f decays in time (asymptoting to 0) taking with it B_z and B_r.

This leaves only B_φ, and it is left to Problem 10.5 to show that this too must diffuse away with time, and an axisymmetric resistive MHD system is always an *anti-dynamo*. □

Note that Theorem 10.1 does *not* say that magnetic induction plays no role in an axisymmetric and resistive MHD problem. What it does say is if a problem in which $\eta \neq 0$ doesn't include a continuous injection of new $\vec{B}$, the system will eventually relax to one in which $\vec{B} \to 0$. It and other theorems like it say that a trivial velocity field won't cut it; MHD dynamos are intrinsically 3-D phenomena and mathematically complicated.

An MHD dynamo begins with a seed magnetic induction and a driver within the fluid that increases or at least maintains $\vec{B}$ against dissipative influences such as viscosity and resistivity; in astrophysics, this driver is usually rotation. Initially, while $\vec{B}$ is still too weak to influence the velocity field driving the dynamo, the dynamo is said to be *kinematic*. Later, as $\vec{B}$ increases to the point where the Lorentz forces start acting back on the velocity driver, the dynamo is said to be *non-linear*. Kinematic dynamos reveal the mechanism by which the fluid flow is able to augment $\vec{B}$, but it takes the transition to the non-linear regime to determine how the dynamo is quenched; that is, $\vec{B}$ cannot continue to increase forever! Invariably, non-linear dynamos contain an aspect of fluid turbulence, which makes them all the more intractable analytically. By the way, for those who went through Chap. 7 and in particular, §7.3, the magneto-rotational instability is a good example of a dynamo.

A few kinematic MHD dynamos are known analytically (*e.g.*, Ponomarenko, 1973), but all realistic non-linear dynamos are known either experimentally or numerically. The first MHD dynamo to be sustained in the lab wasn't until 1999 by the so-called "Riga experiment" in Latvia (Gailitis *et al.*, 2000), while the first fully 3-D simulations to mimic successfully many of the properties of the earth's dynamo were done by Kageyama & Sato (1995) and Glatzmarer & Roberts (1995). We'll return to a qualitative description of the earth's dynamo at the end of this section.

Theorem 10.1 identifies a condition under which an MHD dynamo *cannot* exist. Conversely, can we identify criteria under which a dynamo could or even would exist? For this, we start with the *magnetic incompressible Navier–Stokes equation* and, given that rotation is often the driver, we do so in a frame of reference rotating at an angular velocity $\vec{\omega}$:

$$\frac{\partial \vec{v}}{\partial t} + (\vec{v} \cdot \nabla)\vec{v} = \frac{D\vec{v}}{Dt} = -\frac{1}{\rho}\nabla p - \nabla \phi + \nu \nabla^2 \vec{v} + \frac{1}{\rho}\vec{J} \times \vec{B} - 2\vec{\omega} \times \vec{v}, \quad (10.53)$$

where $D\vec{v}/Dt$ is the Lagrangian derivative indicating a time rate of change in the co-moving frame of the fluid (Eq. 3.2 in §3.1). Eq. (10.53) is essentially the MHD Euler equation (Eq. 4.14) with two terms added to account for viscous effects ($\nu \nabla^2 \vec{v}$, where ν is the kinematic viscosity; Eq. 8.26 in §8.3) and the Coriolis acceleration $(2\vec{\omega} \times \vec{v})$.[14] While these terms have been included for completeness, they won't make much of a difference in our discussion.

As can be found in any sophomore mechanics text, the *Work-Kinetic theorem* states that the change in kinetic energy of a system is the sum of the work done by all external forces acting on that system: $\Delta K = \sum W$. In fluid mechanics, the instantiation of the W-K theorem comes by taking the inner product of $\rho \vec{v}$ with Eq. (10.53):

$$\rho \vec{v} \cdot \frac{D\vec{v}}{Dt} = \frac{Dk}{Dt} = -\vec{v} \cdot \nabla p - \rho \vec{v} \cdot \nabla \phi + \mu \vec{v} \cdot \nabla^2 \vec{v} + \vec{v} \cdot \vec{F}_L, \qquad (10.54)$$

where $k = \frac{1}{2}\rho v^2$ is the kinetic energy density of the fluid, and $\mu = \rho \nu$ is the shear viscosity first introduced by Eq. (8.8). The terms on the RHS of Eq. (10.54) represent, from left to right, the rate at which work is done per unit volume by the pressure gradient, gravity, viscous stresses, and the Lorentz force. Note that since the Coriolis force is proportional to $\vec{\omega} \times \vec{v}$, it can only change the direction of $\vec{v}$ and not its magnitude and thus does no work.

Of particular interest to a dynamo is the last term of Eq. (10.54), $\vec{v} \cdot \vec{F}_L$, representing the rate at which magnetic and kinetic energies are exchanged. When this term is positive, magnetic energy is converted to kinetic energy and we have an anti-dynamo. A dynamo is therefore realised when $\vec{v} \cdot \vec{F}_L < 0$.

To complete the picture, let us take the dot product of $\vec{B}/\mu_0$ with the resistive induction equation, Eq. (10.34), to get,

$$\frac{1}{\mu_0}\vec{B} \cdot \partial_t \vec{B} = \frac{1}{\mu_0}\vec{B} \cdot [\nabla \times (\vec{v} \times \vec{B})] - \frac{\eta}{\mu_0}\vec{B} \cdot (\nabla \times \vec{J})$$

$$\Rightarrow \quad \partial_t e_M = (\vec{v} \times \vec{B}) \cdot \frac{1}{\mu_0}\nabla \times \vec{B} + \frac{1}{\mu_0}\nabla \cdot \overbrace{[(\vec{v} \times \vec{B}) \times \vec{B}]}^{0}$$

$$- \eta \vec{J} \cdot \frac{1}{\mu_0}\nabla \times \vec{B} - \eta \nabla \cdot (\vec{J} \times \vec{B}),$$

where $e_M = B^2/2\mu_0$ is the magnetic energy density and where Identity (A.8) was used on both terms on the RHS. Thus,

$$\partial_t e_M = (\vec{v} \times \vec{B}) \cdot \vec{J} - \eta J^2 - \eta \nabla \cdot \vec{F}_L = -\vec{v} \cdot \vec{F}_L - \eta J^2 - \eta \nabla \cdot \vec{F}_L, \qquad (10.55)$$

using Identity (A.1). Of significance, the rate at which work is done per unit volume by the Lorentz force, $\vec{v} \cdot \vec{F}_L$, appears in both Eq. (10.55) and (10.54) but with opposite sign. In Eq. (10.55), $\vec{v} \cdot \vec{F}_L < 0$ means kinetic energy density is converted to magnetic energy density, the same conclusion drawn from Eq. (10.54). Thus, a

---

[14] See App. G. The reader remembering their sophomore mechanics may wonder why the centrifugal acceleration, $-\vec{\omega} \times \vec{\omega} \times \vec{r}$, has been omitted. For systems such as the earth where $\omega^2 r \ll g$ for most distances r of interest, this term is usually ignored.

necessary (but not sufficient) condition for a dynamo is,

$$\boxed{\vec{v} \cdot \vec{F}_{\mathrm{L}} < 0.}$$

A sufficient condition for a dynamo can be gleaned directly from Eq. (10.55). Integrated over the system volume, Eq. (10.55) becomes,

$$\int_V \partial_t e_{\mathrm{M}} dV = \partial_t E_{\mathrm{M}} = -\int_V (\vec{v} \cdot \vec{F}_{\mathrm{L}} + \eta J^2) dV - \eta \oint_S \vec{F}_{\mathrm{L}} \cdot \hat{n} d\sigma,$$

where E_{M} is the magnetic energy of the system, and where Gauss' theorem (Eq. A.30) was used on the second term on the RHS. Since V contains the entire system, the surface integral is zero and a sufficient condition for a dynamo ($\partial_t E_{\mathrm{M}} > 0$) is,

$$\vec{v} \cdot \vec{F}_{\mathrm{L}} + \eta J^2 < 0 \quad \Rightarrow \quad \boxed{\vec{v} \cdot \vec{F}_{\mathrm{L}} < -\eta J^2.} \qquad (10.56)$$

As ηJ^2 is positive-definite, the boxed Eq. (10.56) implies $|\vec{v} \cdot \vec{F}_{\mathrm{L}}| > \eta J^2$. But,

$$|\vec{v} \cdot \vec{F}_{\mathrm{L}}| \leq v F_{\mathrm{L}} \leq v J B < V J B,$$

where the $\leq$ inequalities arise from setting the cosine or sine of the angles involved to 1, and where V is the "characteristic velocity" of the system which can be thought of as the maximum speed anywhere. Further,

$$\eta J^2 = \eta J \frac{1}{\mu_0} \nabla \times \vec{B} > J \mathcal{D}_{\mathrm{M}} \frac{B}{L},$$

where L is the "characteristic length scale" of the system, and can be thought of as the maximum distance over which B varies significantly. Thus, any spatial derivative of B will be greater than B/L, whence the $>$ inequality.

Bringing these two threads together, we have,

$$V \cancel{J} \cancel{B} > \mathcal{D}_{\mathrm{M}} \frac{\cancel{J} \cancel{B}}{L} \quad \Rightarrow \quad \boxed{\frac{VL}{\mathcal{D}_{\mathrm{M}}} = \mathcal{R}_{\mathrm{M}} > 1,}$$

and a necessary condition for a dynamo is the magnetic Reynolds number (Eq. 10.24) must be greater than unity. This turns out to be a rather weak condition, since for most numerically derived dynamos, $\mathcal{R}_{\mathrm{M}} \gtrsim 20$ (*e.g.*, Nore et al., 1997).

Earth's dynamo

That the earth acts like a giant magnet with well-defined north and south magnetic poles was first proposed by William Gilbert in 1600, although it would be more than two centuries later when a team led by James Clark Ross in 1831 discovered the "north" magnetic pole on the west coast of the Boothia Peninsula in modern-day Nunavut.[15] Each subsequent exploration found the pole to be in a different location from the previous and, in the late 19th, early 20th centuries, it came to be known that the north magnetic pole drifts over the earth's surface by a few km/yr.

---

[15] See www.wikipedia.org/wiki/North_magnetic_pole.

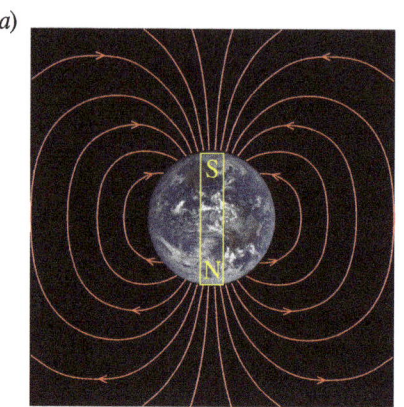
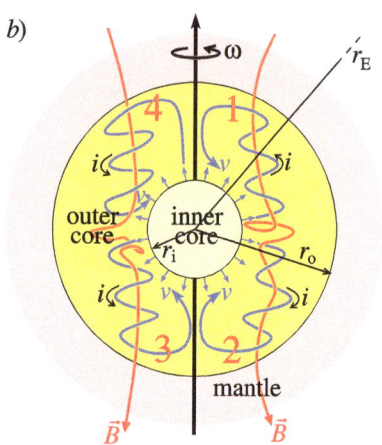

Figure 10.8. *a*) Whatever complicated mechanism may be at the earth's core generating its magnetic field, the appearance from afar (photo credit: NASA) is a dipole field much like an ordinary bar magnet whose magnetic south pole is near the earth's geographic north pole. *b*) A (too) simple model of the earth's dynamo. The earth's core is primarily Ni + Fe where the hotter inner core (6,000 K) is solid, and the cooler outer core (4,000 K) is liquid (say what? see footnote 17 on page 405). The outward temperature gradient drives an outward radial flow which, because of the Coriolis acceleration ($2\vec{v} \times \vec{\omega}$), circulates in a left-handed helix as it makes its way to the inner surface of the mantle (quadrants 1, 3, and 4) or from the mantle to the inner core (quadrant 2), also in a left-handed helix. Since the liquid Ni + Fe is hot enough to be weakly (but sufficiently) ionised, the flow transports a positive current which, by the law of Biot and Savart, induces a magnetic field pointing down, as shown. For a number of reasons, this model is naïve, not least of which because it can't explain polarity reversals of the earth's magnetic field that occur once every $\sim$ half-million years. See text for details.

For reasons poorly understood, the movement of the pole has recently accelerated and, at the time of this writing, seems to be making a bee-line for the Siberian coast at a speed of 50–60 km/yr. In fact, according to the best computer modelling, the north magnetic pole left the Canadian Arctic just at the turn of the 21st century for the first time since at least 1600, when Gilbert first suggested its existence.

Before Ampère, Faraday, and Maxwell brought magnetism into the realm of physics in the 19th century, the earth's magnetic field was known by navigators using "lodestone compasses" where the tip of the needle pointing towards geographical north was, naturally enough, labelled the "north pole". We now know, of course, that opposite poles attract and that the lodestone needle was simply aligning itself with the local dipole magnetic field of the earth and pointing towards its *south* pole. Thus, the magnetic pole currently on its way to Siberia is, in fact, the south magnetic pole putting the actual north magnetic pole somewhere in the Antarctic; Fig. 10.8*a* is so-labelled. Still, you'll find numerous scholarly references to the "north magnetic pole" in the Arctic ocean, and this usage is generally accepted to mean the north*ern* magnetic pole, *i.e.*, the one found within the geographic north.

It is also known that the earth's dipole field (Fig. 10.8*a*) is a residual of a much more complex multi-polar magnetic field configuration deep within the planet generated by a dynamo that's been in place for at least 3.5 billion years (Buffett, 2000). This is remarkable in so many ways, not least of which we wouldn't be here talking about it if it weren't there! Acting like a shield out of *Star Wars*, the earth's magnetic field prevents most solar wind particles from getting close enough to the planet's surface to strip its atmosphere away. This is what happened to Mars when its core solidified some 3.5 to 4 billion years ago, dropping its "magnetic shield" and sealing its fate as the vast, barren desert it has since become.

The broad strokes of how the earth's dynamo functions are fairly well understood, but some important "details" are missing. What's known is the earth's core consists of a solid Fe + Ni "inner core" with a radius $r_i \sim 1{,}200$ km and a temperature $\sim 6{,}000$ K[16] surrounded by a liquid "outer core" also largely Fe + Ni with an outer radius $r_o \sim 3{,}500$ km and a temperature $\sim 4{,}000$ K.[17] Surrounding the outer core is the mantle, a semi-solid semi-molten mishmash of silicates, magnesium oxide, and various "impurities" that reaches within a few tens of km of the earth's surface. We therefore have, at the earth's core, a system in which molten, slightly ionised metal (an incompressible MHD system) is trapped between two solid surfaces and forced to circulate in a rotating system. This is what drives the earth's dynamo.

Figure 10.8 illustrates a rather naïve model of this dynamo. The outward-pointing temperature and pressure gradients drive a slow but steady outflow of liquid Fe and Ni from the inner–outer core boundary. Consider first the blue streamline labelled '1' in Fig. 10.8b. It originates near the equatorial plane where flow is initially driven mostly outwards ($\propto \hat{r}$) and slightly upwards ($\propto \hat{z}$) so that the Coriolis acceleration (last term in Eq. 10.53) is $-2\vec{\omega} \times \vec{v} \propto -\hat{\varphi}$, and thus clockwise. The z-component of motion then extrudes the clockwise circulation into a left helix,[18] forming what is known as a *Taylor column*. Presuming the flow to transport positive ions, a current moving upwards along a left helix would, by Biot and Savart, induce a magnetic induction in the $-\hat{z}$ direction, as shown in Fig. 10.8b.

Streamlines 2 and 3 also originate from near the equatorial plane and these too are driven clockwise by the Coriolis force. However, being in the southern hemisphere, the vertical component of their motion is downward ($\propto -\hat{z}$) extruding the clockwise circulations into right helices. Still, positive current flowing *downwards* along a *right* helix also induces a magnetic induction $\propto -\hat{z}$.

Finally, streamline 4 in Fig. 10.8b is chosen to originate from near the rotational axis. In this case, flow is initially driven upwards with $\vec{v} \propto \hat{z}$ and thus unaffected by the Coriolis force. However, as flow approaches and is deflected by the mantle, the

---

[16] Although at the time of this writing, even this may be in doubt: see Pham & Tkalčić (2023).

[17] It may seem strange that the hotter inner core is solid while the cooler outer core is liquid, especially with their similar compositions. The difference is the tremendous pressure at the earth's centre which solidifies the inner core but, 1,200–3,500 km out, is insufficient to solidify the cooler yet still very hot outer core.

[18] A "left-helix" is one in which, when held vertically and viewed front-on, the near sides of each coil rise from bottom *right* to top *left*. Further, wrapping your right fingers around a left helix so they slide upwards along the coils forces your thumb to point *down*. Conversely, a "right helix" is one whose front edges go from lower *left* to upper *right*, and your right fingers following the coils upwards forces your thumb to point *up*.

acquired outward flow subjects it to – once again – a clockwise Coriolis acceleration, while its downward motion extrudes the clockwise circulation into a right helix. Thus, like streamlines 2 and 3, the downward-directed current induces $\vec{B} \propto -\hat{z}$.

So it seems no matter where on the inner–outer core a fluid streamline originates, with a current of positive ions the Coriolis effect always generates a magnetic induction $\propto -\hat{z}$ and anti-parallel to $\vec{\omega}$. That is to say, this model predicts that the earth's rotation and magnetic axes should point in opposite directions so that the magnetic south pole is near the geographical north pole, and vice versa.

Which it is, so we're all done, right?

Alas, life is never that simple for it is well known geologically that the polarity of the earth's magnetic field has flipped innumerable times. Once every half million years give or take a few hundred thousand, the earth's dynamo undergoes a still-poorly understood process by which, after many convulsions and machinations, the magnetic north and south poles switch places. These flips occur over periods between two and twelve thousand years and it is widely thought that we may be entering such a period now; the last polarity flip happened more than 750,000 years ago which means we're rather overdue. Further, there is no evidence over the æons of a polarity favouritism; $\vec{B}$ is aligned with $\vec{\omega}$ about as often as it's anti-aligned.[19]

Glatzmaier & Roberts (1995) pioneered the first computer simulations (fully 3-D viscid and resistive MHD with very complicated boundary conditions) to demonstrate a polarity flip in an environment that models the earth's core. This and many simulations since have shown that in an extremely complex physical environment of a rotating metallic non-ideal MHD fluid, a dynamo is inevitable and further, polarity flips can happen. What has not been made clear from these studies is the exact mechanism by which these reversals occur. It's one thing to set up a numerical experiment in which certain phenomena can be mimicked. It's entirely another for the numerics to provide a clear explanation of why.

10.4 The Hall effect

Around the time Benjamin Franklin (1706–1790)[20] was flying his kites into storm clouds and coaxing sparks out of dangling keys to demonstrate the electrical nature of lightning, he was also doing laboratory experiments on electrical circuits, trying to determine if there could be any possible practical use for this "electrical fluid". He wasn't the first to perform such experiments, but it was he who designated this "fluid" as being "positive" or "negative" to distinguish that which accumulates on glass rubbed with silk (positive) and on amber rubbed with fur (negative). Now,

---

[19]There are numerous references for this but as is often the case, the most approachable is a Wikipedia page. Try www.wikipedia.org/wiki/Geomagnetic_reversal where the interested reader can follow the numerous references therein.
[20]www.wikipedia.org/wiki/Benjamin_Franklin

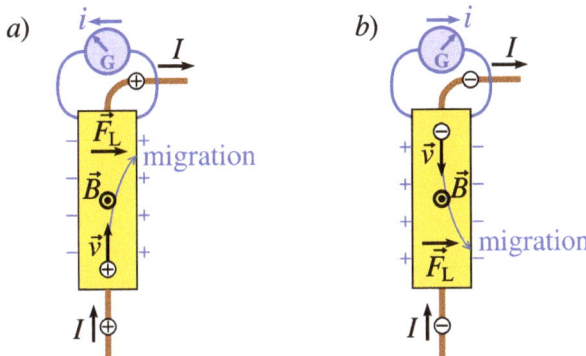

Figure 10.9. In Hall's original experiment, a brass wire carried a current, I, shown here to be directed *upwards*, to a rectangular sheet of gold-leaf with $\vec{B}$ pointing out of the page. *a)* If $q > 0$, q moves *upwards* and the Lorentz force $\vec{F}_L = q\vec{v} \times \vec{B}$ points to the right. Positive charges migrate to the right side of the gold leaf (curved blue arrow), leaving an excess of negative charge on the left. Connecting the two sides, a galvanometer, G, is deflected by a current, i, leftward (direction of positive charge flow). *b)* If $q < 0$, q moves *downwards* but the Lorentz force $\vec{F}_L = q\vec{v} \times \vec{B}$ still points to the right. Negative charges now migrate to the right leaving an excess of positive charge on the left and G deflects rightwards (opposite to negative charge flow). Thus, $q > 0$ if G deflects left, $q < 0$ if G deflects right; a very simple discriminant!

whether it was the positive or negative "fluid" that "flowed" through Franklin's circuits, he could not tell, and so he arbitrarily designated "electrical current" through a wire as the "flow" of "positive fluid".[21]

To most students of physics, the *Hall effect* is how we know the sign of a charge carrier in a metal conductor. Named for its discoverer, Edwin Hall (1855–1938),[22] the effect exploits a sign asymmetry in the Lorentz force that allowed Hall – at the tender age of 24 – to determine that charges conducted in a wire are negative (Fig. 10.9). Still, despite the fact Hall's discovery was in 1879, Franklin's convention that assigns current direction in a circuit to the direction positive charges would flow persists to this day.

In Hall's experiment (see caption of Fig. 10.9 for a quick review), charge migration is driven by the Lorentz force which, on a single charge, q, is given by,

$$\vec{F}_L = q\vec{v}_d \times \vec{B},$$

where $\vec{v}_d$ is the *drift velocity* of the charge, and $\vec{B}$ is the background magnetic induction. Now, as we know today, charge carriers in a wire are *valence electrons* (those most loosely bound to atoms) which, with an electric field (potential difference) along a wire, drift among the stationary atoms in the crystalline structure of

---

[21] I'm putting all these "fluid-analogous" words in quotes *a)* to emphasise their frequency and *b)* even though we no longer think of electricity as a "fluid", to point out the extent to which words like "current" and "flow" remain pervasive in our vernacular when describing electricity!

[22] www.wikipedia.org/wiki/Edwin_Hall

the metal. This is the current I indicated in Fig. 10.9. However, in a metal where electrons are forever colliding with atoms, the actual drift velocity is puny – perhaps 0.2 mm/s – and it is for this reason the Hall effect is difficult to demonstrate experimentally. For example, Hall's use of gold leaf wasn't a display of *richesse*, but rather the need for a medium that could be extruded thinly enough (a few dozen atoms thick) so that migrating charges are sufficiently concentrated at the edges to produce a detectable "Hall-current" (i in Fig. 10.9).

Let n_e be the number density (m^{-3}) of valence electrons ($q = -e$) in a metal conductor. Then, the *current density* (units A m^{-2}) is evidently,

$$\vec{J} = -en_e \vec{v}_d, \qquad (10.57)$$

and the Lorentz force driving the Hall effect is,

$$\vec{F}_L = -e\vec{v}_d \times \vec{B} = \frac{1}{n_e} \vec{J} \times \vec{B}$$

$$\Rightarrow \quad \vec{E}_H \equiv \frac{\vec{F}_L}{e} = \frac{1}{en_e} \vec{J} \times \vec{B}, \qquad (10.58)$$

the so-called *Hall electric field*.

Recognising that the drift velocity in a metal wire is just the velocity of the (drifting) negative charges relative to the (stationary) positive charges, we can generalise all of this to a plasma (Fig. 10.10) where both negative and positive charges (electrons and ions) are mobile, and where the relative (drift) velocity is,

$$\vec{v}_d = \vec{v}_e - \vec{v}_i.$$

Substituting this into Eq. (10.57), we get,

$$\vec{J} = en_e(\vec{v}_i - \vec{v}_e). \qquad \text{Eq. (10.14)}$$

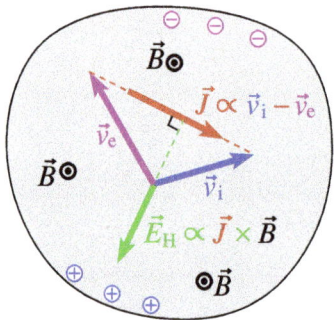

Figure 10.10. In a plasma element where electrons and ions move with different velocities, $\vec{v}_e$ and $\vec{v}_i$, the current density, $\vec{J}$ (red), is $\propto \vec{v}_i - \vec{v}_e$. If $\vec{B}$ points out of the page, the Hall field, $\vec{E}_H \propto \vec{J} \times \vec{B}$ (green), drives charge migration so that charges ⊖ (magenta) and ⊕ (blue) move in opposite directions along $\vec{E}_H$.

Thus, as demonstrated in Fig. 10.10, charges also migrate in a plasma because of a Hall electric field (Eq. 10.58) with a current density given this time by Eq. (10.14). This is the justification for identifying $-\vec{J} \times \vec{B}/en_e$ in Eq. (10.22) as the "Hall effect" in the one-fluid non-ideal induction equation.

10.4.1 The case of a completely ionised fluid

Consider a fully ionised sample of hydrogen where $Z = 1$, $n_i = n_e = n$, $n_n = 0$, and all subscripts 'i' are replaced with 'p' to reflect the fact that in this case, ions are bare protons. Then, Eq. (10.2) and (10.3) written as Euler equations omitting gravity and interactions with a neutral component become:

$$m_p n \left[\partial_t \vec{v}_p + (\vec{v}_p \cdot \nabla) \vec{v}_p \right] = -\nabla p_p + en(\vec{E} + \vec{v}_p \times \vec{B}) + \vec{f}^{\,a}_{e,p}; \qquad (10.59)$$

$$m_e n\left[\partial_t \vec{v}_e + (\vec{v}_e \cdot \nabla)\vec{v}_e\right] = -\nabla p_e - en(\vec{E} + \vec{v}_e \times \vec{B}) + \vec{f}^{\,a}_{p,e}. \tag{10.60}$$

While the RHS of the two equations are comparable, the LHS are proportional to m_p and m_e respectively. Thus, in adding Eq. (10.59) and (10.60), we can safely drop the LHS of the latter to get,

$$m_p n\left[\partial_t \vec{v}_p + (\vec{v}_p \cdot \nabla)\vec{v}_p\right] = -\nabla \underbrace{(p_p + p_e)}_{p} + \underbrace{en(\vec{v}_p - \vec{v}_e)}_{\vec{J}} \times \vec{B} + \underbrace{\vec{f}^{\,a}_{e,p} + \vec{f}^{\,a}_{p,e}}_{0,\text{ Newton's 3}^{\text{rd}}}$$

$$\Rightarrow \quad \rho\left[\partial_t \vec{v} + (\vec{v} \cdot \nabla)\vec{v}\right] = -\nabla p + \vec{J} \times \vec{B}, \tag{10.61}$$

where,

$$\rho = (m_p + m_e)n \approx m_p n \quad \text{and} \quad \vec{v} = \frac{m_p \vec{v}_p + m_e \vec{v}_e}{m_p + m_e} \approx v_p, \tag{10.62}$$

are the average density and bulk velocity of the two-component system with $m_e \ll m_p$, presuming the protons and electrons move with comparable speeds ($v_p \sim v_e$). Equation (10.61) is the one-fluid approximation of the Euler equation for the two-component system where the inertia is dominated by one component (protons).

Of greater interest to the present discussion, however, is what we get when $m_p \times$ Eq. (10.60) is subtracted from $m_e \times$ Eq. (10.59):

$$m_e m_p n\left[\partial_t(\vec{v}_p - \vec{v}_e) + (\vec{v}_p \cdot \nabla)\vec{v}_p - (\vec{v}_e \cdot \nabla)\vec{v}_e\right] = -\overset{\sim 0}{\cancel{m_e \nabla p_p}} + m_p \nabla p_e$$

$$+ e\underbrace{n(m_e + m_p)}_{\rho}\vec{E} + en(m_e \vec{v}_p + m_p \vec{v}_e) \times \vec{B} - (\overset{\sim 0}{\cancel{m_e}} + m_p)\vec{f}^{\,a}_{p,e}, \tag{10.63}$$

using the first of Eq. (10.62) for the underbrace. The LHS does not compress well, and so we'll dispense with it by assuming small deviations [in which case the non-linear terms $(\vec{v}\cdot\nabla)\vec{v} \to 0$] in an otherwise pseudo-steady state (in which case $\partial_t \to 0$). Thus, we're going to bunt and take the LHS to be zero.[23]

As for the RHS, the relevance of the terms can be made more transparent by a few considerations:

1. Assuming $\nabla p_e \sim \nabla p_p$ in a two-component system, then $m_e \nabla p_p \ll m_p \nabla p_e$ and we drop the former term.

2. Next,
$$m_e \vec{v}_p + m_p \vec{v}_e = m_e \vec{v}_p + m_e \vec{v}_e - m_e \vec{v}_e + m_p \vec{v}_e + m_p \vec{v}_p - m_p \vec{v}_p$$

$$= \underbrace{m_p \vec{v}_p + m_e \vec{v}_e}_{\rho \vec{v}/n} - \underbrace{(m_p - \overset{\sim 0}{\cancel{m_e}})(\vec{v}_p - \vec{v}_e)}_{\vec{J}/en} = \frac{\rho \vec{v}}{n} - \frac{m_p}{en}\vec{J},$$

using Eq. (10.62) and (10.14) for the underbraces.

---

[23] See, however, Problem 10.8 for a, perhaps, more-convincing "plausibility argument" that the LHS can be taken as zero.

3. Finally, from Eq. (10.4),

$$\vec{f}^{a}_{p,e} = \gamma_{pe}n^2 m_e m_p \underbrace{(\vec{v}_p - \vec{v}_e)}_{\vec{J}/en} = \frac{\gamma_{pe}nm_e m_p}{e}\vec{J}.$$

Thus, Eq. (10.63) becomes,

$$0 \approx m_p \nabla p_e + e\rho \vec{E} + e\rho \vec{v} \times \vec{B} - m_p \vec{J} \times \vec{B} - \frac{\gamma_{pe}nm_e m_p^2}{e}\vec{J},$$

which we divide through by $e\rho \approx enm_p$ and solve for $\vec{E}$ to get,

$$\vec{E} = \underbrace{-\vec{v} \times \vec{B}}_{\text{induction}} + \underbrace{\frac{1}{en}(\vec{J} \times \vec{B} - \nabla p_e)}_{\text{Hall effect}} + \underbrace{\eta_{p,e}\vec{J}}_{\text{resistive}}. \quad (10.64)$$

Here,

$$\eta_{p,e} \equiv \frac{\gamma_{pe}m_e m_p}{e^2}, \quad (10.65)$$

is the proton–electron resistivity, analogous to Eq. (10.21) for electron–neutral interactions found when interactions between ions and electrons were ignored because of their low concentration compared to neutrals. In the present context, the two-component system is nothing but protons and electrons, and thus any fluid resistivity must come from γ_{pe} (which can be assembled from Eq. 10.5–10.7) as given by Eq. (10.65). For convenience, I've included the ∇p_e term as a correction to the Hall effect, since both contribute to charge separation.

Finally, we "solve" Eq. (10.64) for $\vec{J}$, and find for when the Hall term dominates,

$$\boxed{\vec{J} = \frac{1}{\eta}(\vec{E} + \vec{v} \times \vec{B}) - \frac{1}{en\eta}(\vec{J} \times \vec{B} - \nabla p_e),} \quad (10.66)$$

where $\eta = \eta_{p,e}$ is the fluid resistivity, and where each driver of current density is exposed. Obviously, Eq. (10.66) doesn't actually "solve" Eq. (10.64) for $\vec{J}$, since $\vec{J}$ still appears on the RHS in a cross product. However, this won't deter us in the following example.

Example 10.1. Consider a stagnant ($\vec{v} = 0$) sample of ionised hydrogen in which $\nabla p_e = 0$. Find and describe the current density for $\vec{E} = E\hat{j}$ and $\vec{B} = B\hat{k}$.

Solution: With neither an induction term nor a pressure gradient, Eq. (10.66) becomes,

$$\vec{J} = \frac{1}{\eta}\vec{E} - \frac{1}{en\eta}\vec{J} \times \vec{B} = \frac{E}{\eta}\hat{j} - \frac{1}{en\eta}(-J_x B\hat{j} + J_y B\hat{i}).$$

Equating components on the left- and right-hand sides, we get:

$$J_x = -\frac{B}{en\eta}J_y; \qquad J_y = \frac{E}{\eta} + \frac{B}{en\eta}J_x; \qquad J_z = 0.$$

The Hall effect

Solving (truly, this time!) the first two expressions for J_x and J_y, we get,

$$J_x = -\frac{\zeta}{\eta(1+\zeta^2)}E \equiv -\frac{E}{\eta_\perp} \quad \text{and} \quad J_y = \frac{1}{\eta(1+\zeta^2)}E \equiv \frac{E}{\eta_\parallel}, \quad (10.67)$$

where,

$$\zeta \equiv \frac{B}{en\eta} = \underbrace{\frac{Be}{m_e}}_{\omega_c}\underbrace{\frac{m_e m_p}{e^2\eta}}_{1/\gamma_{pe}}\frac{1}{nm_p} = \frac{\omega_c}{\gamma_{pe}\rho},$$

and where ω_c is the cyclotron (gyration) frequency of an electron immersed in a magnetic field. Note from Eq. (10.67) that with the crossed electric and magnetic fields, the effective resistivity of the fluid is *anisotropic*, with the resistivity parallel to the electric field, $\eta_\parallel$, rather different from the resistivity perpendicular to $\vec{E}$, $\eta_\perp$, and in the direction of Hall migration. Evidently, $\eta_\parallel > \eta$, and the current driven parallel to $\vec{E}$ is reduced while a new component perpendicular to $\vec{E}$ is created.

However, including both components of $\vec{J}$, its magnitude is,

$$|\vec{J}| = \sqrt{J_x^2 + J_y^2} = \frac{E}{\eta},$$

the same as the current density driven by $\vec{E} = E\hat{j}$ when $\vec{B} = 0$. Thus, as shown in the inset, the effect of the magnetic induction and therefore Hall migration is to rotate the direction of the current density relative to the electric field by an angle,

$$\psi = \tan^{-1}\frac{|J_x|}{J_y} = \tan^{-1}\zeta,$$

without changing its magnitude. However, resistive dissipation,

$$\vec{J}\cdot\vec{E} = J_y E = \frac{E^2}{\eta}\frac{1}{1+\zeta^2},$$

is reduced by a factor $(1+\zeta^2)^{-1}$.

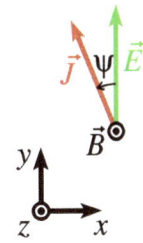

While the Hall effect modifies resistive power dissipation, there is no power dissipation (*e.g.*, Eq. B.15) directly from the Hall electric field itself (Eq. 10.58) since $\vec{J}\perp\vec{E}_H$. □

10.4.2 Magnetic reconnection, revisited

What distinguishes "resistive MHD" from "Hall MHD" is the collisional nature of the fluid. As Fig. 10.3 illustrates, resistive dissipation is the dominant non-ideal term for high number density (n) where the resistivity, η, comes about precisely because of particle–particle collisions between electrons and neutrals (Eq. 10.21). However, as n falls and especially at higher temperatures where the ionisation level increases, electron–neutral interactions are less important and the fluid enters a "Hall regime". Here, the fluid is said to be "collisionless"[24] and, as developed in

---

[24] By this we mean particle–particle interactions are mostly Coulomb in nature, and not direct collisions.

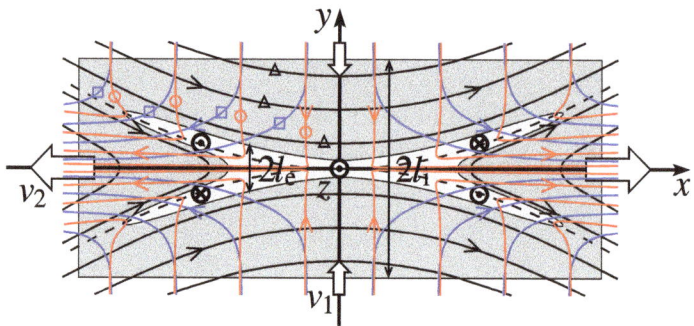

Figure 10.11. A schematic of the Hall MHD model described by Yamada *et al.* (2006), showing how, interior to the reconnection zone of the Sweet–Parker model (Fig. 10.6), ions and electrons decouple because of the Hall term. Ions following the blue □ streamlines demagnetise and diffuse within the *ion diffusion zone* (grey rectangle of width $2l_i$) while electrons following the red ○ streamlines remain coupled to the magnetic induction (black △ solid lines). When electrons enter the *electron diffusion zone* (light grey stretched X-shaped region of average width $2l_e$), they make a very sharp turn away from the y-axis at the *separatrix* (dashed black lines) and momentarily decouple from the magnetic field, allowing magnetic reconnection to occur. This electron flow pattern generates a strong magnetic quadrupole moment (indicated by $\odot$ and $\otimes$) which deflects ions away from the electron diffusion zone.

§10.4.1, the plasma must be described in terms of two co-existing and somewhat independent fluids; one describing the electrons, the other the ions. That these fluids have some independence means, among other things, their velocity fields differ and thus *both* can no longer be coupled to the magnetic field. Typically, the smaller-mass electrons remain "flux-frozen", while the more massive positive ions decouple and become *demagnetised*. The different velocity field also gives rise to the Hall electric field (Eq. 10.58).

It is widely believed that such a transition occurs within the reconnection zone of the Sweet–Parker model (Fig. 10.6); see, for example, Shay *et al.*, (2001), Breslau & Jardin, (2003), and Yamada *et al.*, (2006). As fluid is forced into the reconnection zone, its temperature rises, pressure increases, and fluid is rarified. In this environment, fluid becomes collisionless and crosses into the Hall regime where electrons and ions decouple. Within the *ion diffusion zone* (grey rectangle of width $2l_i$ in Fig. 10.11), the ions demagnetise and veer away from the y-axis (blue □ streamlines in Fig. 10.11) independently of the electrons which follow the red ○ streamlines. Meanwhile, the still flux-frozen electrons continue to drag lines of induction (black △ lines) towards the X-point or, more generally, the *separatrix* (dashed black lines in Fig. 10.11) separating regions of incoming lines of induction before reconnection from outgoing lines of induction after.

Enclosing much of the separatrix is the *electron diffusion zone*, depicted in Fig. 10.11 as a light-grey stretched X-shaped region with an average width of $2l_e$. Once electrons are within this zone, at least two important effects are observed. First, upon reaching the separatrix, electrons are abruptly turned (diffused) away from the y-axis. The strong acceleration of negative charge associated with this abruptness generates, by Biot and Savart, a magnetic induction in the z-direction. Using the upper right quadrant in Fig. 10.11 as an example, the right-hand rule and the negative electron charge require the induced magnetic induction to point into the page ($-\hat{z}$-direction), as indicated by the $\otimes$ symbol in the figure. Repeating this reasoning in the remaining three quadrants one finds that a *magnetic quadrupole moment* ($\odot$ indicating $+\hat{z}$-direction) is generated, entirely driven by the sharp diffusion of electrons at the separatrix. An important consequence of the quadrupole moment is it generates a Lorentz force that diverts positive ions away from the electron diffusion zone, helping to maintain separation of the two fluids. The quadrupole moment is considered to be a *signature* of Hall MHD in this context, since no such moment is possible in the Parker–Sweet model which assumes the electron and ion components of the fluid remain coupled.

Second, once inside their diffusion zone, electrons momentarily decouple from the magnetic induction allowing lines of induction to break, reconnect, and then recouple with the electrons as they are collectively ejected from the electron diffusion zone along the x-axis. As the remagnetised electrons pass through then leave the ion diffusion zone, they recouple with the ions into a single resistive fluid, no longer dominated by the Hall term and where the resistivity, η, is once again determined by electron–neutral collisions.

Notably, where the Hall effect dominates resistance, the magnetic diffusion, $\mathcal{D}_\mathrm{M}$, plays no role in determining the reconnection time scale. In our discussion on the Sweet–Parker model based on resistive MHD (§10.3.3), flux-frozen fluid enters the reconnection zone (dimensions $2L \times 2l$) with speed,

$$v_1 \sim \frac{\mathcal{D}_\mathrm{M}}{l} = \frac{\mathcal{D}_\mathrm{M} v_2}{v_1 L} = \frac{\mathcal{D}_\mathrm{M}}{v_1} \frac{a_1}{L}, \qquad (\text{Eq. 10.48})$$

where a_1 is the Alfvén speed of the fluid as it enters the zone. In Hall MHD, the Sweet–Parker length scale $\mathcal{D}_\mathrm{M}/v_1$ is replaced with l_i (half-width of the ion diffusion zone; Fig. 10.11) which simulations (*e.g.*, Yamada *et al.*, 2006) suggest is of order $0.1L$. Thus, for Hall MHD,

$$v_1 \sim l_\mathrm{i} \frac{a_1}{L} \sim 0.1\, a_1 \quad \Rightarrow \quad \tau_\mathrm{Hall} = \frac{L}{v_1} \sim \frac{L}{0.1\, a_1} \sim 10\,\mathrm{s}, \qquad (10.68)$$

for L and a_1 at the base of a solar flare given in the discussion immediately before Eq. (10.50). This is much more in keeping with observed time scales for solar flares than the Sweet–Parker time scale of 10^7 s (Eq. 10.50).

The reader shouldn't take this seemingly perfect conclusion as evidence that solar flares are now completely understood; they are not. Even in understanding how Eq. (10.68) comes about, the mechanism by which electrons decouple from the magnetic field in their diffusion zone and why it is l_i that the Hall model replaces $\mathcal{D}_\mathrm{M}/v_1$ with are not well understood. Turbulence is probably key, and with turbulence comes a very complicated, fully 3-D problem on which only numerical investigations can shed further light.

10.5 Ambipolar diffusion

10.5.1 Overview and motivation

Through Chap. 9, our discussion was limited to ideal MHD where the plasma is assumed to be 100% ionised. For the first time in §10.2, we considered the possibility that the plasma may not be fully ionised and, in so doing, uncovered the "three players" of non-ideal MHD: resistive dissipation (§10.3), the Hall effect (§10.4), and ambipolar diffusion (AD; this section). For the sake of mathematical expediency, our approach in §10.2.1 was restricted to a *weakly* ionised *isothermal* fluid, still avoiding the more general case of a non-isothermal plasma with *any* level of ionisation. Yet it is here where much of astrophysics lies!

For many astrophysical systems, assuming the ionisation to be *either* complete *or* very low is dodgy at best. The temperature required to ionise hydrogen completely is in excess of 10^5 K and the first ionisation of Helium is complete at a temperature double that. In star-forming regions, temperatures in cloud cores are typically at or under 10^5 K and the plasma is moderately ionised. In stellar outflows, temperatures can range from $\sim 10^8$ K where the jet is launched, to ~ 100 K in the entrained ambient medium (what astronomers refer to as the "second wind"). Therefore, not only is a stellar outflow non-isothermal, its ionisation level can vary with position from 100% to negligible. Even in a planetary disc, we've seen that the temperature in the innermost region can well exceed 10^4 K and, at this location from where matter accretes onto the protostar, the fluid is again moderately ionised.

Evidently, within some astrophysical systems, comparable quantities of ionised and neutral gases coexist, commingle, interact, and exchange as neutrals ionise or ions recombine with the rise and fall of temperature. In such a system, the ionised component – which is coupled to the magnetic induction – collides with and entrains the neutral component in such a way that the motion of the neutrals is modified, indirectly but significantly, by magnetism. As such, the neutral component of a fluid responds to a permeating magnetic induction in a way that a completely neutral fluid would not. Similarly and again via collisions, the neutral component modifies the motion of the ions in competition with the Lorentz force coupling it to $\vec{B}$. As such, a portion of the ionised component can diffuse across lines of magnetic induction in a way that would not occur in a completely ionised fluid.

This essentially describes the "diffusion" part of AD, whose effect on the plasma is highly dependent upon temperature and ionisation level. When introduced in §10.2.1, AD first appeared as a term in the electric field (Eq. 10.20):

$$\vec{E}_{\text{AD}} = -\beta_{\text{i,n}} \vec{f}_{\text{L}} \times \vec{B}, \tag{10.69}$$

which entered the induction equation via Faraday's law as a non-ideal term. When fully spelled out, this term looks like:

$$\partial_t \vec{B}_{\text{AD}} = \frac{1}{\mu_0} \nabla \times \left(\beta_{\text{i,n}} ((\nabla \times \vec{B}) \times \vec{B}) \times \vec{B} \right), \tag{10.70}$$

a truly awe-inspiring construct made worse by the prospect that $\beta_{i,n}$ depends upon the spatially varying density! As impossible as $\partial_t \vec{B}_{\text{AD}}$ may be to fathom, the nicety of the one-fluid, isothermal model is that at least AD is confined to the induction equation; no other MHD equation is affected.

Relaxing the assumptions of weak ionisation and isothermality requires us to go to a so-called *two-fluid model*, where the ions and neutrals are considered as separate fluids, but all the while existing co-spatially. This approach introduces a number of complications that require both care and mathematical fortitude in deriving a self-consistent and useful set of expressions. To give a heads-up and provide a checklist for the issues we'll encounter, consider the following:

1. a separate set of fluid equations is needed for each of the neutrals and ions including continuity, momentum, and an energy equation (to relax isothermality);

2. since ionisation level is a function of temperature, we'll need (for the first time!) *source terms* in the continuity equations to account for transitions between ions and neutrals;

3. as neutrals ionise or ions recombine, momentum and energy are transferred between sub-populations;

4. while isothermality is being relaxed, the fluid components must still remain in thermal equilibrium with each other (otherwise we're really up the crick!); and

5. effects of AD will not be confined to the induction equation.

Most of these complications arise from relaxing the assumption of isothermality and so unsurprisingly, most references you'll find in the literature to a two-fluid AD model explicitly assume an isothermal fluid (*e.g.*, Li, McKee, & Klein, 2006; Chen & Ostriker, 2012; Burkhart *et al.*, 2015; Brandenburg, 2019, to list a few). Examples in the literature of a two-fluid, *non*-isothermal approach are much rarer, owing to concerns of whether ions can even be treated as a fluid in thermodynamical equilibrium (*e.g.*, Draine, 1986), and computational issues such as time step control (Falle, 2003). Despite these hurdles, both Falle (2003) and Tilley, Balsara, & Meyer (2012) have developed computational schemes including bench-mark tests to do non-isothermal, two-fluid AD calculations.

The approach I take here is purely mathematical and goes further than the references above by its application of *strict* thermodynamic equilibrium. Not only do I assume each component – neutrals and ions – is separately in thermodynamic equilibrium, I also assume they are in thermodynamical equilibrium *with each other* (*i.e.*, at any given location, the temperatures of the ions and neutrals are the same). This, after all, is implicit in a one-fluid approach and, as conditions where the one-fluid model is valid (low ionisation) change to where a two-fluid approach is called for, the assumption of strict thermodynamical equilibrium ought to carry forward, at least to some extent.

Finally, the only mechanism I consider to ionise neutrals or recombine ions is changes in temperature. In the dusty, dirty, irradiated environment of a planetary disc, cosmic rays, dust grains, shock waves, UV radiation, *etc.* all contribute to the ionisation level (*e.g.*, Draine, 1986; Falle, 2003), and simple temperature changes in a fluid element may not be as significant. Still, and in keeping with the opening paragraphs of this chapter, using only the temperature – and thus the Saha equation – to determine the ionisation level has the advantage of self-consistency, requiring no additional physics or chemistry than has already been introduced.

And so with all that, let us proceed...

10.5.2 A two-fluid, non-isothermal model for AD

We now begin the most challenging *mathematical* task of the text.[25] Now, I hasten to add, were it not for the miraculous fact that in this universe, $m_e \ll m_i$, we'd be faced with tracking *three* sets of fluid equations, not "just" two! As it is, we'll find our two-fluid model cumbersome enough and, as we're wallowing in the quagmire of two-fluid mathematics, perhaps we can take some solace knowing that at least we aren't dealing with three!

Start by writing down the fluid equations for each of the three components: neutrals; ions; and electrons,[26] omitting for the time being the induction equation:

$$\partial_t \rho_n + \nabla \cdot (\rho_n \vec{v}_n) = \sigma_{i,n} + \sigma_{e,n}; \tag{10.71}$$

$$\partial_t \vec{s}_n + \nabla \cdot (\vec{s}_n \vec{v}_n) = -\nabla p_n - \rho_n \nabla \phi + \vec{f}_{i,n} + \vec{f}_{e,n}; \tag{10.72}$$

$$\partial_t e_n + \nabla \cdot (e_n \vec{v}_n) = -p_n \nabla \cdot \vec{v}_n + \varpi_{i,n} + \varpi_{e,n}; \tag{10.73}$$

$$\partial_t \rho_i + \nabla \cdot (\rho_i \vec{v}_i) = \sigma_{e,i} + \sigma_{n,i}; \tag{10.74}$$

$$\partial_t \vec{s}_i + \nabla \cdot (\vec{s}_i \vec{v}_i) = -\nabla p_i - \rho_i \nabla \phi + Zen_i(\vec{E} + \vec{v}_i \times \vec{B}) + \vec{f}_{e,i} + \vec{f}_{n,i}; \tag{10.75}$$

$$\partial_t e_i + \nabla \cdot (e_i \vec{v}_i) = -p_i \nabla \cdot \vec{v}_i + \varpi_{e,i} + \varpi_{n,i}; \tag{10.76}$$

$$\partial_t \rho_e + \nabla \cdot (\rho_e \vec{v}_e) = \sigma_{n,e} + \sigma_{i,e}; \tag{10.77}$$

$$\partial_t \vec{s}_e + \nabla \cdot (\vec{s}_e \vec{v}_e) = -\nabla p_e - \rho_e \nabla \phi - en_e(\vec{E} + \vec{v}_e \times \vec{B}) + \vec{f}_{n,e} + \vec{f}_{i,e}; \tag{10.78}$$

$$\partial_t e_e + \nabla \cdot (e_e \vec{v}_e) = -p_e \nabla \cdot \vec{v}_e + \varpi_{n,e} + \varpi_{i,e}, \tag{10.79}$$

where, in addition to all the variables defined after Eq. (10.1)–(10.3), we have the following:

- e_n, e_i, e_e are, respectively, the internal energy densities of the neutrals, ions, and electrons, where $p_1 = (\gamma_1 - 1)e_1$ ('1' represents 'n', 'i', or 'e');

- $\sigma_{1,2}$ is the *exchange term* representing the mass per unit volume per unit time

---
[25] Not *computational*! That honour goes to the MHD Riemann solver in Chap. 6!
[26] Not to worry, the electron component will soon go away!

that fluid component '1' converts to '2' (negative if '2' converts to '1'). Thus, $\sigma_{2,1} = -\sigma_{1,2}$;

- $\vec{f}_{1,2}$ is the force density fluid component '1' exerts on '2', where there are now two terms: $\vec{f}_{1,2} = \vec{f}^{\,a}_{1,2} + \vec{f}^{\,x}_{1,2}$. $\vec{f}^{\,a}_{1,2}$ is the ambipolar term from §10.2.1 (Eq. 10.4) while $\vec{f}^{\,x}_{1,2}$ is the *exchange force density*, a new term accounting for momentum transfer when fluid component '1' converts to '2'. As before, $\vec{f}_{2,1} = -\vec{f}_{1,2}$;

- $\varpi_{1,2} = \varpi^{a}_{1,2} + \varpi^{x}_{1,2}$ is the power density delivered from fluid component '1' to '2'. $\varpi^{a}_{1,2}$ is the rate at which the ambipolar force density does work, and $\varpi^{x}_{1,2}$ is the power density delivered from fluid component '1' to '2' (or the reverse if negative) during the exchange to preserve thermal equilibrium among the fluid components.

Our task is now to distil these nine equations into a useful and explicit model for non-isothermal MHD with AD. Relaxing the assumption of low ionisation means we can no longer rely upon $n_i \ll n_n$ and $n_e \ll n_n$ as we did in §10.2.1 to make short-shrift of some of the equations. However, we can and will use liberally the fact that $m_e \ll m_i \sim m_n$ – and in a three-component environment, $p_e \ll p_i \sim p_n$ – which will convert these equations describing three fluid components into the two-fluid, non-isothermal model we seek.

Before tackling that, however, let's first examine the new source terms introduced by our assumption of non-isothermality.

Density source terms

In an isothermal fluid, the fractional ionisation remains constant everywhere and for all time and there is no need to account for the spatially and temporally dependent exchange between neutrals and ions. Relaxing this constraint, then, opens a whole can of worms which starts with accounting for losses and gains in the neutral and ionised sub-components of the fluid, and continues with the corresponding transfers of momentum and thermal energy. Even in an isothermal fluid, conversions between neutrals and ions occur, but equally in both directions. It's only when the temperature increases or decreases that there will be a *net* increase or decrease in the ion number density with a commensurate decrease or increase in the neutral number density as a result of the exchange.

Consistent with the fluid model is the assumption of thermal equilibrium, which is very different from the assumption of isothermality. For our three-component system, this means that each component can be described with the same temperature, T, which, because the system is not isothermal, can still vary with location and time. Thermal equilibrium requires that once perturbed, each fluid component returns to thermal equilibrium within itself and with the others – perhaps to a new common temperature – on a time scale short compared to all other physically relevant time scales for the system.

	H	He	He$^+$
$2g_{s^+}/g_s$	1	4	1
ε_s (eV)	13.54	24.48	54.17

Table 10.1. Statistical weights and ionisation energies for H, He, and He$^+$.

With thermodynamic equilibrium, it is the Saha equation (*e.g.*, MacDonald, 2015),

$$\frac{n_e n_{s^+}}{n_s} = \frac{2g_{s^+}}{g_s}\frac{e^{-\varepsilon_s/k_B T}}{\lambda_e^3(k_B T)} \equiv \nu_s(k_B T), \qquad (10.80)$$

that governs the fractional ionisation of each species everywhere in the fluid as a function of $k_B T$, where k_B is Boltzmann's constant. Here, n_s is the number density of species s (which may, itself, be partially ionised), n_{s^+} is the number density of species s once ionised, n_e is the number density of electrons, g_s and g_{s^+} are the state degeneracies of species s and s$^+$, and ε_s is the ionisation energy of species s (*e.g.*, Table 10.1 for H and He). Further,

$$\lambda_e(k_B T) = \frac{h}{\sqrt{2\pi m_e k_B T}},$$

is the so-called *thermal de Broglie wavelength* of an electron where h is Planck's constant and m_e is the mass of an electron. Finally, $\nu_s(k_B T)$ is the *critical number density* for species s which evidently has a strong dependence on $k_B T$.

The simplest case is pure hydrogen where only one invocation of Eq. (10.80) is required. Let $n_{s^+} = n_{H^+} = n_e$, the number density of ionised hydrogen (n_{H^+}) and free electrons (n_e), and let $n_s = n_H$, the number density of neutral hydrogen. Then, Eq. (10.80) becomes,

$$n_{H^+}^2 = \nu_H n_H. \qquad (10.81)$$

Since the conserved quantity is the total number density, $n = n_{H^+} + n_H$, we set $n_H = n - n_{H^+}$ and Eq. (10.81) becomes,

$$n_{H^+}^2 = \nu_H(n - n_{H^+}) \quad \Rightarrow \quad n_{H^+} = \frac{1}{2}\left(\sqrt{\nu_H^2 + 4n\nu_H} - \nu_H\right). \qquad (10.82)$$

From this, it is easy to show that the ionised and neutral number densities are equal ($n_{H^+} = n_H = n/2$) when $n_{H^+} = \nu_H$, whence the designation of ν_H as a *critical density*.

With this, we can evaluate the exchange term, $\sigma_{n,i}$, for n = H and i = H$^+$. Since $\sigma_{n,i}$ is the rate at which neutrals are converted to ions per unit volume and thus the rate of change of ρ_i, we can write by the chain rule,

$$\sigma_{n,i} = \frac{d\rho_i}{dt} = m_H \frac{dn_{H^+}}{dt} = m_H \frac{dn_{H^+}}{d\nu_H}\frac{d\nu_H}{d(k_B T)}\frac{d(k_B T)}{dt}, \qquad (10.83)$$

where the first two derivatives can be evaluated by differentiating Eq. (10.82) and (10.80) directly, and where the third derivative can be determined from Eq. (10.74) and (10.76) by grace of the ideal gas law, $k_B T = m_H p_i/\rho_i$. The details are left to Problem 10.9.

However, computationally, it is much more practical to evaluate $\sigma_{n,i}$ as the difference in ionisation densities after and before the Saha calculation is performed over a time step δt:

$$\sigma_{n,i} = \frac{\delta \rho_i}{\delta t} = \frac{\rho_{i,x} - \rho_i}{\delta t}, \quad (10.84)$$

where $\rho_{i,x} = m_H n_{H^+}$ is the ionised mass density after the exchange, n_{H^+} is the ionised number density evaluated directly from Eq. (10.82) after the change in temperature, and ρ_i is the ionised mass density before the exchange. Evidently, since mass is conserved,

$$\boxed{\sigma_{i,n} = \frac{\rho_{n,x} - \rho_n}{\delta t} = -\sigma_{n,i} = -\frac{\rho_{i,x} - \rho_i}{\delta t}.} \quad (10.85)$$

Eq. (10.83)–(10.85) illustrate how one might tie n_n and n_i to the changing temperature of the fluid assuming pure hydrogen. It is astonishing, actually, how much more complicated the problem becomes just by adding one more species to the mix. For the mathematically hardy, Problem 10.10 takes the reader through the process of finding n_n and n_i as the temperature changes for a mixture of just two elements, H and He, with no mention of the numerous other elements, molecules, and dust one finds in a realistic astrophysical soup!

Momentum source terms

In addition to the *ambipolar force density* derived in §10.2.1,

$$\vec{f}^{\,a}_{n,i} = \gamma_{n,i} \rho_n \rho_i \vec{v}_{n,i}, \quad \text{Eq. (10.4)}$$

the relaxation of isothermality means that any neutrals being ionised or ions recombining will transfer from their former component to the new whatever momentum they may have possessed. I refer to this transfer of momentum density over a specific period of time, δt, as the *exchange force density*.

Consider a small volume, δV, of fluid in which a net mass,

$$\delta m_{n,i} = (\rho_{i,x} - \rho_i)\delta V > 0,$$

of neutrals is converted to ions in a time δt. Such an exchange delivers a net momentum,

$$\delta \vec{S}_{n,i} = \delta m_{n,i}(\vec{v}_n - \vec{v}_i) = \delta m_{n,i} \vec{v}_{n,i},$$

from the neutrals to the ions, which corresponds to a net exchange force,

$$\vec{F}^{\,x}_{n,i} = \frac{\delta \vec{S}_{n,i}}{\delta t} = \frac{\rho_{i,x} - \rho_i}{\delta t} \vec{v}_{n,i} \delta V = \sigma_{n,i} \vec{v}_{n,i} \delta V,$$

using Eq. (10.84), and thus an exchange force density,

$$\vec{f}^{\,x}_{n,i} = \frac{\vec{F}^{\,x}_{n,i}}{\delta V} = \sigma_{n,i} \vec{v}_{n,i}. \quad (10.86)$$

This is an effective force density exerted by the neutrals on the ions as a result of neutrals converting to ions. In the co-moving frame of the neutrals, $\vec{f}_{n,i}^{\,x} \propto -\vec{v}_i$, and it can be thought of as a "drag force", one that works to reduce the relative speed between the two fluid components.

By Newton's third law, the reaction exchange force density exerted by the ions on the neutrals is evidently,

$$\vec{f}_{i,n}^{\,x} = -\vec{f}_{n,i}^{\,x} = -\sigma_{n,i}\vec{v}_{n,i} = -\sigma_{i,n}\vec{v}_{i,n}, \tag{10.87}$$

since $\sigma_{i,n} = -\sigma_{n,i}$ (Eq. 10.85) and $\vec{v}_{n,i} = -\vec{v}_{i,n}$. Note that Eq. (10.86) and (10.87) are derived assuming a net conversion of neutrals to ions ($\sigma_{n,i} > 0$, $\sigma_{i,n} < 0$).

If, instead, there were a net conversion of ions to neutrals and,

$$\delta m_{i,n} = (\rho_{n,x} - \rho_n)\delta V > 0,$$

then by swapping the indices i and n in Eq. (10.86) and (10.87), the force densities become,

$$\vec{f}_{i,n}^{\,x} = \sigma_{i,n}\vec{v}_{i,n} \quad \Rightarrow \quad \vec{f}_{n,i}^{\,x} = -\sigma_{i,n}\vec{v}_{i,n} = -\sigma_{n,i}\vec{v}_{n,i}, \tag{10.88}$$

this time for $\sigma_{i,n} > 0$ and $\sigma_{n,i} < 0$. Therefore, we can combine Eq. (10.86)–(10.88) to get our final expressions for the exchange force densities,

$$\vec{f}_{n,i}^{\,x} = |\sigma_{n,i}|\vec{v}_{n,i} \quad \text{and} \quad \vec{f}_{i,n}^{\,x} = |\sigma_{i,n}|\vec{v}_{i,n}, \tag{10.89}$$

where their "equal and opposite" nature is self-evident.

Combining Eq. (10.89) with (10.4) gives us expressions for the total force densities neutrals and ions exert on each other:

$$\boxed{\vec{f}_{n,i} = \vec{f}_{n,i}^{\,x} + \vec{f}_{n,i}^{\,a} = (|\sigma_{n,i}| + \gamma_{n,i}\rho_n\rho_i)\vec{v}_{n,i} \equiv \mathcal{Q}_{n,i}\vec{v}_{n,i};} \tag{10.90}$$

$$\boxed{\vec{f}_{i,n} = \vec{f}_{i,n}^{\,x} + \vec{f}_{i,n}^{\,a} = (|\sigma_{i,n}| + \gamma_{i,n}\rho_i\rho_n)\vec{v}_{i,n} \equiv \mathcal{Q}_{i,n}\vec{v}_{i,n} = -\vec{f}_{n,i},} \tag{10.91}$$

where I'm defining $\mathcal{Q}_{n,i} = \mathcal{Q}_{i,n}$ as the combined *exchange-ambipolar coefficient*.

Energy source terms

The first power density source term we'll look at is the simplest of the two. The *ambipolar power density* is the rate at which the ambipolar force density does work, and is therefore the dot-product of $\vec{f}_{n,i}^{\,a}$ and the relative velocity between the two fluid components:

$$\varpi_{n,i}^{a} = \vec{f}_{n,i}^{\,a} \cdot \vec{v}_{n,i} = \gamma_{n,i}\rho_n\rho_i v_{n,i}^2, \tag{10.92}$$

using Eq. (10.4), a positive-definite quantity. Therefore, $\varpi_{n,i}^{a}$ is dissipative and always *increases* the internal energy density of the ions so long as there remains a non-zero relative velocity between the ionised and neutral components. This is consistent with our interpretation of $\vec{f}_{n,i}^{\,a}$ as a drag force in the previous segment.

Further,

$$\varpi_{i,n}^{a} = \vec{f}_{i,n}^{\,a} \cdot \vec{v}_{i,n} = \gamma_{i,n}\rho_i\rho_n v_{i,n}^2 = \varpi_{n,i}^{a}, \tag{10.93}$$

and the internal energy of both fluid components increases by the same amount at the expense of their relative velocity. The analogy I like to use here is that of rubbing hands together. So long as your hands are in contact and there is a relative speed between them, one hand doesn't get warm while the other cools; *both* hands warm at the same rate.

The *exchange power density*, $\varpi_{n,i}^x$, is a bit trickier to evaluate as it bears the responsibility of maintaining thermodynamic equilibrium between the ions and neutrals. The principle is simple. Without specifying the mechanism of energy transfer (*e.g.*, radiation, adiabatic expansion/compression, mixing, whatever), the assumption of thermodynamic equilibrium between the two fluid components presupposes that something, somehow restores their *specific energies* to equilibrium on a time scale short compared to all other processes. As the exchange of particles between ions and neutrals is the only vehicle we have for such an energy exchange, we task $\varpi_{n,i}^x$ with transferring from one component to the other precisely that amount of energy required to maintain thermal equilibrium.

Starting with the ideal gas law, we can enforce thermal equilibrium between the neutrals and ions by setting,

$$k_B T = m_n \frac{p_n}{\rho_n} = m_i \frac{p_i}{\rho_i} \quad \Rightarrow \quad p_n = \tilde{m} \frac{\rho_n}{\rho_i} p_i \tag{10.94}$$

$$\Rightarrow \quad e_i = \frac{p_i}{\gamma_i - 1} = \frac{1}{\tilde{m}\tilde{\gamma}} \frac{\rho_i}{\rho_n} e_n, \tag{10.95}$$

where,

$$\tilde{m} = \frac{m_i}{m_n} \quad \text{and} \quad \tilde{\gamma} = \frac{\gamma_i - 1}{\gamma_n - 1}, \tag{10.96}$$

are defined for convenience. For the present system where ions are created from the neutral particles, $\tilde{m} \sim 1$. By contrast, for the low ionisation system considered in §10.2.1 where ions are generated from "pollutants" such as Na and K in an otherwise H + He gas, we found $\tilde{m} \sim 13$. As for $\tilde{\gamma}$, for most non-relativistic astrophysical applications, $\gamma = 5/3$ (monatomic gas) whether neutral or ionised, and $\tilde{\gamma} \sim 1$. However, for generality, we'll carry it and $\tilde{m}$ through our derivations, at least until they start getting in the way!

To "inform" the fluid equations that thermal equilibrium is being enforced, substitute Eq. (10.95) into Eq. (10.76) (ignoring $\varpi_{e,i}$ as inconsequential) to get,

$$\frac{1}{\tilde{m}\tilde{\gamma}} \left[\partial_t \left(\frac{\rho_i}{\rho_n} e_n \right) + \nabla \cdot \left(\frac{\rho_i}{\rho_n} e_n \vec{v}_i \right) \right] = -p_i \nabla \cdot \vec{v}_i + \varpi_{n,i}^a + \varpi_{n,i}^x, \tag{10.97}$$

where $\varpi_{n,i}^a$ is given by Eq. (10.92) and $\varpi_{n,i}^x$ is the exchange power density we wish to evaluate. To this end, substitute:

$$\partial_t \left(\frac{\rho_i}{\rho_n} e_n \right) = \frac{\rho_i}{\rho_n} \partial_t e_n + e_n \partial_t \left(\frac{\rho_i}{\rho_n} \right);$$

$$\nabla \cdot \left(\frac{\rho_i}{\rho_n} e_n \vec{v}_i \right) = \frac{\rho_i}{\rho_n} \nabla \cdot (e_n \vec{v}_i) + e_n \vec{v}_i \cdot \nabla \left(\frac{\rho_i}{\rho_n} \right),$$

into Eq. (10.97) and multiply through by $\tilde{m}\tilde{\gamma}\rho_n/\rho_i$ to get,

$$\partial_t e_n + \nabla \cdot (e_n \vec{v}_i) + e_n \frac{\rho_n}{\rho_i} \partial_t \left(\frac{\rho_i}{\rho_n}\right) + e_n \vec{v}_i \cdot \left[\frac{\rho_n}{\rho_i} \nabla \left(\frac{\rho_i}{\rho_n}\right)\right]$$
$$= \tilde{m}\tilde{\gamma}\frac{\rho_n}{\rho_i} \left(-p_i \nabla \cdot \vec{v}_i + \varpi_{n,i}^a + \varpi_{n,i}^x\right), \quad (10.98)$$

taking note that the second term on the LHS has $\vec{v}_i$, not $\vec{v}_n$. Next, from Eq. (10.73) we have,

$$\partial_t e_n = -\nabla \cdot (e_n \vec{v}_n) - p_n \nabla \cdot \vec{v}_n + \varpi_{i,n}^a + \varpi_{i,n}^x. \quad (10.99)$$

Further,

$$\frac{\rho_n}{\rho_i} \partial_t \left(\frac{\rho_i}{\rho_n}\right) = \frac{\rho_n}{\rho_i} \left(\frac{1}{\rho_n} \partial_t \rho_i - \frac{\rho_i}{\rho_n^2} \partial_t \rho_n\right) = \frac{1}{\rho_i} \partial_t \rho_i - \frac{1}{\rho_n} \partial_t \rho_n$$
$$= \frac{1}{\rho_i} \left(-\nabla \cdot (\rho_i \vec{v}_i) + \sigma_{n,i}\right) - \frac{1}{\rho_n} \left(-\nabla \cdot (\rho_n \vec{v}_n) + \sigma_{i,n}\right)$$

$$\Rightarrow \quad \frac{\rho_n}{\rho_i} \partial_t \left(\frac{\rho_i}{\rho_n}\right) = -\nabla \cdot \vec{v}_i - \vec{v}_i \cdot \frac{\nabla \rho_i}{\rho_i} + \nabla \cdot \vec{v}_n + \vec{v}_n \cdot \frac{\nabla \rho_n}{\rho_n} + \sigma_{n,i}\left(\frac{1}{\rho_i} + \frac{1}{\rho_n}\right), \quad (10.100)$$

using Eq. (10.71) and (10.74), and since $\sigma_{i,n} = -\sigma_{n,i}$. Then, substituting Eq. (10.99), (10.100), and,

$$\frac{\rho_n}{\rho_i} \nabla \left(\frac{\rho_i}{\rho_n}\right) = \frac{1}{\rho_i} \nabla \rho_i - \frac{1}{\rho_n} \nabla \rho_n,$$

into Eq. (10.98), we get, after a little algebra,

$$-\vec{v}_{n,i} \cdot \underbrace{\left(\nabla e_n - e_n \frac{\nabla \rho_n}{\rho_n}\right)}_{\rho_n \nabla (e_n/\rho_n)} - p_n \nabla \cdot \vec{v}_n + e_n \sigma_{n,i}\left(\frac{1}{\rho_i} + \frac{1}{\rho_n}\right) + \varpi_{i,n}^a + \varpi_{i,n}^x$$
$$= -\tilde{\gamma}\tilde{m}\underbrace{\frac{\rho_n}{\rho_i}p_i}_{p_n} \nabla \cdot \vec{v}_i + \tilde{m}\tilde{\gamma}\frac{\rho_n}{\rho_i}\left(\varpi_{n,i}^a + \varpi_{n,i}^x\right), \quad (10.101)$$

where the second underbrace follows from Eq. (10.94).

For the non-conservative ambipolar power density, we found that $\varpi_{i,n}^a = \varpi_{n,i}^a$. Conversely, the exchange power density is conservative (what one fluid component gives up, the other receives) and thus $\varpi_{i,n}^x = -\varpi_{n,i}^x$. With this final piece of insight, we can solve Eq. (10.101) for $\varpi_{n,i}^x$ to get,

$$\varpi_{n,i}^x = \frac{\rho_i}{\rho_i + \tilde{m}\tilde{\gamma}\rho_n}\left[e_n \sigma_{n,i}\left(\frac{1}{\rho_i} + \frac{1}{\rho_n}\right) + (\rho_i - \tilde{m}\tilde{\gamma}\rho_n)\gamma_{n,i}\rho_n v_{n,i}^2 \right.$$
$$\left. - p_n \nabla \cdot (\vec{v}_n - \tilde{\gamma}\vec{v}_i) - \rho_n \vec{v}_{n,i} \cdot \nabla\left(\frac{e_n}{\rho_n}\right)\right], \quad (10.102)$$

using Eq. (10.92) and after some straight-forward algebra.

At first glance, Eq. (10.102) may not seem antisymmetric in the exchange of its indices i and n, as it must be for $\varpi_{i,n}^x = -\varpi_{n,i}^x$. The proof that Eq. (10.102) is, in fact, antisymmetric is left as an exercise for the reader. [*Hint*: As can be seen from Eq. (10.96), switching i and n forces $\tilde{m} \to 1/\tilde{m}$ and $\tilde{\gamma} \to 1/\tilde{\gamma}$. Liberal use of Eq. (10.94) and (10.95) then completes the proof.]

Growing a little weary now of $\tilde{m}$ and $\tilde{\gamma}$, let's set them both to 1 so that Eq. (10.102) can be reduced to, again after a little algebra:

$$\varpi_{n,i}^x = -\varpi_{i,n}^x = \frac{\sigma_{n,i}}{\rho_n} e_n + \frac{\rho_i - \rho_n}{\rho} \gamma_{n,i} \rho_n \rho_i v_{n,i}^2 \\ - \frac{\rho_i p_n}{\rho} \nabla \cdot \vec{v}_{n,i} - \frac{\rho_i \rho_n}{\rho} \vec{v}_{n,i} \cdot \nabla \left(\frac{e_n}{\rho_n} \right), \quad (10.103)$$

where $\rho = \rho_n + \rho_i$ is the total density of the fluid. To interpret the four terms on the RHS of Eq. (10.103), first note that because of strict thermal equilibrium (Eq. 10.94 and 10.95), $\rho_i p_n = \rho_n p_i$ and $e_n/\rho_n = e_i/\rho_i \propto T$. With this, the four terms considered left to right can be seen to represent:

1. the portion of internal energy density transported with the neutrals as they become ions (or *vice versa*);

2. a correction to the heating caused by the ambipolar power density – initially applied equally to ions and neutrals – so that each subpopulation is heated the same amount *per unit mass*;

3. a correction to the adiabatic heating/cooling in the same spirit as above; and

4. a term proportional to the temperature gradient, and thus the heat conduction needed to compensate for the differing heat transport of the ions and neutrals.

Combining Eq. (10.92), (10.93), and (10.103), we arrive at our final expressions – again after a little algebra – for the combined *exchange-ambipolar power density* each subpopulation delivers to the other:

$$\boxed{\begin{aligned} \varpi_{n,i} &= \varpi_{n,i}^x + \varpi_{n,i}^a \\ &= \frac{e_i}{\rho_i} \sigma_{n,i} + \frac{\rho_n}{\rho} \left[2\gamma_{n,i} \rho_i^2 v_{n,i}^2 - p_i \nabla \cdot \vec{v}_{n,i} - \rho_i \vec{v}_{n,i} \cdot \nabla \left(\frac{e_i}{\rho_i} \right) \right]; \end{aligned}} \quad (10.104)$$

$$\boxed{\begin{aligned} \varpi_{i,n} &= \varpi_{i,n}^x + \varpi_{i,n}^a \\ &= \frac{e_n}{\rho_n} \sigma_{i,n} + \frac{\rho_i}{\rho} \left[2\gamma_{i,n} \rho_n^2 v_{i,n}^2 - p_n \nabla \cdot \vec{v}_{i,n} - \rho_n \vec{v}_{i,n} \cdot \nabla \left(\frac{e_n}{\rho_n} \right) \right]. \end{aligned}} \quad (10.105)$$

Generalised two-fluid MHD

And so now for the final push: extracting from Eq. (10.71)–(10.79) the kernel of mathematics containing the physics of the two-fluid, non-isothermal model for AD-MHD.

Let's start by assessing what we can throw away. First, assuming the inertia and internal energy density of the electron component to be negligible, we can discard Eq. (10.77) and (10.79). Second, the exchange terms involving electrons ($\sigma_{n,e} = -\sigma_{e,n}$ and $\sigma_{i,e} = -\sigma_{e,i}$) are equally negligible and discarded. Third, while the internal energy densities of the neutrals and ions are assumed to be comparable, we still only need to retain one of their energy equations, say that of the ionised component (Eq. 10.76), and replace the energy equation for neutrals with the requirement that the system remain in thermal equilibrium (Eq. 10.94 and 10.95). Already, these three considerations have simplified matters enormously.

But we must be careful not to throw out any of the baby with the bathwater! The electron momentum equation, Eq. (10.78), for example, cannot be so easily dismissed. In comparing Eq. (10.75) with Eq. (10.78), we see they have at least one force density – $\pm e n_e (\vec{E} + \vec{v}_{i/e} \times \vec{B})$ – in common (since $Z n_i = n_e$). Further, while we've already argued that $f_{n,e} \ll f_{n,i}$ (Eq. 10.17 in §10.2.1), $f_{i,e}$ is not equally ignorable, since, in this case, the ions and electrons interact with the full Coulomb force density, and not just the residual ambipolar force density between the neutrals and electrons as described on page 381.

So, proceeding carefully with Eq. (10.78), let's add it to Eq. (10.75) to get,

$$\partial_t \vec{s}_i + \nabla \cdot (\vec{s}_i \vec{v}_i) + \underbrace{\partial_t \vec{s}_e}_{\sim 0} + \underbrace{\nabla \cdot (\vec{s}_e \vec{v}_e)}_{\sim 0}$$

$$= -\nabla(p_i + \underbrace{p_e}_{\sim 0}) - (\rho_i + \underbrace{\rho_e}_{\sim 0})\nabla\phi + \underbrace{e n_e (\vec{v}_i - \vec{v}_e)}_{\vec{J}} \times \vec{B} + \underbrace{\vec{f}_{e,i}}_{} + \vec{f}_{n,i} + \underbrace{\vec{f}_{n,e}}_{\sim 0} + \underbrace{\vec{f}_{i,e}}_{},$$

since $\vec{f}_{i,e} = -\vec{f}_{e,i}$. Thus,

$$\boxed{\partial_t \vec{s}_i + \nabla \cdot (\vec{s}_i \vec{v}_i) = -\nabla p_i - \rho_i \nabla \phi + \vec{f}_L + \mathcal{Q}_{n,i} \vec{v}_{n,i},} \qquad (10.106)$$

using Eq. (10.90). This is the two-fluid momentum equation for ions. Note that the Lorentz force density, $\vec{f}_L = \vec{J} \times \vec{B}$, appears in this equation instead of the momentum equation for neutrals (Eq. 10.72), whereas in the one-fluid model described in §10.2.1, the Lorentz force appears in the neutral momentum equation (Eq. 10.16).

Let us now subtract $m_i \times$ Eq. (10.78) from $Z m_e \times$ Eq. (10.75):

$$Z m_e (\partial_t \vec{s}_i + \nabla \cdot (\vec{s}_i \vec{v}_i)) - m_i (\partial_t \vec{s}_e + \nabla \cdot (\vec{s}_e \vec{v}_e))$$
$$= -Z m_e \nabla p_i + m_i \nabla p_e - \cancel{Z m_e \rho_i \nabla \phi} + \cancel{m_i \rho_e \nabla \phi}$$
$$+ Z m_e e n_e (\vec{E} + \vec{v}_i \times \vec{B}) + m_i e n_e (\vec{E} + \vec{v}_e \times \vec{B}) \qquad (10.107)$$
$$- (Z m_e + m_i) \vec{f}_{i,e} + Z m_e \vec{f}_{n,i} - m_i \vec{f}_{n,e},$$

where the two gravity terms that cancel outright (no approximation) do so because,

$$Z m_e \rho_i = Z m_e m_i n_i = m_e m_i n_e = m_i \rho_e.$$

Numerous other terms in Eq. (10.107) can be dropped as negligible. All terms on the LHS, both remaining terms in the second line, and the first term on the third line are proportional to either m_e or p_e and summarily dismissed. On the fourth line, the first term proportional to m_e can be dropped (as it is being added directly to m_i), but not the second term, for in this case,

$$\frac{m_i f_{n,e}}{Z m_e f_{n,i}} \sim \frac{6.4}{Z} \sqrt{\frac{T}{10^3 \,\mathrm{K}}},$$

using Eq. (10.17) and setting m_i to the mass of a proton; hardly justification to drop either term! Thus, Eq. (10.107) whittles down to,

$$\vec{E} = -\vec{v}_e \times \vec{B} + \frac{Zm_e}{en_e m_i}\vec{f}_{n,i} + \frac{1}{en_e}(\vec{f}_{i,e} + \vec{f}_{n,e}). \tag{10.108}$$

Now, in the one-fluid model discussed in §10.2.1 where the inertia of both the electrons and ions were negligible, three velocities arose that held significant meaning: $\vec{v}_n$, the induction velocity; $\vec{v}_{i,e} \propto \vec{J}$; and $\vec{v}_{i,n} \propto \vec{f}_L$ (see page 384). In the current two-fluid setting, we'll find that induction is tied to the *ion* velocity rather than the neutral velocity, and that $\vec{v}_{i,n}$ has no special significance. Only $\vec{v}_{i,e}$ plays the same role here as it did in the one-fluid model, and only because $m_e \ll m_p$.

Using Eq. (10.90) and (10.91) to replace the force densities in Eq. (10.108) and inserting $\vec{v}_i - \vec{v}_i$ in two strategic places, we get,

$$\vec{E} = -(\vec{v}_e - \vec{v}_i + \vec{v}_i) \times \vec{B} + \frac{1}{en_e}\frac{Zm_e}{m_i}\mathcal{Q}_{n,i}\vec{v}_{n,i}$$
$$+ \frac{1}{en_e}\left(\mathcal{Q}_{i,e}\vec{v}_{i,e} + \mathcal{Q}_{n,e}(\vec{v}_n - \vec{v}_e + \vec{v}_i - \vec{v}_i)\right)$$
$$= -\vec{v}_i \times \vec{B} + \vec{v}_{i,e} \times \vec{B} + \frac{1}{en_e}\left(\frac{Zm_e}{m_i}\mathcal{Q}_{n,i} + \mathcal{Q}_{n,e}\right)\vec{v}_{n,i} + \frac{\mathcal{Q}_{i,e} + \mathcal{Q}_{n,e}}{en_e}\vec{v}_{i,e}$$

$$\Rightarrow \vec{E} = \underbrace{-\vec{v}_i \times \vec{B}}_{\text{induction}} + \underbrace{\frac{1}{en_e}\vec{f}_L}_{\text{Hall}} + \mathcal{B}\vec{v}_{n,i} + \underbrace{\eta_2 \vec{J}}_{\text{resistive}}, \tag{10.109}$$

where the induction, Hall, and resistive terms are immediately recognisable from Eq. (10.20), where there is no apparent ambipolar diffusion term, and where the term $\mathcal{B}\vec{v}_{n,i}$ is new to the two-fluid case. Note that the coefficient,

$$\mathcal{B} \equiv \frac{1}{en_e}\left(\frac{Zm_e}{m_i}\mathcal{Q}_{n,i} + \mathcal{Q}_{n,e}\right), \tag{10.110}$$

has units of magnetic induction, whence our use of the symbol $\mathcal{B}$.

The resistivity, η_2, is a two-fluid generalisation of Eq. (10.21) to include the dissipative interaction of electrons with both ions and neutrals:

$$\eta_2 \equiv \frac{\mathcal{Q}_{i,e} + \mathcal{Q}_{n,e}}{(en_e)^2} = \frac{\gamma_{i,e}\rho_i\rho_e + \gamma_{n,e}\rho_n\rho_e}{(en_e)^2} \approx \frac{m_e}{e^2 n_e}\left(\langle\sigma u\rangle_{i,e}n_i + \langle\sigma u\rangle_{n,e}n_n\right), \tag{10.111}$$

where m_e is set (approximately) to the reduced masses, $\mu_{i,e}$ and $\mu_{n,e}$, $\langle\sigma u\rangle_{i,e}$ is given by Eq. (10.7), and where $\langle\sigma u\rangle_{n,e}$ is given by the first of Eq. (10.6).

The term $\mathcal{B}\vec{v}_{n,i}$ is new to this analysis. Now, in the one-fluid model, $\vec{v}_{n,i} \propto \vec{f}_L$ and there we argued that such a term was negligible compared to what was to become the Hall term. Here, even though $\vec{v}_{n,i}$ is no longer proportional to $\vec{f}_L$, we'll refer to $\mathcal{B}\vec{v}_{n,i}$ as "Hall-like" and ask if it too is negligible compared to the Hall term.

First thing is to look at the coefficient, $\mathcal{B}$. Using Eq. (10.90) and (10.91) for

$\mathcal{Q}_{n,i}$ and $\mathcal{Q}_{n,e}$ and neglecting the exchange term $|\sigma_{n,e}|$, Eq. (10.110) becomes,

$$\mathcal{B} = \underbrace{\frac{1}{en_e}\frac{Zm_e}{m_i}|\sigma_{n,i}|}_{\mathcal{B}^x} + \underbrace{\frac{1}{en_e}\left(\frac{Zm_e}{m_i}\gamma_{n,i}\rho_n\rho_i + \gamma_{n,e}\rho_n\rho_e\right)}_{\mathcal{B}^a}.$$

For illustration, let's set aside the exchange term, $\mathcal{B}^x$, and consider only the ambipolar term, $\mathcal{B}^a$. Since $Zm_e\rho_i/m_i = \rho_e$, we have,

$$\mathcal{B}^a = \frac{\rho_e\rho_n}{en_e}(\gamma_{n,i} + \gamma_{n,e}) = \frac{m_e m_n n_n}{e}\left(\frac{\langle\sigma u\rangle_{n,i}}{m_n + m_i} + \frac{\langle\sigma u\rangle_{n,e}}{m_n + m_e}\right)$$

$$\Rightarrow \quad \mathcal{B}^a \approx \frac{m_e n_n}{e}\left(\tfrac{1}{2}\langle\sigma u\rangle_{n,i} + \langle\sigma u\rangle_{n,e}\right), \tag{10.112}$$

using Eq. (10.5) for the second equality, and setting $m_n \approx m_i$ and $m_n \gg m_e$ for the approximation.

Now, from Eq. (10.6), $\langle\sigma u\rangle_{n,e} > \langle\sigma u\rangle_{n,i}$ and even more so $\tfrac{1}{2}\langle\sigma u\rangle_{n,i}$. Thus, let's compare the second term in Eq. (10.112) with the actual Hall term, $\vec{E}_H = \vec{v}_{i,e} \times \vec{B} = \vec{f}_L/en_e$ by defining the ratio,

$$R_{\mathcal{B}H} = \frac{\mathcal{B}^p_{n,e}v_{n,i}}{|\vec{v}_{i,e} \times \vec{B}|} = \frac{m_e n_n v_{n,i}\chi_{n,e}}{ev_{i,e}B\sin\psi}\sqrt{\frac{T}{T_0}},$$

where ψ is the angle between $\vec{v}_{i,e}$ (i.e., $\vec{J}$) and $\vec{B}$, and where we'll assign $\sin\psi = 2/\pi$ as the average value for $\psi \in [0,\pi]$. As a reminder, $T_0 = 10^3$ K is the fiducial temperature chosen in §10.2.2 for a planetary disc. Then, following our *M.O.* in §10.2.2, we have,

$$R_{\mathcal{B}H} = \frac{\pi m_e \chi_{n,e}}{2e\sqrt{k_B T_0 \gamma \mu_0 \alpha_0}} \overbrace{\frac{v_{n,i}}{v_{i,e}}}^{\sim 1} \sqrt{n_n}\underbrace{\frac{\sqrt{\rho_n \mu_0}}{B}}_{1/a}\underbrace{\sqrt{\frac{\gamma k_B T}{m_n}}}_{c_s}\sqrt{\alpha_0} = \sqrt{\frac{n_n}{n_{\mathcal{B}H}}\frac{\alpha_0}{\alpha}},$$

where I've taken the leap that the relative velocities $v_{n,i}$ and $v_{i,e}$ are, on average, comparable. As usual, $\alpha = a^2/c_s^2$ is the MHD-alpha, $\alpha_0 = 0.01$ is the fiducial value chosen in §10.2.2 for a planetary disc, $\gamma = \tfrac{5}{3}$ for a monatomic gas, and $n_{\mathcal{B}H} \sim 5.28 \times 10^{21}$ m^{-3} is the critical number density evaluated from all the constants. Note that $R_{\mathcal{B}H}$ is almost identical to the relative importance of resistive losses to the Hall effect, R_{RH}, found in Eq. (10.27). I leave it as an exercise to compare the first term in Eq. (10.112) with the Hall term where one should find a similar expression to R_{HA}, the relative importance of the one-fluid Hall and ambipolar diffusion effects (Eq. 10.28). It is therefore not at all apparent that this new term, $\mathcal{B}\vec{v}_{n,i}$, can be neglected, at least not compared to the resistive and Hall terms.

Returning our focus to Eq. (10.109), we use Faraday's law (App. B) to derive the two-fluid induction equation,

$$-\nabla \times \vec{E} = \boxed{\partial_t \vec{B} = \nabla \times \left(\vec{v}_i \times \vec{B} - \eta_2 \vec{J} - \frac{1}{en_e}\vec{f}_L - \mathcal{B}\vec{v}_{n,i}\right),} \tag{10.113}$$

with all "significant" non-ideal terms included. I place *significant* in quotes because in the two-fluid regime ($> 10^5$ K say) where ionisation of H and He are the dominant sources of ions and the electron number density is high, resistive losses and the Hall effect are quite negligible compared to ambipolar diffusion (Fig. 10.3), and certainly negligible compared to the induction term. For that reason, I carry forward from Eq. (10.113) only the induction term into *Equation Set 11*, our final set of equations for an adiabatic, two-fluid, MHD system:

Equation Set 11:

$$\partial_t \rho_n + \nabla \cdot (\rho_n \vec{v}_n) = \sigma_{i,n}; \qquad \text{Eq. (10.71)}$$

$$\partial_t \rho_i + \nabla \cdot (\rho_i \vec{v}_i) = \sigma_{n,i}; \qquad \text{Eq. (10.74)}$$

$$\partial_t \vec{s}_n + \nabla \cdot (\vec{s}_n \vec{v}_n) = -\nabla p_n - \rho_n \nabla \phi + \vec{f}_{i,n}; \qquad \text{Eq. (10.72)}$$

$$\partial_t \vec{s}_i + \nabla \cdot (\vec{s}_i \vec{v}_i) = -\nabla p_i - \rho_i \nabla \phi + \vec{f}_L + \vec{f}_{n,i}; \qquad \text{Eq. (10.106)}$$

$$\partial_t e_i + \nabla \cdot (e_i \vec{v}_i) = -p_i \nabla \cdot \vec{v}_i + \varpi_{n,i}; \qquad \text{Eq. (10.76)}$$

$$\partial_t \vec{B} = \nabla \times (\vec{v}_i \times \vec{B}), \qquad \text{Eq. (10.113)}$$

where $p_i = (\gamma - 1) e_i$ and $p_n = \tilde{m} \rho_n p_i / \rho_i$ ($\tilde{m} = 1$ for $m_i = m_n$) preserve thermodynamic equilibrium, $\sigma_{i,n}$ ($\sigma_{n,i}$) is the recombination (ionisation) rate (mass exchange rates) given by Eq. (10.83) or (10.85), $\vec{f}_{i,n} = -\vec{f}_{n,i}$ is the combined exchange-ambipolar force density exerted by the ions on the neutrals (neutrals on the ions) given by Eq. (10.90) and (10.91), $\vec{f}_L = \vec{J} \times \vec{B}$ is the Lorentz force density, and where $\varpi_{n,i}$ is the combined exchange-ambipolar power density given by Eq. (10.104). For the two-fluid *isothermal* AD model as used by many in the literature (*e.g.*, Li, McKee, & Klein, 2006; Chen & Ostriker, 2012; Burkhart *et al.*, 2015; Brandenburg, 2019), one needs only set $\sigma_{i,n} = 0$, $\vec{f}_{i,n} = \vec{f}^{\,a}_{i,n}$, and drop Eq. (10.76) in Eq. Set 11.

A few comments before we close shop. First, it is evident that the effect of ambipolar diffusion, now contained within the exchange-ambipolar coefficients, $\mathcal{Q}$, is in the momentum equations and not the induction equation as in the one-fluid model, and one can see almost by inspection the "diffusive nature" of AD. Evidently, only the ions are directly influenced by the Lorentz force density, $\vec{f}_L$. However, the ions and neutrals push on each other via equal and oppositely directed forces, $\vec{f}_{i,n} = \mathcal{Q}_{i,n} \vec{v}_{i,n}$ and $\vec{f}_{n,i} = \mathcal{Q}_{n,i} \vec{v}_{n,i}$, and it is by this mechanism that the neutrals are somewhat and indirectly coupled to the magnetic induction. It's the "somewhat" part that allows material in an imperfectly ionised medium to "slip" or diffuse through lines of induction, whence the 'D' in AD.

Second, in the two-fluid model, it is $\vec{v}_i$ and not $\vec{v}_n$ that appears in the induction term ($\vec{v}_i \times \vec{B}$) and thus induces magnetic induction from the flow of charge. In the one-fluid model, we found somewhat paradoxically that $\vec{v}_n$ played that role; indeed, the Lorentz force appeared directly in the neutral momentum equation, despite the fact the neutrals have no charge. I invite the reader to review the discussion after Eq. (10.16) if the significance of this point has been forgotten or overlooked.

And third, as the temperature increases to the point where ionisation is virtually complete, $\rho_n \to 0$, Eq. (10.71) and (10.72) drop out, $\sigma_{n,i}$, $\mathcal{Q}_{n,i}$, and $\varpi_{n,i}$ all go to zero, leaving us with a one-fluid, non-isothermal MHD model with no non-ideal or exchange terms remaining. Once again, we find ourselves in the realm of ideal MHD, bringing us full circle to our opening discussion in Chap. 4.

And to my dear, late father I give the last word. At such a juncture where a suitable "wrap-it-up phrase" was called for, Dad could be relied upon to declare: "And Bob's your uncle!"

$\sim$

Problem Set 10

10.1** In this problem, I'll take you through the steps to find the rate coefficient between ions and electrons, $\langle \sigma u \rangle_{i,e}$, where you will show that,

$$\langle \sigma u \rangle_{i,e} = \frac{4}{3}\sqrt{\frac{2\pi}{m_e}} \left(\frac{Ze^2}{4\pi\epsilon_0}\right)^2 \frac{\ln \Lambda}{(k_B T)^{3/2}}, \qquad (10.114)$$

and where the only quantity new to the reader should be $\ln \Lambda$, the so-called *Coulomb logarithm*. This problem falls solidly within the realm of plasma physics and so I've broken it up into parts with explanations that should be digestible to readers of this text. As a result, the problem presents long but it's shorter than it looks!

Assuming thermodynamic equilibrium, we seek the force on an electron passing through a "sea" of positive ions $(+Ze)$ partially shielded by other electrons (Fig. 10.12). In deflecting the electron, this force – the cumulative effect of numerous Coulomb interactions – decreases its forward velocity component and thus can be thought of as a "drag force" exerted by the ions on the electrons. Evaluating this will lead to an expression for $\langle \sigma u \rangle_{i,e}$ and, as in any complex physics problem, the key to solution is breaking it up into bite-sized pieces. So here we go...

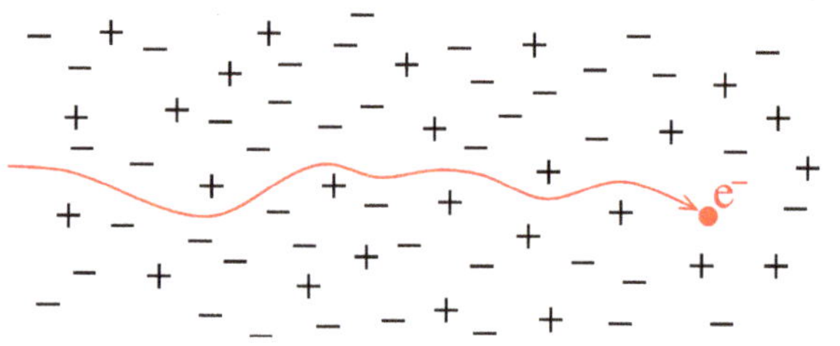

Figure 10.12. (Problem 10.1) An electron (red e^-) makes its way through a sea of positive and negative charges.

a) Consider a single electron with mass m_e and initial velocity $\vec{v}_0 = v_0 \hat{z}$ approaching a single ion with impact parameter b (inset). The Coulomb force of attraction causes the electron to follow a hyperbolic trajectory resulting in an angular deflection, δ, from its original path.

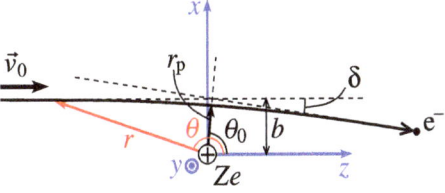

Your second-year mechanics course probably included a unit on central forces where you would have found that the trajectory of such a particle coming in from $z = -\infty$ and $x = b$ is given by,[27]

$$r(\theta) = \frac{a}{1 + \epsilon \cos(\theta - \theta_0)}, \qquad (10.115)$$

where, for a Coulomb interaction,

$$a \equiv \frac{4\pi\epsilon_0}{Ze^2} m_e b^2 v_0^2, \qquad (10.116)$$

is the *latus rectum*, and $\epsilon = \sqrt{1 + \frac{a^2}{b^2}} > 1$ is the *eccentricity* of the hyperbolic orbit.

Assuming small-angle deflections ($\delta \ll \pi/2$), show that: $1 - \cos\delta \approx \frac{2b^2}{a^2}$.

Hint: Given expressions for $r(\theta)$, a, and ϵ, this problem is more geometry than physics. Note that $r \to \pm\infty$ when the denominator of Eq. (10.115) is zero. Setting $r = -\infty$ when $\theta = -\pi$ allows you to evaluate $\cos\theta_0$, where $\theta = \theta_0$ when the electron is at the *pericentre* (point of closest approach; r_p in the inset). Then, set $r = \infty$ at $\theta = -\delta$ to find $\cos\delta$.

b) In solving part a), you should have had to assert that $a \gg b$. Give a physical reason why this is necessary for small-angle deflections.

c) Using your results from part a), show that the change in the component of velocity in the direction of the initial velocity is,

$$\delta v_z = -\frac{2Z^2 e^4}{(4\pi\epsilon_0)^2 m_e^2 b^2 v_0^3}. \qquad (10.117)$$

d) Equation (10.117) is the effect on an electron by a single ion encounter. To find the cumulative effect of all ions the electron passes, consider the problem in the electron's frame of reference as illustrated in the inset on the next page. From its frame, the electron sees a stream of ions passing it with velocity $-v_z \hat{z}$

---

[27] The expressions for $r(\theta)$, a, and ϵ are straight-forward applications of Newton's second law with a Coulomb central force. I'm deliberately avoiding the distraction of taking you through this part of the problem, and encourage any reader for whom this may be unfamiliar to consult just about any sophomore mechanics text (*e.g.*, Fowles & Cassiday or Marion & Thornton, any edition) with a unit on central forces.

and, in a time Δt, the number of ions streaming by within a radial distance $b_{\max}$ is $n_i \pi b_{\max}^2 v_z \Delta t$, where n_i is the ion-number density, $v_z = v_0 \cos\vartheta$,[28] and $b_{\max}$ is the furthest an ion can be from the electron before other electrons in the plasma effectively shield it from the ion – the so-called "Debye-length" (pronounced "d'by") – given by,

$$b_{\max} = \lambda_D = \sqrt{\frac{\epsilon_0 k_B T}{n_e e^2}}. \qquad (10.118)$$

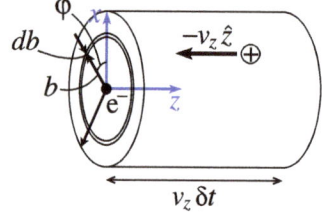

Using the inset as a guide, set up the appropriate integral to show that the effective force exerted on the electron by ions streaming in the z-direction is,

$$F_z(v_0, \vartheta) = -\frac{Z^2 e^4 \cos\vartheta}{4\pi\epsilon_0^2 m_e v_0^2} n_i \ln\left(\frac{b_{\max}}{b_{\min}}\right), \qquad (10.119)$$

where $b_{\min}$ is the minimum impact parameter which you'll evaluate in part e). Since we're considering only small scattering angles, the impact parameter must be greater than a certain $b_{\min}$ lest the electron be scattered out of the cylinder in the inset. Such impacts are rare and neglecting them won't affect the results by much.

Caution: Don't overthink this part; it's actually a two-liner!!

e) The quantity,

$$\ln\Lambda \equiv \ln\left(\frac{b_{\max}}{b_{\min}}\right) = \ln\left(\frac{\lambda_D}{b_{\min}}\right),$$

is known as the *Coulomb logarithm*, and its value depends upon how we choose $b_{\min}$. Fortunately, $b_{\min}$ appears in a logarithm, and how it's set affects things rather weakly.

Small deflections mean $\delta \ll \pi/2$ and thus a hard upper limit is $\delta_{\max} = \pi/2$. For the purists, we could set a smaller $\delta_{\max}$ but again, $b_{\min}$ appears in a logarithm and its actual value matters little. As you'll see, $\delta_{\max} = \pi/2$ happens to be a convenient value.

Show that $\delta_{\max} = \pi/2$ corresponds to,[29]

$$b_{\min} = \frac{Ze^2}{4\pi\epsilon_0 m_e v_0^2}.$$

Then, setting v_0 to the electron thermal velocity,[30] $\frac{1}{2}m_e v_0^2 = \frac{3}{2}k_B T$, show that the Coulomb logarithm is given by,

---

[28] After numerous small deflections, δ, the net deflection angle, ϑ, can become significant and the z-component of velocity becomes $v_z = v_0 \cos\vartheta$.

[29] This is the so-called *classical limit* valid for $T <$ few $\times 10^5$ K. For higher temperatures, $b_{\min}$ is set using quantum mechanical criteria; see, *e.g.*, Chap. 2 in Callen (2006).

[30] This is a result from statistical mechanics for systems in thermal equilibrium that follow a classical Maxwell–Boltzmann distribution.

$$\ln \Lambda = \ln\left(\frac{9N_\mathrm{D}}{Z}\right),$$

where N_D is the number of electrons within a *Debye sphere* (sphere of radius λ_D).

f) For $n_\mathrm{e} = 2 \times 10^9\,\mathrm{m}^{-3}$, $T = 10^3\,\mathrm{K}$, and $Z = 1$,[31] evaluate $\ln \Lambda$. For most (astro)physical plasmas, $12 \lesssim \ln \Lambda \lesssim 17$; your result should land within this range.

g) Equation (10.119) is the net force on a *single* electron with a speed v_0 passing through partially shielded ions. We want the *average* force felt by an electron of *any* speed.

For an ensemble of N_0 particles with differing velocities, the *fractional distribution function*, $f(\vec{v})$, is defined such that the fraction of particles with velocities between $\vec{v}$ and $\vec{v} + d\vec{v}$ is,

$$\frac{dN}{N_0} = f(\vec{v})\,d^3\vec{v}.$$

Evidently,

$$\int f(\vec{v})\,d^3\vec{v} = \frac{1}{N_0}\int dN = 1,$$

where integrating to 1 is the hallmark of a *fractional* distribution function. Note that $d^3\vec{v}$ is a 3-D differential, usually written in spherical polar coordinates in velocity-space: $d^3\vec{v} = v^2 \sin\vartheta\,dv\,d\vartheta\,d\varphi$. Then, for a given particle property, $q(\vec{v})$ (*e.g.*, energy, speed, whatever), the ensemble average of q weighted by the distribution function $f(\vec{v})$ is:

$$\langle q \rangle = \int q(\vec{v}) f(\vec{v})\,d^3\vec{v}.$$

Undoubtedly, the most famous fractional distribution function in physics is the Maxwell–Boltzmann distribution first worked out by Maxwell in 1860 based largely on heuristic arguments, then by Boltzmann in 1877 based on his new theory of statistical thermodynamics. It describes an ensemble of particles in *thermodynamic equilibrium* (*i.e.*, one which can be described by a single temperature, T), and is given by,

$$f_\mathrm{MB}(\vec{v}) = \left(\frac{m}{2\pi k_\mathrm{B} T}\right)^{3/2} e^{-m(\vec{v}-\vec{u})^2/(2k_\mathrm{B}T)}, \qquad (10.120)$$

where m is the mass of a particle, $\vec{u}$ is the centre of the velocity distribution, and T is the temperature describing the ensemble.

Now in the frame of reference of the ions, $\vec{u} = v_{\mathrm{e,i}}\hat{z}$ is the "drift velocity" of the electrons which is small compared to the "thermal velocity", $\vec{v}$ ($v_{\mathrm{e,i}} \ll v$). Show that the ensemble average of the effective force acting on a single electron

---

[31] Consistent with the ballpark figures for an inner planetary disc given on page 388 of the text.

(Eq. 10.119) weighted by the Maxwell–Boltzmann distribution (Eq. 10.120) is given by,

$$\langle F_z \rangle = -\frac{Z^2 e^4}{6\pi \epsilon_0^2 k_B T} \sqrt{\frac{m}{2\pi k_B T}} \ln \Lambda \, n_i v_{e,i}. \qquad (10.121)$$

Hint: The exponential in Eq. (10.120) can be approximated as,

$$e^{-m(\vec{v}-\vec{u})^2/(2k_B T)} \approx e^{-mv^2/2k_B T} \left(1 + \frac{mv_z v_{e,i}}{k_B T}\right),$$

where $v_z = v\cos\vartheta$, and ϑ is the accumulated deflection of the electron from the z-axis.

h) And now for the finish. Equation (10.121) is the average force exerted by the ions *per electron*. To get the force *per unit volume* (force density) exerted by the ions collectively on the electrons, we must multiply $\langle F_z \rangle$ by the electron number density,

$$f_{i,e} = n_e \langle F_z \rangle. \qquad (10.122)$$

With this final piece of insight and the definitions of ambipolar force density and the coupling coefficient (Eq. 10.4 and 10.5 in the text), show that the rate coefficient, $\langle \sigma u \rangle_{i,e}$, is given by Eq. (10.114) at the beginning of the problem.

For $Z = 1$ and using your value for $\ln \Lambda$ from part f, confirm the numerical value for $\chi_{i,e}$ given just after Eq. (10.7) of the text.[32]

To those who get through all of this, congratulations! Who knew? With a little classical mechanics, statistical mechanics, electrodynamics, fluid mechanics, and even a little quantum mechanics thrown in for good measure, you too can be a plasma physicist!

10.2* Assuming a constant resistivity, η, derive Eq. (10.37) in the text, namely,

$$\partial_t e_T^* + \nabla \cdot \left[(e_T + p)\vec{v} + \frac{1}{\mu_0}\vec{B} \times (\vec{v} \times \vec{B}) + \frac{\eta}{\mu_0}\vec{J} \times \vec{B}\right] = 0,$$

the total energy equation for a resistive fluid.

To get a head-start on this, you might consult §4.6 where the total energy equation for ideal MHD, Eq. (4.19), is derived, though you should fill in all the details left out by the text (*e.g.*, flesh out what is meant by "using numerous vector identities from App. A)".

---

[32] A note on temperature units. Most plasma physicists and many astronomers express T in eV, an *energy* unit. As an mks purist, I eschew this practice, though it has the advantage of not having to carry the Boltzmann constant. If you need to compare a numerical result derived in kelvins (*e.g.*, parts f, h) with a corresponding result derived in eV (*e.g.*, Callen, 2006), the rule of thumb is simple: replace $T(K)$ with $T(eV) \times 1.1604 \times 10^4$; or replace $T(eV)$ with $T(K) \times 8.6175 \times 10^{-5}$, the constants being either e/k_B or k_B/e.

10.3

a) Show that in the absence of the resistive and Hall terms, the non-ideal induction equation for the one-fluid model, Eq. (10.22), can be written as,

$$\partial_t \vec{B} = \nabla \times (\vec{v}_n \times \vec{B} - \eta_{AD} \vec{J}_\perp), \qquad (10.123)$$

where,

$$\eta_{AD} = \frac{B^2}{\gamma_{i,n} \rho_i \rho_n},$$

is the *ambipolar resistivity* and where $\vec{J}_\perp$ is the vector component of the current density, $\vec{J}$, perpendicular to $\vec{B}$.

b) Compare the resistive induction equation, Eq. (10.34) in the text, with Eq. (10.123) above. If the resistivity, η, in the former works towards dissipating the current density, $\vec{J}$,[33] what do you suppose the ambipolar resistivity, η_{AD}, does in the latter?

One way to interpret the action of η on $\vec{J}$ in Eq. (10.34) is by eliminating $\vec{J}$ outright, resistive dissipation renders the magnetic induction "force-free" ($\vec{f}_L = \vec{J} \times \vec{B} \to 0$). What is the analogous statement for the ambipolar resistivity, η_{AD}? Does all of $\vec{J}$ have to be dissipated away in order to make $\vec{B}$ force free?

10.4 Section 8.5 of the text started off by *scaling* Euler's equation (Eq. 8.36) and showing it to be *scale-free*.[34] Conversely, once the viscid term, $\nu \nabla^2 \vec{v}$, was included, the resulting incompressible Navier–Stokes equation was found *not* to be scale-free, and it was here that the *Reynolds number*, $\mathcal{R} = VL/\nu$, emerged (Eq. 8.40 which uses R instead of L).

We're going to do the same thing here for the *magnetic Reynold's number*, $\mathcal{R}_M$.

a) Define the "scaling laws":

$$r = Lr'; \qquad \vec{v} = V\vec{v}'; \qquad \vec{B} = B_0 \vec{B}', \qquad (10.124)$$

where quantities on the LHS are the original variables, the primed quantities on the RHS are unitless (scale-free) versions, and L, V, and B_0 are constant scaling factors that carry all the units (*e.g.*, m, m s^{-1}, and T respectively) sometimes referred to as *fiducial values*.

Using Eq. (10.124), show that the ideal induction equation,

$$\partial_t \vec{B} = \nabla \times (\vec{v} \times \vec{B}), \qquad \text{Eq. (4.4)}$$

---

[33] *Rappel*: resistive dissipation works to smooth out $\vec{B}$ and make it uniform everywhere which, among other things, means $\nabla \times \vec{B} = \mu_0 \vec{J} \to 0$.

[34] The reader who has not gone through Chapter 8 yet might read the first two pages of §8.5 (which does not depend unduly on knowledge of viscosity or the Navier–Stokes equation) to acquaint themselves with what it means to "scale an equation" and for an equation to be or not to be (always wanted to use that quotation in a real setting!) "scale-free".

is "scale-free". That is, after substituting Eq. (10.124) into Eq. (4.4), all scaling factors should cancel leaving an equation in primed quantities identical to the original ideal induction equation.

Hint: Think carefully about how you scale the operators, ∂_t and ∇ (*e.g.*, see Eq. 8.38).

b) Next, by scaling the resistive induction equation,

$$\partial_t \vec{B} = \nabla \times (\vec{v} \times \vec{B} - \eta \vec{J}), \qquad \text{Eq. (10.34)}$$

show that you get,

$$\partial_{t'} \vec{B}' = \nabla' \times \left(\vec{v}' \times \vec{B}' - \frac{1}{\mathcal{R}_M} \vec{J}' \right), \qquad (10.125)$$

where,

$$\mathcal{R}_M = \frac{\mu_0 V L}{\eta},$$

is the *magnetic Reynolds number* (Eq. 10.24 in the text). Note that Eq. (10.125) is *not* scale-free since the constant $\mathcal{R}_M$ depends upon scaling factors chosen *a priori*.

Think really carefully about how you scale the current density, $\vec{J} = (\nabla \times \vec{B})/\mu_0 = J_0 \vec{J}'$. In particular, should $\vec{J}'$ be set to $(\nabla' \times \vec{B}')/\mu_0$, or just $\nabla' \times \vec{B}'$? Give a reason for your choice, other than 'one gets the right answer and the other doesn't'!

10.5 Complete the proof of Theorem 10.1 in the text by showing that for an axisymmetric system and no injection of B_φ, B_φ diffuses away in time according to,

$$\frac{d}{dt}\left(\frac{B_\varphi}{r}\right) = \vec{B} \cdot \nabla \left(\frac{v_\varphi}{r}\right) + \frac{\mathcal{D}_M}{r^2} \widetilde{\nabla}^2 (r B_\varphi),$$

where the poloidal components of the magnetic induction, B_z and B_r, diffuse away because of Eq. (10.52).

Thus, no velocity profile can sustain a dynamo in a resistive, axisymmetric MHD system.

Hint: Take note that this problem needs to be done in *cylindrical* coordinates.

10.6* In this problem, we'll use our naïve model based on Taylor columns for the earth's dynamo (Fig. 10.8*b*) to make an even more naïve estimate of the expected strength of the magnetic induction on the earth's surface at the equator.

a) In Eq. (10.53) of the text, assume steady state ($D/Dt = 0$), hydrostatic equilibrium within the earth's core, and ignore viscosity. This leaves just the Coriolis and Lorentz accelerations to balance out. Use this plus whatever else you can find to eliminate J to derive an expression for B_{oc} (the magnetic induction strength somewhere in the middle of the outer core) in terms of ρ_{oc} (density of the molten Fe and Ni mixture), η_{oc} (resistivity of this mixture), and $\omega_\oplus$ (earth's rotation rate). Note that by "even more naïve", I'm inviting you to replace nasty things like $\vec{J} \times \vec{B}$ with JB, etc.

b) Just for fun, find B_{oc} in terms of ρ_{oc}, η_{oc}, and $\omega_\oplus$ again, but this time using dimensional analysis. (*Rappel*: forget what this is? See §2.4, page 51 of the text.)

c) Using $\rho_{oc} \sim 10^4 \, \text{kg m}^{-3}$, $\eta_{oc} \sim 10^{-7} \, \Omega\,\text{m}$ (highly conductive!), and $\omega_\oplus \sim 7.3 \times 10^{-5} \, \text{rad s}^{-1}$, what is your estimate for B_{oc}?

d) Given that a dipole magnetic induction falls of as r^{-3} in a direction perpendicular to the dipole axis, what is your estimate of B on the earth surface at the equator? (Measured values vary, but their average is $\sim 3.0 \times 10^{-5}$ T.)

10.7 Consider the current sheet in Fig. 10.6 in the text. Suppose across this sheet, the magnetic induction is described by,

$$\vec{B} = B_x \hat{i} = B_1 \hat{i} \begin{cases} 1, & y \geq l; \\ \dfrac{y}{l}, & -l < y < l; \\ -1, & y \leq -l, \end{cases} \quad (10.126)$$

where $l > 0$ is half the thickness of the current sheet. Thus, as shown in the inset, B_x has equal but oppositely directed constant values outside the current sheet, and varies linearly within.

a) Find the current density, $\vec{J}$, across the current sheet.

b) Find the Lorentz force density, $\vec{f}_L = \vec{J} \times \vec{B}$, across the current sheet. Does the Lorentz force density tend to drive layers of opposing magnetic induction apart or pinch them together?

c) Let the pressure outside the current sheet ($|y| \geq l$) be p_0, a constant. Assuming a stationary state ($\partial_t = 0$, $\vec{v} = 0$), what is the pressure profile across the current sheet?

d) Using a computer plotting package or doing it by hand, generate a stacked plot of profiles for $p(y)$ (top), $f_y(y)$ (middle), and $J_z(y)$ (bottom). How would you characterise the relationship of each plot with the one immediately above or below it?

10.8 The LHS of Eq. (10.63) in the text was rather arbitrarily set to zero in order to arrive at Eq. (10.64) for the electric field, $\vec{E}$, for a completely ionised fluid.

Let us suppose we can selectively replace $\vec{v}_p$ and $\vec{v}_e$ with the average fluid velocity,

$$\vec{v} = \frac{n_p \vec{v}_p + n_e \vec{v}_e}{n_p + n_e},$$

such that,

$$(\vec{v}_p \cdot \nabla)\vec{v}_p - (\vec{v}_e \cdot \nabla)\vec{v}_e = (\vec{v} \cdot \nabla)(v_p - v_e). \tag{10.127}$$

(We can't, not really, but let's just suppose we can). Show that under the assumption of charge conservation in the form of,

$$\partial_t \vec{J} + \nabla \cdot (\vec{J}\vec{v}) = 0, \tag{10.128}$$

the LHS of Eq. (10.63) is identically zero.

10.9* Find an expression for the exchange term, $\sigma_{n,i}$, using Eq. (10.83) from the text carrying out all the differentiations. Don't expect anything too tidy; it's not going to be pretty! You should aim for something like:

$$\sigma_{n,i} = \frac{m_H}{3}\left(\frac{2\pi m_e}{h^2}\right)^{3/2}\left(\frac{\nu_H + 2n}{\sqrt{\nu_H^2 + 4n\nu_H}} - 1\right)e^{-\varepsilon_H/k_B T}\frac{\sqrt{k_B T}}{n}\left(\frac{\varepsilon_H}{k_B T} + \frac{3}{2}\right)$$
$$\times \left(2\gamma_{n,i}\rho_n\rho_i v_{n,i}^2 - p_i \nabla \cdot \vec{v}_i - p_n \nabla \cdot \vec{v}_n\right),$$

where the subscripts 'n' and 'i' refer to the neutral (H) and ionised (H$^+$) components of the pure hydrogen gas respectively, and where ε_H is the ionisation energy of H (Table 10.1).

Hint: The derivatives of n_{H^+} and ν_H are straight-forward enough, but you may appreciate a nudge on what to do with the $k_B T$ derivative. Start by considering,

$$\partial_t(k_B T) = m_H(\gamma - 1)\partial_t\left(\frac{e_i}{\rho_i}\right) = \cdots,$$

where $\gamma = \frac{5}{3}$ for a monatomic gas, and use Eq. (10.74) and (10.76) from the text. Then, take,

$$\frac{d(k_B T)}{dt} = \partial_t(k_B T) + \vec{v} \cdot \nabla(k_B T),$$

where,

$$\vec{v} = \frac{n_n \vec{v}_n + n_i \vec{v}_i}{n_n + n_i},$$

is the bulk (average) velocity of the fluid and n_n, n_i are number densities. It was quite amazing to see how all the terms managed to combine and cancel to get the temperature derivative!

10.10* Consider a more "realistic" astrophysical fluid of total number density, n, and fractional abundances, f and $1 - f$, of H and He respectively (*e.g.*, $f \sim 0.75$). A partially ionised mix of H and He will have six different number densities to account for: n_H, n_{H^+}, n_{He}, n_{He^+}, $n_{He^{++}}$, and n_e, the number density of electrons. In thermodynamic equilibrium, these are constrained by the Saha equation:

$$n_{H^+} n_e = \nu_H n_H; \qquad (10.129)$$

$$n_{He^+} n_e = \nu_{He} n_{He}; \qquad (10.130)$$

$$n_{He^{++}} n_e = \nu_{He^+} n_{He^+}, \qquad (10.131)$$

where ν_s is the critical number density for species 's', as defined in Eq. (10.80) of the text. These deceptively simple-looking equations are coupled non-linearly by the number density of electrons, now given by,

$$n_e = n_{H^+} + n_{He^+} + 2n_{He^{++}}, \qquad (10.132)$$

along with two conservation constraints:

$$n_H + n_{H^+} = fn; \quad \text{and} \quad n_{He} + n_{He^+} + n_{He^{++}} = (1-f)n. \qquad (10.133)$$

a) Find an expression for n_e^2 in terms of n_H, n_{He}, and n_{He^+} along with their corresponding critical densities.

b) Find independent expressions for n_H, n_{He}, and n_{He^+} in terms of n and n_e (along with f and any critical densities that may come for the ride).

c) Combine your results from parts a) and b) to show that n_e is given by the quartic:

$$\begin{aligned} n_e^4 &+ (\nu_H + \nu_{He})n_e^3 + \left[\nu_{He}(\nu_H + \nu_{He^+}) - n(f\nu_H + (1-f)\nu_{He})\right]n_e^2 \\ &+ \nu_{He}\left[\nu_H(\nu_{He^+} - n) - 2\nu_{He^+}n(1-f)\right]n_e \\ &- \nu_H \nu_{He} \nu_{He^+} n(2-f) = 0, \end{aligned} \qquad (10.134)$$

where all coefficients are made up of known quantities.

d) Given a physical solution to Eq. (10.134), outline how you would go about finding the exchange term, $\sigma_{n,i}$, using Eq. (10.84) in the text.

e) Show that in the limit as $f \to 1$ (pure H), one recovers Eq. (10.82) in the text from Eq. (10.134).

10.11* Show that in the limit where $\rho_i \ll \rho_n$, the two-fluid adiabatic equations in Eq. Set 11 reduce to the rather more standard "weakly ionised one-fluid adiabatic AD-MHD equations":

$$\partial_t \rho + \nabla \cdot (\rho \vec{v}) = 0;$$

$$\partial_t \vec{s} + \nabla \cdot (\vec{s}\vec{v}) = -\nabla p - \rho \nabla \phi + \vec{f}_L;$$

$$\partial_t e + \nabla \cdot (e\vec{v}) = -p\nabla \cdot \vec{v} + 2\beta_{i,n} f_L^2;$$

$$\partial_t \vec{B} = \nabla \times \left[(\vec{v} + \beta_{i,n}\vec{f}_L) \times \vec{B}\right],$$

where:

$$\rho = \rho_n + \rho_i \approx \rho_n; \quad p = p_n + p_i \approx p_n; \quad \vec{v} = \frac{\rho_n \vec{v}_n + \rho_i \vec{v}_i}{\rho} \approx \vec{v}_n, \quad (10.135)$$

and where the ionisation number density, ρ_i (needed for $\beta_{i,n}$), can be given by, for example, Eq. (10.23) in the text. Try to avoid using the approximations in Eq. (10.135) as much as you can. In my solution, I had to make an approximation four times; everywhere else my development was exact.

A little commentary on the internal energy and induction equations. On the former, the factor of two in the last term may be a bit curious, and is addressed in Problem 10.13. On the latter, the term proportional to $\beta_{i,n}$ is the one that blows up to the impossible-looking quadruple cross product in Eq. (10.70). Here, I've arranged it in the fashion I have to reveal, once again, the "diffusive nature" of AD. As written, the magnetic induction is not tied to the flow velocity, $\vec{v}$, but to the flow velocity adjusted by a "slip velocity", $\beta_{i,n}\vec{f}_L$, which is perpendicular to both the current density, $\vec{J}$, and the magnetic induction, $\vec{B}$. It is this "slip" that allows $\vec{B}$ to diffuse away from matter, making AD-MHD qualitatively different from ideal MHD where the magnetic induction is truly "flux-frozen" into the fluid.

10.12**

a) Show that the adiabatic two-fluid *total* energy equations are:

$$\partial_t e_{T_n} + \nabla \cdot \left[(e_{T_n} + p_n)\vec{v}_n\right] = (e_{T_n} - e_n)\frac{\sigma_{i,n}}{\rho_n} + \vec{v}_n \cdot \vec{f}_{i,n} + \varpi_{i,n}; \quad (10.136)$$

$$\partial_t e_{T_i}^* + \nabla \cdot \left[(e_{T_i} + p_i)\vec{v}_i + \vec{S}_P\right] = (e_{T_i} - e_i)\frac{\sigma_{n,i}}{\rho_i} + \vec{v}_i \cdot \vec{f}_{n,i} + \varpi_{n,i}. \quad (10.137)$$

where $\vec{S}_P$ is the MHD Poynting vector given by Eq. (4.15) in the text, where the total energy densities for the neutral and ionised components are:

$$e_{T_n} = e_n + \tfrac{1}{2}\rho_n v_n^2 + \rho_n \phi; \quad e_{T_i} = e_i + \tfrac{1}{2}\rho_i v_i^2 + \rho_i \phi; \quad e_{T_i}^* = e_{T_i} + \frac{B^2}{2\mu_0},$$

and where all other symbols have their meanings from §10.5.2.

Hint: Start with e_{T_i} and follow the derivation of the total energy equation in §4.6 (Eq. 4.19), this time carrying all the ambipolar and exchange terms from §10.5.2. The expression for e_{T_n} follows trivially from that.

b) The closest I've found to Eq. (10.136) and (10.137) in the literature are Eq. (8) and (9) in Duffin & Pudritz (2008) who do not include the exchange terms. Show that in the absence of exchange terms, the RHS of Eq. (10.136) and (10.137) reduce to $\vec{v}_i \cdot \vec{v}_{i,n}/\beta_{i,n}$ and $\vec{v}_n \cdot \vec{v}_{n,i}/\beta_{n,i}$ respectively, where $\beta_{1,2}$ is given by Eq. (10.8) in the text.[35]

---

[35] For anyone inclined to compare this result to D&P, note that an unfortunate typo swapped $\vec{v}_i$ and $\vec{v}_n$ in their Eq. (8) and (9).

c) Finally, show that in the limit where $\rho_i \ll \rho_n$, Eq. (10.136) and (10.137) reduce to the weakly ionised one-fluid adiabatic total energy equation, given by,[36]

$$\partial_t e_T^* + \nabla \cdot \left[(e_T^* + p^*) \vec{v} - \frac{1}{\mu_0}(\vec{v} \cdot \vec{B})\vec{B} + \frac{\beta_{i,n}}{\mu_0} B^2 \vec{f}_L \right] = \beta_{i,n} f_L^2, \qquad (10.138)$$

where:

$$e_T^* = e_{T_n} + e_{T_i}^* \approx e_{T_n} + \frac{B^2}{2\mu_0}; \quad p^* = p_n + p_i^* \approx p_n + \frac{B^2}{2\mu_0}; \quad \vec{v} = \frac{\rho_n \vec{v}_n + \rho_i \vec{v}_i}{\rho_n + \rho_i} \approx \vec{v}_n,$$

and where e_T^* and p^* are the MHD total energy and pressure (Eq. 4.18 and 4.26).

d) Physically, how would you interpret the RHS of Eq. (10.138)?

10.13 In determining the resistive power density, p_R, our approach in §10.3.2 was to take the dot product between the current density, $\vec{J}$, and the electric field, $\vec{E}_\eta$, driving the current density against the resistivity, η (Eq. 10.35 in the text). The one-fluid internal energy equation including resistive heating (Eq. 10.36) was then determined by adding the resistive power density to the RHS of the ideal internal energy equation, namely Eq. (1.34).

a) For a non-isothermal system, why is there no power density for the Hall effect to include in the one-fluid internal energy equation? You may assume ∇p_e is negligible.

b) In the same way that the resistive power density was found, find the power density for ambipolar diffusion, p_{AD}, and add this to the RHS of the ideal internal energy equation, namely Eq. (1.34). If you proceed as I have imagined you would, you should arrive at,

$$\partial_t e + \nabla \cdot (e\vec{v}) = -p\nabla \cdot \vec{v} + \beta_{i,n} f_L^2. \qquad \text{(wrong)}$$

c) Yet, in Problem 10.11, the one-fluid correction to the internal energy equation is $2\beta_{i,n} f_L^2$. Where does that factor of two come from?

---

[36] Again for those inclined to check the references, be aware that D&P define their total energy, E, without the gravitational energy density, and take the unusual step of carrying terms $\propto \nabla \cdot \vec{B}$. These account for the apparent differences between their Eq. (23) and my Eq. (10.138).

Appendices

A Essentials of Vector Calculus

A.1 Vector identities

Let $\vec{A}, \vec{B}, \vec{C}$, and $\vec{D}$ be four arbitrary vectors. Then:

$$\vec{A} \cdot (\vec{B} \times \vec{C}) = \vec{B} \cdot (\vec{C} \times \vec{A}) = \vec{C} \cdot (\vec{A} \times \vec{B}); \tag{A.1}$$

$$\vec{A} \times (\vec{B} \times \vec{C}) = (\vec{A} \cdot \vec{C})\vec{B} - (\vec{A} \cdot \vec{B})\vec{C}; \tag{A.2}$$

$$(\vec{A} \times \vec{B}) \cdot (\vec{C} \times \vec{D}) = (\vec{A} \cdot \vec{C})(\vec{B} \cdot \vec{D}) - (\vec{A} \cdot \vec{D})(\vec{B} \cdot \vec{C}). \tag{A.3}$$

Let f and g be two arbitrary scalar functions of the coordinates, and let $\vec{A}$ and $\vec{B}$ be two arbitrary vector functions of the coordinates. Then:

$$\nabla(fg) = f\nabla g + g\nabla f; \tag{A.4}$$

$$\nabla(f/g) = \frac{g\nabla f - f\nabla g}{g^2}; \tag{A.5}$$

$$\nabla(\vec{A} \cdot \vec{B}) = (\vec{B} \cdot \nabla)\vec{A} + (\vec{A} \cdot \nabla)\vec{B} + \vec{B} \times (\nabla \times \vec{A}) + \vec{A} \times (\nabla \times \vec{B}); \tag{A.6}$$

$$\nabla \cdot (f\vec{A}) = f\nabla \cdot \vec{A} + \vec{A} \cdot \nabla f; \tag{A.7}$$

$$\nabla \cdot (\vec{A} \times \vec{B}) = \vec{B} \cdot (\nabla \times \vec{A}) - \vec{A} \cdot (\nabla \times \vec{B}); \tag{A.8}$$

$$\nabla \times (f\vec{A}) = f\nabla \times \vec{A} + \nabla f \times \vec{A}; \tag{A.9}$$

$$\nabla \times (\vec{A} \times \vec{B})^1 = (\vec{B} \cdot \nabla)\vec{A} - (\vec{A} \cdot \nabla)\vec{B} - \vec{B}(\nabla \cdot \vec{A}) + \vec{A}(\nabla \cdot \vec{B}); \tag{A.10}$$

$$\nabla \times (\nabla f) = 0; \tag{A.11}$$

$$\nabla \cdot (\nabla \times \vec{A}) = 0. \tag{A.12}$$

For $\nabla \times (\nabla \times \vec{A})$, see identity (A.27).

Identities (A.13)–(A.15) below are particularly useful for working with the MHD equations and can be derived from the more fundamental identities above:

$$(\vec{A} \cdot \nabla)\vec{B} = \frac{1}{2}\Big[\nabla(\vec{A} \cdot \vec{B}) + \vec{A}(\nabla \cdot \vec{B}) - \vec{B}(\nabla \cdot \vec{A}) - \nabla \times (\vec{A} \times \vec{B}) \\ - \vec{A} \times (\nabla \times \vec{B}) - \vec{B} \times (\nabla \times \vec{A})\Big]; \tag{A.13}$$

---

[1] $\nabla \times (\vec{A} \times \vec{B})$ may also be expressed in terms of perfect divergences; see identity (A.22).

$$(\vec{A}\cdot\nabla)f\vec{B} = f(\vec{A}\cdot\nabla)\vec{B} + \vec{B}(\vec{A}\cdot\nabla f); \tag{A.14}$$

$$(\vec{A}\cdot\nabla)\vec{A} = \tfrac{1}{2}\nabla A^2 - \vec{A}\times(\nabla\times\vec{A}). \tag{A.15}$$

A.1.1 Identities involving dyadics

Constructs such as $\vec{A}\vec{B}$ (the *dyadic product* of the vectors $\vec{A}$ and $\vec{B}$) as well as $\nabla\vec{A}$ (the gradient of the vector $\vec{A}$) are examples of *dyadics* (rank 2 tensors; matrices) and appear frequently in the MHD equations. In Cartesian coordinates, these look like:

$$\vec{A}\vec{B} = |A\rangle\langle B| = \begin{bmatrix} A_x \\ A_y \\ A_z \end{bmatrix} \begin{bmatrix} B_x & B_y & B_z \end{bmatrix} = \begin{bmatrix} A_x B_x & A_x B_y & A_x B_z \\ A_y B_x & A_y B_y & A_y B_z \\ A_z B_x & A_z B_y & A_z B_z \end{bmatrix}; \tag{A.16}$$

$$\nabla\vec{A} = \begin{bmatrix} \partial_x A_x & \partial_x A_y & \partial_x A_z \\ \partial_y A_x & \partial_y A_y & \partial_y A_z \\ \partial_z A_x & \partial_z A_y & \partial_z A_z \end{bmatrix}. \tag{A.17}$$

The *colon product* (double contraction) of two dyadics $\mathsf{M} = \vec{A}\vec{B}$ and $\mathsf{N} = \vec{C}\vec{D}$ is defined as:

$$\begin{aligned} \mathsf{M}:\mathsf{N} &\equiv \sum_{ij} M_{ij} N_{ij} = \sum_{ij} A_i B_j C_i D_j \\ &= \sum_i A_i C_i \sum_j B_j D_j = (\vec{A}\cdot\vec{C})(\vec{B}\cdot\vec{D}). \end{aligned} \tag{A.18}$$

Thus, the colon product[2] is often referred to as the *double dot product*.

Following are several useful identities involving the tensors $\nabla\vec{A}$, $\vec{A}\vec{B}$, and T, the latter not necessarily a dyadic. The superscript $^\mathrm{T}$ denotes the *transpose* of a rank 2 tensor where, in its matrix representation, the rows of the tensor form the columns of the transpose. Thus, $(\vec{A}\vec{B})^\mathrm{T} = \vec{B}\vec{A}$. The first equality in identity (A.21) is analogous to Eq. (A.7) while the next two follow from Eq. (A.13). Identities (A.22) and (A.23) then follow from Eq. (A.21):

$$\vec{A}\cdot(\nabla\vec{A}) = (\vec{A}\cdot\nabla)\vec{A} = \tfrac{1}{2}\nabla A^2 - \vec{A}\times(\nabla\times\vec{A}); \tag{A.19}$$

$$\vec{A}\cdot(\nabla\vec{A})^\mathrm{T} = \tfrac{1}{2}\nabla A^2; \tag{A.20}$$

$$\left.\begin{aligned} \nabla\cdot(\vec{A}\vec{B}) &= (\vec{A}\cdot\nabla)\vec{B} + \vec{B}(\nabla\cdot\vec{A}) \\ &= \tfrac{1}{2}\Big[\nabla(\vec{A}\cdot\vec{B}) + \vec{A}(\nabla\cdot\vec{B}) + \vec{B}(\nabla\cdot\vec{A}) - \nabla\times(\vec{A}\times\vec{B}) \\ &\quad - \vec{A}\times(\nabla\times\vec{B}) - \vec{B}\times(\nabla\times\vec{A})\Big] \\ &= (\vec{B}\cdot\nabla)\vec{A} + \vec{A}(\nabla\cdot\vec{B}) - \nabla\times(\vec{A}\times\vec{B}); \end{aligned}\right\} \tag{A.21}$$

---

[2]Some authors define the colon product as $\mathsf{M}:\mathsf{N} = \sum_{i,j} M_{ij} N_{ji} = (\vec{A}\cdot\vec{D})(\vec{B}\cdot\vec{C})$, which is more in keeping with the usual rules of matrix multiplication. This definition gives the same result as identity (A.18) if the second tensor is replaced with its transpose.

$$\nabla \cdot (\vec{A}\vec{B}) = \nabla \cdot (\vec{B}\vec{A}) - \nabla \times (\vec{A} \times \vec{B}); \tag{A.22}$$

$$\left.\begin{aligned}\nabla \cdot (\vec{A}\vec{A}) &= (\vec{A} \cdot \nabla)\vec{A} + \vec{A}(\nabla \cdot \vec{A}) \\ &= \tfrac{1}{2}\nabla A^2 + \vec{A}(\nabla \cdot \vec{A}) - \vec{A} \times (\nabla \times \vec{A});\end{aligned}\right\} \tag{A.23}$$

$$\nabla \cdot (f\mathsf{T}) = \mathsf{T} \cdot \nabla f + f \nabla \cdot \mathsf{T}; \tag{A.24}$$

$$\nabla \cdot (\mathsf{T} \cdot \vec{A}) = \mathsf{T} : \nabla \vec{A} + (\nabla \cdot \mathsf{T}) \cdot \vec{A}; \tag{A.25}$$

$$\nabla \vec{A} : \nabla \vec{A} = (\nabla \times \vec{A}) \cdot (\nabla \times \vec{A}) + \nabla \vec{A} : (\nabla \vec{A})^\mathsf{T}; \tag{A.26}$$

$$\nabla \cdot (\nabla \vec{A}) = \nabla^2 \vec{A} = \nabla(\nabla \cdot \vec{A}) - \nabla \times (\nabla \times \vec{A}); \tag{A.27}$$

$$\nabla \cdot (\nabla \vec{A})^\mathsf{T} = \nabla(\nabla \cdot \vec{A}). \tag{A.28}$$

A.1.2 Vector derivatives of $\vec{r}$

Occasionally, the displacement vector, $\vec{r}$, appears directly in a vector derivative. While easiest to prove for Cartesian coordinates, these identities are true for all coordinate systems:

$$\nabla \cdot \vec{r} = 3; \qquad \nabla \times \vec{r} = 0; \qquad \nabla \vec{r} = \mathsf{I}, \tag{A.29}$$

where I is the identity matrix (rank 2 tensor; see Eq. A.17).

A.2 Theorems of vector calculus

Let f and g be arbitrary scalar functions and let $\vec{A}$ be an arbitrary vector function of the coordinates. Let V represent an arbitrary volume and let S represent the closed surface containing V. Let Σ represent an arbitrary open surface and let C represent the circumference of Σ (a closed path). Finally, let dV, $\hat{n}\,d\sigma$, and $d\vec{l}$ represent elements of volume, surface, and displacement respectively. Then...

Three flavours of Gauss' theorem:

$$\int_V \nabla \cdot \vec{A}\, dV = \oint_S \vec{A} \cdot \hat{n}\, d\sigma; \tag{A.30}$$

$$\int_V \nabla f\, dV = \oint_S f\hat{n}\, d\sigma; \tag{A.31}$$

$$\int_V \nabla \times \vec{A}\, dV = -\oint_S \vec{A} \times \hat{n}\, d\sigma. \tag{A.32}$$

Two flavours of Green's theorem:

$$\int_V (f\nabla^2 g + \nabla f \cdot \nabla g)\, dV = \oint_S f\nabla g \cdot \hat{n}\, d\sigma; \tag{A.33}$$

$$\int_V (f\nabla^2 g - g\nabla^2 f)\, dV = \oint_S (f\nabla g - g\nabla f) \cdot \hat{n}\, d\sigma. \tag{A.34}$$

Coordinate system	(x_1, x_2, x_3)	h_1	h_2	h_3
Cartesian	(x, y, z)	1	1	1
cylindrical	(z, r, φ)	1	1	r
spherical polar	$(\varrho, \vartheta, \varphi)$	1	ϱ	$\varrho \sin\vartheta$

Table A.1. Scale factors for the most commonly used orthogonal coordinate systems.

Three flavours of Stokes' theorem:

$$\int_\Sigma \nabla \times \vec{A} \cdot \hat{n}\, d\sigma = \oint_C \vec{A} \cdot d\vec{l}; \tag{A.35}$$

$$\int_\Sigma \nabla f \times \hat{n}\, d\sigma = -\oint_C f\, d\vec{l}; \tag{A.36}$$

$$\int_\Sigma (\hat{n}\, d\sigma \times \nabla) \times \vec{A} = -\oint_C \vec{A} \times d\vec{l}. \tag{A.37}$$

A.3 Orthogonal coordinate systems

Consider an orthogonal coordinate system, (x_1, x_2, x_3), in which a differential length is given by,

$$ds^2 = h_1^2 dx_1^2 + h_2^2 dx_2^2 + h_3^2 dx_3^2,$$

and whose *scale factors* h_1, h_2, and h_3 are, in general, functions of the coordinates. Table A.1 gives the scale factors for the most commonly used orthogonal coordinate systems, namely Cartesian, cylindrical, and spherical polar coordinates.

Let f and $\vec{A}$ be arbitrary scalar and vector functions of the coordinates. Then, the gradient, divergence, curl, and Laplacian are given by:

$$\nabla f = \hat{x}_1 \frac{1}{h_1} \frac{\partial f}{\partial x_1} + \hat{x}_2 \frac{1}{h_2} \frac{\partial f}{\partial x_2} + \hat{x}_3 \frac{1}{h_3} \frac{\partial f}{\partial x_3};$$

$$\nabla \cdot \vec{A} = \frac{1}{h_1 h_2 h_3} \left(\frac{\partial(h_2 h_3 A_1)}{\partial x_1} + \frac{\partial(h_3 h_1 A_2)}{\partial x_2} + \frac{\partial(h_1 h_2 A_3)}{\partial x_3} \right);$$

$$\nabla \times \vec{A} = \hat{x}_1 \frac{1}{h_2 h_3} \left(\frac{\partial(h_3 A_3)}{\partial x_2} - \frac{\partial(h_2 A_2)}{\partial x_3} \right) + \hat{x}_2 \frac{1}{h_3 h_1} \left(\frac{\partial(h_1 A_1)}{\partial x_3} - \frac{\partial(h_3 A_3)}{\partial x_1} \right)$$
$$+ \hat{x}_3 \frac{1}{h_1 h_2} \left(\frac{\partial(h_2 A_2)}{\partial x_1} - \frac{\partial(h_1 A_1)}{\partial x_2} \right);$$

$$\nabla \cdot (\nabla f) = \nabla^2 f = \frac{1}{h_1 h_2 h_3} \left[\frac{\partial}{\partial x_1} \left(\frac{h_2 h_3}{h_1} \frac{\partial f}{\partial x_1} \right) + \frac{\partial}{\partial x_2} \left(\frac{h_3 h_1}{h_2} \frac{\partial f}{\partial x_2} \right) \right.$$
$$\left. + \frac{\partial}{\partial x_3} \left(\frac{h_1 h_2}{h_3} \frac{\partial f}{\partial x_3} \right) \right],$$

where $\hat{x}_1$, $\hat{x}_2$, $\hat{x}_3$ are unit vectors in the x_1-, x_2-, and x_3-directions, and where $\hat{x}_i \cdot \hat{x}_j = \delta_{ij}$. From these expressions, identity (A.13) may be expanded out as:

$$(\vec{A} \cdot \nabla)\vec{B} =$$
$$\hat{x}_1 \left(\vec{A} \cdot \nabla B_1 - \frac{A_2 B_2}{h_2 h_1} \frac{\partial h_2}{\partial x_1} - \frac{A_3 B_3}{h_3 h_1} \frac{\partial h_3}{\partial x_1} + \frac{A_1 B_2}{h_1 h_2} \frac{\partial h_1}{\partial x_2} + \frac{A_1 B_3}{h_1 h_3} \frac{\partial h_1}{\partial x_3} \right)$$
$$+ \hat{x}_2 \left(\vec{A} \cdot \nabla B_2 - \frac{A_3 B_3}{h_3 h_2} \frac{\partial h_3}{\partial x_2} - \frac{A_1 B_1}{h_1 h_2} \frac{\partial h_1}{\partial x_2} + \frac{A_2 B_3}{h_2 h_3} \frac{\partial h_2}{\partial x_3} + \frac{A_2 B_1}{h_2 h_1} \frac{\partial h_2}{\partial x_1} \right) \quad (A.38)$$
$$+ \hat{x}_3 \left(\vec{A} \cdot \nabla B_3 - \frac{A_1 B_1}{h_1 h_3} \frac{\partial h_1}{\partial x_3} - \frac{A_2 B_2}{h_2 h_3} \frac{\partial h_2}{\partial x_3} + \frac{A_3 B_1}{h_3 h_1} \frac{\partial h_3}{\partial x_1} + \frac{A_3 B_2}{h_3 h_2} \frac{\partial h_3}{\partial x_2} \right).$$

For the coordinate systems listed in Table A.1, these expressions become:

Cartesian: (A.39)

$$\nabla f = \hat{x} \frac{\partial f}{\partial x} + \hat{y} \frac{\partial f}{\partial y} + \hat{z} \frac{\partial f}{\partial z};$$

$$\nabla \cdot \vec{A} = \frac{\partial A_x}{\partial x} + \frac{\partial A_y}{\partial y} + \frac{\partial A_z}{\partial z};$$

$$\nabla \times \vec{A} = \hat{x} \left(\frac{\partial A_z}{\partial y} - \frac{\partial A_y}{\partial z} \right) + \hat{y} \left(\frac{\partial A_x}{\partial z} - \frac{\partial A_z}{\partial x} \right) + \hat{z} \left(\frac{\partial A_y}{\partial x} - \frac{\partial A_x}{\partial y} \right);$$

$$\nabla^2 f = \frac{\partial^2 f}{\partial x^2} + \frac{\partial^2 f}{\partial y^2} + \frac{\partial^2 f}{\partial z^2};$$

$$(\vec{A} \cdot \nabla)\vec{B} = \hat{x}(\vec{A} \cdot \nabla B_x) + \hat{y}(\vec{A} \cdot \nabla B_y) + \hat{z}(\vec{A} \cdot \nabla B_z),$$

cylindrical: (A.40)

$$\nabla f = \hat{z} \frac{\partial f}{\partial z} + \hat{r} \frac{\partial f}{\partial r} + \hat{\varphi} \frac{1}{r} \frac{\partial f}{\partial \varphi};$$

$$\nabla \cdot \vec{A} = \frac{\partial A_z}{\partial z} + \frac{1}{r} \frac{\partial (r A_r)}{\partial r} + \frac{1}{r} \frac{\partial A_\varphi}{\partial \varphi};$$

$$\nabla \times \vec{A} = \hat{z} \frac{1}{r} \left(\frac{\partial (r A_\varphi)}{\partial r} - \frac{\partial A_r}{\partial \varphi} \right) + \hat{r} \left(\frac{1}{r} \frac{\partial A_z}{\partial \varphi} - \frac{\partial A_\varphi}{\partial z} \right) + \hat{\varphi} \left(\frac{\partial A_r}{\partial z} - \frac{\partial A_z}{\partial r} \right);$$

$$\nabla^2 f = \frac{\partial^2 f}{\partial z^2} + \frac{1}{r} \frac{\partial}{\partial r} \left(r \frac{\partial f}{\partial r} \right) + \frac{1}{r^2} \frac{\partial^2 f}{\partial \varphi^2};$$

$$(\vec{A} \cdot \nabla)\vec{B} = \hat{z}(\vec{A} \cdot \nabla B_z) + \hat{r} \left(\vec{A} \cdot \nabla B_r - \frac{A_\varphi B_\varphi}{r} \right) + \hat{\varphi} \left(\vec{A} \cdot \nabla B_\varphi + \frac{A_\varphi B_r}{r} \right),$$

spherical polar: (A.41)

$$\nabla f = \hat{\varrho}\frac{\partial f}{\partial \varrho} + \hat{\vartheta}\frac{1}{\varrho}\frac{\partial f}{\partial \vartheta} + \hat{\varphi}\frac{1}{\varrho\sin\vartheta}\frac{\partial f}{\partial \varphi};$$

$$\nabla \cdot \vec{A} = \frac{1}{\varrho^2}\frac{\partial(\varrho^2 A_\varrho)}{\partial \varrho} + \frac{1}{\varrho\sin\vartheta}\frac{\partial(\sin\vartheta\, A_\vartheta)}{\partial \vartheta} + \frac{1}{\varrho\sin\vartheta}\frac{\partial A_\varphi}{\partial \varphi};$$

$$\nabla \times \vec{A} = \hat{\varrho}\frac{1}{\varrho\sin\vartheta}\left(\frac{\partial(\sin\vartheta\, A_\varphi)}{\partial \vartheta} - \frac{\partial A_\vartheta}{\partial \varphi}\right) + \hat{\vartheta}\frac{1}{\varrho}\left(\frac{1}{\sin\vartheta}\frac{\partial A_\varrho}{\partial \varphi} - \frac{\partial(\varrho A_\varphi)}{\partial \varrho}\right)$$
$$+ \hat{\varphi}\frac{1}{\varrho}\left(\frac{\partial(\varrho A_\vartheta)}{\partial \varrho} - \frac{\partial A_\varrho}{\partial \vartheta}\right);$$

$$\nabla^2 f = \frac{1}{\varrho^2}\frac{\partial}{\partial \varrho}\left(\varrho^2\frac{\partial f}{\partial \varrho}\right) + \frac{1}{\varrho^2\sin\vartheta}\frac{\partial}{\partial \vartheta}\left(\sin\vartheta\frac{\partial f}{\partial \vartheta}\right) + \frac{1}{\varrho^2\sin^2\vartheta}\frac{\partial^2 f}{\partial \varphi^2};$$

$$(\vec{A}\cdot\nabla)\vec{B} = \hat{\varrho}\left(\vec{A}\cdot\nabla B_\varrho - \frac{A_\vartheta B_\vartheta}{\varrho} - \frac{A_\varphi B_\varphi}{\varrho}\right) + \hat{\vartheta}\left(\vec{A}\cdot\nabla B_\vartheta - \frac{A_\varphi B_\varphi}{\varrho\tan\vartheta} + \frac{A_\vartheta B_\varrho}{\varrho}\right)$$
$$+ \hat{\varphi}\left(\vec{A}\cdot\nabla B_\varphi + \frac{A_\varphi B_\varrho}{\varrho} + \frac{A_\varphi B_\vartheta}{\varrho\tan\vartheta}\right).$$

A.4 Euler's and the momentum equations

Without the gravitational and magnetic terms, Euler's equation (*e.g.*, Eq. 1.36) and the momentum equation (*e.g.*, Eq. 1.27) are, respectively:

$$\frac{\partial \vec{v}}{\partial t} + (\vec{v}\cdot\nabla)\vec{v} = -\frac{1}{\rho}\nabla p;$$

$$\frac{\partial \vec{s}}{\partial t} + \nabla\cdot(\vec{s}\vec{v}) = -\nabla p.$$

The vector constructs $(\vec{v}\cdot\nabla)\vec{v}$ and $\nabla\cdot(\vec{s}\vec{v})$ can be perplexing to the uninitiated reader. However, with the vector identities worked out in this appendix, it is a straight-forward task to spell these expressions out in a general orthogonal coordinate system.

In particular, from Eq. (A.38), the three components of Euler's equation are given by:

$$\frac{\partial v_1}{\partial t} + \vec{v}\cdot\nabla v_1 = -\frac{1}{\rho h_1}\frac{\partial p}{\partial x_1} + \frac{v_2^2}{h_2 h_1}\frac{\partial h_2}{\partial x_1} + \frac{v_3^2}{h_3 h_1}\frac{\partial h_3}{\partial x_1}$$
$$- \frac{v_1 v_2}{h_1 h_2}\frac{\partial h_1}{\partial x_2} - \frac{v_1 v_3}{h_1 h_3}\frac{\partial h_1}{\partial x_3};$$
(A.42)

$$\frac{\partial v_2}{\partial t} + \vec{v}\cdot\nabla v_2 = -\frac{1}{\rho h_2}\frac{\partial p}{\partial x_2} + \frac{v_3^2}{h_3 h_2}\frac{\partial h_3}{\partial x_2} + \frac{v_1^2}{h_1 h_2}\frac{\partial h_1}{\partial x_2}$$
$$- \frac{v_2 v_3}{h_2 h_3}\frac{\partial h_2}{\partial x_3} - \frac{v_2 v_1}{h_2 h_1}\frac{\partial h_2}{\partial x_1};$$
(A.43)

$$\frac{\partial v_3}{\partial t} + \vec{v} \cdot \nabla v_3 = -\frac{1}{\rho h_3}\frac{\partial p}{\partial x_3} + \frac{v_1^2}{h_1 h_3}\frac{\partial h_1}{\partial x_3} + \frac{v_2^2}{h_2 h_3}\frac{\partial h_2}{\partial x_3}$$
$$- \frac{v_3 v_1}{h_3 h_1}\frac{\partial h_3}{\partial x_1} - \frac{v_3 v_2}{h_3 h_2}\frac{\partial h_3}{\partial x_2}. \qquad (A.44)$$

The terms proportional to v_i^2 are "centrifugal acceleration" terms, while those proportional to $v_i v_j$, $i \neq j$ are "Coriolis acceleration" terms.

Next, with $\vec{A} = \vec{v}$ and $\vec{B} = \rho \vec{v} = \vec{s}$, Eq. (A.21) becomes:

$$\nabla \cdot (\vec{s}\vec{v}) = \tfrac{1}{2}\Big[\nabla(\vec{s}\cdot\vec{v}) + \vec{s}(\nabla\cdot\vec{v}) + \vec{v}(\nabla\cdot\vec{s}) - \vec{s}\times(\nabla\times\vec{v}) - \vec{v}\times(\nabla\times\vec{s})\Big].$$

Expanding out the gradients, divergences, and curls in this expression, we find the three components of the momentum equation to be:

$$\frac{\partial s_1}{\partial t} + \nabla\cdot(s_1\vec{v}) = -\frac{1}{h_1}\frac{\partial p}{\partial x_1} + \frac{s_2 v_2}{h_2 h_1}\frac{\partial h_2}{\partial x_1} + \frac{s_3 v_3}{h_3 h_1}\frac{\partial h_3}{\partial x_1}$$
$$- \frac{s_1 v_2}{h_1 h_2}\frac{\partial h_1}{\partial x_2} - \frac{s_1 v_3}{h_1 h_3}\frac{\partial h_1}{\partial x_3}; \qquad (A.45)$$

$$\frac{\partial s_2}{\partial t} + \nabla\cdot(s_2\vec{v}) = -\frac{1}{h_2}\frac{\partial p}{\partial x_2} + \frac{s_3 v_3}{h_3 h_2}\frac{\partial h_3}{\partial x_2} + \frac{s_1 v_1}{h_1 h_2}\frac{\partial h_1}{\partial x_2}$$
$$- \frac{s_2 v_3}{h_2 h_3}\frac{\partial h_2}{\partial x_3} - \frac{s_2 v_1}{h_2 h_1}\frac{\partial h_2}{\partial x_1}; \qquad (A.46)$$

$$\frac{\partial s_3}{\partial t} + \nabla\cdot(s_3\vec{v}) = -\frac{1}{h_3}\frac{\partial p}{\partial x_3} + \frac{s_1 v_1}{h_1 h_3}\frac{\partial h_1}{\partial x_3} + \frac{s_2 v_2}{h_2 h_3}\frac{\partial h_2}{\partial x_3}$$
$$- \frac{s_3 v_1}{h_3 h_1}\frac{\partial h_3}{\partial x_1} - \frac{s_3 v_2}{h_3 h_2}\frac{\partial h_3}{\partial x_2}. \qquad (A.47)$$

The "inertial" force densities that have been revealed correspond term for term to the inertial accelerations that appear in Eq. (A.42)–(A.44). A little further manipulation of Eq. (A.45)–(A.47) yields the following:

$$\frac{\partial h_1 s_1}{\partial t} + \nabla\cdot(h_1 s_1 \vec{v}) = -\frac{\partial p}{\partial x_1} + \frac{s_1 v_1}{h_1}\frac{\partial h_1}{\partial x_1} + \frac{s_2 v_2}{h_2}\frac{\partial h_2}{\partial x_1} + \frac{s_3 v_3}{h_3}\frac{\partial h_3}{\partial x_1}; \qquad (A.48)$$

$$\frac{\partial h_2 s_2}{\partial t} + \nabla\cdot(h_2 s_2 \vec{v}) = -\frac{\partial p}{\partial x_2} + \frac{s_1 v_1}{h_1}\frac{\partial h_1}{\partial x_2} + \frac{s_2 v_2}{h_2}\frac{\partial h_2}{\partial x_2} + \frac{s_3 v_3}{h_3}\frac{\partial h_3}{\partial x_2}; \qquad (A.49)$$

$$\frac{\partial h_3 s_3}{\partial t} + \nabla\cdot(h_3 s_3 \vec{v}) = -\frac{\partial p}{\partial x_3} + \frac{s_1 v_1}{h_1}\frac{\partial h_1}{\partial x_3} + \frac{s_2 v_2}{h_2}\frac{\partial h_2}{\partial x_3} + \frac{s_3 v_3}{h_3}\frac{\partial h_3}{\partial x_3}. \qquad (A.50)$$

By introducing h_i next to s_i on the left-hand side, the two "Coriolis" terms on the right-hand side are eliminated and one additional centrifugal-like term is added. Note that for Cartesian-like coordinates, $h_i = 1$ and is unitless, whence $h_i s_i = s_i$. For angular coordinates, h_i is the moment arm about the axis of rotation, and $h_i s_i$ can be interpreted as the angular momentum component about that axis. Since angular momentum is the conserved quantity for angular coordinates, expressing the momentum equations in terms of $h_i s_i$ makes sense.

For the coordinate systems listed in Table A.1, Eq. (A.42)–(A.44) and (A.48)–(A.50) become:

Cartesian: (A.51)

$$\frac{\partial v_x}{\partial t} + \vec{v}\cdot\nabla v_x = -\frac{1}{\rho}\frac{\partial p}{\partial x}; \qquad \frac{\partial s_x}{\partial t} + \nabla\cdot(s_x\vec{v}) = -\frac{\partial p}{\partial x};$$

$$\frac{\partial v_y}{\partial t} + \vec{v}\cdot\nabla v_y = -\frac{1}{\rho}\frac{\partial p}{\partial y}; \qquad \frac{\partial s_y}{\partial t} + \nabla\cdot(s_y\vec{v}) = -\frac{\partial p}{\partial y};$$

$$\frac{\partial v_z}{\partial t} + \vec{v}\cdot\nabla v_z = -\frac{1}{\rho}\frac{\partial p}{\partial z}; \qquad \frac{\partial s_z}{\partial t} + \nabla\cdot(s_z\vec{v}) = -\frac{\partial p}{\partial z},$$

cylindrical: (A.52)

$$\frac{\partial v_z}{\partial t} + \vec{v}\cdot\nabla v_z = -\frac{1}{\rho}\frac{\partial p}{\partial z}; \qquad \frac{\partial s_z}{\partial t} + \nabla\cdot(s_z\vec{v}) = -\frac{\partial p}{\partial z};$$

$$\frac{\partial v_r}{\partial t} + \vec{v}\cdot\nabla v_r = -\frac{1}{\rho}\frac{\partial p}{\partial r} + r\omega^2; \qquad \frac{\partial s_r}{\partial t} + \nabla\cdot(s_r\vec{v}) = -\frac{\partial p}{\partial r} + \rho r\omega^2;$$

$$\frac{\partial v_\varphi}{\partial t} + \vec{v}\cdot\nabla v_\varphi = -\frac{1}{\rho r}\frac{\partial p}{\partial \varphi} - v_r\omega; \qquad \frac{\partial (rs_\varphi)}{\partial t} + \nabla\cdot(rs_\varphi\vec{v}) = -\frac{\partial p}{\partial \varphi},$$

where $\omega = v_\varphi/r$ is the angular speed about the z-axis;

spherical polar: (A.53)

$$\frac{\partial v_\varrho}{\partial t} + \vec{v}\cdot\nabla v_\varrho = -\frac{1}{\rho}\frac{\partial p}{\partial \varrho} + \varrho(\Omega^2 + \omega^2 \sin^2\vartheta);$$

$$\frac{\partial v_\vartheta}{\partial t} + \vec{v}\cdot\nabla v_\vartheta = -\frac{1}{\rho\varrho}\frac{\partial p}{\partial \vartheta} + \varrho\omega^2 \sin\vartheta \cos\vartheta - v_\varrho\Omega;$$

$$\frac{\partial v_\varphi}{\partial t} + \vec{v}\cdot\nabla v_\varphi = -\frac{1}{\rho\varrho\sin\vartheta}\frac{\partial p}{\partial \varphi} - v_\varrho\omega\sin\vartheta - \varrho\Omega\omega\cos\vartheta;$$

$$\frac{\partial s_\varrho}{\partial t} + \nabla\cdot(s_\varrho\vec{v}) = -\frac{\partial p}{\partial \varrho} + \rho\varrho(\Omega^2 + \omega^2 \sin^2\vartheta);$$

$$\frac{\partial(\varrho s_\vartheta)}{\partial t} + \nabla\cdot(\varrho s_\vartheta\vec{v}) = -\frac{\partial p}{\partial \vartheta} + \rho\varrho^2\omega^2 \sin\vartheta \cos\vartheta;$$

$$\frac{\partial(\varrho \sin\vartheta\, s_\varphi)}{\partial t} + \nabla\cdot(\varrho \sin\vartheta\, s_\varphi\vec{v}) = -\frac{\partial p}{\partial \varphi},$$

where $\omega = v_\varphi/(\varrho\sin\vartheta) = v_\varphi/r$ is the angular speed about the z-axis, and $\Omega = v_\vartheta/\varrho$ is the meridional speed.

A.5 The Lorentz force

The Lorentz force term in either the momentum or Euler's equation (Eq. 4.13 or 4.14) is proportional to $(\nabla \times \vec{B}) \times \vec{B}$ which, for the general orthogonal coordinate system defined in §A.3, is given by:

$$(\nabla \times \vec{B}) \times \vec{B} =$$
$$-\hat{x}_1 \left[\frac{B_3}{h_3 h_1} \frac{\partial(h_3 B_3)}{\partial x_1} + \frac{B_2}{h_2 h_1} \frac{\partial(h_2 B_2)}{\partial x_1} \right] - \hat{x}_2 \left[\frac{B_1}{h_1 h_2} \frac{\partial(h_1 B_1)}{\partial x_2} + \frac{B_3}{h_3 h_2} \frac{\partial(h_3 B_3)}{\partial x_2} \right]$$
$$-\hat{x}_3 \left[\frac{B_2}{h_2 h_3} \frac{\partial(h_2 B_2)}{\partial x_3} + \frac{B_1}{h_1 h_3} \frac{\partial(h_1 B_1)}{\partial x_3} \right] + \hat{x}_1 \left[\frac{B_3}{h_1 h_3} \frac{\partial(h_1 B_1)}{\partial x_3} + \frac{B_2}{h_1 h_2} \frac{\partial(h_1 B_1)}{\partial x_2} \right]$$
$$+\hat{x}_2 \left[\frac{B_1}{h_2 h_1} \frac{\partial(h_2 B_2)}{\partial x_1} + \frac{B_3}{h_2 h_3} \frac{\partial(h_2 B_2)}{\partial x_3} \right] + \hat{x}_3 \left[\frac{B_2}{h_3 h_2} \frac{\partial(h_3 B_3)}{\partial x_2} + \frac{B_1}{h_3 h_1} \frac{\partial(h_3 B_3)}{\partial x_1} \right],$$

where the first three $(-)$ terms are the "compressional Lorentz forces" (that behave much like thermal pressure), and the last three $(+)$ terms are the "transverse Lorentz forces" which govern the transmission of Alfvén waves. I find this expansion to be particularly useful for computational purposes.

For pen-and-paper calculations, a more useful expansion comes from Eq. (A.15),

$$(\nabla \times \vec{B}) \times \vec{B} = (\vec{B} \cdot \nabla)\vec{B} - \tfrac{1}{2}\nabla B^2,$$

where expressions for $(\vec{A} \cdot \nabla)\vec{B}$ in §A.3 and Identity (A.7) (since $\nabla \cdot \vec{B} = 0$) can be used to complete the expansions.

Thus, for the coordinate systems listed in Table A.1, these two flavours of $(\nabla \times \vec{B}) \times \vec{B}$ become:

Cartesian: (A.54)

$$(\nabla \times \vec{B}) \times \vec{B} = -\hat{x}\frac{1}{2}\frac{\partial(B_y^2 + B_z^2)}{\partial x} - \hat{y}\frac{1}{2}\frac{\partial(B_z^2 + B_x^2)}{\partial y} - \hat{z}\frac{1}{2}\frac{\partial(B_x^2 + B_y^2)}{\partial z}$$
$$+ \hat{x}\left[B_z\frac{\partial B_x}{\partial z} + B_y\frac{\partial B_x}{\partial y}\right] + \hat{y}\left[B_x\frac{\partial B_y}{\partial x} + B_z\frac{\partial B_y}{\partial z}\right]$$
$$+ \hat{z}\left[B_y\frac{\partial B_z}{\partial y} + B_x\frac{\partial B_z}{\partial x}\right]$$
$$= \nabla \cdot (B_x \vec{B})\hat{x} + \nabla \cdot (B_y \vec{B})\hat{y} + \nabla \cdot (B_z \vec{B})\hat{z} - \frac{1}{2}\nabla B^2;$$

cylindrical: (A.55)

$$(\nabla \times \vec{B}) \times \vec{B} = -\hat{z}\frac{1}{2}\frac{\partial(B_r^2 + B_\varphi^2)}{\partial z} - \hat{r}\frac{1}{2}\left[\frac{1}{r^2}\frac{\partial(rB_\varphi)^2}{\partial r} + \frac{\partial B_z^2}{\partial r}\right] - \hat{\varphi}\frac{1}{2r}\frac{\partial(B_z^2 + B_r^2)}{\partial \varphi}$$

$$+ \hat{z}\left[\frac{B_\varphi}{r}\frac{\partial B_z}{\partial \varphi} + B_r\frac{\partial B_z}{\partial r}\right] + \hat{r}\left[B_z\frac{\partial B_r}{\partial z} + \frac{B_\varphi}{r}\frac{\partial B_r}{\partial \varphi}\right]$$

$$+ \hat{\varphi}\left[\frac{B_r}{r}\frac{\partial (rB_\varphi)}{\partial r} + B_z\frac{\partial B_\varphi}{\partial z}\right]$$

$$= \nabla \cdot (B_z \vec{B})\hat{z} + \left(\nabla \cdot (B_r \vec{B}) - \frac{B_\varphi^2}{r}\right)\hat{r} + \frac{1}{r}\nabla \cdot (rB_\varphi \vec{B})\hat{\varphi} - \frac{1}{2}\nabla B^2;$$

spherical polar: (A.56)

$$(\nabla \times \vec{B}) \times \vec{B} = -\hat{\varrho}\frac{1}{2\varrho^2}\frac{\partial[\varrho^2(B_\vartheta^2 + B_\varphi^2)]}{\partial \varrho} - \hat{\vartheta}\frac{1}{2\varrho}\left[\frac{1}{\sin^2\vartheta}\frac{\partial(B_\varphi\sin\vartheta)^2}{\partial \vartheta} + \frac{\partial B_\varrho^2}{\partial \vartheta}\right]$$

$$- \hat{\varphi}\frac{1}{2\varrho\sin\vartheta}\frac{\partial(B_\varrho^2 + B_\vartheta^2)}{\partial \varphi} + \hat{\varrho}\frac{1}{\varrho}\left[\frac{B_\varphi}{\sin\vartheta}\frac{\partial B_\varrho}{\partial \varphi} + B_\vartheta\frac{\partial B_\varrho}{\partial \vartheta}\right]$$

$$+ \hat{\vartheta}\frac{1}{\varrho}\left[B_\varrho\frac{\partial(\varrho B_\vartheta)}{\partial \varrho} + \frac{B_\varphi}{\sin\vartheta}\frac{\partial B_\vartheta}{\partial \varphi}\right]$$

$$+ \hat{\varphi}\frac{1}{\varrho}\left[\frac{B_\vartheta}{\sin\vartheta}\frac{\partial(B_\varphi\sin\vartheta)}{\partial \vartheta} + B_\varrho\frac{\partial(\varrho B_\varphi)}{\partial \varrho}\right]$$

$$= \left(\nabla \cdot (B_\varrho \vec{B}) - \frac{B_\vartheta^2 + B_\varphi^2}{\varrho}\right)\hat{\varrho} + \frac{1}{\varrho}\left(\nabla \cdot (\varrho B_\vartheta \vec{B}) - B_\varphi^2\cot\vartheta\right)\hat{\vartheta}$$

$$+ \frac{1}{\varrho\sin\vartheta}\nabla \cdot (\varrho\sin\vartheta B_\varphi \vec{B})\hat{\varphi} - \frac{1}{2}\nabla B^2.$$

Note that in the second expansions for each of cylindrical and spherical polar coordinates, the "Coriolis-like" terms have been absorbed into the divergences (similar to what was done for Eq. A.49 and A.50) where the appearance of an extra factor of the radial coordinate (r or ϱ) is a consequence.

B Essentials of Electrodynamics

B.1 Maxwell's equations

Let $\vec{E}$ ($\vec{D}$) be the electric field (displacement), and let $\vec{H}$ ($\vec{B}$) be the magnetic field (induction). In free space, $\vec{D} = \epsilon_0 \vec{E}$, $\vec{B} = \mu_0 \vec{H}$ where, in mks units, $\epsilon_0 = 8.8542 \times 10^{-12} \, \text{A}^2 \, \text{s}^4 \, \text{kg}^{-1} \, \text{m}^{-3}$ ($\text{C}^2 \, \text{N}^{-1} \, \text{m}^{-2}$) is the *permittivity of free space*, and $\mu_0 = 4\pi \times 10^{-7} \, \text{kg} \, \text{m} \, \text{A}^{-2} \, \text{s}^{-2}$ ($\text{N} \, \text{A}^{-2}$) is the *permeability of free space*. Note that $\epsilon_0 \mu_0 = c^{-2}$ where $c = 2.9979 \times 10^8 \, \text{m} \, \text{s}^{-1}$ is the speed of light.

If $\vec{v}$ is the velocity of a fluid element and ρ_q is the local charge density, then the current density, $\vec{J} = \rho_q \vec{v}$, is such that the current enclosed, i_{enc}, by an arbitrary loop surrounding an open surface Σ is given by:

$$i_{\text{enc}} = \int_\Sigma \vec{J} \cdot \hat{n} \, d\sigma,$$

where $\hat{n}$ is a unit vector normal to the surface element $d\sigma$.

With all these definitions stated, we write James Clerk Maxwell's (1831–1879)[1] famous set of equations governing electrodynamics in the following tables.

Maxwell's equations in differential form:

	general	free space
Gauss, electric	$\nabla \cdot \vec{D} = \rho_q$	$\nabla \cdot \vec{E} = \dfrac{\rho_q}{\epsilon_0}$
Gauss, magnetic	$\nabla \cdot \vec{B} = 0$	$\nabla \cdot \vec{B} = 0$
Faraday	$\nabla \times \vec{E} = -\dfrac{\partial \vec{B}}{\partial t}$	$\nabla \times \vec{E} = -\dfrac{\partial \vec{B}}{\partial t}$
Ampère–Maxwell	$\nabla \times \vec{H} = \dfrac{\partial \vec{D}}{\partial t} + \vec{J}$	$\nabla \times \vec{B} = \dfrac{1}{c^2}\dfrac{\partial \vec{E}}{\partial t} + \mu_0 \vec{J}$

(B.1)

---

[1] www.wikipedia.org/wiki/James_Clerk_Maxwell

Maxwell's equations in integral form:

	general	free space	
Gauss, $\vec{E}$	$\oint_S \vec{D} \cdot \hat{n}\, d\sigma = q_{\text{enc}}$	$\oint_S \vec{E} \cdot \hat{n}\, d\sigma = \dfrac{q_{\text{enc}}}{\epsilon_0}$	
Gauss, $\vec{B}$	$\oint_S \vec{B} \cdot \hat{n}\, d\sigma = 0$	$\oint_S \vec{B} \cdot \hat{n}\, d\sigma = 0$	(B.2)
Faraday	$\oint_C \vec{E} \cdot d\vec{l} = -\dfrac{d\Phi_B}{dt}$	$\oint_C \vec{E} \cdot d\vec{l} = -\dfrac{d\Phi_B}{dt}$	
A–M	$\oint_C \vec{H} \cdot d\vec{l} = \dfrac{d\Phi_D}{dt} + i_{\text{enc}}$	$\oint_C \vec{B} \cdot d\vec{l} = \dfrac{1}{c^2}\dfrac{d\Phi_E}{dt} + \mu_0 i_{\text{enc}}$	

where the enclosed charge, q_{enc}, and fluxes, Φ_Q with $Q = B, E, D$, are given by:

$$q_{\text{enc}} = \int_\tau \rho_q\, d\tau; \qquad \Phi_Q = \int_\Sigma \vec{Q} \cdot \hat{n}\, d\sigma, \tag{B.3}$$

where $d\tau$ is a volume differential.[2] Complementing Maxwell's equations is the electromagnetic force equation,

$$\vec{F}_{\text{EM}} = q(\vec{E} + \vec{v} \times \vec{B}), \tag{B.4}$$

where $q\vec{E}$ and $q\vec{v} \times \vec{B}$ are, respectively, the electric and Lorentz forces acting on a charge q moving with a velocity $\vec{v}$ in the vicinity of $\vec{E}$ and $\vec{B}$.

B.2 Electric energy density

Consider a volume, τ, with a charge density $\rho_q(\vec{r})$ and a potential function $V(\vec{r})$, and consider the task of modifying that charge density everywhere by $\delta\rho_q$. Then the work required to add an infinitesimal charge $dq = \delta\rho_q d\tau$ to the volume element $d\tau$ is,

$$dW = V\, dq = V\, \delta\rho_q\, d\tau.$$

Thus, over the entire volume, the work required to change the charge density by $\delta\rho_q$ is,

$$\delta W = \int_\tau V \delta\rho_q\, d\tau = \int_\tau V \delta(\nabla \cdot \vec{D})\, d\tau = \int_\tau V \nabla \cdot (\delta \vec{D})\, d\tau, \tag{B.5}$$

where the second equality invokes the first of Maxwell's equations (Eq. B.1) and the third equality invokes the linearity of the *nabla* operator, ∇. From Eq. (A.7) in App. A, we have:

$$V \nabla \cdot (\delta \vec{D}) = \nabla \cdot (V \delta \vec{D}) - \delta \vec{D} \cdot \nabla V.$$

Substituting this into Eq. (B.5), we get:

$$\delta W = \int_\tau \nabla \cdot (V \delta \vec{D})\, d\tau - \int_\tau \delta \vec{D} \cdot \nabla V\, d\tau = \oint_S V \delta \vec{D} \cdot \hat{n}\, d\sigma + \int_\tau \vec{E} \cdot \delta \vec{D}\, d\tau,$$

---

[2] I'm using τ here instead of V to represent volume since I'll soon need V to represent potential.

where Gauss' theorem (Eq. A.30) and the relationship between electric field and the potential function, $\vec{E} = -\nabla V$, have been used.

Assuming that the charge distribution, ρ_q, has a finite extent, the volume integral will be independent of the volume, τ, chosen so long as τ includes all the charge. The same is not true of the surface integral. As $r \to \infty$, the "far-field" approximations to the electric displacement and potential function, namely Coulomb's law, applies, and $V\,\delta\vec{D} \to r^{-3}$. Since the surface area only increases as r^2, the surface integral itself falls off as r^{-1} and thus vanishes as $r \to \infty$. Since we are free to choose any surface that encloses τ, we choose the surface at infinity thereby eliminating the surface integral and leaving us with,

$$\delta W = \int_\tau \vec{E} \cdot \delta \vec{D}\, d\tau.$$

Thus, the *work density* (work per unit volume) is given by $\delta w = \vec{E} \cdot \delta \vec{D}$ and represents the increment of work necessary to change the electric displacement by $\delta \vec{D}$ in a background electric field $\vec{E}$. By the conservation of energy, this must also represent the amount of additional energy stored in the electric field. Therefore, the total energy density stored in the electric field, e_E is given by,

$$\boxed{e_E = w = \int \vec{E} \cdot d\vec{D} = \frac{1}{2}\epsilon_0 E^2,} \tag{B.6}$$

where the last equality assumes the constitutive equation for free space, namely $\vec{D} = \epsilon_0 \vec{E}$. Finally, the power density required to modify the electric displacement is obtained by,

$$\boxed{p_E = \frac{dw}{dt} = \vec{E} \cdot \frac{d\vec{D}}{dt} = \vec{E} \cdot \dot{\vec{D}}.} \tag{B.7}$$

B.3 Magnetic energy density

For magnetic induction, the analogue of the work done in moving an incremental charge, δq, through a potential, V, is the work done in incrementing a given current loop by an amount δi immersed in a magnetic induction, $\vec{B}$ (Fig. B.1). To this end, we begin with the integral form of Faraday's law (Eq. B.2) and write down an expression for the induced *emf*,

$$\mathcal{E} = \oint_C \vec{E} \cdot d\vec{l} = -\frac{d\Phi_B}{dt}, \tag{B.8}$$

where Φ_B is given by Eq. (B.3), and the negative sign acknowledges Lenz' law. Thus, the work done in pushing an incremental charge, δq, *against* this *emf* (potential) is,

$$\delta W = -\mathcal{E}\,\delta q = -\mathcal{E}\,i\,\delta t = i\,\delta \Phi_B. \tag{B.9}$$

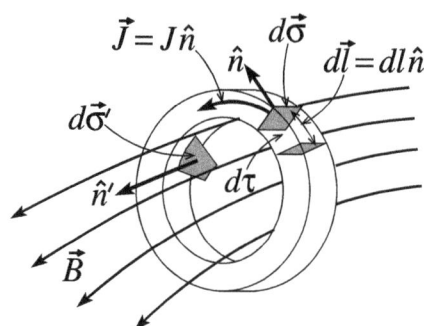

Figure B.1. According to Faraday's law, time-varying magnetic flux density directed along $\hat{n}'$, induces a change in the current density directed along $\hat{n}$, and around a loop enclosing the magnetic flux. Differentials in length ($\vec{dl} = \hat{n}\, dl$), area ($\hat{n}\, d\sigma$ and $\hat{n}'\, d\sigma'$), and volume ($d\tau = \vec{dl} \cdot \hat{n}\, d\sigma = dl\, d\sigma$) defined in the text are indicated.

Now,

$$\delta \Phi_B = \int_{\Sigma'} \delta \vec{B} \cdot \hat{n}'\, d\sigma' = \int_{\Sigma'} \nabla \times (\delta \vec{A}) \cdot \hat{n}'\, d\sigma' = \oint_C \delta \vec{A} \cdot \hat{n}\, dl, \quad (B.10)$$

where $\vec{A}$ is the *vector potential* ($\vec{B} = \nabla \times \vec{A}$, introduced in §4.8), where the first flavour of Stokes' theorem (Eq. A.35) is invoked, and where C is the closed contour (perimeter) of Σ'. The distinction between the two unit vectors is important. As indicated in Fig. B.1, $\hat{n}'$ is a unit vector parallel to the axis of the loop (normal to surface element $d\vec{\sigma}'$ of open surface Σ'), whereas $\hat{n}$ is a unit vector tangential to the loop (normal to surface element $d\vec{\sigma}$ of open surface Σ) and in the direction of current flow. Thus,

$$i = \int_\Sigma \vec{J} \cdot \hat{n}\, d\sigma = \int_\Sigma J\, d\sigma, \quad (B.11)$$

for $\vec{J} \parallel \hat{n}$, as in Fig. B.1. Substituting Eq. (B.10) and (B.11) into Eq. (B.9) yields:

$$\delta W = \int_\Sigma J\, d\sigma \oint_C \delta \vec{A} \cdot \hat{n}\, dl = \int_\Sigma \oint_C \delta \vec{A} \cdot (J\hat{n})\, dl\, d\sigma = \int_\tau \delta \vec{A} \cdot \vec{J}\, d\tau,$$

where, as evident from Fig. B.1, $d\sigma\, dl = d\tau$, a volume element. In the absence of an electric field, Ampère–Maxwell's law (Eq. B.1) requires $\vec{J} = \nabla \times \vec{H}$ and thus:

$$\delta W = \int_\tau \delta \vec{A} \cdot (\nabla \times \vec{H})\, d\tau. \quad (B.12)$$

From vector identity (A.8), we have,

$$\delta \vec{A} \cdot (\nabla \times \vec{H}) = \vec{H} \cdot (\nabla \times \delta \vec{A}) - \nabla \cdot (\delta \vec{A} \times \vec{H}),$$

and Eq. (B.12) becomes:

$$\delta W = \int_\tau \vec{H} \cdot (\nabla \times \delta \vec{A})\, d\tau - \int_\tau \nabla \cdot (\delta \vec{A} \times \vec{H})\, d\tau$$

$$= \int_\tau \vec{H} \cdot \delta \vec{B} \, d\tau - \int_{\Sigma''} (\delta \vec{A} \times \vec{H}) \cdot \hat{n}'' d\sigma'',$$

where $\delta \vec{B} = \nabla \times (\delta \vec{A})$ and where Gauss' theorem (Eq. A.30) was used. Note that Σ'' is a surface surrounding the entire volume containing the magnetic field (induction) distribution, and is different again from either Σ' and Σ used above.

Just as in the discussion in §B.2, in the far-field limit the surface integral disappears,[3] leaving us with,

$$\delta W = \int_\tau \vec{H} \cdot \delta \vec{B} \, d\tau.$$

Thus, the incremental work density is given by $\delta w = \vec{H} \cdot \delta \vec{B}$, and the total energy density stored in the magnetic field becomes:

$$\boxed{e_M = w = \int \vec{H} \cdot d\vec{B} = \frac{B^2}{2\mu_0},} \quad (B.13)$$

where the last equality assumes the constitutive equation for free space, namely $\vec{B} = \mu_0 \vec{H}$. Finally, the power density necessary to modify the magnetic inductance for a given magnetic field is,

$$\boxed{p_M = \frac{dw}{dt} = \vec{H} \cdot \frac{d\vec{B}}{dt} = \vec{H} \cdot \dot{\vec{B}}.} \quad (B.14)$$

B.4 Resistive energy density

The power delivered by the electromagnetic fields to a charged fluid element moving at velocity $\vec{v}$ is given by,

$$\mathcal{P} = \vec{f} \cdot \vec{v} = q(\vec{E} + \vec{v} \times \vec{B}) \cdot \vec{v} = q\vec{E} \cdot \vec{v}.$$

Therefore, the power density, p_R, is given by,

$$\boxed{p_R = \rho_q \vec{v} \cdot \vec{E} = \vec{J} \cdot \vec{E}.} \quad (B.15)$$

For $\vec{J} \cdot \vec{E} > 0$, electric power is being converted to other forms of energy (dissipation, kinetic energy), and the energy density of the electromagnetic fields is being drained. For $\vec{J} \cdot \vec{E} < 0$, other forms of energy are being converted to electromagnetic energy, and the energy density of the electromagnetic is being augmented.

---

[3]In this case, $H \sim r^{-3}$ and $A \sim r^{-2}$, and so the surface integral vanishes as r^{-3} as $r \to \infty$.

B.5 The Poynting vector

In 1884, John Henry Poynting (1852–1914) made an important discovery from Maxwell's equations which showed how energy is transported by electromagnetic fields.

The total electromagnetic power into or out of a given volume, τ, is given by:

$$\mathcal{P}_{\text{EM}} = \int_\tau (p_E + p_M + p_R)\,d\tau = \int_\tau \left(\vec{E}\cdot(\dot{\vec{D}} + \vec{J}) + \vec{H}\cdot\dot{\vec{B}}\right)d\tau. \tag{B.16}$$

But, from Maxwell's equations (Eq. B.1), $\dot{\vec{D}} + \vec{J} = \nabla\times\vec{H}$ and $\dot{\vec{B}} = -\nabla\times\vec{E}$, and Eq. (B.16) becomes:

$$\mathcal{P}_{\text{EM}} = \int_\tau (\vec{E}\cdot\nabla\times\vec{H} - \vec{H}\cdot\nabla\times\vec{E})\,d\tau = \int_\tau \nabla\cdot(\vec{H}\times\vec{E})\,d\tau,$$

where vector identity (A.8) has been used. The cross product $\vec{E}\times\vec{H}$ is so important to the theory of electrodynamics that it is given its own symbol and name:

$$\boxed{\vec{S}_{\text{P}} = \vec{E}\times\vec{H},} \tag{B.17}$$

where $\vec{S}_{\text{P}}$ is the *Poynting vector*,[4] named for the first person to identify its importance. With units $\mathrm{J\,m^{-2}\,s^{-1}}$, it can be interpreted as an *energy flux density* that measures the amount of electromagnetic energy passing through a unit area in a unit time. Thus,

$$\mathcal{P}_{\text{EM}} = -\int_\tau \nabla\cdot\vec{S}_{\text{P}}\,d\tau = -\oint_S \vec{S}_{\text{P}}\cdot\hat{n}\,d\sigma = -\Phi_S, \tag{B.18}$$

where Φ_S is the *Poynting flux*, the total power (energy per unit time) transported by the electromagnetic fields across a given closed surface, S. For $\Phi_S > 0$, $\mathcal{P}_{\text{EM}} < 0$ and there exists a net drain of electromagnetic energy from the volume τ. Conversely, for $\Phi_S < 0$, $\mathcal{P}_{\text{EM}} > 0$ and there exists a net gain of electromagnetic energy by the volume τ.

The *Poynting power density*, p_S, is another useful quantity that describes the electromagnetic power gained (>0) or lost (<0) per unit volume, and is given by:

$$\boxed{p_S = \frac{d\mathcal{P}_{\text{EM}}}{d\tau} = -\nabla\cdot\vec{S}_{\text{P}} = -\nabla\cdot(\vec{E}\times\vec{H}).} \tag{B.19}$$

---

[4] The Poynting vector is normally designated simply as $\vec{S}$. However, since $\vec{S}$ is already used for the total momentum, I've added the subscript 'P' to aid in their distinction.

C. The "Conics" of PDEs

Consider a second-order partial differential equation of the form:
$$(\alpha\partial_{tt} + \beta\partial_{tx} + \beta\partial_{xt} + \gamma\partial_{xx} + \delta\partial_t + \epsilon\partial_x + \zeta)f(x,t) + \eta = 0, \qquad (C.1)$$
where,
$$\partial_{tx} = \frac{\partial}{\partial x}\frac{\partial}{\partial t} = \frac{\partial^2}{\partial x\,\partial t},$$
etc. The *discriminant*, $\mathcal{D}$, is given by the determinant of the 2×2 matrix formed from the coefficients of the second derivatives:
$$\mathcal{D} = \begin{vmatrix} \alpha & \beta \\ \beta & \gamma \end{vmatrix} = \alpha\gamma - \beta^2. \qquad (C.2)$$

If $\mathcal{D} > 0$, Eq. (C.1) is said to be *elliptical*, $\mathcal{D} = 0$, *parabolic*, and $\mathcal{D} < 0$, *hyperbolic*.

The designation elliptical/parabolic/hyperbolic is borrowed from its usage in *conics*, in which the three types of curves in 2-D (ellipse, parabola, hyperbola) can be obtained by taking planar slices through a right cone (*conic sections*), as shown in Fig. C.1. In general, a conic section can be described by the equation:
$$ay^2 + 2bxy + cx^2 + dy + ex + f = 0,$$
where one gets a circle/ellipse for $ac - b^2 > 0$, a parabola for $ac - b^2 = 0$, and a hyperbola for $ac - b^2 < 0$, reminiscent of the discriminant in Eq. (C.2).

As an example, the 1-D wave equation,
$$\partial_{tt}p = c_{\text{s}}^2 \partial_{xx}p \quad \Rightarrow \quad \partial_{tt}p - c_{\text{s}}^2 \partial_{xx}p = 0, \qquad (C.3)$$

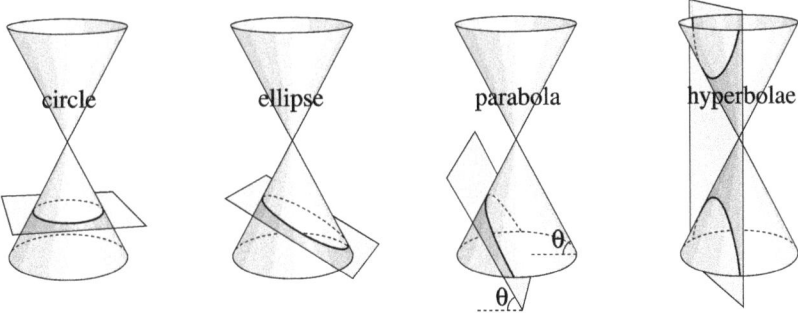

Figure C.1. *Conic sections* showing, from left to right, a circle taken from a horizontal slice through a right cone; an ellipse taken from a slice inclined at an angle less than the conic angle, θ; a parabola taken from a slice inclined at θ, and a pair of hyperbolæ taken from a slice inclined at an angle greater than θ.

is a hyperbolic equation since $\mathcal{D} = \alpha\gamma - \beta^2 = (1)(-c_s^2) - 0 < 0$. Similarly, the 2-D Poisson equation,

$$\partial_{xx}\phi + \partial_{yy}\phi = 4\pi G\rho,$$

is elliptical, since $\mathcal{D} = \alpha\gamma - \beta^2 = (1)(1) - 0 > 0$. With a suitably modified definition for the discriminant for equations of three and four variables, the wave equation is still hyperbolic in 3-D and Poisson's equation is still elliptical.

A second-order hyperbolic PDE may be transformed into a hyperbolic system of two first-order PDEs. Thus and for example, we saw in §2.1 that the second-order wave equation in p (Eq. C.3) can be derived from two first-order PDEs which, in the Lagrangian (co-moving) frame of reference, are given by:

$$\partial_t p = -\rho_0 c_s^2 \partial_x v;$$

$$\partial_t v = -\frac{1}{\rho_0}\partial_x p.$$

Note that a second-order wave equation in v may also be derived from these two equations. In the Eulerian (lab) frame, these equations are:

$$\partial_t p + v_0 \partial_x p + \rho_0 c_s^2 \partial_x v = 0;$$

$$\partial_t v + \frac{1}{\rho_0}\partial_x p + v_0 \partial_x v = 0,$$

which, as shown in §2.1.2, can be written more compactly as:

$$\partial_t |q\rangle + \mathsf{J}\partial_x |q\rangle = 0, \tag{C.4}$$

where,

$$|q\rangle = \begin{bmatrix} p \\ v \end{bmatrix} \quad \text{and} \quad \mathsf{J} = \begin{bmatrix} v_0 & \rho_0 c_s^2 \\ 1/\rho_0 & v_0 \end{bmatrix}.$$

One can show that for a hyperbolic system of first-order equations, the eigenvalues of the Jacobian, J, are real and its eigenvectors are linearly independent of each other. If all eigenvalues are distinct, the system is said to be *strictly hyperbolic*. Conversely, if there is a possibility some or all of the eigenvalues are degenerate (equal), the system is said to be *not strictly hyperbolic*.

One is not limited to just two first-order PDEs in a hyperbolic system. The full 1-D hydrodynamics problem consists of three first-order PDEs (when continuity is brought in; see Eq. 3.20) and the 1-D MHD primitive Eq. (5.1)–(5.7) represent a hyperbolic system of *seven* first-order PDEs. Each of these systems may be written in the form of Eq. (C.4) where, for the 1-D HD problem, $|q\rangle$ and J are given by Eq. (3.22) and, for the 1-D MHD problem, by Eq. (5.9). The fact that all eigenvalues are real (Eq. 3.33 and 5.27) attests to the hyperbolicity of these systems of equations.

The wave approach to solving a system of differential equations is unique to hyperbolic systems of first-order PDEs. Because the eigenvalues are real, factors such as e^{ikut} – where u is one of the real eigenvalues (characteristic speeds) and k the wavenumber – are oscillatory, whence the wave approach. By the same token, because their eigenvalues are, in general, complex, parabolic and elliptical systems of equations do not lend themselves well to a wave decomposition, and other methods must be employed.

The eigenvalues of the 1-D hydrodynamic equations, namely $v \pm c_s$ and v, are clearly distinct since $c_s > 0$. Thus, these equations represent a strictly hyperbolic system of three first-order PDEs. On the other hand, it is shown in §5.2.3 that degeneracy among the MHD characteristics is possible, and thus the seven 1-D MHD equations do not constitute a strictly hyperbolic system. This degeneracy has implications on how numerical MHD algorithms are designed, since differences between characteristic speeds frequently appear in denominators of critical quantities. As shown in §6.2, this problem can be mediated by a suitable normalisation of the eigenvectors.

D | The Secant Method

D.1 Univariate root finder

For a single transcendental equation in one unknown,

$$f(x) = 0,$$

one is often faced with the task of solving for x; that is, finding the root(s) of $f(x)$. The most straight-forward numerical method, known as *bisection*, starts by finding x_1 and x_2 such that the product $f(x_1)f(x_2) < 0$ guaranteeing that the root, x, lies somewhere between x_1 and x_2. One then evaluates:

$$x_3 = \frac{x_1 + x_2}{2},$$

(thus bisecting the original interval $[x_1, x_2]$), and tests whether the root lies in $[x_1, x_3]$, and thus $f(x_1)f(x_3) < 0$, or $[x_3, x_2]$, and thus $f(x_3)f(x_2) < 0$. If the former (latter) is true, set $x_2 = x_3$ ($x_1 = x_3$), and repeat the process until $f(x_3)$ is, by whatever measure, sufficiently close to zero. Bisection is an example of an algorithm with *first-order convergence*, and typically will give a root to several digits of accuracy after 30 to 40 iterations.

With knowledge of the function's first derivative, $f'(x)$, one can easily construct a method with *second-order convergence* which is just as simple to code as bisection, yet converges to the root with several digits of accuracy in several iterations. Such algorithms are so simple, so fast, and so robust that it is rarely necessary to go beyond what is discussed in this appendix for finding almost every root of almost every univariate function, though the interested reader is referred to the excellent and definitive discussion in *Numerical Recipes* (Press et al., 1992) for alternatives.

Consider a two-term Taylor expansion of $f(x)$ about the point x_0 which, in this context, would be our first guess at a root:

$$f(x) \approx f(x_0) + (x - x_0)f'(x_0). \tag{D.1}$$

In practice, it will not matter much what value we choose for x_0, so long as it is "closer" to the root we seek than other roots $f(x)$ may have. For continuous and monotonic functions which can only have one root, it won't matter at all what our initial value for x_0 is; second-order root finders will just about always converge on the unique root in fewer than ten iterations.

If, in Eq. (D.1), x is the putative root, $f(x) = 0$ and we solve for x to get our

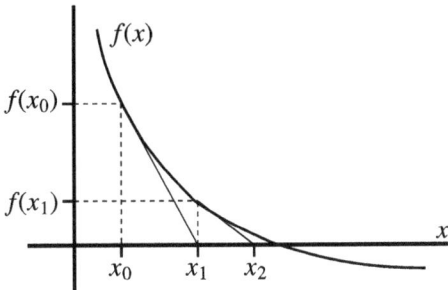

Figure D.1. Finding the root of a function $f(x)$ using a second-order root finding method in which one extrapolates from the previous guess, $(x_0, f(x_0))$ to $(x_1, 0)$ using a straight line of slope $f'(x_0)$.

refined guess at the root, x_1, namely,

$$x_1 = x_0 - \frac{f(x_0)}{f'(x_0)}. \tag{D.2}$$

This is represented graphically in Fig. D.1, where it can be seen that x_1 is necessarily a better estimate of the root than x_0 so long as we are restricted to a monotonic portion of the function near the actual root, and thus where $f(x)$ crosses the x-axis. One then sets $x_0 = x_1$ and repeats the process given by Eq. (D.1) and (D.2).

So what do we do about $f'(x)$? If the derivative function is known, we simply include in the computer program two functions: one each to describe $f(x)$ and $f'(x)$. Such a variation is known as the *Newton–Raphson method*, named for its inventors. If, on the other hand, $f'(x)$ is not a known function (or we simply can't be bothered programming the derivative function!), an *equally* good method to Newton–Raphson is the *secant method*.

In the secant method, two "first guesses" are made, $f(x_{-1})$ and $f(x_0)$ say, where x_{-1} is "close" to x_0, and then the derivative, $f'(x_0)$ is approximated by a secant, namely,

$$f'(x_0) \approx \frac{f(x_0) - f(x_{-1})}{x_0 - x_{-1}}. \tag{D.3}$$

One then uses Eq. (D.2) to find x_1, resets $x_{-1} = x_0$ and $x_0 = x_1$, re-evaluates the derivative using Eq. (D.3), and generates the next guess using Eq. (D.2). This cycle is repeated until sufficient convergence on the root has been achieved, generally after no more than several iterations.

A *FORTRAN* function named `secant` is given in the following pages that can be freely copied into any computer program requiring a secant finder. You will need to provide your own function that returns the value of $f(x)$ for any value of x passed to it. Then, to find the root of a particular function, you issue the statement:

```
root = secant ( x0, fofx )
```

in your program, where `root` is the variable set to the root of the function, `x0` is your current guess (x_0), and `fofx` is the name of your function describing $f(x)$.

```
c=================================================================================
c
      function secant ( xinit, f )
c
c     written by: David Clarke
c     date      : February, 2002
c     modified 1:
c
c     PURPOSE: This routine performs the secant method to find the root of
c     the input function, f.
c
c     INPUT VARIABLES:
c       xinit        initial guess; a close initial guess is good, but not
c                    always necessary.
c       f            external function for which root is being sought.
c
c     OUTPUT VARIABLES:
c       secant       returned as the root when the improvement over the
c                    previous iteration is less than maxerr, a parameter
c                    set by this routine.
c
c---------------------------------------------------------------------------------
c
      implicit    none
c
      integer     maxiter
      real*8      maxerr  , small
      parameter   ( maxiter=30, maxerr=1.0d-8, small=1.0d-99 )
c
      integer     iter
      real*8      xinit   , xnm1     , xn       , fxn      , fxnm1
     1                    , err      , ferr
      real*8      f       , secant
c
c---------------------------------------------------------------------------------
c
c     Initialise variables.
c
      iter = 0
      xn   = xinit
      xnm1 = 1.1d0 * xn
      if (xn .eq. 0.0d0) xnm1 = 0.1d0
      fxn   = f(xn)
      fxnm1 = f(xnm1)
c
c----- Top of secant loop. -------------------------------------------------------
c
10    continue
         iter  = iter + 1
c
c     Evaluate the next Secant guess for the root (secant).
c
         secant = xn - fxn * ( xn - xnm1 ) / ( ( fxn - fxnm1 ) + small )
c
c     Estimate error.
c
```

```
              err   = abs ( secant - xn )
              ferr  = err / ( abs (secant) + small )
c
c     If maximum number of iterations has been exceeded, issue warning
c     and return to calling routine.
c
              if ( iter .gt. maxiter ) then
                 write (6, 2010) maxiter, xn, 100.0d0*ferr, fxn
                 go to 20
              endif
c
c     If the fractional error of "xn" is less than or equal to
c     "maxerr", the root has been found and execution is returned to the
c     calling routine.
c
              if ( ferr .le. maxerr ) go to 20
c
c     Otherwise, perform another iteration.
c
              xnm1   = xn
              fxnm1  = fxn
              xn     = secant
              fxn    = f(xn)
              go to 10
 20        continue
c
c----- Bottom of secant loop. -----------------------------------------
c
 2010     format('SECANT  : Maximum number of iterations ',i3,' exceeded.'
     1         ,/
     2         ,'SECANT  : xr (best guess) = ',1pg12.5,' +/- ',g9.2
     3         ,' %.  f(xr) = ',g12.5,'.')
c
           return
           end
c
c=======================================================================
```

D.2 Multivariate root finder

Consider now two transcendental equations in two unknowns:
$$f_1(x_1, x_2) = 0 \quad \text{and} \quad f_2(x_1, x_2) = 0. \tag{D.4}$$

Suppose $\vec{r}_0 = (x_1, x_2)$ is our current guess for the roots of Eq. (D.4), but do not solve Eq. (D.4) to within our stated tolerance. We wish to find a $\delta \vec{r} = (\delta x_1, \delta x_2)$ such that $\vec{r} = \vec{r}_0 + \delta \vec{r}$ comes closer to satisfying Eq. (D.4) than $\vec{r}_0$ does.

To this end, consider first-order Taylor expansions of f_1 and f_2 in the direction $\delta \vec{r}$:

$$f_1(x_1+\delta x_1, x_2+\delta x_2) \approx f_1(x_1, x_2) + \nabla f_1 \cdot \delta \vec{r} = f_1(x_1, x_2) + \partial_1 f_1 \, \delta x_1 + \partial_2 f_1 \, \delta x_2;$$

$$f_2(x_1+\delta x_1, x_2+\delta x_2) \approx f_2(x_1, x_2) + \nabla f_2 \cdot \delta \vec{r} = f_2(x_1, x_2) + \partial_1 f_2 \, \delta x_1 + \partial_2 f_2 \, \delta x_2,$$

where $\partial_i f_j = \partial f_j / \partial x_i$. In matrix form, this becomes:

$$|f(\vec{r})\rangle \approx |f(\vec{r}_0)\rangle + \underbrace{\begin{bmatrix} \partial_1 f_1 & \partial_2 f_1 \\ \partial_1 f_2 & \partial_2 f_2 \end{bmatrix}}_{\mathsf{J}} |\delta r\rangle, \qquad (\text{D.5})$$

where J is the Jacobian matrix, first introduced in §2.1.2. Taking $|f(\vec{r})\rangle \approx |0\rangle$ ($\vec{r}$ is closer to the root we seek than $\vec{r}_0$), we set Eq. (D.5) approximately to zero to find,

$$\boxed{\mathsf{J}|\delta r\rangle \approx -|f(\vec{r}_0)\rangle,} \qquad (\text{D.6})$$

which we solve for $|\delta r\rangle$. Since this is an approximate equation, $\vec{r} = \vec{r}_0 + \delta\vec{r}$ still won't satisfy Eq. (D.4), but $|f(\vec{r})\rangle$ should be closer to zero than $|f(\vec{r}_0)\rangle$ was, and thus we iterate until a set of values, (x_1, x_2), have been found to satisfy Eq. (D.4) to within our stated tolerance. Note that while our discussion has been restricted to two equations in two unknowns, Eq. (D.6) is just as applicable to systems of n equations (constraints) in n unknowns (parameters).

Formally, the solution to Eq. (D.6) is:

$$|\delta r\rangle = -\mathsf{J}^{-1}|f(\vec{r}_0)\rangle,$$

where J^{-1} is the inverse of the Jacobian. In practice, however, it is more efficient to do a so-called LU decomposition on J, and solve for the components of $|\delta r\rangle$ by forward and backward substitution. The interested reader will find guidance in widely available resources (*e.g.*, Press *et al.*, 1992).

Finally, if the derivatives of the various functions (constraints) are known, using them directly in the Jacobian would make this a *multivariate Newton-Raphson root finder*. Without the derivatives, one can estimate them by taking "secants" such as,

$$\partial_j f_i \approx \frac{\delta f_i}{\delta x_j} = \frac{f_i(x_1,\ldots,x_n) - f_i(x_1,\ldots,x_j-\delta x_j,\ldots,x_n)}{\delta x_j}, \qquad (\text{D.7})$$

for J_{ij}, the $(i,j)^{\text{th}}$ element of the Jacobian. This describes a *multivariate secant root finder*, and is the method used in the algorithm for the MHD Riemann solver (§6.4.3).

E   Roots of a Cubic

With the quadratic formula known since antiquity, Niccolò Tartaglia (1500?–1557) revealed to a colleague, Gerolamo Cardano, in 1530 his formula for the roots of a cubic in terms of its coefficients. He did so on condition Cardano not publish it and – as was not uncommon in those days – it was written entirely in verse! Some time later, Cardano came upon an independently derived and unpublished solution by Scipione del Ferro and, as this was dated before Tartaglia's poem, felt this released him from his promise not to publish it. He then included it – along with the solution to the quartic derived in 1540 by his student, Lodovico Ferarri – in his book *Ars Magna* published in 1545. Even though he gave ample credit to both mathematicians in his text, Tartaglia felt betrayed and publicly chastised Cardano for a decade, maintaining one of the most storied feuds in the history of mathematics.

Today, the formula for the roots of a cubic is known as the Cardano–Tartaglia formula, derived in a time when most mathematicians did not acknowledge, never mind understand, the existence and properties of imaginary numbers! Thus, the solution presented here is a very modern version of the CT formula, one that neither man would likely recognise.

The three roots to a general cubic,

$$f_3(x) = a_0 + a_1 x + a_2 x^2 + x^3 = (x - x_1)(x - x_2)(x - x_3) = 0,$$

are given by,

$$x_k = -\frac{1}{3}\left(a_2 + \omega^k \zeta + \frac{\delta_0}{\omega^k \zeta}\right); \quad k = 1, 2, 3,$$

where:

$$\omega = \frac{1}{2}\left(-1 + \sqrt{3}\,i\right) = \sqrt[3]{1}; \qquad \delta_0 = a_2^2 - 3a_1;$$

$$\zeta = \sqrt[3]{\delta_1 + \sqrt{\delta_1^2 - \delta_0^3}}; \qquad \delta_1 = a_2^3 - \tfrac{9}{2} a_2 a_1 + \tfrac{27}{2} a_0.$$

In general, one can choose any root or cube root one wants; doing so simply permutes the subscript index, k, on the roots. The one exception is if $\delta_0 = 0$, one chooses $\sqrt{\delta_1^2} = \delta_1$, thus preserving the sign of δ_1 so that $\zeta \neq 0$.

Finally, the *discriminant* is given by:

$$\Delta = \frac{4}{27}(\delta_0^3 - \delta_1^2) \begin{cases} > 0 & \text{three distinct real roots;} \\ = 0 & \text{three real but degenerate roots; or} \\ < 0 & \text{one real root and two complex conjugate roots.} \end{cases} \qquad (E.1)$$

F Sixth-Order Runge–Kutta

Consider a function, $y(x)$, whose values are known only at discrete points separated by the *step size*, h (Fig. F.1a), and for which we wish to estimate its derivative, $y'(x)$. If we take a Taylor expansion from x to $x + h$, then,

$$y(x+h) = y(x) + hy'(x) + \frac{h^2}{2}y''(x) + \frac{h^3}{3!}y'''(x) + \cdots$$

$$\Rightarrow \quad y'_f(x) \equiv \frac{y(x+h) - y(x)}{h} = y'(x) + \underbrace{\frac{h}{2}y''(x) + \frac{h^2}{3!}y'''(x) + \cdots}_{\text{error terms}},$$

which defines the *forward derivative* (estimated from the function evaluated at x and the point *forwards*, $x + h$; red line in Fig. F.1a), whose leading error term is evidently $\propto h$ [*i.e.*, of order h, written as $\mathcal{O}(h)$]. That is, if one were to halve the step size, h, one would also halve the leading, and normally most onerous, error term, $hy''(x)/2$.

Similarly, we can take a Taylor expansion from x to $x - h$:

$$y(x-h) = y(x) - hy'(x) + \frac{h^2}{2}y''(x) - \frac{h^3}{3!}y'''(x) + \cdots$$

$$\Rightarrow \quad y'_b(x) \equiv \frac{y(x) - y(x-h)}{h} = y'(x) - \underbrace{\frac{h}{2}y''(x) + \frac{h^2}{3!}y'''(x) - \cdots}_{\text{error terms}},$$

to define the *backward derivative* (estimated from the function evaluated at x and the point *backwards*, $x - h$; blue line in Fig. F.1a), whose leading error term is also $\mathcal{O}(h)$. Note that this time, however, the error terms form an *alternating* series.

Both forward and backward derivatives are similarly accurate [$\mathcal{O}(h)$], and there is nothing to choose between them in this regard. However, we can define a *central derivative* (green line in Fig. F.1a) by averaging the two, in which case we get:

$$y'_c(x) \equiv \frac{y'_f(x) + y'_b(x)}{2}$$

$$= \frac{y'(x) + y'(x)}{2} + \frac{h}{2}\frac{y''(x) - y''(x)}{2} + \frac{h^2}{3!}\frac{y'''(x) + y'''(x)}{2} + \cdots$$

$$= y'(x) + \underbrace{\frac{h^2}{3!}y'''(x) + \frac{h^4}{5!}y^{(v)}(x) + \cdots}_{\text{error terms}},$$

where the leading error term is now $\mathcal{O}(h^2)$. That is, halving h *quarters* the error

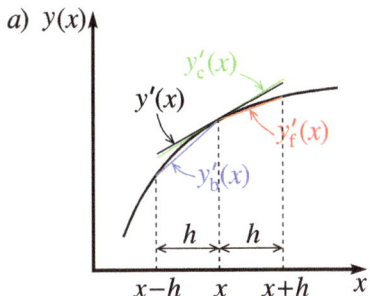 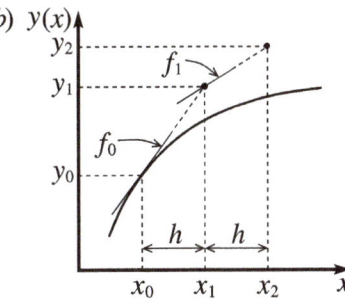

Figure F.1. *a*) Forward (y'_f, red), backward (y'_b, blue), central (y'_c, green) and "actual" (black) derivatives at x. Of the three coloured slopes, the central (green) most closely mimics the black. *b*) First-order Euler solver illustrated over two intervals.

committed in making the central derivative estimate of $y'(x)$. Further, all error terms of odd order cancel, and the next error term is $\mathcal{O}(h^4)$, followed by $\mathcal{O}(h^6)$, etc. Evidently, the central derivative is more accurate than the forward and backward derivatives, as seen in Fig. F.1*a* where, of the three derivative estimates pictured, the actual derivative (black line) is best approximated by the central one.

So, with this mini-tutorial on the accuracy of numerical derivatives under our belts, let us now consider a general first-order ODE:

$$y'(x) = f(x, y(x)), \tag{F.1}$$

whose solution, $y(x)$, we seek at discrete points, $x_j = x_0 + jh$, $j \geq 0 \in \mathbb{Z}$, with h once again the step size, for a specified boundary condition, $y(x_0) = y_0$. To advance the solution from the one known point, (x_0, y_0), we perform a Taylor expansion from x_0 to x_1:

$$y(x_1) \equiv y_1 = y(x_0) + hy'(x_0) + \cdots$$
$$\approx y_0 + hf_0,$$

where, from Eq. (F.1), $y'(x_0) = f(x_0, y_0) \equiv f_0$ is a quantity we can calculate.

Now that we have y_1, we calculate the slope of the function at x_1, namely $y'(x_1) = f(x_1, y_1) \equiv f_1$, and Taylor-expand from x_1 to the next point, x_2:

$$y(x_2) \equiv y_2 = y(x_1) + hy'(x_1) + \cdots$$
$$\approx y_1 + hf_1,$$

and so it goes.

This describes a so-called *first-order Euler solver*, with the first two steps illustrated in Fig. F.1*b*. Evidently, each step is advanced by a backward derivative and, as such, this method is *first-order accurate* [$\mathcal{O}(h)$]. First-order methods can require *extremely* small step sizes to maintain the solution to within even a modest tolerance, say 1%, over a useful domain in x. As a case in point, the step size in Fig. F.1*b* is clearly too large since, just after two steps, the estimated value for y_2 has deviated significantly from the target function represented by the solid curve.

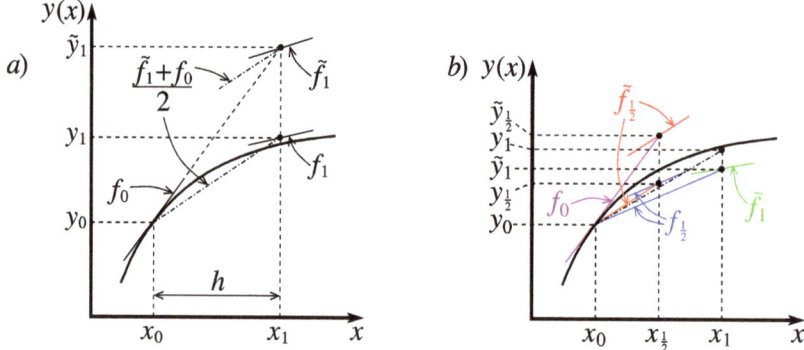

Figure F.2. *a*) Second-order predictor-corrector solver illustrated over one interval. *b*) Fourth-order Runge–Kutta solver illustrated over two half-intervals.

Indeed it can be a losing battle, where the step size needed to maintain accuracy is so small that the number of steps required to integrate over the specified domain can make the calculation prohibitive. In a word, first-order methods can be *useless*.

To devise a second-order method, we need a central derivative, and one way to achieve this is a method known as *Predictor–Corrector* (PC; Fig. F.2*a*). Here, we *predict* values for y_1 and the slope, f_1, at the end of the interval using the first-order Euler solver:

$$\tilde{y}_1 = y_0 + hf_0; \quad \tilde{f}_1 = f(x_1, \tilde{y}_1),$$

(tildes ~ indicate "predicted values"), then *correct* the value for y_1 by starting again from x_0 this time using the average of the slopes f_0 and $\tilde{f}_1$. Once we get y_1, we compute the "corrected slope" there too:

$$y_1 = y_0 + h\frac{f_0 + \tilde{f}_1}{2}; \quad f_1 = f(x_1, y_1), \tag{F.2}$$

to be used in integrating across the next interval. Using a slope from effectively the middle of the interval renders this algorithm second-order accurate and requires far fewer intervals than a first-order method to maintain a given accuracy over the specified domain. As Fig. F.2*a* shows, the deviation of y_1 from the putative $y(x)$ (solid curve) is far less severe than that of $\tilde{y}_1$.

But wait! We're just getting started!! An embarrassingly clever technique in numerical estimation known as *Richardson extrapolation* exploits our knowledge of the order of the error terms. Since PC is second-order accurate where halving the step size quarters the error, integrating $y(x)$ from x_0 to x_1 in two successive steps of step size $h/2$ (to obtain $y_{1,\frac{h}{2}}$) should commit a quarter of the error as integrating from x_0 to x_1 in a single step of step size h (to obtain $y_{1,h}$). Thus, in the quantity,[1]

$$y_1 = \frac{4y_{1,\frac{h}{2}} - y_{1,h}}{3}, \tag{F.3}$$

---

[1]Think of y_1 as being formed from four times the estimate of $y(x_1)$ constructed by taking two successive half steps, $y_{1,\frac{h}{2}}$, and subtracting from that one estimate of $y(x_1)$ constructed by taking one full step, $y_{1,h}$. Thus, the numerator represents three estimates altogether, and the denominator $(3 = 4 - 1)$ is the required normalisation.

the error term of $\mathcal{O}(h^2)$ is eliminated, leaving a leading error term of $\mathcal{O}(h^4)$ (recall central differences have no error terms of odd order), and y_1 in Eq. (F.3) is *fourth-order* accurate.

Using PC, we can work out a closed-form expression for a fourth-order accurate y_1 as follows. Integrating Eq. F.1 with two successive half steps of PC yields:

$$\tilde{y}_{\frac{1}{2},\frac{h}{2}} = y_0 + \frac{h}{2}f_0; \quad y_{\frac{1}{2},\frac{h}{2}} = y_0 + \frac{h}{2}\frac{f_0 + \tilde{f}_{\frac{1}{2}}}{2}, \tag{F.4}$$

where $f_0 = f(x_0, y_0)$ and $\tilde{f}_{\frac{1}{2}} = f(x_{\frac{1}{2}}, \tilde{y}_{\frac{1}{2},\frac{h}{2}})$, and where:

$$\tilde{y}_{1,\frac{h}{2}} = y_{\frac{1}{2},\frac{h}{2}} + \frac{h}{2}f_{\frac{1}{2}}; \quad y_{1,\frac{h}{2}} = y_{\frac{1}{2},\frac{h}{2}} + \frac{h}{2}\frac{f_{\frac{1}{2}} + \tilde{f}_1}{2}, \tag{F.5}$$

where $f_{\frac{1}{2}} = f(x_{\frac{1}{2}}, y_{\frac{1}{2},\frac{h}{2}})$ and $\tilde{f}_1 = f(x_1, \tilde{y}_{1,\frac{h}{2}})$. Now, substitute the second of Eq. (F.4) into the second of Eq. (F.5) to get,

$$y_{1,\frac{h}{2}} = y_0 + \frac{h}{4}(f_0 + \tilde{f}_{\frac{1}{2}} + f_{\frac{1}{2}} + \tilde{f}_1). \tag{F.6}$$

Finally, substituting Eq. (F.6) and the first of Eq. (F.2) into Eq. (F.3), we find,

$$\boxed{y_1 = y_0 + \frac{h}{6}(f_0 + 2\tilde{f}_{\frac{1}{2}} + 2f_{\frac{1}{2}} + \tilde{f}_1),}$$

a fourth-order estimate of y_1 given that all four slopes are first order.

This is the essence of the *fourth-order Runge–Kutta method*, named for the German mathematicians Carl Runge and Martin Kutta who developed the technique at the turn of the 20th century. Thus, with the various slopes illustrated in Fig. F.2b, the method requires the following calculations to advance the solution a single step, h, from (x_0, y_0) to (x_1, y_1):

$$\left. \begin{array}{ll} f_0 = f(x_0, y_0); & \tilde{y}_{\frac{1}{2}} = y_0 + \dfrac{h}{2}f_0; \\[6pt] \tilde{f}_{\frac{1}{2}} = f(x_{\frac{1}{2}}, \tilde{y}_{\frac{1}{2}}); & y_{\frac{1}{2}} = y_0 + \dfrac{h}{2}\tilde{f}_{\frac{1}{2}}; \\[6pt] f_{\frac{1}{2}} = f(x_{\frac{1}{2}}, y_{\frac{1}{2}}); & \tilde{y}_1 = y_0 + h f_{\frac{1}{2}}; \\[6pt] \tilde{f}_1 = f(x_1, \tilde{y}_1); & y_1 = y_0 + \dfrac{h}{6}(f_0 + 2\tilde{f}_{\frac{1}{2}} + 2f_{\frac{1}{2}} + \tilde{f}_1). \end{array} \right\} \tag{F.7}$$

The eight expressions in Eq. (F.7) comprise what I refer to as "the RK step", which can be easily programmed for a known boundary condition, (x_0, y_0), and a known derivative function, $y'(x) = f(x, y)$. Further, it should be apparent that nowhere was it required that y and f be scalars. Indeed, if we were to write Eq. (F.1) as,

$$|y'(x)\rangle = |f(x, y(x))\rangle, \tag{F.8}$$

where the kets are n-dimensional, *nothing* in the previous discussion would change. Thus, a Runge–Kutta routine can be written for one first-order ODE in one unknown, or for n coupled first-order ODEs in n unknowns; for a programmer, the difference amounts to nothing more than putting key expressions inside a do-loop.

Adaptive step size

Fourth-order accuracy is good, so sixth order must be better, but the reason for *sixth-order Runge–Kutta* is much more profound than just upping the accuracy to the next level. Indeed, the fact that sixth-order Runge–Kutta is sixth-order accurate is actually a by-product of a much more important property bestowed upon it: adaptive stepping.

An algorithm's order of accuracy is just one aspect of error control; it is equally important to ensure the solution remains within a prescribed tolerance. While an nth order scheme means that the solution improves by a factor of 2^n if the step size is halved, it cannot, by itself, control what the absolute error is. Indeed, in a region where the solution, $y(x)$, varies rapidly with small changes in x, one can imagine that a very much smaller step size is required to achieve the same absolute error than in an asymptotic region where $y(x)$ barely varies at all even for large changes in x.

First, a word on the nature of error accumulation in Runge–Kutta methods. It should give the reader some cause for pause that one starts at a single known point of the solution, namely the boundary condition (x_0, y_0), then effectively extrapolates all the way across the domain from there, knowing only what the first derivative is at each point. Further, if the derivative depends upon y, then the use of errant values of y will make even the derivative known only to within some – possibly unknowable – uncertainty. No matter how good the scheme, errors will accumulate along the way, and requiring a tolerance of, say, 10^{-8} only means that Δy across the next step will be accurate to within that level, not the value of $y(x)$ itself. If it takes 10^4 steps to cross the domain, then, in the best case scenario where the errors combine linearly and statistically, the error at the end of the domain may be as high as 10^{-6}. If the accumulation of errors is nonlinear (*e.g.*, y' depends explicitly on y), error accumulation could be much worse.

With this caveat in mind, we need a measure of the error committed in Δy across the current interval in order to know whether it is within our intended tolerance. With a fourth-order scheme, we know that halving the step size drops the error by a factor of $2^4 = 16$. Thus, consider integrating across the current interval twice, first with one RK step using a full step size, h, to get $y_{1,h}$, then again with two RK steps each using half a step size, $h/2$, to get $y_{1,\frac{h}{2}}$.[2] First, this allows us to take another Richardson extrapolation,

$$y_1 = \frac{16 y_{1,\frac{h}{2}} - y_{1,h}}{15}, \tag{F.9}$$

whose leading error term is now $\mathcal{O}(h^6)$, whence the moniker *sixth-order Runge–Kutta*. Second and more importantly, it allows us to estimate the error of y_1 by

---

[2] I appreciate I am reusing notation from just two pages back where we extended second-order PC by one extrapolation step to fourth order. However, there are only so many subscripts, superscripts, tildes, primes, *etc.*, one can impose upon a label before the notation becomes completely inscrutable!

comparing it to the next best guess we have, namely $y_{1,\frac{h}{2}}$:

$$\varepsilon_{\text{RK}} = \frac{y_1 - |y_{1,\frac{h}{2}}|}{\max\left(|y_1|, |y_{1,\frac{h}{2}}|\right)}. \tag{F.10}$$

If the error, ε_{RK}, is greater than the tolerance, we halve the step size ($h \to h/2$) and start the current step again. If ε_{RK} is, say, less than a tenth of the stated tolerance, we keep our current step knowing its accuracy is even better than we require, but increase h by, say, 50% for the next step. Finally, if ε_{RK} is less than the stated tolerance but greater than a tenth of it, we leave well enough alone and retain h for the next step. Evidently, an estimate of the error allows us to adapt the step size to the local nature of the solution, an attribute of the scheme whose importance cannot be overstated.

Sixth-order RK algorithm

Following is a skeletal algorithm for sixth-order Runge–Kutta, written assuming n coupled first-order ODEs as represented by Eq. (F.8). Text in `typewriter font` represents pseudo-code or pseudo-code variables. Variable names ending in `left` or `right` indicate values at the left/right side of the current interval. Variables `x` and `y` gather discrete values of the independent and dependent variables as RK integrates across the domain. The right arrow ($\to$) indicates the quantity on the left is replaced with that on the right. Text in a rectangular box indicates subroutines whose algorithms follow the main one. Text in boxes with rounded corners indicates targets, similar to those used by the old *FORTRAN* "go to" statements.

1. Initialisation.

    - set error tolerance, `errmax`, and domain, (`x0`, `xmax`).

    - set initial step size: `h = 0.1 * errmax * (xmax-x0)`.

    - set boundary conditions:

        `step = 1;`

        `x(step) = xleft = x0;`

        `y(i,step) = yleft(i) = y0(i), i=1,...,n.`

A. (top of adaptive RK loop)

2. Integrate solution across current interval with RunKut twice:

    - in one step with step size `h`: `yh(i)`;

    - in two successive steps with step size `h/2`: `yh2(i)`.

3. Evaluate yright(i) from Eq. (F.9): $\text{yright}(i) = \dfrac{16 * \text{yh2}(i) - \text{yh}(i)}{15}$.

4. Estimate error from Eq. (F.10): $\text{err}(i) = \dfrac{|\text{yright}(i) - \text{yh2}(i)|}{\max(|\text{yright}(i)|, |\text{yh2}(i)|)}$;

   - err = max[err(i)];

   - if err > errmax, h → h/2, go to A ;

   - else, if err < 0.1 * errmax, h → 1.5*h.

5. Integration across current interval is within error tolerance; prepare for next step:

   - step → step + 1;

   - xleft → xleft + h; x(step) = xleft;

   - ∀ i=1,n, y(i,step) = yright(i);

   - if xleft < xmax, go to A for next step; else, go to B .

B . (bottom of adaptive RK loop; exit)

Niceties of the algorithm include setting a value hmin below which the step size is not allowed to fall to avoid spiralling into an infinite loop should the integration, for whatever reason, fail to converge. Similarly, one can set a maximum number of integration steps that the loop is allowed to take. It's rare for sixth-order Runge–Kutta to require more than 10^3 steps to integrate across a typical domain for a typical function, and it could be a sign of trouble if more steps than this are needed.

Subroutine RunKut

To integrate across a single interval, this routine is called thrice. The first call integrates across the entire interval in one RK step starting at xleft with interval size h. The second and third calls use a step size h/2 with the second call starting at xleft and the third at xleft+h/2.

Here we shall refer to the starting point and interval size as xbeg and Δx respectively. The middle of the interval is xmid = xbeg + Δx/2, while the values of the n dependent variables at the beginning, middle, and end of the interval are ybeg(i), ymid(i), and yend(i), i=1,...,n, respectively.

Each time this routine is called, its task is to evaluate Eq. (F.7), which requires four evaluations of $y'(x)$.

6. Evaluate first and second of Eq. (F.7). All variables beginning with y or f are n-dimensional vectors; 't' at the end of a variable name means "twiddle":

$$\text{fbeg} = \boxed{\text{yprime}}(\text{xbeg}, \text{ybeg}); \quad \text{ymidt} = \text{ybeg} + \dfrac{\Delta x}{2} * \text{fbeg}.$$

7. Evaluate third and fourth of Eq. (F.7):
$$\text{fmidt} = \boxed{\text{yprime}}(\text{xmid}, \text{ymidt}); \quad \text{ymid} = \text{ybeg} + \frac{\Delta x}{2} * \text{fmidt}.$$

8. Evaluate fifth and sixth of Eq. (F.7):
$$\text{fmid} = \boxed{\text{yprime}}(\text{xmid}, \text{ymid}); \quad \text{yendt} = \text{ybeg} + \Delta x * \text{fmid}.$$

9. Evaluate seventh and eighth of Eq. (F.7):
$$\text{fendt} = \boxed{\text{yprime}}(\text{xend}, \text{yendt});$$
$$\text{yend} = \text{ybeg} + \frac{\Delta x}{6} * (\text{fbeg} + 2 * \text{fmidt} + 2 * \text{fmid} + \text{fendt}).$$

Function $\boxed{\text{yprime}}$

10. Evaluate Eq. (F.8) from input values of x and y.

    This is the only place in the routine where the differential equations are specified; Steps 1–9 are completely generic in that regard.

G Coriolis' Theorem

We are all familiar with the experience of being "pushed back" into the seat of an aggressively accelerated car, or of being "pressed into" the car door if the driver takes a sharp corner too quickly. In this context, the phrases "pushed back" and "pressed into" have entered the vernacular since our common experience is these are somehow *forces* being exerted upon us as the car accelerates. However, as physicists we know that these forces are not "real" (however hard we may struggle to counter them), but rather "inertial" that arise by virtue of being in an accelerating reference frame. In this appendix, I remind the reader of *Coriolis' theorem* which identifies four distinct types of inertial forces arising in such a frame.

Figure G.1a on the next page shows two Cartesian coordinate systems, O (black) and O' (blue) each observing a point P (red) and whose origins are separated by a displacement $\vec{R}$. O, defined by unit vectors $\hat{\imath}$, $\hat{\jmath}$, and $\hat{k}$ in the x-, y-, and z-directions respectively, is an *inertial* (non-accelerating) frame of reference while O', with unit vectors $\hat{\imath}'$, $\hat{\jmath}'$, and $\hat{k}'$ is an accelerating frame of reference because of its rotation about a fixed axis, A,[1] and possibly because $\ddot{\vec{R}} \neq 0$. Without loss of generality, let $\hat{k}$ be aligned with A.

The question we wish to address is as follows. If O observes P to be at displacement $\vec{r}$ moving with velocity $\vec{v}$ and acceleration $\vec{a}$, how are these related to the corresponding kinematical quantities observed by O'?

The first is easy. Evidently:

$$\vec{r} = \vec{R} + \vec{r}', \tag{G.1}$$

where $\vec{r} = x\hat{\imath} + y\hat{\jmath} + z\hat{k}$ and $\vec{r}' = x'\hat{\imath}' + y'\hat{\jmath}' + z'\hat{k}'$.

For the velocity, differentiate Eq. (G.1) with respect to time to get,

$$\vec{v} = \frac{d\vec{r}}{dt} = \frac{d}{dt}(\vec{R} + \vec{r}') = \vec{V} + \frac{d}{dt}(x'\hat{\imath}' + y'\hat{\jmath}' + z'\hat{k}')$$

$$\Rightarrow \quad \vec{v} = \vec{V} + \underbrace{\dot{x}'\hat{\imath}' + \dot{y}'\hat{\jmath}' + \dot{z}'\hat{k}'}_{\vec{v}'} + x'\dot{\hat{\imath}}' + y'\dot{\hat{\jmath}}' + z'\dot{\hat{k}}', \tag{G.2}$$

where $\vec{V} = \dot{\vec{R}}$ is the relative velocity between the origins O and O'.

Two things here are worthy of note. First, unlike $\hat{\imath}$, *etc.*, of the inertial frame O whose orientations remain fixed, the directions of $\hat{\imath}'$, *etc.*, in the rotating frame O' are always changing, and thus have time derivatives which we evaluate below. Second, the velocity of P as measured by O', namely $\vec{v}'$, is completely specified by

---

[1] Rotation is in a "tidally locked" fashion. Should an observer at O' be facing the rotation axis, A, that observer remains facing A as O' rotates.

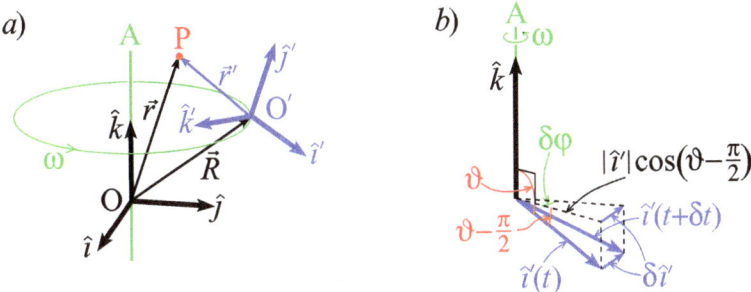

Figure G.1. *a)* Two coordinate systems, O (black) and O' (blue), displaced from each other by $\vec{R}$ both observe point P. O is inertial and O' is accelerating and/or rotating about $\hat{k}$. *b)* The unit vector $\hat{i}'$ from rotating frame O' in panel *a* rotates by a small angle about the $\hat{k}$ axis from inertial frame, O. Since $\hat{i}' \not\perp \hat{k}$ in general, it's the *projection* of $\hat{i}'$ onto the plane perpendicular to $\hat{k}$ that rotates by angle $\delta\varphi = \omega\delta t$.

the underbrace: the vector sum of the time-rate-of-change of the coordinates and their unit vector ($\dot{x}'\hat{i}'$, etc.). One would no more include the terms proportional to $\dot{\hat{i}}'$, etc., in $\vec{v}'$ than one would worry about how the rotation of the earth changes the absolute direction of a highway. If I'm driving on a straight road at 100 kph due east, as far as I'm concerned I'm always driving due east regardless of how that direction may change relative to an inertial observer hovering over the planet. Accordingly, the last three terms in Eq. (G.2) are not included as part of $\vec{v}'$.

Of course, in the inertial frame O, terms in Eq. (G.2) proportional to $\dot{\hat{i}}'$, etc., do matter, and so we look now to evaluate them. Figure G.1*b* illustrates how the unit vector $\hat{i}'$ changes relative to the inertial frame, O, after a small rotation, $\delta\varphi = \omega\delta t$, about $\hat{k}$. As $\hat{i}'(t)$ rotates to $\hat{i}'(t+\delta t)$, the magnitude of $\delta\hat{i}'$ is evidently given by,

$$|\delta\hat{i}'| = |\hat{i}'|\cos\left(\vartheta - \tfrac{\pi}{2}\right)\delta\varphi = \omega\delta t \sin\vartheta,$$

where ϑ is the angle between $\hat{i}'$ and $\vec{\omega} = \omega\hat{k}$. This and the fact that $\delta\hat{i}'$ is perpendicular to both $\hat{i}'$ and $\vec{\omega}$ for $\delta\varphi \ll 2\pi$ require that,

$$\delta\hat{i}' = (\vec{\omega}\delta t) \times \hat{i}' \quad \Rightarrow \quad \dot{\hat{i}}' = \vec{\omega} \times \hat{i}'.$$

Similarly, $\dot{\hat{j}}' = \vec{\omega} \times \hat{j}'$, $\dot{\hat{k}}' = \vec{\omega} \times \hat{k}'$, and Eq. (G.2) becomes:

$$\vec{v} = \vec{V} + \vec{v}' + x'\vec{\omega} \times \hat{i}' + y'\vec{\omega} \times \hat{j}' + z'\vec{\omega} \times \hat{k}'$$
$$= \vec{V} + \vec{v}' + \vec{\omega} \times \vec{r}'.$$

For the acceleration, we differentiate once again to find,

$$\vec{a} = \frac{d\vec{v}}{dt} = \vec{A} + \underbrace{\frac{d}{dt}(\dot{x}'\hat{i}' + \dot{y}'\hat{j}' + \dot{z}'\hat{k}')}_{\vec{a}' + \vec{\omega} \times \vec{v}'} + \frac{d\vec{\omega}}{dt} \times \vec{r}' + \vec{\omega} \times \underbrace{\frac{d}{dt}(x'\hat{i}' + y'\hat{j}' + z'\hat{k}')}_{\vec{v}' + \vec{\omega} \times \vec{r}'},$$

where $\vec{A} = \ddot{\vec{R}}$ is the acceleration of the origin of O' relative to O, $\vec{a}'$ is the acceleration of point P relative to O' (the quantity we've been seeking), and where the

underbraces are evaluated by repeating the steps previous. Thus, we find,
$$\vec{a} = \vec{A} + \vec{a}\,' + \dot{\vec{\omega}} \times \vec{r}\,' + 2\vec{\omega} \times \vec{v}\,' + \vec{\omega} \times (\vec{\omega} \times \vec{r}\,').$$
Note that $\vec{\omega} = \omega \hat{k}$, and thus its derivative does not liberate a term proportional to the time-derivative of a unit vector.

Solving for $\vec{a}\,'$, we get,
$$\vec{a}\,' = \vec{a} - \vec{A} - \dot{\vec{\omega}} \times \vec{r}\,' - 2\vec{\omega} \times \vec{v}\,' - \vec{\omega} \times (\vec{\omega} \times \vec{r}\,'), \tag{G.3}$$
and then multiplying through by m, the mass of the object at P, we arrive at *Coriolis'* theorem:
$$\boxed{\sum \vec{F}\,' = m\vec{a}\,' = \sum \vec{F} + \vec{F}_{\text{tr}} + \vec{F}_{\perp} + \vec{F}_{\text{Cor}} + \vec{F}_{\text{cent}},} \tag{G.4}$$
where, in addition to the "real" forces $(\sum \vec{F})$ such as gravity,[2] electromagnetic forces, *etc.*, four separate *inertial forces* have been revealed.

1. $\vec{F}_{\text{tr}} = -m\vec{A}$ is the *translational force* on m caused by the acceleration of origin O' relative to O. This is what "pushes you into the car seat" as your driver accelerates aggressively.

2. $\vec{F}_{\perp} = -m\dot{\vec{\omega}} \times \vec{r}\,'$ is the *transverse force* on m acting perpendicular to its displacement from O' should $\dot{\vec{\omega}} \neq 0$.

3. $\vec{F}_{\text{Cor}} = -2m\vec{\omega} \times \vec{v}\,'$ is the *Coriolis force* on m by virtue of its motion relative to O' and, most famously, is responsible for hurricanes. This is the term that Coriolis discovered by his analysis, and why this subject bears his name.

4. $\vec{F}_{\text{cent}} = -m\vec{\omega} \times (\vec{\omega} \times \vec{r}\,')$ is the *centrifugal force* on m and is what "pushes you into the car door" as your driver takes a turn too quickly.

Taking these four inertial forces into account, one can relate forces and acceleration in the accelerating frame of reference as Newton would have us do in an inertial frame of reference, namely $\sum \vec{F}\,' = m\vec{a}\,'$.

However, for the discussions in §7.3, 9.2, and 10.3.4, we need only the Coriolis, centrifugal, and translational *accelerations* which, as identified in Eq. (G.3), are evidently:
$$\boxed{\vec{a}_{\text{Cor}} = -2\vec{\omega} \times \vec{v}\,'; \quad \vec{a}_{\text{cent}} = -\vec{\omega} \times (\vec{\omega} \times \vec{r}\,'); \quad \text{and} \quad \vec{a}_{\text{tr}} = -\vec{A}.}$$

Anticipating there may be a reader who might appreciate an example of how Coriolis' theorem is used in practice to help complete their "memory refresh", I finish this appendix with the classic example of a "bead on a rod". In addition to being a good exemplar of this type of problem, it also has astrophysical implications near

---

[2] However, students of general relativity know that Einstein regarded gravity as another "inertial force", on the same par as the Coriolis and centrifugal forces uncovered here.

Coriolis' Theorem

and dear to my heart, as it turns out to be a model for an MHD mechanism by which astrophysical jets are launched! (Blandford & Payne, 1982; Spruit, 1996.) This is discussed in some detail in §9.2–9.4 where some familiarity with this example would be beneficial.

Example G.1. A bead of mass m is released from rest at a distance r_0 from the end of a frictionless rod of length $l > r_0$ about which the rod rotates at a constant angular speed ω in the horizontal plane.

a) Find the displacement of the bead along the rod as a function of time.

b) At what angle, φ_{rod}, does the bead reach the end of the rod and what is its speed?

c) Find the normal force exerted by the rod on the bead as a function of position along the rod.

d) Describe this problem in the inertial (non-rotating) frame of reference. In particular, how does the frictionless rod – which can only exert a *normal* force on the bead (and thus perpendicular to the rod) – manage to impart an apparent *radial* acceleration to the bead (thus parallel to the rod) as it rotates?

Solution: a) Define a rotating coordinate system, O', such that $\hat{\imath}'$ always points along the rod. Further and as shown in Fig. G.2, let the origins of O' and the inertial coordinate system, O, coincide (thus, $\vec{R} = 0$) on the rotation axis fixed at one end of the rod. Then:

acceleration of O' relative to O, $\quad \vec{A} = 0$;

angular velocity of O' relative to O, $\quad \vec{\omega} = \omega \hat{k}'$;

angular acceleration of O' rel. to O, $\quad \dot{\vec{\omega}} = 0$;

position of m relative to O', $\quad \vec{r}' = x' \hat{\imath}'$;

velocity of m relative to O', $\quad \vec{v}' = \dot{x}' \hat{\imath}'$;

acceleration of m relative to O', $\quad \vec{a}' = \ddot{x}' \hat{\imath}'$.

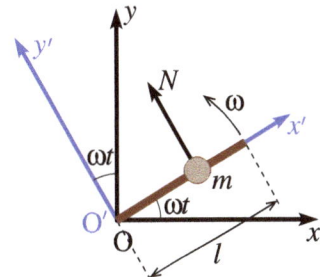

Figure G.2. A bead, m, slides along a frictionless rotating rod.

Next, evaluate the inertial forces:

transverse force, $\quad -m\dot{\vec{\omega}} \times \vec{r}' = 0$;

Coriolis force, $\quad -2m\vec{\omega} \times \vec{v}' = -2m\omega \dot{x}' \hat{k}' \times \hat{\imath}' = -2m\omega \dot{x}' \hat{\jmath}'$;

centrifugal force, $\quad -m\vec{\omega} \times (\vec{\omega} \times \vec{r}') = -m\omega^2 x' \, \hat{k}' \times (\hat{k}' \times \hat{\imath}') = m\omega^2 x' \hat{\imath}'$.

Finally assemble the forces. From Fig. G.2, the only real force acting in the x'–y'

plane is the normal force: $\vec{F} = N\hat{j}'$. Substituting this and the inertial forces into Coriolis' theorem (Eq. G.4), we get,

$$\vec{F}' = m\vec{a}' = m\ddot{x}'\hat{i}' = N\hat{j}' - 2m\omega\dot{x}'\hat{j}' + m\omega^2 x'\hat{i}'. \tag{G.5}$$

Considering for the moment the $\hat{i}'$ component of Eq. (G.5), we get,

$$\ddot{x}' = \omega^2 x' \quad \Rightarrow \quad x'(t) = Ae^{\omega t} + Be^{-\omega t} \quad \text{and} \quad \dot{x}'(t) = A\omega e^{\omega t} - B\omega e^{-\omega t}. \tag{G.6}$$

Imposing boundary conditions,

$$\left.\begin{array}{ll} x'(0) = r_0 & \Rightarrow \quad A + B = r_0 \\ \dot{x}'(0) = 0 & \Rightarrow \quad A - B = 0 \end{array}\right\} \quad \Rightarrow \quad A = B = \frac{r_0}{2},$$

we find,

$$\boxed{x'(t) = \frac{r_0}{2}\left(e^{\omega t} + e^{-\omega t}\right) = r_0 \cosh \omega t.} \tag{G.7}$$

b) Let the time when the bead reaches the end of the rod be t_{rod}. Then,

$$x'(t_{\text{rod}}) = l = \frac{r_0}{2}\left(e^{\omega t_{\text{rod}}} + e^{-\omega t_{\text{rod}}}\right) \quad \Rightarrow \quad r_0 e^{2\omega t_{\text{rod}}} - 2l e^{\omega t_{\text{rod}}} + r_0 = 0$$

$$\Rightarrow \quad e^{\omega t_{\text{rod}}} = \frac{l + \sqrt{l^2 - r_0^2}}{r_0} = \lambda + \sqrt{\lambda^2 - 1}, \tag{G.8}$$

where $\lambda \equiv l/r_0$. Here, we've chosen the '+' root so that $e^{\varphi_{\text{rod}}} > 1$ and $\varphi_{\text{rod}} > 0$. Thus,

$$\boxed{\omega t_{\text{rod}} = \varphi_{\text{rod}} = \ln\left(\lambda + \sqrt{\lambda^2 - 1}\right),} \tag{G.9}$$

is the angle at which the bead leaves the rod. Note that φ_{rod} depends only on λ, the ratio of the rod's length, l, to the bead's starting position, r_0, and not on the angular speed, ω.

To find the speed of the bead at the end of the rod, differentiate Eq. (G.7) with respect to time and evaluate it at $t = t_{\text{rod}}$ to get,

$$v'(t_{\text{rod}}) = \frac{\omega r_0}{2}\left(e^{\omega t_{\text{rod}}} - e^{-\omega t_{\text{rod}}}\right) = \frac{\omega r_0}{2}\left(\lambda + \sqrt{\lambda^2 - 1} - \frac{1}{\lambda + \sqrt{\lambda^2 - 1}}\right),$$

using Eq. (G.8). Thus,

$$\boxed{v'(t_{\text{rod}}) = \omega r_0 \sqrt{\lambda^2 - 1} = \omega\sqrt{l^2 - r_0^2},}$$

after a little algebra. Note that the final speed *does* depend on ω.

c) For the normal force exerted by the rod on the bead, start with the y' component of Eq. (G.5):

$$N_{y'}(t) = 2m\omega\dot{x}' = 2m\omega^2 r_0 \sinh \omega t. \tag{G.10}$$

Since $\sinh \omega t = \sqrt{\cosh^2 \omega t - 1}$ and $\cosh \omega t = x'/r_0$ (Eq. G.7),

$$\sinh \omega t = \sqrt{\frac{x'^2}{r_0^2} - 1} \quad \Rightarrow \quad N_{y'}(x') = 2m\omega^2 \sqrt{x'^2 - r_0^2}.$$

To complete the answer, we include the z'-component of the normal force (not shown in Fig. G.2) which must balance the weight of m:

$$N_{z'} = mg.$$

Thus, the normal force exerted by the rod on the bead as a function of x' is,

$$\boxed{\vec{N}(x') = 2m\omega^2\sqrt{x'^2 - r_0^2}\,\hat{j}' + mg\hat{k}'.}$$

d) To determine what real forces drive the bead, this problem must be examined in a Cartesian inertial frame of reference which, as we'll see, leads to a rather nasty differential equation. For this reason, most on-line and textbook solutions to this problem typically "bunt", and describe it in polar coordinates. While certainly a valid approach and one where the correct answer is found easily, it fails to identify convincingly the real force(s) driving the motion.

To wit, in polar coordinates (r, φ), the kinematical quantities are given by:

$$\vec{r} = r\hat{r}; \qquad \vec{v} = \dot{r}\hat{r} + r\dot{\varphi}\hat{\varphi}; \qquad \vec{a} = (\ddot{r} - r\dot{\varphi}^2)\hat{r} + (r\ddot{\varphi} + 2\dot{r}\dot{\varphi})\hat{\varphi},$$

where the additional terms in $\vec{v}$ and $\vec{a}$ arise from the time dependence of the unit vectors $\hat{r}$ and $\hat{\varphi}$.[3] And so, since the normal force, $\vec{N} = N\hat{\varphi}$, is the only real force acting in the horizontal plane, we can write,

$$\vec{F} = m\vec{a} \quad \Rightarrow \quad N\hat{\varphi} = m(\ddot{r} - r\dot{\varphi}^2)\hat{r} + m(r\ddot{\varphi} + 2\dot{r}\dot{\varphi})\hat{\varphi},$$

which, broken up into its $\hat{r}$ and $\hat{\varphi}$ components, leads to,

$$\ddot{r} = r\omega^2 \quad \text{and} \quad N = 2m\omega\dot{r}, \tag{G.11}$$

for $\dot{\varphi} = \omega = $ constant. These are identical to the first of Eq. (G.6) and last of Eq. (G.10) derived from the rotating reference frame, O', and thus their solutions,

$$r = r_0 \cosh\omega t \quad \text{and} \quad N = 2m\omega^2 r_0 \sinh\omega t,$$

also solve Eq. (G.11). However, to my taste this doesn't come close to explaining how it can be the normal force that accelerates the bead in the r-direction since N doesn't even appear in the first of Eq. (G.11)!

And so we tackle this problem as we should have in the first place: in Cartesian coordinates. From Fig. G.2, we resolve N into its x- and y-coordinates and write Newton's second law as,

$$-N\sin\omega t = m\ddot{x} \quad \text{and} \quad N\cos\omega t = m\ddot{y}.$$

Already we see in these coordinates that the normal force accelerates m in both the x- and y-directions. Dividing the first of these equations by the second, we get,

$$\ddot{x} = -\ddot{y}\tan\omega t. \tag{G.12}$$

---

[3] For example, with a little reflection the reader should be able to confirm that $\dot{\hat{r}} = \dot{\varphi}\hat{\varphi}$ which leads to the tangential component ($\propto \hat{\varphi}$) of the velocity, $r\dot{\varphi}$. Similarly, $\dot{\hat{\varphi}} = -\dot{\varphi}\hat{r}$.

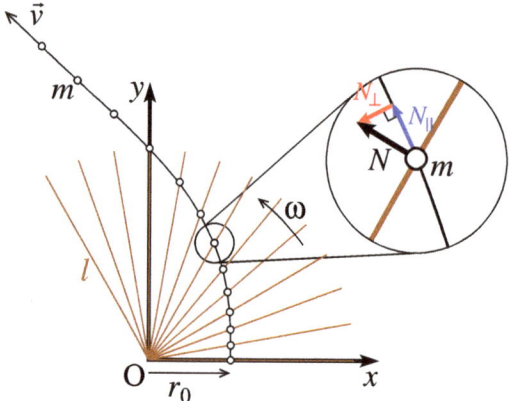

Figure G.3. Illustrated are 13 snapshots of a bead sliding along a frictionless rod rotating in a horizontal plane at a constant angular speed. The curved black line shows the trajectory of the bead as viewed from an inertial frame of reference, O. See text for details.

Further, this is a constrained problem: whatever x and y may be, the ratio y/x must always be the tangent of the angle $\varphi = \omega t$ through which the rod has swung. Thus,

$$y = x \tan \omega t \quad \Rightarrow \quad \dot{y} = \dot{x} \tan \omega t + x\omega \sec^2 \omega t$$

$$\Rightarrow \quad \ddot{y} = \ddot{x} \tan \omega t + 2\dot{x}\omega \sec^2 \omega t - 2x\omega^2 \tan \omega t \sec^2 \omega t,$$

after a little algebra. Substituting this into Eq. (G.12), we get,

$$\ddot{x} = -2\dot{x}\omega \tan \omega t - 2x\omega^2 \tan^2 \omega t, \tag{G.13}$$

again, after a little algebra. Similarly, the equation of motion for the y-direction is,

$$\ddot{y} = 2\dot{y}\omega \cot \omega t - 2y\omega^2 \cot^2 \omega t. \tag{G.14}$$

Now, who would like to try solving these? Certainly not me!

So, to reassure ourselves that these are indeed the correct equations of motion for the bead on a rod problem in an inertial Cartesian frame of reference, one can substitute the known solutions from Eq. (G.7) (where we identify $x' = r$), namely:

$$\left. \begin{array}{l} x(t) = r(t) \cos \omega t = r_0 \cosh \omega t \cos \omega t; \\ y(t) = r(t) \sin \omega t = r_0 \cosh \omega t \sin \omega t, \end{array} \right\} \tag{G.15}$$

in to Eq. (G.13) and (G.14) respectively.

I'll let the sceptical reader do that. The point is, doing the problem from an inertial Cartesian reference frame clearly identifies the normal force as the real force accelerating the bead in each of the x- and y-directions.

Figure G.3 shows the rod (thin brown lines, length l) at thirteen positions in $10°$ increments for $0° \leq \varphi = \omega t \leq 120°$. Choosing for illustration $\lambda = l/r_0 = 2.5$, the small circles indicate the bead's location (using Eq. G.15) along the rod until it reaches $\varphi_{\text{rod}} = 89.8°$ (Eq. G.9). Once it leaves the rod, the bead carries on in a straight line with its final velocity, $\vec{v}$, whose components are given by differentiating Eq. (G.15) with respect to time. Its overall trajectory is indicted in Fig. G.3 by the curved black line.

At a typical point along the rod, the blow-up in Fig. G.3 shows how the normal force on the bead can be broken up into its parallel ($N_\parallel$; blue) and perpendicular ($N_\perp$; red) components relative to the bead's trajectory. Evidently, $N_\parallel$ accelerates the bead along its direction of motion, while $N_\perp$ acts as a centripetal force, causing the curvature in the bead's path. A more "realistic" depiction of Fig. G.3 is given in Fig. 9.3 on page 353.

And for those who *still* haven't had their fill on Coriolis' theorem, I refer you to the two classic texts by Marion & Thornton (1995) and Fowles & Cassiday (2004).

H The Diffusion Equation

If a few tablespoons of bleach are poured onto a saucer and placed at one end of a closed room, it does not take long for the smell of bleach to fill the air. Evidently, the bleach spontaneously evaporates and distributes itself uniformly throughout the room with no intervention required. Of course, the reverse process in which all the bleach vapour spontaneously reassembles itself back into a liquid and onto the saucer never occurs; not because it would violate any conservation law such as energy or particle number, but because it is just exceedingly unlikely. This is the essence of the second law of thermodynamics: the macrostate of a system – how it is observed – is the one that is most likely to occur. Flip a coin 1,000 times, and there is just one way for there to be 1,000 heads. However, there are numerous ways more for there to be ~ 500 heads, and so that is the macrostate of coin tosses one observes.

First written down by Adolf Fick in 1855, the diffusion equation describing how the smell of bleach permeates a room is a direct consequence of the second law. If the total number of particles of a certain substance in a volume V is conserved, then the equation governing how its number density, n, is transported within V is a continuity equation (*e.g.*, Eq. 1.19),

$$\partial_t n + \nabla \cdot \vec{f}_n = 0 \quad \Rightarrow \quad \partial_t n = -\nabla \cdot \vec{f}_n, \tag{H.1}$$

where $\vec{f}_n$ is the *particle flux density* entering or leaving a volume differential, δV, across its surface. Particles enter or leave a given δV simply by their random (Brownian) motions. The most likely state – where n is uniform everywhere – is one where as many particles enter a given δV as leave it, rendering the net particle flux zero and thus the value of n unchanged. This is the state in which the smell of bleach is the same everywhere in the room.

However, while the bleach is still evaporating from the saucer, n is not uniform throughout the room. In this case, two neighbouring pockets of air with differing concentrations of bleach will not experience net zero particle flux across their mutual surface. Rather, if pocket A has more bleach molecules than pocket B, more molecules are knocked from pocket A to B than from pocket B to A, and there is a net migration of bleach particles from A to B. Fick reasoned that the net particle flux should point away from the direction of increasing number density and thus opposite to ∇n (*Fick's first law of diffusion*). That is,

$$\vec{f}_n = -\mathcal{D}\nabla n, \tag{H.2}$$

where the *diffusion coefficient* $\mathcal{D}$ could, in principle, be a function of position, $\vec{r}$, and of the number density, n, itself. Combining Eq. (H.2) with Eq. (H.1) gives us the so-called *diffusion equation*,

$$\partial_t n = \nabla \cdot (\mathcal{D} \nabla n) = \mathcal{D} \nabla^2 n, \qquad (\text{H.3})$$

where the last equality holds only if $\mathcal{D}$ is constant. Written in 1-D as $\partial_t n - \mathcal{D} \partial_{xx} n$ and comparing this to Eq. (C.1) in App. C, we see that the discriminant is zero, and thus the diffusion equation is *parabolic* in nature.

The reader may already be familiar with the *heat equation*,

$$\partial_t T = \alpha \nabla^2 T,$$

where T is the temperature and α is the *heat diffusivity* of the medium. This was first written down by Joseph Fourier himself in 1822 to describe how heat (what was then known as a mysterious fluid called *caloric*) "moves" from a warmer object to a cooler one. And while the heat equation is mathematically identical to the diffusion equation, it wasn't actually linked to the idea of the diffusion of matter until Fick. The particulate nature of matter – even air – still wasn't appreciated widely enough.

Another famous diffusion equation *of a sort* is Schrödinger's wave equation for free particles (potential $V = 0$),

$$\partial_t \Psi = \frac{i\hbar}{2m} \nabla^2 \Psi,$$

written in a rather non-standard form to emphasise its similarity to Eq. (H.3). In this case, the "diffusion coefficient" is imaginary, which is why I described this as a diffusion equation "of a sort". Still, it is well known to students of quantum mechanics that a free wave packet whose position is initially known with great precision (*e.g.*, where the norm of its wavefunction, $\Psi^*\Psi$, is a narrow Gaussian) spreads out – diffuses – as it propagates through space and time.[1] This is a direct consequence of the diffusive nature of Schrödinger's wave equation.

---

[1] See, for example, www.wikipedia.org/wiki/Wave_packet.

Variable Glossary

The following tabulates an incomplete list of variables used throughout the text with their definition(s). They are listed in alphabetical order, with Roman character 'A' following Greek character 'ω', upper case before lower case, scalars before vectors.

α	"MHD-alpha" $= a^2/c_s^2 = 2/\gamma\beta$; constant of convenience (§3.6); angular acceleration (Chap. 8)
$\alpha_{f,s}, \alpha_x, \alpha_\perp$	"MHD-alpha" variations: relative to fast/slow speeds, $a_{f,s}^2/c_s^2$; relative to B_x, a_x^2/c_s^2; relative to $B_\perp$, $a_\perp^2/c_s^2$
β	"plasma-beta" $= 2\mu_0 p/B^2$; occasionally used for other purposes (Eq. 3.56, 7.2)
$\beta_{1,2}$	ambipolar coefficient (Chap. 10)
Γ	fluid circulation (vorticity flux)
γ	ratio of specific heats of a gas
$\gamma_{1,2}$	coupling coefficient (Chap. 10)
ΔA	$= (\Delta l)^2$ or $(\Delta x)^2$, face area of a (small) cubic fluid element
ΔV	$= (\Delta l)^3$, volume of a (small) fluid element
Δl	length of a fluid element such that $\delta l \ll \Delta l < \mathcal{L}$
δ_{ij}	Kronecker delta; 1 if $i = j$, 0 otherwise
δ_x	$= \alpha_x - 1$ (Eq. 6.22)
δl	mean free path of a fluid particle between collisions
ϵ	perturbation label whose formal value is 1 (§2.1, 5.2.2); angle between $\vec{B}$ and disc surface (§9.2)
ϵ_0	permittivity of free space $= 8.8542 \times 10^{-12}$ C^{2}J^{-1}m^{-1}
ε	specific internal energy $= e/\rho \propto T$, J kg^{-1}
ε_s	ionisation energy for species 's' (§10.5.2)
ζ	downwind to upwind pressure ratio (Eq. 3.40, 5.96); $\chi/(v_{rms}k)$ (§7.4)
η	ratio of downwind to upwind density (§5.3.5); mass load, kg m^{-2}s^{-1}T^{-1} (§9.1); resistivity, kg m^3 s^{-1} C^{-2} (Chap. 10)
θ	angle between $\vec{B}$ and $\hat{x}$ (Chap. 5); inclination angle (Chap. 8)

ϑ	meridional spherical polar coordinate, $(\varrho, \vartheta, \varphi)$
κ	adiabatic gas law constant; $p = \kappa\rho^\gamma$; $d\Omega^2/d(\ln r)$ at $r = r_0$ (§7.3); unitless wave number, kL (§7.4)
λ	perturbation wavelength (Chap. 7); ratio $v_{\rm p}/B_{\rm p}$ (Chap. 9)
$\lambda_{\rm e}$	thermal De Broglie wavelength (§10.5)
μ	normalisation of fast and slow eigenkets (Eq. 6.26); shear viscosity $= \nu\rho$, N m^{-2} s (Chap. 8)
μ_0	permeability of free space $= 4\pi \times 10^{-7}$ kg m C^{-2}
$\mu_{1,2}$	reduced mass of species 1 and 2, $m_1 m_2/(m_1 + m_2)$ (§10.2)
μ_ρ	"magneto-rotational number" (§7.3)
ν	frequency (§2.1.1); normalisation of fast and slow eigenkets (Eq. 6.27); kinematic viscosity $= \mu/\rho$, m^2 s^{-1} (Chap. 8, §10.3.4)
ξ_i	co-moving coordinate of i-family wave, $\xi_i = x - u_i t$, with characteristic speed u_i; unitless vertical coordinate, $\xi = kz$ (§7.4)
$\vec{\xi}, \xi_x, \xi_y$	displacement perturbation (§7.3)
$\varpi_{1,2}, \varpi^{\rm a}_{1,2}, \varpi^{\rm x}_{1,2}$	$\varpi_{1,2} = \varpi^{\rm a}_{1,2} + \varpi^{\rm x}_{1,2}$, ambipolar + exchange power densities (Chap. 10)
ϱ	radial spherical polar coordinate, $(\varrho, \vartheta, \varphi)$
ρ	matter density, kg m^{-3}
$\rho_{\rm A}$	density at Alfvén point (Chap. 9)
ρ_0, p_0, v_0	unperturbed values of ρ, p, and v (§2.1.1, 7.4)
$\rho_{\rm d}, p_{\rm d}, v_{\rm d}$	downwind values of ρ, p, and v (§3.5)
$\rho_{\rm l}, p_{\rm l}$	values in lower layer of RTI (§7.2)
$\rho_{\rm n}, \rho_{\rm i}, \rho_{\rm e}$	matter density of neutrals, ions, electrons (Chap. 10)
$\rho_{\rm p}, p_{\rm p}, v_{\rm p}$	perturbations to ρ, p, and v (§2.1.1, 7.4)
ρ_q	charge density
$\rho_{\rm u}, p_{\rm u}, v_{\rm u}$	upwind values of ρ, p, and v (§3.5); values in upper layer of RTI (§7.2)
Σ	open integration surface; source term (§1.3)
$\vec{\Sigma}, \Sigma$	cross-sectional area (§9.1)
σ	differential area element; specific source term (§1.3); cross-sectional area (§4.1, 10.5)
$\sigma_{1,2}$	exchange term between species 1 and 2 (§10.5)
$\sigma_{\rm M}, \sigma_{\rm CR}$	$p_{\rm M}/p$ and $p_{\rm CR}/p$ respectively (§7.4)
$\langle\sigma u\rangle_{1,2}$	rate (collisional) coefficient (Chap. 10)
τ	volume elements (Chap. 3, App. B); time scales (§8.6.1, 10.3.3)
$\tau_{\rm X}$	X = KH, RT, MR, P; fluid instability e-folding time (Chap. 7)

$\vec{\tau}, \tau$	torque, torque density (Chap. 8, §9.1)
Φ	flux of an arbitrary vector field, $\vec{\phi}$ (§4.3)
Φ_B	magnetic flux
Φ_S	Poynting flux
φ	cylindrical (z, r, φ) and spherical polar $(\varrho, \vartheta, \varphi)$ azimuthal coordinate
ϕ	gravitational potential, $\mathrm{J\,kg^{-1}}$ or $\mathrm{m^2\,s^{-2}}$
ϕ_{eff}	effective grav. potential in rotating reference frame (§9.2)
$\vec{\phi}$	arbitrary vector field (Chap. 4)
χ	imaginary frequency $(-i\omega) \equiv$ inverse e-folding time $(1/\tau)$ (Chap. 7)
$\chi, \chi_{\mathrm{L,R}}$	orientation angle of $\vec{B}_\perp$ (§6.4.1)
$\chi_{1,2}$	constants for the rate coefficients (Chap. 10)
$\chi_{\mathrm{f,s}}$	tentative normalisations for fast and slow eigenkets (§6.2.1)
χ_i	Riemann invariants (characteristics of the flow, §3.5)
$\Psi(y)$	amplitude of stream function, $\vec{\psi}$ (§7.2)
ψ	magnetic pitch angle, $\tan^{-1} B_\varphi/B_{\mathrm{p}}$ (§9.3)
$\psi_{\mathrm{f,s}}$	tentative normalisations for fast and slow eigenkets (§6.2.1)
ψ_i	$i = 1, 6$, parameters for multivariate secant solver (§6.4)
$\vec{\psi}, \psi$	stream function, $\vec{v} = \nabla \times \vec{\psi}$ (§7.2)
Ω, Ω_0	angular speed of rotation (§7.3, Chap. 9)
ω	angular frequency, speed; normal mode frequencies (Chap. 7)
ω_{A}	Alfvén frequency (§7.3)
$\vec{\omega}$	vorticity $= \nabla \times \vec{v}$; angular velocity (App. G)
$\mathcal{A}$	Atwood number (§7.2)
A_{p}	poloidal Alfvén number, $v_{\mathrm{p}}/a_{\mathrm{p}}$ (§9.1)
A_x	longitudinal Alfvén number, v_x/a_x (§5.3.5)
$A_\perp$	transverse Alfvén number, $v_\pi/a_\perp$ (§5.3.5)
$\mathcal{A}^\pm$	Alfvén Riemann invariants
$\vec{A}$	vector potential; $\vec{B} = \nabla \times \vec{A}$
a	Alfvén speed $= B/\sqrt{\mu_0 \rho}$; occasionally acceleration (Chap. 1)
$a_\varphi, a_{\mathrm{p}}$	azimuthal and poloidal Alfvén speed: $B_\varphi/\sqrt{\mu_0 \rho}$; $B_{\mathrm{p}}/\sqrt{\mu_0 \rho}$
$a_{\mathrm{Cor}}, a_{\mathrm{cent}}, a_{\mathrm{tr}}$	Coriolis, centrifugal, translational accelerations (§7.3, App. G)
$a_{\mathrm{f,s}}$	fast and slow magnetosonic speeds (Eq. 5.23, 5.25)
$a_x, a_\perp$	longitudinal and perpendicular Alfvén speed: $B_x/\sqrt{\mu_0 \rho}$; $B_\perp/\sqrt{\mu_0 \rho}$

Variable Glossary

$\mathcal{B}, \mathcal{B}_M$	Bernoulli function (§2.4, Chap. 9)
B_φ	azimuthal component of $\vec{B}$ (§7.3, Chap. 9, §10.3.4)
B_p	poloidal component of $\vec{B}$ (Chap. 9)
B_x	longitudinal component of $\vec{B}$ (§5.3)
$\vec{B}$	magnetic induction
$\vec{B}_\perp$	component perpendicular to longitudinal (x) direction (§5.3)
b	ratio of downwind to upwind $B_\perp$ (Eq. 5.95)
$b_\perp$	ratio of downwind $B_\perp$ to B_x (Eq. 5.109)
C	contour of integration (§4.7, A.2); "Parker's constant" (§7.4)
$\mathcal{C}^{\pm,0}$	characteristic paths for ideal HD
$\mathcal{C}^{\pm}_{f,s,x}$	fast, slow, Alfvén characteristic paths for ideal MHD
$c^{\pm,0}$	characteristic speeds for ideal HD ($v \pm c_s$, v)
c_iso	$= \sqrt{p/\rho}$ = isothermal sound speed
c_s	$= \sqrt{\gamma p/\rho}$ = adiabatic sound speed
D	fast and slow discriminant (Eq. 6.23); constant for PI (§7.4)
$\mathcal{D}$	diffusion coefficient (App. H)
$\mathcal{D}_M$	magnetic diffusion coefficient (§10.3)
D_v/Dt	Lagrangian derivative (Eq. 3.2)
$D_t^{\pm}$	abbreviated Leibniz notation for Lagrangian derivative (§5.2.1)
$\vec{D}$	electric displacement
d	$= D/c_s^2$ (Eq. 6.25)
E	internal energy of a fluid volume
$\mathcal{E}$	induced emf $= \oint \vec{E} \cdot \vec{dl}$
E	symmetrised strain tensor (Chap. 8)
E_{ij}	the $(i,j)^{\text{th}}$ element of E (Chap. 8)
E_T	total HD energy (thermal + kinetic + gravitational) of a fluid
$\vec{E}$	electric field
$\vec{E}_\text{ind}$	induced electric field $= -\vec{v} \times \vec{B}$
e	internal energy density $\propto p$, J m^{-3}
e_n, e_i, e_e	internal energy density of neutrals, ions, electrons (Chap. 10)
e_T	total HD energy density, $e + \rho\phi + \rho v^2/2$, J m^{-3}
e_T^*	total MHD energy density, $e_T + B^2/2\mu_0$, J m^{-3}
$\mathcal{F}$	Froude number (§7.2, Problem 8.6)
$\vec{F}_\text{EM}$	electromagnetic force $= q(\vec{E} + \vec{v} \times \vec{B})$
f	flux function (rA_φ) that labels lines of $\vec{B}$ in φ-symmetry (Chap. 9, §10.3.4)

$\lvert f \rangle$	ket formed from flux densities of (M)HD primitive variables
$f_i(\psi_j)$	functions whose value $\to 0$ when MHD Riemann problem converges (§6.4)
$\vec{f}_{1,2}, \vec{f}^{\,a}_{1,2}, \vec{f}^{\,x}_{1,2}$	$\vec{f}_{1,2} = \vec{f}^{\,a}_{1,2} + \vec{f}^{\,x}_{1,2}$, ambipolar + exchange force densities (Chap. 10)
$\vec{f}_{\text{ext}}$	external force densities (N m^{-3}) acting on a fluid element
$\vec{f}_{\text{L}}$	Lorentz force density $= \vec{J} \times \vec{B}$ (N m^{-3})
$\vec{f}_L$	angular momentum flux density (Eq. 7.60)
G	Newton's gravitational constant $= 6.6743 \times 10^{-11}\ \text{N m}^2\,\text{kg}^{-2}$
g_{gal}	acceleration of gravity at galactic plane (§7.4)
$\vec{g}, g$	acceleration of gravity at the earth's surface $= 9.81\ \text{m s}^{-2}$
$\mathcal{H}$	the "hydrodynamical term" (Eq. 5.104)
$\vec{H}$	magnetic field
h	enthalpy (of gas), $\gamma p/(\gamma - 1)\rho$; $\text{m}^2\,\text{s}^{-2}$
I	identity tensor (matrix)
i	$\sqrt{-1}$; current (§9.1, 10.4, App. B)
J	impulse (Chap. 1)
J	Jacobian matrix (subscripts "p" and "c" $\Rightarrow$ "primitive" and "conservative")
$\mathcal{J}^{\pm}$	sonic Riemann invariants for ideal HD
J_{ij}	matrix element of J
$\vec{J}$	current density
K	kinetic energy (per particle); alternate wave number (§7.4)
k	wave number, $2\pi/\lambda$; kinetic energy density (Eq. 10.54)
k_{B}	Boltzmann constant, $= 1.3807 \times 10^{-23}\ \text{J K}^{-1}$
L	scale height (Chap. 7)
$\mathcal{L}$	smallest scale of interest (§1.1); differential operator (§7.4)
L	matrix of eigenbras (§3.5.2)
$\vec{L}$	angular momentum (§7.3)
l	specific angular momentum, $\text{m}^2\,\text{s}^{-1}$ (Chap. 9); length element (§8.6.5, App. B)
ℓ	angular momentum density, $\text{kg m}^{-1}\text{s}^{-1}$ (§7.4, 8.6.5)
$\langle l_i \rvert$	the ith left eigenvector (bra)
l_{ij}	the jth component of the ith left eigenvector
M	Mach number, v/c_{s}; occasionally total mass of a fluid volume
$\mathcal{M}$	lab frame Mach number (§2.2.3); mass flux density (ρv; §5.3)

Variable Glossary

M_A	Mach number at Alfvén point (where $a_x = c_s$; Problem 5.23)
$M_{f,s}$	fast and slow magnetosonic numbers; $v/a_{f,s}$ (§5.3, 9.4)
$M_\pm$	Mach number at fast and slow points (where $a_{f,s} = c_s$; Problem 5.23)
m	mass of a single fluid particle
$\mathcal{N}$	number of fluid particles inside a fluid volume
n_n, n_i, n_e	number density of neutrals, ions, electrons (Chap. 10)
$\mathcal{P}_{app}$	rate at which work is done on fluid volume by neighbouring fluid volumes ($J\,s^{-1}$)
$P(y)$	y-dependent amplitude of the pressure perturbation (§7.1)
p	thermal pressure, $(\gamma - 1)e$
p_{CR}	cosmic ray pressure (§7.4)
p_M	magnetic pressure, $B^2/2\mu_0$
p_R	resistive power density (§10.3, B.4)
p_S	Poynting power density (§4.5, B.5)
p_{app}	applied power density ($\mathcal{P}_{app}$ per unit volume; $J\,s^{-1}m^{-3}$)
p_n, p_i, p_e	thermal pressure of neutrals, ions, electrons (Chap. 10)
p^*	thermal plus magnetic pressure, $p + p_M$
Q, q	an arbitrary quantity; occasionally used for particle charge
$\mathcal{Q}_{1,2}$	combined exchange-ambipolar coefficient (Chap. 10)
$|q\rangle$	ket formed from the (M)HD primitive variables
$|\tilde{q}\rangle$	ket formed from the initial conditions of the (M)HD primitive variables
R	matrix of eigenkets (§3.5.2)
$\mathcal{R}$	Reynolds number, VR/ν
$\mathcal{R}_M$	magnetic Reynolds number $= \mu_0 VL/\eta$ (§10.2)
$R_{1,2}$	ratio of non-ideal electric fields (§10.2)
r	radial cylindrical coordinate, (z, r, φ)
r_A	radial cylindrical coordinate at Alfvén point (Chap. 9)
r_L	Larmor radius, mv/qB
$|r_i\rangle$	i^{th} right eigenvector (ket)
r_{ij}	j^{th} component of ith right eigenvector
S	entropy of a fluid volume ($J\,K^{-1}$); closed integration surface
$\mathcal{S}$	unitless entropy per particle $= ms/k$
$\mathscr{S}$	Lundquist number $= \mu_0 aL/\eta$ (§10.2)
S	viscid portion of stress tensor (Chap. 8)

$\mathcal{S}^0$	entropy Riemann invariant for ideal (M)HD
S_{ij}	$(i,j)^{\text{th}}$ element of S (Chap. 8)
$\vec{S}$	total momentum of a fluid volume
$\vec{S}_{\text{P}}$	MHD Poynting vector
s	specific entropy (per unit mass; $\text{J K}^{-1}\text{kg}^{-1}$); coordinate along line of induction (Chap. 9)
s_i	generalised coordinate; $ds_i = w_i(q)d\xi_i$ (§3.5.3, 6.2)
$\vec{s}$	momentum density ($\text{kg m s}^{-1}\text{m}^{-3}$); momentum per particle
$\hat{s}$	unit vector along line of induction (§9.2)
$\vec{s}_{\text{n}}, \vec{s}_{\text{i}}, \vec{s}_{\text{e}}$	momentum density of neutrals, ions, electrons (Chap. 10)
$\hat{s}_{\text{p}}$	unit vector along poloidal component of line of induction (§9.2)
T	(as a superscript) indicates tensor (matrix) transpose
T	thermal temperature
T	stress tensor (Chap. 8)
T_{ij}	$(i,j)^{\text{th}}$ element of stress tensor (Chap. 8)
t	time
U	diagonal matrix of eigenvalues u_i (§3.5.2)
u_i	eigenvalues/eigenspeeds of Jacobian (§3.5.3, 5.2)
$\mathcal{V}$	shock or bore speed, usually in lab frame (Chap. 2, §3.6, 6.4.3); scalar gauge potential (related to $\vec{A}$) (§4.8, 7.4.2, 9.1)
v_{φ}	azimuthal component of velocity (§7.3, 8.6, Chap. 9)
v_{p}	poloidal component of velocity (Chap. 9); perturbed velocity (§2.1)
v_{rms}	root-mean-square velocity of an ensemble of particles
$\vec{v}$	fluid velocity
$\vec{v}_{1,2}$	velocity of species 1 relative to 2, $\vec{v}_1 - \vec{v}_2$ (Chap. 10)
$\vec{v}_{\text{d}}$	drift velocity, $\vec{v}_{\text{e}} - \vec{v}_{\text{n}}$ (§10.4)
$\vec{v}_{\text{n}}, \vec{v}_{\text{i}}, \vec{v}_{\text{e}}$	velocity of neutrals, ions, electrons (Chap. 10)
$\vec{\vec{v}}_{\perp}$	component perpendicular to longitudinal (x) direction (§5.3)
$w_i(q)$	scaling functions for eigenkets (§3.5.3, 6.2)
x	first Cartesian coordinate, (x,y,z)
$Y(x,t)$	function giving shape of a boundary layer (§7.1, 7.2)
y	second Cartesian coordinate, (x,y,z)
Z	ionisation number (Chap. 10)
Z_0	fluid impedance, $c_s\rho$ (§2.1.2)
z	third Cartesian coordinate, (x,y,z); axial cylindrical coordinate, (z,r,φ)

References

Alfvén, H. 1942, *Existence of electromagnetic-hydrodynamic waves*, Nature, v. 150, p. 405.

Appenzeller, I. 1971, *Observational evidence for Parker's instability of the interstellar gas and magnetic field*, Astron. Astrophys., v. 12, p. 313.

Arfken, G. B., Weber, H. J., and Harris, F. E. 2013, *Mathematical Methods for Physicists*, 7th ed., Chap. 9 (Oxford: Academic Press), ISBN 978-0-12-384654-9.

Armitage, H. J. 2011, *Dynamics of protoplanetary disks*, Annu. Rev. Astron. Astrophys., v. 49, p. 195.

Avenhaus, H. et al. 2018, *Disks around T Tauri stars with SPHERE (DARTTS-S). I. SPHERE/IRDIS Polarimetric imaging of eight prominent T Tauri disks*, Astrophys. J., v. 863, p. 44.

Balbus, S. A. 2009, *Magnetohydrodynamics of protostellar disks*, in "Physical Processes in Circumstellar Disks Around Young Stars", ed. P. Garcia (Chicago: University of Chicago Press), arXiv:0906.0854 [astro-ph.SR].

Balbus, S. A. and Hawley, J. F. 1991, *A powerful local shear instability in weakly magnetized disks*, Astrophys. J., v. 376, p. 214.

Balbus, S. A. and Terquem, C. 2001, *Linear analysis of the Hall effect in protostellar disks*, Astrophys. J., v. 552, p. 235.

Batchelor, G. K. 2000, *An Introduction to Fluid Dynamics* (Cambridge: Cambridge University Press) ISBN 0-521-66396-2.

Beck, R. 2007, *Galactic Magnetic Fields*, www.scholarpedia.org/article/Galactic_magnetic_fields (Scholarpedia).

Beck, R., Brandenburg, A., Moss, D., Shukurov, A., and Sokiloff, D. 1996, *Galactic magnetism: recent developments and perspectives*, Annu. Rev. Astron. Astrophys., v. 34, p. 155.

Black, D. C. and Scott, E. H. 1983, *A numerical study of the effects of ambipolar diffusion on the collapse of magnetic gas clouds*, Astrophys. J., v. 263, p. 696.

Blandford R. D. and Payne D. G. 1982, *Hydromagnetic flows from accretion discs and the production of radio jets*, Mon. Not. R. Astron. Soc., v. 199, p. 883.

Bradley, G. 1975, *A Primer of Linear Algebra* (New Jersey: Prentice Hall), ISBN 0-13-700328-5.

Brandenburg, A. 2019, *Ambipolar diffusion in large Prandtl number turbulence*, Mon. Not. R. Astron. Soc., v. 487, p. 2673.

Breslau, J. A. and Jardin, S. C. 2003, *Global extended magnetohydrodynamic studies of fast magnetic reconnection*, Phys. Plasmas, v. 10, p. 1291.

Bridle, A. H., Perley, R. A., and Henriksen, R. N. 1986, *Collimation and polarization of the jets in 3C219*, Astronom. J., v. 92, p. 534.

Brio, M. and Wu, C. C. 1988, *An upwind differencing scheme for the equations of ideal magnetohydrodynamics*, J. Comput. Phys., v. 75, p. 400.

Buether *et al.* (editors) 2014, *Protostars and Planets VI* (Tucson: University of Arizona Press), ISBN 978-0-8165-3124-0.

Buffett, B. A. 2000, *Earth's core and the geodynamo*, Science, v. 288, p. 2007.

Bullard, E. C. 1955, *The stability of a homopolar dynamo*, Proc. Cambridge Phil. Soc., v. 51, p. 744.

Burkhart, B., Lazarian, A., Balsara, D., Meyer, C., and Cho, J. 2015, *Alfvénic turbulence beyond the ambipolar diffusion scale*, Astrophys. J., v. 805, p. 118.

Burns, J. O., O'Dea, C. P., Gregory, S. A., and Balonek, T. J. 1986, *Observational constraints on bending the wide-angle tailed radio galaxy 1919+479*, Astrophys. J., v. 307, p. 73.

Callen, J. D. 2006, *Fundamentals of Plasma Physics* (Madison: University of Wisconsin Press), www.cptc.wisc.edu/course-materials/.

Carroll, B. W. and Ostlie, D. A. 2017, *An Introduction to Modern Astrophysics*, 2nd ed. (Cambridge: Cambridge University Press), ISBN 978-1-108-42216-1.

Chandrasekhar, S. 1960, *The stability of non-dissipative Couette flow in hydromagnetics*, Proc. Natl. Acad. Sci., v. 46 (2), p. 253.

Chen, C.-Y. and Ostriker, E. C. 2012, *Ambipolar diffusion in action: transient C-shock structure and prestellar core formation*, Astrophys. J., v. 744, p. 124.

Chen, F. F. 1984, *Introduction to Plasma Physics and Controlled Fusion*, 2nd ed. (New York: Plenum Press), ISBN 0-306-41332-9.

Clarke, D. A. 1996, *A consistent method of characteristics (CMoC) for multidimensional MHD*, Astrophys. J. v. 457, p. 291.

Clarke, D. A., MacDonald, N. R., Ramsey, J. P., and Richardson, M. 2008, *Astrophysical jets*, Physics in Canada, v. 65, p. 47.

Clarke, D. A., Norman, M. L., and Burns, J. O. 1986, *Numerical simulations of a magnetically confined jet*, Astrophys. J., v. 311, p. L63.

Clarke, D. A., Norman, M. L., and Burns, J. O. 1989, *Numerical observations of a simulated radio jet with a passive helical magnetic field*, Astrophys. J., v. 342, p. 700

Courant, R., Friedrichs, K., and Lewy, H. 1928, *Über die partiellen differenzengleichungen der mathematischen Physik*, Math. Ann. (in German), v. 100 (1), p. 32.

Draine, B. T. 1986, *Multicomponent, reacting MHD flows*, Mon. Not. R. Astron. Soc., v. 220, p. 133.

Draine, B. T., Roberge, W. G., and Dalgarno, A. 1983, *Magnetohydrodynamical shock waves in molecular disks*, Astrophys. J., v. 264, p. 485.

Duffin, D. F. and Pudritz, R. E. 2008, *Simulating hydromagnetic processes in star formation: introducing ambipolar diffusion into an adaptive mesh refinement code*, Mon. Not. R. Astron. Soc., v. 391, p. 1659.

Evans, C. E. and Hawley, J. F. 1988, *Simulation of magnetohydrodynamical flows: a constrained transport method*, Astrophys. J., v. 332, p. 659.

Faber, T. 1995, *Fluid Dynamics for Physicists* (Cambridge: Cambridge University Press) ISBN 0-521-41943-3.

Falle, S. A. E. G. 2002, *Rarefaction shocks, shock errors, and low order of accuracy in ZEUS*, Astrophys. J., 577, L123.

Falle, S. A. E. G. 2003, *A numerical scheme for multifluid magnetohydrodynamics*, Mon. Not. R. Astron. Soc., v. 344, p. 1210.

Feng, H. and Wang, J. M. 2008, *Observations of a $2 \rightarrow 3$ type interplanetary intermediate shock*, Solar Phys., v. 247, p. 195.

Feng, H., Wang, J. M., and Chao, J. K. 2009, *Observations of a subcritical switch-on shock*, Astron. Astrophys., v. 503, p. 203.

Ferriere, K. 2001, *The interstellar environment of our galaxy*, Rev. Mod. Phys., v. 73 (4), p. 1031; see also www.wikipedia.org/wiki/Inter-stellar_medium.

Fiedler, R. A. and Mouschovias, T. C. 1993, *Ambipolar diffusion and star formation: formation and contraction of axisymmetric cloud cores*, Astrophys. J., v. 415, p. 680.

Fricke, K. 1969, *Stability of rotating stars II. The influence of toroidal and poloidal magnetic fields*, Astron. Astrophys., v. 1, p. 388.

Fowles, G. and Cassiday, G. 2004, *Analytic Mechanics*, ed. 7 (Belmont: Thomson Brooks/Cole), ISBN 0-534-49492-7.

Goedbloed, H., Keppens, R., and Poedts, S. 2019, *Magnetohydrodynamics of Laboratory and Astrophysical Plasmas* (Cambridge: Cambridge University Press), ISBN 978-1-107-12392-2.

Gailitis, A., Lielausis, O., Dement'ev, S., Platacis, E., Cifersons, A., Gerbeth, G., Gundrum, T., Stefani, F., Christen, M., Hänel, H., and Will, G. 2000, *Detection of a flow-induced magnetic field eigenmode in the Riga Dynamo Facility*, Phys. Rev. Lett., v. 84, p. 4365.

Galtier, S. 2016, *Introduction to Modern Magnetohydrodynamics*, §10.3 (Cambridge: Cambridge University Press), ISBN 978-1-107-15865-8.

Glatzmaier, G. A. and Roberts, P. H. 1995, *A three-dimensional self-consistent computer simulation of a geomagnetic field reversal*, Nature, v. 377, p. 203.

Hada, T. 1994, *Evolutionary conditions in the dissipative MHD system: stability of intermediate MHD shock waves*, Geophys. Res. Lett., v. 21, p. 2275.

Hagen, H. J. and Helmi, A. 2018, *The vertical force in the solar neighbourhood using red clump stars in TGAS and RAVE*, Astron. Astrophys., v. 615, p. A99.

Halliday, D., Resnick, R., and Walker, J. 2003, *Fundamentals of Physics*, 6th ed. (New York: Wiley), ISBN 0-471-22862-1. (Chap. 1: Chap. 20 in HRW; Chap. 9: Chap. 30, 31; Chap. 10: p. 622)

Harten, A. 1983, *High resolution schemes for hyperbolic conservation laws*, J. Comput. Phys., v. 49, p. 357.

Inoue, T. and Inutsuka, S. 2007, *Evolutionary conditions in dissipative MHD systems revisited*, Prog. Theor. Phys., v. 118, p. 47.

Jackson, J. D. 1975, *Classical Electrodynamics*, 2nd ed. (New York: Wiley), ISBN 0-471-43132-X.

Kageyama, A. and Sato, T. 1995, *Computer simulation of a magnetohydrodynamical dynamo, II*, Phys. Plasmas, v. 2, p. 1421.

Landau, L. D. and Lifshitz, E. M. 1987, *Fluid Mechanics*, 2nd ed. (Amsterdam: Elsevier), ISBN 978-0-08-033933-7.

Lesur, G., Kunz, M. W., and Fromang, S. 2014, *Thanatology in protoplanetary discs: the combined influence of Ohmic, Hall, and ambipolar diffusion on dead zones*, Astronomy & Astrophys., v. 566, p. A56.

Li, P. S., McKee, C. F., and Klein, R. I. 2006, *The heavy-ion approximation for ambipolar diffusion calculations for weakly ionised plasmas*, Astrophys. J., v. 653, p. 1280.

Liberman, M. A. and Velikovich, A. L. 1986, *Physics of Shock Waves in Gases and Plasmas*, Springer Series in Electrophysics, v. 19 (Berlin: Springer).

Lorrain, P. and Corson, D. 1970, *Electromagnetic Fields and Waves*, 2nd ed. (San Francisco: Freeman), ISBN 0-7167-0331-9.

Lynden-Bell, D. 1996, *Magnetic collimation by accretion discs of quasars and stars*, Mon. Not. R. Astron. Soc., v. 279, p. 389.

MacDonald, J. 2015, *Structures and Evolution of Single Stars*, (San Rafael: Morgan and Claypool).

Marion, J. B. and Thornton, S. T. 1995, *Classical Dynamics of Particles and Systems*, 4th ed. (Fort Worth: Saunders College Publishing), ISBN 0-03-097302-3.

Markovski, S. A. 1998, *Non-evolutionary discontinuous MHD flows in a dissipative medium*, Phys. Plasmas, v. 5, p. 2596.

McKee, C. F. and Ostriker, E. C. 2007, *Theory of star formation*, Annu. Rev. Astron. Astrophys., v. 45, p. 565; see also www.wikipedia.org/wiki/Star_formation.

Nore, C., Brachet, M. E., Politano, H., and Pouquet, A. 1997, *Dynamo action in the Taylor–Green vortex near threshold*, Phys. Plasmas, v. 4, p. 1.

Paris, D. T. and Hurd, F. K. 1969, *Basic Electromagnetic Theory*, 2nd ed. (New York: McGraw-Hill), ISBN 07-048470-8.

Parker, E. N. 1957, *Sweet's mechanism for merging magnetic fields in conducting fluids*, J. Geophys. Res., v. 62, p. 509.

Parker, E. N. 1958, *Dynamics of the interplanetary gas and magnetic fields*, Astrophys. J., v. 128, p. 664.

Parker, E. N. 1966, *The dynamical state of the interstellar gas and field*, Astrophys. J., v. 145, p. 811.

Perley, R. A., Dreher, J. W., and Cowan, J. J. 1984, *The jet and filaments in Cygnus A*, Astrophys. J., v. 285, p. L35.

Petschek, H. E. 1964, *Magnetic field annihilation*, NASA Special Publication, No. SP 50, p. 425.

Pham, T. and Tkalčić, H. 2023, *Up-to-fivefold reverberating waves through the earth's centre and distinctly anisotropic innermost inner core*, Nature Communications, v. 14, article 754.

Pinkney, J., Burns, J. O., and Hill, J. M. 1994, *1919–479: big WAT in a poor cluster*, Astronom. J., v. 108, p. 2031.

Ponomarenko, Y. B. 1973, *Theory of the hydromagnetic generator*, J. Appl. Mech. Tech. Phys., v. 14, p. 775.

Press, W. H., Vetterling, W. T., Teukolsky, S. A., and Flannery, B. P. 1992, *Numerical Recipes* (Cambridge: Cambridge University Press), ISBN 0-521-43064-X.

Ramsey, J. P. and Clarke, D. A. 2011, *Simulating protostellar jets simultaneously at launching and observational scales*, Astrophys. J., v. 728, p. L11.

Ramsey, J. P. and Clarke, D. A. 2019, *MHD simulations of the formation and propagation of protostellar jets to observational length-scales*, Mon. Not. R. Astron. Soc., v. 484, p. 2364.

Reynolds, O. 1883, *An experimental investigation of the circumstances which determine whether the motion of water shall be direct or sinuous, and of the law of resistance in parallel channels*, Philos. Trans. R. Soc., v. 174, p. 935.

Roe, P. L. and Balsara, D. S. 1996 *Notes on the eigensystems of magnetohydrodynamics*, SIAM J. Appl. Math., v. 56, p. 57.

Ryu, D. and Jones, T. W. 1995, *Numerical magnetohydrodynamics in astrophysics: algorithm and tests for one-dimensional flow*, Astrophys. J., v. 442, p. 228.

Ryu, D., Jones, T. W., and Frank, A. 2000, *The magnetohydrodynamic Kelvin–Helmholtz instability: a three-dimensional study of nonlinear evolution*, Astrophys. J., v. 545, p. 475.

Saripalli, L., Subrahmanyan, R., and Udaya Shankar, N. 2003, *Renewed activity in the radio galaxy PKS B1545-321: Twin edge-brightening beams within diffuse radio lobes*, Astrophys. J., v. 590, p. 181.

Shibata, K., Tajima, T., Matsumoto, R., Horiuchi, T., Hanawa, T., Rosner, R., and Uchida, Y. 1989, *Non-linear Parker instability of isolated magnetic flux in a plasma*, Astrophys. J., v. 388, p. 471.

Shu, F. H. 1974, *The Parker instability in differentially-rotating discs*, Astron. Astrophys., v. 33, p. 55.

Sod, G. 1978, *A survey of several finite difference methods for systems of nonlinear hyperbolic conservation laws*, J. Comput. Phys., v. 27, p. 1.

Solanki, S. K., Inhester, B., and Schüssler, M. 2006, *The solar magnetic field*, Rep. Prog. Phys., v. 69, p. 563.

Shay, M. A., Drake, J. F., Rogers, B. N., and Denton, R. E. 2001, *Alfvénic collisionless magnetic reconnection and the Hall term*, J. Geophys. Res., v. 106, p. 3759.

Spruit, H. C. 1996, *Magnetohydrodynamical jets and winds from accretion disks*, NATO Advanced Science Institutes Series C, v. 477, ed. Wijers, R. A. M. J., Davies M. B., and Tout C. A., p. 249 (Dordrecht: Kluwer Academic).

Stone, J. M., Hawley, J. F. Gammie, C. F., and Balbus, S. A. 1996, *Three-dimensional magnetohydrodynamical simulations of vertically stratified accretion disks*, Astrophys. J., v. 463, p. 656.

Sweet, P. A. 1958, *The neutral point theory of solar flares*, in "Proceedings of the IAU Symposium #6 on Electromagnetic Phenomena in Cosmical Physics", Stockholm, Sweden, 1956, ed. Bo Lehnert, p. 123 (Cambridge: Cambridge University Press).

Takahashi, K. and Yamada, S. 2013, *Regular and non-regular solutions of the Riemann problem in ideal MHD*, J. Plasma Phys., v. 79, p. 335 (TY13).

———. 2014, *Exact Riemann solver for ideal MHD that can handle all types of intermediate shocks and switch-on/off waves*, J. Plasma Phys., v. 80, p. 255 (TY14).

Tilley, D. A., Balsara, D. S., and Meyer, C. 2012, *A numerical scheme and benchmark tests for non-isothermal two-fluid ambipolar diffusion*, New Astronomy, v. 17, p. 368.

Velikhov, E. P. 1959, *Stability of an ideally conducting liquid flowing between cylinders rotating in a magnetic field*, J. Exptl. Theoret. Phys., v. 36, p. 1398.

Weber, E. J. and Davis Jr., L. 1967, *The angular momentum of the solar wind*, Astrophys. J., v. 148, p. 217.

Yamada, M., Ren, Y., Ji, H., Breslau, J. A., Gerhardt, S. P., Kulsrud, R. M., and Kuritsyn, A. 2006, *Experimental study of two-fluid effects on magnetic reconnection in a laboratory plasma with variable collisionality*, Phys. Plasmas, v. 13, p. 052119.

Index

$1\frac{1}{2}$-D flow, definition, 34, 123

abbreviated Leibniz notation, 123
accretion discs. *See* planetary discs
Alfvén, Hannes, ii, xiii, 102, 127–128
Alfvén number, **156**, 157, 165, 344
Alfvén point, **165**, 166–170, 179, 344, 346, 349, 352, 360, 371
Alfvén speed, a, a_x, **127**, 133, 145, 147, 148, 156, 186, 344, 360, 387, 413
  trans-Alfvénic, 155
Alfvén waves. *See* wave families, MHD
Alfvén's theorem, 105–106, 157, 342, 393, 396
ambipolar diffusion, 414–427
  ambipolar electric field, $\vec{E}_{AD}$, 414
  ambipolar resistivity, η_{AD}, 433
  coefficients
    ambipolar, $\beta_{1,2}$, 382, 386
    coupling, $\gamma_{1,2}$, 382
    exchange-ambipolar, $\mathcal{Q}_{i,n}$, 420, 425, 427
    rate, $\langle \sigma u \rangle_{1,2}$, 382, 426, 428–432
  density source terms, 417–419
    exchange, $\sigma_{1,2}$, 417–419, 424, 436, 437
  energy source terms, 420–423
    ambipolar, $\varpi^a_{1,2}$, 417, 420–421
    exchange, $\varpi^x_{1,2}$, 417, 421–423
  momentum source terms, 419–420
    ambipolar, $\vec{f}^a_{1,2}$, 380–385, 417, 419
    exchange, $\vec{f}^x_{1,2}$, 417, 419–420
  two-fluid, isothermal model, 415
  two-fluid, non-isothermal model, 410–427
  two-fluid resistivity, η_2, 425
Ampère's law, 106, **454**, 456
anti-curl, 112, 118
applied power, $\mathcal{P}_{app}$, 7, 16
  ambipolar power density, p_{AD}, 439
  density, p_{app}, 15, 16
  electric power density, p_E, 455
  electromagnetic power, $\mathcal{P}_{EM}$, **458**
  magnetic power density, p_M, **457**
  Poynting power density, p_S, **458**
  resistive power density, p_R, 392, **457**
astrophysical jets, 62–68, 100, 366–371
  Mach disc (hot spot), 62
  restarting jet, 63–65
  wide-angle tailed sources (WATs), 65–68
AZEuS, 366–371

barotropic gas, 20, 23, 24, 358
Bay of Fundy, 42
  lunar resonance, 59
bead-on-a-rod problem, 352–353, **478–483**
  inertial reference frame, 481–483
  non-inertial reference frame, 479–481
Bernoulli levitation, 52–54
  air gap, 53
  maximum supportable mass, 62
Bernoulli's theorem, HD, **49**, 47–55
  gas, 49
  and Kelvin–Helmholtz instability, 252
  liquid, 49
Bernoulli's theorem, MHD, **352**
  Bernoulli function, $\mathcal{B}_M$, 349–372
    critical points, 359–360, 362–364, 375
    as function of ρ and s, 358–359
    inertial frame derivation, 372
    unitless, 362, **364**
    value for stellar winds, 361
  as a driver for outflow, 352
  effective potential, ϕ_{eff}, 356–357, 374
  role of $\vec{B}$, 352–354, 372
Biot and Savart, law of, 400, 405, 412
bores, 42–47
  foaming, 43, 45, 46
  lab frame, 46–47
  sluice gate, 61
  standing (hydraulic jump), 43, 61
  tidal, 42
  undulating, 43, 45, 46
  velocity jump, 45
boundary conditions
  axisymmetry, 334
  Cauchy, 330, 334
  fluid–solid (no slip), **327**, 327–330, 336
  free boundary, 330
bra-ket notation, $\langle \, | \, \rangle$, 30
  eigenbras $\langle \, |$, 84–85, 96, 175, 236
  eigenkets $| \, \rangle$, 83–85, 96
Brio & Wu problem, 160, 161, 183
broad-crested weir, 49–52
  flowrate, 51

500

Cardano–Tartaglia formula, 163, **467**
  properties of cubic roots, 169
characteristic paths, **74**, 76, 78, 80,
    128–129, 135, 137–139, 177, 202–203,
    210
characteristic speeds, 71, 73–74, 86, 461
  1-D MHD, 128, 186
  Alfvén waves, 135
  magnetosonic waves, 148
characteristics
  Alfvén waves, $\mathcal{A}^{\pm}$, 135, 136, 177
  calculating, 85
  entropy wave, $\mathcal{S}^0$, **73–74**, 75, 76, 86
  fast/slow waves, $\mathcal{F}^{\pm}$, $\mathcal{S}^{\pm}$, 201
  sound waves, $\mathcal{J}^{\pm}$, **73–74**, 76, 201
  where $\mathcal{J}^{\pm}$, $\mathcal{S}^0$ are constant, 86, 97
circulation, Γ, 24, 117
colon product of matrices, A : B, 321, **444**
combined laws of thermodynamics, 18, 38
conics of PDEs, 459–461
conservation laws for MHD, 106
  energy, 7
  magnetic flux, 106
  mass, 7
  Newton's second law, 7
conservative variables, 22
contact discontinuity, HD, 35, 258
  isothermal, 57
  polytrope, 58
contact discontinuity, MHD, 154, 202
continuity equation, **15**
  incompressible flow, 43, 258
  linearised, 293
  steady state, 48, 342
control volume (steady-state HD), 43
cosmic rays, 100, 286–287
  pressure equation, 292
    linearised, 294
  pressure, p_{CR}, 288, 289, 305
Crab nebula, 257
current density, $\vec{J}$, 106, **143**, **383**, 408,
    424, 435, 456
current sheet, 395, 435

de Laval nozzle, 54–55, 62
  choke point, 55
  de Laval's equation, 55
  in radio source 1919+479, 66
difference theory, 150–151, 178
diffusion equation, 327, **484–485**
  diffusion coefficient, 327, 485
  Fick's first law, 484
diffusion time scale, 327
  magnetic, 393, 395
dimensional analysis, 51–52
discharge rate
  flow between plates, 329, 339

Hagen–Poiseuille flow, 334
  open channel flow, 331
discontinuous flow, 22, 90
downwind, definition, 34
dyadic product, 22, 111, **444**
dynamos, **399**, 399–406
  anti-dynamos, 400–402, 434
  Bullard dynamo, 399–400
  Earth's dynamo, 403–406, 434
    magnetic pole drift, 403
    polarity flips, 406
    Taylor column, 405
  kinematic vs. non-linear, 401
  necessary and sufficient conditions, 403

Earth's magnetic field, **3**, 403–406, 434
eigenvalues, HD, 85
  sound waves, 31
eigenvalues, MHD, **127–128**, 172, 174, 186
  Alfvén waves, 132
  magnetosonic waves, 140–141
eigenvectors, HD
  rarefaction wave, 86, 95
  sound waves, 31
eigenvectors, MHD
  Alfvén waves, 132, 186, 235
  entropy wave, 186, 235
  magnetosonic waves, 141–145, 186–190
    normalisation, ψ_f, ψ_s, 188, 236
    scaling factors, μ, ν, 190–192, 195,
      220, 236, 238
electric energy density, e_E, 455
electric field, $\vec{E}$, **453**
  ambipolar diffusion, $\vec{E}_{AD}$, 414
  Hall, $\vec{E}_H$, 408
  non-ideal MHD fluid, 385
  resistive, $\vec{E}_\eta$, 392
  static vs. induced, 102
  supported by a conducting medium, 102
electromagnetic force, $\vec{F}_{EM}$, 101, 391, 454
  density, $\vec{f}_{EM}$, 381
elliptical equations, 18, **459**
energy. See internal, total, or magnetic
    energy
enthalpy, h, 49, 351, 356
entropy, S, 18, 38, 73
  per particle, $\mathcal{S}$, 19, 23
  specific, s, 19
equations of HD, 18, 20, 21, 71
  conservative form, **18**, 80
    1-D steady state, 34
  differences between forms, 20–22
  Eulerian form, **71**
  Lagrangian form, **71**
  primitive form, **21**, 80
    1-D general solution, 84, 96
    in 1-D, 81

equations of MHD (ideal), **109**, 111
  conservative form, **111**
    1-D steady state, 149
  in 1-D, **124**, 172
  in most compact form, 112
  for Parker instability, 292
  primitive form, 122
    1-D linearised, 140, 173
  in 1-D, **124**, 172
  steady-state, 342
equations of MHD (non-ideal)
  neutrals, ions, electrons, 416
  one-fluid, isothermal model, 386
    limitations, 390
  one-fluid, non-isothermal model, 438
  three non-ideal terms, 379
    their comparison, 387–391
  two-fluid, isothermal model, 427
  two-fluid, non-isothermal model, 427
equilibrium, stable vs. unstable, 243
Euler number, $\mathcal{E}$, 324
Eulerian reference frame, 71, 280
  space-time diagrams, 74, 76, 77
Euler's equation, HD, **20**, 22, 26, 322
  1-D, 72
  linearised, 27, 29, 247, 260
  orthogonal coordinate systems, 448–450
  scaled version, 322–323, 338
Euler's equation, MHD, 107, 177
  steady state, 342, 398
evolutionary vs. non-evolutionary, 160
extensive vs. intensive var., 12, 103
  relationship between, 12

Faraday's law, 102, 385, 392, 426, **454**
fast point, **165**, 169, 180, 354, 360, 362–364, 375
fast speed, a_f. See magnetosonic speeds
flowline, 48
fluid, definition, 2, **7–9**
  impedance, Z_0, 32
  inviscid, 12, 25
  viscid, 12
fluid dynamics, definition, 2
fluid mechanics, definition, 2
flux, flux density, 103
  definitions, 14
  HD flux densities, 22
    linearised, 30
  MHD flux densities, 125
flux-freezing, 105, 157, 342, 396
flux function, f, 342–345, 358, 400
  coordinate, s, 350
  as lines of induction, $\vec{B}$, 342
  twisting lines of induction, 344–345
flux-linking, 115
  on solar surface, 116

flux loop, 114
flux theorem, 103–105, 117
flux tube, 105, 347–348
force densities, $\vec{f}_{\text{ext}}$, 17
  ambipolar, $\vec{f}^{\,a}_{1,2}$, 380
  exchange, $\vec{f}^{\,x}_{1,2}$, 420
  gravity, $\vec{f}_\phi$, 17
  Lorentz, $\vec{f}_L$, 107
  pressure gradient, $\vec{f}_p$, 17
  viscous stresses, $\vec{f}_T$, 318
force-free condition, 357, 374
forces
  applied vs. external, 7
  collisional, 7–10
Froude number, $\mathcal{F}$, 263, 338

gas dynamics, definition, 2
Gauss' law, **454**
Gauss' theorem, 14, 21, 44, 397, 403, **445**, 457
generalised Ohm's law. See electric field, $\vec{E}$; non-ideal MHD fluid
Green's theorem, **445**

Hall MHD, 406–414, 435
  Hall current, 410
  Hall effect (lab), 407–408
  Hall effect (plasma), 408
  Hall electric field, $\vec{E}_H$, 408, 426
  magnetic reconnection, 411–414
  proton–electron resistivity, $\eta_{p,e}$, 410
  quadrupole magnetic moment, 413
  two component model, 408–410
helicity. See magnetic helicity
helicity flux, $\vec{\mathcal{F}}_h$, 114
hydrodynamics
  definition, 2
  ideal, 12
hyperbolic equations, 82, 83, 128, **459**
  strictly vs. not strictly, 82, 128, 202

ideal gas law, 10–12, 289, 421
induction equation
  ideal, **102**, 109–111, 177
    cf. vorticity equation, 118
    linearised, 294
    steady state, 342
  non-ideal, **385**
    ambipolar diffusion, 386
    Hall, 386, 408, 426
    resistive, 386, 391–392, 394, 400, 402
    two-fluid, 426

instabilities. *See* KHI, RTI, Kruskal, MRI, Parker
intensive var. *See* extensive *vs.* intensive var
intermediate point, **166**, 169, 170
internal energy, E, 10–11
  density, e, 11
  specific, ε, 19, 73
internal energy equation
  adiabatic, **20**, 23
  isothermal, 23
  one-fluid, ambipolar, 439
  resistive, 393
  viscid form, 321, 337
interstellar medium (ISM), 286–288

Jacobian matrix, HD
  conservative, 95
  primitive, 82
  sound waves, 30
Jacobian matrix, MHD
  Alfvén waves, 131
  conservative, 125–126, 172
  magnetosonic waves, 140
  primitive, 124, 127, 172, 186
  Riemann problem. *See* Riemann problem, MHD; Jacobian
Jupiter's Great Red Spot, 253–254

Kelvin–Helmholtz instability (KHI), 245–255, 325
  and Bernoulli's theorem, 252
  cat's eyes, 252
  condition for instability, 249
  dispersion relation, 248–249
  growth rate, 250
  linear *vs.* non-linear theory, 251–252
  normal mode analysis, 246–250
  numerical analysis, 251–252, 254–255
  slab jet, 305
Kelvin's circulation theorem, 24, 117, 258
kinetic energy
  density, k, 402
  flux, $\vec{\mathcal{K}}$, 375
Kruskal–Schwarzschild instability, 267
  dispersion relation, 267

Lagrangian derivative, 71, 73, 135, 177, 401
Lagrangian reference frame, 71, 280
  space-time diagrams, 74, 75
Lagrangian velocity, 71
laminar flow, 325–336
Laplace's equation, 25, 259, 374
  pseudo-Laplacian operator, $\widetilde{\nabla}^2$, 400
Larmor radius, r_L, 100
liquid, definition, 2

Lorentz force, $\vec{F}_L$, 3, 99, **106**, 289, 396, 400, 408
  density, $\vec{f}_L$, 106, 143, 145, 270, 373, 383, 424
  longitudinal terms, 111
  orthogonal coordinate systems, 451–452
  transverse terms, 111
LU decomposition, 213–214
Lundquist number, S, 387, 398

Mach number, M, **36**, 55, 180
  downwind of shock, 37
  transonic point, 37, 40, 55
  upwind of shock, 157
magnetic diffusion, 393
magnetic diffusivity, $\mathcal{D}_M$, 393, 403
magnetic energy, E_M, 403
  density, e_M, 111, 402, **457**
magnetic field, $\vec{H}$, 3, **453**
  in astrophysics, 99–100
  potential field, 374
magnetic flux, Φ_B, **105–106**, 110
  conservation of, **105–106**, 115, 157, 361
magnetic helicity, H_A, 113–114
  as a conserved quantity, 114
  cross helicity, $h_\times$, 122
  density, h_A, 113–114
  evolution equations, 113–114
  value in flux loop(s), 115
magnetic induction, $\vec{B}$, 3, **453**
magnetic reconnection, 393–399, 411–414
  Hall regime, 412
    quadrupole magnetic moment, 413
    reconnection time scale, 413
  Sweet–Parker model, 395–399
    reconnection time scale, 398
  X-point, 395, 398, 399, 412
magnetic topology, 114–116
magnetic torque/moment, 347–348
  torque density, 347
magneto-acoustic waves, 145, 149, 201, 202
  speed, a_M, 145, 195
magnetohydrodynamics (MHD)
  definition, 2
  ideal, definition, 101
magneto-rotational instability (MRI), 268–286
  angular momentum transport, 278–307
    $\vec{L}$ transported per revolution, 283
  Balbus & Hawley, Shaw Prize, 268, 278
  comparison to KHI, 274, 285
  condition for instability, 274–275
  diffusion coefficient, 307
  dispersion relation, 273
  dynamical equations, 273
  growth rate, 277–278, 306

magneto-rotational instability (MRI) (cont.)
  normal mode analysis, 272–274
  numerical analysis, 283–286
  physical model, 275–277
magnetosonic numbers, M_f, M_s, 165, 169, 368
magnetosonic speeds, a_s, a_f, **127**, 141, 148, 186, 360
  identities, 173
  inequalities, 128, 173
  limits, 174
Maxwell's equations
  differential form, 453
  integral form, 453–454
mean free path, definition, 2
method of characteristics (MoC), 76–78
  applied to Alfvén waves, 135–176
  applied to Riemann problem, 78–80
  as a numerical scheme, 77, 94
MHD-alpha, α, 106, **130**, 157, 179, 236, 237, 388, 426
  fast, slow, $\alpha_{f,s}$, 180, **189**, 190, 195
  identities, 188
momentum equation, HD, **17**, 310
  orthogonal coordinate systems, 449–450
momentum equation, MHD
  Hall, 408
  ideal, **107**, 111, 383
  ions, two-fluid, 424
  linearised, 293, 294
  neutrals, ions, electrons, 380
momentum, $\vec{S}$, 7
  density, $\vec{s}$, 17

Navier–Stokes equation, 317–320
  compressible, 319
  incompressible, 320
    scaled version, 323
  inertial term, $\nabla \cdot (\vec{s}\vec{v})$, 318
  magnetic, 401
  stress force density, $\vec{f}_T$, 317–318
  viscid momentum equation, 318
  viscid term, $\nabla \cdot (\mu S)$, 318
Newtonian fluids, 314–315, 318
non-inertial reference frame, 269–350, 352, 401, **476–483**
  Coriolis theorem, 478
  inertial accelerations, 270–271, 350, 405–406, **478**
normal mode analysis
  Alfvén waves, 131–134, 175
  ball on a mound, 305
  explained, 247
  KHI, 246–250
  magnetosonic waves, 140–141
  MRI, 272–274

Parker instability, 298–301
rarefaction fan, 82
RTI, 257–262
sound waves, 30–33
numerical considerations
  convergence, 217, 219, 237, 238, 468–469
  preserving precision, 180, 191, 208–209, 212–216, 218, 220–221
  scaling, 262–264
  suppressing pressure perturbations, 284
numerical MHD, 119, 139, 366–371

Ohmic resistance. *See* resistive MHD
outflow mechanisms
  bead-on-a-rod (BRM), 353–357
  critical angle (60°), 355–357, **374**
  energy flux, 375
  magnetic tower (MTM), 354, 373
  and MHD Bernoulli theorem, 352

parabolic equations, 326, 327, **459**, 485
Parker instability, 286–305
  2-D equations of MHD, 292
  comparison to RTI, 291
  condition for instability, 289, 301–303
  dynamical equation, 298
  growth rate, 288, 290, 301–302
  interstellar clumps, 287–288, 290, 303
  normal mode analysis, 298–301
  perturbation analysis, 292–294
  qualitative description, 286–291
  quantitative description, 291–305
particle path, 47
Pascal's law, 318
PdV term, 16
planetary discs, 268, 426
  anomalous viscosity, 268
  artist's conception, 379
  formation, 354–355
  number density, 388
  T Tauri IM Lup, 378
  temperature, 379, 388
plasma physics, definition, **3**, 99
plasma-beta, β, **129**
Poisson's equation, 18
polytropic gas, 57
power. *See* applied power
Poynting
  flux, Φ_S, 107, **458**
  power density, p_S, 107, 109, **458**
  vector, $\vec{S}_P$, 107, 375, **458**
pressure
  cosmic ray, p_{CR}, 288, **289**, 305
  magnetic, p_M, **111**, 142–145, 270, 288
  MHD, p^*, **111**, 271, 279
  thermal, p, **17**
    collisional, 8

isotropic, 8–10
pressure equation, **20**, 23
  1-D, 72
  barotropic, 26
  cosmic rays, 292
  linearised, 27, 29, 294
pressure head, 44, 259
primitive variables, 22, 34
principle of equipartition, **11**, 100, 303

quasi-steady state. *See* steady state

Rankine–Hugoniot, HD, **35**, 80, 90
  isothermal, 57
  polytrope, 57
Rankine–Hugoniot, MHD, **150**, 153, 157
rarefaction fans, HD, 80
  generalised coordinate, s_i, 86
  profiles as function of s_i, 87
  profiles as function of u_i, 88, 89, 97
  strength/width, 87
  transition, 87, 89
rarefaction fans, MHD, 185, **192–202**, 239
  fast fans, **193**
    fast Euler fans, **193**, 196, 200, 201, 204, 207
    saturation, **193**, 196, 201, 207, 219
    switch-off fans, **193**, 196, 200, 203, 236
  fast/slow differences, 192, **200–202**
  generalised coordinate, s_i, 186
  profiles as function of s_i, 186, 218
  profiles as function of u_i, 196–200
  similarity to HD fans, 192, 195, 200
  slow fans, **193–195**
    asymptotic limits, 193, 200
    slow Euler fans, **193**, 195, 200, 201, 204
    switch-on fans, **194–195**, 198–201
  strength/width, 186, **193**, 200, 207
ratio of specific heats, γ, 11
Rayleigh–Taylor instability (RTI), 256–267
  Atwood number, 261
  condition for instability, 261
  dispersion relation, 259–261
  growth rate, 261
  link with KHI, 262, 265, 266
  normal mode analysis, 257–262
  numerical analysis, 262–267
resistive MHD, 391–406
  energy dissipation, 392–393
  resistive electric field, $\vec{E}_\eta$, 392
  resistivity, η, 385
Reynolds number, $\mathcal{R}$, **324**, 322–326
  inertial *vs.* viscous dominance, 325
  magnetic, $\mathcal{R}_M$, 387, 403
Riemann, Bernhard, 69
Riemann invariants. *See* characteristics

Riemann problem, HD
  defined, 69
  solution, 90–93, 98
Riemann problem, MHD
  defined, 183–185
  exact solver, 204–221, 239
    algorithm, 210–221
    constraints, 206, 210–212
    fast shock, 215–216
    Jacobian, 209–210, 212–214
    parameters, 205–208, 210
    rarefaction fans, 218–221
    slow shock, 216–218
    strategy, 208–209
  solutions, 221–235
  uniqueness, 160, 204
rms speed, $v_{\rm rms}$, **10**, 12, 29, 289, 304
rotational discontinuity, **155**, 165, 167, 178, 181
Runge–Kutta, sixth order, **468–475**
  algorithm, 473–475
  derivation, 468–473
  MHD rarefaction fans, 196, 218

Saha equation, 304, 416, **417–418**, 437
  thermal de Broglie wavelength, λ_e, 418
scale height, L, **289**, 290, 304
secant root finder, 462–466
  multivariate, 205, 210–211, **465–466**
  univariate, 92, 93, 205, 364, **462–465**
    FORTRAN 77 listing, 463–465
shear layer, 245
shock tube, 33
  general MHD, 149
  Sod, 93
shock waves, HD, **37–42**, 80
  entropy condition, **38–40**, 87
  general frame, 58
  hypersonic limit, 37
  lab frame, 40–42
  shock strength, 58
  variable jumps, 37
shock waves, MHD, **155–172**, 181, 182
  entropy condition, **156**, 158, 159, 163, 169–171, 181
  Euler branch, 164–166, **167**, 170–172, 180
  evolutionary condition, **159**, 163, 170, 172
  fast shocks, **158–161**, 164–172, 181
    switch-on shocks, 157, **164–165**, 166, 167, 170–172, 179, 180, 203
  $i \to j$ designation scheme, **159**, 167, 168
  intermediate shocks, **159–161**, 165–172, 181
  shock types, **159**, 166, 170, 171, 180

shock waves, MHD (cont.)
  slow shocks, **158–161**, 165–172, 179, 181
    switch-off shocks, **167**, 167
  strength, 157, 207
  variable jumps, **163**, 207
slow point, **165**, 169, 170, 180, 360, 375
slow speed, a_s. *See* magnetosonic speeds
smooth flow, **22**, 72, 80, 82, 90
solar flares, 395, 398, 413
  anomalous resistivity, 399
sound speed, c_s, 27, 127
  adiabatic, 28
  astrophysical values, 29
  isothermal, c_{iso}, 28, 388
  value in dry air at STP, 29
sound waves, 26–33
  frequency, ω, 28
  linear algebra solution, 29–33
  perturbation analysis, 26
  secular equation, 31
  solution to wave equation, 26–29
  wave vector, $\vec{k}$, 28
space-time diagrams, **74–76**, 77, 128
  1-D MHD, 128, 202
  Alfvén waves, 135, 137
  event, 74
  footprints, 76–78, 136
  magnetosonic waves, 148
  Riemann problem, 78–80, 90
  sonic cone, 76
  worldline, 74, 75
steady state
  definition, 34, 341
  quasi, 329, 331
stellar winds (Weber–Davis), 361–366
  additional assumptions, 361
  asymptotic behaviour, 365–366, 376
  boundary conditions, 357
  profiles for ρ, v_p, ψ, 364–365, 376
Stokes' theorem, 104, **446**, 456
strain tensor, E, 316
  strain components, $\partial_j v_i$, 314
streakline, 47
stream function, ψ, **258**, 259, 291
streaming motion, 9
streamline, 47
streamtube, 48
stress tensor, T, 311–317
  components, T_{ij}, 311–317
    cylindrical coordinates, 334–336
    compressive stresses, 310, 312, 316
    shear stresses, 310
    trace, tr(T), 312–313, 337
      relation to thermal pressure, p, 318
superfluids, 325
surface-conserved quantity, 104, 109, 110
synchrotron emission, 286

tangential discontinuity, HD, 35
tangential discontinuity, MHD, 36, 154
  as limit to slow and Alfvén waves, 195
Theorem of hydrodynamics, **13–15**, 103, 117
total energy equation, HD
  inviscid form, 17
  scaled version, 337
  viscid form, 320, 337
total energy equation, MHD, **109**, 110, 119
  differenced 1-D steady state, 152, 178
  resistive, 393, 432
  steady state, 342, 372
  two-fluid, non-ideal, 438
total energy, HD
  E_T, 7, **15**
  density, e_T, **15**
total energy, MHD
  density, e_T^*, **108**, 393
triple umbilic. *See* wave families, MHD
turbulence, 243–245, 253, 262, 283, 309, **324–326**
  super-Alfvénic, 244

upwind, definition, 34

vector derivatives, 446–448
  Cartesian coordinates, 447
  cylindrical coordinates, 447
  spherical polar coordinates, 447
vector identities, 443–445
  with dyadics, 444–445
vector potential, $\vec{A}$, **112**, 112–113, 118, 292, 293, 400–401, 456
  contours as lines of $\vec{B}$, 292
  evolution equation, 112
viscometer, 334, 336, 340
viscosity
  kinematic, ν, **319**, 324, 393, 402
  shear, μ, **314**, 318, 319, 328
viscous dissipation, 321
viscous flow
  Couette, 334–336, 340
    torque, 336, 340
  forced between co-axial cylinders, 339
  forced between plates, 328–329, 339
  Hagen–Poiseuille, 333–334
  open channel, 329–332
  plane laminar, 327–328
volume-conserved quantity, 14, 18, 109, 110
vorticity, $\vec{\omega}$, 23, **105**, 254, 309, 319
  comparison with $\vec{B}$, 24, 339
vorticity equation, 23

wave equation
  Alfvén waves, 131
  sound waves, 27, 246

wave families, HD, 86, 130, 202
wave families, MHD, 130, **128–149**, 202–204, 206
   Alfvén waves, 129, **130–139**, 175
      compressional, 145, 149
      linear algebra solution, 131–134, 175
      properties, 131, 134, 146–149, 174
      torsion, 177, 354, 368
      wave equation, 131
   compound wave, **160**, 200, 203, 204
   degeneracy, 145–149
   entropy wave, 129, 202
   magnetosonic waves, 129, 139–146
      fast *vs.* slow waves, 142–145
      linear algebra solution, 140–141
      perturbation analysis, 139–140
      properties, 141, 146–149, 174
   triple umbilic, 148, **190**, 191–195, 198, 200–204, 209
weakly ionised medium, 380, 383
   interpretation of velocities, 384
Weber–Davis constants, **342–349**, 372
   angular speed, Ω_0, 343–344, 369
   mass load, η, 345–346, 369
   specific angular momentum, l, 347–349, 369, 371
   steady-state axisymmetry, 342
work-kinetic theorem, 402

Zemplén's theorem. *See* entropy condition; shock waves
ZEUS-3D, 63, **77**, 119, 160, 244, 251–252, 254–257, 262–265, 283–285, 366–371

For EU product safety concerns, contact us at Calle de José Abascal, 56–1°, 28003 Madrid, Spain or eugpsr@cambridge.org.

www.ingramcontent.com/pod-product-compliance
Lightning Source LLC
LaVergne TN
LVHW072011060526
838200LV00010B/329